8th International Conference on Compressors and their Systems

---

DEDICATION

These proceedings are dedicated to the memory of Geoffrey Soar. Starting from the bottom of the ladder, as a machinist, he rose to the top as Engineering Director of his company and was a pioneer in the introduction of computer aided design to the compressor industry. Geof was an active member of the Fluids Machinery Group of the IMechE for many years and a keen supporter and a major participant in the organisation of this series of conferences since they were first conceived in 1997. Although he knew that his time was limited he bore his illness bravely and was active in participating in the conference organisation until the last weeks of his life. He was a true traditional English Gentleman and is sadly missed.

Organising Committee

Ian Smith *(Chairman)*	City University London
Amelia Brunt	The Institution of Mechanical Engineers
Graeme Cook	Howden Compressors
Chris Holmes	PTG Advanced Developments Ltd
Guy Hundy	Aleph Zero
Ahmed Kovacevic	City University London
Ivor Rhodes	ACE Cranfield
Melanie Rigby	Corac
Nikola Stosic	City University London
Sham Rane	City University London
Louise Gordon	City University London
Emma J Leaver	City University London

International Liaison Committee

Ian Bennett	Shell Global Solutions, UK
Cesar DeChamps	University of Santa Catarina, Brazil
Mitsuhiro Fukuta	Shizuoka University, Japan
Yuri Galerkin	St Petersburg University, Russia
Amin Haghjoo	Ingersoll Rand, Germany
Liansheng Li	Xi'an Jiaotong University, China
Erich Machu	Consultant, Austria
Kim TiowOoi	Nanyang Technology University, Singapore
Jack Sauls	Ingersoll Rand/Trane, USA

8th International Conference on Compressors and their Systems

9–10 September 2013
City University London, UK

in conjunction with

Oxford   Cambridge   Philadelphia   New Delhi

Published by Woodhead Publishing Limited
80 High Street, Sawston, Cambridge CB22 3HJ, UK
www.woodheadpublishing.com
www.woodheadpublishingonline.com

Woodhead Publishing, 1518 Walnut Street, Suite 1100, Philadelphia, PA 19102-3406, USA

Woodhead Publishing India Private Limited, G-2, Vardaan House, 7/28 Ansari Road, Daryaganj, New Delhi – 110002, India
www.woodheadpublishingindia.com

First published 2013, Woodhead Publishing Limited
© The author(s) and/or their employer(s) unless otherwise stated, 2013
The authors have asserted their moral rights.

This book contains information obtained from authentic and highly regarded sources. Reprinted material is quoted with permission, and sources are indicated. Reasonable efforts have been made to publish reliable data and information, but the authors and the publisher cannot assume responsibility for the validity of all materials. Neither the authors nor the publisher, nor anyone else associated with this publication, shall be liable for any loss, damage or liability directly or indirectly caused or alleged to be caused by this book.

Neither this book nor any part may be reproduced or transmitted in any form or by any means, electronic or mechanical, including photocopying, microfilming and recording, or by any information storage or retrieval system, without permission in writing from Woodhead Publishing Limited.

The consent of Woodhead Publishing Limited does not extend to copying for general distribution, for promotion, for creating new works, or for resale. Specific permission must be obtained in writing from Woodhead Publishing Limited for such copying.

Trademark notice: Product or corporate names may be trademarks or registered trademarks, and are used only for identification and explanation, without intent to infringe.

British Library Cataloguing in Publication Data
A catalogue record for this book is available from the British Library.

Library of Congress Control Number: 2013946843

ISBN 978 1 78242 169 6 (print)
ISBN 978 1 78242 170 2 (online)

Produced from electronic copy supplied by authors.
Printed in the UK and USA.
Printed in the UK by 4edge Ltd, Hockley, Essex.

CONTENTS

FOREWORD 1

KEYNOTE PAPERS

C1390/219 What does the future hold for compressor manufacturers? 5
A B Pearson, Star Refrigeration Ltd, UK

C1390/107 Recent research of novel compression concepts for vapor 15
compression heat pumping, air conditioning and refrigeration
systems
E A Groll, O Kurtulus, Purdue University, USA

C1390/220 Sliding vane rotary compressor technology and energy saving 27
R Cipollone, University of L'Aquila, Italy

COMPRESSOR SYSTEMS

C1390/143 Experimental study of compressor operating characteristics 57
and performance when using refrigerants R32, R1234yf, and
two new low GWP developmental refrigerants as drop-in
replacements for R410A
*L Cremaschi, X Wu, A Biswas, P Deokar, Oklahoma State
University, USA*

C1390/100 Performance and suitability comparisons of some R22 possible 67
substitute refrigerants
*A Subiantoro, TUM CREATE; K T Ooi, A Z Junaidi, Nanyang
Technological University, Singapore*

C1390/117 Oil level measurement in oil-injected screw compressor 77
packages used in the petroleum, petrochemical, refrigeration
and fuel gas markets
*W J Milligan, D I Muir, Howden Process Compressors;
D K Harrison, Glasgow Caledonian University, UK*

C1390/133	Simulation of a cold climate heat pump furnished with a scroll compressor with multiple vapour injection lines *I H Bell, University of Liège, Belgium; E A Groll, J E Braun, W T Horton, Purdue University, USA*	87
C1390/179	Compressor for exhaust treatment of non-road vehicles *R Sachs, W Asal, Gardner Denver Schopfheim GmbH, Germany*	103
C1390/114	Screw pulsation generation and control: a shock tube mechanism *P X Huang, S Yonkers, Hi-Bar MC Technologies, LLC; D Hokey, D Olenick, GE Oil & Gas, USA*	113
C1390/166	Numerical analysis of unsteady behaviour of a screw compressor plant system *E Chukanova, N Stosic, A Kovacevic, City University London, UK*	129
C1390/112	Optimization of integrated energy process in China industrial compressed air system *T Ma, H Li, Beijing University of Technology; Z Jia, Beijing Huayuanyitong Heating Development Co., Ltd, China; X Zhang, X Zhao, University of Hull, UK*	141
C1390/128	Investigation on working characteristics of micro compressed air energy storage system *Q C Yang, Y Y Zhao, L S Li, Z G Qian, Hefei General Machinery Research Institute, China*	151

ENERGY EFFICIENCY

C1390/115	Compressor and system energy efficiency improvements through lubricant optimization *J Karnaz, CPI Engineering Services, USA*	163
C1390/193	Energy saving in sliding vane rotary compressors *R Cipollone, G Bianchi, University of L'Aquila; G Valenti, T Calvi, Politecnico di Milano; S Murgia, G Contaldi, Ing. Enea Mattei S.p.A., Italy*	173
C1390/194	Energy recovery using sliding vane rotary expanders *R Cipollone, G Bianchi, University of L'Aquila; G Contaldi, S Murgia, Ing. Enea Mattei SpA, Italy*	183

SCREW COMPRESSORS

C1390/212	Calculation of discharge pressure pulsations of a screw compressor using the one-dimensional method of characteristics A Linkamp, A Brümmer, TU Dortmund University, Germany	197
C1390/207	Discussion of actual profile clearances' calculation method in rotary compressors in the absence of rotor timing units R R Yakupov, T N Mustafin, M S Khamidullin, I G Khisameev, Kazan National Research Technology University; V N Nalimov, NIIturbocompressor n.a. V.B.Shnepp, Russian Federation	209
C1390/153	Theoretical analysis of loads on the gate rotor bearings in the single screw compressor T Li, Z L Wang, R Huang, W F Wu, Q K Feng; Xi'an Jiaotong University, China	219
C1390/206	Investigation of heat exchange in the working chamber of rotary compressors I I Sharapov, A G Saifetdinov, A M Ibraev, M S Khamidullin, I G Khisameev, Kazan National Research Technological University, Russian Federation	227
C1390/162	Experimental validation of a geometry model for twin screw machines D Buckney, Howden Compressors Ltd; A Kovacevic, N Stosic, City University London, UK	237
C1390/140	Thermodynamic simulation of multi-stage screw compressors using chamber-based screw model J Hauser, M Beinert, S Herlemann, GHH Rand Schraubenkompressoren GmbH, Germany	247
C1390/210	Effect of variable volume index on performance of single screw compressor M Masuda, H Ueno, T Inoue, K Hori, M A Hossain, Daikin Industries Ltd, Japan	257
C1390/127	Investigations of deformation failure of the slide valve in single screw refrigeration compressors F Liu, Z L Wang, W F Wu, Q K Feng, Xi'an Jiaotong University, China	265

SCROLL COMPRESSORS

C1390/177	Assessment of reed valve dynamic behavior in a scroll compressor through visualization A J P Zimmermann, University of Illinois at Urbana-Champaign; P S Hrnjak, University of Illinois at Urbana-Champaign and CTS - Creative Thermal Solutions Inc., USA	277
C1390/149	Fatigue design and safety factor for scroll compressor wraps D Gross, DANFOSS Commercial Compressors, France	285
C1390/168	Simulation model to predict temperature distribution along scroll wraps M C Diniz, C J Deschamps, Federal University of Santa Catarina; E L L Pereira, Embraco, Brazil	301
C1390/176	Oil flow at discharge valve in a scroll compressor A J P Zimmermann, University of Illinois at Urbana-Champaign; P S Hrnjak, University of Illinois at Urbana-Champaign and CTS - Creative Thermal Solutions Inc., USA	311

NOVEL COMPRESSORS AND MANUFACTURING TECHNOLOGIES

C1390/187	Analysis and development of a new compressor device based on the new finned piston M Heidari, A Rufer, École Polytechnique Fédérale de Lausanne (EPFL), Switzerland	323
C1390/163	Oil-flooded screw vacuum pumps with a novel flexible discharge port design Y Tang, Kaishan Compressor, China	335
C1390/218	Spool compressor tip seal design considerations C R Bradshaw, Torad Engineering LLC, USA	341
C1390/118	Generating grinding in rotor production – KAPP rotor grinding machine RX 120 A Köhler, J Heyder, F Wölfel, KAPP Werkzeugmaschinen GmbH, Germany	353

| C1390/129 | Experimental study of noise and vibration reduction in a medium-size oil-flooded twin-screw compressor by the application of helix relief
K Matsuo, Mayekawa Mfg. Co., Ltd, Japan; C S Holmes, PTG Advanced Developments, Holroyd, UK | 363 |

RECIPROCATING COMPRESSORS

C1390/136	Survey of factors influencing reciprocating compressor efficiencies and discharge gas temperatures E H Machu, Consulting Mechanical Engineer, Austria	375
C1390/201	An investigation of the heat transfer phenomena between the hermetic reciprocating compressor components A R Ozdemir, E Oguz, Arcelik Research and Development Center; S Onbasioglu, ITU Mechanical Faculty, Turkey	385
C1390/145	Study of aerodynamic noise in hermetic refrigerator compressor S Lin, Z He, J Guo, Xi'an Jiaotong University, China	397

EXPANDERS

C1390/161	Physics of a dry running unsynchronized twin screw expander J Hütker, A Brümmer, TU Dortmund, Germany	407
C1390/160	3D CFD analysis of a twin screw expander A Kovacevic, S Rane, City University London, UK	417
C1390/174	Sizing models and performance analysis of volumetric expansion machines for waste heat recovery through organic Rankine cycles on passenger cars L Guillaume, A Legros, S Quoilin, S Declaye, V Lemort, University of Liège, Belgium	431
C1390/150	Simulation of expansion process in positive displacement expander K M Ignatiev, M M Perevozchikov, Emerson Climate Technologies Inc., USA	445
C1390/172	Investigation on a scroll expander for waste heat recovery on internal combustion engines A Legros, PSA Peugeot Citroen, France and University of Liège, Belgium; M Diny, PSA Peugeot Citroen, France; L Guillaume, V Lemort, I Bell, S Quoilin, University of Liège, Belgium	453

TURBO MACHINES

C1390/182	Oil-free centrifugal refrigeration compressors: from HFC134a to HFO1234ze(E) *J J Brasz, Danfoss Turbocor Compressors Inc., USA*	467
C1390/203	The application of the Universal Modeling Method to development of centrifugal compressor model stages *Y B Galerkin, K V Soldatova, TU Saint-Petersburg, Russia*	477
C1390/209	Flow and performance investigation of the specially designed channel diffuser of centrifugal compressor *M Kalinkevych, A Skoryk, Sumy State University, Ukraine*	489
C1390/121	Investigation of gas flow with injection in vaneless diffuser of centrifugal compressor *M Kalinkevych, O Shcherbakov, V Ihnatenko, Sumy State University, Ukraine*	501
C1390/200	Experimental study of radial low specific speed turbocompressor running in reverse as turboexpander *M Arjeneh, K R Pullen, City University London; S Etemad, Dynamic Boosting Systems (DBS) Ltd, UK*	511
C1390/123	Shape optimization of a centrifugal compressor impeller *S Khalfallah, Ecole Militaire Polytechnique; A Ghenaiet, University of Sciences and Technology USTHB, Algeria*	523
C1390/204	Centrifugal compressor efficiency types and rational application *Y B Galerkin, A Drozdov, K V Soldatova, TU Saint-Petersburg, Russia*	533

VALVES

C1390/135	Historic review of attempts to model valve dynamics *E H Machu, Consulting Mechanical Engineer, Austria*	545
C1390/169	Numerical analysis of seat impact of reed type valves *F C Lajús Junior, C J Deschamps, Federal University of Santa Catarina; M Alves, University of São Paulo, Brazil*	555
C1390/156	The improved valve assembly of swash plate compressor for vehicle air conditioning system *G-H Lee, Doowon Technical College; T-J Lee, S-W Lee, H-J Kim, Doowon R&D Center, Korea*	565

C1390/195	Transient and dynamic numerical simulation of the fluid flow through valves based on large eddy simulation models O Estruch, J Rigola, A Oliva, C D Pérez-Segarra, Universitat Politècnica de Catalunya (UPC); O Lehmkuhl, Universitat Politècnica de Catalunya (UPC) and TermoFluids S.L., Spain	577

COMPUTATIONAL FLUID DYNAMICS

C1390/213	Use of computational fluid dynamics to develop improved one-dimensional thermodynamic analyses of refrigerant screw compressors J Sauls, S Branch, Ingersoll Rand, USA	591
C1390/139	CFD grid generation and analysis of screw compressor with variable geometry rotors S Rane, A Kovacevic, N Stosic, M Kethidi, City University London, UK	601
C1390/148	CFD modelling of scroll compressor intermediate discharge ports B Angel, Renuda; P Ginies, D Gross, C Ancel, Danfoss Commercial Compressors, France	613
C1390/141	CFD analysis of pressure pulsation in screw compressors – Combine theory with practice J Hauser, M Beinert, GHH Rand Schraubenkompressoren GmbH, Germany	625
C1390/124	CFD analyses of a radial inflow turbine M Cerdoun, Ecole Militaire Polytechnique; A Ghenaiet, University of Sciences and Technology USTHB, Algeria	635
C1390/125	Numerical and experimental investigation of the efficiency of vaned diffuser of centrifugal compressor O Obukhov, A Smirnov, PJSC «Sumy Frunze NPO»; O Gysak, Sumy State University, Ukraine	649
C1390/184	Simulation and validation of the compressor stage of a turbocharger using OpenFOAM M Heinrich, R Schwarze, Technical University Bergakademie Freiberg, Germany	659
C1390/146	Influence of the suction arrangement and geometry of the inlet port on the performance of twin screw compressors M Pascu, M Heiyanthuduwage, S Mounoury, G Cook, Howden Compressors Ltd, UK	669

MODELLING

C1390/196	A new dynamic heat pump simulation model with variable speed compressors under frosting conditions N Park, J Shin, B Chung, LG Electronics, Korea	681
C1390/144	Developing simulation tools for design of low charge vapour compression refrigeration systems G L Ding, T T Wang, J D Gao, Shanghai Jiao Tong University; Y X Zheng, Y F Gao, J Song, International Copper Association Shanghai Office, China	697
C1390/147	Modeling of small-size turbocharger compressors' performance curves K V Soldatova, TU Saint-Petersburg, Russia	707
C1390/131	Development of a generalized steady-state simulation framework for positive displacement compressors and expanders I H Bell, V Lemort, University of Liège, Belgium; E A Groll, J E Braun, W T Horton, Purdue University, USA	717
C1390/192	A parallel object oriented code framework for numerical simulation of reciprocating compressors – introduction of solid parts modeling J López, J Rigola, A Oliva, Universitat Politècnica de Catalunya (UPC); O Lehmkuhl, Universitat Politècnica de Catalunya (UPC) and Termo Fluids S.L., Spain	731
C1390/126	Analysis of the basic geometrical parameters influence on the efficiency of the Roots-type compressor on the basis of thermodynamic processes simulation A M Ibraev, S V Vizgalov, I G Khisameev, Kazan National Research Technological University, Russian Federation	739
C1390/101	A comprehensive simulation model of the dynamics of the revolving vane machine A Subiantoro, TUM CREATE; K T Ooi, Nanyang Technological University, Singapore	749

AUTHOR INDEX

SPONSORS

City University London and the Institution of Mechanical Engineers would like to thank the following sponsors:

Platinum sponsor

Gold sponsor

Silver sponsor

ABSOLUTE PRECISION IN ROTOR DESIGN

PTG is proud to once again be the platinum sponsor of the International Conference on Compressors and their Systems.

Since City's Centre for Positive Displacement in Compressors was established in 1995, we have collaborated to advance the design of compressors and expanders through research and development.

Today, our Precision Components division is recognised as leading the way in screw and rotor compressor manufacture and works closely with customers to design, prototype and produce all types of helical forms, rotors, superchargers, blowers, pumps and vacuum screws.

At the same time, Holroyd Precision Limited, part of our machine tool design and build division, leads the world in the development of high-precision machine tools – such as the Zenith 400 helical profile grinder – for the manufacture of ultra-precise helical components.

To find out more about our technologies and capabilities:

Tel +44 (0)1706 526590
Email info@holroyd.com

www.holroyd.com

Part of the CQME Group of Companies

ABSOLUTE PRECISION MAKES ALL THE DIFFERENCE

WORLD PIONEERS OF **ROTARY TWIN SCREW COMPRESSORS**

SINCE PIONEERING THE FIRST TWIN SCREW COMPRESSOR IN THE 1930's, HOWDEN HAS SUPPLIED OVER 35,000 SCREW COMPRESSORS WORLDWIDE.

Setting the industry standard for both gas and refrigeration applications, Howden-designed and manufactured screw compressors are used for a variety of demanding process gas compression and industrial refrigeration duties. We also supply units for gas boosting, such as feeding gas turbines with fuel gas pressure, and for a range of other applications. We produce a wide and versatile range of bare-shaft screw compressor units that are supplied globally to our compressor packaging customers. This allies our engineering expertise with local knowledge, and enables strong support to be provided to end-user clients.

MANUFACTURING EXCELLENCE
Equipped with some of the world's most advanced rotor and casing machining equipment, our screw compressor centre of excellence facility in Glasgow, UK, ensures the quality and reliability of our products every time. This factory holds **triple certification** to international standards:

Quality management: ISO 9001:2008
Environmental management: ISO 14001:2004
Health & Safety management: OHSAS 18001

For further information contact:

Howden Compressors
133 Barfillan Drive, Glasgow G52 1BE
Scotland, United Kingdom
Tel: +44 141 882 3216   Fax: +44 141 882 8648
Email: hcl.sales@howden.com to discuss twin screw compressors
Email: hcl.aftersales@howden.com to request after-sales support

CHOOSE HOWDEN SCREW COMPRESSORS FOR:
- ADVANCED TECHNOLOGY
- OIL INJECTED AND OIL FREE
- PROVEN RELIABILITY AND DURABILITY
- MANUFACTURING EXCELLENCE
- BUILT TO API STANDARD
- GLOBAL SUPPORT
- LIFETIME COMMITMENT

 Find out more:
www.howden.com/en/compressors

Howden

© Howden Ltd. All rights reserved. 2013

KAPP RX 120

New Technology for Rotor Grinding

KAPP NILES is a global market leader in manufacturing machines and tools for finishing gears and profiles. Technologies from KAPP NILES guarantee both precision and cost-effectiveness when manufacturing sophisticated components. In this way, KAPP NILES sets their customers in precise motion – on land, in the water and in the air.

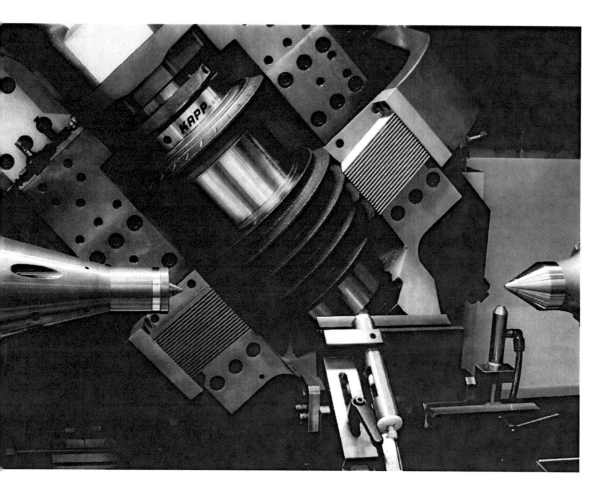

Enquire about designing your own machining concept. We will gladly advise you!
KAPP Werkzeugmaschinen GmbH • Callenberger Str. 52 • 96450 Coburg • GERMANY
Tel: +49 9561 866-0 • E-Mail: sales@kapp-niles.com • Internet: www.kapp-niles.com

FOREWORD

Following our last event, two years ago, it is my pleasure, again, to welcome you to this, our 8th International Conference on Compressors and their Systems, to be held at the City University London on 9th and 10th September 2013.

As for its predecessors, the conference is conducted by the School of Engineering and Mathematical Sciences at City University in conjunction with the Institution of Mechanical Engineers (IMechE) and, as for the 7th conference, the administration has been carried out mainly by City University London staff.

This year we received a record number of submissions and hence, in order not to hold more than two sessions in parallel at any time, while retaining the conference within the planned two day schedule, we were compelled to limit some of the last submissions to poster display only. This, in no way detracts from our perceived merit of these papers and I hope that all of you will take advantage of the breaks in the sessions to visit these exhibits and discuss them with their authors. Naturally, all papers, these included, are to be found in the following pages of these proceedings.

I would like to draw your attention to a new feature of this event, namely that the short course, that we hold preceding it, has been increased in scope to become a forum on the application of Computational Fluid Dynamics (CFD) analysis to all types of compressors. An international team of eleven experts will be giving a series of lectures on a wide range of uses for this form of analysis and I hope that as many of you as possible will have taken advantage of this event, while coming for the conference itself.

On this occasion, I must thank all those who have helped to make this event. Special thanks go to Ahmed Kovacevic, who introduced us to ConfTool, thereby making the conference organisation much easier, to Louise Gordon and Emma J Leaver, of the University's events team, who did most of the work and to all members of the organising committee, the international committee and all those who participated in refereeing the papers. On a sadder note, I must record the loss of Geoff Soar, who played a key role in the organisation of these conferences from the time that they were first planned in 1997, but who died earlier this year.

Finally, I must, again, give special thanks to our sponsors, whose generosity, has enabled us to present this event at a reasonable cost to the delegates.

Please do not forget to complete your assessment of the conference before leaving, and indicate how you think it can be improved next time. All the members of the Conference Committee hope that you find the event to be both worthwhile and enjoyable and we look forward to welcoming you to City University London.

Professor Ian Smith
Conference Chairman
School of Engineering and Mathematical Sciences,
City University London, UK

KEYNOTE PAPERS

What does the future hold for compressor manufacturers?

A B Pearson
Star Refrigeration Ltd, UK

ABSTRACT

This paper explores some future opportunities in refrigeration for compressor designers and manufacturers. Some of the most interesting concepts are in expanders and in particular how to integrate them into systems, how to target suitable applications and how to convince non-technicians that this is necessary and worthwhile. In some cases the compressor may hold the key to a problem experienced in another part of the system; at other times the solution to a compressor design challenge may be found by making changes elsewhere.

1   INTRODUCTION

Industrial refrigeration covers a wide variety of applications and sizes of system. The United Nations report on Refrigeration Technical Options (1) takes the range of sizes for industrial systems as 10 kW to 10 MW of refrigerating effect, at cooling temperatures ranging from -50 °C to +20 °C, with the added criterion that failure of the cooling system would jeoparise the operation of the facility that it serves. For example an office air-conditioning system may not be essential to continued operation but the cooling plant serving a data centre is "mission critical". Commercial refrigeration covers some of the same ground, usually in a temperature band from -30 °C to +5 °C and a capacity range from 5 kW to 500 kW in point-of-sale facilities, for example supermarkets, mini-markets and local shops. The compressors that currently serve these market sectors are usually of the piston, screw or scroll type with suction swept volumes from about 1 m^3 hr^{-1} up to 10,000 m^3 hr^{-1}. They are most often direct drive, with the small to medium sized ones of a "semi-hermetic" construction. The drive is usually a 2-pole, 4-pole or 6-pole motor (2950 rpm, 1450 rpm or 975 rpm on a 50 Hz supply) but there is a growing trend in variable speed drive, either using a frequency inverter on the direct drive motor or with a permanent magnet motor. The working fluid is typically a halogenated hydrocarbon, with an increase in the use of hydrocarbons in small systems, carbon dioxide in small to medium-sized systems and ammonia in medium to large-sized systems.

Much of the development effort in refrigeration compressors over the last twenty years has been centred on the transition from chlorinated hydrocarbons. This imperative has to an extent diverted attention away from more radical forms of development, although there has been a flurry of recent innovation in compressor type which bodes well for the future. For example Orosz et al (2) describe a novel rotary compressor based on an oscillating spool, Teh and Ooi (3) show a form of rotary vane compressor where the cylinder casing rotates eccentrically with the rotor thereby creating a variable volume, and Wissink (4) uses a torsional effect on a free piston compressor to improve efficiency. Other researchers have applied improvements to more long-established technologies, for example the oil-flooded scroll compressor in Bell et al (5), the application of permanent magnet motors to swing compressors in Sekiguchi et al (6) and the implementation of a water-cooled hermetic motor to an ammonia reciprocating compressor in Boone (7) go well beyond the incremental refinement of existing designs.

Future development needs will be driven by more than the transition from chlorinated hydrocarbons. Three interlinked drivers were identified by the UK Government report "The Future of Food and Farming" under the Foresight (8) programme. They are population growth/migration, energy demand and food security. By 2050 the world population will have increased to about 9.3 Billion, of which about 75% will live in cities according to the Institution of Mechanical Engineers (9) – this means that the population of cities by then will match the total population at the present time; more than double the present population of cities. This rapid increase will place unprecedented demands on energy supply and the food chain. Resultant threats include uncontrollable climate change, energy price volatility and food deficiencies; meaning that there may be sufficient food but insufficient nutritional value, or in the wrong place or at the wrong price.

Converting this overview into a set of priorities for the refrigeration world is not easy. Pearson (10) outlines some of the implications. Transition from high global warming potential (high GWP) working fluids must be achieved without causing increased energy or water consumption. Investment decisions must be based on life cycle climate performance not lowest capital cost, and equipment must be easy to maintain in peak operating condition without skilled intervention.

2   TRENDS IN REFRIGERATION

Vapour compression accounts for the great majority of refrigeration systems in the world, using variants of the Perkins cycle of compression, condensation, expansion and evaporation. The compressor's role is to raise the pressure of dry gas (usually superheated gas) from a pressure at which the working fluid evaporates from liquid when heat flows to it from the surroundings. The gas is compressed to a higher pressure which is sufficient to allow heat rejection to ambient (or to some process which derives benefit from the heat input) either by desuperheating and condensing the gas back to liquid or, in a few cases, by the removal of sensible heat from a gas in the supercritical state, for example in transcritical CO_2 systems. The working fluids are generally stable chemicals or blends of chemicals which have been mixed in order to give more appropriate properties for a given operating condition. Most compressors are oil lubricated and the oil is used for a number of ancillary functions in addition to lubricating the moving parts in bearings, cylinders, vanes, volutes or rotors. Oil is used to seal the compression path and improve efficiency, it is used as hydraulic fluid to drive actuated components such as capacity control gear, it feeds the drive shaft seal keeping it lubricated and cool and it is used as a heat transfer fluid to cool the compression process. The oil might be miscible with the working fluid, which poses some challenges in the compressor and lubrication circuit, but keeps life simple in the evaporator, or it may be immiscible which gives more stable oil condition to the compressor but makes return from the low side of the system more difficult. In a few tough cases, for example high temperature heat pumps using ammonia with polyalfaolefin or hydrocracked mineral oil, the oil may be miscible in the bearings and lubricant circuit, but immiscible in the evaporator – the worst of both worlds.

Recent developments in compressors have included various efficiency improvements. In piston compressors this has been achieved through material selection, valve design and segregation of high and low temperatures as described by Bon (11). In screw compressors optimisation of screw profile and orientation of the economiser and oil injection ports have been the most common developments. In both types of compressor the improvements are incremental as the basic designs are by now very well understood. Other developments have centred on raising the maximum speed, lowering the minimum speed and ensuring that the machine operates across the full range without excessive resonance. Screw compressors running up to 6,000 rpm and piston compressors up to 3,000 rpm have been recently introduced in larger sizes than previously seen at those speeds and the availability of cost-effective inverters in larger sizes have enabled efficient part-load operation to be achieved.

A medium-sized centrifugal compressor designed for HFC-134a was introduced about ten years ago by Conry (12) and has recently been redesigned for the unsaturated HFC-1234ze(E) as described by Pearson (13). Similar compressors are now on offer from several manufacturers, with capacities at water chilling conditions ranging from 200 kW of cooling up to 2,000 kW. These machines use either magnetic or gas bearings to achieve oil-free operation and usually have the speed control electronics integrated into the compressor assembly. This offers the possibility of far greater monitoring and diagnostic capability than is available with more traditional compressor types. One manufacturer of small piston compressors has followed suit and included a diagnostic module with their compressor, but the full capability of this package is not yet in widespread use.

The term "Not in kind refrigeration" is used to capture a wide range of technologies that offer an alternative to vapour compression. These include magneto-caloric refrigeration, thermoelectric (Peltier effect) and thermoacoustic as well as various forms of absorption and adsorption. Apart from the traditional absorption systems used in heat-powered chillers and camping fridges the only one of these technologies to be used in mainstream commercial products is the Peltier effect which is used for drinks coolers and small portable fridges. Absorption (where the refrigerant vapour is absorbed into a liquid and pumped to high pressure) has been used for over a century in industrial systems. Adsorption (where the vapour is adsorbed into a solid material) has gained popularity more recently and is not as widely commercialised. It seems to be less suitable for application to large cooling loads, so it is perhaps less of a threat to compressor manufacture, at least in the industrial and commercial sectors. The thermophysical effects are unlikely to be scaled up to industrial size due to low inherent efficiency and high capital cost. Other gas cycles such as the Stirling cycle and the Brayton cycle offer opportunities for manufacturers, but are unlikely to be used for industrial systems. In the Stirling cycle the heating and cooling occurs on opposite ends of the "engine", so a secondary fluid would be required to transfer the cooling capability to the heat source (in the freezer or cold room). The cost of incorporating suitable heat exchangers on an industrial scale into the Stirling engine would be extremely high; it would be more apropriate to think in terms of heat exchangers with Stirling engines built into them. The Brayton cycle (also often known as air cycle refrigeration) requires a compressor and expander combination, frequently done in the style of a turbocharger. These devices are potentially cheap to make, but they require relatively expensive and bulky air to air heat exchangers to make the efficiency anywhere close to acceptable and they are most suited to applications where both the heat source and heat sink are served by a large temperature change which can be arranged in counterflow to the airstream. This is also difficult and expensive and does not fit the normal requirement for refrigeration, where the product temperature is to be kept at as steady a temperature as possible.

It therefore seems probable that the not-in-kind technologies do not present any threat to the manufacturers of compressors for industrial refrigeration. Indeed one recent development, sometimes called the hybrid cycle offers an intriguing crossover between the two camps. A two component solution, usually comprising water and ammonia, is used in an absorption system with generator and desorber, but the evaporator outlet gas is compressed in a typical refrigeration compressor, while the weak solution is pumped to high pressure and then mixed with the compressor outlet. This system is attractive for heatpumps as it allows high temperatures to be reached in the heat sink circuit, but without the high pressures that would be required in a Perkins cycle ammonia heat pump. For example to heat water to 90 °C requires an ammonia discharge pressure of about 5 MPa, whereas the hybrid cycle would operate at about half that pressure, well within the range of a standard machine. A disadvantage of the hybrid compression-absorption cycle is that, like the air cycle, there is a large temperature change in the heat source and sink as the cycle progresses, so to take full advantage requires an unusual heat source profile and an expensive pure counterflow evaporator.

3   ADDITIONAL CHALLENGES

Addressing the drivers identified in the Foresight report will require more than just incremental change to existing technology. It is useful to take a step back from current technology and consider what we would ideally like a compressor to be (or not to be)! Systems must use less energy and less water, but must be highly reliable without requiring specialist intervention. To achieve this we require compressors to achieve more complex tasks, but to be simpler than current technology. They should have higher efficiency but be lower cost. They should be integrated into the system operation rather than being a stand-alone component.

Oil free operation has been achieved in the medium sized centrifugal compressors described earlier, but they are presently restricted to a few working fluids and only for operation in chill conditions (evaporating temperature above -10 °C). Piston, screw and scroll compressors have been available for many years in other applications, but not for refrigeration. However the problems introduced on the low pressure side of a refrigeration plant by the accumulation of oil, not least the environmental impact of inefficient operation and the hazard associated with oil draining from a live system, mean that providing oil-free refrigeration machines would be a major benefit.

Wet operation is also a desireable goal, since it would deliver benefits not only to the compressor but to the rest of the system too. Traditionally operating with liquid in the suction has been problematic; the liquid may cause damage to valves or cylinder heads and can also wash lubricant away from bearings. It also may reduce the discharge temperature to the point where more liquid forms during the compression process. However liquid in the gas stream would provide at least two of the functions currently provided by oil; sealing and cooling, so may help towards the goal of oil free operation. Liquid refrigerant can also in some cases provide bearing lubrication itself. The challenge would be to get the right amount of liquid in the right place at the right time. It could also increase the suction gas mass flow by ensuring maximal density at the compressor inlet. The benefit to the system of a compressor that is tolerant of some liquid is that evaporator flow control can be made much simpler, evaporating temperatures can be raised (helping to improve efficiency) and suction separation vessels could be made a bit smaller (and cheaper) if the standard of separation were to be relaxed.

Hermetic operation for ammonia compressors would also be a significant factor in facilitating the use of ammonia in a wider range of applications. At present a few machines have been demonstrated, including piston compressors in Germany, screw compressors in Germany and a scroll compressor in Japan, but none of these have been made available to the wider market and have tended to suffer from low efficiency. The recent announcement by Boone (7) of a water cooled piston compressor brings new hope of a significant breakthrough in this field, and the greater use of permanent magnet motors may also offer new ways to reach this goal. Wider use of ammonia is a relatively simple route to higher efficiency systems, particularly if efficient heat exchangers (for example microchannel aluminium panels) are also used.

4   ADDITIONAL OPPORTUNITIES

The expansion process in the Perkins cycle was once described by Lorentzen as "the internal haemhorrhage of the refrigeration process" (14). Expanders have been discussed over many years, but there is a strong disincentive to their adoption, as they are as complicated as a compressor (the most sophisticated mechanical component in the cycle) but replace a component that may be as rudimentary as a small orifice plate and is never more complicated than a ballcock. There needs to be a strong imperative to install the complexity, cost, risk and maintenance overhead inherent in an expander. The means to employ the recovered work is also not well defined. It could be converted to electric power in an alternator, but then there is the question of how to connect to the mains supply; the safety devices to ensure synchronisation and to protect the expander from a sudden loss of mains connection are expensive but necessary. Alternatively the expander could be directly connected to the compressor shaft, either within a single "compander" or on the opposite end of a double shaft motor so that the expander work output reduces the motor power input. The latter arrangement is more readily achieved in the short term but rather goes against the previously described goal of achieving hermetic operation. The compounded machine (either two devices on a single shaft within a common housing or else with both compression and expansion occurring on a common device such as the "expressor" described, for example, by Hansen et al (15) and Brasz (16).

In these compound machines the compression and expansion processes take place in the same location, which does not suit the traditional arrangement of many large refrigeration systems where the compressors are typically in a machinery room and the expansion takes place at the surge drum, which might be at the other side of the machinery room, or outdoors beside the condenser, or out on the plant, close by the evaporators. This presents a further problem; large plants typically have multiple compressors and a requirement to function at low part load for much of the time. A fixed capacity expander on the back of one of the compressors will not match the part load capacity of the whole plant, and may not run at all if its partner compressor is switched off.

One solution to the part load problem would be to pipe high pressure liquid (perhaps highly subcooled to avoid flash gas in the liquid line) out to the multiple evaporators on the plant and have an expressor at each evaporator, sized to match the load requirement of that unit alone. The energy recovered from the pressure reduction of the subcooled liquid would not be enough to raise the evaporator outlet gas to condensing pressure, but it would get part way there, thus reducing the power requirement of the main compressors in the machinery room. The expressor would only run when the evaporator control called for cooling, so part load would be managed by switching individual units on and off. It ought to be possible to deliver sufficient energy from subcooled high pressure liquid to raise the outlet pressure of cold store evaporators to match the suction pressure of the high stage compressors serving chill rooms, thus removing booster compressors from a two-stage system. It would also be possible in this way to avoid the wasteful practice of running cold store coolers at very low evaporating pressure when freezers are in operation.

It is likely that expanders will either be two-phase devices, in which case there are a number of possible platforms (including all of the compressor variants mentioned previously) or else they will draw in subcooled liquid which produces little or no flash gas as it expands. These devices could not be positive displacement machines as there is very little volume change to drive them. The type of expander required to power an organic Rankine cycle power generation system is also different, as it operates in the superheated gas area of the pressure-enthalpy map. Expanders can therefore be categorised according to whether they have liquid, vapour or both at the inlet and whether they have liquid, vapour or both at the outlet. The most attractive opportunities in both refrigeration and power generation lie in the areas that require devices handling both liquid and vapour. If wet expansion is necessary then perhaps wet compression should be revisited at the same time, although the challenge of incompressible liquid in the machine is greater.

A further opportunity for compressor manufacturers lies in the chance to pump heat from a low temperature to a higher one. Of course all vapour compression refrigeration systems do this, but the output at the high temperature end is usually called "waste heat" and is literally thrown away. In future, as the energy demand outlined in the Foresight report begins to bite, venting heat may become as unacceptable as venting refrigerant. There is a higher demand for heating than for refrigeration in any developed society so the wastefulness inherent in our failure to connect the two processes is unsustainable. The main barrier to this connected approach is logistical; the heating and cooling demands are not coincident (in time or space). As we build new cities over the next thirty years in order to double the urban population they will be less haphazard and more organised than our current urban sprawls which have evolved over centuries to their current size. Industrial parks can contain heating and cooling utility loops with smart meters to enable individual companies operating within the park to draw off or feed in heating and cooling as their processes require it. Provided there is a mix of industry on the estate the loads will be reasonably balanced, so that the data centre can reject its heat into a heating circuit which serves an office block or hospital, or an industrial laundry running a heat pump on its water supply can sell cooling to a drinks bottler. For a variety of reasons the most suitable fluids for use in these types of system are ammonia and carbon dioxide, but they both require compressors capable of running higher pressures than we have been used to in the refrigeration world in the past. 75 bar is the design benchmark for ammonia and sub-critical carbon dioxide, and if the carbon dioxide system discharges heat from supercritical gas a design pressure of 100 bar to 140 bar is likely to be needed.

5 NEW TECHNOLOGIES

5.1 Sensors
The development of raw computing power delivered by the continued expression of Moore's Law has already resulted in huge leaps in sensing capability in recent years. In future it will be feasible to provide sufficient pressure, temperature and acceleration probes on a dynamic machine to monitor its condition cost effectively in real time. This will enable more sophisticated compressor control to be implemented, perhaps including the injection of oil at specific points of the compression cycle in measured doses that are the bare minimum to achieve the required effect, rather than flooding the compression chamber as we do at present. They might also permit the use of measured doses of liquid refrigerant in just the right spots for cooling, sealing and lubricating, eliminating the need for oil

completely. Greater use of powerful magnets, perhaps even in ambient superconductors, will allow far greater use of actuators within machines, enabling more of the parameters affecting the compression process to be carefully controlled, in the manner that combustion is now more carefully controlled in a car engine. It will be possible to vary capacity, pressure ratio and discharge temperature by modulating the size and shape of the compression chamber, whether it is a piston, screw or scroll machine – or one of the new generation currently called "novel".

There is now a capacitance based sensor on the market that can measure the relative proportions of vapour and liquid in a refrigerant pipe. If the compressor were liquid-tolerant then this sensor would eliminate the need for superheat control of thermostatic expansion valves, thereby raising the suction condition in a typical application by about 5K and improving the system CoP by about 15%. The expansion valve would be set to ensure a 2% - 5% overfeed from the evaporator rather than 6K superheat as is current practice. This would also reduce evaporator size by about 15%, the heat transfer surface typically required to provide the superheat.

5.2 Diagnostics
The proliferation of sensors will enable an entirely new approach to compressor maintenance, based upon real time measurement of the condition of all the moving parts in the machine. Significantly improved software will be required to maximise the benefit, and as Ron Conry said of the Turbocor compressor, it will be better to think of it as a computer that pumps gas rather than as a compressor with a lot of controls built in. The compressor will be able to monitor its performance and advise when preventive remedial action is required not only to avoid expensive breakdowns but more importantly to keep it running at peak efficiency. In effect the compressor becomes its own calorimeter, enabling system efficiency to be accurately calculated in real time based on a small number of refrigerant state measurements. Real time sensing could also be used to greatly reduce vibration levels by fine-tuning compressor ports to match the cylinder and discharge manifold pressures more accurately.

5.3 Materials
Plastics are already being widely used in some elements of compressors, including valve gear, piston rings and seals. They will become increasingly used in structural elements such as housings, cylinder linings and ports, including "memory materials" which change shape at different temperatures and could be used to ensure much tighter fits (and hence better efficiency without oil) over a wider temperature range. They could also make compressors far lighter, revolutionising the way in which they are built into systems.

5.4 Production techniques
When plastic materials are more widely adopted for the wetted parts of a compressor they will enable machines to be produced in entirely new ways. For example, intricate interlinked pieces can already be made in 3-D printers that are only as sophisticated as a dot-matrix printer was in the 1980s. They already print in a variety of plastics (in full colour), bronze and stainless steel. As these machines become more refined the quality of their output will improve to the point that pieces can be taken straight from the printer to the assembly line.

5.5 New concepts
3-D printing would also open up the possibility of making novel types of compressor that are not cost effective to machine at present, such as the "cylindrical-cylindrical" single screw compressor (as opposed to the more common "cylindrical-planar" arrangement) described by Heidrich (17). The linear-torsional and spool

compressors mentioned earlier would also benefit from these advanced production techniques, as would more traditional scroll compressors.

6   JUSTIFICATION FOR THE CHANGE

The main imperative for making this change will be energy efficiency, driven in part by increased energy cost and cost volatility. Financial assessment will become more realistic, requiring an assessment of full life energy cost that factors in the balance between cost of maintenance and cost of performance degradation. Life cycle costing will not assume as-new efficiency for the duration. The winning technology will therefore be the one that does maintain its "out-the-box" performance and does so without significant manual intervention, but also is seen to do so and therefore can validate the energy cost savings claimed for it. The level of instrumentation, engine management software and diagnostics found under the bonnet of an average family car would almost be enough to enable this to be done today, so it is not such a big leap to foresee it in refrigeration compressors in ten or twenty years time. Those that do not provide this level of provable performance will not sell, no matter how cheap they are.

7   CONCLUSIONS

There are exciting times ahead for compressor manufacturers. A demand for higher operating pressures and temperatures to service the needs of a renewed heat pump market, coupled with a requirement for low-maintenance systems and energy efficiency will drive the development of oil-free, liquid tolerant, hermetically driven compressors and expanders (sometimes combined into a single unit). The continued advances expected in sensor technology will enable a raft of energy saving control and monitoring techniques that revolutionise the way that we interact with these machines, enabling them to report on system efficiency and advise precise intervention intervals.

All this needs to be done at a price that is significantly lower than the present day and in a manner that enables the compressor manufacturers to make a healthy profit from their expertise.

REFERENCE LIST

(1) UNEP, "Assessment Report of the Refrigeration, Air-Conditioning and Heat Pumps Technical Options Committee", Nairobi, 2010
(2) Orosz, J., Kemp, G., Bradshaw, C. and Groll, E., Performance and Operating characteristics of a Novel Rotating Spool Compressor, International Compressor Engineering Conference, Purdue, 2012
(3) Teh, Y.L., Ooi, K.T., Analysis of internal leakage across radial clearance in the improved revolving vane (RV-i) compressor, International Compressor Engineering Conference, Purdue, 2008
(4) Wissink, E., Dedicated Compressor Technology for a next generation domestic heat pump – free piston with oil free torsion drive, GL2012 IIR Conference, Delft, 2012
(5) Bell, I., Groll, E., Braun, J., Horton, T., Experimental Testing of Oil-Flooded Hermetic Scroll Compressor, International Compressor Engineering Conference, Purdue, 2012
(6) Sekiguchi, T., Development of Lightweight and High Efficiency Swing Type Compressor using New Interior Permanent Magnet Synchronous Motor, International Compressor Engineering Conference, Purdue, 2012

(7) Boone, J., Ammonia Chillers in different industrial plants in Switzerland, International Institute of Refrigeration conference "Ammonia Refrigeration Technology", Ohrid, 2013
(8) Foresight, The Future of Food and Farming (2011) Final Project Report. The Government Office for Science, London, 2011.
(9) Institution of Mechanical Engineers "Population: One planet, too many people?" London, 2011
(10) Pearson, A., The role of refrigeration in the future of food and farming, 42nd Congress on HVAC&R, KGH, Belgrade, 2011
(11) Bon, G., New high efficiency piston compressors for ammonia, GL2012 IIR Conference, Delft, 2012
(12) Conry, R., A brief overview of the Turbocor compressor – the Road to Discovery, Proc Inst Ref, London, 2009
(13) Pearson, A., R-1234ze for variable speed centrifugal chillers, Proc Inst Ref, London, 2013
(14) Lorentzen, G., Throttling, the internal haemhorrhage of the refrigeration process, Proc Inst Ref, London, 1983
(15) Hansen, T., Smith, I., Stosic, N., Combined Industiral Cooling and Heating with Transcritical CO_2 Heat Pumps Utilising the Work of Expansion, GL2004 IIR Conference, Glasgow, 2004
(16) Brasz, J.J., Single Rotor Expressor as Two-Phase Flow Throttle Valve Replacement, US Patent, n.006185956, 2001
(17) Heidrich, F., Water Flooded Single Screw (SSP) Compressor Technology, International Compressor Engineering Conference, Purdue, 1996

Recent research of novel compression concepts for vapor compression heat pumping, air conditioning and refrigeration systems

E A Groll, O Kurtulus
Purdue University, School of Mechanical Engineering, USA

ABSTRACT

The past phase-out of CFC and HCFC refrigerants and the upcoming phase-out of HFC refrigerants, combined with advances in compressor technology, such as reduced noise and vibration and mechanical capacity control, has motivated the development of novel compression concepts for use in vapor compression cycles. This paper provides an overview of several new developments for refrigeration, air-conditioning and heat pumping applications, including three novel compressors called bowtie compressor, rotary spool compressor, and z-compressor. The bowtie compressor offers an integrated method of capacity modulation for use in domestic refrigerators/freezers. It modulates the cooling capacity by changing the piston stroke without changes of the clearance volume for better thermodynamic efficiency. The bowtie compressor received its name due to its two sector-shaped, opposing compression chambers forming a bowtie. The novel rotary spool compressor combines various aspects of rotary and reciprocating devices. Studies to date have shown that the rotary spool compressor can achieve high efficiencies at low manufacturing costs. The z-compressor is a dual-chamber hermetic rotary compressor with opposing upper and lower compression chambers separated by z-shaped blade. The compressor offers significantly lower noise and vibration levels than a conventional rolling piston compressor. In addition to presenting these new compressor types, recent research on linear compressors and diaphragm compressors will also be presented.

1 INTRODUCTION

The HVAC&R industry has been a maturing industry for more than 100 years. In its infancy, prior to the Second World War, there was virtually no air-conditioning in residential homes or automobiles, and for the most part food preservation relied on cooling with ice. Today, refrigeration has become essential to ensure the maintenance of food supplies throughout the world, and air-conditioning is relied on to support the comfort and standard of living to which contemporary society has become accustomed. Various manufacturing industries have also developed an increased dependence on refrigeration technology. The electronics and chemical industries, for example, require this technology for the production of computer chips and for the manufacturing of synthetics. In recent years, there have been significant challenges for the compressor industry mainly because of the need for more efficient components and systems in order to reduce energy consumption but also to address the ever changing landscape of refrigerants and the need to provide reliable products that are in-expensive to manufacture.

For example, in electronics cooling, larger cooling capacities are needed, which cannot be achieved by using air-cooling methods alone. Thus, manufacturers started to look into miniature refrigeration systems to be able to provide increasing cooling demand. To date, there are no suitable compressors available for this application matching the require size constraints and operating conditions.

This paper describes some of the research that has been conducted in the area of compressors to address relevant design parameters which affect the performances of compressors. In particular, the topics of bowtie compressor, rotary spool compressor, z-compressor, linear compressor and diaphragm compressor will be discussed in greater detail.

2   BOWTIE COMPRESSOR

A novel method of capacity modulation of positive displacement compressors has been proposed by the author's research group (1). The new method is called bowtie compressor and modulates the cooling capacity by changing the piston stroke without changes in the clearance volume for better thermodynamic efficiency. The new method is based on a unique off-center-line mechanism so that the piston stroke can be varied without changes in motor rotation. The new compressor concept is shown in Figure 1. The design is based on the Beard-Pennock Variable-Stroke Compressor (2), but includes significant modifications. It may provide an ideal solution to achieve mechanical capacity modulation. The crankshaft rotates counter-clockwise, and the piston rod makes the piston reciprocate axially not linearly. Thus, two compressions take place at the same time. The crankshaft, piston rod, and piston are arranged in such a way that the piston rod is perpendicular to the crankshaft when the piston reaches the top-dead center as shown in the first diagram of Figure 1. The cylinder is placed in the sliding chamber with mechanical springs and pressurized on both ends by the suction and the discharge gases. The cylinder is allowed to slide based on the difference of the two pressure forces. When more capacity is demanded, the difference between the two pressure forces increases. Thus, the force on the right side of the cylinder increases. As a result, the cylinder moves to the left creating a longer stroke which increases the compression volume as shown in the second diagram of Figure 2. It has to be noted that this does not create an extra clearance volume. Thus, the capacity modulation is achieved solely by changing the piston stroke.

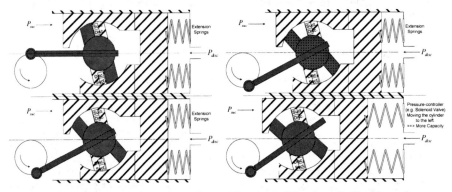

Figure 1: Diagrams of the bowtie compressor design

Figure 2: Diagram indicating change of piston stroke of bowtie compressor

A simulation model of the bowtie compressor was developed to predict the cooling capacity, COP, volumetric efficiency, leakage flow rate, mass flow rate, overall isentropic compressor efficiency, and the compressor power input. The predictions of the compressor mass flow rate and the overall isentropic compressor efficiency as a function of the swept volume are shown in Figure 3 at the operating conditions of ANSI-ASHRAE Standard 23-1993 (condensing temperature of 54.4°C, evaporating temperature of -23.3°C, and compressor suction temperature of 32.2°C) for a bowtie compressor with a 2.5 μm leakage clearance and a ratio of 1.25 of vane radius to vane height.

At these operating conditions, the model of the bowtie compressor predicts a 52% reduction in compressor mass flow rate from 1.5 g/sec to 0.72 g/sec with an approximately 20% decrease in overall isentropic compressor efficiency from 60 to 48%, as shown in Figure 4. To achieve the capacity modulation, the swept volume decreases from 4.70 cm³ to 3.04 cm³. One reason for the decrease in overall isentropic compressor efficiency is the fact that the motor runs off its optimum efficiency at the low capacity conditions.

The pressure-volume diagrams for five swept volumes are shown in Figure 4. The figure indicates that the swept volume increases with an increase of the control length, x, from the reference point, where the minimum swept volume occurs. Figure 4 also shows that the over-compression during the discharge process gradually decreases as the swept volume decreases. This is mainly because a decrease in swept volume results in a lower refrigerant mass flow rate, which in turn creates less flow restriction through the same discharge port.

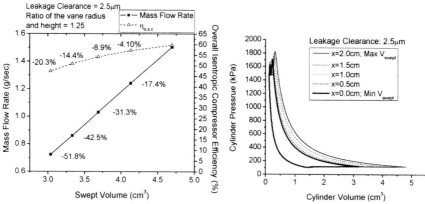

Figure 3: Diagrams of the bowtie compressor design

Figure 4: Compression cycles in a pressure-volume diagram

The results indicate that it is feasible to reduce the cooling capacity by half by using an on-off solenoid valve. Moreover, the cooling capacity varies linearly with changes in control length. This implies that a continuous cooling capacity modulation is possible when a linear controller is used instead of the solenoid valve.

3   ROTARY SPOOL COMPRESSOR

The rotating spool compressor is a novel rotary compressor mechanism most similar to the sliding vane compressor. Primary differences are described by Kemp et al. (3), (4), and include three key differences from a sliding vane compressor.
- The vane is constrained by means of an eccentric cam allowing its distal end to be held in very close proximity to the housing bore (typically less than 0.30 mm) while never contacting the bore.

- The rotor has affixed endplates that rotate with the central hub and vane forming a rotating spool.
- The practical use of dynamic sealing elements to minimize leakage between the suction and compression pockets as well as between the process pockets and the compressor containment.

These differences are shown in Figure 5 which presents a cutaway view of a rotating spool compressor with the key geometric features highlighted.

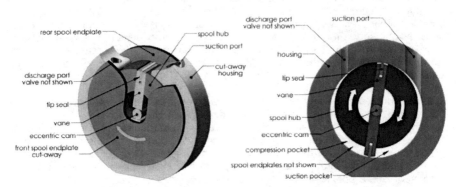

Figure 5: Cutaway view of rotating spool compressor mechanism with key components highlighted

More demand for higher efficiency components has resulted in a renewed interest in detailed compressor modeling to predict the performance of novel compressors. A recent approach to compressor modeling, called the comprehensive approach, has provided a complete analysis of positive displacement compressors. Comprehensive model approach is utilized to predict the performance of rotating spool compressor. This approach has been presented in detail previously for other positive displacement compressors (5, 6, 7, and 8). The approach relies on a governing set of equations, which are derived from a mass and energy balance within the working volumes of a positive displacement compressor. A detailed model is available in Kemp et al., (9). The model results were compared with experimental data from the prototype scroll compressor, which has a displaced volume of 23.9 cm^3. The compressor was operated at rotational speeds from 1750 to 3250 RPM and suction conditions at 905 kPa and 11 °C pressure and superheat, respectively. R410A was used as the working fluid. The compressor discharge pressure was varied to achieve pressure ratios between 2.1 and 2.9.

Figure 6 shows volumetric efficiencies of the rotary spool compressor prototype compared to other compressors in market. Peak volumetric efficiency at a pressure ratio of 2.5 is 2% less than the scroll compressor while loss per unit of pressure is 3.9%. When compared to the rolling piston type compressor, the slope of the pressure loss is comparable while the volumetric efficiency is 4 points higher. Figure 7 shows a comparison of the overall isentropic efficiency of the same compressors that were compared in Figure 6 for the volumetric efficiency. The absolute values for the efficiency show that the rotary spool prototype compressor is 7.5 point higher than a rotary compressor and 5 points lower than a scroll compressor at a pressure ratio of 2.5.

The spool compressor is a novel device that combines various design attributes of other technologies currently in production to achieve high efficiencies while maintaining low production costs. Orosz et al. (10) present a work which characterizes the performance of the spool compressor, identifies the areas of loss and, identifies the benefits as they relate to the current market demands. This is accomplished through the iteration of prototype spool compressor designs, which explore a variety of geometric configurations and operating conditions. A dimensionless number called the Zsoro number is presented by Orosz et al. (10), which relates the effective frictional load to the total displaced volume of the compressor. It should be noted that a Zsoro number of less than two provides more reasonable results and can serve as a design indicator for a compressor that can be further improved by optimizing the other performance related features. Achieving a Zsoro number of less than two is limited by two factors, the ability to manufacture the compressor L/D larger and the ability of the longer internal tip seal to function properly.

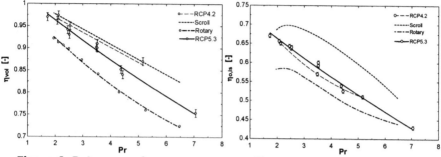

Figure 6: Rotary spool compressor prototypes volumetric efficiency vs. commercial technologies

Figure 7: Rotary spool overall isentropic efficiency vs. other technologies

4　Z-COMPRESSOR

A novel rotary dual-chamber hermetic compressor, termed the z-compressor, has been developed for use with the refrigerant R410A. In one rotation of the shaft, the compressor undergoes two compression processes shifted by one-half revolution. Figure 8 shows a cutaway view showing the upper and lower suction and compression chambers of the z-compressor. For both upper and lower sets of chambers, separation between the suction and compression chambers is provided by a vane on one side and the section of the blade that is in contact with the upper or lower bearing on the other side. The position of the lower bearing (sub-bearing) is shown in Figure 9 and the upper bearing is positioned on the other side of the shaft. The refrigerant enters the suction chamber through a port that faces both the upper and lower suction chambers and does not require any valves. A reed-style discharge valve is employed at the discharge port and opens when the pressure within the compression chamber exceeds the pressure within the discharge port. Discharge gases from the compression chambers enter the shell of the compressor. A mathematical description of the geometry of this compressor has been presented by Jovane et al. (11).

A study was conducted to understand the factors that affect the energy performance of the z-compressor and to identify design improvements. For this purpose, a detailed model was developed. The model predictions were compared to experimental results. The model predicts the mass flow rate and the power input of the compressor within 7% and the discharge temperature within 5 °C of the measured values. The model results show that for suction pressures ranging from 0.6 to 1MPa and discharge pressures from 2 to 3.4 MPa, the volumetric efficiency of the z-compressor varies between 0.84 and 0.89 while its overall isentropic efficiency varies between 0.53 and 0.59 as reported by Jovane (12).

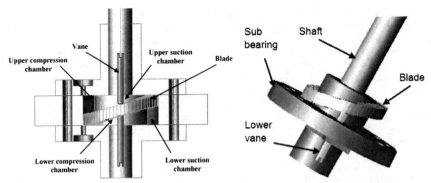

Figure 8: Cutaway view of the z-compressor

Figure 9: Shaft, blade, vane, and sub-bearing of the z-compressor

There are three dimensions of the z-compressor that affect the volume of the chambers of the compressor. These dimensions are the maximum chamber height (h_b), the inner blade radius (r_i), and the outer blade radius (r_o) (as indicated in Figure 10 and Figure 11). The leakage and frictional interactions of the blade-cylinder contact are proportional to the outer radius of the blade. The first strategy to improve the efficiency of the z-compressor consisted of changing the ratio between the outer diameter of the blade and the maximum chamber height while keeping the inner blade radius constant. This is done in such a way that the swept volume of the chambers is kept constant. Figure 12 shows the volumetric efficiency of the z-compressor as a function of the height-to-diameter ratio (h/D). Figure 12 show that there is a slight increase in the volumetric efficiency up to an h/d ratio between 0.30 and 0.35, but the improvement is not significant. In addition, the leakage interactions are not decreased from an overall point of view. This can be observed in Figure 13, where the leakage interactions of the blade-cylinder contact are shown along with the leakage interactions associated with the vane. It can be seen from the figure that the interactions associated with the vane increase as the leakage interactions of the blade-cylinder contact decrease. This is due to the increase in the maximum chamber height that affects the area of leakage associated with the vane.

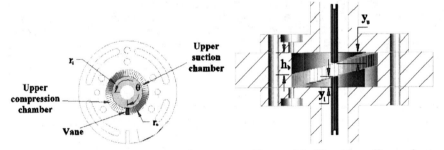

Figure 10: Blade dimensions

Figure 11: Chamber dimensions

Clearly, variations in the h/D ratio affect the leakage flow paths. However, variations in the h/D ratio also affect the frictional interactions of the blade-cylinder contact. The decrease in the frictional interactions of these two contacts is reflected in the overall isentropic efficiency where the improvements are slightly greater than the improvement in the volumetric efficiency. The isentropic efficiency follows the same trend as the volumetric efficiency but the improvement in frictional interactions result in an approximately 2% improvement in the isentropic efficiency. This is four times the improvement of the volumetric efficiency.

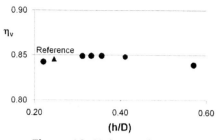
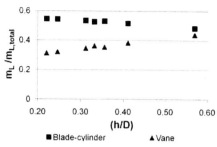

Figure 12: Volumetric efficiency as a function of the h/D ratio

Figure 13: Leakage interactions as a function of the h/D ratio

Although the improvement in isentropic efficiency is more significant than the improvement in volumetric efficiency, this efficiency still remains below the efficiency reported for conventional rolling piston rotary compressors.

5 LINEAR COMPRESSOR

A comprehensive simulation model of a miniature-scale linear compressor was recently developed by Bradshaw et al. (5). Figure 14 depicts the major components and design parameters of a linear compressor. The geometry of the piston is directly related to both the friction and leakage of a compressor. Therefore, for a fixed displacement volume, some piston diameter and stroke combinations will provide higher efficiencies than others. The impact of changes to these parameters proves useful when designing a linear compressor, and warrants further investigation.

A method for calculating the resonant frequency of a linear compressor has been developed by Bradshaw et al. (5). An approach to numerical control the piston stroke is also provided that ensures compressor operation at the desired stroke. A series of sensitivity studies have been carried out using the simulation model, which highlight the sensitivity in performance with respect to the leakage gap and eccentricity as well as the piston geometry (13).

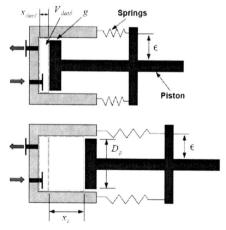

Figure 14: Schematic diagram of linear compressor at Top Dead Center (TDC, top) and Bottom Dead Center (BDC, bottom) with primary linear compressor components and design parameters highlighted

The model presented in Bradshaw et al. (5) utilizes a two-degree-of-freedom system to describe the piston motion. These two degrees of freedom are the desired piston translation and the undesired rotation of the piston within the cylinder. However, it has been observed that the resonant frequency of the linear compressor is predicted accurately using a one degree of freedom approach (14; 15). The approach developed by Bradshaw et al. (5) estimates the damping and stiffness of the system when the stroke is at a desired condition. By linearizing these equations, the dependence on instantaneous stroke is removed. Therefore, the resonant frequency can be calculated external to the comprehensive linear compressor model.

The study highlighted four parameters that significantly impact the overall performance metrics of the compressor: motor efficiency, η_{motor}, dry friction coefficient, f, spring eccentricity, ε, and leakage gap g. For simplicity, changes to η_{motor} are not considered, as changes in this parameter would provide little insight into compressor design; the higher the linear motor efficiency, the better is the compressor overall performance. Thus, the efficiency is fixed at 90% for the rest of this work. The operating conditions are set at 20 °C, 40 °C, and 5 °C for evaporating, condensing, and superheat temperatures, respectively. This operating condition represents a typical electronics cooling application. In addition, the clearance gap between the piston at TDC and the valve plate, x_{dead}, as illustrated in Figure 14, is fixed at 3 mm.

The dry friction coefficient and spring eccentricity both relate to friction between the piston and cylinder. These two parameters are coupled by the normal force acting on the piston. For eccentricity, the range of values considered spans from 0.1 cm to 0.9 cm, where the upper limit represents an extreme situation. The leakage gap ranged from 1 μm to 23 μm, which spans a realistic range of values for compressor leakage gaps (1, 10). In addition, the stroke is fixed at 2.54 cm (1 inch), as presented by Bradshaw et al. (13)

The geometry of a compressor piston can have a large impact on the overall performance of the device (1, 16). The impact of scaling of the linear compressor is also of interest to the goal of miniaturizing the compression technology. To explore the impact of scaling, three compressor displacement volumes of 2, 3, and 6 cm^3 are examined. For each displaced volume the piston diameter is varied from 0.8 to 1.7 cm. Fixing the compressor displacement volume and varying the piston diameter requires modification of the compressor stroke. Therefore, the stroke-to-diameter ratio is investigated for three displacement volumes.

The leakage area is determined by the piston diameter, with larger piston diameters corresponding to larger leakage areas. When the stroke-to-diameter ratio is small, piston period is also small, but the leakage area is large. When the stroke-to-diameter ratio is large, on the other hand, the piston period is large but the leakage area is small. This trade-off is largely dominated by piston period but does exhibit an optimum at a stroke-to-diameter ratio of between 3 and 4. While the leakage losses tend to increase sharply after a stroke-to- diameter ratio of between 3 and 4, this does not have a corresponding effect that is apparent in the volumetric efficiency. The reason for this is that the magnitude of the leakage losses is relatively small even at the maximum. Therefore, there is little impact on the volumetric efficiency.

The sensitivity studies conducted show that the linear compressor is highly sensitive to changes in the leakage gap between the piston and cylinder as well as the spring eccentricity; both parameters should be minimized for optimal performance. Therefore, it is important to quantify and control these parameters in any compressor design that is mass-produced to maximize performance.

The recent research study illustrates the ability of the linear compressor to be readily scaled to smaller capacities. The ability to handle large amounts of dead volume without performance degradation could also allow this technology to be used to control the capacity of the refrigeration system. Capacity control is a critical need for high-performance refrigeration systems in electronics cooling and should be further investigated.

6   DIAPHRAGM COMPRESSOR

An electrostatically actuated diaphragm compressor offers promise for the miniature cooling system application because of its potential for high efficiency, compactness and scalability. The diaphragm compressor, schematically represented in Figure 15, consists of a flexible circular diaphragm clamped at its circumference, enclosed by two identical halves of a conformal chamber. Gas is admitted into the chamber through the suction ports along the circumference; while the discharge valves control discharge flow and pressure rise. Metallic electrode layers are deposited on the diaphragm and on the chamber surfaces and dielectric layers are deposited on the top of the metallic electrodes to prevent electrical shorting when the diaphragm touches the chamber surface. The principle of operation of the diaphragm compressor is based on progressive electrostatic zipping of the diaphragm towards the chamber when a DC potential difference is applied across them. An analytical model for such a diaphragm compressor was developed by the Sathe et al., (17), and validated against results from the literature as well as against experimental results conducted using a custom test setup. The diaphragm compressor model and comparisons were limited to a specific set of geometric parameters, and the effects of variation of the chamber dimensions on the overall performance of the diaphragm compressor were not considered. For a potential application in electronics cooling, an optimized design of the diaphragm compressor that offers the best performance at the lowest compression power is desired.

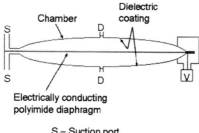

S – Suction port
D – Discharge port and valve
V – DC voltage power supply

Figure 15: Schematic of an electrostatically actuated diaphragm compressor

While the dielectric constant of the dielectric layer on the chamber surface limits the maximum pull-down voltage that can be imposed before dielectric breakdown, the maximum pressure rise achieved in the chamber is limited by the geometric and elastic properties of the diaphragm. Volume flow rate are also related to chamber geometry.

Sathe et al. (18) conducted a study to understand the behavior of diaphragm compressors. A novel diaphragm compressor model was developed. Figure 16 shows where the chamber pressure rise and refrigerant flow rate are plotted as a function of the chamber radius. Any design strategies focused on enhancing the pressure rise would result in a decrease in volume flow rate and vice versa. Since these two parameters are independent of each other, a design optimization that would maximize the performance of the compressor is needed.

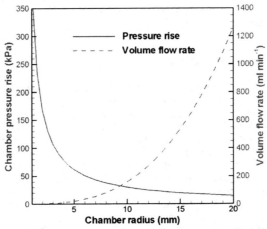

Figure 16: Chamber pressure rise and refrigerant volume flow rate versus chamber radius

As stated above, the most important performance parameters of the compressor, pressure rise and volume flow rate are physically independent of each other. To account for both of these parameters, the compression power has been chosen as an optimization variable. While it cannot be considered as the actual power input to the compressor, it is argued that minimizing the compression power will make the compressor thermodynamically more efficient.

For an electrostatically actuated diaphragm compression, another important design parameter, in addition to the pressure rise and the volume flow rate, is the required diaphragm pull-down voltage. Higher operating voltages are undesirable because of the required increase in the dielectric layer thickness. A detailed analytical approach for calculating the pull-down voltage for the given chamber geometry and the chamber pressure rise is described by Sathe et al. (17).

Calculation of the pull-down voltage at radius R_1 shows that the pull-down voltage is 387.7 V, which is 0.3% higher than the pull-down voltage at radius R_2. On the other hand, calculating the thermodynamic compression power at radius R_2 gives 31.8 mW, which is 5.2% higher than that at radius R_1. Hence, radius R_1 (= 8.5 mm) is selected as the optimized chamber radius at which both variables the theoretical power and the pull-down voltage are minimum. For an aspect ratio of 100, the maximum chamber depth is estimated as 85 µm. Design parameters for optimized diaphragm listed in Table 1.

Table 1: Design parameters for the optimized diaphragm compressor

Chamber radius	8.5 mm
Maximum chamber depth	85 µm
Maximum pressure rise	35.6 kPa
Volume flow rate	97 ml min^{-1}
Pull-down voltage	387 V

As an example, to achieve the pressure rise and refrigerant volume flow rate required for an 80 W cooling application, an array of the diaphragm compressor has to be used, where multiple units of diaphragm compressor are arranged in parallel and series to achieve the desired volume flow rate and the desired pressure rise, respectively. Cabuz et al. (19) also proposed a 3-D array of dual diaphragm pumps for enhancing the pumping rate. Similar arrangements for increasing the pressure rise have been proposed by Yoon (20) and by Sathe et al. (18).

According to the study by Sathe et al. (17) for 80 W cooling capacity, 3-D compressor arrays of 172,104 and 126 diaphragm compressors are needed to make up the required pressure rise and volume flow rate using refrigerants R134a, R236fa and R245fa, respectively. The calculated volumes of the diaphragm compressor arrays are compared with the available volume for the compressor which shown in Figure 17. The comparison indicates that it is theoretically possible to fit 3-D compressor arrays using any of these refrigerants within the volume constraints of 32 cm^3.

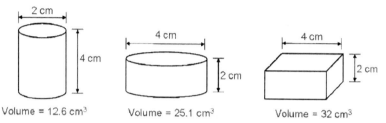

Figure 17: Possible compressor shapes and dimensions for 80 W notebook computer cooling application

While the fabrication complexities and associated cost for such a compressor array have not been considered in this analysis, it is believed that with rapid advances in the silicon micro-fabrication techniques, the diaphragm compressor holds promise.

7   SUMMARY

This paper provided a review of some of the latest developments and research activities that have been conducted with respect to novel compressors for vapor compression refrigeration, air conditioning, and heat pumping systems. The bowtie compressor, rotary spool compressor, Z-compressor, linear compressor and diaphragm compressor have been introduced and the parameters that affect their performance (efficiency, capacity, etc.) are described. In particular, it is outlined which parameters are increasing the compressor performance. In addition, two different compressor technologies for small-scale electronics cooling applications, namely linear compressor and diaphragm compressor, have been discussed in detail and important parameters for design and modeling have been presented.

8   REFERENCES

1. Kim, J.-H., and Groll, E.A., 2007. "Feasibility study of a bowtie compressor with novel capacity modulation," *Int'l J. Refrigeration*, Vol. 30, No. 8, pp. 1427-1438.
2. Pennock, G. R., and Beard, J. E., 1988, "The Beard-Pennock Variable-Stroke Compressor," *Proc. Int'l Compressor Eng. Conf.*, Purdue University, West Lafayette, IN, USA, pp. 599-607.

3. Kemp, G., Garrett, N., Groll, E., 2008. Novel Rotary Spool Compressor Design and Preliminary Prototype Performance. In: Proceedings of the International Compressor Engineering Conference. Purdue University, West Lafayette, IN USA. No. 1328.
4. Kemp, G., Elwood, L., Groll, E., 2010. Evaluation of a Prototype Rotating Spool Compressor in Liquid Flooded Operation. In: Proceedings of the International Compressor Engineering Conference. Purdue University, West Lafayette, IN USA. No. 1389.
5. Bradshaw, C.R., Groll, E.A., Garimella, S.V., 2011, "A comprehensive model of a miniature-scale linear compressor for electronics cooling". Int. J. Refrigeration, 34(1), pp.63-73.
6. Bell, I. 2011. Theoretical and Experimental Analysis of Liquid Flooded Compression in Scroll Compressors. PhD thesis, Purdue University.
7. Mathison, M., Braun, J., Groll, E. 2008. Modeling of a two-stage rotary compressor. HVAC&R Research, 14(5):719-748.
8. Mathison, M. 2011. Modeling and Evaluation of Advanced Compression Techniques for Vapor Compression Equipment. PhD thesis, Purdue University.
9. Kemp, G, Bradshaw, C.R., Orosz, J., Groll, E.A., 2012. "A comprehensive model of a novel rotating spool compressor", 21st International Compressor Conference, Purdue University, West Lafayette, IN, July 16-19, 1142-10 pages.
10. Orosz, J., Kemp, G., Bradshaw, C.R., 2012, "Performance and operating characteristics of a novel rotating spool compressor", 21st International Compressor Conference, Purdue University, West Lafayette, IN, July 16-19, Ref. Num. 1257-9 pages.
11. Jovane, M.E., Braun, J.E., Groll, E.A., and Lee, S.J., 2006, "Theoretical Analysis of a Novel Rotary Compressor", *Proc. of the 18th Int'l Compressor Eng. Conf. at Purdue*, Purdue University, West Lafayette, IN, July 17-20, 9 pages.
12. Jovane, M.E., 2007. "Modeling and Analysis of a Novel Rotary Compressor. PhD Thesis. School of Mechanical Engineering". Purdue University. West Lafayette, Indiana, USA.
13. Bradshaw, C.R., Groll, E.A., Garimella, S.V., 2012, "A Sensitivity Analysis of a Miniature-Scale Linear Compressor for Electronics Cooling Using a Comprehensive Model", 21st International Compressor Conference, Purdue University, West Lafayette, IN, July 16-19, Ref. Num. 1133-11 pages.
14. Pollak, E., Soedel, W., Cohen, R. & Friedlaender, F., 1979. On the resonance and operational behavior of an oscillating electrodynamic compressor. *J. Sound Vib.*, 67, pp.121-33.
15. Cadman, R. & Cohen, R., 1969. Electrodynamic oscillating compressors: part 2 evaluation of specific designs for gas load. *ASME J. Basic Eng.*, December, pp.664-70.
16. Rigola, J., Perez-Segarra, C. & Oliva, A., 2005. "Parametric studies on hermetic reciprocating compressors." Int. J. Refrigeration, 28(2005), pp.253-266.
17. Sathe, A.A., Groll, E.A., Garimella, S.V., 2008. "Analytical model for an electrostatically actuated miniature diaphragm compressor" Journal of Micromechanics and Microengineering, 18 (3), # 035010.
18. Sathe, A., Groll, E.A., Garimella S.V., 2009, "Optimization of electrostatically actuated miniature compressors for electronics cooling", *International Journal of Refrigeration*, Vol 32, Issue 7, Pages 1517-1525.
19. Cabuz, C., Cabuz, E. I., Herb, W. R., Rolfer, T., and Zook, D., 1999. Mesoscopic sampler based on 3D arrays of electrostatically actuated diaphragms. Proc. of the 10th Int. Conf. on Solid-State Sensors and Actuators, Transducers'99, Sendai, Japan. 1890-1891.
20. Yoon, J. S., 2006. Studies on the micro vapor compressor for the application to a miniature refrigeration system. PhD thesis, School of Mechanical and Aerospace Engineering, Seoul National University, South Korea.

Sliding vane rotary compressor technology and energy saving

R Cipollone
University of L'Aquila,
Department of Industrial and Information Engineering and Economy, Italy

ABSTRACT

Energy saving, CO_2 reduction and energy generation from renewable sources represent the three cornerstones of the energetic and environmental commitments of all the Countries in the World.

These three elements are considered to be able to give a quantitative contribution to the sustainability in an industrial environment. Among those mentioned, the most important and the one which represents a driver in many sectors is the limitation of the CO_2 concentration in the atmosphere: most recent data (2013) from NOAA-ESRL set at 395.55 parts per million (ppm) the CO_2 in the atmosphere and the continuous increasing trend will quickly allow the reaching of the 450 ppm level which is considered as a safeguard limit to avoid irreversible environmental and socio economics problems, [1].

Looking at the energy consumption side, energy saving is a key factor. Compressed air production doesn't escape this requirement and, for the compressor manufacturing industry, this can represent an opportunity with great potential benefits.

Compressed air is produced by electrical energy and the consumption accounts as much as 10 % of industrial consumption of electricity, [2,3,4,5,6]. A lower estimate [7] places at 6 % this share but an additional 12 % is estimated to be associated with the commercial and residential markets (portable tools, air pumps, pneumatic heating, ventilation, air conditioning, etc...), so overall compressor needs are estimated equal to 20 % of the industrial electricity needs. Considering that industrial consumption of electricity represents a given share of the overall electrical energy consumption (it depends on the geographical context, social development, industrial level, etc....), with good approximation, compressed air can be associated to the overall electricity consumption and to primary energy consumption too. So, it can be compared with the others energy alternatives: when the data is reliable and correspondent to real situations, actions to promote energy efficiency in compressed air systems can be identified with their real importance and compared with all the others measures.

From many independent studies [2,4,6] the most important energy saving measures are associated to the: (1) reduction of leakages on the distribution lines, (2) a more appropriate compressed air system design, (3) use of adjustable speed drives, (4) waste heat recovery. All these aspects, in a ten year period of operation, weigh 70-75 % of the overall compressed air costs [7,8,9,11,12]: compressor technology is, therefore, a key factor to reduce energy consumption including in it load control, variable speed operation, compressor sizing, etc... A great potential

© The author(s) and/or their employer(s), 2013

saving is associated with leaks, friction pipes, etc… but these actions are downstream of the compressed air production.

After having discussed some issues concerning the future overall energy consumption and Co2 emissions, considering the development of the electricity market in the World in the near future, and overall energy characteristics of existing machines widely used in the compressed air market, the article goes deep inside a specific compressor technology which is represented by the sliding vanes rotary type. Principal processes inside these machines are discussed in the light of the recent scientific literature advancement of a theoretical and experimental nature.

The general idea was that these machines are not so well known and their use is not so widespread: thanks to a deeper scientific interest over the past few years, these compressors have had a notable performance improvement, meaning this technology has a potentially greater industrial role to play than previously thought. Thanks to some specific intrinsic aspects of these machines some energy issues are presented and premium sectors discussed.

INTRODUCTION

In 2011, in spite of the anomalous events (Arab spring, earthquake and tsunami in Japan, etc…) and of the economic and financial crisis which hit Nations in a way never seen till now, the global energy consumption grew by 2.5 % following the historical trend, but well below the 5.1 % seen in 2010. The most part of this growth continues to shift from the OECD to emerging economies, especially in Asia: China alone counted for 71 % of the mentioned growth. So particular attention must be placed in that area when looking at worldwide energy planning.

Fossil fuels still dominate energy consumption, with a market share of 87 %.

Oil's share is 33 % of global energy consumption reaching 88 million barrels/day (b/d), with an annual increase by 0,6 million b/d, or 0.7 % with respect to 2010, in spite of the outages in Libya and elsewhere.

Natural gas consumption grew by 2.2 % reaching a rate of 8,8 million m^3/d with an annual increase by 0,19 million m^3/d, or 2.2 % with respect to 2010, being responsible of 23.6 % of global energy consumption.

Coal consumption grew by 5.4 % in 2011, reaching a rate of 10,20 million toe/d accounting with an annual increase by 0,52 million of toe/d, for 30.3% of global energy consumption, the highest share since 1969.

The remaining 13 % (for a total of 4.34 million toe/d) is covered by hydro, nuclear and with only 0.534 million toe/d, by renewables accounting for only 2 % of energy consumption globally.

The share of 87 % covered by fossil fuels is equivalent to 29,28 million toe/d: an artistic view of this consumption could be done if one considers that this quantity corresponds to 214.63 million b/d: a stack of barrels each day 215000 km high, which will reach the Moon in less than two days and the Sun in less than two years. Similar comparisons could be done for methane and coal production with similarly surprising conclusions. The unsustainability of this economy should be considered with deeper attention and not only as a *politically correct* concept.

The result of this fossil economy is the CO_2 increase into the atmosphere: the corresponding share (with respect to the emissions of other greenhouses gases) is 65 % which rebounds both from final energy use (energy supply, transport, residential, agriculture, forestry, etc...) and type of species emitted (CO_2, N_2O, CH_4, etc...). The scientific community agrees about a positive feedback between climate change and the carbon cycle: the most important issue is the magnitude of this feedback which is at the center of the scientific debate: based upon current understanding, the stabilization of CO_2 at 450 ppm will likely result in a global equilibrium warming of 1.4 °C to 3.1 °C, with a best guess of about 2.1 °C: this would require a reduction of current annual greenhouse gas emissions by 70-80 % by 2100: so the 450 ppm level was adopted as a political number, in order to inspire political economic planning.

The actual CO_2 concentration ranked at an unfortunate level of 395.55 ppm in 2013, so an urgent intervention should be in the agenda of policy makers.

In spite of this established awareness, CO_2 concentration grew in 2012 having an average concentration in January 2012 equal to 393.84 ppm and in 2011 equal to 391.65 ppm. For the past decade (2003-2012) the average annual increase is 2.1 ppm per year while the average for the prior decade (1993-2002) is 1.7 ppm per year.

So, fossil fuels based economy is still stronger than sustainable development.

In an energy market so unbalanced toward fossil fuels, and in the context in which CO_2 has been committed as emissions, energy efficiency and energy recovery represent the most important technological challenges and opportunities: they are also the easiest interventions to put in place. So, energy is not a private but a common property.

Compressed air makes use of electrical energy and it could be useful to re-read the numbers above in terms of electrical energy.

The electricity consumption in 2009 was 17770,8 TWh [13] whose fossil fuel share is 67 %: this means 11906,4 TWh produced by burning oil, gas and coal. The main use of electricity is for industry, transport, building and others. Table 1 shows the relative share among these sectors, related to different geographical contexts. The data and differences deserve some consideration[1]:

1) The use of electrical energy in industry is more pronounced in Europe and in Japan, ranking 36.3 % and 28.7 % with respect USA, 21.9 %;
2) BRIC Countries have a greater share (China 65,5 %, India 46.6 %, Russia 45.7 %, Brazil 45.7 %) and their economic growth is mainly based on the industry development;
3) Absolute values demonstrate that energy consumption in industry in the BRIC Countries ranks at 6128.2 TWh, 35 % of the overall electricity consumption; China alone represents 61.5 % of this share;
4) The main electricity consumption in USA is in the building sector reaching 77.7 % of the overall consumption: a huge potential saving is expected.

Data in Table 1 are reported in Figure 1 which immediately evidences the differences among different sectors and puts in evidence the most important and effective interventions.

---

[1] Other energy sector covers the use of energy by transformation industries, energy losses in converting primary energy, by gas works, by petroleum refineries, coal and gas transformation and liquefaction. It also includes energy used in coal mines, in oil and gas extraction and in electricity and heat production. This consumption has not been considered in the % values to estimate real consumed energy values.

Table 1. Electricity share among sectors

Area/Country	Industry	Transport	Building	Other	Oth. en.	Total	% (share)	% Industry	% Buildings
USA	802,47	11,63	2837,72	0	558,24	4210,06	23,69	21,97	77,71
EU	988,55	69,78	1616,57	46,52	476,83	3198,25	18,00	36,32	59,40
Asia Oceania	581,5	23,26	976,92	11,63	209,34	1802,65	10,14	36,50	61,31
Japan	267,49	23,26	395,42	244,23	104,67	1035,07	5,82	28,75	42,50
Russia	314,01	81,41	279,12	11,63	290,75	976,92	5,50	45,76	40,68
China	2035,25	34,89	895,51	139,56	628,02	3733,23	21,01	65,54	28,84
India	325,64	11,63	244,23	116,3	209,34	907,14	5,10	46,67	35,00
Middle East	116,3	0	453,57	34,89	151,19	755,95	4,25	19,23	75,00
Africa	232,6	0	267,49	23,26	116,3	639,65	3,60	44,44	51,11
Brazil	186,08	0	197,71	23,26	104,67	511,72	2,88	45,71	48,57
TOTAL						17770,64	100,00		

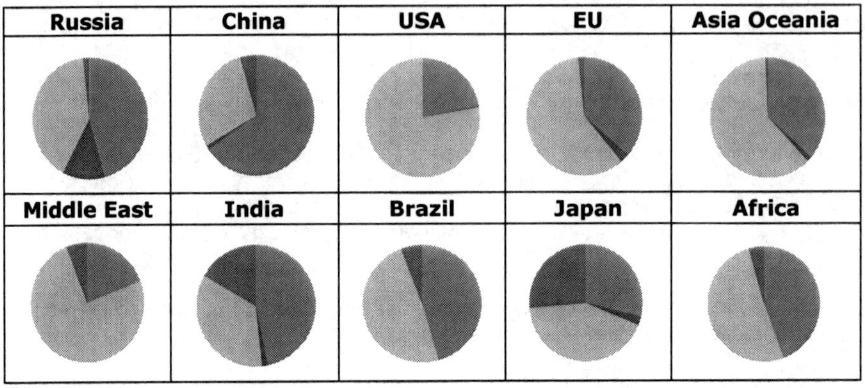

Figure 1. Electricity share among sectors (blue: industry, red: transport; green: building; magenta: other)

When a projection for future consumption is made, according to a reasonable policy scenario[3], electricity consumption will grow according to Figure 2: BRIC Countries in 2020 will have a dominant share (40 %) and China will become the most important electricity user (greater than USA and Europe, close to the sum of these two). Energy saving will have a major importance in these geographical contexts and the overall industrial sector must consider this opportunity (energy saving as a new technological action).

Compressed air production in developed Countries accounts for a mean value around 10 % of the overall electricity consumption in the industrial sector: this figure is almost fully agreed in technical literature, [2,3,4,5,6]. This share is referred to the industrial needs of compressed air; if one adds all the other "compression needs" i.e. the commercial and residential markets (portable tools, air pumps, pneumatic heating, ventilation, air conditioning, personal uses, etc...), the consumption grows to 20 % of the industrial electricity needs [7]. No estimates are available when considering the compressed air for transportation (train, trucks, buses, etc...): sometimes in this case a direct conversion is done by mechanical energy so, no electrical consumptions can be directly associated.

---

2 Defined as "New Policy scenario" as intermediate between Current Policy Scenario and 450 ppm Scenario leading to a 450 ppm the CO2 concentration by 2035.

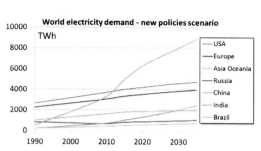 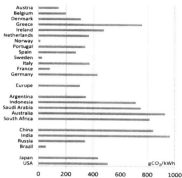

Figure 2. Electricity growth **Figure 3. CO_2 emissions per kWh**

Specific industrial sectors account for lower and greater electricity consumptions shares, [6]: a range from few point percent (leather, printing, stone clay, etc…) to 20 % (chemicals, oil and gas, etc…) can be found: the interest toward energy optimization processes when producing compressed air is relevant.

In developing Countries, many factors related to the:

1. mean technological level of the industrial sectors in developing Countries which is lower than the corresponding productions in the "original" Countries;
2. transfer of the old already used production lines;
3. lack of awareness in the energy saving and waste contest;
4. lower efficiencies in the final energy uses (motors, lighting, etc…);
5. absence of integrated energetic technologies (cogeneration, etc…);
6. longer production cycles;

would justify higher shares of compressed air consumption with respect to the overall electricity consumption in industry. In absence of specific data, the same mean share (10%) has been used.

In 2020, almost 24000 TWh will be consumed: compressed air will account for 2400 TWh. When these numbers are transformed into CO_2 emissions, the driver for the energy technology development, the situation is a concern: when electricity projections are used and specific CO2 emission rate per kWh of different geographical contexts, Figure 3 is used and CO2 emissions related to compressed air can be calculated. Consolidated data in 2009 demonstrates 1.4 $GtCO_2$/Y and predictions in 2020 bring to 2.2 Gt CO_2/Y. In terms of Carbon, during the period 2009-2020 the cumulative emissions could be estimated equal to 20 GtCO2 which corresponds to 5 GtC. According to a mean Carbon airborne ratio equal to 0.5, this would produce an atmospheric CO_2 increase of 1.5 ppm: this is 7-10 % of the overall admitted CO_2 increase into atmosphere if the stabilization at 450 ppm by 2035 is considered as political goal.

Therefore, compressed air represents a relevant sector of applications of technological improvement when CO_2 reduction is considered as a major future concern.

Energy saving in compressed air systems can be done on many sectors and with many interventions: Table 2 shows mean potential savings as referred by literature, [2,4,6].

Table 2. Energy saving actions

Action	Saving, %
High efficiency motors	2
Speed control	15
Upgrading compressors	7
Sophisticated control systems	12
Reducing frictional pressure losses	3
Optimizing end use devices	39
Reducing air leaks	20
Filter replacement	2
Total	100

An additional potential saving (as energy recovered) could be reached considering: (1) re-use of waste heat; (2) better overall system design; (3) new technologies for cooling, drying and filtering.

Real absolute potential saving strongly depends on a combination and application of the actions in the existing operating compressor and this depends on many factors, some not so obvious. According to actual electricity consumption a real potential saving is close to 20-35 % of the electricity consumption, [2, 6]; this saving could be easily increased by adopting simple *upstream* (awareness of compressed air users, implementing maintenance programs, using trained personnel, etc...) and *downstream* (leak detection, optimization of end use, etc...) actions. Easily the potential saving could be doubled with a tremendous energy benefit.

From these considerations, one can share the simple concept that actual production of compressed air is very wasteful process, [12].

Compressor technology and operation accounts for a mean consumption share close to 30-35 %. In this share, electric motors, speed control and the use of sophisticated control systems (part load) are responsible for the most significant part (70-80 %). So, machines more oriented to reduce energy consumption and, most preferably, more oriented to reduce consumption during flow rate modulation (matching between compressed air demands and delivery) should be preferred.

1. SLIDING VANE COMPRESSOR TECHNOLOGY AND ENERGY CONSUMPTION

Figure 4 shows typical costs of the compressed air supply in a typical industrial site, [2, 8, 9, 12]: energy cost accounts for 70-75 % and is the main cost compared with the capital costs (15-20 %) and maintenance costs (7-10 %). Installation costs are negligible (2 %). Data in Figure 4 are referred over a continuous 10 year operation period: so, with a simple exercise concerning the economic correlation between investing more money (to have e more energy efficient compressed air service) and reducing electricity costs has been done. The energetic costs have been referred to the overall share, being related to the compression of the air, even though air is wasted or used in an inefficient way. The purpose of this analysis was to estimate, (starting from "mean" actual technology and compressed air management) how much of a cost increase due to a more efficient compressor, would be justified in order to have in a 10 year continuous operating period, a saving in the discounted overall costs.

In Figure 5 the result of the calculation is presented: "x" represents the investment cost increase, "y" the minimum compressor efficiency increase which is needed to reach the same discounted cost. The result is not dependent by the discounted coefficient. For a given (accepted) investment increase, the "y" value gives the minimum efficiency increase that justifies the additional cost (discounted) over a ten year period. Values above the line produce a net cost reduction. An investment increase of 10 % is justified only if the compressor efficiency increase is greater than 4 % with respect to actual technology; if the cost increase is 25 % it is justified only if the saving is greater than 8 %: the ratio between the two parameters remains constant.

Therefore serious consideration should be given to invest in more energy efficient compressors when the additional capital cost can easily be offset by the cost of the energy saved.

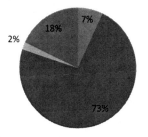

Figure 4. Compressed air costs over a ten year period (red: energy costs, blue: maintenance, green: installation, magenta: capital costs)

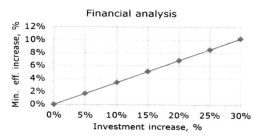

Figure 5. Relationship between investment increase on compressors (choice of a more efficient machine) and minimum efficiency improvement which justifies the investment

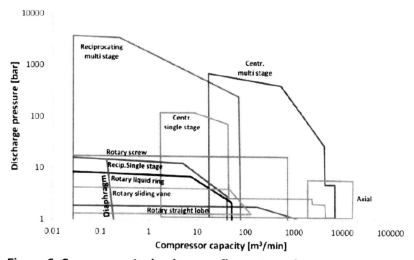

Figure 6. Compressor technology vs. flow rate and pressure delivered

In industrial applications, the most frequently used compressors for producing compressed air are rotary volumetric. Figure 6 shows the area of intervention of the different compressor technologies: centrifugal compressors have, with respect to the rotary ones, major common areas but they are confined to higher pressure ratio, where specific purposed must be fulfilled (compressed air for processes); axial (but also centrifugal) are justified for very high flow rate not usually required for compressed air end use applications. So, in the range 7-12 bar and flow rates less than 1000 m^3/min with an electrical power range from a few kW to several hundred kW, rotary volumetric machines represent the reference technology.

In the rotary type family, screw compressors are the most common and widespread technology. Features like reliability, low cost manufacturing, proven technology, strong scientific and research support, very high knowhow inside manufacturers and, more generally, less maintenance, robustness, capability *to stay on the market* are the main reasons of this success.

Concerning sliding vane rotary compressors, one must observe that the current literature which refers to the corresponding performances doesn't seem to be updated. A maximum power available for these compressors is declared to be 75 kW [8] while in reality SVRC of much greater power were and are in operation; sometimes, the efficiency is declared lower than other type of rotary compressors, [9], while in reality this is not necessarily corresponding to the actual operation of all SVRC. One should observe that only recently (last two decades) SVRC improved in performances and areas of application, thanks to the investment in the sector and also in terms of the research performed.

Since the energy aspect is so important, an analysis of the energy consumption of existing machines has been done, almost exclusively referred to the screw type. An interesting source of data has been the Compressed Air and Gas Institute, CAGI Data Sheet, [10]: original data is available from the manufacturer's sites to which the CAGI's site readdresses. Thousands of data sheets have been processed and a clear overview of the energetic performances of actual compressor technology is presented here.

An important remark concerning the treatment of CAGI's data has to be underlined. The mentioned data have been produced by different manufacturers at different delivery pressures by the compressors (from 5 bar to 13 bar), but, for a given pressure level, raw data as given by CAGI's data sheets cannot be compared. In fact, frequently, the pressure set during measurement was "around" a specific value, within a tolerance too wide to be referred to as a specific value. This produced higher or lower values of the specific energy which were not the result of a better machine but only the result of an operating pressure too distant from the specific value set up. In order to avoid this inconsistency when comparing data, a complex procedure has been applied which limited to +/- 10 % the pressure difference with respect to the reference value (7,8,9,10 bar) and, inside this tolerance, the specific consumption has been exactly recalculated (to the exact pressure reference value) interpolating among machines of the same manufacturer. This allowed an exact comparison even though it excluded almost 25 % of the original data.

CAGI's data, even though not specifically mentioned, should be referred to electric motors at 60 Hz. When a compressor designed with an electric motor operating at a different frequency (for instance, 50 Hz), is tested at 60 Hz, a slightly greater specific consumption is produced and the comparison is not straightforward.

Figures 7, 8, 9 and 10 show the performances of the existing compressors on the market: for sake of privacy, the data is presented anonymously, eliminating the manufacturer's name. Data have been organized as a function of:

a) Pressure delivered at rated and zero flow: the first condition refers to the usual design operation, the second one, frequent in reality, when the load/unload control is used. This control is also known as constant-speed control, allows the motor to run continuously, but unloads the compressor when the discharge pressure is adequate. Data at zero flow have been split among machines running at fixed speed or variable speed in order to outline the differences;

b) Oil cooled by air or water.

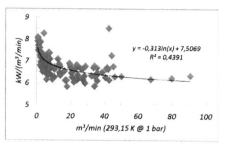

Figure 7a) Specific energy, 7bar@rated flow@air cooled

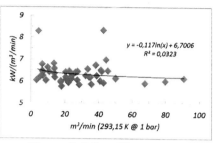

Figure 8a) Specific energy, 7bar@rated flow@water cooled

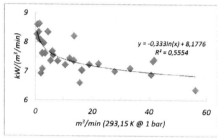

Figure 7b) Specific energy, 8bar@rated flow@air cooled

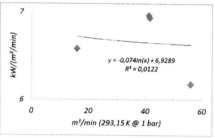

Figure 8b) Specific energy, 8bar@rated flow@water cooled

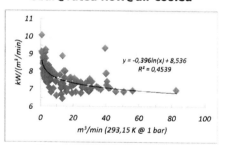

Figure 7c) Specific energy, 9bar@rated flow@air cooled

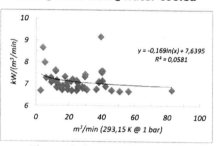

Figure 8c) Specific energy, 9bar@rated flow@water cooled

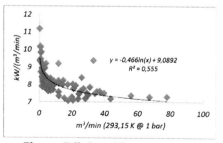

Figure 7d) Specific energy, 10 bar@rated flow@air cooled

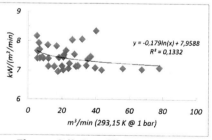

Figure 8d) Specific energy, 10 bar@rated flow@ water cooled

Absolute power can be easily calculated from data shown by multiplying the "x" and the "y" data. Volumetric flow rate has been evaluated at 1 bar, 298.15 K.

The following considerations apply:

a) Specific power decreases in general with respect to flow rate, demonstrating that bigger machines are more efficient. In reality, this is more evident for air cooled machines (with respect to water cooled) and for higher pressure delivered (9-10 bar);

b) Water cooled machines are more efficient than air cooled. This is due to the consumption of the fan which cools the oil: the circulation of the water is produced by an external pump (or pressurized system) and it is negligible in terms of energy consumed;

c) For a delivered pressure equal to 7 bar, air cooled machines have a technological scatter (among manufacturers) very significant depending on flow rate up to 50 m^3/min. For compressors whose rated flow rate is close to 10 m^3/min a mean value close to 6.8 kW/m^3/min is the reference value, with a scatter of 1.5 kW/m^3/min: so, for these conditions the compressors must be chosen properly. When flow rate increases (30 m^3/min), mean value and scatter decrease reaching 6.5 kW and 1 kW/m^3/min. Higher flow rates (> 50 m^3/min) are characterized by a mean value close to 6.2 kW/m3/min which seems to be an asymptotic performance.

For water cooled compressors, Figure 8a), mean specific power decreases and remains more constant when flow rate changes: 6.2 kW/m^3/min is the reference value with a scatter among manufacturers slightly less than 1 kW/m^3/min;

d) For higher delivered pressures the differences between air and water cooled compressors increase, in favor of the water cooled versions, and the scatter remains wider for air cooled machines when compared to water cooled versions. For a delivered pressure of 9 bar, at 20 m^3/min 7.4 kW/m^3/min is a mean value, but with a scatter among different machines close to 1 kW/m^3/min. Water cooled machines have a slightly lower value and a reduced scatter. 7 kW/m^3/min is the asymptotic value for flow rates greater than 50 m^3/min. A correct compressor choice is needed and significant energy quantity (and cost) can be saved;

e) When delivered pressure is further increased (10 bar), a reference value is 8.2 kW/m^3/min for flow rates limited to 40 m^3/min, with a large scatter among machines: similarly as in the previous cases, a correct choice has to be made;

f) Mean values reported are characterized by very high differences among machine manufacturer but some premium machines from an energetic point of view are present on the market: limiting the attention to air cooled machines, noteworthy are:

- 5.7 kW/m^3/min is the best value when rated flow rate is 20 m^3/min for a delivered pressure equal to 7 bar (Figure 7a);
- 6.5 kW/m^3 min is the best value at 8 bar, when flow rate is close to 20 m^3/min (Figure 7b); a premium value is at 60 m^3/min for very high flow rate (60 m^3/min);
- 6.3 kW/m^3/min is best value when rated flow rate is 15-18 m^3/min for a delivered pressure equal to 9 bar, Figure 7c);
- 7 kW/m^3/min is best value when rated flow rate is 20 m^3/min for a delivered pressure equal to 10 bar, Figure 7d); higher values (7.4 kW/m^3/min) characterize higher flow rates.

The specific consumption at zero capacity is a significant part of the rated machine consumption; so, load/unload machine control is a wasteful strategy.

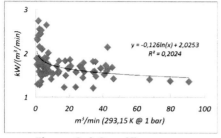

Figure 9a) Specific energy, 7bar@0Flow @fixed speed

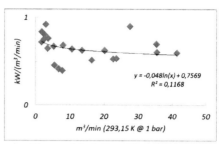

Figure 10a) Specific energy, 7bar@0Flow @variable speed

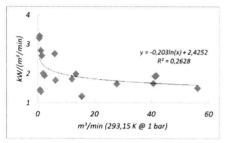

Figure 9b) Specific energy, 8bar@0Flow@fixed speed

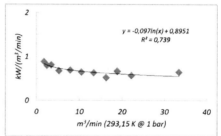

Figure 10b) Specific energy, 8bar@0Flow@variable speed

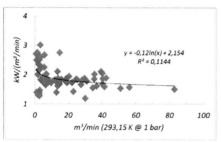

Figure 9c) Specific energy, 9bar@0Flow@fixed speed

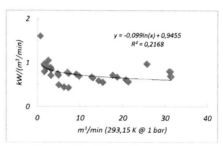

Figure 10c) Specific energy, 9bar@0Flow@variable speed

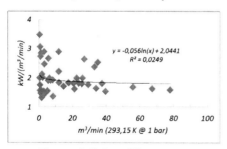

Figure 9d) Specific energy, 10bar@0Flow@fixed speed

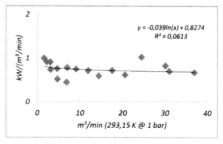

Figure 10d) Specific energy, 10bar@0Flow@variable speed

Figures 9 and 10 refer to the specific consumption according to machines operated at fixed speed (Figure 9) and those at variable speed (Figure 10). Variable speed machines have around one half of the consumption of the fixed speed machines and the scatter among different manufacturers is significantly reduced.

For a delivered pressure of 7 bar, a mean 25 % of the rated power is lost at zero capacity but higher values, up to 40 %, are possible. This share remains almost unchanged when delivered pressure increases (1.8-2 kW/m³/min as mean values at 8 bar, 9 bar and 10 bar).

R^2 is a statistic datum that will give some information about the goodness of fit of a model, i.e., in regression, the R^2 coefficient of determination is a statistical measure of how well the regression model approximates existing energetic performances of machines (real data points). The model assumed is $Y = A \cdot \ln X + B$ being Y the specific energy consumption in kW/m³/min and X the volumetric flow rate in m³/min. R^2 of 1 indicates, as known, that the model approximates (real) data. The low values of R^2 clearly demonstrate that the technology is not aligned to the best standards and that many machines designed many years ago are still in the market and no upgrading effort has been done.

2. SLIDING VANE COMPRESSOR: PRINCIPLE OF OPERATION

The geometry of a SVRC is well known. A cylindrical rotor, placed eccentrically within a corresponding cylindrical stator, during rotation expels a given number of blades arranged inside the rotor slots. This causes the formation of closed cells having a volume which decreases from the intake to the exhaust port during rotation, so compressing the trapped air.

At the contact between blade tip and stator inner wall, a thin oil layer (pressurized by the same compressed air) prevents a dry contact, minimizing friction and losses. Oil also ensures cell sealing, keeping a high volumetric efficiency.

A typical, simplified geometry of an SVRC is shown in Figure 11: the slots inside the rotor can be simply radial or placed at an angle, so offering a degree of freedom in terms of minimizing friction losses. The blade inclination, in fact, changes the reaction forces and, consequently, the power dissipated by friction and the pressure distribution on the oil layer which avoids dry contact.

Main processes concerning the behavior of a SVRC are:

a) Filling and emptying of the vanes;
b) Thermodynamic transformations inside the cells, considering oil injection and interactions (air cooling, oil drops motion and thermal effects. etc...);
c) Motion of the blades inside the slots and reaction forces at the contact points;
d) Pressure stabilization inside the oil film at the tip blade-stator contact;
e) Oil separation from compressed air.

All of these phenomena are complex: if one identifies a main process in terms of "causes" and "effects", the very short characteristic times of the "causes" requires an unsteady treatment which increases complexity and requires a suitable mix between physical representation and approximations, required to solve definitively the mathematical formulation.

Several of these aspects have been object of a recent scientific attention from the Author and colleagues: a greater detail can be found in [14, 15, 16, 17, 18, 19, 20, 21, 22].

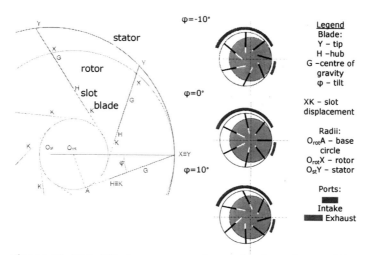

Figure 11. Simplified vane geometry and relevant quantities

2.1 Vane filling and emptying

In order to improve SVRC design the filling and emptying of the cells required particular care in order to predict the mass inducted and expelled in or from the vane.

While the induction process could be considered steady, the exhaust is completely unsteady considering that the cell pressure when it opens toward the discharge usually does not match the line pressure, i.e., the pressure in the discharge volume.

This feature makes the discharge process more difficult and more generally, it has a negative impact in terms of specific compression energy. Considering that the cell-discharge volume pressure difference could be in the order of 5-6 bar, a strong unsteady process occurs during cell voiding as the line pressure is in the range of 7-13 bar. Vane filling and emptying occur through ports which can be built axially or radially, according to the direction of the flow which moves through them, Figure 12. In both cases, the angular position of the beginning or of the end of ports is quite important for a complete filling and for pressure optimization. An example: an early opening of the exhaust port will produce a sudden isochoric compression requiring more work to deliver air at a given pressure.

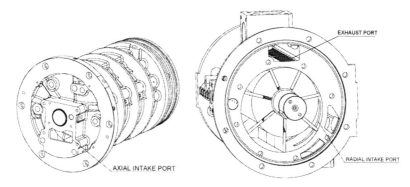

**Figure 12. Ports axonometric views
(from Ing. E. Mattei S.p.A.)**

While vane filling proceeds at ambient pressure (if the control valve is opened or at a steady lower value), a 1D unsteady fluid dynamic treatment is needed at the exhaust: the Author proved a model which was developed for inlet and exhaust manifolds in Internal Combustion Engines, to be very suitable to this case: the theoretical approach is called quasi propagatory model (QPM) [19] and, in a very efficient and fast way, it evaluates the unsteadiness during the exhaust phase. The method makes use of the mass, momentum and energy unsteady equations and solves them in an analytical form catching the inertial, capacitive and propagatory nature of the process (unsteady).

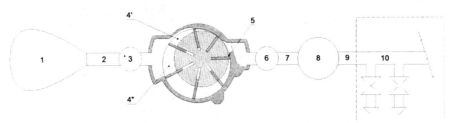

Figure 13. Capacities and equivalent pipes specifying unsteady behavior

QPM considers the SVRC as in Figure 13. An inlet plenum (1) having infinite capacity is connected with the intake vane (4) though an intake manifold (2) and an intermediate capacity (3) which is the one in which an inlet valve is mounted, controlling with its opening the flow rate delivered. Similarly, the exhaust cell (5) discharges toward a capacity (6) represented by the volume which immediately receives the flow. Through an equivalent duct (7), the compressed air (and oil) fill the oil separator (8) which behaves as a capacity and, through a pipe (9) having a given designed length, it reaches the working line (10) which behaves as a plenum of a finite capacity at pressure line. The mass & enthalpy fluxes are calculated in each relevant section (equivalent pipes and capacities) solving an unsteady 1D model: inlet plenum (1), intermediate inlet volume (3), cell which is emptying and voiding (4, 5), intermediate capacity (6), oil separator (8) and line (10) behave as time varying boundary conditions. Each capacity is characterized by boundary conditions (infinite plenum, throttled energy ellipse, static to static conditions, etc....) and waves which propagate inside the discharge line intersect with boundary conditions and produce a returning wave. Similarly when these reach the upstream boundary conditions: residual waves move downstream. The sequences of these waves can be analytically described, [19], and report instantaneous pressure, velocity and density of the air.

2.2 Thermo dynamic cell model

The model is based on a lumped parameter assumption having, uniform thermodynamic properties (temperature, pressure, composition, etc...). Considering the quantity of the oil injected, each cell is considered as a mixture of air (inducted) and oil (injected). Different phase transitions (liquid-vapor) are allowed according to physical tendency of oil evaporation (comparison between its partial pressure and the saturation pressure which changes with the temperature inside the cell).

The cell behaves like an open system (with respect to mass and energy); therefore, energy conservation equation has the form:

$$\frac{dU}{dt} = \sum \dot{m} h + \dot{q} - p\dot{V} - \lambda(T) \left[\frac{dm_{o,vap}}{dt} \right] \tag{1}$$

Eq. (1) requires the knowledge of the input and output terms which derive from the cell voiding and emptying.

Equation (1) included the heat exchanged by the air during compression: in fact, when it is compressed the temperature increases and it would be very beneficiary to cool it in order to decrease energy required for compression. From this point of view, equation (1) considers two terms: (1) forced convection between air and metallic parts of the machine; (2) the heat subtracted by the air due to oil evaporation improved by spraying it.

These two terms have been accounted inside the term "q". In order to take benefits from a positive oil cooling, the oil injection has been recently studied examining the spray produced by a pressure swirl injector and experimentally validated, [18, 21].

2.3 Oil spray injection model
Oil circulates inside the machine naturally, the driving force of this circulation being the oil pressure when it is discharged by the vane. Distributed pressure losses as well as concentrated ones (changes in directions and cross sections) are considered: the last pressure drop is represented by the oil injection rail from which several injections are made.

The spray produced by the orifice has drops of different sizes. The spray pattern is normally a solid cone, in which more closely spaced drops are found when moving from the outer cone surface toward the axis. A Rosin Rammler drop size distribution function has been used as an input where the Sauter Mean Diameter (SMD) has been correlated to the main properties of the oil, spray cone angle, orifice diameter and pressure across injectors. Once drops are generated inside the jet, they are animated by an initial velocity (which respect mass conservation) whose direction is randomly chosen. Each drop is followed in its motion described by forces acting on each drop:

$$m_d \frac{d(\vec{V_d})}{dt} + \vec{V_d}\frac{d(m_d)}{dt} = \vec{F}_{drag} + \vec{F}_{Coriolis} + \vec{F}_{centrifugal}$$

$$\vec{F}_{drag} = -\frac{\pi}{8} d_d^2 \rho_a (\vec{V_d})^2 C_d(R_e)$$

$$\vec{F}_{Coriolis} = -2m\vec{\omega} \times \vec{V_d}$$

$$\vec{F}_{centrifugal} = -m\omega^2 \vec{R} \qquad (2)$$

In this way, oil puddles on the blades as they cross the jet and on the rotor can be calculated: when they are formed, they do not participate to the air cooling. The term which applies in equation (1) can be evaluated according to the equation:

$$\frac{d\left[m_d C_{pd}(T_d - T_r)\right]}{dt} = q = q_{a-d} + q_{ev} \qquad (3)$$

which states the drop temperature increase. The two terms account for molecular diffusion between drop surface and air and droplet evaporation when its temperature reaches the boiling value.

Vane filling and emptying, and closed volume transformations, including oil injection, have been integrated in a comprehensive mathematical model, solved and theoretical results have been compared with experimental data. Most intimate variable i.e. pressure inside the cell has been measured according to sequentially spaced piezoelectric transducers. Figure 14 shows the sensor position mounted in a way to allow the reconstruction of the pressure inside the cell, including intake and exhaust. Figure 15 shows the single pressure traces before reconstruction: piezoelectric transducers, in fact, have a very low response time so they allow the measurement of highly fluctuating pressures in time. Unfortunately, they measure differential pressures and when they continuously monitor the pressure in fixed position, the overall pressure signal must be reconstructed adding sequentially the values measured by the piezoelectric sensors.

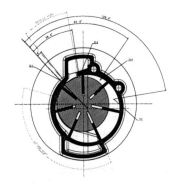

Figure 14. Position of pressure transducers allowing the reconstruction of the pressure inside the cell

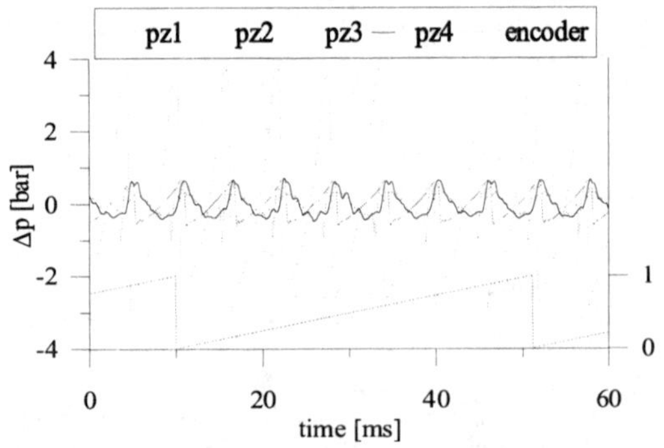

Figure 15. Pressure signals measured by the transducers before theoretical treatment to build the absolute signal

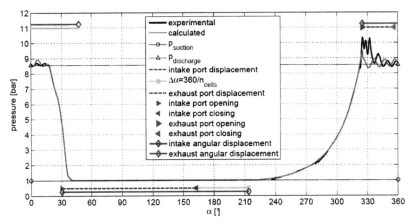

Figure 16. Indicated pressure

Figure 16 shows the p-V diagram (reconstructed from differential pressures given by sensors) comparing theoretical and experimental data: it is evident there is steadiness during intake and that there is strong unsteady behavior during exhaust. The coincidence between an adiabatic (isoentropic) transformation and real pressure trace demonstrates the limited cooling action of the oil, according to data reported in Figure 16.

When a new technology for oil injection is considered (pressure swirled), oil droplets produced by the injectors are small enough to be heated up during their travel (before impinging) as shown in Figure 17. This derived from the high convective heat transfer conditions produced by the high speed drops, Figure 18, and to the high surface to volume ratio which characterizes the drops. The effect of the pressure inside the cell is evident in Figure 19 which demonstrates that when drops are reduced, the cooling effect is enhanced and this slightly reduces the pressure inside the cell but, more importantly, it reduced the air temperature during compression, reducing significantly the compression work.

Air temperature is shown in Figure 20 which clearly states the differences between an oil free compression and an oil injected one, having drops close to 65 µm as SMD. The effect on compression work is immediately evident as Figure 21 reduction of 1.2 kW over an initial 9.6 kW is allowable, representing 12 % reduction.

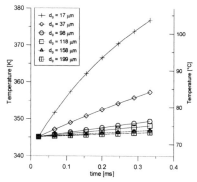

Figure 17. Temperature of the oil drops during injection (traveling time before impingement) as a function of injection's SMD

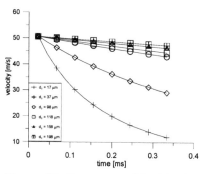

Figure 18. Drops speed inside the cell during injection as a function of injection's SMD

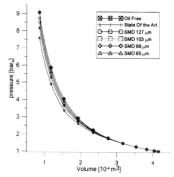

Figure 19. Pressure inside the cell during compression as a function of the injection's SMD

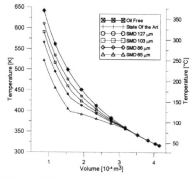

Figure 20. Air temperature inside the cell during compression a function of injection's SMD

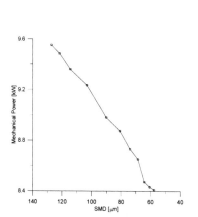

Figure 21. Compression work reduction due to air cooling produced by the oil drops

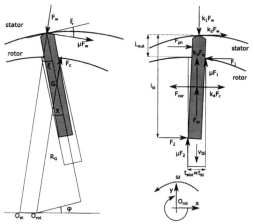

Figure 22. Blade dynamics: forces acting on the blade

2.4 Friction modelling and mechanical power lost

The power dissipated by friction is calculated evaluating the forces exchanged between the blade and the surfaces during their motion inside the slot and during the rotor's rotation. Contact surfaces are at blade tip (in contact with the stator's surfaces) and to the lateral surfaces of the blades, in contact with the slot's surfaces. When equations which express the conservation of momentum are solved, the position of the blade inside the slot can be calculated, according to the four possible states reported in Figure 23: blades periodically move inside the slots assuming different settings.

The four states correspond to four equilibriums in which the blades pass during one full rotation: during rotation, Figure 24, according to the forces acting on it, the blade tilts inside the slots. Rotor moves in counterclockwise direction and opening and closing of the inlet and outlet ports are 30.3°-162.5° and 325°-356°, respectively.

It is evident that, during compression phase (closed cell) the greater pressure inside the vane ahead tilts the blade as in position #2; during discharge and intake the blade stays attached to the surface slot, in position #1 and #3, respectively. After the discharge, the blade has a sudden tilting to position #2 and #4 and definitively rearranges to position #2 during intake.

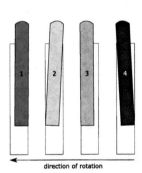
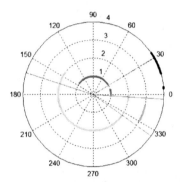

Figure 23. Blade possible settings   **Figure 24. Blade position during rotation**

The calculation of friction coefficient is one of the major concern is SVRC: from indicated power (calculated and/or measured) and from a theoretical approach which determines that friction power is equal to:

$$P_{fr}(\mu) = \mu \cdot \left(F_w \cdot \omega \cdot R_Y + F_1 \cdot v_1 + F_2 \cdot v_2 \right) \tag{4}$$

where v_1 and v_2 are the blade speed at the contact inside the slot.

The friction coefficient can be identified, when shaft power is known from measurements. Reaction forces can be calculated by the equilibrium of the forces acting on the blade (radially and tangentially) and of the torques acting on the blade.

The calculation of friction power allows a geometrical optimization of the compressor in terms of stator-rotor diameter ratio concerning specific energy absorbed.

The variation of the geometrical diameter-length ratio, in fact, modify the reaction forces at the blade tip and, therefore, for the same indicated power (compression of the air), the overall power absorbed, [22]. Starting from an original geometry of an existing industrial 22 kW compressor and keeping the same rotational speed, Figure 25 a) and 25 b) show the effect of the stator and rotor increase on the machine, keeping the same volumetric intake capability.

Friction power increases when increasing the stator diameter due to the higher centrifugal forces (Figure 25 a) producing for a variation of 6 % on stator diameter a 20 % increase of the friction power which provides a 4 % decrease in the mechanical efficiency. The effect of the rotor is inversely proportional (Figure 25b): a rotor increase decreases the mechanical losses due to the power due to friction reduction at the lateral surface between blade and slot: when the rotor increases, eccentricity is reduced and the relative speed between blade lateral surface and slot decreases, leading to an overall reduction of the friction losses. The contribution at the blade tip remains almost unchanged.

When the dimensions of the stator/rotor change (where the volumetric intake vane remains the same, Figure 26 a) in an existing machine, the built in volumetric compression ratio changes, modifying the pressure inside the cell when the discharge port opens. When the line pressure is greater than that pressure, a sudden isochoric compression is done inside the vane, increasing indicated power and decreasing efficiency. In Figure 26b), for three values of the pressure line, the variation in the indicated pressure is done, according to the built in volumetric pressure ratio. For each value of the line pressure, there is an optimum pressure ratio and, moving from the optimum datum, increases of about 5 % can be suffered in terms of indicated power.

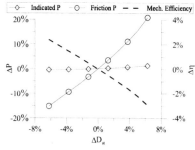

Figure 25a) Effect of the stator's diameter on the indicated and friction power

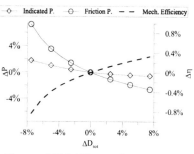

Figure 25b) Effect of the rotor's diameter on the indicated and friction power

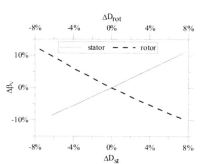

Figure 26a) Variation of the pressure built in pressure ratio

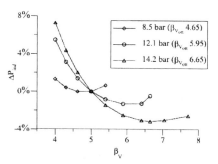

Figure 26b) Indicated power vs built in pressure ratio

A very interesting parameter inside SVRC to decrease the power lost to friction is the rotational speed which can be decreased without significant operational boundaries. From this point of view, SVRC show a better behavior with respect to screw compressors in which the speed of rotation influences the volumetric efficiency. From this point of view, a stable oil film at the blade tip and on the perpendicular covers insures in SVRC volumetric efficiency close to unity, even when, at the ideal contact rotor-stator, the vane is completely squeezed preventing any leaks from exhaust to intake ports.

Figure 27 demonstrates the effect of the rotational speed on the power absorbed by friction: in the 22 kW industrial compressor under analysis, the speed is reduced keeping the same volumetric dimensions of the vane (the simplest way is to increase compressor length). Indicated power remains unchanged and a significant benefit is observed in the reduction in friction losses. It is not without explanation that the best energy saving SVRC operates at lower speeds. In fact, when rotational speed is decreased, the only way to recover flow rate is make the compressor longer. This produces - Figure 27a) - a linear increase of the mass of the blade and consequently a linear increase of the centrifugal forces (radial blades) which, on the other hand, decreases in a quadratic way when speed is decreased. Moreover, relative tip velocity decreases too, linearly with the rotational speed. The net effect is a linear decrement of the friction losses, Figure 27a): for a reduction of about 30 % (from 1500 RPM to 1000 RPM) a similar reduction in friction losses is obtained. The effect on overall compressor efficiency is shown in Figure 27b).

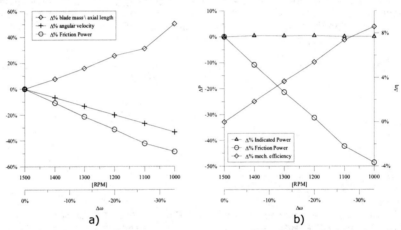

Figure 27. Power lost by friction as a function of revolution speed

Indicated power remains almost constant (the flow rate has been kept constant and PV cycle remains almost unchanged) and the result of a reduction of about 50 % in the friction losses produces an increase in the overall compressor efficiency close to 8 %.

In order to describe in a deeper physical approach the hydrodynamic situation of the oil film between blade tip (with its tip shape) and stator surface, the momentum and mass conservation equations have been solved in a 1D problem described by:

$$\frac{d}{dx}\left(\frac{h^3}{12\mu}\frac{dp}{dx}\right) = \frac{u}{2}\frac{dh}{dx} \qquad (5)$$

where h(x) is the film thickness which depends on the geometry and has a convergent-divergent geometry, p(x) the pressure (uniform on the thickness) and x a 1D spatial coordinate. Incompressible oil and no slip at the walls were most important assumptions; boundary conditions are the values of the pressure upstream and downstream of the oil film, being fixed by the pressures among adjacent cells (from thermodynamic cell model. Unsteady terms are neglected.

The solution, [23], gives the pressure distribution inside the oil film and minimum thickness distribution. The occurrence of situations in which the oil is unable to sustain the blade leading to a dry destructive contact can be identified and avoided by a model based approach (minimum allowed value of h).

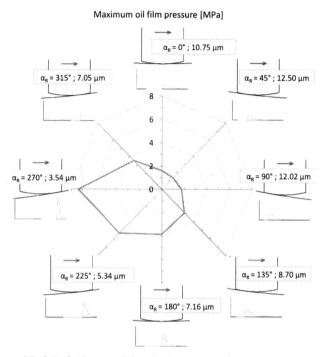

Figure 28a) Relative position between blade and stator and relevant data and minimum oil thickness during rotation

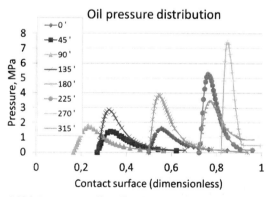

Figure 28b) Pressure distribution of the oil layer between blade tip and stator. Contact surface changes according to the blade position during rotation. Different colors correspond to different positions of the blade during rotation

Figure 28a) shows most relevant quantities during blade motion. The pressure distribution is represented at the blade tip-stator surface contact whose maximum value is reported in the polar drawing. Quantitative data are reported in figure 28b) where the contact surface has been reported in a dimensionless way: it changes during rotation (according to the position of the blade during rotation) and pressure distribution starts at different points. Maximum value is 7 MPa: in this condition, a minimum oil thickness equal to 3.54 µm is found. For each blade position during rotation, minimum oil thickness is reported: when $α_R$ moves from 0 to 360 (ideal contact between rotor and stator with the blade completely inside the slot). Data refer to an industrial 22 kW compressor, 136 mm and 111 mm as stator and rotor diameter, 10.5 mm as vane tip radius, 38 mm as blade height, operating at 1482 RPM delivering at 9.5 bar (absolute pressure).

All the sections previously considered have been managed inside a virtual platform which gives, once the geometry is designed and ambient and line pressure are known, the behavior of the SVRC as a function of the operating conditions. The platform has been continuously under upgrade and represents a very useful design tool and scientific platform.

2. PREMIUM SECTORS OF SVRC

Rotary volumetric compressors have several advantages when compared to the other compressors and, as known, they have reached a dominant position for compressing air for industrial application, Figure 6.

Within the rotary volumetric types, screw compressors play a dominant role, with the overall market share of SVRC being only few percent. The manufacture of SVRC is also limited to just a few Companies unlike the wider range of screw manufacturers and screw compressor packagers in the world.

So, an eventual useful comparison (from an industrial point of view) among these two technologies must keep in mind these relative positions.

Inside this conceptual context, it is in the Author's opinion that the following advantages are found with the SVRC. These aspects must be emphasized and when they are really important for a specific application, SVRC offer a proven and reliable solution. It is the Author's opinion that, often, SVRC are not considered simply because they are unknown or not known as well as screw compressors.

They are:

a) The geometrical simplicity of SVRC with respect to the complex geometries of the rotors in screw compressors. Simpler machines move in favor of reliability, reduced maintenance, absence of overhauls, absence of re-machining, etc...;
b) The reduced rotor mass which characterizes SVRC favors a reduced temperature sensitivity of the machine and less important clearance control: it is known that the bearing precision and mounting is an important aspect of screw compressors. SVRC do not have bearings but use oil bushings with a simple geometry;
c) Smaller masses in rotation mean lower vibrations and quieter machines: there are specific applications in which these features are particularly requested (transportation sector, compressed air for trucks and buses, etc...);
d) Greater masses in rotation mean a longer time to reach steady conditions as happens in screw compressors when compared to SVRC: when these conditions are reached, a correct and sufficient pressure is reached to

lubricate the bearings and the working chamber. This lack of lubrication causes dry contacts and higher air temperature during compression, [24];

e) Dry contact, even though limited to a very small contact area, between male and female rotors, in screw compressors this can't be avoided: the abundant oil injection tries to prevent this occurrence, but the rotation of the two rotors and the massive oil squeezing, results in a dry contact: the situation gets worse with aging and during transient operations;

f) SVRC can be operated at very low speed and benefits from the friction reduction which is proportional to the square of rotational speed; the reduction of the blade masses further participates to the reduction of the friction losses thanks to the reduction of the reaction forces. From this point of view mechanical efficiency of the SVRC can be improved (and it is under actual scientific attention) acting in these two directions (reducing speed and blade masses) while the efficiency characterizing screw compressors seems to have reached an asymptotic value: friction is limited by the fact that screws do not rely on the stator but they are strictly reliant on the accuracy of the bearings which, unfortunately, tends to decrease during time. In SVRC the blades rely on the stator even though the contact is prevented by an oil film. Today, transient conditions in SVRC's, which may imply little or no lubricating oil reaching the stator wall, are prevented by special honing off the stator and the blade tips. This honing traps enough oil from the previous steady state condition to assure safe transient lubrication conditions. Studies are now under way to further improve these operating conditions in an attempt to further reduce friction and improve machine performance;

g) SVRC can reach high volumetric efficiency at low rotational speed. Also in this case, the abundant oil injection seals the clearances between rotor and stator (on the perpendicular planes which "closes" the cylindrical stator) and when the vane is pushed all the way into the rotor (when the vane after having discharged the compressed air, is moving towards a new intake). Screw compressors have difficulty operating at low rotational speed and their volumetric efficiency is higher when rotational speed increases. This is due to the geometrical "blow hole" a fixed leakage path between the high pressure and low pressure sides of a screw compressor, which is inherent to its very design and impossible to eliminate totally. The higher the rotating speed the more volumetric flow from the compressor leading to a dilution of the recirculated air through the blow-hole and, hence, a higher volumetric efficiency;

h) From an energy consumption point of view, Figures 29, 30, 31 and 32 show the position of SVRC with respect to the other machines (mainly screws). Figures labeled as "a" refer to a rated full flow (x-axis); figures labeled as "b" to a zero flow operating conditions. In sequence, the rated pressure is 7 bar, 8bar, 9 bar and 10 bar. Mean values of the overall compression machines (screw compressors and SVRC) have been represented by the continuous line using the model able to reproduce (even with a high scatter) actual technology;

At 7 bar SVRC behave like the best machines up to 40 m^3/min, the maximum actual flow rate delivered. Some "dots" are outside of the scatter range: 5.7 kW/m3/min appears to be the best value al 60 Hz. At the other operating pressures, SVRC stay on mean values even though at 9 bar and small flow rate (up to 10 m^3/min), a slightly better behavior is shown; generally speaking, SVRC have energy performances very close to those characterizing the mean technological values, with the exception of some specific highly dedicated machines, specially designed for energy saving; At 10 bar, SVRC appear more efficient;

i) An interesting feature is shown by Figures 29b), 30b), 31b) and 32b) concerning the energy absorption at zero flow: this situation occurs when a load/unload control is done. SVRC demonstrate at all the pressure levels a much lower energy consumption; so, they are more suitable for this control strategy which has a huge application in practical situations: it is important to remark that in those applications in which flow rates are variable, the load/unload behavior can account for a significant part of the compressor's operating life;

j) SVRC shows additional potential benefits when the rotational speed is further reduced, Figure 27. This degree of freedom, which increases mechanical efficiency due to the peripheral speed reduction at the contact blade tip-stator surface, can't be used in screw compressors (as it can be modified for SVRC). In this case, in fact, speed reduction doesn't modify volumetric efficiency as happens in screw compressors. There are no specific limitations in SVRC to bring the rotational speed down to 500-750 RPM with a huge additional energy benefit, whilst keeping similar very high values for the volumetric efficiency.

Due to the combination of several factors, certain fields of application have chosen the SVRC technology. Two main sectors of applications can be mentioned as a result of an in-depth technological effort.

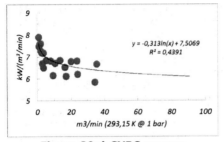

Figure 29a) SVRC energy consumption, rated conditions, 7 bar

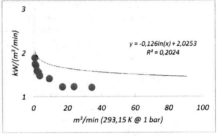

Figure 29b) SVRC energy consumption, zero flow, 7 bar

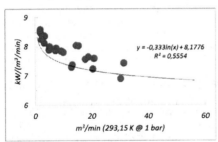

Figure 30a) SVRC energy consumption, rated conditions, 8 bar

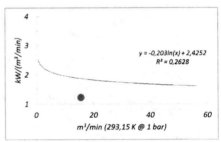

Figure 30b) SVRC energy consumption, zero flow, 8 bar

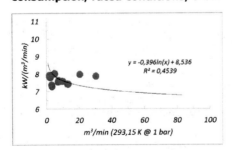

Figure 31a) SVRC energy consumption, rated conditions, 9 bar

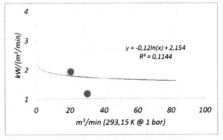

Figure 31b) SVRC energy consumption, zero flow, 9 bar

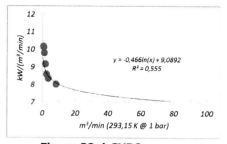
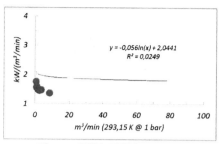

Figure 32a) SVRC energy consumption, rated conditions, 10 bar

Figure 32b) SVRC energy consumption, zero flow, 10 bar

In the first case, [25], the compressed air production is done in a very severe environment. In the aluminium smelting business the compressed air is continually used to tap the sludge crusts that from on top of the molten aluminum during the refining process and also to move the material using the production of vacuums. Having to run the compressor over 1km long molten aluminum pools with very fine aluminum dust and 80°C ambient temperatures, has led the aluminium smelter crane manufacturers to look for the most reliable, sturdy, but at the same time efficient compressor solutions. The SVRC is particularly adept at dealing with such hot and dusty environments since the large quantities of oil injected into the compressor do a very good job at keeping the rotor/stator/blade assembly clean and cool. Also, using oil bushes with direct oil injection and dedicated oil filters, leads to great reliability on the rotating shaft. This is not the case with oil flooded rotating bearings typically found in screw compressors. The high temperature operating conditions also favor SVRC for their insensitivity to these and particularly when compared with screw compressors with respect to this parameter.

In the second case, [26], in the transit industry, the most important factor for a compressor operating the brakes is reliability. Although the retrofit market players have known for many years that a better compressor solution existed than the ones being installed by the train industry OEMs (pistons and screws), it is only recently that a major manufacturer has decided to include, in their first installation, an SVRC. Once again, the vibrations involved in a moving vehicle make the bushes involved in an SVRC a winner. Precise screw compressor shaft bearings cannot withstand the high vibrations and the hot and dusty environment present on the underside of the train. This had led train operating companies to often perform a preventative complete substitution of screw compressors in underfloor train systems, sometimes even up to 3 times a year. With SVRCs this environment does not lead to shortness of product life and the compressor, if serviced correctly, can live up to its industrial equivalent of a couple of hundred thousand operating hours. The very compact compressor size combined with a high degree of customization that can be carried out, add flexibility to the application.

A recent application specifically conceived for reducing energy consumption, [27] has demonstrated the potentiality of SVRC for very large compressed air consumption. The activity was referred to a very large compressed air installation with 3.5 MW of installed electric power. In the compressor station, centrifugal compressors, screw oil flooded and screw oil free compressors were in operation: the substitution with the most advanced SVRC machine in terms of efficiency produced an energy saving close to 30%. Since this specific industrial sector is committed in terms of greenhouse gas emissions reduction, the equivalent reduction in CO_2 emission is of great importance.

Additional consumption benefits have been predicted reducing leakages and optimizing end use tools: an estimate of an additional 20% saving has been calculated, reaching an overall 45 % energy saving. From an economic point of view, the saving is even greater if one considers the value of the white certificates which are recognized for new compressor installations specifically designed around energy saving targets.

3. CONCLUSIONS

Compressed air in industry represents an important part of the energy consumption. According to different sectors, the mean share of this consumption ranges from few percentage points (paper, stone clay and glass, printing and publishing, etc...) to values close to 10 % (primary metal industries, rubber and plastics, electronics, lumber and woods, etc...) reaching greater values for chemicals, petroleum, etc.. up to 20 %. More efficient machines and, in general, a more efficient compressed air management is strongly needed, considering the mandatory commitments adopted almost everywhere in the World concerning energy reduction, saving and CO_2 emissions. The potential saving in the sector is huge, reaching 30-40 % considering all the potential measures; the consequential carbon emission reduction is an added valued which today represent the new driver in many industrial sectors.

Rotary volumetric machines are the technology reference in compressed air production for industrial needs, and among those, screw compressors for many reasons have a dominant and universally recognized role. This was also thanks to the huge research and development efforts which sustained this type of compressor providing training for specifically oriented engineers and technicians as well as academic environments.

Sliding vane rotary compressors (SVRC) in the last decade have reached unforeseen performances which offer to the market a potential alternative: this was also thanks to research efforts done internally by the manufacturers and by a renewed interest from the scientific and academic world.

Today a comprehensive modeling of the processes occurring inside a SVRC is available, having different levels of complexity: from a lumped parameter form modeling describing blade motion and dynamics and thermodynamic properties inside closed vanes during compression, to a one dimensional unsteady vane filling and emptying to a full 3D fluid dynamic approach where pressure inside the oil film between stator and blade tip needs to be calculated. To correct and improve some aspects which can't be fully represented through a mathematical approach, a specific experimental activity has been developed and it offers a very suitable engineering support. This activity offers a tool to optimize SVRC thanks to a model based design with produces a great economy in terms of time to market and costs.

The paper highlights some specific features of these machines summarizing the main theoretical and experimental results achieved during the recent years.

Some particularly advanced machines have reached specific energy consumptions which have the lowest values among the rotary volumetric compressors.

Compressed air for transportation sector (truck, trains, etc...) appreciates the extremely durable, reliable and robust design of SVRCs, the possibility to optimize compressor shaping, to reduce dimensions, and to have extremely light weight solutions. The vibrations involved in a moving vehicle make the bushes involved in an SVRC a very safe and reliable and maintenance free solution. Precise screw compressor shaft bearings cannot withstand the high vibrations or the hot and dusty environment present on the underside of the train. Moreover, in the metal (aluminum) melting processes where compressed air is used to tap the sludge crusts that form on top of the molten aluminum during the refining process and also to move the material using the production of vacuums, the SVRC is particularly adept at dealing with such hot and dusty environments since the large quantities of oil injected into the compressor do a very good job at keeping the rotor/stator/blade assembly clean and cool.

A final remark concerns energy saving: SVRC are, at some pressure levels (7 bar), the most efficient machine in the market. Flow rates required by the system can be matched by increasing the number of compressors (where space permits), so ensuring the best overall energy saving. An additional potential saving can be reached and is under investigation by reducing machine speed which produces a direct reduction in friction losses. This action, which can't be done in other positive displacement machines, does not produce any volumetric efficiency change; this remains very high (close to unity), thanks to the stable oil film layers between moving and fixed surfaces in relative motion.

ACKNOWLEDGEMENTS

The Author acknowledges Ing. E. Mattei S.p.A. for continuous research funding and support. Dr. Giulio Contaldi, CEO of the Company, is particularly acknowledged. The Author is also grateful to the co-partners of the European SAVE Research Project entitled "Compressed Air systems in the European Union" with which shared a fundamental experience in the sector, [2].

REFERENCES

[1] IPCC Fourth Assessment Report: Climate Change 2007 (AR4), http://www.ipcc.ch/publications_and_data, 2007
[2] P. Radgen, E. Blaustein (Eds), Compressed Air systems in the European Union. Energy, Emissions, Saving, Potential and Policy actions, LOG_X, ISBN 3-932298-16-0
[3] Efficient Vermont – guide to Savings helping Vermont Businesses saving energy and money – Compressed air systems
[4] GPG 385 Energy efficient compressed air systems – Carbon Trust by British Compressed Air Society Ltd, February 2005
[5] Control techniques – A Guide to Energy Saving with Compressed Air, Stafford Park 4, Telford, Shropshire, TF3 3BA www.controltechniques.com, Part No. 0175- 0305 01/01, 1999
[6] Assessment of the market for Compresses Air Efficiency Services Office of energy Efficiency and renewable energies, US Department of Energy with the technical support provided by CAC (Compressed Air Challenge®), June 2001DOE – GO 102001 – 1197
[7] Improving Compressed Air System Performance: a sourcebook for industry – U.S. Department of Energy's Office of Energy Efficiency and Renewable Energy (EERE) Best Practices and the Compressed Air Challenge®, DOE/GO-102003-1822, November 2003
[8] Compressed air: opportunities for the business – ©The Carbon Trust 2012, January 2012
[9] F. Da Cunha, Compressed air: energy efficient reference guide, CEA Technology Inc. (CEATI) Customer Energy Solution Interest Group (CESIG), 2007
[10] www.cagi.org
[11] Norgren How to improve energy efficiency – Energy saving in compressed air systems – how Norgren is helping to improve energy efficiency

[12] Compressed air energy input and useful energy (Adapted from North West Energy Alliance), ©Copyright 2007 CEA Tech. Inc
[13] World Energy Outlook 2011 by IEA, International Energy Agency http://www.worldenergyoutlook.org/publications/weo-2011/
[14] Cipollone R., Contaldi G., Sciarretta A., Tufano R., A comprehensive model of a sliding vane rotary compressor system, IMechE International conference on compressors and their systems, 2005
[15] Cipollone R., Contaldi G., Villante C., Tufano R., A theoretical model and experimental validation of a sliding vane rotary compressor, 18th International Compressor Engineering Conference at Purdue, July 17-20, 2006
[16] Cipollone R., Contaldi G., Valente R., Tufano, R. An integrated two stage rotary vane compressor: theoretical optimization, IMechE International conference on compressors and their systems, 2007
[17] Cipollone R., Contaldi G., Capoferri A., Valente R., Theoretical and experimental study of the p-V diagram for a sliding vane rotary compressor, ImechE International conference on compressors and their systems, 2009
[18] Cipollone, R., Bianchi, G., G. Contaldi, Sliding vane rotary compressor Energy optimization, Proceedings of the ASME 2012 International Mechanical Engineering Congress & Exposition, November 9-15, 2012, Houston, Texas, USA, IMECE2012-85955
[19] Cipollone R., A. Sciarretta, The Quasi-Propagatory Model: a New Approach for Describing Transient Phenomena in Engine Manifolds, SAE Technical paper no. 2001-01-0579
[20] G. Valenti, L. Colombo, S. Murgia et al., "Thermal effect of lubricating oil in positive-displacement air compressors", Applied Thermal Engineering, vol. 51, no. 1-2. pp.1055-1066, Mar., 2013
[21] R. Cipollone, G. Bianchi, G. Contaldi, Ottimizzazione energetica di compressori a palette, Congresso ATI 2012. Trieste, Italy, September 2012 (in Italian)
[22] R. Cipollone, G. Contaldi, G. Bianchi, D. Di Battista, A. Capoferri, S. Murgia, Energy optimization in air compression: theoretical and experimental research activity on a sliding vane rotary compressors, Motor Driven Systems Conference, Birmingham 8-9, 11, 2011, UK
[23] A. V. Olver et al. Tip blade hydro-dynamic behavior in SVRC, Private communication, Department of Mechanical Engineering, Imperial College, London, September 2004
[24] E. Chukanova, N. Stosic, A. Kovacevic, S. Rane, Identification ad quantification of start up process in oil flooded screw compressors, IMECE 2012-86497, Proc. of the ASME 2012 International Mechanical Engineering Congress and Exposition, November 9-5, 2012, Houston Texas, USA
[25] R. Cipollone et al. – Theoretical analysis concerning the use of SVRC on the Aluminium Industry for ECL (Rio Tinto Alcan) – Private internal report produced under the Research Contract between Ing. E. Mattei S.p.A – Department of Mechanical Engineering Energy and Management, University of L'Aquila, Italy, April 2006
[26] R. Cipollone et al. – SVRC Optimization in the transportation sector for Bombardier Transportation Internal report produced under the Research Contract between Ing. E. Mattei S.p.A – Department of Mechanical Engineering Energy and Management, University of L'Aquila, Italy, June 2010
[27] R. Cipollone et al. – SVRC Optimization in a large compressor station for the Pilkington glass industry: potential of recovery and testing activity, Internal report produced under the Research Contract between Ing. E. Mattei S.p.A – Department of Mechanical Engineering Energy and Management, University of L'Aquila, Italy, April 2013

COMPRESSOR SYSTEMS

Experimental study of compressor operating characteristics and performance when using refrigerants R32, R1234yf, and two new low GWP developmental refrigerants as drop-in replacements for R410A

L Cremaschi, X Wu, A Biswas, P Deokar
Oklahoma State University, School of Mechanical and Aerospace Engineering, USA

ABSTRACT

This paper documents the operating characteristics and performances of two compressors when refrigerant R410A was replaced with refrigerants R32, R1234yf, and two new developmental refrigerants that have GWP of about 300 and 500. The thermal and volumetric efficiencies were measured for a broad range of ambient temperatures in air conditioning applications. One compressor was for residential applications while the second compressor was for commercial applications. R32 had the highest discharge temperature in both compressors. When the lubrication circulation through the compressors was intensified then lower discharge temperatures were measured at the expense of augmented compressor pumping power and pressure ratio.

NOMENCLATURE

CFCs	Chlorofluorocarbons	MFM	Mass flow meter
COMP	Compressor	OCR	Oil circulation ratio
COP	Coefficient of Performance	P	Pressure
DR	Developmental Refrigerant	POE	Polyolester
GWP	Global Warming Potential	P_r	Compressor pressure ratio
HCFCs	Hydro chlorofluorocarbons	T	Temperature
HFCs	Hydro fluorocarbons	Ts	Surface Temperature
HT	High Temperature	η_T	Compressor thermal efficiency
HVAC	Heating, ventilating and air conditioning	η_V	Compressor volumetric efficiency
LGWP	Low Global Warming Potential		

1 INTRODUCTION

Improving energy efficiency of air conditioning systems has high priority in the HVAC industry [1, 2]. To meet the terms of the Montreal Protocol, CFCs and HCFCs have been gradually phased out and they have been replaced by new blends that have zero ozone depletion potential. However, HFCs and new blends have global warming potential (GWP) that might still be of concern from an environmental perspective in case of leakage or in case of improper charge management. For example, refrigerant R410A had GWP of 2088 [3] and several researchers investigated refrigerants that could potentially retrofit R410A in air conditioning systems. Refrigerant R32 has been proposed in mini-split systems and primarily in

© The author(s) and/or their employer(s), 2013

China and Japan [4]. R32 has a GWP of 675 [3] but its flammability characteristics pose some concerns in case of leakage or in case of failure of equipment. Few studies on refrigerants that have zero ozone depletion potential and GWP less than 500 are available in the literature. Minor and Spatz [5] and McLinden [6] provided some guidelines of how low GWP refrigerants can be implemented into existing equipment. Some experimental studies for retrofitting R410A in AC systems with new low GWP developmental refrigerants can be found in the literature [7, 8], and in authors previous work [9-11]. Preliminary findings from these studies suggested that new development refrigerants were viable options. System COP and capacity were measured in air conditioning systems and heat pumps but there is not much information on the compressor operating characteristics and compressor efficiencies when the new refrigerants were used as drop-in R410A replacements.

This paper presents new data of compressors operating characteristics and compressor performance of four refrigerants, which were used as drop-in R410A replacements in residential and commercial air-source air-conditioning systems. The experimental investigation aimed to document the drop-in behaviour of refrigerants R32, R1234yf, and two new developmental refrigerants, referred to as DR-4 and DR-5 throughout the present paper (DR- stands for developmental refrigerant). The new development refrigerant DR-5 is a mixture of R32 and R1234yf in concentration of 72.5 and 27.5 by weight [12]. It has a GWP of about 500 and a temperature glide of 1°C (1.8°F) during phase change from saturated liquid to saturated vapour [8]. It is compatible with POE lubricant, chemically stable, not corrosive, and had flammability characteristics of class 2L refrigerants [13]. The new developmental refrigerants DR-4 has GWP of about 300 and had a temperature glide of 5°C (9°F) [8]. It is compatible with POE lubricant, chemically stable, not corrosive and has flammability characteristics of class 2L [8, 14]. The cycle parameters of the tested refrigerants were described in details in author's previous work [9-11].

2 EXPERIMENTAL METHODOLOGY

Two compressors were investigated in the present work and their specifications are summarized in Table 1. Both compressors were designed for refrigerant R410A and for air conditioning applications. Compressor 1 was selected from a commercially available air-source split system for ducted residential applications while compressor 2 was selected from an air-source rooftop unit for light commercial building applications. It should be noticed that both compressors were scroll type but were intentionally from different manufacturers and of different brand. The compressor 1 was pre-charged with 1.57 litres (53 oz.) of POE lubricant by the manufacturer. An additional amount of 0.65 litres (22oz.) of lubricant was charged into the compressor to compensate for the oil being retained in the suction line of split system; this additional lubricant amount was recommended by the manufacturer installation guidelines. Figure 1(a) shows compressor 1 as installed in the air-source split system and the corresponding instrumentation for the performance measurements. The refrigerant pressures were taken by absolute piezo-transducers with accuracy of ±0.13% of full scale. The refrigerant temperatures were measured by T-type thermocouples with ±0.1°C (0.2°F) accuracy. The discharge in-line thermocouple and pressure transducer, shown by T_1 and P_1 in Figure 1(a) were installed right after the 4-way reversing valve, at about 1 m (3 feet) from the compressor discharge port. The 4-way reversing valve and the pipelines were well insulated to minimize heat losses between the discharge port and the location of the temperature sensor T_1. The suction pressure was directly measured by using the outdoor unit access valve, indicated as P_2, which was at only few inches from the compressor suction port. Surface thermocouples were installed right at the compressor discharge and suction ports to correct the measurements of the discharge and suction temperatures given by the in-line temperature sensors T_1 and T_2. A Coriolis-type mass flow meter (indicated as MFM) was installed on the liquid line of the air conditioning split system to measure the refrigerant flow rate.

Table 1 Specifications of the two compressors tested in the present work

Comp	System and application	Nominal capacity [HP]	voltage [Vac]	frequency [Hz]	speed [RPM]	Swept Volume [cm³/rev]	displacement volume [m³/h]
1	Heat pump split system for residential applications	4	208–230	60	3600	47.4	9.9
2	Rooftop unit for commercial building applications	32	460–480	60	3500	77.2	16.21

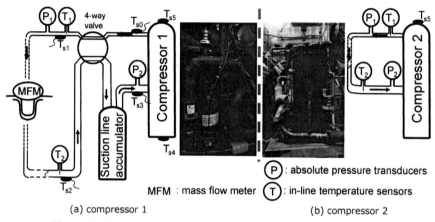

(a) compressor 1 (b) compressor 2

Figure 1 Compressors and corresponding instrumentation

Compressor 2 was also a scroll type compressor and it was installed in a dedicated compressor test set up in laboratory, shown in Figure 1(b). The compressor was pre-charged with 1.77 litre (60 oz) of POE lubricant by the manufacturer and two oil separators (not shown in Figure 1(b)) were installed at the compressor discharge to promote oil return to the compressor suction line. Figure 1(b) shows a photo and the schematic of compressor 2 and corresponding instrumentation. The dedicated compressor test set up consisted of a water-cooled condenser and a water-to-refrigerant evaporator (not shown in Figure 1(b)) that control the high side and low side saturation pressures, respectively. The in-line temperature sensors were located at about 31 cm (~1 foot) away from the compressor suction and discharge ports. The pipelines were well insulated and the in-line sensors were close enough to the compressor ports and surface temperature sensors were installed on the top shell of the compressor 2 to monitor the housing temperature at the top discharge section. The mass flow meter, pressure transducers and temperature sensors in the test set up for compressor 2 were of the same model and had the same accuracy as that for compressor 1.

3 RESULTS AND DISCUSSION

Compressor 1 was installed in an air-source heat pump system that was run in cooling mode inside a climate controlled psychometric chamber. The operating conditions were measured at AHRI design cooling conditions with outdoor temperatures of 27.8°C (82°F) and 35°C (95°F), namely B-test and A-test. Additional tests were conducted at extreme high ambient temperature conditions of

43°C and 46°C (110°F and 115°F), referred to as HT-1 and HT-2 tests respectively. The indoor conditions were constant at 26.7°C (80°F) and 50% R.H. Compressor 2 was tested in the dedicated compressor test set up and the measurements were taken at evaporation saturation temperature that ranged from 2.8 to 15°C (37 to 58°F). Refrigerant R410A was run first as baseline fluid. Then refrigerant R32, R1234yf, DR-4, and DR-5 were tested in drop-in tests. The results are presented next and were grouped for compressor 1, that is, residential applications, and for compressor 2 for light commercial building applications.

3.1 Analysis of compressor 1 operating characteristics in air-source air conditioning split systems for residential applications

Compressor discharge temperature and pressure might have an impact on compressor reliability. For example, excessive discharge temperature might cause metal fatigue of the valves and thermal stress of the lubricant. Figure 2 shows the discharge temperatures for all the refrigerants during the tests of compressor 1 when it was run in the air-source air conditioning split system. The temperatures were normalized with respect to the discharge temperatures of R410A at the same ambient temperature and testing conditions. Refrigerant R32 had 17 to 30°C (30 to 55°F) higher discharge temperature than that of R410A. DR-5 had slightly increased discharge temperature by 3 to 5°C (5.4 to 9°F). For R1234yf and DR-4, the compressor discharge temperatures were lower by about 11 to 30°C (20 to 55°F) and 5 to 9°C (9 to 16.2°F), respectively, in comparison to that of R410A.

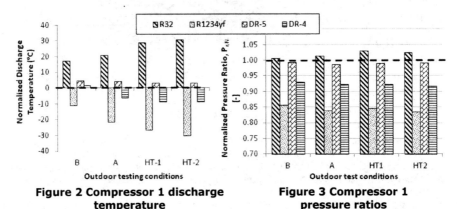

Figure 2 Compressor 1 discharge temperature

Figure 3 Compressor 1 pressure ratios

Figure 3 provides the compressor pressure ratios of all the refrigerants normalized ($P_{r,N}$) with respect to that of R410A. The normalized pressure ratios were calculated using Equation (1) and (2) below.

$$P_r = \frac{Compressor\ discharge\ pressure}{Compressor\ suction\ pressure} = \frac{P_{discharge}}{P_{suction}} \quad (1)$$

$$P_{r,N} = \frac{P_{r,ref}}{P_{r,R410A}} \quad (2)$$

In Figure 3 the refrigerants R32 and DR-5 had compressor pressure ratios similar to that of R410A for the B test conditions. At A test conditions and at extreme high ambient temperatures R32 had up to 3% higher pressure ratio in comparison to R410A while DR-5 pressure ratio was lower than that of R410A. For R1234yf and DR-4, the pressure ratios were about 15% and 7% lower, respectively, than that of R410A when the system was run at the same ambient temperatures and same indoor test conditions.

The volumetric efficiency of the compressor 1 was calculated from the pressure, temperature, and refrigerant flow rate measurements. The ratio of actual and ideal mass flow rates were estimated using Equation (3), and then normalized ($\eta_{v,N}$) with respect to R410A using Equation (4).

$$\eta_v = \frac{Actual\ measured\ mass\ flow\ rate}{Ideal\ mass\ flow\ rate} = \frac{Actual\ measured\ mass\ flow\ rate}{(suction\ density) \cdot \dot{V}_{comp}} \quad (3)$$

$$\eta_{v,N} = \frac{\eta_{v,ref}}{\eta_{v,R410A}} \quad (4)$$

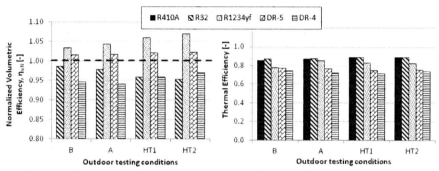

Figure 4 Compressor 1 volumetric efficiencies

Figure 5 Compressor 1 thermal efficiencies

The suction density was calculated from the pressure and temperature measurements at the suction port of the compressor while the ideal volumetric flow rate of the compressor, $\dot{V}_{comp}$, was determined from manufacturer technical data. Figure 4 shows the normalized volumetric efficiencies for all the refrigerants with respect to R410A for the entire range of outdoor temperatures. For R32 the volumetric efficiency was 1 to 2% lower than that of R410A at design A and B operating conditions. For extreme high ambient temperatures the volumetric efficiency of R32 decreased to 5% lower than R410A. DR-4 volumetric efficiency was also 3 to 6% lower in comparison to R410A. R1234yf yielded to higher volumetric efficiency by up to 7% with respect to that of R410A. Refrigerant DR-5 had similar to 2% higher volumetric efficiency than that of R410A at the same ambient temperatures.

Thermal efficiency of the compressor (η_T) was defined as shown in Equation (5)

$$\eta_T = \frac{Isentropic\ work\ of\ compressor}{Actual\ work\ of\ compressor} = \frac{h_{comp,discharge,isentropic} - h_{comp,suction}}{h_{comp,discharge,actual} - h_{comp,suction}} \quad (5)$$

Where $h_{comp,dis,actual}$ and $h_{comp,dis,isentropic}$ are the actual and isentropic enthalpies at compressor discharge and $h_{comp,suc}$ is the compressor suction enthalpy. The isentropic work was calculated from the measurements of suction temperature and pressure and discharge pressure and the actual work was calculated based on the compressor suction and discharge temperatures and pressures. The actual heat losses from the compressor shell were not accounted in the estimation of the compressor thermal efficiency. However, the compressor 1 was enclosed in an insulated box inside the unit. Since the boundaries at the compressor shell might not be perfectly adiabatic the authors refer to this quantity as thermal efficiency rather than isentropic efficiency of the compressor. Figure 5 provides the thermal efficiency data for all the refrigerants during the tests on compressor 1 when it was run in the air-source split system. It should be emphasized that the data of thermal efficiency in Figure 5 were not normalized and R410A is shown by the solid black bars. Refrigerant R32 has similar thermal efficiency as that of R410A and it ranged from 0.87 to 0.89 for the range of ambient temperatures from 27.8°C (82°F, B test) to 46°C (115°F, HT2 test). For R1234yf the thermal efficiency was slightly lower and DR-5 and DR-4 had thermal efficiencies that ranged from 0.75 to 0.77 and from 0.71 to 0.74, respectively. This result is expected since the scroll compressor was designed and optimized for R410A and the lower compressor thermal efficiency of DR-5 and DR-4 were due to their thermodynamic properties and heat transfer characteristics of the refrigerant gases during the compression process. It should be emphasized that since the 4-way valve was present during the measurements, heat is exchanged inside the 4-way valve from the hot discharge vapor line to the cold suction vapor line. The higher the temperature difference between these two streams, the higher is the heat exchange. For this reason, the 4-way valve introduced an uncertainty in the actual refrigerant temperature measurements and it was eliminated in the dedicated compressor set up used for compressor 2.

3.2 Analysis of compressor 2 of air-source rooftop unit for light commercial buildings air conditioning applications

The refrigerants were tested in a second compressor for air-source rooftop unit for light commercial buildings air conditions applications. A dedicated compressor test set up was used for these tests. Figure 6 shows the discharge temperatures of compressor 2 with refrigerants R32, DR-4 and DR-5. The temperatures were normalized with respect to the discharge temperatures of R410A at similar evaporating saturation temperatures. It should be emphasized that the data for R410A in compressor 2 were measured in presence of high oil circulation ratio (OCR) in the discharge flow rates. Two test series were conducted on R32 to investigate the effect of lubricant on the compressor discharge temperatures. The results are shown in Figure 6; in case of OCR less than 1% in weight, R32 had higher discharge temperatures of about 20 to 27°C (36 to 48.6°F) with respect to R410A. This result was consistent with the behaviour of R32 in compressor 1 discussed earlier. If the OCR increased then R32 discharge temperature was about 10°C (18°F) higher than R410A at similar evaporating saturation temperatures. In this second series of tests with R32 very high oil circulation ratio (OCR) in the system were observed and they are reported as "R32 with oil" in Figure 6. For these tests the OCR was measured by using a sampling method and resulted in the range from 10 to 20% in weight. This range is typically considered too high for actual systems. However, the laboratory experiments of "R32 with oil" reported in this section was used to provide some information about the effect of the lubricant in circulation on the R32 operating characteristics in scroll compressors. For DR-5 and DR-4 the amount of oil circulation ratio in the refrigerant flow was also measured and resulted less than 1% in weight. DR-5 had from 13 to 17°C (23.4 to 30.6°F) higher discharge temperature than R410A. DR-4 discharge temperature was about 12 to 15°C (21.6 to 27°F) higher than R410A. The difference in behaviour with respect to compressor 1 (see discharge temperature results seen in Figure 2) is due to the fact the data for R410A in Figure 6(a) were measured in presence of high OCR, which ultimately diminished the discharge temperature of the baseline test by few degrees.

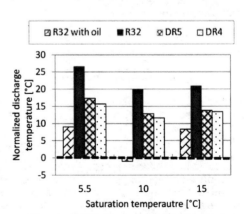

Figure 6 Compressor 2 discharge temperatures

The compression ratios of compressor 2 for R410A, R32, DR-4 and DR-5 at different evaporation saturation temperatures are shown in Figure 7. With increasing evaporation saturation temperature and at constant condensing temperature, the compression ratio decreased. Similarly to the behaviour observed for compressor 1, the pressure ratios of DR-5 in compressor 2 were about 30% lower than R410A, which had the highest compression ratio in the range of evaporation saturation temperatures tested. The pressure ratio of DR-4 and R32 were fairly close to each other throughout the range of the evaporation saturation temperatures. One may also notice that the amount of lubricating oil in circulation with the refrigerant flow impacted the compression ratio of the compressor. In the case of R32 with oil, 15 to 20% of oil circulation ratio (OCR) increased the compression ratio by about 22-31%, indicating that more work was required to achieve the same pressure lift. Thus, higher OCR reduced the discharge temperature but increased the pressure ratio and the corresponding pumping power of the compressor 2.

In Figure 8, the thermal efficiency of scroll compressor 2 is given when it was tested with R32, DR-5 and DR-4 in the case of low oil circulation ratio in the test set up. The thermal efficiency for all the refrigerants ranged from 0.6 to 0.7, which was about 10% to 20% lower than the thermal efficiency measured in compressor 1. Similar to compressor 1, refrigerant R32 thermal efficiency in compressor 2 was about 1 to 2% higher than DR-5 and 3 to 4% higher than DR-4 when the saturation temperature was lower than 10°C (50°F). However, at higher saturation temperature of 15°C (59°F), DR-4 showed the highest thermal efficiency of 0.75, which was about 15% higher than R32. It should be noted that, the data for compressor 1 were gathered during system level testing while data for compressor 2 were recorded from specific compressor tests by using a dedicated set up. Even though the results for compressor 1 are reported with reference to the air side conditions (i.e. reference to A, B, HT1, and HT2 conditions) and the results for compressor 2 are reported based on saturation temperatures, both series of data basically reported the same type of information with regard to the compressors operating characteristics at various suction and condensing pressures.

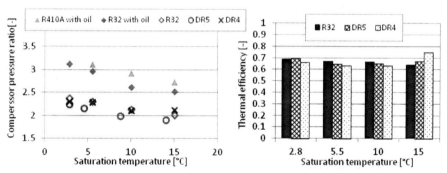

Figure 7 Compressor-2 pressure ratios

Figure 8 Compressor-2 thermal efficiencies

3.3   Analysis of lacking of proper lubrication inside compressor 2

As discussed by Hall [15], flood start, flood back, contamination, improper charging; overcharge and undercharge, and operation of the compressor outside the design envelope conditions are main reasons that could lead to mechanical failure of the compressor. Failure of the compressor 2 was observed during the experimental campaign presented in this work. The reasons of failure are unknown but severe bearing wear was observed, as indicated in Figure 9(a). The POE oil recovered from the failed compressor appeared quite dark and dense with small particles as shown Figure 9(b). These symptoms led to the conclusion that excessive wear occurred due to lacking of proper lubrication inside the compressor 2. The measurements of the electrical current indicated that the current consumption increased by 50% before failure of the compressor 2 occurred. In addition, the discharge temperature increased by 27.2°C (50°F) at similar tests conditions. These measurements suggested that oil was carried over with the refrigerant in the test set up but did not return to the compressor. It was estimated that the nominal charge of lubricant inside the compressor crankcase would be depleted in about 15 minutes of operation if no oil returned to the compressor during that period. Since the mechanical failure of compressor 2 was observed after one to two weeks of continuous testing, the authors speculated that some of the oil returned to the compressor while some oil was trapped in the test set up. After a visual analysis of the interior of the compressor, shown in Figure 9(a), it was noticed severe wearing on the bearings and scratches on the top and bottom surfaces of the scrolls. Future research is needed to clarify whether this mechanical failure was caused by a systematic oil trap in the system, by issues with the POE oil transport with the various refrigerants during drop-in tests, or by the fact that the compressor had unusual high oil concentration ratio during some of the experiments. An analysis of the refrigerant thermodynamics with the system operating characteristics was proposed in [8, 13, 14]. The experimental work of this paper build upon their results and the new data confirmed that their simulations showed similar trends as the present experiments do. However, a more detailed modelling approach of the refrigerant thermal and fluid dynamic processes inside the compressors is required to improve the accuracy of the existing models in the literature and in order to link the present results with the actual refrigerant properties and compressor operating characteristics.

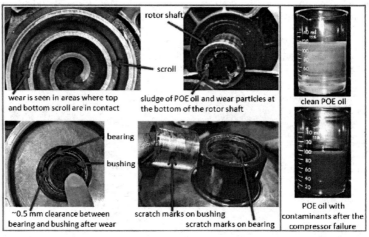

(a) Cut outs of the scroll compressor   (b) oil color comparison

Figure 9 Cut outs of the compressor 2 after mechanical failure and analysis of the lubricant before and after compressor failure

4   CONCLUSIONS

Data of compressor operating characteristics and performance were presented for refrigerants R32, R1234yf, and two new developmental refrigerants, referred to as DR-4 and DR-5. These refrigerants were used as drop-in R410A replacements candidates in two compressors, which were designed for R410A air conditioning systems: one compressor was for residential applications while the second compressor was for light commercial buildings applications.

At the same testing conditions R32 had the highest discharge temperature for both compressors. The discharge temperatures of DR-5 were slightly higher than R410A while R1234yf and DR-4 had lower discharge temperatures. Increasing the oil in circulation with the refrigerant stream decreased the discharge temperature. The compressor pressure ratio for R32 was about 5% higher than DR-5 for both compressors and it was measured that an increase of the amount of lubricating oil in circulation with the R32 refrigerant flow increased the compressor pressure ratio by as much as 31%. The volumetric efficiency of DR-5 was similar to that of R410A, while R32 and DR-4 were about 2 to 5% lower. The compressor thermal efficiencies of R32 and R410A were similar, whereas the ones from DR-4 and DR-5 drop-in tests were lower.

During the experimental campaign mechanical failure of one of the compressor tested was observed and it was due to lacking of proper lubrication inside the compressor. Future research is needed to clarify whether this failure was caused by a systematic oil trap in the system, by issues with POE oil transport with the various refrigerants during drop-in tests, or by the fact that the compressor had unusual high oil concentration ratio during some of the experiments.

ACKNOWLEDGMENT

The authors thankfully acknowledge E.I. du Pont de Nemours and Company for supporting this work and for providing the refrigerant samples.

REFERENCES

[1] M. Moezzi, "Decoupling energy efficiency from energy consumption", *Energy and Environment,* vol. 11, pp. 521-537, 2000.
[2] J. Althof, et al., "The HVAC response to the energy challenge", *ASHRAE Journal,* vol. 43, pp. 40-43, 2001.
[3] P. Forster, et al., Changes in Atmospheric Constituents and in Radiative Forcing. In: Climate Change 2007: The Physical Science Basis. Contribution of Working Group I to the Fourth Assessment Report of the Intergovernmental Panel on Climate Change. Cambridge, United Kingdom Cambridge University Press, 2007.
[4] H. Pham, "Next generation refrigerants: Standards and climate policy implications of engineering constraints", in *The 2010 ACEEE summer study on energy efficiency in buildings,* Pacific Grove, CA, USA, 2010.
[5] B. Minor and M. Spatz, "HFO-1234yf Low GWP Refrigerant Update", in *12th International Refrigeration and Air Conditioning conference at Purdue, Paper No.2349, July 14, 2008 - July 17, 2008,* West Lafayette, Indiana, United States 2008.
[6] M. O. McLinden, "Property data for low-GWP refrigerants", in *Seminar 6-- Removing Barriers for Low-GWP Refrigerants, ASHRAE Winter Meeting, January 30, 2011,* Las Vegas, NV, 2011.
[7] S. Yana Motta, et al., "Analysis of LGWP alternatives for small refrigeration (plugin) applications" in *12th International Refrigeration and Air Conditioning conference at Purdue, Paper No.2499, July 14, 2008 - July 17, 2008,* West Lafayette, Indiana, United States 2010.
[8] T. J. Leck, "Property and performance measurements of low GWP fluids for AC and heat pump applications", in *The 23rd IIR international confress of refrigeration, Paper No.610, Prague, Czech Republic, Aug 21-26, 2011,* Prague, Czech Republic, 2011.
[9] A. Barve and L. Cremaschi, "Drop-in Performance of Low GWP Refrigerants in a Heat Pump System for Residential Applications", in *14th International*

Refrigeration and Air Conditioning conference at Purdue, Paper No.2197, July 16, 2012 - July 19, 2012, West Lafayette, Indiana, United States 2012.

[10] A. Biswas and L. Cremaschi, "Performance and capacity comparison of two new LGWP refrigerants alternative to R410a in residential air conditioning applications", in *14th International Refrigeration and Air Conditioning conference at Purdue, Paper No.2196, July 16, 2012 - July 19, 2012*, West Lafayette, Indiana, United States 2012.

[11] A. Biswas, *et al.*, "An experimental study of the performance of new low global warming potential (LGWP) refrigerants at extreme high temperature ambient conditions in residential AC ducted split systems", in *2013 ASHRAE Winter Conference, Paper ID. DA-13-C013. January 26, 2013 - January 30, 2013*, Dallas, TX, United states, 2013.

[12] K. Schultz and S. Kujak, "AHRI test Report #1, System drop-in test of R410A alternative fluids (ARM-mode), in AHRI low-GWP alternative refrigerant evaluation program", AHRI, Arlington, VA, availabe online at http://www.ahrinet.org/App_Content/ahri/files/RESEARCH/AREP_Final_Reports/AHRI%20Low-GWP%20AREP-Rpt-001.pdf2012.

[13] T. J. Leck and Y. Yamaguchi, "Development and Evaluation of Reduced GWP AC and Heating Fluids" in *JRAIA International Symposium on New Refrigerants, December 2, 2010-December 3, 2010*, Kobe, Japan, 2010.

[14] T. J. Leck, "New high performance, low GWP Refrigerants for stationary AC and refrigeration", in *13th International Refrigeration and Air Conditioning conference at Purdue, Paper No.2160, July 12, 2010 - July 15, 2012*, West Lafayette, Indiana, United States 2010.

[15] J. R. Hall, "Q&A: Compressor Problems and Solutions", *Air Conditioning, Heating & Refrigeration News,* vol. 245, pp. 10-,12, 2012.

Performance and suitability comparisons of some R22 possible substitute refrigerants

A Subiantoro[1], K T Ooi[2], A Z Junaidi[2]
[1] TUM CREATE (Technische Universität München - Campus for Research Excellence And Technological Enterprise), Singapore
[2] School of Mechanical and Aerospace Engineering, Nanyang Technological University, Singapore

ABSTRACT

All the R22 substitutes from the US EPA list together with some popular alternative refrigerants were studied. From the ODP and GWP criteria, only seven substitute refrigerants (R134a, R407C, R437A, R744, R1234yf and R290) were accepted. The environmental impact, compatibility with existing R22 systems and system performance of each refrigerant were compared. The system performances were studied in systems without and with internal heat exchangers. It was found that R744 is the most environmentally friendly refrigerant. R290 and R407C are the most suitable to retrofit existing systems. R290 and R134a have superior system performances as compared to other alternative refrigerants.

NOMENCLATURE

ρ density (kg/m³)
COP   coefficient of performance (-)
h   specific enthalpy (kJ/kg)
$\dot{m}$ mass flow rate (kg/s)
$\dot{Q}$ cooling capacity (W)
$\dot{V}$ volume flow rate (m³/s)
VFR   volumetric flow ratio (-)
$\dot{W}$ compressor work (W)

Subscript
I   Case 1 (without HEX)
II   Case 2 (with 10°C sub-cooling and corresponding superheating)
III   Case 3 (with 10°C superheating and corresponding sub-cooling)
refr   refrigerant being studied

1 INTRODUCTION

The refrigerant HFCF-22 or R22 was first introduced in 1930s but its usage is only getting more popular much later. Today, it has been used for more than 40 years in air-conditioning and heat pump systems for residential and plants as well as in food refrigeration industry. However, due to its relatively high ozone depleting potential (ODP) of 0.05 and high global warming potential (GWP) of 1700 [1], it was agreed in the Montreal Protocol 2007 to phase out R22. Today, in all the developed countries, no new products are introduced with R22. Some replacements have been proposed, including those by the US Environmental Protection Agency (EPA) [2]. Many of these substitutes are short term. The world is still in the process of finding refrigerants which are good enough not only from the environmental perspective but also from financial, energy and safety consideration in their applications [3-6].

This paper compares the direct environmental impact (through global warming potential), compatibility with existing systems and system performance of R22 alternative fluids. Refrigerants considered are from the US Environmental Protection Agency (EPA) list [2] and some of the current most popular refrigerants outside of the list. In this study, a system with a standard refrigerant cycle and another with an internal heat exchanger (HEX) (see Figure 1) are investigated. To ease the comparisons, scoring scheme is used to rank the refrigerants in each category.

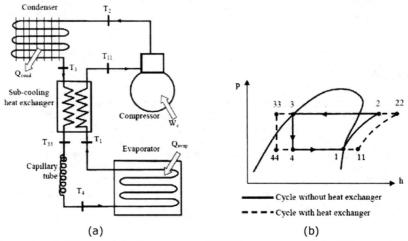

(a)   (b)

Figure 1 – (a) Schematic diagram and (b) Pressure-enthalpy diagram of a vapour compression system with a sub-cooling heat exchanger (HEX) [7]

2 COMPARISONS OF DIRECT ENVIRONMENTAL IMPACT

R22 has ODP of 0.05 and GWP of 1700. It is reasonable to expect the substitute to have better or at least comparable environmental impacts to R22. Therefore, all alternative refrigerants with ODP of more than 0 or GWP of more than 1700 are not considered. From the EPA list [2], only R134a, R407C and R437A are acceptable. Among the popular refrigerants, R744 (CO_2), R1234yf, R32 and R290 (propane) are included in this analysis. All of these refrigerants have zero ODPs. The GWPs are listed in Table 1.

Table 1 – Global warming potentials (GWPs) of various refrigerants [8-11] (1 to 7 – worst to best)

	R22	R134a	R407C	R437A	R744	R1234yf	R32	R290
GWP	1700	1300	1600	1684	1	4	675	20
Score	NA	3	2	1	**7**	6	4	5

Table 1 shows that the most environmentally friendly R22 substitute refrigerant is R744. It has the lowest GWP of all. Moreover, it is natural and not flammable. R1234yf also has low GWP. However, it is synthetic and recent analysis shows that is slightly flammable [12]. R290 is natural and has low GWP, but it is very flammable.

It should be noted that ODP and GWP only indicate the direct environmental impact of a refrigerant. To investigate the more comprehensive environmental impact, many other factors should be considered. These include the life cycle analysis of the electrical energy needed to run the compressor, the energy and materials used to produce the refrigerants, etc. These are not considered in this article.

3 COMPARISONS OF COMPATIBILITY WITH EXISTING R22 SYSTEMS

Another important consideration for the selection of suitable R22 substitutes is the compatibility with the existing R22 systems. It is desirable to have an alternative refrigerant that can be easily used for retrofitting without redesigning the compressor, in particular, and the entire system, in general. Three good indicators of the retrofitting potential of an alternative refrigerant are the similarities of the evaporating pressure, condensing pressure and volumetric flow rate of the compressor suction with those of R22.

The comparison of evaporator and condenser pressures at various evaporating and condensing temperatures are shown in Figure 2. The thermophysical properties are obtained from REFPROP [13]. The condenser pressures of R744 are not shown as at the corresponding condensing temperatures, the state of the refrigerant is supercritical and the cycle is a transcritical cycle. A gas cooler, instead of a condenser, is used in this case and the pressure is usually in the range of 90 bar and above, way higher than the typical R22 condenser pressures.

As shown in Figure 2, the refrigerant with the closest evaporator pressure profile to R22 is R290. The refrigerants with the closest condenser pressure profile are R407C and R290, depending on the condensing temperature.

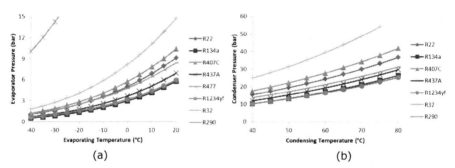

Figure 2 – (a) Evaporator and (b) condenser pressures of various refrigerants at different evaporating and condensing temperatures

To compare the similarities of the compressor suction volumetric flow rates of various refrigerants with that of R22 at the same cooling capacity, $\dot{Q}$, a dimensionless parameter called the "volumetric flow ratio (VFR)" is introduced. The definitions are shown in Equations (1) and (2). A refrigerant with a good retrofitting potential should have VFR values close to unity. It is noted that when designing a new system, and not retrofitting an existing system, a low VFR is actually more desirable as it indicates that the refrigerant requires a smaller compressor for the same cooling capacity.

$$\dot{Q} = \dot{m}(h_1 - h_4) = \rho_1 \dot{V}_1 (h_1 - h_4) \tag{1}$$

$$VFR = \dot{V}_{1,refr}/\dot{V}_{1,R22} = [\rho_{1,R22}(h_{1,R22} - h_{4,R22})]/[\rho_{1,refr}(h_{1,refr} - h_{4,refr})] \quad (2)$$

It is assumed that the refrigeration cycle follows that of the "Cycle without heat exchanger" in Figure 1, where at the expansion device inlet the refrigerant is saturated liquid and at the compressor inlet it is saturated vapour. The condensing temperature is fixed at 54.4°C while the evaporating temperature is varied.

The VFR data are shown in Figure 3. R744 is again excluded from the comparison due to the requirement of a transcritical cycle at the corresponding condensing temperature. It can be seen that R290 and R407C have the closest volumetric flow rate at the compressor suction to R22.

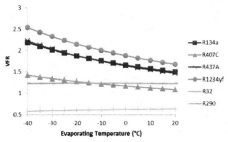

Figure 3 – Volumetric flow ratio (VFR) of various refrigerants at various evaporating temperatures

To summarize the compatibility comparison of various R22 substitutes in this section, the scores of each of the fluids considered in the three parameters studied are tabulated in Table 2. From the total score, it can be seen that R290 is the R22 substitute with the highest retrofitting potential. R407C is a close second. They operate at similar pressure conditions. Moreover, the volumetric flow rates at the compressor suction are also similar to that of R22, indicating a similar compressor size requirement. The least compatible substitute is R744 due to its absolutely different properties with R22.

Table 2 – Compatibility scores of various R22 substitute refrigerants (1 to 7 – worst to best)

Scores	R134a	R407C	R437A	R744	R1234yf	R32	R290
Evaporator pressure	3	6	5	1	4	2	7
Condenser pressure	4	6.5	5	1	3	2	6.5
Volumetric flow rate	3.5	6.5	3.5	1	2	5	6.5
TOTAL	10.5	19	13.5	3	9	9	20

4 COMPARISONS OF SYSTEM PERFORMANCES

To compare the system performances of the refrigerants, three (3) cases were simulated (see Figure 1):
1. Refrigeration system without an internal heat exchanger (HEX)
    It is assumed that the refrigerant at the evaporator outlet and compressor inlet (point 1) is at saturated vapour state. At the condenser outlet and expansion device inlet (point 3), it is at saturated liquid state.

2. **Refrigeration system with an internal heat exchanger (HEX) #1**
   It is assumed that the HEX causes 10°C sub-cooling after the condenser outlet and the corresponding superheating after the evaporator outlet. No heat loss exists in the HEX, so all the heat from the hot refrigerant is transferred to the cold refrigerant.
3. **Refrigeration system with an internal heat exchanger (HEX) #2**
   It is assumed that the HEX causes 10°C superheating after the evaporator outlet and the corresponding sub-cooling after the condenser outlet. No heat loss exists in the HEX, so all the heat from the hot refrigerant is transferred to the cold refrigerant.

In all cases, it is assumed that the compression process is isentropic, the expansion process in the expansion device is isenthalpic and there is no pressure loss in the pipes. The condensing temperature is uniform at 54.4°C except for the R744 system where the system is assumed to be operated in a transcritical cycle with a discharge pressure of 90 bar. The temperature of the R744 fluid entering the expansion device in Case 1 is 54.4°C and in Case 2 is 44.4°C. To have a fairer comparison, the size and operating speed of the compressor are fixed, resulting in uniform volumetric flow rates at the compressor suctions across all the systems. To ease the comparison further, the flow rates at the compressor suction are arbitrarily set as 0.001 m³/s.

Four parameters were studied to characterize the system performances:
1. Cooling capacity
2. Coefficient of Performance (COP)
3. Compressor discharge pressure
4. Compressor discharge temperature

4.1 Cooling capacity

Cooling capacity indicates the amount of heat the system can remove from the refrigerated space over time. It is equal to the change in the specific enthalpy of the refrigerant in the evaporator caused by the refrigeration load multiplied by the mass flow rate of the refrigerant. It is expressed in Equations (3a) and (3b) for Cases 1, 2 and 3. The cooling capacities of Cases 2 and 3 are based on the enthalpy of point 1, not 11, because the heat from the refrigeration load only increases the enthalpy from point 44 to point 1. The further change from point 1 to point 11 is caused by the HEX and therefore, should not be included in the cooling capacity calculation.

$$\dot{Q}_I = \dot{m}(h_1 - h_4) = \rho_1 \dot{V}_1 (h_1 - h_4) \tag{3a}$$
$$\dot{Q}_{II \text{ or } III} = \dot{m}(h_1 - h_{44}) = \rho_{11} \dot{V}_{11} (h_1 - h_{44}) \tag{3b}$$

The values of the cooling capacities of various refrigerants at various evaporating temperature of all cases are plotted in Figure 4. As mentioned above, the volumetric flow rates at the compressor inlet are set as 0.001 m³/s. From the figure, it can be seen that R32 consistently produces the largest cooling capacity. It is also the only R22 substitute studied here that consistently produces higher cooling capacity than R22. This is consistent with the data shown in Figure 3, where R32 is also the only R22 substitute with a VFR less than unity which indicates a higher cooling capacity generation.

Figure 4 also shows the interesting behaviour of R744. It produces very small cooling capacities when used in a system without HEX. When the evaporating temperature is too high, it cannot even produce any positive cooling capacity. However, when a HEX with a 10°C sub-cooling is used, its cooling capacity is increased significantly. If a HEX with a 10°C superheating is used instead, the increase is less significant. This is because the amount of heat transferred through the HEX in Case 2 is more than that in Case 3, resulting in a more efficient system in Case 2.

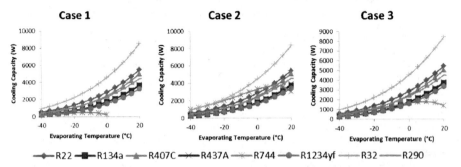

Figure 4 – Cooling capacity of various refrigerants at various evaporating temperatures for cases without and with HEX

4.2 Coefficient of Performance (COP)

The main characteristic of an energy efficient refrigeration system is the Coefficient of Performance (COP). COP is defined as the cooling capacity produced per unit of work required. It is expressed in Equations (4a) and (4b) for Cases 1, 2 and 3. For simplicity, the work requirement is assumed to be equal to compression work, neglecting any mechanical or thermodynamics losses. A good refrigeration system should have a high COP.

$$COP_I = \frac{\dot{Q}_I}{\dot{W}_{I,comp}} = \frac{(h_1 - h_4)}{(h_2 - h_1)} \tag{4a}$$

$$COP_{II \text{ or } III} = \frac{\dot{Q}_{II}}{\dot{W}_{II,comp}} = \frac{(h_1 - h_{44})}{(h_{22} - h_{11})} \tag{4b}$$

The COP values of various refrigerants at various evaporating temperatures for all cases are plotted in Figure 5. It can be seen that most of the COPs of the R22 substitutes studied are somewhat comparable to that of R22 except for R744. Upon closer inspection, R22 is still superior, R134a is a close second while R744 and R407C consistently exhibit the lowest COPs of all the refrigerants studied.

It is interesting to recall that R32 consistently produces the largest cooling capacity in Figure 4. However, its COP is not superior here, indicating high compressor work requirements.

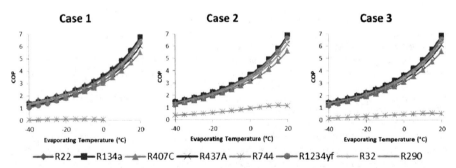

Figure 5 – COP of various refrigerants at various evaporating temperatures for cases without and with HEX

4.3 Compressor discharge pressure

Compressor discharge pressure (which is equal to the condenser pressure) is dictated by the condensing temperature, except for R744. From design point of view, it is more preferred to have a system with a lower compressor discharge pressure to minimize the wall thickness requirement. From Figure 2(b) shown in Section 3, it can be seen that R1234yf is the R22 substitute with the lowest corresponding discharge pressure. R744 (not shown) is the highest of all, reaching about 90 to 100 bar due to the requirement to operate at transcritical conditions.

4.4 Compressor discharge temperature

It is desirable to have a system with a low compressor discharge temperature as it will cause less strain on the compressor, in which will lead to a longer compressor life. Assuming isentropic compression process and condensing temperature of 54.4°C or 90 bar condenser pressure for R744, the compressor discharge temperatures of various refrigerants in all cases can be computed. These are shown in Figure 6.

It can be seen that R32 consistently produces the highest discharge temperature followed by R744. The other R22 substitutes have lower compressor discharge temperatures than R22 with R1234yf always produces the lowest temperature. However, one caution is noted here where the discharge temperature of R1234yf in Case 1 is always 54.4°C. This means that the compression process involves compression of a vapour-liquid mixture, which is not desirable. Therefore, R1234yf has to be operated with some degree of superheating at the compressor suction.

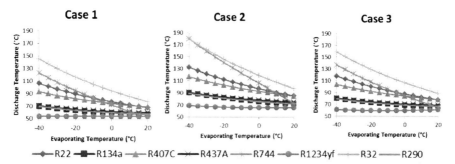

Figure 6 – Compressor discharge temperature of various refrigerants at various evaporating temperatures for cases without and with HEX

4.5 Summary of performance comparison

With all the four parameters studied, the performances of the R22 alternative refrigerants studied are summarized in Table 3. It can be seen that from the total score, R290 has the most superior general performance as compared to the other R22 substitutes. It is interesting to note however, that it may not be the most superior in each category. For example, in terms of cooling capacity production, R32 and R407C are better than R32.

The worst performing refrigerant is R744. However, it should be noted that R744 operates at a transcritical cycle in the assumed operating conditions in this study. The cycle conditions are not optimized and hence, may not give the optimal results.

Table 3 – Performance scores of various R22 substitute refrigerants
(1 to 7 – worst to best)

Category	Case	R134a	R407C	R437A	R744	R1234yf	R32	R290
Cooling capacity	1	3	6	4	1	2	7	5
	2	2	5	3	6	1	7	4
	3	3	6	4	1	2	7	5
COP	1	7	2	3.5	1	3.5	5.5	5.5
	2	7	2	3	1	5	4	6
	3	7	2	3	1	4	5	6
Discharge pressure		6	3	5	1	7	2	4
Discharge temperature	1	5.5	3	4	2	NA	1	5.5
	2	5	3	4	2	7	1	6
	3	5	3	4	2	7	1	6
TOTAL		50.5	35	37.5	18	38.5	40.5	53

5 CONCLUSIONS

Seven R22 possible substitute refrigerants (R134a, R407C, R437A, R744 (CO_2), R1234yf and R290 (propane)) were investigated. The environmental impact (particularly ODP and GWP), compatibility with existing R22 systems and system performance of each refrigerants were compared.

To study the system performances, three cases were simulated: 1) without an internal heat exchanger (HEX), 2) with a 10°C sub-cooling and corresponding superheating and 3) with a 10°C superheating and corresponding sub-cooling. The base condensing temperature was 54.4°C. For the R744 system, the gas cooler pressure was set as 90 bar. The volumetric flow rate at the compressor suction was kept constant and uniform. Four parameters were of main consideration to characterize the system performance, namely the cooling capacity, the coefficient of performance (COP), compressor discharge pressure and compressor discharge temperature.

A scoring system was used to simplify the comparison. The summary of the ranks of each refrigerant in the three comparison categories are tabulated in Table 4.

Table 4 – Ranks of various R22 substitute refrigerants in various aspects
(I to VII – best to worst)

Rank	R134a	R407C	R437A	R744	R1234yf	R32	R290
Environmental impact	V	VI	VII	I	II	IV	III
Compatibility	IV	II	III	VII	V	V	I
Performance	II	VI	V	VII	IV	III	I

The findings are:
1. In terms of direct environmental impact through global warming potential (GWP), R744 is the most preferred refrigerant while R437A is the least preferred. R1234yf and R290 are 2nd and 3rd, respectively. However, one main weakness of R1234yf and R290 is that they are flammable and hence, less safe as compared to the other refrigerants.
2. In terms of compatibility with existing R22 systems, R290 is the most preferred followed by R407C. R744 is the least preferred as it requires extremely high pressure conditions. It has the most similar pressure conditions at various evaporating and condensing pressures to R22. It also has the most similar

volumetric flow rate at the compressor suction to R22 at the same cooling capacity. These allow the use of the same pipes and other system components as those of R22.
3. In terms of general system performance, R290 is also the most superior of all the R22 substitutes followed by R134a. R744 is the lowest ranked.
4. More detailed analysis of the system performances comparisons also show the followings:
    a. R32 consistently produces the largest cooling capacities, but it requires high discharge pressure and temperature.
    b. R134 consistently produces the highest COPs, but it produces comparably smaller cooling capacity at the same flow rate.
    c. R1234yf has the lowest discharge pressure and temperature at the same operating conditions, but it produces comparably smaller cooling capacity as compared to other refrigerants.

From the findings, it can be concluded that:
1. For the very short term, R290 or R407C are most suited for retrofitting of existing R22 systems.
2. For the medium term, R290 and R134a are more preferred for development of new systems.
3. For the long term, depending on the state of the technological development of the system, R744 should be used for future refrigeration systems.

ACKNOWLEDGMENT

This work was financially supported in part by the Singapore National Research Foundation under its Campus for Research Excellence And Technological Enterprise (CREATE) program. The views expressed herein are solely the responsibility of the authors and do not necessarily represent the official views of the Foundation.

REFERENCE LIST

[1] Refrigerants - Ozone Depletion (ODP) and Global Warming Potential (GWP), http://www.engineeringtoolbox.com/Refrigerants-Environment-Properties-d_1220.html (accessed April 29, 2013).
[2] Acceptable Substitutes in Household and Light Commercial Air Conditioning, http://www.epa.gov/ozone/snap/refrigerants/lists/homeac.html (accessed April 29, 2013).
[3] Fatouha, M., Ibrahim, T.A., Mostafa, A., 2010, Performance assessment of a direct expansion air conditioner working with R407C as an R22 alternative. Applied Thermal Engineering, 30(2–3):127–133.
[4] Cabello, R., Torrella, E., Navarro-Esbri,' J., 2004, Experimental evaluation of a vapour compression plant performance using R134a, R407C and R22 as working fluids. Applied Thermal Engineering, 24(13):1905–1917.
[5] Douglas, J.D., Braun, J.E., Groll, E.A., Tree, D.R., 1999, Cost-based method for comparing alternative refrigerants applied to R-22 systems. International Journal of Refrigeration, 22(2):107–125.
[6] La Rocca, V., Panno, G., 2011, Experimental performance evaluation of a vapour compression refrigerating plant when replacing R22 with alternative refrigerants. Applied Energy, 88(8):2809–2815.
[7] Bolaji, B.O., Huan, Z., 2012, Comparative analysis of the performance of hydrocarbon refrigerants with R22 in a sub-cooling heat exchanger refrigeration system. Journal of Power and Energy, 226(7):882-891.

[8] National Refrigerants – R437A Safety Data Sheet, http://www.nationalref.com/PDF's/MSDS/Refrigerants/MSD437A.pdf (accessed April 29, 2013).

[9] R407F Refrigerant - http://www.actrol.com.au/Products/Refrigerant-Oils--Gases/Refrigerant/HFC/R407F-Refrigerant/ (accesed April 29, 2013).

[10] NIST Property data for low-GWP refrigerants: - http://tc31.ashraetcs.org/pdf/Property%20Data%20Low%20GWP%20Refrigerants.pdf (accessed April 29, 2013).

[11] Honeywell / Dupont Joint Collaboration, HFO-1234yf A Low GWP Refrigerant for MAC - http://www.1234facts.com/wp-content/uploads/2012/06/sae_world_congress.pdf (accessed April 29, 2013).

[12] ACR News, Daimler shuns new refrigerant R1234yf - http://www.acr-news.com/news/news.asp?id=3002&title=Daimler+shuns+new+refrigerant+R1234yf (accessed April 30, 2013).

[13] Lemmon, E.W., Huber, M.L., McLinden, M.O., 2007, NIST Standard Reference Database 23: Reference Fluid Thermodynamic and Transport Properties - REFPROP. Gaithersburg, National Institute of Standards and Technology, Standard Reference Data Program.

Oil level measurement in oil-injected screw compressor packages used in the petroleum, petrochemical, refrigeration and fuel gas markets

W J Milligan, D I Muir
Howden Process Compressors, UK

D K Harrison
Glasgow Caledonian University, UK

ABSTRACT

Screw compressors used in the petroleum, petrochemical, refrigeration and fuel gas tend to feature lubrication, shaft-sealing and oil-control systems designed to the general requirements of API-614. These systems require oil level measurement for process control, alarm and trip functions. This paper focuses on the level measurement within the package's primary separator. Oil injected screw compressor packages inject oil not only into the drive train bearings but also into the compressors working chamber. During the compression process the gas becomes entrained within the oil and the primary separator vessel is used to separate the two mediums. The level of the oil within the separator requires to be monitored as too high a level would increase the amount of oil carry over into the process and too low a level could eventually cause a process trip. The process and internal design of the separator vessel does not lend itself to easy measurement such that the oil to be measured can be at a low or high temperature, the dielectric constant varies with the amount and type of gas entrained within the oil and the viscosity of the oil to be measured varies with its temperature. Add to this the oil and gas mixture discharged from the compressor is being deliberately blasted against a diffuser to separate the two mediums makes for difficult measurement.

This paper discusses each of the different parameters in turn with a view to aligning them with measurement technologies before concluding with the most suitable technology for this application.

1 OIL-INJECTED SCREW COMPRESSOR LUBRICATION OIL SYSTEM

1.1 Overview
The lubrication oil system of an oil-injected screw compressor employed in petroleum, petrochemical, refrigeration and fuel gas applications and designed in general accordance with API-614 are quite complex. Before starting on the overview it is important to understand that the package lubrication system used in these applications essentially follow the same design with only the physical parameters (pressures, temperatures, flows and dimensions) involved vary for each individual package with respect to the process it has been deployed in. The first item to note is the difference between an oil-injected compressor's lubrication oil system and an oil-free compressor's lubrication oil system. The difference being an oil-injected system features separation vessels that double as reservoirs whereas an oil-free compressor has a more basic reservoir and is completely separate from the gas

© The author(s) and/or their employer(s), 2013

system. From the primary separator vessel, two oil pumps supply oil through separate pressure control and temperature control systems before arriving at the oil filter. Post filter is a selection of valves that allow the oil to circulate freely or be injected directly into the compressor bearings and working chamber. If the compressor is running the oil returns to the primary separator mixed with the process gas to be separated or, if the compressor has stopped, any excess oil in the chamber can return via the drain valve. This is shown in Figure 1 below. Control of the lubrication oil system is discussed later in the paper.

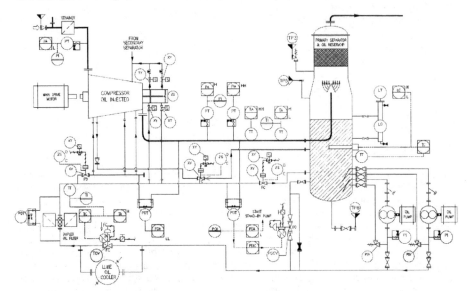

Figure 1, the piping and instrumentation diagram of a lubrication oil system for an oil-injected screw compressor.

From Figure 1, it can be seen that the system monitors both the pressure and temperature of the lubrication oil.

1.2 Primary Separator
The primary separator is a pressure vessel and must be designed to a suitable design code such as ASME VIII Division 1 or PD5500. As mentioned previously, the primary separator on oil-injected packages has two functions, one is to separate the gas and oil mixture that is discharged from the compressor and the other is to act as a reservoir for the oil pumps. The size of the vessel, both diameter and height, are sized so both functions can be carried out.

1.2.1 As a separation process
The sketch shown in Figure 2 highlights the main parts of the vessel. The inlet device, or diffuser, is connected to the end of the compressor discharge piping and situated in the centre of the vessel. The diffuser is carefully designed to reduce the velocity and to allow slow moving droplets of oil to separate from the gas stream under gravity. The diffuser effectively disengages the bulk of the oil, reducing the overall load on the demister.

After the bulk of the oil has been knocked out of the gas using the diffuser, the process gas passes through a high performance demister as the next part of the separation process. The demister is self-cleaning in operation, have a very small pressure drop (150Pa), require no maintenance and can achieve separation levels down to about 100ppm by weight in isolation.

After the process gas exits the primary separator some packages, depending on the application, have a secondary separator installed to further reduce the oil entrained in the gas to only a few parts per million. The oil collected by the secondary separator is returned back to the lubrication oil system.

1.2.2 As an oil reservoir

The oil capacity of the primary separator is such that it will contain two minutes retention capacity where retention capacity is defined as the total volume below the minimum operating level (1). The mimimum volume of oil in the vessel therefore requires to be two times the normal flow of oil to the compressor where normal flow is defined as the total amount of luid required by equipment components such as bearings, seals, couplings and controls excluding transient flow for controls or fluid bypassed directly back to the reservoir (1).

The oil level within the separator at start-up must be below the rotors, this is to prevent the situation where the compressor is trying to start-up and compress oil, resulting in a much increased starting torque. The compressor rotor shaft on the package will be at such a height to ensure sufficient oil will drain from the compressor casing drain back to the separator.

Oil level indication is required on the separator. This could be a sight glass, level switch or continuous measurement transmitter. The range of the indication in the vessel should span from 50mm below the lowest operating level to 50mm above the rundown level where the rundown level is defined as the highest level that the oil reservoir can reach when the entire system is shut down. When the system shuts down, the operating level while the compressor is running may have been at 60% but as all the oil in the system drains back to the reservoir after it has shut down the rundown oil level may rise substantially. It is essential that if a level gauge and a level transmitter are both fitted they should both have the same range to prevent operators reading one level on the gauge but the control system reading another.

Primary Separator Operation Description:

Demister
- Knitted Mesh Pad.
- Oil mist condenses.
- High efficiency oil removal from process gas.

Gas with <200ppm Oil exits vessel to secondary separator.

Diffuser
- Inlet Device.
- Forces Oil/Gas to change direction.
- Slows down.
- Oil drops out gas into reservoir.

Gas/Oil mixture enters from compressor discharge.

Oil Reservoir
- Store required Oil for compressor operation.
- Sized for Oil Level control of system.
- Allow rundown of system after shutdown.

Figure 2, a sketch of a primary separator detailing the demister, diffuser and reservoir.

It is important to understand that the oil level indication, whether it be a gauge or transmitter, is only used for alarm and indication: no tripping function is required to protect the package or indeed any fiscal measuring or custody transfer. There is no need for high accuracy. If the oil level fell such that the pump discharge pressure fell, the transmitters monitoring the oil to gas differential pressure would detect this and shut the package down. A direct comparison can be drawn between the level instrument on the primary separator and the fuel gauge on a motor car, it will alert you to a low level but will continue running until empty.

From the nozzles on the primary separator an additional chamber, sometimes referred to as a bridle, is attached to house the level instruments. The reason for the separate chamber is to avoid inaccurate readings as a result of the chaos caused by the compressor gas and oil mixture discharging against the diffuser. A common arrangement, minus isolation valves, is shown in Figure 3.

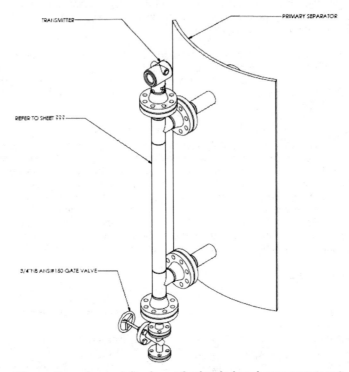

Figure 3, an isometric view of a basic level measurement chamber attached to a primary separator.

The bridle shown in Figure 3 is a very simple design featuring one transmitter. More elaborate bridle designs, featuring multiple transmitters and gauges in separate chambers have also been installed. The available measurement technologies are discussed later.

1.2.3 System Control
The reader should note that because this paper is focusing solely on the oil system surrounding the primary separator many other parts of the functionality of the control system, such as the gas system and the slide valve, have been omitted for clarity.

Primary separators are fitted with oil heaters either to prevent attempts being made to start and operate the compressor with oil so cold and viscous that it will not circulate properly. This is particularly important for packages in cold environments. The heater also prevents the build up of liquid refrigerant (condensate) in the separator during shutdown. The heater should be fitted below the minimum oil level, have a safety high temperature cut-out and have a maximum rating of

$2W/cm^2$ to prevent burning the oil. The heater should be sized appropriately to heat the oil within the separator within twelve hours (2).

During normal operation the oil level within the separator will gradually reduce as a result of the oil carry over. To prevent having to stop the package in order to top up the oil level a complex piping arrangement featuring multiple non-return valves, top-up tank and pump can be installed to facilitate replenishing the oil while the package is still running. The top-up pump discharge pressure must be higher than the compressor discharge pressure in order to force the oil into the separator. The starting and stopping of the top-up pump is a manual function which can be done via a local control station or the human machine interface (HMI).

The lubrication oil system, including the primary separator, is controlled by the package control system but before any automatic functions concerning the lubrication oil system can be carried out the end user must first: select a duty lubrication oil pump; select a lubrication oil temperature setpoint and set an oil to gas differential pressure setpoint. During the power-up sequence of the package the programmable logic controller (PLC) or client's distributed control system (DCS), whichever is controlling the package, will carry out a number of automatic functions including enabling the separator oil heater control and setting the valves to the correct position. Low oil to gas differential pressure alarms and trips are inhibited to prevent needless alarms. During the power-up sequence the conditions within the primary separator are static with only the heater energised but the oil still at a low temperature and the oil free of any process gas.

In order to start the pre-start sequence certain conditions must first be met. The primary separator level must be above the low alarm limit and the lubrication oil temperature within the separator must be above 10^0C. When these parameters are satisfied the lubrication oil system can be started to allow the system to get up to pressure and temperature prior to the compressor start. During the pre-start sequence the duty lubrication oil pump pumps oil from the separator, through the oil to gas differential pressure control system, through the temperature control system, through the filter before bypassing the compressor and returning to the separator via an oil re-circulation valve and separate nozzle. Once the oil manifold temperature, different from the separator temperature, has maintained a steady $30\sim32^0C$ the lubrication oil system is ready for a compressor start. During the pre-start sequence the conditions within the primary separator are relatively static with the oil still free of any process gas, still at a relatively low temperature and entering the vessel from conventional nozzles.

During start-up of the actual compressor the drain valve is closed, the oil-re-circulation valve is closed and the compressor manifold isolation valve is opened allowing oil to be sent to the compressor bearings and working chamber. After these valves have travelled and the oil/gas differential pressure control valve has re-established a healthy pressure the main drive motor of the compressor is started. During the start-up sequence the conditions within the primary separator are the same as the pre-start conditions.

When the compressor is running the conditions inside the primary separator can only be described as chaotic. The oil and gas mixture discharged from the compressor is blasted off the diffuser sending it everywhere, the heat generated by the compressor sends the separator temperature soaring and the demister pads are also dropping oil back into the reservoir before being sucked back out of the separator by the lubrication oil pump to go round again.

The stopping sequence, concerning the lubrication oil system, is a reversal of the start-up sequence where after the main driver has been stopped, the manifold isolation valve is closed, the re-circulation valve is opened, and the drain valve is opened. This allows any oil within the chamber to drain back to the separator but also allows the oil to continue to circulate in the event that the system has to be re-started within a short time frame. During the pre-start sequence the conditions within the primary separator are relatively static with the oil still free of any process gas, still at a relatively low temperature and entering the vessel from conventional nozzles. During the stopping sequence the conditions within the primary separator return to a relatively static condition similar to the pre-start condition with the oil still at a relatively low temperature and entering the vessel from conventional nozzles. The difference being the oil is now entrained with process gas for a retention time. Retention time is the time allowed for disengagement of entrained gas (1).

During emergency stopping of the package, all power is removed from the main drive motor, auxiliary drive motors (pumps), heaters and valves returning the primary separator to a static condition.

1.2.4 Process Dynamics
During the System Control section the differences in conditions within the primary separator during the various running conditions was described. Taking the temperature first, during a pre-start sequence the oil temperature could start off at -20^{0}C, warm up to 32^{0}C during pre-start before continuing all the way up to around 100^{0}C after the compressor has been running for a short period of time. As the oil temperature changes, so does its viscosity (3)(4).

The density of the oil also changes within the separator between pre-start and running. The large increase in temperature makes the oil expand and as density is defined as the mass of fluid per unit volume, the mass stays the same and the volume increases thus reducing the density (5). Further to this, as the process gas gets entrained in the oil the density reduces further.

Some level measurement technologies require the dielectric constant of the media it has to measure. The dielectric constant is widely assumed to be another form for relative permittivity or the amount of energy that can be stored in a material. The dielectric constant of a material is dimensionless because it is a ratio of the permittivity of the material under discussion to the permittivity of vacuum (6). The dielectric constant of a material can be impacted by a number of factors such as the measurement frequency, pressure, temperature, humidity, concentration, contamination and polarisation (7).

So we can conclude that during the course of a few hours the oil temperature, viscosity, density and dielectric constant of the oil that is to be measured can change while bearing in mind the chaotic conditions within the primary separator itself.

The process gas composition can also be dynamic. The packages deployed in petroleum, petrochemical, refrigeration and fuel gas markets are designed to last a minimum of 25 to 30 years. The gas composition over the life of the package can change considerably. The make up of the gas within an oil and gas reservoir for instance is a function of position it is extracted from within the reservoir. As the reservoir is formed over many years, processes such as oil migration, fluid convection and diffusion as well as the effects of horizontal and vertical temperature gradients all contribute to the make varying gas compositions within the one reservoir(8). It has also been found that as wells are depleted the overall composition differs. At the beginning of its life the well produces mainly dry gas made up of methane, nitrogen and carbon-dioxide. As this is depleted wet gas is extracted this time composed up of methane but also light hydrocarbons. The gas condensate phase is next producing gas and heavy hydrocarbons so we can conclude that during the course of a few years, let alone the lifespan of the package, the gas composition can change thus adversely affecting the oil properties for level measurement as it is entrained in the oil.

2 LEVEL MEASUREMENT TECHNOLOGIES

2.1 Guided Wave Radar

Guided wave radar (GWR) transmitters are mounted at the top of the tank or chamber with a probe extending down to the bottom of the measurement range. A low energy pulse of microwaves is sent down the probe. At the point of the liquid level a large proportion of the waves are reflected back up the probe to the transmitter. The transmitter measures the time between the transmitted signal and the received echo and calculates the level. As a proportion of the waves continue down the probe, a second echo can be detected from an interface between two different liquids. Changes in density will not impact on measurement accuracy and with no moving parts they are very low maintenance (9).

GWR level transmitters offer many advantages for level measurement within the primary separators such as: low maintenance; highly reliable and unaffected by changes in density. However, if we bypass the manufacturers marketing we find that it can be affected by pressure, temperature, humidity, concentration, contamination all of which occur within the primary separator. Therefore in order to use GWR transmitters confidently and accurately within primary separators the measurement inaccuracies caused by the change in dielectric properties of the oil must first be quantified and engineered out as far as reasonably practicable.

2.2 Pressure

Pressure transmitters are a common and well understood technology especially within the petroleum, petrochemical, refrigeration and fuel gas markets and they are extremely economical compared to other technologies. If a level is to be made in an open vessel a pressure transmitter (PT) is placed at the bottom and the head pressure caused by the weight of the liquid can be used to calculate the height of the liquid. Differential pressure transmitters can be used on pressurised vessels using the same principles (9).

Level measurement using pressure transmitters cannot be deployed in primary separators as the change in temperature will impact the measurement, the change in density of the oil during operation will impact the measurement and any fluctuation of the oil/gas differential pressure control valve or pressure disturbances would also impact on the reading.

2.3 Magnetostrictive

This technology measures the intersection of two magnetic fields, one in a float and the other in a guide tube. The float is free to travel up and down the guide tube as the liquid level changes. The transmitter sends a low current down the guide tube and when it reaches the magnetic field generated by the float a torsional twist is initiated creating a sonic wave which is detected and timed with the level then calculated (9).

Magnetostrictive transmitters have a few limitations also. The moving parts, over the period of many years, may eventually foul and as such if they were to be deployed within the level measurement chamber would need scheduled into a maintenance programme where they are removed and cleaned every few years. The float would need to be designed to work over a range of densities also as a float designed only to work in unsaturated oil at 32^0C would quickly sink when the compressor started running.

2.4 Ultrasonic

An ultrasonic level transmitter will house both the transmitter and receiver alongside some control electronics. A sound pulse transmitted from the unit is reflected back off the surface of the liquid and received as an echo by the receiver. The measured propagation time is proportional to the distance between the instrument and the liquid surface. A major disadvantage ultrasonic level measurement has is the transmitter takes a short period of time to change from transmit to receive mode meaning there is a minimum distance the transmitter should be placed above the maximum level. The measurement is also influenced by temperature change so most units have a temperature sensor incorporated to compensate for any change in temperature (9).

Ultrasonic level measurement must also be discounted as the mist created in the separator will impact on the measurement and the physical dimensions of the chamber requirements, provided it was to be situated in a separate chamber, would prevent it from being fitted to most chambers. Interference from nozzles, mist, droplets and the diffuser prevent it from being mounted within the main separator.

Figure 4, a primary separator with test chamber and reference gauge used to trial level devices.

3 CONCLUSIONS

End users within the petroleum, petrochemical, refrigeration and fuel gas markets tend to favour the technology with the least moving parts as they reason that with no moving parts they cannot foul and will require less maintenance. This is because unplanned maintenance on a transmitter may involve shutting a package down and in turn this may shut a particular process down or cause a gas turbine to come offline or any number of extremely expensive consequences.

Another consideration the end user has is the amount of spare transmitters they keep on site. This is not so much a problem for on-shore refineries or poly-olefin plants but off-shore rigs where space is at a premium they cannot keep a whole variety of spares.

As a result of these considerations, when designing packages for these markets the specifications tend to point to a particular manufacturer and also a particular technology. As GWR transmitters are used throughout refineries and billed as being lower maintenance with greater reliability than their peers with moving parts, the specifications will continue to point in their direction. Future work in the area of level measurement within primary separators should focus on the acceptance of magnetostrictive transmitters by the end users and quantifying the changes in the dielectric properties of the oil within the separator when the package is running to facilitate the reliable use of guided wave radar.

4   REFERENCES

(1) API 614, American Petroleum Institute (2008) Lubrication, Shaft-sealing and Oil-control Systems and Auxiliaries (API STANDARD 614), 5th ed, Washington: API Publishing Services, 1-p58
(2) API 614, American Petroleum Institute (2008) Lubrication, Shaft-sealing and Oil-control Systems and Auxiliaries (API STANDARD 614), 5th ed, Washington: API Publishing Services, 2-p7
(3) Exxon Mobil DTE Names Series Datasheet, [Online] Available at: http://www.mobil.com/Hungary-English/Lubes/PDS/EUXXENINDMOMobil_DTE_Named.aspx [Accessed: 10 February, 2013]
(4) Shell Polyco 32 Industrial Gas Compressor Oil Datasheet, [Online] Available at: http://www.epc.shell.com/Docs/GPCDOC_GTDS_Polyco_32.pdf [Accessed: 10 February, 2013]
(5) Stauss T. (2004) Flow Handbook, 2nd ed, Reinach: Endress+Hauser Flowtec AG, p.34, ISBN 3-9520220-4-7
(6) Emerson Process Management. (2011) The Engineer's Guide to Level Measurement, 1st ed, USA: Rosemount, p.21, Lit Ref: 00805-0100-1034
(7) Clarke B. (2003) A Guide to characterisation of dielectric materials at RF and microwave frequencies, 1st ed, London: Institute of Measurement and Control, p.3, ISBN 0 904457 38 9
(8) Thomas O. (2007) Reservoir Analysis Based on Compositional Gradients, [Online] Available at: https://pangea.stanford.edu/ERE/pdf/pereports/PhD/Thomas07.pdf [Accessed: 10 February, 2013]
(9) Emerson Process Management. (2011) The Engineer's Guide to Level Measurement, 1st ed, USA: Rosemount, p30-37, Lit Ref: 00805-0100-1034

Simulation of a cold climate heat pump furnished with a scroll compressor with multiple vapor injection lines

I H Bell
University of Liège, Belgium

E A Groll, J E Braun, W T Horton
Purdue University, USA

ABSTRACT

A detailed open-source scroll compressor simulation has been developed as described in a companion paper. In this work, the compressor simulation code is used to investigate refrigerant vapor injection in scroll compressors for cold climate heat pumps. The required analysis is developed for vapor injection in scroll compressors with one or two injection lines. The improvement in heating-mode efficiency at a -20°C evaporation temperature with one injection line is as much as 10 % while with two injection lines the increase in efficiency can be as much as 16 %.

NOMENCLATURE

COP_h	Coefficient of Performance (-)	$\dot{Q}_{evap}$	Evaporator input (kW)
h_s	Scroll wrap height (m)	$\dot{W}_{comp}$	Compressor power (kW)
$h_1, h_2, ..$	Enthalpy at state (kJ kg^{-1})	r_a	Radius of discharge arc (m)
$h'_{f,1}$	Sat. liq. enth. flash #1 (kJ kg^{-1})	r_b	Base circle radius (m)
$h'_{f,2}$	Sat. liq. enth. flash #2 (kJ kg^{-1})	r_o	Orbiting radius (m)
$h''_{f,1}$	Sat. vap. enth. flash #1 (kJ kg^{-1})	t_s	Scroll wrap thickness (m)
$h''_{f,2}$	Sat. vap. enth. flash #2 (kJ kg^{-1})	x	Cartesian coordinate (m)
$\dot{m}_{cond}$	Condenser mass flow (kg s^{-1})	X_1	Injection ratio flash #1 (-)
$\dot{m}_{evap}$	Evaporator mass flow (kg s^{-1})	X_2	Injection ratio flash #2 (-)
$\dot{m}_{f,1}$	Vap. gen. flash #1 (kg s^{-1})	y	Cartesian coordinate (m)
$\dot{m}_{f,2}$	Vap. gen. flash #2 (kg s^{-1})	θ	Crank angle (rad)
$\dot{m}_{inj,1}$	Mass flow inj. #1 (kg s^{-1})	φ	Involute angle (rad)
$\dot{m}_{inj,2}$	Mass flow inj. #2 (kg s^{-1})	φ_0	Initial involute angle (rad)
$\dot{p}_{f,1}$	Press. flash #1 (kg s^{-1})	φ_{i0}	Inner wrap initial angle (rad)
$\dot{p}_{f,2}$	Press. flash #2 (kg s^{-1})	φ_{is}	Inner wrap starting angle (rad)
$\dot{p}_{inj,1}$	Press. inj. line #1 (kg s^{-1})	φ_{ie}	Inner wrap ending angle (rad)
$\dot{p}_{inj,2}$	Press. inj. line #2 (kg s^{-1})	φ_{o0}	Outer wrap initial angle (rad)
$\dot{Q}_{cond}$	Condenser output (kW)	φ_{os}	Outer wrap starting angle (rad)
		φ_{oe}	Outer wrap ending angle (rad)

© The author(s) and/or their employer(s), 2013

1 INTRODUCTION

With the continuing emphasis on efficiency of heating and cooling systems, new technologies must be investigated to further improve system efficiency. For heat pump systems operating over large temperature lifts, refrigerant vapor injection is one technology that has been used in the past as a means of improving cycle efficiency.

Systems with refrigerant vapor injection exhibit similar behavior to systems with two-stage compression and economization. The limiting case of refrigerant vapor injection is that the injection occurs instantaneously at the injection pressure. This is equivalent to the compression of the suction gas to an intermediate pressure, instantaneous mixing at constant pressure, and finally the compression of the mixture of the two streams to the discharge pressure.

The advantage of refrigerant vapor injection over conventional two-stage compression is that only one compressor is required, saving capital cost. In order to save manufacturing costs it is also possible to carry out two-stages of compression within one machine (1). In either case, the costs for the two-stage system are greater than that for the conventional heat pump cycle.

Other options are available within one compressor to achieve economizing, like parallel compression economization (2) with a reciprocating compressor.

While two-stage compression and refrigerant vapor injection share many similar features from a cycle perspective, the analysis of vapor injection is significantly more complex as in a real system the injection process is tightly coupled with the cycle behavior. For instance, the amount of vapor generated at the intermediate pressure must be balanced by the amount of vapor that the compressor will accept at a given injection pressure. The further development of this theme is presented below.

There are a number of different types of compressors that are well-suited to refrigerant vapor injection. In particular, the compressor geometry must be amenable to the addition of injection ports which take a finite space and must be open to the refrigerant stream for a specified range of the crank angle. From these standpoints, rotary, spool, scroll and screw compressors are good candidates.

Refrigerant vapor injection in spool compressors with two injection ports has been previously investigated (3), for which the predicted increase in cooling mode COP for two ports is as much as 20%. Vapor injection in scroll compressors has also been studied, (4; 5; 6; 7; 8; 9), and the authors have found theoretical and experimental benefits from vapor injection with one set of injection ports. All the authors have found a significant benefit to cycle performance with refrigerant vapor injection, generally the benefit to cooling-mode COP is greater than 10% at extreme operation conditions.

In addition, a number of authors (10; 11; 9; 12) have considered liquid-refrigerant injection into scroll compressors as a means of decreasing the discharge temperature of the compressor for large pressure ratio applications. The analysis developed in this paper is strictly limited to refrigerant vapor injection.

A detailed simulation code has been developed as described in a companion paper (13; 14). This simulation code can be used to analyze the steady-state performance of a wide range of volumetric machines, including compressors and expanders.

The target application for cold-climate heat pumps are climate zones in which heat pumps do not currently find wide application and where less efficient systems are currently in use. For instance in the USA, as of 2009, 34% of the households use electric heat (15, Table 2.7). To achieve the same primary energy efficiency as natural gas condensing boilers, the seasonal COP of the heat pump must be greater than approximately 3.0.

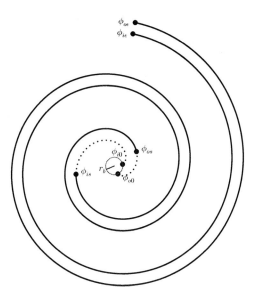

Figure 1: Involute angles for the fixed scroll wrap

This paper begins with a development of the requisite analysis for vapor injection in a scroll compressor. Then the cycle analysis is carried out, and finally, the compressor model and the cycle models are coupled together.

2 COMPRESSOR ANALYSIS AND SIMULATION

2.1 Scroll Compressor Geometry

The detailed analytical description of scroll compressor geometry, forces and other geometrically related terms is beyond the scope of this paper, but can be found in Bell (16). Nevertheless, the scroll compressor's geometry is fundamental to both an understanding of the scroll compressor's efficiency as well as the integration of vapor injection. An abbreviated description of the geometry is presented here, the reader is directed to Bell (16) for further information.

The most common configuration of a scroll compressor is composed of two scroll wraps, one of which is fixed, the other of which orbits. The orbiting motion of the motive scroll wrap traps crescent-shaped pockets of vapor which decrease in volume as they move towards the center of the compressor. The high-pressure vapor in the pockets is then discharged.

One wrap of the scroll compressor is formed of two involutes; an involute is formed by the unwrapping of a circle. Each involute is unwrapped from a base circle with radius r_b, which yields the Cartesian coordinates of a point on the involute curve by

$$x = r_b (\cos \varphi + (\varphi - \varphi_0) \sin \varphi)$$
$$y = r_b (\sin \varphi - (\varphi - \varphi_0) \cos \varphi) \tag{1}$$

where φ takes on the values from φ_s to φ_e. The innermost portion of the involute from φ_0 to φ_s is not considered as part of the involute, rather in this area, different curves are used to join the involute curves. Analytical solutions for the most common sets of these geometries are provided in Bell (16).

The inner and outer involutes of the scroll wrap have initial involute angles of φ_{i0} and φ_{o0}, respectively. This yields two involute curves with a constant distance between them; these curves form the outer walls of the scroll wrap, as seen in Figure 1. The thickness of the scroll is given by

$$t_s = r_b (\varphi_{i0} - \varphi_{o0}). \tag{2}$$

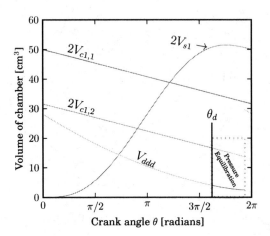

Figure 2: Volumes of the compressor chambers for the compressor under study

Figure 2 presents the volumes of each of the chambers over the course of a rotation. The volumes of the suction pocket V_{s1} initially increases during the suction process, which draws vapor into the compressor. The suction pocket is sealed off after one rotation and becomes a compression chamber $V_{c1,1}$. This compression chamber decreases in volume over the course of a few rotations (during which part of the time it is defined as $V_{c1,2}$) until the discharge angle is reached, at which point it merges with the refrigerant vapor remaining in the discharge region. After pressure equilibrium, the refrigerant in the merged discharge chamber V_{ddd} flows out into the discharge line.

At the beginning of the rotation ($\theta = 0$), the scroll wraps are in contact at the end of the scrolls. Over the course of one rotation (2π radians), all of the contact points move 2π radians towards the center of the compressor.

It should be mentioned here that this analysis is only valid for scroll compressors with constant wrap thickness and completely symmetric wraps. The analysis of variable wall thickness scroll wraps can be found in literature (17; 18)

2.2 Compressor Sizing

The pressure ratios for cold climate heat pump applications are quite large. For an adiabatic compression of propane from 244.5 kPa to 1476.7 kPa, with a compressor inlet superheat of 11.1 K, the ratio of outlet to inlet densities is equal to 5.52 which is equivalent to the ideal built-in volume ratio. Irreversibilities in the compression process will tend to decrease the outlet density due to the larger outlet temperature and yield a smaller required volume ratio. In addition, injection of refrigerant vapor will increase the pressure of the fluid in the working chambers, further decreasing the required volume ratio when vapor injection is applied. For these reasons, a built-in volume ratio of 3 was selected for this application. If the selected volume ratio is too large, there will be large over-compression losses in the compressor when operating at high evaporation pressures due to the mal-adjustment of the large volume ratio for the

Table 1: Parameters of the scroll compressor studied

Geometric Parameters	
Volume ratio [-]	3.0
r_b [mm]	2.228
r_o [mm]	4.000
h_s [mm]	52.346
t [mm]	3.000
φ_{i0} [rad]	0.000
φ_{is} [rad]	3.142
φ_{ie} [rad]	21.096
φ_{o0} [rad]	-1.346
φ_{os} [rad]	0.300
φ_{oe} [rad]	21.096
r_a [mm]	6.426
Discharge port diameter [mm]	9.000
Symmetric Scrolls	Yes
Flow Parameters	
Leakage gap width [µm]	15
Flank gap width [µm]	15
Leakage flow model	(19)

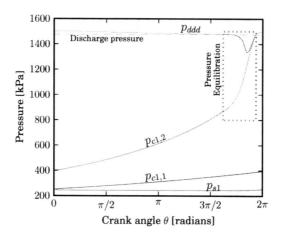

Figure 3: Pressures in compressor evaporating at 244.5 kPa [-20°C saturated] and condensing at 1476.7 kPa [43.3°C saturated]

small imposed pressure ratio. When working with vapor injected scroll compressors, it is generally better to err on the side of smaller volume ratios in order to avoid the potentially large overcompression effects at evaporation pressures above the design point.

The remaining scroll compressor geometric parameters were selected in order to yield an efficient and robust compressor; all the parameters needed to simulate the compressor are outlined in Table 1. Furthermore, additional parameters required to simulate the machine are provided in the simulation code [1]. A pressure versus crank-angle plot is provided in Figure 3 which shows that the volume ratio is well-selected at the rating point as there are reasonable under-compression losses.

---

[1] Posted in the University of Liège repository at http://orbi.ulg.ac.be/handle/2268/147945

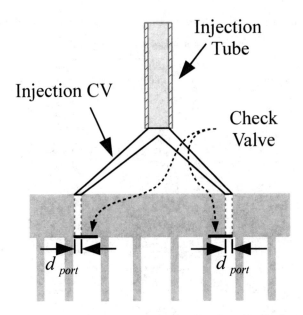

Figure 4: Schematic of the components used in the simulation for one pair of injection ports

2.3 Injection Ports and Lines

The location and sizing of the injection ports has a profound impact on the working process of the vapor injected compressor. In this analysis it is assumed that each of the injection ports is symmetric with respect to the center of the compressor and is added to the fixed scroll. Wang (5; 6) provides a quite thorough analysis of the optimal locations and characteristics of the injection ports when one injection line is used.

When there is one set of injection lines, the first pair of injection ports is located such that the ports are only in contact with the outermost compression chambers so that there is no flow of injected fluid into the suction chamber. Injection of vapor directly into the suction chamber can be thought of as a leakage term that tends to reduce the amount of volume in the suction chamber that can be used to accept suction vapor.

In the simulation code, a few simplifying assumptions are made about the injection process. It is assumed that there is no heat transfer in the injection line between the injected refrigerant and the ambient, and that the pressure drop is also zero in this line.

The system of injection line and ports is decoupled into a network of components as shown in Figure 4. The refrigerant enters into the injection tube which is treated as being a quasi-steady tube as described in Bell (13). The steady-state mean value for the mass flow rate through the tube (which is updated after each cycle of the compressor simulation) is used to calculate the pressure drop through the line.

The injection tube is connected to a time-variant control volume that allows for the dynamics of the pressure in the injection lines to be modeled due to outflow from the control volume into the compression pocket of the compressor. Finally, an isentropic nozzle model is used to model the flow rate between the injection line control volume and the compression chambers.

As is noted by Wang (5; 6), check valves can be useful to avoid backflow of refrigerant into the injection lines under high evaporation pressure conditions. Simplified check

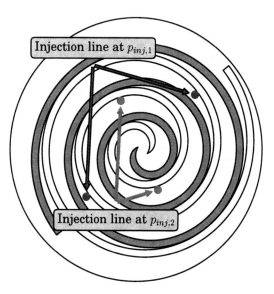

Figure 5: Scroll compressor with injection ports shown at a crank angle of $\theta = \pi/2$ radians

valves have been implemented in the simulation by assuming idealized check valve behavior. That is, if the pressure in the injection control volume is less than the pressure in the compression chamber it is connected to, no mass is allowed to flow back into the injection control volume. The additional pressure drop associated with the check valve has not been included.

The description of the dimensions of the ports and lines considered can be found in Table 1. The ports themselves are located at an involute angle as close to the suction as possible, which yields the injection port configuration as shown in Figure 5. The inner set of injection ports at $p_{inj,2}$ (as described in a following section) are only used when both pairs of injection ports are active.

The injection port is considered to be fully open to the chamber that it is connected to, and the mixing of the injected refrigerant is assumed to happen instantaneously in the compression pocket.

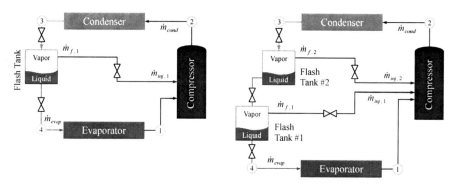

Figure 6: Schematic of one- and two-line vapor injection systems

3 CYCLE ANALYSIS

In order to more fully study the compressor behavior with vapor injection it is necessary to develop the appropriate cycle models.

3.1 No Injection Lines (Conventional Operation)

When no injection lines are used, the performance of the cycle is that of the conventional four-component heat pump cycle. The working fluid vapor exits the evaporator slightly superheated (state point # 1) and then enters into the compressor where it is compressed to the discharge pressure at state point # 2. The working fluid is condensed in the condenser to a subcooled state (state point # 3), then expanded in the expansion device to the evaporating pressure at state point # 4, and finally evaporates in the evaporator to a superheated state.

The mass flow rate through the system is the same through all the components and is given by $\dot{m}_{evap}$. The rate of heat input to the evaporator is given by

$$\dot{Q}_{evap} = \dot{m}_{evap}(h_1 - h_4) \qquad (3)$$

and the heat rejected by the condenser (the useful output of the heat pump) is given by

$$\dot{Q}_{cond} = \dot{m}_{cond}(h_2 - h_3) \qquad (4)$$

where $\dot{m}_{cond} = \dot{m}_{evap}$ and the sign of $\dot{Q}_{cond}$ is selected to be positive. The electrical power input to the compressor $\dot{W}_{comp}$ and the mass flow rate through the evaporator $\dot{m}_{evap}$ are given from the compressor simulation code. $\dot{W}_{comp}$ is in general greater than $\dot{m}_{evap}(h_2 - h_1)$ due to ambient heat loss in the compressor. The heating-mode coefficient of performance is given by

$$COP_h = \frac{\dot{Q}_{cond}}{\dot{W}_{comp}} \qquad (5)$$

3.2 One Injection Line Cycle Analysis

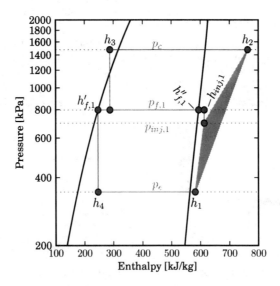

Figure 7: Pressure versus enthalpy for one injection line system

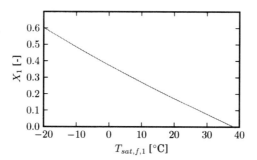

Figure 8: Injection fraction for propane evaporating at 244.5 kPa [-20°C saturated] and condensing at 1476.7 kPa [43.3°C saturated] with 5 K subcooling at the outlet of the condenser

When one injection line is employed, a few components need to be added to the conventional heat pump system. In the single-injection configuration, the primary expansion valve flashes the refrigerant to an intermediate pressure $p_{f,1}$, at which point the phases are separated in a flash tank at constant pressure. The saturated vapor is sent to the compressor, and the saturated liquid is further expanded to the evaporating pressure. The two-step expansion yields a lower inlet enthalpy to the evaporator, which results in higher rate of heat input, and therefore also a higher rate of heat output in the condenser.

The mass and energy balances on the flash tank can be expressed as

$$\begin{aligned} \dot{m}_{cond} &= \dot{m}_{evap} + \dot{m}_{f,1} & \text{(Mass)} \\ \dot{m}_{cond} h_3 &= \dot{m}_{evap} h'_{f,1} + \dot{m}_{f,1} h''_{f,1} & \text{(Energy)} \end{aligned} \qquad (6)$$

where $h'_{f,1}$ and $h''_{f,1}$ are the saturated liquid and saturated vapor enthalpies for a given flash tank pressure $p_{f,1}$. This yields the ratio of the mass flow rates of vapor to liquid generated in the flash tank of

$$X_1 = \frac{\dot{m}_{f,1}}{\dot{m}_{evap}} = \frac{h_3 - h'_{f,1}}{h''_{f,1} - h_3} \qquad (7)$$

and the condenser mass flow rate is given by $\dot{m}_{cond} = \dot{m}_{evap}(1+X_1)$. The capacity and COP of the system are as given in Equations 4 and 5, where in this case, h_4 is equal to $h'_{f,1}$. The electrical power input to the compressor W_{comp} is given by the compressor simulation code, as will be further described below.

As an example, Figure 8 shows the value of X_1 as a function of the flash tank saturation temperature. Mass flow fractions below zero are non-physical and represent impossible operation states. The highest flash tank saturation temperature that can yield vapor generation is 38.3°C under these conditions. This is a characteristic curve for the fluid that is independent of the compressor. The lower limit on the flash tank saturation temperature is the evaporation saturation temperature.

It should be mentioned that this analysis is intended only for pure working fluids. More care is required to use this analysis with zeotropic blends with temperature glide or azeotropic blends (like R404A or R410A) with nearly no temperature glide. In particular, for mixtures, the dew and bubble temperatures of the mixture are not equal for a given pressure.

3.3 Two Injection Line Cycle Analysis

In the two injection line case, the analysis is quite similar to that of the one injection line case. A system of equations is set up like Equation 6; there is one pair of equations

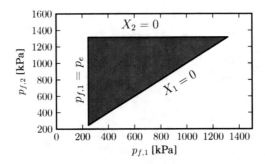

Figure 9: Range of validity of flash tank pressures for two lines for propane evaporating at 244.5 kPa [-20°C saturated] and condensing at 1476.7 kPa [43.3°C saturated] with 5 K subcooling at the outlet of the condenser

for each flash tank which arrive from mass and energy balances:

$$\dot{m}_{cond} = \dot{m}_{f,2} + \dot{m}_{f,1} + \dot{m}_{evap}$$
$$\dot{m}_{cond} h_3 = (\dot{m}_{evap} + \dot{m}_{f,1}) h'_{f,2} + \dot{m}_{f,2} h''_{f,2} \quad (8)$$

$$\dot{m}_{cond} - \dot{m}_{f,2} = \dot{m}_{evap} + \dot{m}_{f,1}$$
$$(\dot{m}_{cond} - \dot{m}_{f,2}) h'_{f,2} = \dot{m}_{evap} h'_{f,1} + \dot{m}_{f,1} h''_{f,1}$$

which yields the ratio of vapor generated to liquid generated in each flash tank of

$$X_2 = \frac{\dot{m}_{f,2}}{\dot{m}_{evap} + \dot{m}_{f,1}} = \frac{h_3 - h'_{f,2}}{h''_{f,2} - h_3} \quad (9)$$

$$X_1 = \frac{\dot{m}_{f,1}}{\dot{m}_{evap}} = \frac{h'_{f,2} - h'_{f,1}}{h''_{f,1} - h'_{f,2}} \quad (10)$$

As with the one-injection-line analysis, there are a few thermodynamic constraints on the possible flash tank pressures. One constraint is that $p_{f,2}$ must be greater than $p_{f,1}$, and a second constraint is that both $\dot{m}_{inj,1}$ and $\dot{m}_{inj,2}$ must be positive. For a given operating condition, it is therefore possible to map the range of possible flash tank pressures, as shown in Figure 9.

4 COMPRESSOR SIMULATION WITH ONE PAIR OF INJECTION PORTS

The analysis of vapor injection in compressors integrated into the heat pumps can be decoupled into two separate sub-models. In the first sub-model, the compressor is simulated for a given injection pressure and injection port area, independent of the cycle.

For a range of compressor injection pressures, it is then possible to back-calculate the flash tank pressures that would be required to yield the same injection mass flow rate fraction as the simulation code predicts. In other words, the simulation code will provide predictions of $\dot{m}_{evap}$ and $\dot{m}_{inj,1}$ for a given value of $p_{inj,1}$. Equation 7 can then be iteratively solved to find the $p_{f,1}$ that yields the same value of X_1. Finally, Figure 10 shows the results of this analysis; the flash tank saturation pressure is calculated as a function of the pressure of the injected vapor and the diameter of the injection port. Only the simulation results that yield a physical solution ($p_{f,1} > p_{inj,1}$) are retained.

These results show that for a given injection port diameter there is a single solution that yields the same pressure between the injection line and the flash tank, denoted by

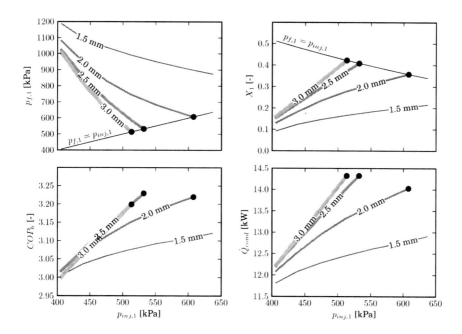

Figure 10: Flash tank pressure, system COP, heating capacity and injection fraction as a function of injection pressure to compressor and injection port diameter (evaporating at -20°C saturated and condensing at 43.3°C saturated, 0 K subcooling)

the solid black circles. For a given injection pressure, as the port diameter is increased, the injection mass flow rate is increased, while $\dot{m}_{evap}$ is effectively constant, resulting in a larger X_1. Thus as the port diameter is increased, the intersection of the system- and compressor X_1 curves moves to lower injection pressures.

This is analogous to the relationship between pump and system curves when sizing pumps for HVAC systems. The stable operation point is the intersection of the compressor (pump) curve and the system curve.

If there is a finite pressure drop between the flash tank and the injection line, the stable operation point can be found that yields the given difference in pressure.

For a given injection port diameter, as the compressor injection pressure is increased, the COP increases monotonically, up to the point where the flash tank and injection pressures are equal. Therefore, the optimal system COP will always be found when the pressure drop between the flash tank and the compressor injection port is zero. Any additional pressure drop between the flash-tank and injection pressure generates irreversibilities.

There is an optimal injection port diameter that maximizes the system COP for the case where $p_{f,1} = p_{inj,1}$. As the port diameter decreases, the flash tank pressure increases, which in turn decreases the capacity as well as increasing the throttling irreversibilities in the primary expansion valve. As the port diameter increases, the capacity increases, though the injection process irreversibility generation also increases due to the increased injection mass flow rate, which results in a fall-off of COP at higher injection port diameters.

On the other hand, from the standpoint of capacity, there is no penalty to compressor capacity at higher injection port diameters. The increase in heating capacity with

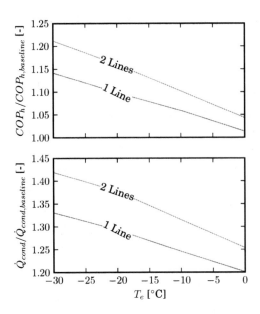

Figure 11: Optimized compressors for a condensing temperature of 43.3°C

vapor injection is largely due to the fact that the mass flow rate passing through the condenser is increased by a multiplicative factor of $1 + X_1$ while the outlet of the condenser is anchored. There is a small reduction in the outlet temperature of the compressor at higher injection mass flow rates, but the increase in capacity due to the increase in condenser mass flow rate is the dominant physical effect.

In the ideal case that there is no pressure drop between the flash tank and the compressor injection port, designing the compressor and the system for maximum efficiency yields nearly the maximum capacity as well. Therefore, in the analysis that follows, the injection port of diameter equal to 0.0025 m [2.5 mm] was selected as the injection port diameter.

5 OPTIMIZATION OF VAPOR-INJECTED COMPRESSORS WITH ONE AND TWO INJECTION LINES

As the analysis in the previous section showed, the optimal COP with one injection line is a function of both injection port diameter as well as injection pressure.

Analogously, with two injection lines, the optimal performance is achieved for a compressor that has no pressure drop between each of the injection pressures and the respective flash tank pressure.

Thus with two injection lines it is possible to find the set of injection pressures that result in equivalent flash tank pressures. This is achieved through the use of a multi-dimensional non-linear system of equations solver.

For given injection pressures $p_{inj,1}$ and $p_{inj,2}$, the simulation code will then yield predictions of $\dot{m}_{inj,1}$, $\dot{m}_{inj,2}$, as well as $\dot{m}_{evap}$. Based on these predicted flow rates it is possible to numerically solve Equation 9 to find the flash tank pressures $p_{f,1}$ and $p_{f,2}$. The residuals to be driven to zero by the numerical solver are the pressure drops $p_{f,1} - p_{inj,1}$ and $p_{f,2} - p_{inj,2}$ by adjusting the parameters $p_{inj,1}$ and $p_{inj,2}$.

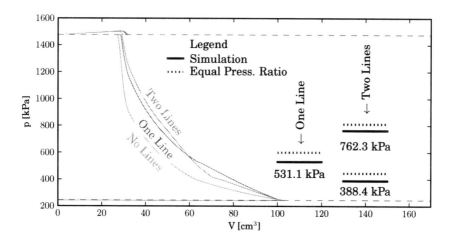

Figure 12: Pressure-volume plot for compressor with no, 1 and 2 injection lines and diagram of injection pressures (evaporating at -20°C saturated and condensing at 43.3°C saturated)

Figure 11 shows the COP and capacity for the systems utilizing compressors with one and two injection lines. The injection pressure that yielded no pressure drop in the injection line was obtained using a numerical solver. These results show that the performance with two injection lines is always better than that with one injection line, which is in turn always better than the baseline system. At lower evaporation temperatures, the increase in system COP is greater than 20% with two injection lines. In addition, as the evaporation saturation temperature decreases, the improvement in heating capacity relative to the baseline improves monotonically.

Furthermore, Figure 12 shows a pressure-volume plot of the compressor operating at an evaporation pressure of 244.5 kPa [-20°C saturated] at the optimal COP_h point. In adding one injection line, the net impact is to shift the pressure curve upwards. In adding a second injection line, the pressure-volume curve is warped during the compression process. The addition of the second injection line results in a lower first injection pressure of 386.6 kPa which results in less injection during the initial part of the compression. The second injection pressure is higher than the injection pressure with one line. The net result is that the pressure increases once the second port opens, and at the end of the compression, the pressure in the working chamber is slightly greater than that in the one-injection-line case.

CONCLUSIONS

Scroll compressors with multiple injection lines offer the prospect of significant improvements in system efficiency. When optimizing the system design for vapor injection, it is necessary to carefully select the injection port diameter in order to obtain the maximum cycle efficiency. When the compressor is well-matched with the application, the increase in heating-mode COP can be greater than 20% with two injection lines.

Further work is ongoing to provide experimental validation of the results presented here.

REFERENCES LIST

[1] M. M. Mathison, James E. Braun, and Eckhard A. Groll. Modeling of a two-stage rotary compressor. *HVAC&R Research*, 14:719--748, 2008.

[2] Ian Bell. Performance increase of carbon dioxide refrigeration cycle with the addition of parallel compression economization. In *Proc. IIR Gustav Lorentzen Conf. Natural Working Fluids, Glasgow, UK.*, 2004.

[3] Margaret M. Mathison, James E. Braun, and Eckhard A. Groll. Modeling of a novel spool compressor with multiple injection ports. In *International Compressor Engineering Conference at Purdue, July 16-19, 2012*, 2012.

[4] Baolong Wang, Xianting Li, Wenxing Shi, and Qisen Yan. Design of experimental bench and internal pressure measurement of scroll compressor with refrigerant injection. *Int. J. Refrig.*, 30(1):179 -- 186, 2007.

[5] Baolong Wang, Wenxing Shi, Linjun Han, and Xianting Li. Optimization of refrigeration system with gas-injected scroll compressor. *Int. J. Refrig.*, 32(7): 1544 -- 1554, 2009.

[6] Baolong Wang, Wenxing Shi, and Xianting Li. Numerical analysis on the effects of refrigerant injection on the scroll compressor. *Appl. Therm. Eng.*, 29(1):37 -- 46, 2009.

[7] Baolong Wang, Wenxing Shi, Xianting Li, and Qisen Yan. Numerical research on the scroll compressor with refrigeration injection. *Appl. Therm. Eng.*, 28(5-6): 440 -- 449, 2008.

[8] Youn-Cheol Park, Yongchan Kim, and Honghyun Cho. Thermodynamic analysis on the performance of a variable speed scroll compressor with refrigerant injection. *Int. J. Refrig.*, 25:1072--1082, 2002.

[9] Eric L. Winandy and Jean Lebrun. Scroll compressors using gas and liquid injection- experimental analysis and modelling. *Int. J. Refrig.*, 25:1143--1156, 2002.

[10] Asit K. Dutta, Tadashi Yanagisawa, and Mitsuhiro Fukuta. An investigation of the performance of a scroll compressor under liquid refrigerant injection. *Int. J. Refrig.*, 24:577--587, 2001.

[11] S. Ayub, J. W. Bush, and D. K. Haller. Liquid refrigerant injection in scroll compressors operating at high compression ratios. In *International Compressor Engineering Conference at Purdue University*, pages 561--567, 1992.

[12] Honghyun Cho, Jin Taek Chung, and Yongchan Kim. Influence of liquid refrigerant injection on the performance of an inverter-driven scroll compressor. *Int. J. Refrig.*, 26(1):87 -- 94, 2003.

[13] Ian H Bell, Vincent Lemort, Eckhard A. Groll, and James E. Braun. Development of a generalized steady-state simulation framework for positive displacement compressors and expanders. In *International Conference on Compressors and their Systems 2013*, 2013.

[14] Ian Bell. Positive Displacement Machine Simulation (PDSIM): http://pdsim.sf.net, 2012.

[15] U.S.A. Energy Information Administration. Annual energy review. Technical report, U.S.A. Energy Information Administration, 2012.

[16] Ian Bell. *Theoretical and Experimental Analysis of Liquid Flooded Compression in Scroll Compressors*. PhD thesis, Purdue University, 2011.

[17] Bryce R. Shaffer and Eckhard A. Groll. Parametric representation of scroll geometry with variable wall thickness. In *International Compressor Engineering Conference at Purdue, July 16-19, 2012*, 2012.

[18] Jens Gravesen and Christian Henriksen. The Geometry of the Scroll Compressor. *SIAM Review*, 43(1):113--126, 2001.

[19] Ian H. Bell, Eckhard A. Groll, James E. Braun, and W. Travis Horton. A compu-

tationally efficient hybrid leakage model for positive displacement compressors and expanders. *Int. J. Refrig.*, 2013.

Compressor for exhaust treatment of non-road vehicles

R Sachs, W Asal
Gardner Denver Schopfheim GmbH, Development Engineering, Germany

ABSTRACT

Diesel particulate filter systems are applied to meet emission standards for engines in non-road mobile machinery. A burner with an air supply unit can ensure the regeneration of the filter without machine stoppages. A low cost unit with a constant performance and a low flow pulsation is therefore needed: a so-called COR®-compressor (1) consisting of two axial rotors with a trochoidal shaped gearing is proposed. The rotor material of this dry running positive displacement machine needs to combine a high geometrical accuracy with a good tribological behaviour. Hence plastic rotors are utilised which are produced using a two-component thermoset injection moulding process.

1   INTRODUCTION

Exhaust particulates are becoming more and more the focus of legislators. Also, the emission standards for Diesel powered non-road mobile machinery are getting stricter (2, 3, 4). One can choose from a number of options to meet the resultant requirements. Other than a direct combustion calibration a selective catalytic reduction, exhaust gas recirculation, or a Diesel particulate filter are also applicable to reduce exhaust pollution. A combination of these techniques is also possible.

To operate a Diesel particulate filter a residue-free incineration of the soot inside the filter is needed to control the internal flow resistance. This active regeneration has to be done to control an increasing particulate load while the Diesel engine is still in operation. The Diesel injected burner system applied for this purpose requires a compressor for the air supply. The air supply unit has to provide the burner with the necessary air during the incineration process.

The given specification of this air supply unit leads to the choice of a robust, uncomplicated, maintenance free and a reasonably priced design. The characteristics of the compressor should ensure a gas flow independent of its pressure difference as well as a low flow pulsation. The volume flow after switch off has to go down quickly without applying a brake. In use the compressor has to run in a 20 minutes time on and 6 hours time off cycle. The required total runtime in its life cycle is 1500h. Furthermore, the mechanical connection to the exhaust system should be possible in any position. The power supply is connected to the 12/24V mains of the vehicle.

© The author(s) and/or their employer(s), 2013

2 COR® TECHNOLOGY

A COR® machine is a positive displacement machine with a complex 3-dimensional front toothed trochoidal gearing (1). Figure 1 shows an example of the two rotors.

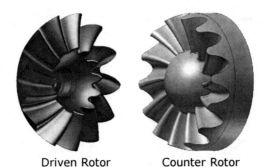

Driven Rotor   Counter Rotor

Figure 1: COR® rotors

The rotors run in a housing and the "driven rotor" (named by the inventor) is powered by an external drive train, e.g. an electric motor. The "counter rotor" is moved by the driven rotor in a direct contact. To minimize the wear caused by friction a low speed difference between the rotors is necessary. This is achieved by having one tooth less on the driven rotor than on the counter rotor.

The rotor axles form a tilt angle α. In connection with their trochoidal design the rotors can form closed working chambers; Figure 2. It should be noted that all tooth flanks are in contact the whole time. This provides a high internal tightness and it is an advantage when compared to 2-shaft machines with synchronising gears and rotors that are not in contact. The turning of the rotors causes a decreasing and increasing of the chamber size from zero to maximum volume during one rotation. The volume flow of the machine is dependent on the diameter of the rotors, the tilt angle, the number of teeth and the speed.

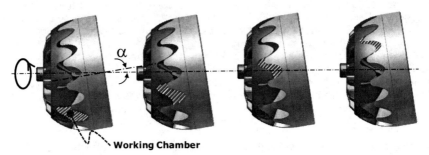

Figure 2: Compression via decreasing working chambers

The inlet and outlet ports in the housing control the fluid flow and define the internal compression of the machine; Figure 3.

A critical influence on the wear of the rotors is the contact force between them. Therefore, the counter rotor needs to be moveable in its axial direction. Thus the gap between the rotors and hence the pressure from one rotor to the other is adjustable. There are different ways to implement this function: a mechanical compensation is done by a spring behind the counter rotor; a pneumatical compensation uses a bypass to the high pressure at the backside of the counter rotor (5). In this way the abrasion of the rotors is reduced and the ability of the machine to handle small particles in the fluid flow is improved. Furthermore, this mechanism enables a compensation of the axial drift of the rotors by their abrasion during its lifetime. This attribute is a difference when compared to most other positive displacement machines and it allows the use of abrading rotor materials.

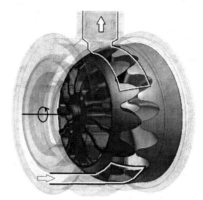

Figure 3: Control of compression via inlet and outlet ports

3   COR® COMPRESSOR

The COR® principle is at present mostly used in liquid pumps. An air compressor of this size was presented for the first time in 2012 (6). To start with it has to be noted that this compressor is a dry running machine. The rotors described above are made of a plastic material and are produced using a two-component thermoset injection moulding process. This approach makes it feasible to produce these complex rotors at reasonable costs.

3.1   Design

In Figure 4, the design of the compressor used in exhaust treatment of non-road vehicles is presented. The driven rotor is attached to the shaft of the brushless direct current motor. This shaft is seated in ball bearings. One of the bearings is installed in the motor housing that also encloses the motor itself, the electronic device and the inlet air flow system. The other bearing is installed in the compressor housing that includes the internal air flow system, the outlet and the two compressor rotors. The counter rotor runs on two more ball bearings, which are connected to the axis at the top cover. This design prevents external leakage because the machine is closed hermetically.

One of the requirements of this compressor is the ability to run without external cooling air. The electronics are the most temperature-sensitive part of the machine. The cool inlet air is therefore guided along the surface next to the electronics at first; see arrows in Figure 4. This area is also in the range of the highest flow velocity and vorticity outside the working chamber and this improves the effectiveness of the heat transfer to the air. The slightly warmer air then cools the motor housing. Finally, the air is compressed in the working chamber, gets much hotter and goes to the outlet.

The external contours are simple leading to an easy mounting with two brackets in any position in the vehicle, Figure 5. This advantage is also made possible because there is no oil filling to be considered. In addition, the blower housing and the motor housing, which are the inlet and outlet, can be easily twisted against each other. This increases even more the variability for integration in any location.

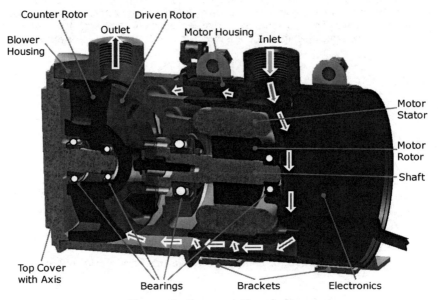

Figure 4: Cross sectional view

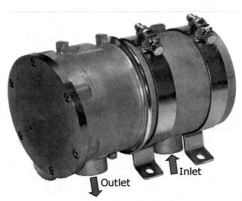

Figure 5: Full view

3.2 Characteristics

In the course of development several modifications are done in order to optimize the compressor for the present application. The main factors for a good isentropic efficiency and performance are the tilt angle, the internal gaps i.e. leakage, the friction between the rotors and an adjusted internal compression. One factor that had to be adapted is the volume flow. By increasing the tilt angle the gearing silhouette changes and the working chambers get larger. But the available energy to drive the compressor, given by the maximum amperage in 12/24 V electrical vehicle systems, has to be taken into consideration. As a result off this the inner compression, depending on the shape and position of the inlet and the outlet ports was improved. The final tilt angle α of this compressor equals now to 9°. Its nominal speed is 7000 rpm and the gearing ratio is 13/14.

The performance of the compressor is illustrated in Figure 6. The outlet pressure is normalized by the maximum pressure of the application, the volume flow by the maximum flow of the compressor and the isentropic efficiency by the maximum coupled power of the compressor. The isentropic efficiency is the ratio of the isentropic power output requirement to the coupled power input (7).

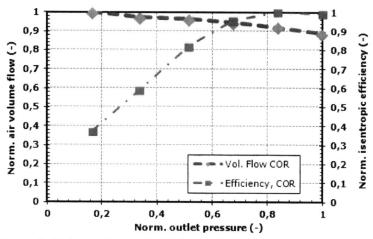

Figure 6: Volume flow and efficiency (based on measured values)

The burner air supply requires a relatively constant flow. Both amplitude and frequency of the air flow pulsation should be appropriate. A fluid flow machine would have big advantages here and another positive displacement machine with e.g. 2 working chamber big disadvantages. The present compressor with 13 working chambers is a good compromise. Figure 7 clarifies the pressure pulsation of the flow which is normalized by the maximum pressure of the application. In contrast to a COR® a fluid flow machine would need a long time to run down after being switched off. But the machine has to decrease the volume flow from 100% to 10% in less than 2 seconds in order to not affect the burner. The deceleration time of the compressor is short enough for this application; see figure 7.

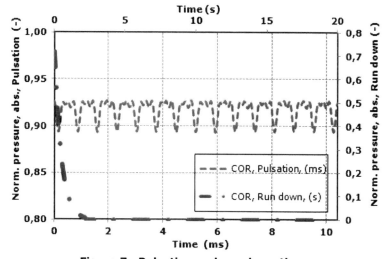

Figure 7: Pulsation and run down time

107

4 DEVELOPMENT OF PLASTIC ROTORS FOR THE COR® COMPRESSOR

The production and use of plastic rotors for positive displacement machines has become the focus of industry for a long time. This development has been driven mainly by the technical (e.g. lower noise) and cost advantages associated with their use. Beside physical, mechanical and material issues one has to cope with sophisticated production tasks. One of the issues for a compressor is the high operating temperature. Using water or oil for lubricating and cooling the rotors can help (8). Another challenge is the production of plastic rotors with a complex geometry and a sufficient strength and resilience. To overcome this, the rotor can be made from plastic, but the shaft still from metal (9). Last but not least, the tribological behaviour of the material is important, if the rotors are in contact. This is a broad field with many setbacks. For example, even the use of graphite fillers in polyimide, a thermoplastic material, does not always significantly reduce the wear (10).

4.1 Challenges

The complex rotor geometry does not allow a cutting manufacturing process in the serial production for cost reasons. For sufficiently large quantities an injection moulding process can be taken into consideration.

For the purpose of verifying the rotor layout, the rotor geometry and checking the CFD calculation, the test rotors are made of a Phenol-Formaldehyde material. The rotors are not injection moulded but are milled from solid. The rotor lifetime, thermal expansion and the ability of the rotor to run in a fail safe mode are not the focus of this preliminary investigation. But these have to be considered for the second rotor generation in this application which is injection moulded. Alongside a cost-effective production the rotors have to withstand mechanical and thermal loading in operation. Particularly important is the tribological suitability and a high dimensional accuracy and stability.

4.2 Choice of materials

Table 1: Different thermosetting plastic for rotors

No.	Generic Identification ISO 11469:200	Tribological properties	Tensile strength/density R_m/ρ $[\times 10^3 \ m^2/s^2]$	Young's modulus/ density E/ρ $[\times 10^6 \ m^2/s^2]$	Material price
1	UP-(MD59+GF20)	no	20	6	low
2	PF-(GB+GF)75	no	60	12	low
3	PF-CD65	yes	33	10	low
4	PF-CF50	yes	97	24	high
5	PF-CDxx	yes	30	9	medium

Index:

Base material		Fillers			
UP	Unsaturated Polyester	GB	Glass Beads	CF	Carbon Fiber
PF	Phenol-Formaldehyde	CD	Carbon Powder	GF	Glass Fiber
				MD	Mineral Powder

Different materials were investigated to ensure that they comply with the demands mentioned above. The basic choice is a thermosetting plastic. With this preference the requirements set above as well as a comparatively high wall thickness are realizable. Also, the shrinkage in the casting tool is controllable and the cycle time for moulding is short due to rapid hardening. Another advantage for the project is the existing experience with highly accurate parts in the injection moulding industry (e.g., throttle body housings). Investigations of the thermoset injection moulding process substantiate the use of two components: a "perform" with a layer; see section 4.3. This gives the necessary geometrical exactness needed and better possibilities of setting the tribological behaviour.

The following materials were taken into account, Table 1. No. 1 and No.2 show materials for the rotor raw part. They have no tribological properties but a low price. The parameter "Tensile strength/density" for No. 1 is on the borderline. FEM calculations have shown an extremely high load. Also, the parameter "Young's modulus/density", which gives information about the influences of the centrifugal force, indicates that No. 1 is the weaker material. For the tribologically active layer the favourites are Nos. 3 to 5. As a result of the tests mentioned in section 4.5, the choice is No. 4. The material combination No. 2 for the raw part and No. 4 for the layer has the best properties, the best adhesion of the layer and the best strength.

4.3 Two-component thermoset injection moulding process

The choice of an injection moulding to manufacture the rotors is justified by economical and technical reasons (dimensional and shape accuracy). But the rotors have to be off-tool parts, i.e., they must not be reworked after moulding due to the large cost involved. At the beginning some flow simulations are done. This gives valuable information about the details of the design of the thermoset injection moulding tool. The filling process is calculated separately as a function of pressure and as a function of time; for the latter see Figure 8. The simulation gives answers about the places for ventilation bores and about the active temperature and pressure control. Also, the formation of joint lines can be predicted. With this information recommendations for the geometry of the tool and the part can be given.

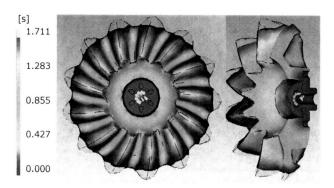

Figure 8: Mould filling process simulated as a function of time

The tool for the two-component thermoset injection moulding process is used in two steps and two types; firstly, in moulding the preform and secondly, in moulding the layer. In the first step the raw part material is injected and the machine has to keep the moulding pressure high. As the pressure sensor is monitoring the actual status, the compression core is moved into the preform. This action helps to manage the shrinkage and moulding accuracy. The reason for this can be attributed to a change from hydrodynamic pressure of the normal injection moulding process to a nearly hydrostatic behaviour (11). For the second step the process continues as stated in step one to apply the wear resistant layer. Essential for a successful production (e.g., a high geometrical accuracy) is an exact and constant process management. The parameters to be considered are temperature, pressure, injection speed, injection quantity, hold time and a constant cycle time. Beside these factors a constant quality of the raw material is fundamental.

4.4 Verification of rotor geometry

In Figure 9, the rotor geometry modified for the injection moulding process can be seen. The significant changes when compared to the milled rotor are the draft angles and nearly constant wall thickness.

Driven rotor

Figure 9: Rotor-CAD-model modified for injection moulding

The first challenge is to match this geometry with the moulded parts. The individual factors of influence mentioned above result in several production steps, which leads to an increase understanding of the exact conditions and corrections regarding the tool and the procedure.

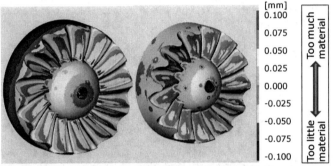

Rotor from first tool   Rotor from optimized tool

Figure 10: Surface measurement of counter rotor

To help further our understanding, a comparison between the 3D-CAD model and the real rotor is done. With the help of an optical 3-D measurement system the differences becomes visible; Figure 10. The diameter of the rotor and the shape of the gearing are improved by reworking the tools and by using optimized process parameters.

4.5 Endurance tests

After a successful adjustment of all parameters rotors with the correct geometry can be produced. In a very first test a slight mechanical impact causes a delamination; see Figure 11. The reason is because of a weak connection between the preform and the layer. After adjusting the process parameters the adhesion of the layer is so tough that a strong stroke would not peel off the layer but break the tooth.

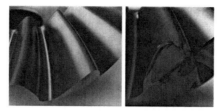

Figure 11: Delamination of coating by mechanical impact

If the inner sphere of the counter rotor is too large it touches the driven rotor. In contrast to the situation at the gears this is a permanent contact of the same surface. For this reason there is no cooling gas in contact with this surface at any time. The material overheats and blisters arise on the outside as shown in Figure 12.

Figure 12: Blistering caused by overheating in a test run

Cracks emerged during a test run of rotors made of the layer material 3 or 5 (see table 1) and the raw part material 2; see Figure 13. The connection of the layer to the preform was good enough to pass the mechanical impact test. It can be assumed that the reasons for the cracks are different thermal expansion coefficients of the applied materials which leads to high tensions at the operating temperature in the compressor.

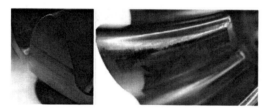

Figure 13: Cracks due to test run

Figure 14: Driven rotor after the required lifetime test run

Figure 14 shows the result of a successful test. The material combination is No. 2 for the raw part and No. 4 for the layer. The driven rotor looks good after a lifetime run of 1500 h. The conditions are a constant speed at maximum pressure in the laboratory. The application of COR® rotors produced in a two-component thermoset injection moulding process looks promising. The real challenge is the control of all parameters that are influencing the production.

5   CONCLUSION

The presented COR® compressor for burner systems in Diesel particulate filter arrangements meets the requirements. The main advantages are its constant performance, high efficiency and dynamic behaviour. The installation in any space and any position in the vehicle is easy due to compact design, low weight and non-use of oil. A two-component thermoset injection moulding process is established for the manufacturing of the rotors. The important properties are excellent tribology, high geometrical accuracy and low material costs. As shown in this paper, the design, the rotor material selection and the verification of the rotor and compressor layout are successfully done.

6   OUTLOOK

The next step for the COR® compressor is validation under realistic conditions. All necessary requirements for this procedure are fulfilled. The future task would be the stabilisation of the moulding process. The possibilities of a combination of different layers on driven and counter rotor can be investigated to increase the lifetime.

7   REFERENCE LIST

(1) Patentschrift - Drehkolbenmaschine: Arnold, Felix, DE59805024.3, 20.08.1998
(2) Achtundzwanzigste Verordnung zur Durchführung des Bundes-Immissionsschutzgesetzes (Verordnung über Emissionsgrenzwerte für Verbrennungsmotoren) vom 20. April 2004 (BGBl. I S. 614, 1423), die zuletzt durch Artikel 1 der Verordnung vom 8. April 2011 (BGBl. I S. 605) geändert worden ist
(3) http://www.dieselnet.com/standards/us/nonroad.php#tier4
(4) http://www.epa.gov/nonroaddiesel/2004fr.htm
(5) Patentschrift - Spaltverluststromsteuerung. Amesoeder, Dieter et al., DE102004026048.6, 25.05.2004
(6) R. Sachs, I.Nowitzky: Burner Air Supply Unit for an Active Regeneration of Diesel Particulate Filter (DPF)-Systems, In: International Rotating Equipment Conference – Compressors Users International Forum 2012, Proceedings, pp. 257-268; Düsseldorf/Germany 27th-28th September 2012, © Compressors, Compressed Air and Vacuum Technology Association within VDMA e.V.
(7) W. Beitz, K.-H. Grote (Hrsg.): Taschenbuch für den Maschinenbau / Dubbel, 20. neu bearbeitete und erweiterte Auflage, Springer Verlag, Berlin ff., 2001
(8) http://www.opcon.se/web/Water_lubricated_compressors.aspx
(9) Patentschrift – Verfahren zur Herstellung von Kunststoff-Rotoren: M. Sundström, K. Timuska – Svenska Rotor Maskiner AB, Stockholm, DE60016796T2, 24.11.2005
(10) Abschlußbericht SFB 316 - Metallische und metallkeramische Verbundwerkstoffe: H.-D. Steffens, F.-W. Bach (Hrsg,), 1. Auflage, Dortmund 1999, ISBN 3-9806415-2-x
(11) P. Thienel, A. Lück, et al.: Das Spiel mit dem Licht, Kunststoffe 6/2010, Carl Hanser Verlag, München, 2010

Screw pulsation generation and control: a shock tube mechanism

P X Huang, S Yonkers
Hi-Bar MC Technologies, LLC, USA

D Hokey, D Olenick
GE Oil & Gas, USA

ABSTRACT

Screw pulsations commonly exist in HVACR, automotive, energy and other processing industries and are believed to be a major source for system vibrations, noises and fatigue failures. It has been widely accepted that screw pulsations mainly take place at discharge side of the compressor and their magnitudes are especially significant at off-design conditions of either an under-compression (UC) or over-compression (OC).

However, due to the transient nature of pulsation phenomena and associated non-linear PDEs, some fundamental questions still remain to be unanswered even today, such as: What is the physical nature of screw pulsation and the conditions that trigger its happening? Where and when is it generated? How do we estimate its magnitude and travel direction? And is there a different way to tackle screw pulsation from insight gained from the above questions?

This paper attempts to answer these questions by applying the classic shock tube theory to the transient process of an UC or OC. A shock tube analogy is established with the hypothesis (implied from experimental observations) that the instant cavity opening to different outlet pressure of a screw compressor triggers pulsation generation. In light of this theory, it is revealed that the nature of a screw pulsation is a composition of strong bi-directional pressure waves AND induced unidirectional fluid flow in an inseparable CW-IFF-EW formation when initiated. Moreover, the pulsation level is directly proportional to the square root of the pressure ratio of UC or OC. Pulsation Rules are deduced to predict the location of initiation, maximum magnitude, travelling directions and speed from the design parameter and operating condition of the compressor. Based on this insight, a pro-active control strategy called pulsation trap is devised, targeting the unique characteristics of CW-IFF-EW and tackling them right at the predicted source. A 100% UC prototype testing indicates 20 dB pulsation reductions under different load and speed conditions without any major adverse effects.

NOMENCLATURE

c	speed of sound	Lp	sound pressure level
CW	compression waves pulsation component in screw compressor, shock wave in shock tube	OC	over compression
		p	absolute gas pressure or rms value for SPL
EW	expansion waves pulsation component in screw compressor and shock tube	Q	Screw inlet flow rate
		R	gas constant
		T	absolute temperature
IFF	induced fluid flow pulsation component	UC	under compression

© The author(s) and/or their employer(s), 2013

ΔU   contact surface velocity in shock tube, IFF velocity in screw compressor
W   shockwave velocity in shock tube, CW velocity in screw compressor
ρ   gas density
γ   ratio of specific heats

Subscripts
1   initial low pressure in shock tube, or in cavity or outlet in screw compressor during UC or OC
2   pressure after shockwave in shock tube, pressure after CW in screw compressor
3   pressure after expansion wave in shock tube, pressure after EW in screw compressor
4   initial high pressure in shock tube or in cavity or outlet in screw compressor during UC or OC
cavity   screw compressor cavity
inlet   screw compressor inlet
outlet   screw compressor outlet

1. INTRODUCTION

1.1 Screw Under-Compression, Over-Compression and Sources of Gas Pulsations

A rotary screw compressor uses two helical screws, known as rotors, to compress the gas. To demonstrate its operating principle, a complete cycle of a screw compressor is illustrated in Figures 1a-1e by following one flow cell in a typical 4x6 lobe configuration. Gas with flow rate Q at pressure p_{inlet} first enters the suction side and moves through the threads trapped as the screws rotate. Then the internal trapped volumes between the threads decrease and the gas is compressed to designed pressure p_{cavity}. The cavity then opens to outlet pressure p_{outlet} as shown in Figure 1d which is connected to a discharge dampener in series. It is essentially a positive displacement mechanism but using rotary screws instead of reciprocating motion so that displacement speed can be much higher. The result is a more continuous and smoother stream of flow with more compact size.

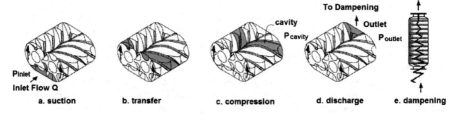

a. suction   b. transfer   c. compression   d. discharge   e. dampening

Figure 1(a-e): Conventional screw compression cycle with a serial dampener at discharge

At design condition, cavity pressure p_{cavity} is the same as the outlet pressure p_{outlet} of the screw. However, when the operating conditions differ from its design condition, there exists either an under-compression or over-compression. An under-compression, or UC, happens when the pressure at the outlet is greater than the pressure of the cavity just before the opening. This results in a rapid backflow of the gas into the cavity. The pulsed backflow for each compression cycle is widely believed, according to the conventional theory, to be the main source of the gas pulsations distinguished with high amplitude and cavity passing frequency. All fixed pressure ratio compressors suffer to a degree from under-compression due to mismatch of varying system back pressure with fixed design pressure. An extreme case is the Roots type blower where there is no internal compression at all, or the under-compression is 100%. On the other hand, an over-compression, or OC, takes place when the pressure at the outlet is lower than the pressure of the compressed gas within the cavity, causing a rapid forward flow of the gas into the discharge for each cycle, another source of the gas pulsations.

The conventional theory as exemplified by (1, 2), models the thermodynamic process of an UC or OC as isochoric, deviating abruptly from the adiabatic compression curve at sudden discharge opening as represented by a vertical rise or fall of pressure on P-V diagram as illustrated in Figure 2. The compressor efficiency becomes worse than the design point as indicated by additional work (horn areas) consumed.

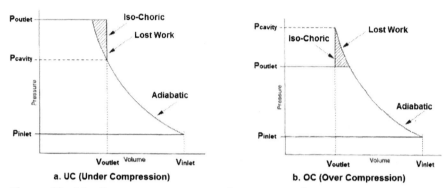

a. UC (Under Compression)   b. OC (Over Compression)

Figure 2(a-b): Screw compressor under compression or over compression processes on P-V diagram

1.2 Screw Pulsation Adverse Impact and Present Control Methods

It has long been observed that screw pulsations take place mainly at discharge side of the compressor with cavity passing frequency and its amplitude is especially significant under high operating pressures or at off-design conditions of either an UC or OC, as reported by (3, 4, 5, 6 and 7). For Roots type (100% UC), gas pulsation constantly exists and pulsation magnitude is directly proportional to the pressure rise from blower inlet to outlet.

Screw pulsations are transient in nature and gas borne. They travel through the downstream piping system and if left uncontrolled, could potentially damage pipe line in-line equipments and excite severe vibrations and noises as reported in (8, 9, 10 and 11). For this reason, screw compressors are often cited unfavorably with high pulsation induced NVH and low efficiency at off-design when compared with dynamic types like the centrifugal compressor. At the same time, the ever stringent NVH regulations from the government and growing public awareness of the comfort level in residential and office applications have given rise to an urgent need for quieter screw compressors.

Currently, an orifice plate as reported by Peters (12) or a large size reactive type pulsation dampener is required at the discharge side of a screw compressor as shown in Figure 1e. They are capable of reducing pressure pulsations by 5-10 folds, or 10-20 dB. However, both methods suffer in efficiency (pressure loss) and often result in fatigue failure of the dampener itself. Various efforts have been attempted to reduce screw pulsations without using a serially connected dampener or orifice plate at discharge. The most widely used method is based on a flow feedback principle as disclosed by (11, 13). The idea is to feed back a portion of the compressed gas from a modified outlet port shape or through a pre-opening slot or holes, thereby gradually equalizing gas pressure inside the cavity and reducing discharge pressure spikes compared with an abrupt lobe opening at discharge.

Figure 3 is an example of a pre-opening design with a modified outlet shape from (11). However, its effectiveness for pulsation attenuation is typically limited to less than 2-fold pulsation reduction, or 5-6 dB.

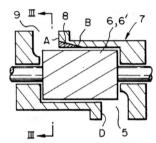

Figure 3: A screw compressor with a modified outlet port from (11)

1.3 Screw Pulsation Levels and Limitation of Acoustic Theory

Screw pulsation levels are measured as fluctuating pressures using dynamic pressure transducer and their magnitudes are typically ranging from 0.02 – 2 bar (0.3 - 30 psi) or equivalent to 160-200 dB in terms of sound pressure level (SPL) as defined by Equation (1), where p stands for the RMS value of the time domain dynamic pressure and p_{ref} is equal to 20 µPa. In comparison, pressure fluctuations of acoustic waves are typically less than 0.0002 bar (0.003 psi) or equivalent to 120 dB according to Beranek (14). Table 1 lists the corresponding values of gas pulsations and acoustic pressures in bar and dB. It can be seen that pressure levels of screw pulsations are several orders of magnitude higher and well beyond the small disturbance assumption in classic acoustics. Moreover, screw pulsation has a distinguished cavity passing frequency, and pulsation wave form rises and falls much faster than the linearized harmonic waves in the Acoustics.

$$L_p = 10 \log_{10} \left(\frac{p^2}{p_{ref}^2} \right) = 20 \log_{10} \left(\frac{p}{p_{ref}} \right) \text{ dB}$$

The implication is challenging: the linearized wave equation that allows the superposition of solutions based on small perturbations can no longer be accurately used to model screw pulsation.

Table 1: Magnitude of acoustic wave in comparison with gas pulsation from screw compressor

	Acoustic Wave			Gas Pulsations in RED			
Pressure Pulsation, bar	0.000002	0.00002	0.0002	0.002	0.02	0.2	2
Sound Pres. level, L_p, dB	80	100	120	140	160	180	200

1.4 Fundamental Questions and a Transient Approach

Due to the difficulty in modeling and solving mathematically the transient behavior and associated non-linear PDEs of screw pulsation, there seems to be a lack of theory even today that could satisfactorily explain the generation mechanism of the screw pulsation. Fundamental questions still remain unanswered, such as: What is the physical nature of screw pulsation? Is it a wave or a fluid particle movement or both? What exactly triggers the transient process? Where and when are pulsations generated, and how do we estimate their magnitude and travel directions? More importantly, how can we devise a more effective pulsation control strategy than the current methods? It is believed that an effective pulsation control method is only possible as a result of correctly answering these questions. This paper will focus on the mechanism of screw pulsation generation and control while the topic of under (or over) compression mechanism can be found in paper (15).

In order to answer these questions while avoiding any non-linear differential equations, a different approach is desired that could explain the observed physical phenomena, reflect the transient nature of pulsation and estimate its quantitative magnitude/travelling directions. Searching for answers, we came across a powerful analytical tool, the shock tube, which is simple to use yet well established for studying transient behavior of a screw discharge process.

2. SCREW PULSATION GENERATION MECHANISM: A SHOCK TUBE THEORY

2.1 Introduction to Shock Tube Theory

The shock tube, invented in 1899 by French scientist Pierre Vieille, is a device used to study the transient aerodynamic phenomena under a wide range of temperatures and pressures for a variety of gases. It has been studied extensively by researchers from the start of the last century and has effectively served the developments of the supersonic and hypersonic flights and a range of transient devices, such as pulse-jet engines and a special supercharger called Comprex, which is summarized in (16). However, the physical phenomena observed in the shock tube and the well established shock tube theory have thus far not served for examining, hence determining the gas pulsation mechanism of a screw compressor in general.

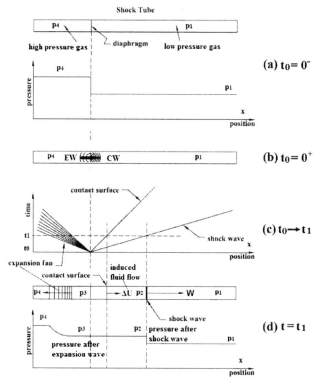

Figure 4(a-d): Diaphragm-triggered shock tube wave diagram and pressure distribution

An ideal diaphragm-triggered shock tube, as shown in Figure 4a, from (17, 18, 19), is a tube in which gases at low and high pressure regions are separated by a diaphragm. This diaphragm then suddenly bursts open, shown in Figure 4b, which produces a series of compression waves (CW), each increasing the speed of sound behind them so that they quickly coalesce into a shock wave propagating into the low pressure gas, as illustrated in "position vs. time" wave diagram and pressure distribution in Figures 4c-4d. The shock wave increases the temperature and pressure of the low-pressure gas adiabatically when sweeping through at shockwave velocity W and also induces a fluid flow in the same direction but with a much slower velocity ΔU. Simultaneously, a fan of rarefaction (expansion) waves (EW) travel back into the high-pressure gas, decreasing its pressure and temperature. The interface separating low and high-pressure gases is referred to as the contact surface, which travels at the same velocity ΔU as the induced fluid flow between the shock and expansion wave.

The analytical solutions for shock wave and shock tube were available more than a century ago, thanks to the independent efforts of Rankine, Taylor and Hugoniot. The key results were re-derived and summarized by Anderson (18) as below:

$$\frac{p_4}{p_1} = \frac{p_2}{p_1} \left[1 - \frac{(\gamma_4 - 1)(c_1/c_4)(p_2/p_1 - 1)}{\sqrt{2\gamma_1[2\gamma_1 + (\gamma_1 + 1)(p_2/p_1 - 1)]}} \right]^{-2\gamma_4/(\gamma_4 - 1)}$$
(2)

$$p_2 - p_1 = \rho_1 \, W \, \Delta U$$
(3)

Where γ is the ratio of specific heats, c is the speed of sound that is equal to $\sqrt{\gamma RT}$.

Equation (2), known as the Shock Tube Equation, gives the incident shock strength p_2/p_1 as an implicit function of the diaphragm pressure ratio p_4/p_1 for known gas types. While Equation (3), known as the Rankine-Hugoniot Equation, relates the abruptly arisen pressure (p_2-p_1), a measure of shock strength, to shockwave speed W, initial gas density ρ_1 and induced flow velocity ΔU. Moreover, if the gas types on two sides of the diaphragm are same, say air, and pressure ratio $p_4/p_1 < 6$, the magnitudes of the simultaneously generated Compression Waves (CW) and Expansion Waves (EW) can be approximated by:

$$p_2/p_1 = p_4/p_2 = (p_4/p_1)^{1/2}$$
(4)

Note that the wave magnitude of CW or EW is defined as the pressure difference (p_2-p_1) or (p_4-p_2) respectively. It should also be pointed out that the CW and EW values obtained from equations (2) - (4) are the maximum possible magnitude that could be achieved assuming zero diaphragm bursting time. Moreover, they are independent of shock tube length.

In summary, two conditions are sufficient to trigger a shock tube for a transition from static state to dynamic: the existence of a pressure difference (p_4-p_1) and sudden opening of a valve (diaphragm is a special onetime valve) that separates the gases with that pressure difference.

2.2 Screw Lobe as Fast Rotary Valve
A traditional mechanical open/close valve is a device that starts or stops the flow of a fluid media inside a pipe by opening or closing its passageway. It has long been observed that fast opening or closing valves inside a pipe could trigger transient phenomena such as gas hammer and water hammer, which are characterized by non-linear waves and induced fluid flow. The rotary lobe inside a screw compressor is essentially a rotary type mechanical valve that is timed to open or close the inlet and outlet by rotary motion in a screw compression cycle, as shown in Figure 1. Unlike a recip type, a rotary valve opens and closes in a non-stop manner with timing entirely determined by the speed of the rotor and casing geometry, not by the actuation from a pressure difference like a reed valve.

As shown in Figure 1 of a traditional screw compression cycle, the suction process begins very early, right after finishing discharge, and lasts until the cavity is closed by lobes and casing wall. The pressure inside the cavity is slightly less than the inlet pressure so that gas can gradually fill the cavity. Then the trapped internal volumes inside the cavity gradually decrease and the gas is being compressed to its designed pressure $p_{cavity} = p_{outlet}$ during the so called internal compression in a perfect match. However, when the operating conditions of a screw differ from its design condition, an UC or OC is resulted as soon as the cavity is opened suddenly to a mis-matched discharge pressure.

In other words, the screw UC or OC process is characterized by the sudden exposure of two gases with a pressure difference of Δp (=$p_{outlet} - p_{cavity}$), equivalent to the suddenly opened diaphragm to Δp_{41} for a shock tube as shown in Figure 4b. This likening immediately suggests a possible shock tube mechanism for the initiation of screw pulsation. This insight is also consistent with test observations that screw pulsations take place mainly at discharge, and its magnitude is directly proportional to the degree of UC or OC.

2.3 Screw Pulsation Mechanism: A Shock Tube Analogy
So an analogy between the shock tube and screw compression is established at the moment just before and after the sudden diaphragm opening as illustrated in Figures 5a and 5b, and the analogous equivalents are correlated in Table 2. Hence, a tentative screw pulsation theory can be proposed by a single postulate as follows:

> Screw cavity with a low gas pressure p_1 (high gas pressure p_4 for OC) is instantly opened to a higher outlet pressure p_4 (low gas pressure p_1 for OC) in the under compression phase (Over Compression phase).

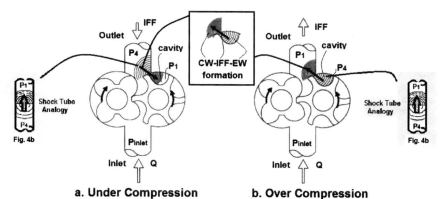

a. Under Compression   b. Over Compression

Figure 5(a-b): Compression at off-design of a screw compressor according to the shock tube theory

Table 2: Analogy - Shock Tube vs. Screw Outlet

Device: shock tube	Device: screw outlet
moment: before ($t = 0^-$) and after ($t = 0^+$) the diaphragm opening	moment: before ($t = 0^-$) and after ($t = 0^+$) the cavity opening in UC or OC
focusing location: near front and back of the diaphragm with Δp_{41}	focusing location: divider between the cavity and outlet the cavity is opened to
high pressure region: p_4	outlet side pressure (UC): p_4
low pressure region: p_1	cavity side pressure (UC): p_1

This postulate is fundamental and significant here not only because it establishes the link between the conditions of the screw discharge and the shock tube just before the discharge opening, but also because it implies that what happens thereafter can be applied to the screw discharge due to the same dynamic forces of the suddenly exposed pressure difference ($p_4 - p_1$). Therefore, the well established results of the shock tube theory accumulated over the past 110 years, though degraded somehow by the ideal gas and one dimensional assumption, can be readily applied to reveal the screw pulsation generation mechanism without the need to solve any non-linear PDEs.

Now let's take a new look at the transient process of an UC for a screw compressor again, as shown in Figure 5a, in light of the new screw pulsation mechanism. The suddenly opened cavity at screw discharge, as shown in Figure 5a, resembles the moment just after the diaphragm bursting of a shock tube shown in Figure 4b. A series of compression waves (CW) would be generated into the cavity at the wave speed. Simultaneously, a fan of expansion waves (EW) would be generated downstream, reversing the downstream fluid flow instantly as they sweep through at wave speeds, which results in a unidirectional backflow (IFF) between the CW and EW fronts. Hence, the induced fluid flow (IFF), or the backflow as in conventional UC theory, is the result of CW and EW wave actions, according to the new theory.

Since all screw compressors mechanically divide the incoming continuous gas stream into parcels of cavity size for compression and release at discharge, they inherently generate pulsations with lobe opening frequency at the moment of the cavity opening to the outlet whenever there is a pressure difference between the cavity and outlet.

2.4 Pulsation Rules for Screw Compressor

For the convenience of effectively applying the analytical results of the shock tube theory to the screw compressor, especially for a layman to use, the following Pulsation Rules are summarized as a simplified way to determine the following: Where is the source of gas pulsation, and when does it happen? What are the sufficient conditions to trigger its generation, and how do we predict quantitatively its magnitude and travel direction at its source? In principle, these rules, if validated by future experiments, are applicable to different gases and for gas pulsations generated by any PD (Positive Displacement) type gas machinery, such as engines, expanders, pressure compressors and vacuum pumps.

1. Rule I: For any two divided compartments (either moving or stationery) with different gas pressures p_1 and p_4, there will be no or little gas pulsations generated if the two compartments stay divided;
2. Rule II: If, at an instant, the divider between the high pressure gas p_4 and the low pressure gas p_1 is suddenly removed in the direction of divider surface, gas pulsations are instantaneously generated at the location of the divider and instant of the removal as a composition of a fan of Compression Waves (CW) or a quasi-shock wave, a fan of Expansion Waves (EW) and an Induced Fluid Flow (IFF) with magnitudes as follows:

$$CW = p_2 - p_1 = p_1\,[(p_4/p_1)^{1/2} - 1] = (p_4 \times p_1)^{1/2} - p_1 \tag{5}$$

$$EW = p_4 - p_2 = CW * (p_4/p_1)^{1/2} = p_4 - (p_4 \times p_1)^{1/2} \tag{6}$$

$$\Delta U = (p_2 - p_1)/(\rho_1 \times W) = CW/(\rho_1 \times W) \tag{7}$$

Where ρ_1 is the gas density at low pressure region, W is the speed of the lead compression wave, ΔU is the velocity of Induced Fluid Flow (IFF);

3. Rule III: Pulsation component CW is the action by the high pressure (p_4) gas to the low pressure (p_1) gas while pulsation component EW is the reaction by low pressure (p_1) gas to high pressure (p_4) gas in the opposite direction, and their magnitudes are such that they approximately equally divide the initial pressure ratio p_4/p_1 [that is: $p_2/p_1 = p_4/p_2 = (p_4/p_1)^{1/2}$]. At the same time, CW and EW pair together to induce the third pulsation component: a unidirectional fluid flow IFF in a fixed relationship of CW-IFF-EW.

Rule I implies that there would be little or no gas pulsations during the suction, transport and internal compression phases of a screw cycle because of the absence of either a pressure difference or an abrupt opening. The focus instead should be placed upon the sudden discharge phase, especially at the instant when the cavity is suddenly opened to a different outlet pressure during the off-design conditions in the form of UC or OC.

Rule II indicates specifically that the **moment** of pulsation generation is the instant the lobe separating p_4 and p_1 suddenly opens and that the **location** is at the opening lobe. Moreover, it defines the **two sufficient conditions** for pulsation generation as follows:

a) The existence of a pressure difference Δp_{41};
b) The sudden opening of the divider separating the pressure difference Δp_{41}.

Because all screw compressors have a "sudden" opening to Δp_{41} under UC or OC for each cycle, both sufficient conditions are satisfied at the moment of the cavity opening to the discharge, thus the reason gas pulsations are inherent and strong for screw at off-design conditions.

It should be noted that Rule II specifies that the divider sudden opening is, as an ideal case, in the same direction of the divider surface, and there is no velocity component in the IFF direction. This is an approximation to the lobe opening of a screw or scroll compressor, which acts essentially like a gate valve in contrast to a reed valve that has a dominant normal component in the IFF direction.

Pulsation Rule II predicts that the nature of screw pulsation is not just a flow phenomenon, as theorized today, but a composition of strong bi-directional waves AND induced fluid flow in an inseparable CW-IFF-EW formation when initiated. These waves are non-linear in characteristics with changing waveform during propagation. This is in direct contrast to the linear acoustic waves and unchanging wave fronts during propagation, and do not induce a mean through flow. It is interesting to note the wholeness of the three different pulsation components in an inseparable formation as CW-IFF-EW, which is generated simultaneously, and one cannot be produced without the others. This implies that noises are generated in the form of CW and EW simultaneously as the induced fluid flow IFF and represented as a loud bang due to the steep wave shape and large magnitude. It further suggests that gas pulsations would be very difficult to control because not one but all three components (CW-IFF-EW) need to be handled. It should be pointed out that the pulsation strength predicted by Rule II is the ideal maximum possible magnitude generated at the source by assuming instant opening time.

Rule III shows further that the interactions between gases at different pressures are mutual and in pairs so that for every CW pulsation into or out of the cavity, there is always an equal but opposite EW pulsation going or coming downstream in terms of the pressure ratio (approximately equal: $p_2/p_1 = p_4/p_2$). Paired together collectively, they induce, by the pulling force of CW at front and pushing force of

EW from behind, a unidirectional backflow (IFF) in an inseparable formation CW-IFF-EW.

The challenge is: could the Pulsation Rules be used as a guideline to devise a new pulsation control method for the screw compressor such that all three components of pulsation (CW-IRFF-EW) can be dealt with at the same time.

3. PULSATION CONTROL: PULSATION TRAP METHOD

3.1 Application of Pulsation Rules for Pulsation Control: A Pulsation Trap Method

For pulsation control purpose, two possibilities could be reasoned from the two sufficient conditions for pulsation generation: either make the sudden opening at the under or over compression phase gradual or reduce the pressure difference Δp_{41} of an UC or OC at the discharge in such a way so that it becomes zero at discharge. Because the velocity of the lobe opening is mainly determined by the rotor speed and flow rate, the size of the compressor would become very large if RPM is reduced significantly for the same amount of flow rate.

The alternative is to change the port shape into a gradual opening at discharge or to have a pre-opening slot before discharge shown in Figure 6a, as suggested and tested in (11, 13), but this route has been proven not very effective, only achieving a 2-fold pulsation reduction. A clue to this deficiency can be traced back to the fundamental questions raised in paragraph 1.4 of this paper: What is the true nature of screw pulsation? Is it just a fluid flow phenomenon illustrated in Figure 6a as assumed by the current UC or OC theory, or a combined wave and fluid action in a CW-IFF-EW formation according to the newly proposed shock tube theory in Figure 6b? The answer to this question determines the strategy of screw pulsation control. As can be seen from Figures 5a and 6b, which illustrates the same under compression phase but interpreted by the new pulsation generation mechanism, the EW pulsation is left loose at the outlet even though the slot size of pre-opening makes the opening more gradual. This could be the reason for the limited success (just 2-fold reduction) of the pre-opening method. So a modification is needed that could not only eliminate the pressure difference Δp_{41} before discharge but also control the EW pulsation at the same time. This new method is called pulsation trap in this paper and relevant references (20, 21).

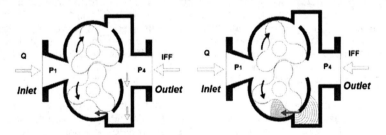

a. backflow compression   b. CW-IFF-EW compression

Figure 6(a-b): Comparison of pulsation generation mechanism between Pulsation Rules and backflow theory for a Roots blower

As a brief introduction to the principle of a pulsation trap, Figure 7 shows again a screw compressor at UC for a 4x6 lobe combination but with an addition of a pulsation trap near the compression cavity. In principle, pulsation trap is used to trap *and* to attenuate all three components of pulsation CW-IFF-EW at the same time. For comparison, the traditional discharge dampener is connected in series with the cavity (compression chamber) AFTER the discharge port, as shown in Figure 1e and 8a, and through which both cavity flow Q and pulsation flow IFF pass, while the pulsation trap is connected in parallel with the cavity BEFORE the discharge port through which only pulsation flow IFF passes, shown in Figure 8b. The phases of flow suction, transfer and internal compression are still the same as those shown in Figures 1a to 1c. But during the discharge phase say at UC as illustrated in Figure 7, instead of waiting for the lobe opening at discharge as a conventional screw compressor, the flow cavity is pre-opened to an injection port (or trap inlet). The trap inlet is connecting, before discharge, the cavity to the pulsation trap in parallel, which in turn is also communicating with the compressor outlet through a feedback port (trap outlet). Between the trap inlet and trap outlet and within the pulsation trap, there is a pulsation dampening means to control the EW pulsation component that was "left loose" in the pre-opening method by (11, 13).

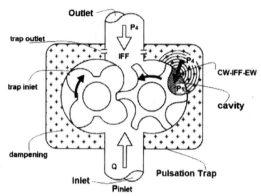

Figure 7: Compression phase of a screw compressor with a Pulsation Trap for an Under Compression

As shown in Figure 7 for the case of UC, both CW-EW waves and IFF flow are generated as soon as the cavity is opened to the trap inlet due to the pressure difference between the pulsation trap (relates to outlet pressure p_4) and cavity (relates to cavity pressure p_1). The suddenly generated CW waves travel into the cavity, compressing the gas inside with the assistance of IFF. At the same time, the simultaneously generated EW waves at the trap inlet are propagating into the pulsation trap, and therein are being contained *and* attenuated. Because waves travel at the speed of sound, about 5-20 times faster than the rotor tip speed, the CW and EW pulsation could be well settled after a few reflections before the lobe tip reaches the outlet, hence discharging a pulsation-free flow. Thus, there is no more need for a serial dampener.

In principle, the trap inlet of a pulsation trap can be designed so that the screw compressor will operate under a partial internal compression mode and partial Under-Compression mode, but never under Over-Compression mode in order to maximize average system efficiency and minimize pulsations and noises.

3.2 Pulsation Trap: A Parallel Dampener
For pulsation trap, the gas Q inside the cavity being compressed and pulsation flow IFF going through the trap being controlled are parallel with each other as shown in Figure 8b, instead of in series as in a conventional screw compressor shown in Figure 8a.

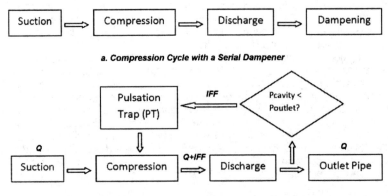

Figure 8(a-b): Comparison of Pulsation Trap with a conventional serial dampener for Under Compression

There are several advantages associated with the parallel pulsation trap compared to the traditional serially connected pulsation dampener. First of all, pulsation flow IFF attenuation is separately handled from the main cavity flow Q so that an effective IFF dampening will not affect the main cavity flow Q, resulting in both higher system efficiency (no discharge pressure loss associated with a serial dampener) and attenuation efficiency. In a traditional serially connected dampener, both pulsating flow IFF and cavity flow Q travel mixed together through the dampener where a better attenuation always comes at a cost of higher pressure losses. So a compromise is often made in order to reduce losses by sacrificing the degree of pulsation dampening or employing a very large volume and costly dampener in a serial setup.

Secondly, the parallel pulsation trap attenuates pulsation closer to the pulsation generation source and in a shape more conformal and compact than a serial one. Pulsation trap can be designed and built as an integral part of the compressor casing as shown in Figure 7. The noise source is encased inside, resulting in much less noise radiation surfaces, hence smaller size and less weight.

The key question is: will the actual test support the pulsation trap concept designed according to the Pulsation Rules rather than the backflow theory?

3.3 Prototype Testing and Results

For preliminary testing, the special case of the 100% UC is prototyped using a 75 HP Roots blower with a pulsation trap designed and built based on Pulsation Rules targeting pulsation formation of CW-IFF-EW. Base tests are first conducted for two traditional designs: a Roots blower with a serial dampener at discharge and a Roots blower with a pre-opening injection port called WhispAir (22). Then for consistency, the same test setup and instruments are used again to test and measure the newly designed and built prototype with a pulsation trap. The test setup follows ASME PTC-9 specification and a Himmelstein torque meter is used for the shaft power measurement. Pressure pulsations are measured using PCB dynamic pressure transducers at the blower inlet and discharge, before and after the dampener for the whole range of design pressure and speed. Both CSI signal analyzer and National Instruments NI-DAQ system are used and compared for data acquisition of dynamic pressures. A sample discharge pulsation measurement data in Figure 9 show the time traces, large magnitude (pulsation unit as psi) and the dominant pulsation at cavity passing frequency.

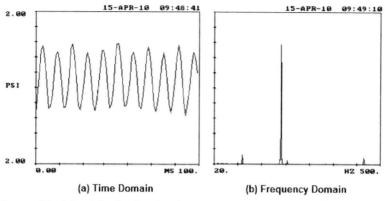

(a) Time Domain   (b) Frequency Domain

Figure 9(a-b): A typical pulsation measurement at Roots discharge with data taken by a CSI analyzer

The discharge pulsations under different loads and speeds are presented in Figure 10 for pulsation trap in comparison with other methods. Firstly, it confirms the observed results that gas pulsation from an under compression takes place mainly at discharge with magnitude ranging from 0.01 – 0.1 bar (0.15 – 1.5 psi in Figure 10) and is directly proportional to the pressure difference of a UC (up to 1 bar, or 15 psi), while the inlet side pulsation is at least one order of magnitude lower, less than 0.01 bar (0.15 psi). Secondly, it demonstrates the effective dampening by a traditional serially connected dampener, 5-10 fold pulsation reduction (or 10-20 dB) at discharge, but with a dampener loss ranging from 0.01-0.03 bar (0.15-0.45 psi) depending on the size of dampener. Thirdly, it further confirms the level of dampening by a pre-opened feedback design or WhispAir: about 2 fold pulsation reduction at discharge, or 5-6 dB. Lastly, 10 fold pulsation reduction is achieved (at high end of pressure rise) with a pulsation trap design without using a discharge dampener, hence resulting in no dampener related losses. It should be pointed out that the above pulsation results are fairly uniform and stable for the whole working range of pressure rises and operating speeds of the blower, while the temperature rise and shaft power are measured and found not be affected adversely.

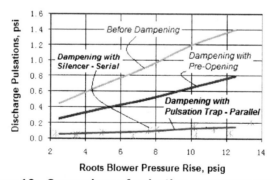

Figure 10: Comparison of pulsation measurements for different dampening methods under UC mode

The data showing effective pulsation attenuation by the Pulsation Trap can also be interpreted as a partial validation for the CW-IFF-EW pulsation theory because it addresses EW pulsation in addition to CW and IFF pulsation components. It is the additional control specifically targeting the CW and EW (wave control in addition to flow control) that differentiates the pulsation trap with the pre-opening method.

4. CONCLUSIONS AND DISCUSSIONS

It is noted that the instant cavity opening to different outlet pressure as we have postulated for screw under or over compression immediately suggests a possible shock tube mechanism for the initiation of screw pulsation.

The attempted application of the classic shock tube theory to the transient process of a screw under or over compression reveals several surprising results. Firstly, it is shown that screw pulsation is not just a pure fluid flow phenomenon as theorized today but a composition of strong bi-directional pressure waves AND induced unidirectional fluid flow in an inseparable CW-IFF-EW formation when initiated. Moreover, it is revealed that the instant flow IFF is induced by the pushing force CW and the pulling force EW as they sweep through the fluid flow. Secondly, the pulsation phenomena are directly triggered by the sudden exposure of the pressure difference at screw discharge, and the pulsation magnitude can be estimated quantitatively and is proportional to the square root of the pressure ratio of UC or OC, i.e.: $(p_4/p_1)^{1/2}$. Furthermore, the location of initiation, travelling directions and speed can all be estimated based on the design parameters and operating condition of the screw compressor.

For the convenience of practical applications, three Pulsation Rules are summarized to predict the maximum magnitude of the three pulsation components, CW-IFF-EW, and to obtain the sufficient conditions for their generation. Moreover, it has not escaped our notice that the three Pulsation Rules somehow resemble the three Newton's Laws of Motion. It could be just a coincidence or can also be interpreted as one form of the manifestation of Newton's Laws of Motion in its application to the Shock Tube and its subsequent application to gas pulsation phenomena. No matter what the reasons are, Pulsation Rules are a simplified way for a layman to use.

Based on insight gained from the Pulsation Rules, a new pro-active control strategy called the Pulsation Trap is devised, which tackles the CW-IFF-EW formation simultaneously and right at the predicted source. Prototype testing for the case of 100% UC demonstrates 20 dB pulsation reductions under different load and speed conditions without a serially connected discharge dampener and associated dampener losses.

It should be pointed out that though illustrations and description in this paper are devoted to a screw compressor, the principle of Pulsation Rules and pulsation trap can also be applied to scroll compressor or other PD (Positive Displacement) type gas machinery, such as engines, expanders, pressure compressors or vacuum pumps.

Finally, we hope, in our continued search to understand screw compression and pulsation generation mechanism, further experiments can directly confirm the existence of CW and EW components of screw pulsation and validate the formula for predicting each pulsation component. A prototype based on an actual screw compressor incorporating a pulsation trap should be built and tested under both UC and OC conditions. With more research and development resources devoted by both academia and industry, it is anticipated that future generation of screw compressors can be designed to be even simpler in structure, smaller in size and smoother in running than those used today.

ACKNOWLEDGEMENTS

The authors would like to thank all the reviewers for their time and professional suggestions that have made this paper better. We would also like to pay special tribute to Dr. Joost Brasz whose influence and advice have always proved to be valuable. Most importantly, a special grant from GE Oil and Gas that makes the prototype testing during the last few years possible is greatly appreciated by the authors.

REFERENCES

1. Stoecker, W., *Industrial Refrigeration Handbook, 2nd Edition*, McGraw-Hill, p. 125-139, 2004.
2. Xing, Z., *Screw Compressors: Theory, Design and Applications*, Machinery Industry Publisher, p. 89-92, 2000.
3. Mujic, E., Kovacevic, A., Stosic, N., Smith, I., *The influence of port shape on gas pulsations in a screw compressor discharge chamber*, International Conference on Compressors and their Systems 2007, London, UK, p. 213-223, 2007.
4. Koai, K., Soedel, W., *Determination of port flow and interpretation of periodic volume source*, The 1990 International Compressor Engineering Conference at Perdue, p. 369-377, 1990.
5. Sangfors, B., *Computer simulation of gas-flow noise from twin-screw compressors*, International Conference on Compressor and Their Systems 1999, London, p.707-716, 1999.
6. Wu, H., Xin, Z., Peng, X., Shu, P., *Simulation of discharge pressure pulsation within twin screw Compressors*, International Conference on Compressors and their Systems 2007, London, UK, 2004.
7. Gavric, L., Badie-Cassagnet, A., *Measurement of gas pulsations in discharge and suction lines of refrigerant compressors*, The 2000 International Compressor Engineering Conference at Perdue, p. 627-634, 2000.
8. Price, S., Smith, D., *Sources and Remedies of Hi-Frequency Piping Vibration and Noise*, Proceedings of The 28th Turbomachinery Symposium, p. 189-212, 1999.
9. Tweten, D., Nored, M., Brun, K., *The Physics of Pulsations*, Gas Machinery Conference, 2008.
10. Lucas, L., *Pulse-free blower*, US Patent# 4,215,977, 1984.
11. Yanagisawa, K., Maeda, M., *Screw Compressor*, US Patent# 5,051,077, 1989.
12. Peters, R., *Damping of low frequency noise in piping systems by means of perforated plates*, International Conference on Compressor and Their Systems, London, p.545-558, 2003.
13. Weatherston, L., *Pulse-free blower*, US Patent# 4,215,977, 1984.
14. Beranek, L., *Noise & Vibration Control*, Inst. of Noise Control Engineer, New York, p. 396, 1988.
15. Huang, P., *Under Compression: An Isochoric or Adiabatic Process?* The 2012 International Compressor Engineering Conference at Perdue, 2012.
16. Mueller, N., *A Review of Wave Rotor Technology and Its Applications*, ASME J. of Engineering for Gas Turbines and Power, vol. 128, p. 717-731, 2006.
17. Rudinger, G., *Non-steady Duct Flow: Wave Diagram Analysis*, Dover, New York, 1969.
18. Anderson, J., *Modern Compressible Flow*, McGraw-Hill Book Company, New York, p. 172–205, 1982.
19. Weber, H. E., *Shock Wave Engine Design*, John Wiley & Sons, Inc., New York, 1995.
20. Huang, P., *Rotary Lobe Pump or Vacuum Pump with Shunt Pulsation Trap*, US2011/0300014 A1, 2009.

21. Huang, P., *Screw Compressor with a Shunt Pulsation Trap*, US2012/171069 A1, 2011.
22. Roots, Dresser Industries, Inc., *Noise Control for Rotary Blowers*, Connersville, Indiana, 1980.

Numerical analysis of unsteady behaviour of a screw compressor plant system

E Chukanova, N Stosic, A Kovacevic
City University London,
Centre for Positive Displacement Compressor Technology, UK

ABSTRACT

In order to analyse the performance of screw compressors operating under varying load conditions, an unsteady one dimensional model of the compressor process was modified to include all plant components, including tanks and connecting piping. This was based on the differential equations of conservation of mass and energy. The results derived from this model have been verified by experiment in order to obtain a reliable tool that can simulate a variety of scenarios which may occur in everyday compressor plant practice.

Keywords: Mathematical modelling, Screw compressor process, Compressor plant

NOMENCLATURE

A	– cross section area of discharge valve	T_{in}	– temperature of the gas entering the tank
M	– mass	T_{out}	– temperature of the gas leaving the tank, equal to T_2
m_{in}	– mass flow entering the tank	Δt	– time step
m_{out}	– mass flow leaving the tank	U	– internal energy
p_0	– atmospheric pressure	u	– specific internal energy
p_1	– pressure in the compressor tank	V	– volume of the plant containing tank and pipes
p_2	– pressure in the tank at the next time step		
R	– gas constant	μ	– flow coefficient
T_2	– Temperature in the tank	ρ_2	– density of gas in the tank

1. INTRODUCTION

Design tools for the design of screw compressors and the prediction of their performance under steady operating conditions, are now widely available. However, despite the fact that such machines frequently and, in some cases even continuously operate under unsteady conditions, there are few published studies of how this affects their performance, while it is quite possible that the need to operate under such conditions may influence the compressor design. The following summarises known studies.

Jun and Yezheng (1),(2) carried out experimental studies on the effects of working fluid migration during the start-up and shut-down cycles of a refrigeration system with a reciprocating compressor. They developed a program to estimate energy losses and how to calculate how much they are due to this effect, with the aim of reducing energy consumption.

Fleming, Tang and You (3) published a paper on simulation of shutdown processes in screw compressor driven refrigeration plant. Their idea was to use a reverse rotation brake instead of a suction non-return valve, which prevented reverse rotation, leading to a significant decrease of the compressor backflow. As a consequence, a reduction of the shutdown torque occurred. However, only the mathematical model was presented and no experiment data has followed to validate it. A disadvantage of the reverse rotation brake is that it might trigger failure if there is significant rotor backlash in the compressor.

Li and Alleyne (4) investigated transient processes in the start-up and shutdown of vapour compression cycle systems with semi-hermetic reciprocating compressors. They established a model of a moving boundary heat exchanger and validated it experimentally. Ndiaye and Bernier (5) developed a dynamic model for a reciprocating compressor in on-off cycle operation and validated it as part of an experiment to justify water-to air heat pump models. A recent paper, by Link and Deschamps (6), deals with the numerical methodology and experimental validation of start-up and shutdown transients in reciprocating compressors.

This paper is the third of a series intended to give an insight into screw compressor transient behaviour. Previous papers described compressor start-up as a transient process and were presented in two papers by Chukanova, Stosic and Kovacevic, (7),(8). Experimental results and their analysis were presented and simulations included inertia effects during start up.

The work now described, covers the numerical simulation of unsteady behaviour of a screw compressor within a compressor plant, including the filling and emptying of the plant tank and associated connecting pipes during different types of compressor starts. This model has been integrated with SCORPATH (Screw Compressor Optimal Rotor Profiling and Thermodynamics), an existing compressor design program, developed in house. The model is written in FORTRAN and is based on the differential equations of mass and energy conservation, developed and tested in earlier work. It is sufficiently general to take into account dry and oil flooded compressors and various plant tanks connected by gas pipes in different combinations providing that they are characterized by one volume and one exit valve.

An interface was written to couple the compressor and plant model elements for this purpose and has been used to show how the tank pressure is affected by the gas mass flow rate, the compressor discharge gas temperature, and the volume of the tank and communicating pipes. The tank pressure is then used to calculate the compressor performance in the next time step. The sequence is repeated for the whole compressor plant system until the specified time is reached.

The model was verified by comparison of predictions obtained from it with measurement results obtained in a series of tests performed on a compressor test rig. A detailed description of the experiments is given in Chukanova, Stosic and Kovacevic (7),(8). A part related to the model verification is presented in section 3 of this paper.

2. MATHEMATICAL MODEL OF THE SCREW COMPRESSOR PLANT

Screw compressor modelling combines the analysis of thermodynamic and fluid flow processes. Both are dependent on the screw compressor geometry and combining them in a mathematical model as a complex process.

The algorithm of the thermodynamics and flow processes in a screw compressor, described here, is based on a mathematical model, defined by a set of equations which describe the physics of the complete process in a compressor. The equation set consists of the equations for the conservation of energy and mass continuity together with a number of algebraic equations defining the flow phenomena in the fluid suction, compression and discharge processes together with the differential kinematic relationship which describes the instantaneous operating volume and its change with rotation angle or time. In addition, the model accounts for a number of 'real-life' effects which may influence the final performance of a real compressor and make the model valid for a wider range of applications.

In the past, these equations have often been simplified in order to achieve a more efficient and economical numerical solution of the set. In this case, where all the terms are included, the effect of such simplifications on the solution accuracy can be assessed.

2.1 Equations governing screw compressor process

The working chamber of a screw machine together with the suction and discharge plenums can be described as an open thermodynamic system in which the mass flow varies with time and for which the differential equations of conservation laws for energy and mass are derived using Reynolds Transport Theorem.

A feature of the model is the use of the unsteady flow energy equation to compute the effect of profile modifications on the thermodynamic and flow processes in a screw machine in terms of rotational angle, or time.

The following conservation equations have been employed in the model.

The conservation of internal energy:

$$\omega\left(\frac{dU}{d\theta}\right) = \dot{m}_{in} h_{in} - \dot{m}_{out} h_{out} + \dot{Q} - \omega p \frac{dV}{d\theta}$$

where θ is angle of rotation of the main rotor, $h=h(\theta)$ is specific enthalpy, $\dot{m} = \dot{m}(\theta)$ is mass flow rate $p=p(\theta)$, fluid pressure in the working chamber control volume, $\dot{Q} = \dot{Q}(\theta)$, heat transfer between the fluid and the compressor surrounding, $V = V(\theta)$ local volume of the compressor working chamber.

Flow through the suction and discharge port is calculated from the continuity equation. The suction and discharge port fluid velocities are obtained through the isentropic flow equation. The computer code also accounts for reverse flow.

Leakage in a screw machine is a substantial part of the total flow rate and affects the compressor delivery, i.e. volumetric efficiency and the adiabatic efficiency, the gain and loss leakages are considered separately. The gain leakages come from the discharge plenum and from the neighbouring working chamber with a higher pressure. The loss leakages leave the chamber towards the discharge plenum and to the neighbouring chamber with a lower pressure.

The leakage velocity through the clearances is considered to be adiabatic Fanno-flow through an idealized clearance gap of rectangular shape and the mass flow of leaking fluid is derived from the continuity equation. The effect of fluid-wall friction is accounted for by the momentum equation with friction and drag coefficients expressed in terms of Reynolds and Mach numbers for each type of clearance.

The injection of oil or other liquids, for lubrication, cooling or sealing purposes, modifies the thermodynamic process substantially. The same procedure can be used to estimate the effects of injecting any liquid but the effects of gas or its condensate mixing and dissolving in the injected fluid or vice versa should be accounted for separately.

In addition to lubrication, the main purpose for injecting oil into a compressor is to cool the gas. The solution of droplet energy equation in parallel with the momentum equation yields the amount of heat exchange with the surrounding gas.

The equations of energy and continuity are solved to obtain U(θ) and m(θ). Together with V(θ), the specific internal energy and specific volume u=U/m and v=V/m are now known. T and p, or x can then be calculated. All the remaining thermodynamic and fluid properties within the machine cycle are derived from the pressure, temperature and volume directly. Computation is repeated until the solution converges.

For an ideal gas, the internal thermal energy of the gas-oil mixture is given by:

$$U = (mu)_{gas} + (mu)_{oil} = \frac{mRT}{\gamma - 1} + (mcT)_{oil} \qquad T = (\gamma - 1)\frac{U - (mcT)_{oil}}{mR}$$

Hence, the pressure or temperature of the fluid in the compressor working chamber can be explicitly calculated by the equation for the oil temperature T_{oil}.

For the case of a real gas the situation is more complex, because the temperature and pressure cannot be calculated explicitly. However, since the equation of state $p=f_1(T,V)$ and the equation for specific internal energy $u=f_2(T,V)$ are decoupled, the temperature can be calculated numerically from the known specific internal energy and the specific volume obtained from the solution of differential equations, whereas the pressure can be calculated explicitly from the temperature and the specific volume by means of the equation of state.

In the case of a phase change for a wet vapour during the compression process, the specific internal energy and volume of the liquid-gas mixture are:

$$u = (1-x)u_f + xu_g \qquad v = (1-x)v_f + xv_g$$

where u_f, u_g, v_f and v_g are the specific internal energy and volume of liquid and gas and are functions of saturation temperature only. The equations require an implicit numerical procedure which is usually incorporated in property packages. As a result, temperature T and dryness fraction x are obtained. These equations are in the same form for any kind of fluid, and they are essentially simpler than any others in derived form. In addition, the inclusion of any additional phenomena into the differential equations of internal energy and continuity is straightforward. A full account of the compressor model used in this work can be found in Stosic, Smith and Kovacevic (9).

2.2 The unsteady process in a lumped volume of the plant reservoirs and connecting pipes

A two tank plant model was investigated which enables closed systems to be simulated, such as refrigeration, air-conditioning and heat pump plants, as well as plants which operate under power cycles, like Joule, Rankine and Organic Rankine cycles to be investigated. In fact, since the one tank model is a special case of this and if volume of the compressor inlet tank is left very large or infinity, it may be used to simulate atmosphere. Thus, the developed two tank model can be used to obtain the one tank model results.

A two tank plant is presented in Figure 1. Gas from Tank 1 goes to suction of screw compressor, then it discharged to Tank 2 and through throttle valve goes back to Tank 1.

All connecting pipes in the compressor plant are considered to be short enough that their volume, together with the reservoir volumes to be summed up into one lump tank volume. This assumes that all the thermodynamic properties are uniform within such a control volume. Thus the conservation equations of continuity and energy already used in the compressor model may be utilized for the tank calculations. The tank filling/ emptying equations for that analysis are as follows.

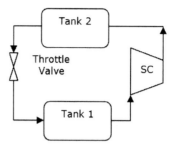

Figure 1: Two Tanks schematics

$$m_2 u_2 - m_1 u_1 = (\dot{m}_{in} h_{in} - \dot{m}_{out} h_{out}) \Delta t$$
$$m_2 - m_1 = (\dot{m}_{in} - \dot{m}_{out}) \Delta t$$

where indices 1 and 2 denote start and end time of filling/emptying respectively and Δt is time difference between these.

In the ideal gas case, the finite difference equations can be written as:

$$p_2 = p_1 + \frac{\gamma R \Delta t}{V}(\dot{m}_{in} T_{in} - \dot{m}_{out} T_{out})$$

$$\dot{m}_{out} = \mu A \sqrt{2 \rho_2 (p_2 - p_0)} \qquad \rho_2 = \frac{m_2}{V} \qquad T_2 = \frac{p_2}{R \rho_2}$$

To estimate the unsteady behaviour of a compressor plant system, the tank equations are coupled with the compressor model equations and solved in sequence to obtain a series of results for each time step. When the pressure p_2 in the tank at each time step is known, the flow and temperature m_{in} and T_{in} at the compressor discharge can be calculated. These derived values are then taken as the input parameters for the next time step. When the tank pressure p_2 is calculated. m_{out} is either known, or calculated as the flow through the exit throttle valve to pressure p_0 and T_2 becomes T_{out} in the next time step. The calculation was repeated until the final time was reached.

Mass inflow and outflow was calculated as a pipe flow with restrictions which comprised line and local losses therefore defining pressure drops within the plant communications. Since the tanks are of far higher volume than the communications, which results in far lower gas velocities, the losses in the tanks are far lower than the pipe losses.

Two levels of programming were applied, firstly the compressor and plant processes were solved separately. The compressor process was calculated through SCORPATH software and the tank model is processed, with mutual interchange of output and input data. Since this combination appeared to be slow in data transfer, the compressor and tank procedures were programmed together to get instant data exchange. This resulted in a very quick calculation allowing the bulk estimation of the unsteady behaviour of a screw compressor plant under various scenarios.

3. EXPERIMENTAL VERIFICATION OF THE RESULTS OBTAINED FROM THE MATHEMATICAL MODEL

The air compressor test rig with the oil flooded air compressor was used to validate the predicted results. An oil-flooded twin screw compressor was coupled to a 75 kW electric motor and driven by a six-band belt drive which speed is controlled by a frequency converter. A two stage oil separator consists of two separator tanks joined together by a short pipe for which the maximum working pressure is 15 bars. The oil cooler is a shell and tube heat exchanger. In this system, the oil is injected into the compressor by means of the pressure difference between the oil separator and the compressor working chamber. A motor driven throttle valve after the oil separator is used to control the air pressure inside the oil separator.

Apart of the laboratory ambient temperature and pressure, which were manually put into the test rig computer, all measured physical quantities are obtained as electric signals and transferred to an InstruNet data logger. Instantaneous values of pressure, temperature, speed and torque are displayed on the test data monitor. More details about the particular measurements can be found in Chukanova et. al. (7).

Measurement records were collected twice a second and saved in a separate file which was used for further analysis. Before measurements, the compressor and its plant were run for 30 minutes to obtain steady temperature in the compressor casing and to bring the oil temperature to its working level.

The experimental and predicted results of pressure variation during the start up, presented in Figure 2 for the starting receiver pressures 8 and 10 bars, show good agreement.

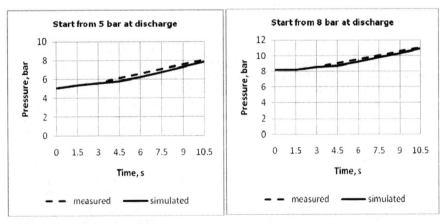

Figure 2: Diagrams of pressure rising in the tank during the compressor start

The tested compressor has a lobe configuration 4/5. The main rotor diameter is d=128mm, while the length to diameter ratio is L/d=1.55. Final speed of the main rotor during the experiment was retained constant and equal to 3000 rpm.

4. PRESENTATION AND ANALYSIS OF THE SIMULATED CASES

As previously stated, the developed model which combines the compressor and plant together gives a good opportunity to simulate various kinds of instabilities which might happen during real compressor plant operation. Several cases were presented and analysed for variety of starting tank pressures, tank volumes and valve areas, all of them for an infinite volume inlet tank, atmosphere. Then a two tank model results were presented which enables to closed cycle systems.

The same compressor was used as the basis for experimental testing, as presented in the previous section, where several cases were considered to check the plant model viability. The results are presented in groups, showing the effects of varying the throttle valve area, the tank volume and the tank pressure, as well as by varying the compressor shaft speed.

4.1 Variation of Valve Area

γ	R J/molK	V m^3	p$_1$ bar	Tin K	Shaft Speed rpm	Δt s	A$_v$ m^2	p$_0$ bar	T$_{out}$ K
1.4	287	0.30	1	350	3000	1	See below	1	350

Case 1 – A$_v$=70mm^2, Case 2 – 50 mm^2, Case 3 – closed valve, Case 4 – 30 mm^2

From the diagrams in Figure 3 it can be seen how the pressure and temperature change for different valve areas. For example for the case of the closed valve pressure in the tank reaching 33 bar in less than 2 minutes and the temperature of the air increasing from 350K (77°C) to 450K (177°C) in just 10 seconds. In fact, this confirms how the valve area can be used to control the discharge pressure.

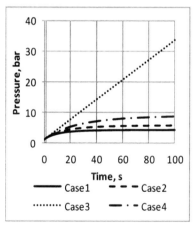 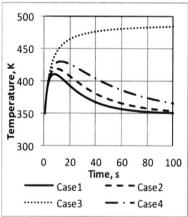

Figure 3: Pressure (left) and Temperature (right) in the tank for cases with different valve areas

4.2 Changing of Tank Volume
The tank volume is varied as follows: Case 1 – V=0.3m^3, Case 2 – 0.03m^3, Case 3 – 0.1m^3, Case 4 – 0.6m^3.

It can be seen from the diagrams in Figure 4 that for a given throttle valve area, the final discharge pressure will be the same for different tank volumes. It is only a question of the time for it to reach its final value: for a tank of 30 litres it will be 2 seconds, for 600 litres about 2 minutes. Similar characteristics apply to the temperature: the less volume the faster temperature reaches its peak (400-420K) and the faster that it returns to its initial value of 350K.

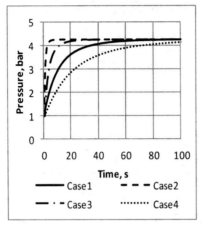 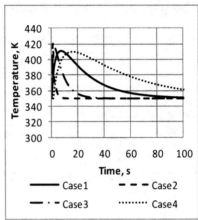

Figure 4: Pressure (left) and Temperature (right) in the tank for cases with different tank volumes

4.3 Changing of Tank Pressure
The pressure in the tank was varied as follows: Case 1 – p_2=1bar, Case 2 – 3bar, Case 3 – 5bar, Case 4 – 7bar.

The diagram in Figure 4 confirms that whatever the starting pressure, it will reach a final value which is determined by the valve area. This is 4.2 bar in all cases, as presented in Figure 5. It is shown that for starting pressure lower than 4.2 bar the pressure will quickly rise together with temperature. Conversely, if pressure if the starting tank pressure is above 4.2 bars, the pressure and temperature will rapidly fall to their final values.

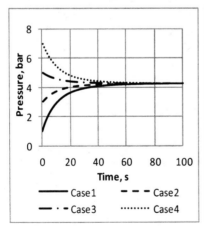 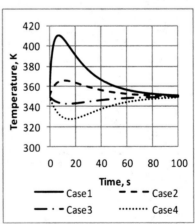

Figure 5: Pressure (left) and Temperature (right) in the tank for cases with different starting pressure in the tank

4.4 Two Tank Case

A two tank plant model was then developed to enable closed systems to be simulated. In fact, the one tank model is a special case of this and if volume of the compressor inlet tank is left very large, or infinity to simulate atmosphere, the one tank model results will be obtained.

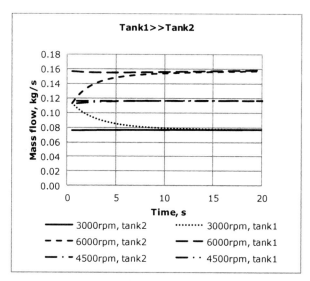

Figure 6: Pressure in the Tanks 1 and 2 for different shaft speeds

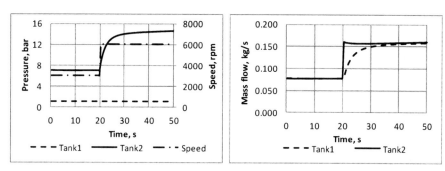

Figure 7: Pressure in the tank, left and mass flow in and out, right for speed variation from 3000 to 6000rpm

Figure 7 left shows the effect of a sudden change in the compressor shaft speed during the plant operation. The pressure in the discharge tank is doubled when the shaft speed is doubled. In this case, the volume of the inlet tank is kept much larger than that of the discharge tank.

As soon as the shaft speed is increased from 3000 to 6000rpm the pressure in Tank 2 starts to rise, but the pressure in Tank 1 remains almost constant because of its larger volume. As a result, the mass flow rate to Tank 2 doubles immediately, but the flow rate from Tank 2 into Tank 1 needs some time to reach this value, as a result of the increase of pressure in Tank 2, as shown in Figure 7 right.

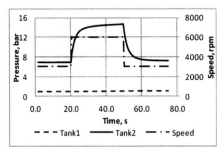 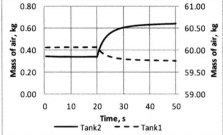

Figure 8: Pressure, left and mass, right in both tanks while changing speed from 3000 to 6000rpm and back to 3000

The resulting variation of mass contained in the tanks with time is presented in Figure 8 right. If the speed increases from 3000rpm to 6000rpm and then is brought back to its original value, the pressure history, as shown in Figure 8 left confirms that the pressure reaches its starting value.

5. CONCLUSION

By including the tank volume and other elements of a compressor plant system, into a well proven mathematical model for estimating screw compressor performance, it was possible to calculate the interaction between compressors and their systems under unsteady conditions. The predicted results, thus obtained agree well with measured results. Thus the simulation procedure has been validated and can be used as a useful and convenient tool for the analysis of unsteady behaviour of screw compressors in their plant. The modelling techniques were developed in a step by step iterative process starting with a simple analytical model and systematically taking more into account factors. This model is a powerful instrument that simulated a variety of scenarios which may occur in everyday compressor plant practice.

REFERENCES

(1) Jun, W., Yezheng, W., 1988. On the On-Off Operation in a Reciprocating Compressor Refrigeration System with Capillary. In International Refrigeration and Air Conditioning Conference. Purdue University.
(2) Jun, W., Yezheng, W., 1990. Start-up and shut-down operation in a reciprocating compressor refrigeration system with capillary tubes. *International Journal of Refrigeration*, 13(3), pp.187–190.
(3) Fleming, J.S., Tang, Y. & You, C.X., 1996. Shutdown process simulation of a refrigeration plant having a twin screw compressor. *International Journal of Refrigeration*, 19(6), pp.422–428.
(4) Li, B. & Allyene, A.G., 2010. A dynamic model of a vapor compression cycle with shut-down and start-up operations. *International Journal of Refrigeration*, 33(3), pp.538–552.
(5) Ndiaye, D. & Bernier, M., 2010. Dynamic model of a hermetic reciprocating compressor in on–off cycling operation. *Applied Thermal Engineering*, 30(8-9), pp.792–799.
(6) Link, R. & Deschamps, C.J., 2011. Numerical modeling of startup and shutdown transients in reciprocating compressors. *International Journal of Refrigeration*, 34(6), pp.1398–1414.

(7) Chukanova, E., Stosic, N., Kovacevic, A., Dhunput, A., 2012. Investigation of Start Up Process in Oil Flooded Twin Screw Compressors. International Compressor Engineering Conference. Purdue University.
(8) Chukanova, E., Stosic, N., Kovacevic, A., Rane, S., 2012. Identification and Quantification of Start Up Process in Oil Flooded Twin Screw Compressors. International Mechanical Engineering Congress & Exposition IMECE 2012.
(9) Stosic N, Smith A.K, Kovacevic A, 2005. Screw Compressors: Mathematical Modeling and Performance Calculation, Monograph, Springer Verlag, Berlin, published June 2005, ISBN: 3-540-24275-9.

Optimization of integrated energy process in China industrial compressed air system

T Ma[1], Z Jia[2], H Li[1], X Zhang[3], X Zhao[3]
[1] School of Environmental and Energy Engineering, Beijing University of Technology, China
[2] Beijing Huayuanyitong Heating Development Co., Ltd, China
[3] School of Engineering, University of Hull, UK

ABSTRACT

This article optimized the energy process of an industrial compressed air system based on the Total Site Integration (TSI) theory. The current circumstance of air compressors in China and their relevant improving approaches were described. A global optimization method was subsequently proposed and applied to analyze the integrated energy process in compressed air system. Relationship between the global and local optimization methods was also discussed. The research results indicated the global optimization method was able to enhance the energy performance of compressed air system, which would further reduce the share of energy consumption in industrial sector and substitute the carbon intensive fuels worldwide.

Keywords: Compressed air system; process optimization; Total Site Integration.

NOMENCLATURE

a	electricity price, 0.55 RMB/kW·h
k	unloaded power consumption rate
p	pressure, Pa
P	power, kW
Pa	absolute air pressure, 0.1MPa
q	air displacement rate, m³/min
q_v	volume flow rate, L/min
Q	electricity, kW·h
S	cost, RMB
t	annual operating time, 8000h/yr
t'	start-up time saved, h
t''	end-up time saved, h
t'''	average operating time saved, h

Greek

η_l	load ratio
η_n	unloaded ratio
μ	specific power ratio, kW/(m³/min)

Subscript

i	optimization node
in	inlet
out	outlet
p	pneumatic
r	rated
s	shaft

1. INTRODUCTION

In 2012, the Chinese government drafted the new roadmap for the national carbon emission reduction that the national energy consumption should drop down to 0.0869 tons/thousand RMB of equivalent standard coal by 2015, which was approximately 32% of the baseline in 2005 (0.1276 tons/thousand RMB) and less than 16% (0.1034 tons/thousand RMB) in 2010. In context of this task, the compressed air system, which occupies the total power consumption of the factory by 10%-20% (even up to 35%) in China's industrial sector, will inevitably become one of the promising targets for energy conservation. In practice, the air compressor is the main source of power consumption, which is about 70% of total energy consumption throughout its life span.

In China's industry, the power consumption of compressor significantly increases every year, which is nearly 180 billion kW·h in 2006, 214 billion kW·h in 2008 and up to 270 billion kW·h in 2010, accounting for about 8.7% of total industry power consumption over the country [1]. The Chinese government has paid much attention to the industrial compressed air system through establishing a series of standards including GB 19153 "Minimum allowable values of energy efficiency and energy efficiency grades for displacement air compressors" [2]. This standard defines displacement air compressor (hereinafter referred to as the air compressor) of energy efficiency level, minimum allowable values of energy efficiency, target minimum allowable values of energy efficiency, evaluation value of energy conservation, test methods and inspect rules. In 2009, the latest version of the standard added energy efficiency level and target minimum allowable values of energy efficiency.

Early in the 1980s, the pinch point technology which was initially proposed by Linnhoff and Tomio Umeda was widely applied in the energy-saving process of the energy intensive industry [3, 4]. From the 1990s, the technology was further developed as "total site integration" (TSI) [5]. Following by twenty-year research and application, the TSI method has been fully investigated by University of Manchester and Professor Bagajewicz of the University of Oklahoma [6-8]. On basis of the TSI theory, the global optimization into the energy process of an industrial compressed air system in China was especially analyzed. The advantages by using the global optimization rather than local approach were then presented.

2. THEORETICAL ANALYSIS

According to the theory of whole life cycle cost analysis [9], based on a 10-year operation, the initial investment and maintenance costs account for about 12% of the whole life cycle cost of the compressed air system, while the energy costs accounts for 76%. A large number of experiments and case study show that it's attainable to achieve 20%-50% or more of the energy saving effect through the appropriate measures to optimize the compressed air system.

The process industry and other related industries are often found such problems in the retrofit: the optimization of separate subsystem (such as a selection of high efficient and energy-saving equipment) doesn't turn out satisfactory to the whole system. This phenomenon is mainly due to the following three reasons:
(1) The whole system is consisted of several relatively independent but interrelated subsystems with their own specific functions. The relationship between the different subsystems will be cut off if one of them is replaced separately.
(2) There exists the contradictory relationship between the optimization targets of total system and subsystem. The optimization of each subsystem separately usually can't meet the requirements of the total system.
(3) The design of process system is usually multi-objective, while the optimization of it cannot meet the demand of each subsystem. The optimization objectives of each subsystem have certain internal limitations.

Hence, it's desired to further investigate the optimization methods by considering the entire system. By using the TSI theory, the maximum amount of energy saving of the compressed air system should be achieved.

2.1 Division of subsystems
A modern industrial compressed air system is composed of several major sub-systems and many sub-components. Figure 1 shows a representative industrial compressed air system and its components including the compressor, prime mover, controls, treatment equipment and accessories and the distribution system.

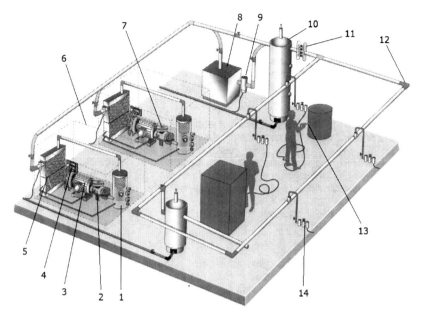

Figure 1 Components of a Typical Industrial Compressed Air System

1-Lubricant/Air separator 2-Compressor air end 3-Motor 4-Control panel 5-Aftercooler and lubricant cooler 6-Compressor package enclosure 7-Air inlet filter 8-Dryer 9-Air filter 10-Air receiver 11-Pressure/Flow controller 12-Distribution system 13-Pneumatic tool 14-Filter, regulator and Lubricator

For simplicity, the compressed air system can be divided into three parts: compressed air manufacturing, transport network and gas terminal, as shown in Figure 2. As there are corresponding gas requirements and operation principles in the design of gas terminal, and manual operation has a significant influence in gas terminal, gas terminal will not be analyzed separately. So in this paper, the whole compressed air system is only divided into two subsystems: compressed air manufacturing and transport network.

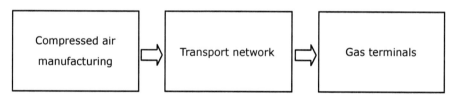

Figure 2 Structure of compressed air system

In accordance to the composition of the compressed air system, the energy waste of the subsystems mainly consists of the following two parts:

(1) Compressed air manufacturing process: Air compressor
Air compressor is the crucial component of compressed air manufacturing process. However, there are mainly two problems existing in daily operation of air compressor, i.e. manual switching and non-automatic shut down.

(2) Transport network
In the transport network of compressed air system, there are several widespread problems including unreasonable supply pressure, pressure loss and gas leakage. Because of the uncertain piping pressure drop, traffic peak in starting compressor and other factors, the supply pressure generated by air compressor is sometimes higher than required by 0.2-0.3MPa, which results in a great waste of energy. And owing to

unreasonable network layout, poor sealing performance, ignorance of leak detection and other reasons, the pressure loss and gas leakage along the pipeline often occupy 20%-30% of gas production [10].

2.2 Global system optimization sequence

It is well known that the subsystems are not absolutely independent but interactive with each other. So the total system can be optimized according to the relationship between the subsystems step by step. In this paper, the optimization sequence is proposed as follows.

(1) **Unreasonable supply pressure:** Determine the parameters of air compressor operation according to the specific needs of gas terminal;
(2) **Pipeline pressure loss and gas leakage:** Reduce network inlet pressure through the network optimization design and leak detection. Network energy-saving has certain effect on the air compressor at the same time;
(3) **Air compressor operation process:** Involve the intelligent control technology for system optimization instead of manual switching. To non-automatic shut down, apply the frequency conversion technology combined with automatic switching.

2.3 Model analysis

The parameters of a selected compressed air system were presented in Table 1. The net total demand of compressed air in the production process was 154.84m^3/min. And 10% margin is taken for pipeline leakage and other unpredictable consumption. The total demand is 154.84×(1+10%+10%)=185.8m^3/min.

Table 1 Parameter table of different gas supply units

Gas unit	Air pressure (MPa)	Air demand (m^3/min)	Remark
Zone A	0.18-0.21	74.5	
	0.28-0.4	52.5	
	0.7	0.8	
	0.6	15.5	
Zone B	0.6	5.8	intermittent use
Zone C	0.4	2.34	
	0.7	3.4	dried and purified

The compressed air system consists of: 7 piston compressors with rated power of 132kW, displacement of 20m^3/min, and exhaust pressure of 0.8MPa; 2 piston compressors with rated power of 65kW, displacement of 10m^3/min, and exhaust pressure of 0.8MPa; 3 screw compressors with rated power of 185kW, displacement of 30.4m^3/min, and exhaust pressure of 0.8MPa. In the manufacturing process, the air compressor exhaust pressure is set at 0.8MPa. After dried and purified process, the compressed air will go through the valve to release certain pressure to different levels for each gas point.

(1) Optimization for unreasonable supply pressure

From Table 1, it's obviously seen that the volume flow rate of compressed air with pressure no more than 0.4MPa is 129.5m^3/min accounts for the largest proportion of demanded air at 83.5%. There exists a large amount of energy waste by reducing the air pressure from 0.8MPa to 0.4MPa through the valve. Based on this situation, the air compressors were divided into several groups according to the air pressure. Separate gas cylinders are used to store compressed air at different pressures. The compressed air manufacturing process is illustrated in Figure 3.

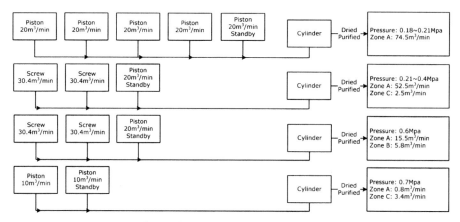

Figure 3 Compressed air system

The amount of energy saving by adjusting the air compressor air supply pressure is defined as ΔQ_1, calculated by

$$\Delta Q_1 = \Delta \mu q t \qquad (1)$$

$\Delta \mu$ stands for the specific power change after adjusting the air supply pressure.

$$\mu = P/q \qquad (2)$$

Energy loss is derived from the testing data provided by the compressor manufacturer, as shown in Table 2 and Table 3.

Table 2 The test data of the piston compressor with rotor code 337

p (MPa)	0.3	0.4	0.5	0.6	0.7	0.8
P (kW)	74.5	83.5	91.5	99.5	107	116
q (m³/min)	20.1	20.0	20.0	19.8	19.8	19.6
μ (kW/(m³/min))	3.7065	4.1750	4.5750	5.0252	5.4040	5.9184

Table 3 The test data of the screw compressor with rotor code 301

p (MPa)	0.5	0.6	0.7	0.8	0.9	1
P (kW)	132.1	145	156	163	174	184
q (m³/min)	29.8	29.7	29.6	29.4	29.3	29.2
μ (kW/(m³/min))	4.4329	4.8822	5.2703	5.5442	5.9386	6.3104

Thus, the corresponding cost saving is ΔS_1 (RMB), writing by

$$\Delta S_1 = \Delta Q_1 a \qquad (3)$$

For the compressed air of 0.18-0.21MPa, the piston compressor exhaust pressure is lowered from 0.8MPa to 0.3MPa:

$\Delta \mu_{1-1} = 5.9184 - 3.7065 = 2.2119 \text{kW}/(\text{m}^3/\text{min})$;
$\Delta Q_{1-1} = 2.2119 \times 74.5 \times 8000 = 1.318 \times 10^6 \text{kW·h}$;
$\Delta S_{1-1} = 1.224 \times 10^6 \times 0.55 = 725$ thousand RMB.

For the compressed air of 0.21-0.4MPa, the screw compressor exhaust pressure is lowered from 0.8MPa to 0.5MPa:

$$\Delta\mu_{1-2}=5.5442-4.4329=1.1091 kW/(m^3/min);$$
$$\Delta Q_{1-2}=1.1091\times55\times8000=4.880\times10^5 kW\cdot h;$$
$$\Delta S_{1-2}=4.880\times10^5\times0.55=268 \text{ thousand RMB.}$$

For the compressed air of 0.6MPa, the screw compressor exhaust pressure is lowered from 0.8MPa to 0.7MPa:

$$\Delta\mu_{1-3}=5.5442-5.2703=0.2739 kW/(m^3/min);$$
$$\Delta Q_{1-3}=1.1091\times55\times8000=4.667\times10^4 kW\cdot h;$$
$$\Delta S_{1-3}=4.667\times10^4\times0.55=26 \text{ thousand RMB.}$$

To sum up

$$\Delta Q_1=\Delta Q_{1-1}+\Delta Q_{1-2}+\Delta Q_{1-3}=1.853\times10^6 kW\cdot h;$$
$$\Delta S_1=\Delta S_{1-1}+\Delta S_{1-2}+\Delta S_{1-3}=1019 \text{ thousand RMB.}$$

(2) Optimization for pipeline pressure loss

The optimization for pressure loss in pipeline is based on the study of pneumatic power calculation [11], which is given as

$$P=pq_v \ln\frac{p}{p_a} \qquad (4)$$

The amount of power consumption here is defined at Q_2

$$Q_2=(P_{in}-P_{out})t\times10^{-3} \qquad (5)$$

The corresponding cost saving is ΔS_2 (RMB), expressed as

$$\Delta S_2=\Delta Q_2 a \qquad (6)$$

Assuming the pressure reduction is 0.01MPa after the transformation of the transport network, the amount of power saving by the optimization of the transport network can be calculated as Table 4 and Table 5:

Table 4 Calculation of pneumatic power before the optimization

p_{out}(MPa)	0.18-0.21	0.21-0.4	0.6	0.7
p_{in}(MPa)	0.3	0.5	0.7	0.8
q_v(L/min)	80000	60000	21300	4200
P(10^4 kW·h)	11.12	11.44	4.89	1.01
Sum		2.85×10^5kW·h		

Table 5 Calculation of pneumatic power after the optimization

p_{out}(MPa)	0.18-0.21	0.21-0.4	0.6	0.7
p_{in}(MPa)	0.29	0.49	0.69	0.79
q_v(L/min)	80000	60000	21300	4200
P(10^4 kW·h)	9.79	10.25	4.39	0.91
Sum		2.53×10^5kW·h		

Thus, the total amount of power and cost saving will be
$$\Delta Q_2=3.2\times10^4 kW\cdot h;$$
$$\Delta S_2=17.6 \text{ thousand RMB.}$$

(3) Optimization for network leakage

According to previous test and analysis results [12], a hole with the diameter of 4 mm can lead to an annual power consumption ΔQ_3 of 5.2×10^4 kW·h, while the electricity cost ΔS_3 caused by the hole can be up to 28.6 thousand RMB under feature condition (0.6MPa of working pressure; 8000h of annual operation time; 0.55 RMB/kW·h of electricity price).

(4) Optimization for manual switching

During the operation process, the power consumption of air compressor is affected by motor rated power, operation time and operation efficiency. The power consumption of the air compressor can be calculated according to the following formulas

$$Q = Pt(\eta_i + \eta_n k) \tag{7}$$

For the manual operation, the advanced technology, such as intelligent control system, can be applied to control system automatically which can avoid the additional energy consumption caused by the delayed manual switching.

The amount of power saving here is given as ΔQ_4

$$\Delta Q_4 = \sum_{i=1}^{n}[(P_i t' + P_i t'' k)] \tag{8}$$

The corresponding cost saving is ΔS_4

$$\Delta S_4 = \Delta Q_4 a \tag{9}$$

It is assumed that the average time for advanced starting can be saved at 78h, while the average time for delayed shut can be saved up to 56h and the unload power consumption rate is recommended at 0.45 [13]. It can be therefore calculated ΔQ_4 at 2.156×10^5 kW·h, and ΔS_4 at 118 thousand RMB.

(5) Optimization for non-automatic shut down

The amount of power saving by applying the frequency conversion technology combined with automatic switching is defined as ΔQ_5, deduced from Formula (7) at

$$\Delta Q_5 = \sum_{i=1}^{n}(P_i t''' k) \quad (i = 1,2,3 \ldots) \tag{10}$$

The corresponding cost saving is ΔS_5

$$\Delta S_5 = \Delta Q_5 a \tag{11}$$

The average operating time saved after the optimization is assumed at 400h and the unloaded power consumption rate is suggested at 0.45 [13]. It can be calculated ΔQ_5 at 1.931×10^5 kW·h and ΔS_5 at 106 thousand RMB.

3. RESULTS AND DISCUSSION

3.1 The optimization results

The final results of energy saving in the compressed air system through the global optimization analysis are given in Table 6.

Table 6 Energy saving summary

Subsystem optimization	Unreasonable supply pressure	Network pressure loss	Network leakage	Manual switching	Non-automatic shut	Sum
ΔQ_i (10^4kW·h, i=1, 2, 3, 4, 5)	185.3	3.2	5.2	21.56	19.31	234.57
ΔS_i (thousand RMB, i=1, 2, 3, 4, 5)	1019	17.6	28.6	118	106	1289.2

Table 7 presents the operation energy consumption before the global optimization:

Table 7 Statistics of annual operating energy consumption

Compressor type	Piston 20m³/min	Piston 10m³/min	Screw 30.4m³/min
Single power (kW)	132	65	185
Number of units	4	1	3
Sum Power (kW)	528	65	555
Total power (kW)		1148	
Annual power consumption (10^4kW·h)		918.4	
Annual electricity cost (thousand RMB)		5051.2	

It can be calculated that the annual power consumption of the compressed air system is 6.838×10^6kW·h, and the annual electricity cost is 3762.3 thousand RMB after the global optimization. The following figure indicates of the energy saving percentage of each subsystem when compared with the amount of energy consumption before the global optimization.

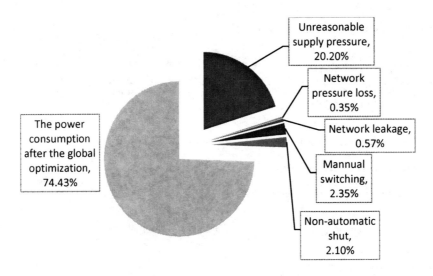

Figure 4 Energy saving percentage of different subsystems

3.2 Relationship between the global and the local optimization

The global optimization is a comprehensive analyzing method based on the integration of each subsystem, whose goal is to reach the maximum of energy saving. The shortcoming of this approach is a large demand of workload. A wide range of professional knowledge background is also required in the global optimization design.

The local optimization focuses on the efficient and energy-saving products on the market, which simply adds or replaces equipment independently from the viewpoint of demand only, ignoring other impacts in the energy-saving of the whole system. However, the compressed air system is divided into several subsystems in the theory of global optimization, where the analysis of each subsystem is based on the local optimization method.

In overall, the global optimization method has more advantages against the local approach when considering the energy performance of whole system. It can assist to enhance the overall energy performance of compressed air system, and would further reduce the share of energy consumption in industrial sector and substitute the carbon intensive fuels worldwide.

4. ACKNOWLEDGEMENTS

The authors wish to express their thanks to National Science and Technology Support Program of China, Code: 2011BAB02B02, for financing our research activities in the field of rational use of energy in the compressed air system.

5. REFERENCES

[1] Shanghai HQ, Mc Kane A. Improving energy efficiency of compressed air system based on system audit. LBNL, 2008.
[2] GB 19153-2009 "Minimum allowable values of energy efficiency and energy efficiency grades for displacement air compressors".
[3] Linnhoff B, et al. User guide on process integration for the efficient use of energy [C].The Inst. of Chem. Engineers, England, 1982.
[4] Linnhoff B. Use pinch analysis to knock down capital costs and emissions [J]. Chem. Eng. Prog., 1994, 90(8): 32-57.
[5] Dhole V R, Linnhoff B. Total site targets for fuel, co-generation, emissions and cooling [J]. Computers and Chem. Eng., 1992, 17: 101-109.
[6] Smith R. State-of-the-art in process integration [J]. Applied Them. Eng., 2000, 20: 1337-1345.
[7] Klemes J, Stehlik P. Recent adavances on heat, chemical and process integration [J]. Applied Them. Eng., 2006, 26: 1339-1344.
[8] Rossiter A P. Succeed at process integration [J]. Chem. Eng. Prog., 2004, 100(1): 58-62.
[9] U.S. Department of Energy. Improving compressed air system performance [M]. Washington D.C.: U.S. Department of Energy's Office of EERE, 2003.
[10] Dai Yande. Energy Efficiency and Market Potential of Electric Motor System in China [M]. Beijing, China Machine Pressure, 2001.
[11] Cai Maolin, Kagawa Toshiharu. Energy consumption assessment and energy loss analysis in pneumatic system [J]. Chinese Journal of Mechanical Engineering, 2007, 9: 69-74.
[12] Qin Hongbo, Hu Shougen. Study on optimization potential of industrial compressed air system [J]. Fluid Machinery, 2010, 38(2): 49-52.
[13] Sun Tieyuan, Cai Maolin. Operating status and energy conservation measures in compressed air system [J]. Machine Tool & Hydraulics, 2010, 38: 108-110.

Investigation on working characteristics of micro compressed air energy storage system

Q C Yang, Y Y Zhao, L S Li, Z G Qian
State Key Laboratory of Compressor Technology,
Hefei General Machinery Research Institute, China

ABSTRACT

Based on the basic working principle of compressed air energy storage (CAES), the theoretical model of the compression process, storage and expansion process was established using thermodynamics theory. The operating characteristics of the micro CAES system under different working conditions were investigated. A multiple throttling method was proposed to further use the compressed air energy and improve the system efficiency. The effect of constant volume and constant pressure storage methods on the performance were analyzed. The results showed that the multiple throttling method could improve the storage efficiency greatly. The proposed analysis model will provide the foundation to design the high efficient cycle of micro CAES.

Keywords: compressed air energy storage; compression; expansion; efficiency

NOMENCLATURE

E	available energy (kJ)	η	efficiency
ED	energy storage density (kJ/m³)	**Subscripts**	
k	adiabatic exponent	c	compression
m	mass (Kg)	e	expansion
n	compression stage	n	polytropic
P	pressure (kPa)	v	valve
Rg	gas constant (kJ/kgK)	s	storage
T	temperature (K)	t	tank
V	volume (m3)	0	ambient
W	work (kJ)		

1 INTRODUCTION

Energy storage systems are becoming more important for load leveling, especially for widespread use of intermittent renewable energy[1]. Compressed air energy storage is one of the promising methods for energy storage, with high efficiency and environmental friendliness[2]. With the development of renewable energy, the combination of compressed air energy storage (CAES) and wind energy is an important method for utilizing the wind energy[3, 4]. In these plants, air is compressed and stored in the form of compressed air in a reservoir during off-peak periods and then used on demand peak periods to generate power with an expander-generator system.

The first commercial scale CAES plant in the world is the 290MW Huntorf, Germany, operated by Nordwest Deutsche Kraftwerke (NDK) since 1978. The Alabama Electric Cooperative Inc. in McIntosh built the second commercial scale CAES plant. The success of these two operated CAES plants shows that the CAES is a reliable energy storage technology. An increasing attention has been paid to CAES and the working characteristics, cycle optimization, system integration of CAES has been investigated theoretically and experimentally by many researchers in recent years. Large scale CAES are dependent on the suitable underground geology. Micro CAES with man-made air vessel is a more adaptable solution for distributed future power networks. For the micro CAES system, the compressed air energy is usually converted to electricity through expander. The compressed air storage temperatue has a large effect on the energy storage efficienty, therefore, the adiabatic CAES system were proposed, in which the compression heat during compression process is stored by thermal energy storage mthod and then was used to preheat the inlet air of expander. The adiabatic CAES system has the advantages of high efficiency. The expansion process is the key process of micro CAES and is relevant to the quantity of power generation of the system. However, there is little work on the expansion process of CAES.

This paper focus on working process of the power generation and operating characteristics of micro CAES system. The theoretical models of the compression process, storage and expansion process were established using thermodynamics theory. A multiple throttling method was proposed to further use the compressed air energy and improve the system efficiency.

2 SYSTEM DESCRIPTION

For large scale CAES system, the compressed air is stored in a large reservoir, typically in an underground geologic formation. The micro CAES can be an above ground air storage system that uses air tank or air piping for the air storage. Such systems are very attractive because they allow CAES plants to be sited almost anywhere, since no underground geologic formation is needed. A simple cycle of micro-CEAs is shown in figure 1. Different types of CAES system can be built according to the choice of compression, expansion process and the heat management methods.

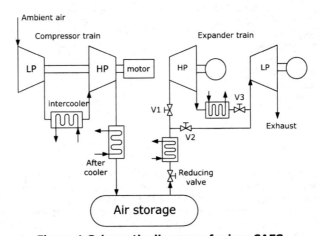

Figure 1 Schematic diagram of micro CAES

Ambient air can be compressed and stored at high pressure, and then use in the working process of a compressed air expander. The multistage compressor with inter cooler and two stage expansion with inter heater was adopted to improve the overall efficiency of the system. The high temperature and high pressure compressed air was further cooled through the aftercooler. Since the ambient air contains the moisture, the driers is need in this system before the air enters the storage tank. In this system, multiple throttling was adopted to make full use of the stored compressed air. During the first throttling period, as shown in figure 1, the valve V1 and V3 is open, V2 is closed, the pressure of compressed air was reduced to a constant value and then expansion with ambient temperature through high pressure (HP) expander and low pressure (LP) expander. During the second throttling period, valve 1 and V3 is closed, V2 is open, the pressure of compressed air is further reduced to a lower level and expansion only by LP expander. The advantage of this multiple throttling method is that both the expander can work under high efficiency, since the inlet pressure is constant.

Two different approaches to storage tank design are adopted: constant volume, where pressure can change during the process, and constant pressure, where the compressed air maintained a constant pressure in the reservoir[5].

2.1 Constant volume air storage
Conventional CAES systems most commonly operate under constant volume conditions[5, 6]. In general, both charging and discharging processes of the high-pressure vessel are unsteady state processes, where the pressure ratios are changing. These varying conditions can result in low efficiencies of compression and expansion due to the deviation from the designed points. So most effects have focused on keeping the inlet pressure of the turbine constant by throttling the upstream air to a fixed pressure because of the increase in turbine efficiency that results from constant inlet pressure operation. It is often optimal to operate a CAES system in this mode (as is the case at both the Huntorf and McIntosh plants) The Huntorf CAES plant is designed to throttle the cavern air to a pressure of 46 bar at the turbine inlet (with caverns operating between 48 and 66 bar) and the McIntosh system similarly throttle the incoming air to 45 bar (operating between 45 and 74 bar). Recently, an EPRI CAES system, with aboveground storage, has been designed to throttle the incoming air to 55 bar (operating between 55 and 103 bar) in order to reduce the cost of vessels or pipeline air storage systems, despite the big throttling loss[7].

2.2 Constant pressure air storage
The release of compressed air energy is critical to the efficiency of the system. In the case of the large scale of CAES plant, it is needed to either the volume of cavern to limit pressure variations or to utilize water column to maintain a constant pressure in the cavern, where water from a surface reservoir displaces the compressed air. The constant pressure air storage can keep the CAES plants high efficiency and decrease the air storage volume. Under this condition, all the stored compressed air can be utilized to generate power by expansion. And the air mass rate is also constant. Some typical design methods were discussed on the constant pressure air storage[8, 9]. It is a storage method that could make maximum use of the available energy of compressed air. However, it is rather difficult to design and construct the constant pressure air storage devices.

3 THERMODYNAMIC ANALYSIS

Figure 2 shows the schematic diagram of micro CAES in the pressure-volume chart. 1-3 denotes an isothermal process, 1-2 is an adiabatic compression process and 3-4 is an adiabatic expansion process. 2-3 is a constant pressure or isobaric transformation and the temperature will be decreaed through the aftercooler. 4-1 is also the isobaric process. The shadow area 1-2-3-4 represents the energy loss of CAES cycle.

The CAES system can be separated into charging and discharging process. During the charging process, the high pressure compressed air was compressed by compressor train that driven by off-peak electricity and then stored in air tank or vessel. During the discharging process, the high pressure compressed air generated power by driving expander. The energy storage efficiency of CAES was defined as the ratio of compression work and expansion work.

$$\eta = W_{c,tot} / W_{e,tot} \tag{1}$$

where $W_{c,tot}$ is the compression work, $W_{e,tot}$ is expansion work.

Figure 2 shows that the energy storage efficiency is equal to 1 when compression and expansion are isothermal process. The energy storage density is related to output power during the expansion process and can be defined as the total output work per unit storage unit.

$$ED = W_{e,tot} / V_s \tag{2}$$

Where ED is energy storage density, $W_{e,tot}$ is total expansion work, V_s is the storage tank volume.

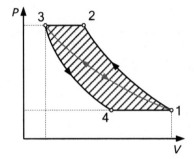

Figure 2 Thermodynamic cycle of compressed air energy storage system

The available energy of compressed air in the storage tank can be defined as the working capacity based on the ambient pressure and temperature[6]. It consists of the pressure energy and temperature energy. Since the temperature of the stored air is close to the ambient temperature, the main available energy of compressed air is pressure energy. It can be calculated as follows equation.

$$E_t = P_s V_s \ln(P_s / P_0) \tag{3}$$

Where E_t is the available energy, P_s is stored pressure, V_s is storage tank volume, P_0 is atmospheric pressure.

For constant volume air storage system, the high pressure compressed air was throttled to a constant pressure and keep the inlet pressure of expander constant and high efficiency. The available energy loss during throttling process can be calculated by following equation[10]:

$$E_v = P_s V_s \ln(P_s / P_v) \tag{4}$$

Where P_v is the pressure after throttling.

3.1 Compression

The high pressure compressed air in storage tank in CAES is compressed by compressor driven by electricity and stored in reservoirs. The power consumption of compressor mainly depend on the adiabatic, isothermal and polytropic process. The power consumption is smallest in the isothermal compression process. In order to improve the compressor efficiency and decreasing the power consumption, the multi-stage with intercooler compressor is preferable selection in the compressed air generation. The power consumption per unit mass compressed air in multi stage compressing process is calculated by

$$W_{cn} = \frac{k}{k-1} n P_1 V_1 \left[\left(\frac{P_2}{P_1} \right)^{(k-1)/nk} - 1 \right] \quad (5)$$

Where, k is adiabatic exponent, n is the compression stages, P_1 and P_2 are the inlet and outlet pressure of compressor.

In this analysis, air is assumed behaving like ideal gases. It obeys the ideal gas law expressed as:

$$m_s = P_s V_s / R_g T_s \quad (6)$$

Where m_s denotes the stored air mass, P_s the stored pressure, V_s the storage tank volume, R_g the gas constant, T the temperature.

3.2 Expansion

As a form of potential energy, the energy contained in a compressed air tank is equal to the work that can be done when gas in that tank expands to ambient pressure. This relationship is described as follows, which shows tank energy as a function of its volume and pressure:

$$E = P_s V_s \ln(P_s / P_0) \quad (7)$$

Where E is tank energy, V_s tank volume, P_s and P_0 are storage pressure and ambient pressure, respectively.

The available energy in compressed air of CAES system is significantly reduced by the thermodynamics of the expansion process. The efficiency of expansion can be improved by separating the process into several stages, allowing the gas to return to ambient temperature in between. In multistage expansion, the pressure between stages grows by a constant ratio. The multistage approach significantly improves the efficiency of the adiabatic process. For this analysis, the system assumed to use two-stage expansion. It is assumed that the expansion process is adiabatic. The work done when gas expands from tank pressure to ambient atmospheric pressure, shown in equation 8, can be described as a function of the tank pressure and volume, atmosphere pressure and the coefficient, which indicated the degree to which the expansion process is adiabatic or isothermal.

$$W_{es} = \frac{k}{k-1} m_s R_g T_1 \left[1 - \left(\frac{P_0}{P_s} \right)^{(1-k)/k} \right] \quad (8)$$

In the limit where $k=1$, the thermodynamic work can be expressed in a simplified form in equation 7, which is equivalent to equation 3, for the energy stored in a tank of compressed air, and is consistent with the definition of tank energy, in which the energy contained in a compressed air tank is equal to the maximum work done when gas in that tank expands to ambient pressure.

Figure 3 shows the effect of inlet temperature of expander on its output power. For all the thermodynamic process, the expansion work is increasing with the increase of the inlet temperature of expander. This is the reason why the inlet air was preheated by the compression heat, solar energy and exhaust heat in most of the CAES systems studied by many researchers.

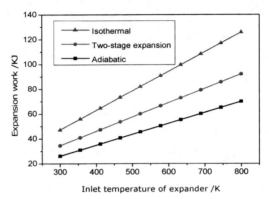

Figure 3 Effect of inlet temperature on output power

4 RESULTS AND ANALYSIS

Based on above thermodynamic analysis of the CAES cycle, the working performance of this system is investigated and the inlet air of expander is non-preheated. Figure 4 shows the energy storage efficiency of CAES under different thermodynamic process. The inlet air of expander was not preheated in this analysis. From this figure, maximum (100%) efficiency is achieved by the isothermal process, in which the temperature of the gas is held constant. Minimum efficiency is achieved by the single stage adiabatic process, in which no heat is transferred into or out of the system. Multistage compression and expansion could improve the system efficiency due to the heat transfer and it reflects the real working process of CAES in practice. In the multi stage working process, the storage efficiency is decreasing with the increasing of storage pressure. In a typical storage pressure range of 5-10MPa, the storage efficiency is changing from 66.4~61.9% and is higher than that (32.7~26.8%) in the single stage working process.

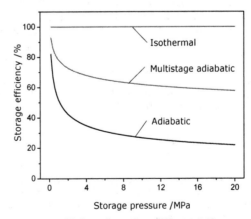

Figure 4 Energy storage efficiency under different thermodynamic process

Figure 5 shows the comparison of the energy storage density of the system between constant volume and constant pressure air storage. The P_{S1}/P_{S2} is the ratio of tank pressure and the inlet pressure of expander. For commercial Huntorf and McIntosh CAES plants, this ratio is 1.4 and 1.6. With the increasing of storage pressure, the energy storage density is also increasing. The density of constant pressure air storage is much higher than that of constant volume air storage. The energy storage density is increasing with the increasing of pressure ratio in its typical changing periods.

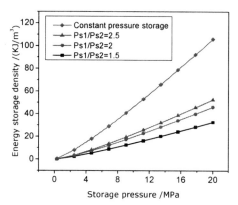

Figure 5 Comparison of energy storage density under constant pressure and constant volume

Although the inlet pressure of expander was adjusted to keep constant by adopting the reducing valve, there exits energy loss during the throttling process. From the figure 6, for given storage pressure, the larger of the pressure difference between storage pressure and the inlet pressure of expander, the greater of throttling energy loss. On the other hand, the amount of compressed air that used to generate power by expansion is increasing when the pressure difference in throttling valve is large. Therefore, the expansion work may be increased according to the comprehensive effect. The energy storage density is determined by the pressure difference in throttling process and the amount of compressed air used for expansion.

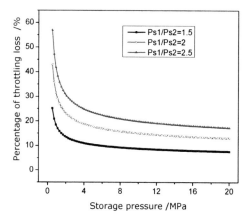

Figure 6 Throttling energy loss under constant volume air storage

Figure 7 shows the changing of storage volume ratio with the storage pressure. The storage volume ratio is the volume needed in constant volume air storage system divided by the volume needed in constant pressure air storage system. For example, when storage pressure is 9 MPa, the ratio of inlet pressure of expander and storage pressure is 1.5, 2 and 2.5, the volume needed in constant volume system is 3.3, 2.4 and 2.1 times larger than that in constant pressure volume system. Therefore, the use of a constant pressure tank requires the smallest volume. The design and construction of the constant pressure storage tank are important for the CAES system to improve the performance.

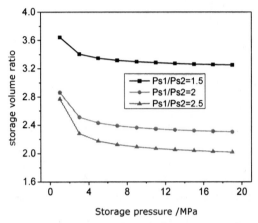

Figure 7 Comparison of storage volume of constant volume and constant pressure air storage

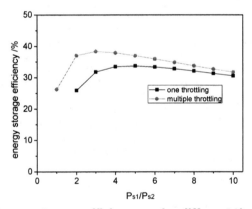

Figure 8 Energy storage efficiency under different throttling mode

The energy storage efficiency of CAES is determined by the expansion process. As shown in above figure 1. The compressed air was throttled to a constant inlet pressure of the high pressure expander and conducted first stage expansion and then the second expansion to ambient atmosphere pressure. In order to further use the stored compressed air energy, the compressed air was throttled second time to a constant pressure that is equal to the middle pressure. Thus the second stage expander could operate under the design working conditions. Figure 8 shows the comparison of energy storage efficiency of CAES system between these two expansion methods. The energy storage efficiency of system has a maximum value with the variation of throttling pressure for a given storage pressure. At the same storage pressure and throttling pressure, the energy storage efficiency could increase from 31.8% to 38.4%. It can be concluded that the output expansion work and energy storage density of the CAES system were improved by multiple throttling while maintaining the high efficiency of expander.

5 CONCLUSION

Based on the basic working principle of compressed air energy storage, the theoretical models of the compression process, storage and expansion process were established using thermodynamics theory. Using energy storage efficiency and energy storage density as evaluate index, the operating characteristics of compressed air energy storage system under the isothermal, adiabatic and multi-stage polytropic working conditions were investigated. A multiple throttling method was proposed to further use the compressed air energy and improve the system efficiency. The effect of constant volume and constant pressure storage methods on the performance were analyzed. The results showed that increasing the storage pressure and utilizing the multi-stage compression or expansion could improve the efficiency and the energy storage density of the system. The proposed analysis model will provide the foundation to design the high efficient cycle of micro CAES.

ACKNOWLEDGEMENTS

The work was supported by the National Basic Research Program of China (2012CB724307) and Anhui Provincial Natural Science Foundation (1208085ME68).

REFERENCES

(1) Beaudin M., Zareipour H., Anthony S., et al. Energy storage for mitigating the variability of renewable electricity sources: An updated review. Energy for Sustainable Development, 2010(14):302-314.
(2) Cavallo A. Controllable and affordable utility-scale electricity from intermittent wind resources and compressed air energy storage (CAES). Energy, 2007, 32(2): 120-127.
(3) Greenblatt J. B., Succar S., Denkenberger D. C., et al. Baseload wind energy: modeling the competition between gas turbines and compressed air energy storage for supplemental generation. Energy Policy, 2007, 35(3):1474-1492.
(4) McDowall J. Integrating energy storage with wind power in weak electricity grids. Journal of Power sources, 2006, 162(2):959-964.
(5) Succar S., Williams R. H. Compressed air energy storage: theory, resources, and applications for wind power. Princeton USA: Princeton Environmental Institute, 2008.
(6) Cai M, Kawashima K., Kagawa T. Power assessment of flowing compressed air. Journal of Fluids Engineering, 2006, 128(3):402-405.
(7) Nakhamkin M., Chiruvolu M., Daniel C. Available compressed air energy storage (CAES) plant concepts. Available online: http://espcinc.com/library/PowerGen_2007_paper.pdf.
(8) Kim Y.M., Shin D.G., Favrat D. Operating characteristic of constant-pressure compressed air energy storage (CAES) system combined with pumped hydro storage based on energy and exergy analysis. Energy, 2011, 36(10): 6220-6233.
(9) Kim Y.M., Favrat D. Energy and exergy analysis of a micro-compressed air energy storage and air cycle heating and cooling system. Energy, 2010. 35(1):213-220.
(10) Liu Hao, Tao Guoliang, Chen Ying, et al. Exergy analysis on power system of air-powered vehicle. Journal of Zhejiang University (Engineering Science), 2006, 40(4):694-698 (in Chinese).

ENERGY EFFICIENCY

Compressor and system energy efficiency improvements through lubricant optimization

J Karnaz
CPI Engineering Services, USA

ABSTRACT

This paper explores various theoretical and experimental techniques used to develop and commercialize the optimal lubricant that will enhance refrigeration and air-conditioning system energy efficiency and reliability. Concepts and techniques such as min/max oil film thickness, miscibility, solubility, volumetric efficiency, calorimeter studies, lubricant properties and lubricant-refrigerant interaction will be investigated and utilized for lubricant optimization.

1 IMPORTANCE OF LUBRICANT OPTIMIZATION

Global energy demand is on the rise and the impact this has on global economics and the environment continues to be an important topic of discussion. Designers and manufacturers of refrigeration, air conditioning and other equipment are constantly looking for ways to improve the energy efficiency of their products through such efforts as: improving heating and cooling losses, better heat transfer, enhancements to electric motors, and so on. A significant number of systems employ compressors that are operated with some type of lubricant that is vital to compressor performance. The lubricant needs to meet the required reliability and performance benefits for compressor operation. Often lubricant can be discharged into the system and is in circulation with the refrigerant throughout the refrigeration and air conditioning cycle. Lubricant in circulation needs to be accounted for to assure proper return back to the compressor and limit any negative effects on heat transfer. Optimization of these lubricants is required in order to improve energy efficiency in a system. Since optimization is tied to the refrigerant used, this becomes an important activity given recent and future changes to refrigerants in the market.

Environmental significance of refrigerants is creating the need to generate and evaluate lower global warming potential (GWP) options. These other refrigerants could carry different challenges when trying to maintain required efficiency, performance, stability and economic costs; so it becomes important to be able to evaluate lubricants with refrigerants that might have different interactive properties. This evaluation can be completed with a systematic approach that will encompass various screening and in some cases more elaborate testing. Once appropriate candidates have been identified, then compressor and concluding system tests can be completed to make sure the lubricant of choice meets all the requirements for reliable and efficient system operation.

2 EVALUATION OF LUBRICANT AND REFRIGERANT INTERACTIONS

An important activity before any lubricant is evaluated in a compressor or system is investigation to how the lubricant interacts with the refrigerant. A number of bench tests and specialized equipment can be used to make an initial determination to the compatibility of the lubricant and refrigerant. I believe it is important to look at these tests and how they are used for evaluation to maintain proper system operation.

2.1 Stability and compatibility tests

Making sure the combination of lubricant, refrigerant and materials used in a compressor and refrigeration system maintains adequate stability is important to establish. This evaluation can be done with a bench screening method that looks at the effects of the lubricant and refrigerant along with parameters such as time, temperature and concentration at an accelerated rate. An industry standard based on ASHRAE 97 "Glass Sealed Tube Tests" [1] has been used for years to evaluate refrigerant, lubricant and material stability and compatibility. The method provides an initial look at interactions at varying conditions. For refrigerant stability this type of testing was developed over sixty years ago to show: how Refrigerant 12 had instability at certain conditions due to weakening carbon – chlorine bonds [2]; more recently showing that HFC refrigerant had strong stability at most normal operations; and today how the unsaturated bond found in newly developed hydrofluoro olefin (HFO) refrigerants may affect overall system stability [3]. These types of tests will remain important with future investigation of new refrigerants in order to maintain not only a reliable system but also an efficient system.

2.2 Miscibility

The ability of lubricant and refrigerant to mix and stay as one phase is typically defined as the miscibility or critical solution temperature (CST) between the lubricant and refrigerant. The miscibility is usually represented at lubricant concentrations in refrigerant between the range of 1 and 40% and recorded at the temperature at which this concentration separates. Temperatures are measured above room temperature to represent operation in a high side region of a system and below room temperature to represent operation in a low side region of a system. The chemistry of both the lubricant and refrigerant will affect how the two mix and maintain separation. Figure 1 shows a diagram that represents a typical critical solution temperature (CST) evaluation [4].

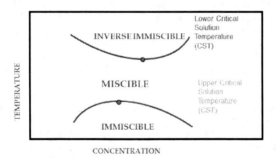

Figure 1: Critical solution temperature (CST)

The proper miscibility evaluation of lubricant and refrigerant is important to designing the most optimized system. Performing an initial bench test evaluation gives a good first look. In general it is beneficial to have a lubricant and refrigerant that maintains miscibility throughout the operating temperature range at concentrations found in circulation. This miscibility allows for adequate oil return back to the compressor and optimized system operation through maximized heat transfer effects. Even though a miscible lubricant-refrigerant combination is preferred in most systems, there are some systems that rely on oil separation so there is ability to take advantage of a combination that is immiscible.

Numerous papers and presentations have shown how optimizing the lubricant and refrigerant properties in a system, such as miscibility, can have benefits to overall system performance [5]. These benefits are usually associated with heat transfer effects and changes of state seen at expansion devices.

2.3 Solubility

Earlier miscibility was described at concentrations that are more refrigerant rich. For solubility the opposite condition of refrigerant poor (lubricant rich) conditions are investigated. The solubility measurement is usually helpful in designing compressor bearing lubrication but also can be a function of proper oil management. Figure 2 is a typical solubility curve of a POE lubricant and HFC refrigerant that relates temperature to pressure of a lubricant – refrigerant mixture.

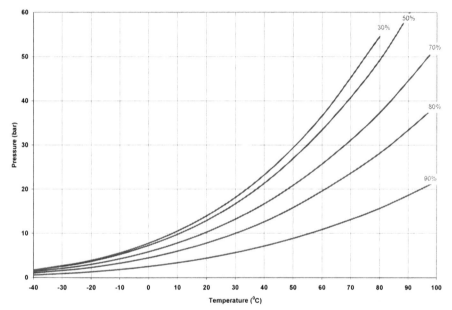

Figure 2: Solubility – pressure vs. temperature curve

When the concentration of lubricant and refrigerant is set, then the pressure associated at this concentration is measure at varying temperatures. The more soluble the refrigerant is into the lubricant, the less pressure will be measured at comparative concentrations. The relative solubility will then dictate the measured viscosity of the lubricant and refrigerant mixture which will be discussed in the next section.

2.4 Pressure-viscosity-temperature

This type of data is described by a number of names: pressure-viscosity-temperature (PVT) curves; vapor-liquid-equilibrium (VLE) curves; or Daniel Plots. They all describe the way the refrigerant interacts with the lubricant by way of the solubility and how this affects the viscosity. This interaction or dilution of the refrigerant into the lubricant is sometimes described as the working viscosity of the lubricant-refrigerant mixture and is valuable when determining lubricant supply to a bearing surface. Figure 3 shows a PVT curve at concentrations between 100% to

70% lubricant to refrigerant. When conditions of operation are known, the working viscosity of the lubricant and refrigerant mixture can be calculated. This value can be used as a preliminary estimate for what measure of film thickness could be expected at a given bearing. Depending on the type of compressor mechanism, quick determination can be made to the likelihood the lubricant minimum or maximum film thickness will be achieved. If needed, changes can be made to the starting lubricant viscosity or a complete chemistry change in order to satisfy the lubrication requirements. These plots and techniques have been cited in numerous references [6].

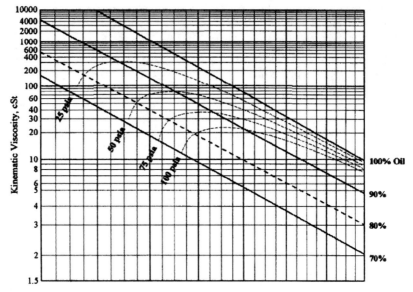

Figure 3: Pressure-viscosity-temperature (PVT) curve

2.5 Compressor studies

Sometimes it is of benefit to gather information regarding how a lubricant might perform inside of a compressor or system. Typically this exercise is left up to the compressor or system manufacturer to specify the lubricant and conduct whatever accelerated life tests are necessary. Commonly included tests are high load, stop/start, flooded start and liquid floodback. Such tests are conducted on a statistical number of compressors/units. For a lubricant manufacturer a few tests based on compressor endurance or calorimeter studies can be helpful to determine the proper lubricant with a given refrigerant. Lessons learned with the bench studies described above can be utilized to determine the direction of study with a given refrigerant. As an example, hydrocarbon refrigerants like R-600a and R-290 which are being investigated because of their low GWP values have low temperature miscibility with a number of refrigeration lubricant chemistries. But even though the miscibility temperatures do not change significantly, the solubility of the hydrocarbon refrigerant in the oil can be different for various lubricant chemistries. This type of difference can be used to benefit the overall performance by affecting the compressor volumetric efficiency. This data has been presented with testing of R-600a with various lubricant chemistries [7]. The data showed an improvement in the energy efficiency ratio (EER) and coefficient of performance (COP) when the solubility of the R-600a refrigerant was changed (reduced) in the lubricant. Reduced solubility effectively increased the volumetric capacity while still maintaining similar power input.

A small number of endurance tests at a short duration of time (500 to 1000 hours) can also give useful information which can help determine the acceptability of a lubricant candidate. More rigorous testing of the lubricant is then done under various conditions and time intervals in order to qualify the lubricant with a given refrigerant, compressor and system designs.

3 LUBRICANT AND REFRIGERANT INTERACTION STUDIES

The testing methods described above can be used to study lubricants that will be needed for the variety of low GWP refrigerants already used or coming to the market.

3.1 Lubricants for R-32 refrigerants

R-32 is a hydrofluorocarbon (HFC) refrigerant but is being investigated as a lower GWP option to R-410A (675 GWP vs. 2100 GWP) [8] in R-22 phase out taking place in developing nations. R-22 and R-410A have been the primary refrigerants used in scroll and rotary compressors for various types of air conditioning compressors. Developed countries which have switched to HFC refrigerants such as R-410A and synthetic lubricants like polyol esters (POE) have developed and optimized system for use based on stability, lubrication and miscibility. For R-32 these same types of synthetic lubricants look like good chemistry candidates but some differences have been discovered which may require some modifications.

3.1.1 Miscibility studies with R-32

For R-410A, lubricants with (viscosity grade) VG 32 have been used with scroll compressors and VG 68 with rotary compressors. When these same lubricant candidates where considered with R-32, it was discovered a noteworthy difference in miscibility temperature. Table 1 shows a comparison in miscibility at 20% lubricant concentration in the described refrigerants.

Table 1: Miscibility data on current lubricants: R-410A and R-32

Lubricant	Refrigerant	Miscibility Temp,°C
VG32 (A) POE	R-410A	-25
VG68 (A) POE	R-410A	-12
VG32 (A) POE	R-32	5
VG68 (A) POE	R-32	Immiscible @ RT

As can be seen in the table, when the same lubricants that were used with R-410A are used with R-32 there is a difference in the miscibility temperature. In some applications this might be an issue, and the lubricant – refrigerant mixture might not be an optimized combination. To remedy the miscibility difference, new lubricant chemistry formulations have been developed to try and meet the same miscibility numbers. These new lubricant miscibility results are found in table 2.

Table 2: Miscibility data on new formulated lubricants: R-32

Lubricant	Refrigerant	Miscibility Temp,°C
VG32 (B) Reformulated	R-32	-30
VG68 (B) Reformulated	R-32	-15

From the data, miscibility parameters of lubricants with R-32 are closely matched with lubricants for R-410A. This should help optimize particular system performance based on miscibility, but this does not mean it is necessarily the proper lubricant to use. Changing from R-410A to R-32 refrigerant will bring about changes to molecular chemistry, solubility, pressures, mass flow, etc.; all of which will cause other affects that need further evaluation.

3.1.2 Bearing lubrication studies with R-32

A key component when looking at the use of R-32 in place of R-410A is what will be the effect to bearing lubrication. R-32 can bring about a few changes that can be predicted such as higher pressures and temperatures over R-410A. This in itself could require the need for different lubricants when compared to R-410A. But in addition, R-32 could have other characteristics that possibly will lead to further lubricant evaluations. Therefore, further lubricant and refrigerant studies were done based on solubility and PVT data. Figure 4 shows data generated from PVT charts of lubricants that have similar miscibility values with each corresponding refrigerant.

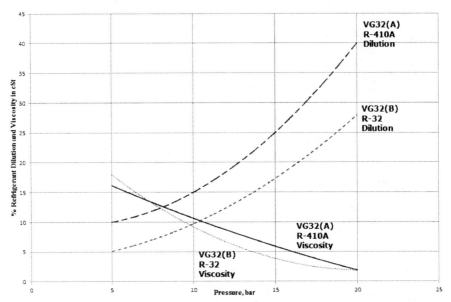

Figure 4: Relationship between dilution and viscosity: R-410A and R-32

The results indicate that even though the mixtures have similar miscibility, a viscosity difference can be seen with the lubricant /R-32 mixture resulting in a viscosity reduction. This combination could have a negative effect on bearing lubrication at certain conditions and a different combination may be necessary.

3.2 Lubricants for hydrocarbon refrigerants

Hydrocarbon refrigerants such as isobutane, R-600a, have been used for a number of years in small appliances. Propane, R-290, is used in industrial applications and is now being evaluated for commercial applications. Usually when used, they have good performance but are not in wide spread use due to their high flammability (A3 as rated by ASHRAE 34 [9]) and limits to charge size. Optimizing lubricants used with hydrocarbon refrigerants is critical to maximizing performance due to high refrigerant solubility and refrigerant charge to unit capacity ratio. Different types of lubricant chemistries can also improve reliability over the operating life of a system.

3.2.1 Compressor performance evaluation with R-600a

Studies have been done showing how lubricant chemistry changes measured in fractional horsepower compressors can affect the overall compressor performance by increasing the volumetric capacity/efficiency [7]. This is very important to this industry which is heavily graded on efficiency. Further efficiency improvements are sometimes accomplished by moving to lower and lower viscosity lubricants. This type of direction for R-600a can be a challenge when typical mineral oil chemistry is used due to high refrigerant dilution which can lead to reliability issues. To minimize this affect while still maintaining performance benefits such as improved efficiency and reliability, a synthetic based lubricant can be applied to the application.

To quantify this improvement, tests were conducted where compressor performance was measured both before and after submitting the compressor to accelerated testing. Table 3 is a snapshot of these results; the compressors which were a nominal capacity of 249 Watts were performance tested before and after a 1000-hour endurance test. Three different lubricant chemistries were evaluated which should provide different outcomes. After completion the compressor bearing components were evaluated.

Table 3: Performance test results and accelerated testing

Lubricant	Pre-Endurance Results			Post-Endurance Results		
	Capacity Watts	Power Watts	COP	Capacity Watts	Power Watts	COP
Product A	247.4	159.2	1.55	255.6	157.4	1.62
Product B	250.0	163.1	1.53	264.1	158.2	1.67
Product C	252.3	163.4	1.64	265.5	158.1	1.68

The data indicates that overall each lubricant could be an acceptable product but Product C stands out. This product will provide the best efficiency. Trending of the results along with observation of the components confirms that Product C will have the superior overall performance (efficiency plus reliability). Further impact on performance could also be achieved through slight lubricant modifications.

3.3 Lubricants for HFO refrigerants
Refrigerants with higher GWP values, like hydrofluorocarbons (HFC), are being scrutinized for their potential negative environmental impact. Hydrofluoro olefin (HFO) refrigerants like the hydrocarbon refrigerants mentioned above have low GWP values (GWP 4-6) [10] but are ASHRAE designated A2 versus A3 for hydrocarbons. One questioned limitation is the stability of the refrigerant due to an unsaturated bond in the molecule. Instability, if it would occur, could have negative effects on the short or long-term operation of the system.

Some HFO chemistries may also have high dilution rates in lubricants when compared to current commercial combinations which will need investigation.

3.3.1 HFO stability studies
Strength of the unsaturated bond in HFO refrigerants and reaction with lubricants was investigated in an Air-conditioning Heating Refrigeration Institute (AHRI) compatibility project [11]. Tests with refrigerants HFO-1234yf, HFO-1234ze(E) and 50/50 blend of HFO-1234ze(E) and R-32 were evaluated in ASHRAE 97 glass sealed tube tests at various levels of contaminants (air and water). Overall, at elevated levels of air and water the tests conducted with HFO refrigerants and lubricants showed higher levels of reactivity when compared to the same conditions with refrigerants R-134a and R-410A with lubricants (test controls). One of the indicators of reactivity was elevating levels of acidity. Low or more normal levels of contaminants did not indicate issues with any of the tested combinations.

One could question whether the higher level contamination would be represented in the field. If instability were to occur, then there is a possible potential for degradation of system performance (efficiency and reliability). To improve this outcome, different lubricants and lubricant formulations can be investigated and compared to the above AHRI results at the elevated contaminant level. Table 4 shows the results of testing different lubricants with HFO-1234yf refrigerant at elevated levels of contaminants with the acidity level (TAN) measured as the indicator.

Table 4: HFO-1234yf stability testing at 175°C and 14 days

Change in Acid Number, mg KOH/g oil	Standard	Formulation 1	Formulation 2
	0.74	0.06	0.05

The results demonstrate an improvement when compared to a lubricant currently used with HFC refrigerants. Also, even when elevated levels of contaminants are present in systems using HFO refrigerants, this reaction can be slowed down or neutralized by formulating the lubricant.

3.3.2 HFO solubility studies

HFO-1234ze(E) refrigerant, due to its low GWP value, is being evaluated as an HFC refrigerant substitute in larger industrial refrigeration systems that typical use larger refrigerant charge amounts. Characteristically some of the compressor equipment used in these applications requires a higher value of viscosity at bearings in order to maintain good performance. When comparing the solubility of HFO-1234ze(E) with an HFC refrigerant like R-134a, at normal operating conditions, it appears that HFO-1234ze(E) will dilute more into a lubricant than would R-134a. Table 5 shows the comparison of POE lubricants that are used with HFC refrigerants and a higher viscosity POE lubricant tested with HFO-1234ze(E). The data was calculated from PVT charts and the values represent two different operating conditions and also a condition after cooling the lubricant at the same dilution levels which is common practice in some applications.

The information in the chart shows that when HFO-1234ze(E) is used at certain operating conditions the refrigerant will dilute more into the lubricant then conventional refrigerants like R-134a. This increased dilution will require additional investigation in order to determine the optimal lubricant and refrigerant combination. This will help assure proper bearing lubrication and overall system performance.

Table 5: Comparison of HFO and HFC Dilution and Viscosity

Condition	ISO68 POE R-134a	ISO120 POE R-134a	ISO170 POE HFO-1234ze(E)
	% Dilution and/or Viscosity	% Dilution and/or Viscosity	% Dilution and/or Viscosity
60°C and 10 bar	20% and 9 cSt	18% and 17 cSt	35% and 3 cSt
Oil Supply at 20°C at Dilution	40 cSt	80 cSt	9 cSt
95°C and 20 bar	22% and 3.5 cSt	18% and 6.5 cSt	30% and 2.4 cSt
Oil Supply at 20°C at Dilution	30 cSt	80 cSt	16 cSt

4 CONCLUSIONS

Lubricants perform many functions within a refrigeration system: lubricating, sealing, heat and debris removal, etc. It is important to understand ways of optimizing the lubricant for increased system performance. The environmental impacts that are associated with some chemicals, are being accessed which includes the global warming potential of refrigerants; this could bring about the need to evaluate a large combination of refrigerants and lubricants. Energy consumption also impacts the environment (in some regions a third of energy consumption can be related to HVACR usage). So it is important to make sure when new investigations are made that energy efficiency and system reliability match existing system performance and moves toward performance improvements.

Test methods have been defined that can allow for study of proper lubricant and refrigerant combinations that address the need for desired overall system performance. These test methods were used to evaluate lubricants for various low GWP refrigerants outlined in this paper allowing for maximum system performance. From some of the studies, it appears a need still exists to define the correct lubricant for various refrigerant candidates through further research.

5 REFERENCES

[1] ASHRAE Standard 97. "Sealed Glass Tube Method to Test the Chemical Stability of Materials for Use within Refrigeration Systems" (2007).
[2] H. Elsey, L. Flowers, J. Kelley. "A Method of Evaluating Refrigeration Oils". Refrigeration Engineering, Vol. 60, No. 7. (1952).
[3] T. Leck. "Evaluation of HFO-1234yf as a Potential replacement for R-134a in Refrigeration Applications". 3rd IIR Conference. Boulder (2009).
[4] J. Karnaz. "So Many Refrigerants – Is One Lubricant Possible?". The International Symposium on New Refrigerants and Environmental Technology. Kobe (2012).
[5] L. Cremaschi. "Lubricant Effects on Heat transfer and Pressure Drops in Heat Exchangers". ASHRAE Winter Conference. Dallas (2013).
[6] J. Karnaz: "Utilizing Lubricant-Refrigerant Interaction Data for Compressor and System Design". Advanced Compressor Modeling (Compressor 102). Purdue Short Course (2012).
[7] J. Karnaz. "Lubricants: Changing the Game". International Conference on Air-conditioning & Refrigeration. Korea (2011).
[8] American Society of Heating, Refrigerating and Air Conditioning Engineers. 2009 Fundamentals Handbook. Section 29.4.
[9] ASHRAE/ANSI Standard 34. "Designation and Safety Classification of Refrigerants" (2007).
[10] M. Spatz: "Latest Developments in Refrigerants for A/C&R Applications". The International Symposium on New Refrigerants and Environmental Technology. Kobe (2008).
[11] AHRTI Report No. 09004-01 "Materials Compatibility & Lubricant research for Low GWP Refrigerants – Phase 1: Thermal and Chemical Stability of Low GWP Refrigerants with Lubricants" (2012).

Energy saving in sliding vane rotary compressors

R Cipollone*, G Valenti**, G Bianchi*, S Murgia***, G Contaldi***, T Calvi**
* University of L'Aquila, Department of Industrial Engineering, Italy
** Politecnico di Milano, Dipartimento di Energia, Italy
*** Ing. Enea Mattei S.p.A., Italy

ABSTRACT

Electrical energy for producing compressed air in industrial contexts represents an important share of the overall electricity consumption: this figure accounts for 4-5 %.

Compressed air is produced by means of rotary volumetric machines which are proven to be more suitable than other types (dynamic, reciprocating, etc...) in terms of pressure and flow rate delivered.

Sliding Vane Rotary Compressors (SVRC) compared to screw type compressors are not as widespread. However, thanks to the technological development made in the last two decades, they are characterized by premium specific energy consumption and demonstrate unforeseen potential in terms of energy saving due to some intrinsic features specifically related to this machine.

The paper focuses the attention on a new technology under development related to the oil injection inside the machine able to cool the air during compression. A comparison between the results of a mathematical model of the new injection oil technology and experimental p-V measured by means of piezoelectric transducers is shown.

The compression work reduction measured on the shaft and observed integrating the p-V cycle gives a strong consistency to the modelling toward a comprehensive physically consistent software platform and to the injection technology.

1. INTRODUCTION

Energy saving will be one of the most important drivers in the next years for many energy consuming users. Energy saving is recognized as equivalent to an energy source, being, with respect to a real energy source, characterized by a much lower cost per unit energy (saved or produced).

Compressed air is universally produced by electrical energy. It is a non-replaceable utility and, based on the global electric energy consumption, it is responsible for a 4-5 % share, [1, 2].

The flow and pressure requirements of typical industrial applications make rotary volumetric compressors the most common kind in the industry: of these, screw type compressors are the most diffused and characterized by a proven and reliable technology. Sliding Vane Rotary Compressors (SVRC) are the second most diffused after screw compressors. It is in the Author's opinion that these machines are not very well known in terms of energy saving potential. In the past two decades, they have been subjected to important technological improvements making them today's most efficient single-stage compressors in the industry.

In recent years, some of the Authors focused their attention on going deep inside the physical processes governing the behaviour of such types of machines: main processes, such as air intake, compression inside the cells, compressed air exhaust, oil circulation and injection, blade motion inside the rotor slots and friction phenomena have been described [3-5]. Thanks to these efforts, a virtual platform has been built with a good correspondence with measured data and it was the conceptual base to improve performances and conceive new arrangements [6]. In parallel, a thorough experimental activity was undertaken: the base of this experimental activity was the measurement of the pressure inside the cell, when intake, compression and exhaust occur. The theoretical treatment of p-V data made possible the identification of the coefficient of friction [6, 7], a better understanding of the cooling action done by the oil during injection [8], and suggested geometrical re-shaping of the machine for optimisation [6]. Also thanks to this activity, the specific energy consumption of some models of SVRC present among the lowest values in the industrial compressor global market: values around 5.4-5.8 kW/m^3/min - ISO 1217 conditions (p intake = 1 bar; T intake = 293.15 K) – at 7 bar as pressure delivered.

Thanks to some theoretical treatment of p-V data, it was possible to discover that the oil spray formed by actual technology (simply calibrated holes) is unable to cool the air effectively during compression, [7,9]. This was caused by some thermodynamic properties of the oil (molecular diffusion coefficient of oil droplets in air and oil saturation properties) but also from the mean drop dimensions produced by the jet: in fact, as it is known, big drops are heated and evaporate with great difficulty during their travel inside the cell. Also it is difficult to subject these to initial momentum erosion and immediately collapse with the metallic walls of the cell (rotor and blade surfaces) keeping their original speed almost unchanged [8]. On the other hand, if oil evaporation would take place, a strong air cooling would result which would decrease compression work but would also result in a work increase due to the compression of the oil vapour which requires additional work (with respect to the oil pressurization in liquid form). Therefore, enough care must be done in order to have the benefits of the air cooling and to avoid the work increase due to the compression of the oil in the vapour phase.

Very recently some of the Authors concentrated their attention on the actual oil injection technology in order to understand under which conditions oil could cool air during injection [8,9]. A complex comprehensive oil injection model was developed in order to understand main processes occurring during jet breakup, particles formation, droplets dynamics, heat transfer between oil and air, oil heating and vaporization, re-condensation due to pressure increase during compression, oil puddles formation due to the interaction between the oil spray and the metallic surfaces; the rotor and blade lateral surfaces. These results were simulated using an existing industrial machine (M111 H, 22 kW at 1500 rpm).

In this paper the Authors present an experimental activity on the same compressor in which proper pressure swirled injectors replaced the conventional injection rail. The injectors were specifically designed for this application and optionally fed by an external pump in order to test higher pressures with respect to those produced by the machine. Pressure increase was required to vary continuously the dimension of the droplets. p-V data were recorded as well as the mechanical shaft power.

2. OIL INJECTION MODELING INSIDE THE VANE

Some of the Authors already developed in [8] a comprehensive physically consistent model which describes main processes produced by a pressure swirled injector inside the cells during compression. The model was integrated in a previous model already referenced in literature [3,10] which evaluates the performances of SVRC subdividing the physical behavior in a set a processes each other integrated. For the quantitative aspects of these models, references [10] and [8] can be addressed. Only a brief description is done in the following.

The comprehensive cell model predicts pressure and temperature inside the cell. It considers:
a) Vane filling and emptying trough intake and exhaust ports suitably described in order to closely match their real shape. The mass transfer makes use of a suitably formulated 1D unsteady approach, considering the transients occurring during discharge;
b) Pressure and temperature inside the cell during time (compression) by means of the energy conservation equation in a lumped parameter form. Energy exchanges between air and oil are considered as previously described as well as the heat exchanged through the stator surfaces.

The oil injection model is different according to the technology used. Current solution is represented by a series of holes of a proper diameter fed by a common rail in which pressurized oil is brought (after separation from the compressed air at the SCRC discharge). The jet which is produced doesn't produce any cooling effect on the air during compression [7]. A technology advancement has been represented by using pressure swirled injectors which demonstrated the capability to cool the air and reduce compression work. The pressure swirled oil injection model predicts:
a. The break up distance from the injector orifice, from which the oil jet starts transforms in a spray. The jet is subdivided in several portions and for each of these, using a Rosin-Rammler drop size density distribution, number of particles for each class of dimensions is calculated. Original injection direction of such particles is defined in a random way inside the spray cone experimentally observed. Initial speed of particles is defined by spray correlations related to the oil flow rate injected;
b. The trajectories of the drops as result of the solution of momentum equation in which aerodynamic forces (drag and shear lift), inertial forces (virtual mass and Bassett history), volume forces (gravity and buoyancy), non-inertial forces (Centrifugal and Coriolis), and pressure forces are considered in order to evaluate droplet motion; oil puddles occurring on rotor surfaces (injectors are mounted on the stator side) and blades as result of jet impingement are considered;
c. Heat transfer between oil drops and air due to forced convection: oil heating and air cooling result till the oil saturation temperature is reached. From this point on, oil in vapor phase is produced, and the latent heat of vaporization is exchange, in proportion with the oil mass evaporated;
d. Oil mass diffusion during droplet motion due to molecular diffusivity in air: this phenomenon modifies the drops' masses and their momentum.

The interactions between cell model and oil injection is represented in Figure 1: the basic concept is that the heat exchange between each drops and air realizes an internal inter-refrigeration for the compression process leading to a lower pressure at the discharge so reducing the amount of energy required.

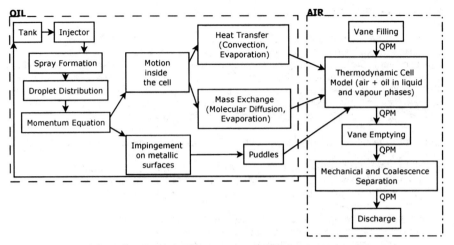

Figure 1 – Interactions among different processes

3. EXPERIMENTAL ACTIVITY

In order to verify experimentally the theory explained above, the injection system of a 22 kW SVRC was modified from the conventional setup in which oil is injected through straight calibrated orifices to an enhanced architecture in which oil is sprayed via pressure-swirled solid-cone nozzles.

A few types of nozzles were specifically designed to be fitted on the compressor either radially or axially. Figure 2 shows the images, taken with a high speed camera, of two different types of nozzles spraying oil at 80 °C and at 6 barg into an ambient reservoir. These pictures show that oil break-up takes place within a short distance in conditions typical for an air compressor, yet it does not lead exclusively to spherical droplets but also to ligaments, ramifications and undefined structures [9].

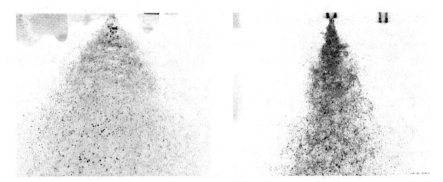

Figure 2 – Images taken with a high speed camera of two different types of nozzles (left: full-cone large angle; right: full-cone narrow angle) spraying oil at 80 °C and 6 barg into an ambient reservoir

A conventional 22 kW SVRC and the modified model featuring the new injection system were tested on a rig (see Figure 3).

Figure 3 – Compressor rig

This experimental rig employed the necessary instrumentation to measure air temperatures and pressures along the process, the volume flow rate at the discharge (from which the mass flow rate was computed), oil temperatures and pressures along the process, the volume flow rate prior to injection and finally shaft torque and rotational speed.

A process flow diagram of the test rig is depicted in Figure 4 illustrating: the air stream (red) with highlighted the rated flange for measuring the volume flow rate on the discharge line; the oil stream (brown) which is split into the oil to shaft bearing and the oil the injection system, both conventional and enhanced; the water stream (cyan) for oil cooling; and the power stream (black) that drives the shaft. The test rig allows to test the conventional and the enhanced compressor while varying: the discharge pressure, the rotational speed, the injected oil temperature and pressure; for the enhanced compressor it allows to change the nozzles that are activated. Four piezoelectric pressure transducers circumferentially placed along one of the end plates were used to measure p versus time data i.e. versus angular displacement i.e. versus cell volume.

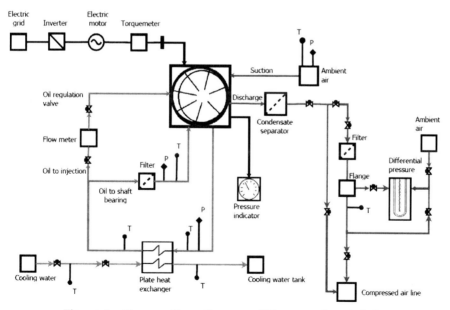

Figure 4 – Process flow diagram of the experimental rig

4. RESULTS

A number of tests have been conducted to create an experimental database that could be used for the validation of the simulation code as well as for understanding the margins of further improving the compressor specific energy.

Because in general the flow rate through a nozzle is much smaller than that through an orifice at same pressure and temperature, the total injected oil was relatively low in the enhanced compressor despite the number of nozzles installed. In order to improve flow rate and make a finer spray, a pump was utilized to boost upstream injector pressure.

The analysis was done in order to:
   a) Further validate the belief that current oil injection technology doesn't produce any significant air cooling effect;
   b) Compare the compressor performances, with respect to the current ones, when the upstream injector pressure is equal to the pressure of the compressed air;
   c) Compare the compressor performances when the upstream injector pressure is increased using an additional pump.

Most part of the comparisons have been made making reference to the p-V data which represent the most intimate information concerning real behaviour. Among the wide testing done, the cases reported in Table 1 have been chosen as most representative for the goal of the analysis.

Case #1 represents a typical condition in which the oil is injected according to the conventional technology: the pressure inside the rail is 7,9 bar_a. In reality, the compressor works with a conventional oil flow rate. Case #2 refers to a pressure swirled injection fed at 20.2 bar_a, assisted by an external pump. Only a similar oil flow rate – slightly lower – has been reached, due to the fact that nozzles require higher pressures to achieve the same flow rate as orifices. Case #3 refers to a pressure swirled injection fed at 8.1 bar_a, without any external pump: this is the main reasons of a reduced oil flow rate with respect to previous values.

Table 1. Experimental cases

Parameter	Unit	Experimental conditions		
		1	2	3
Rotation speed	Rpm	1500	1498	1504
Injection pressure	bar_a	7,9	20.2	8.1
Free Air Delivery	l/min	3984	4001	3848
Air flow rate, dry	kg/s	0.069	0,070	0,068
Air temperature	°C	80,1	73,9	90,8
Oil temperature	°C	67,4	60,0	60,0
Oil flow rate	l/min	37,0	31.0	15.0
End pressure	bar_g	7.5	7.5	7.5
Indicated power	kW	20,90	19,41	19,68
Shaft power	kW	23,08	21,40	21,89
Mechanical efficiency		0,90	0,91	0,90

Figure 5 shows the p-V diagram measured referred to Case #1 and Case #2. Most evident aspects are:
a) the compression during Case #1 closely stays on an adiabatic transformation, so the oil conventionally injected doesn't produce any air cooling;
b) oil injection at high pressure (Case #2) realizes a visible cooling of the air due to the spray which reduces mean drop diameters. The cooling remains effective

during the all injection duration. When this is ended, the pressure trace continues more or less parallel to the adiabatic transformation, slightly lower than the conventional case;
c) The lower oil flow rate and the less effective oil spray, produce in case #3 – Table 1 - a higher air temperature even though a slightly lower indicated power with respect to the conventional case is measured. It doesn't behave too differently from the conventional case, so no additional references are given.

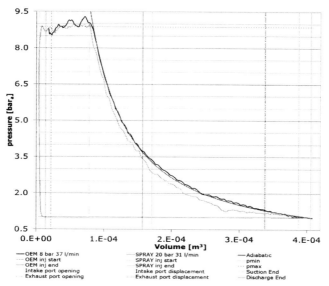

Figure 5 – p-V data, measured and calculated, Case#1 and Case#2

The difference between the pressure trace between case #1 and case #2 integrated in the V direction is equal to the actual heat removed by the spray. From the integration this results equal to 1.50 kW which is very close to the differences found in the shaft power, Table 1.

A theoretical verification of this important datum has been done following the mathematical representation of the interactions between compressed air and spray, [8]. With reference to the conditions of Case #2 test, spray MSD (Mean Sauter Diameter) has been calculated, for an orifice diameter D_{or}, by:

$$SMD = 4.52 \cdot \left(\frac{\sigma_o \mu_o^2}{\rho_a \Delta p}\right)^{0.25} \left(2.7 \cdot \left(\frac{D_{or} \dot{m} \mu_o}{\rho_o \Delta p}\right) \cdot \cos\theta\right)^{0.25} + 0.39 \cdot \left(\frac{\sigma_o \mu_o}{\rho_a \Delta p}\right)^{0.25} \left(2.7 \cdot \left(\frac{D_{or} \dot{m} \mu_o}{\rho_o \Delta p}\right) \cdot \cos\theta\right)^{0.75} \quad (1)$$

being ρ, μ, σ the density, viscosity and surface tension; θ the half spray cone angle, Δp the pressure difference across the injector; "o" and "a" refer to oil and air. $\dot{m}$ is the oil mass flow rate. From the SMD knowledge, the assumption of a Rosin-Rammler drop size distribution gives the number of particles for each class of dimension chosen.

For each droplet, the thermal power exchanged with air $\dot{Q}_{a-d}$ results from:

$$\dot{Q}_{a-d} = \pi \cdot d_d \cdot k_{mix} \cdot N_u^* \cdot (T_a - T_d) \quad (2)$$

being "d" the droplet diameter, k_{mix} the thermal conductivity of the air and oil mixture, T the temperature. N_u^* is the corrected Nusselt number, [11].

From previous studies [8], the thermal power exchanged between jet and air is almost exclusively due to forced convection, having observed that oil vaporization doesn't occur. The overall thermal power exchanged with the air is the sum of all the contributions given by the all particles, whose properties change during motion before impingement.

Equation (1) is reported in Figure 6: for an upstream injection pressure equal to 20,2 bar_a (the mean pressure difference is close to 16-17 bar), and for an orifice diameter close to 0.5-0.7 mm, a SMD in the range of 70-75 μm is calculated.

Figure 7 shows the overall thermal power exchanged vs. spray SMD. The predicted theoretical value corresponds to 2.2 kW. This estimation is almost 1.45 times greater than the experimental measured data.

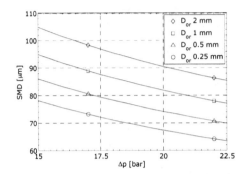

Figure 6 – SMD versus orifice dimension and feeding pressure, Case #2

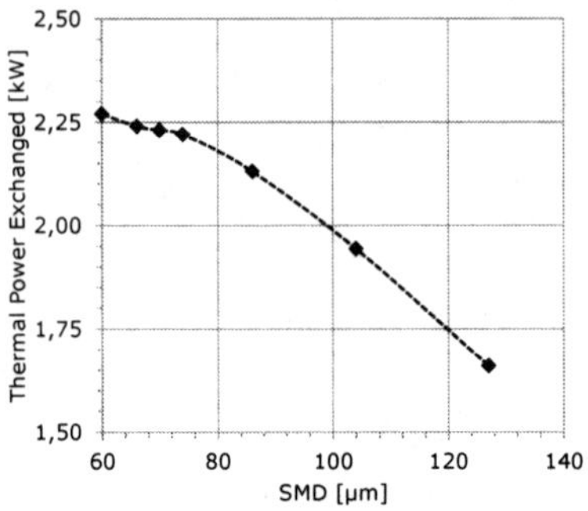

Figure 7 – Thermal power exchanged as a function of SMD, Case #2

This difference can be retained satisfactory, considering all the approximations introduced in the spray modelling: the most important one is related to the fact that each sub injection considered (to reproduce the real process) do not interact. In reality, drop collapses happen, the smaller drops being caught and enveloped by the bigger ones. This reduces heat transfer removed by the air. In this sense, the model appears to be a good compromise between the need of a physical representation (with its high complexity) and the need of having a model which can be run in an engineering software platform.

5. CONCLUSIONS

Oil injection technology in SVRC can further improve specific energy with respect to current values. This is due to the additional effect that oil can introduce, with respect to those related to friction reduction and sealing, related to the air cooling process during compression.

In this paper the Authors present the results obtained making reference to pressure-swirled nozzles substituting the current technology which employs a series of calibrated orifices fed by a oil common rail.

The main goal is the investigation of the cooling effect of sprayed oil on the air which can result in a reduced work of compression, hence in a more efficient compressor.

In order to perform and in depth analysis of this effect, a theoretical model reproducing the performance of a SVRC was presented in literature, and recently was updated with a mathematical model of a pressure swirled oil injection process. The main result was the calculation of the heat removed from the air during compression which resulted from oil drops heating till saturation and subsequent evaporation.

In this paper a detailed experimental SVRC test rig has been built in order to: (a) validate spray modelling in particular concerning the heat removed from the air (cooling); (b) verify compressor performances when the new injection technology is used. A 22 kW existing industrial compressor has been tested at diverse working conditions.

The experimental result demonstrate that with pressure swirled injectors fed at 20 bar_a thanks to an additional pump, the shaft power decreases by about 1,7 kW; this datum corresponds to the treatment of p-V data recorded. The spray modelling applied to the condition tested gives a thermal cooling equal to 2.2 kW while the treatment of p-V data gives 1.5 kW. The difference can still be considered as satisfactory, in spite of the model simplicity.

Currently, the experimentation on the enhanced compressor is continuing. This initial experimentation has shown that the current common rail position may not be ideal for the new pressure swirl injectors. New injector positions on the compressor are being tested and initial results show interesting improvements on specific energy values even without the necessity of an external oil pump.

REFERENCE LIST

1. IEA, "Key world energy statistics". 2011.
2. Peter Radgen and Edgar Blaustein, "Compressed Air Systems in the European Union". 2001.
3. Roberto Cipollone, Giulio Contaldi, Antonio Sciarretta et al., "A comprehensive model of a sliding vane rotary compressor system". IMechE International conference on compressors and their systems. 2005.
4. Tramschek and Mkumbwa, "Mathematical modeling of radial and non-radial vane rotary sliding vane compressors". 1996.
5. Roberto Cipollone, Giulio Contaldi, Antonio Sciarretta et al., "Theoretical Model and Experimental Validation of a Sliding Vane Rotary Compressor". International Compressor Engineering Conference. 2006.
6. R. Cipollone, G. Bianchi, and G. Contaldi, "Ottimizzazione energetica di compressori a palette". 67° Congresso Nazionale ATI. 2012 (in Italian).
7. Roberto Cipollone, Giulio Contaldi, Davide Di Battista et al., "Energy Optimisation in air compression: theoretical and experimental research activity on sliding vane rotary compressors". Motor Driven Systems conference. 2011.
8. Roberto Cipollone, Giuseppe Bianchi, and Giulio Contaldi, "Sliding vane rotary compressor energy optimization". ASME International Mechanical Engineering Congress & Exposition. IMECE 2012.
9. G. Valenti, L. Colombo, S. Murgia et al., "Thermal effect of lubricating oil in positive-displacement air compressors", Applied Thermal Engineering, vol. 51, no. 1-2. pp.1055-1066, Mar., 2013.
10. Roberto Cipollone, Giulio Contaldi, Raffaele Tufano et al., "A Theoretical Model and Experimental Validation of a Sliding Vane Rotary Compressor". 18th International Compressor Engineering Conference. 2006.
11. Abramzon, B., and Sazhin, S., 2006. "Convective vaporization of a fuel droplet with thermal radiation absorption". Fuel, 85(1), pp. 32 – 46.

Energy recovery using sliding vane rotary expanders

R Cipollone*, G Contaldi^, G Bianchi*, S Murgia^
* University of L'Aquila, Department of Industrial and Information Engineering and Economy, Italy
^ Ing. Enea Mattei SpA, Italy

ABSTRACT

Energy recovery from low grade energy sources represents a technological challenge which can significantly contribute to the energy balance between production and consumption. Mechanical and electrical production can be done both from thermal energy usually wasted and low temperature renewable sources. In this sense, the interest is twofold. Many efforts have been made and studies produced, mainly making reference to a well known organic Rankine cycle (ORC) of transformations: in spite of this, the expander technology still represents an open aspect and call for a reliable and low cost development.

This paper presents the experimental validation of a comprehensive mathematical model of a sliding vane rotary expander: the model, fully physically based, has been validated thanks to a test bench in which, through an ORC, a novel expander has been tested and operated.

NOMENCLATURE

n_{cell}	Number of rotating cells	**Subscripts**	
p	Pressure	1,2	Contact points
t	Time	k_1, k_2, k_3, k_4	Coefficients
v	velocity	bl	Blade
F	Force	fr	Friction
P	Power	in	Inertial
O	Center	ind	Indicated
R	Radial distance	Y,W	Blade tip
V	Volume of the cell	Cor	Coriolis
μ	Friction coefficient	cen	Centrifugal
η	Efficiency	sh	Shaft
ω	Rotational speed	rot, st	rotor, stator

INTRODUCTION

Energy transformations in mechanical or electrical form from fossil fuels involve an energy waste whose value, form a thermodynamic point of view, is strictly related to temperature. Often, the quantitative dimension of this waste is greater than the useful energy produced. Among the most energy wasteful sector, the transportation for people and goods is a reference example: mechanical efficiency less than 40 % is already a good value, being the most part of this range below 30 %. Thermal energy is wasted at medium temperature range (< 300°C) and the recovery (form one to two thirds of the input energy) is more a need than a concept and a huge interest is developing.

Moreover, many industrial thermal needs at medium and low power ranges produce energy streams discharged into atmosphere more or less at all temperature levels (< 300-350 °C). Usually components or devices or energy integrated packaged systems require cooling, not related to the thermodynamic cycles, and the energy is wasted. Compressed air (CA) in industrial applications doesn't escape from this *waste phenomenon*: an engineering rule of thumb states that almost the same amount of electrical energy absorbed is exchanged as waste heat into the lubricating oil and towards the environment by convection: temperature levels are low, 80-100 °C, and even the slightest recovery of this energy would be a significant contribution to the CA package efficiency. In fact, electrical energy for producing compressed air in the industrial context represents an important share of the overall electricity consumption: this figure accounts for 4-5 % of global electrical consumption, and a saving of 5-10 % in terms of mechanical or electrical energy would result in a huge contribution. Many forms of renewable energy (geothermal, solar, etc...) behave as low temperature energy sources could make use of innovative waste heat recovery solutions.

For these sectors, the energy recovery into mechanical and electrical form is a challenge. From an economic point of view, it could add substantial consistency to the green economy sector.

Usually, a recovery system using a low grade temperature source is designed considering a proper working fluid (WF) which is subjected to a series of transformations which are represented by a Rankine cycle: the WF in liquid state at a given pressure and temperature (eventually under-cooled) is pumped and, thanks to the energy recovery, it is economized and vaporized (eventually slightly super-heated). So, the vapor is expanded producing mechanical energy. The WF at lower pressure (expander exit) is cooled down to the initial temperature (inlet pump conditions) and this requires, eventually, de-superheating, condensation and under cooling. As it is known, the concept is a question of basic engineering but the availability of the technology (reliable, low cost, safe, etc...) is a complex problem, (1), (2). Main concerns are: (a) WF choice and (b) Expander and Pump Technology (ET & PT). Considering that the PT can benefit from many known technologies (even though there are specific concerns which call for new technology), WF and ET remain the most important items and there has been a focus on these as seen in the technical and scientific literature.

The selection of a pure WF fixes the operating pressures of the Rankine cycle, once the temperatures of the two sources are known: pressure ratio is, therefore, fixed for the pump and the expander. Reference literature for the WF can be found in (3-7): multiple fluids are available and have been tested in literature but when ozone depletion Potential (ODP) and Global Warming Potential (GWP) constraints are included, few of them remain effectively available.

Figure 1 shows ODP vs. GWP for the multitude of WF considered: GWP is the main concern, being the fluid usually considered conceived from the original high ODP.

The most important aspect that today limits a widespread application of low temperature ORCs is the available Expander Technology. Of all the possible machines that have been used and tested, in the power range of dozen to a few hundred kW, rotary volumetric machines are more suitable (8) than turbines or reciprocating types even though both have been investigated and tested.

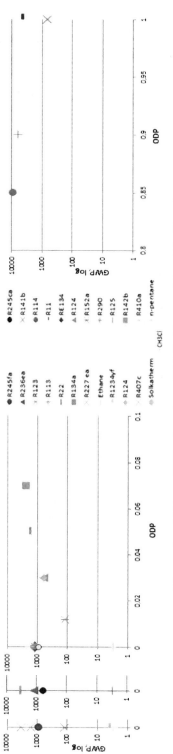

Figure 1 – ODP vs. GWP of various ORC fluids considered in literature

Amongst volumetric rotary machines, scroll (10,11,12), gerotor (13), rotary vanes (14,15) and screw machines (16, 17) have been tested; a dominance for the scroll type must be recognized. Isentropic efficiencies of these expanders are in the range of: 0,45-0.85 for gerotors, 0.3-0.8 for scrolls and 0.5-0.85 for rotary vanes, 05-0.8 for screw expanders.

The differences depend on the power delivered by the machine but they are below 10 kW. A higher power has been reached for screw expanders, up to 30 kW and also above (50 kW): scroll and rotary vanes recently were more referenced, being the screw type characterized by an evident greater complexity.

Rotary vanes machines, from this point of view, are more critical and require a specific know how: so, their potential seems to be more interesting.

In this paper the Authors present a theoretical and experimental activity concerning the design of a sliding vane rotary expander (SVRE) making use of a proven and referenced experience at designing compressors. The design, from the beginning, has been supported by a mathematical model which is under construction as a virtual design platform for expanders. It has been developed thanks to a previous activity which revealed very useful for compressors, (18, 19, 20). The model predicts main expander performances and blade dynamics and involves a friction model: in this paper it has been experimentally validated to behave as a theoretical support for design optimization.

An experimental test bench has been built in which an expander has been tested fed by an organic fluid (R236fa) which operates reproducing a very low maximum temperature Rankine cycle. The testing activity includes a measurement of the pressure inside the vane: a set of piezoelectric transducers allowed for the reconstruction of the p-V cycle and indicated work has been evaluated.

1. MODEL DESCRIPTION

1.1. Geometry

The geometry of the vane has been reproduced according to a basic representation which can consider inclined blades (forward & backward) with respect to the rotor radial direction. Intake and exhaust are represented in their real configuration: their angular displacement represents the most important datum, being responsible of the correct built in volume ratio. The geometry representation is done in a simple parametric way and this allows a very simple modification aiming at an iterative performance optimization.

1.2. Closed volume transformations

A comprehensive model has been implemented able to predict pressure and temperature inside the cell. It calculates:
a) flow rates entering and exiting the cell. The model makes use of a 1D unsteady formulation of the mass and momentum conservation equations: a suitable solution has been developed by one of the Authors and it is called QPM (Quasi Propagatory Model): it reproduces, in a lumped parameter form, the unsteady phenomena occurring when fluid is exchanged between plenums whose boundary conditions characterize their behaviour (plenums of infinite or finite capacity, throttled and un-throttled, static pressure conditions, etc...) particularly intense during discharge. Reference concerning QPM can be found in (20);
b) thermodynamic properties inside the cell (during filling and emptying) and during the expansion process (closed volume): the calculation is done by solving in an explicit way the energy equation. Properties are considered uniform inside the cell and leakages between adjacent vanes can be considered. WF and, eventually, oil (in liquid and vapour phases) are considered as ideal mixtures.

Convection (forced and natural) between WF and stator, and stator and environment are calculated according to known heat transfer convective equations. The presence of the oil usually injected by means of an external pump can be taken into account concerning spray formation and correspondent heat transfer (21).

The model can consider any gases as working fluid as well as vapours (superheated) of organic fluids (OF) or steam.

1.3. Blade dynamics

Blade dynamics represents one of the most critical processes because of the prediction of the mechanical power lost by friction.

During its motion, the blade rotates with the rotor and slides along the slot. This leads to inertia and centrifugal forces as well as the Coriolis force.

At the same time the pressure difference between the cells that the blade separates, generates a side force that acts on the surface outside the slot and tends to tilt the blade in it. Thus, the interactions between the blade and the metallic surfaces of the machine happen in three points, namely W at the blade tip and the stator, and the points 1 and 2 with the side walls of the slot (Figure 2). Moreover, since the materials in contact are the same in all three points, a unique value for the friction coefficient has been assumed.

The unknown forces between the blade and the casing represented in Figure 2, are solved thanks to the Newton's laws (forces along radial and circumferential directions and momentum balance).

Four configurations are possible for the blade positioning inside the slot. Depending on the signs of the forces F_1 and F_2 according to Table 1, the blade can be tilted or pushed on the slot walls, either forward or backwards (Figure 2).

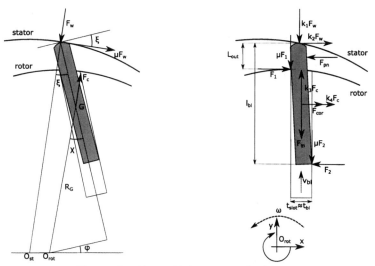

Figure 2 – Force Balance on the blade

Table 1 – Blade configurations with respect to Figure 2

	1-Hard backward	2-Tilt backward	3-Hard forward	4-Tilt forward
F_1	< 0	< 0	> 0	> 0
F_2	> 0	< 0	< 0	> 0

2. EXPERIMENTAL SET UP

An experimental setup has been built in order to evaluate:
a) the expander performances;
b) the pressure inside the vane during rotation in order to evaluate the indicated work;
c) the ORC plant performances using $CF_3CH_2CF_3$, 1,1,1,3,3,3-hexafluoropropane as WF (MW = 152 kg/kmol; T_c = 124.9 °C, P_c = 3.2 MPa, ρ_c = 551.3 kg/m3).

Main specifications of the expander designed are reported in Table 2.

Table 2 – Expander data

Stator diameter	76 mm	Intake port opening	4.4°
Rotor diameter	65 mm	Intake port closing	49.4°
Eccentricity	5.5 mm	Exhaust port opening	200.1°
Number of cells	7	Exhaust port closing	317.4°

Figure 3 shows the ORC plant: main components are indicated. The high temperature source is represented by hot oil (coming out from a compressor) which has to be cooled before reentering the machine.

In Figure 3 a sketch of the ORC plant is shown.

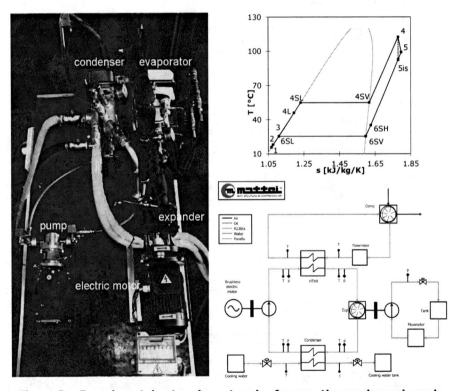

Figure 3 – Experimental setup, layout and reference thermodynamic cycle

This apparatus was used to investigate the main performance of the Mattei-ORC. The system essentially consists of a test bench supporting all the mechanical components and the instrumentation directly involved in the process. The schematic diagram of the proposed ORC system is also displayed. The experimental setup was assembled using both commercial and ad-hoc designed items: the pump and expander used were purpose designed and produced by Ing. Enea Mattei SpA. The main components of the system include four units: (1) *Circulation pump*: a small rotary vane pump coupled with a brushless electric motor was used to circulate the working fluid. The pump is able to circulate a variable volume flow rate (up to 12 l/min at 3000 RPM) of working fluid (2); *Evaporator*: the working fluid was heated in a plate-type heat exchanger, using the hot lubricant from an air compressor as hot source. A flow meter was used to measure the hot oil volume flow rate; (3) *Expander*: a rotary vane expander was used for the working gas expansion: its main geometrical characteristics are reported in Table 2. The expander was directly coupled with an electric motor/alternator operating at 1500 RPM whose electrical efficiency was known in order to determine the mechanical power. (4) *Condenser*: Tap water, which was made to flow into the condenser and out to the drain, was used as cold source. Plate-type heat exchanger was used as the condenser. In order to improve sealing in the expander a 5 % of oil was added to the working fluid; the oil is entrained by the fluid without any separation.

3. EXPERIMENTAL ACTIVITY

An extensive experimental activity has been done in order to investigate main performances of the expander and of the plant. Table 3 summarizes the tested conditions.

Table 3 – Experimental activity

		S1	S2	S3	S4	S5	S6	S7	S8	S9	S10	S11	S12	S13	S14
T [°C]	1	24.2	25.9	23.6	22.3	20.4	20.2	25.7	18.9	17.5	16.5	15.7	17.8	29.5	19.8
	2	25	26.8	24.7	23.2	21.4	21.2	26.5	20	18.8	18.3	18	19.9	30.4	20.5
	4	90.6	96.1	98.3	100.1	106.4	101.6	97.3	104.5	107.6	110.8	112.4	71.7	90.5	78.2
	5	69.4	75.9	79.3	81.3	87.2	84	77	87.5	91.1	94.7	99	57.4	68.5	59.9
p [bar_a]	1	3.4	3.36	3.32	3.28	3.23	3.34	3.66	3.21	3.06	2.97	2.84	2.7	3.67	3.33
	2	12.43	12.7	12.1	11.41	10.8	10.6	13.4	9.92	8.76	7.94	6.69	7.05	12.4	3.37
	4	12.11	12.4	11.8	11.18	10.6	10.4	13.2	9.75	8.59	7.79	6.9	7.11	12.5	3.31
	5	4.58	4.57	4.37	4.15	3.94	3.95	4.78	3.68	3.38	3.18	2.76	3.16	3.82	8.4

The lower cycle pressure is around 3 bar while maximum pressure around 12 bar. At this pressure the vapour is slightly superheated at the expander inlet in order to reduce condensation when the WF touches the metallic walls. Mean maximum temperature of the WF is around 110 °C. In fact, maximum temperature is limited by the source temperature which is represented by the oil exiting a compressor. Figure 4 shows the results in terms of mechanical power recovered on the expander shaft.

Data are referred to 13 testing conditions explored: the x-axis represents the heat transferred to the ORC plant at the evaporator: it ranges from 10 kW to 20 kW. This information allows to associate specific testing conditions to specific hot source thermal capacity (oil from air compressor of various installed powers). The dimension of the circles is proportional to the overall expander efficiency, defined on the base of the adiabatic isentropic expansion work evaluated when the expander inlet conditions are known.

Data reported in Figure 4 can be rearranged in order to evaluate plant efficiency which, definitively, represents the most important datum. Figure 5 summarises this information. A net overall efficiency in the range of 6.5 % - 7.5 % has been achieved which still represent a good result considering the low temperature range of the ORC and the low mechanical power produced.

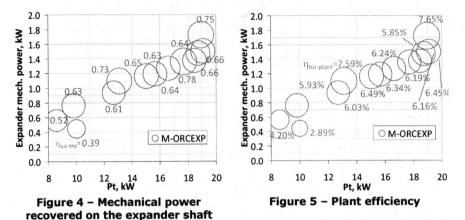

Figure 4 – Mechanical power recovered on the expander shaft

Figure 5 – Plant efficiency

A subsequent mixed theoretical and experimental activity has been done in order to understand how the overall performances could be increased. This was done "mixing" the results from the mathematical model and those coming out from direct pressure measurements inside the cell.

A sequence of three fast response piezoelectric pressure transducers were installed on the "lateral cover" of the expander as shown in Figure 6. The sensors positioning allowed the reconstruction of the pressure evolution inside the cell, from the suction process to the exhaust. Transducers were of the Kistler type 601A with charge amplifier 5064B21.

The ambient pressure has been used as a reference for all the piezoelectric transducers. Therefore, an accurate reconstruction was needed to link the three raw signals: this was done thanks to a given overlapping angular interval of the three sensors. The final experimental pV, together with the numerical predictions of the mathematical modeling for test S13, is represented in Figure 7. From the measured data, the indicated power has been evaluated according to equation (1).

$$P_{ind} = \frac{n_{cells} \oint p \, dV}{t_{cycle}} \quad (1)$$

$$\eta_{vol} = \frac{\dot{m}_{real}}{\dot{m}_{theor}} \quad (2)$$

$$P_{fr} = \mu \left(F_w \omega R_Y + F_1 v_1 + F_2 v_2 \right) \quad (3)$$

$$\mu = \frac{P_{ind} - P_{sh}}{F_w \omega R_Y + F_1 v_1 + F_2 v_2} \quad (4)$$

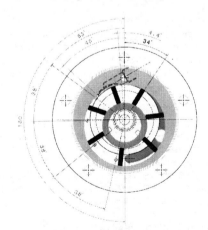

Figure 6 – Sensor positioning for p-V data

The following consideration applies in order to correlate friction ad shaft power as well as volumetric efficiency. As it is known, in positive displacement machines, the estimation of volumetric efficiency is very complex. The indicated power – as it comes from a direct pressure measurement – can be calculated integrating the pressure plot according to Equation (1). It takes into account the real mass expanding inside the cell which differs from the intake flow rate due to internal volumetric losses: part of the intake flow rate fills the cell, the remaining part crosses the machine without producing work. Volumetric efficiency is, therefore, given by Equation (2). Its evaluation is outside of the aims of this paper and requires a complex procedure which is under development and based on the pressure data measured and geometrical characteristics of the machine. The method is based on the conservation of energy during vane discharge and filling and the state equation solved in a way which matches the pressure experimentally measured.

The power lost due to friction is reported in Equation (3) where a mean equivalent friction coefficient can be identified by means of Equation (4) where shaft power (P_{sh}) is known and indicated power P_{ind} is evaluated according to Equation (1). Reaction forces have been calculated by the model solving the equilibrium of the forces and angular momentum (Figure 2).

Figure 7 shows for test number S13, the p-V data as the result of measurements and theoretical predictions.

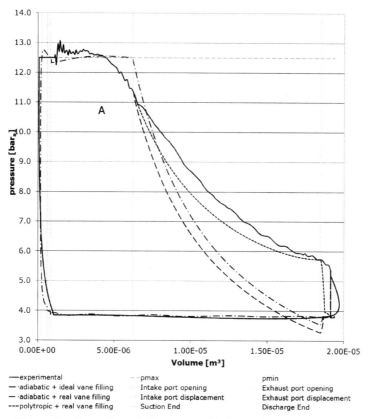

Figure 7 – Experimental vs numerical p-V data (test #S13)

Data allow for the following considerations:
a) During the intake phase, after an initial filling done more or less at constant pressure, the pressure inside the cell starts to decrease when the port is still opened. A correct vane filling (at constant pressure) would require the matching between the flow rate coming from the pump and the one requested by the volume increase of the vane itself (inducted by the expander). In the present testing conditions, the contribution from the pump is lower than the mass flow requested by the vane for a proper filling: this results in a pressure decrease inside the vane. The volumetric flow rate of the machine is a little bit greater than that strictly requested. Part of the potential work is lost with respect to an intake phase done at constant pressure;
b) Following the experimental data, a pressure mismatching when the vane opens toward the exhaust with the exhaust pressure is evident. Since the pressure inside the cell is higher than the one in the discharge plenum, a sudden expansion, theoretically isochoric, balances the gap. However, the finite capacity of the discharge plenum, makes the expansion non instantaneous (strictly isochoric). When the pressure reached the plenum datum, the discharge follows at constant pressure. Part of the work is lost with respect to the case in which the expansion inside the vane would continue to the discharge plenum datum: this would require a greater available volume inside the cell which doesn't match with machine geometry;
c) the mathematical modeling shows different predictions according to the initial pressure inside the vane, at the intake port closing and according to the thermodynamic transformation when the vane is closed. The dash-dotted line corresponds to an adiabatic expansion in which the vane filling extends up to the geometrical suction end, i.e. when the second blade of the vane referred to the direction of rotation closes the intake port. Assuming as initial pressure the value at point A, Figure 7 reports an adiabatic and a polytropic transformation. The former would lead the pressure at a value much smaller than the real one: it is clearly evident that such a transformation does not represent real expansion process which, on the contrary, has been definitively observed for air compressors [19]. Indeed, a partial condensation due to the contact between the WF and the metal surfaces can't be avoided, even in presence of a good thermal insulation. This transfers to the fluid inside the cell a portion of the latent heat of condensation, with a polytropic coefficient equal to 0.63 (from the adiabatic value equal to 1.15) which matches real pressure at the discharge port opening. It would correspond to 473.7 W transferred to the WF.

According to this test, indicated work is 2054.5 W. With a friction coefficient equal to 0.19 (evaluated from equation 4), the power lost by friction is 333.4 W and the shaft power equal to 1721.1 W. The datum concerning indicated work is predicted by a polytropic transformation (1727.9 W) with an error close to 16 %.

Overall plant efficiency, including power absorbed by the pump, is equal to 7.65 %, recovering from the high temperature source almost 19 kW thermal.

The result is really interesting considering the low grade of the high temperature source. Considering a potential scalability, for a 100kW air compressor a net mechanical power recovered would be in the range of 7,65kW. An equivalent decrease in specific energy is to be expected.

4. CONCLUSIONS

A sliding vane rotary expander has been designed and tested in order to recover low grade thermal power from waste heat sources, represented in this paper by the hot oil circulating in an industrial air compressor.

The design was done thanks to a comprehensive mathematical model specifically conceived for representing real processes. The model validation required a custom built experimental test bench with a set of piezoelectric transducers mounted on the expander in order to measure p-V data in real operation. The expander operated with a dry organic fluid ($CF_3CH_2CF_3$) entering slightly superheated inside the machine and performing the expansion in a full dry vapor phase.

A wide testing activity has been done and reliability of the expander demonstrated. With a hot source at 110°C and 19kW thermal, the recovered mechanical power was 1.7 kW with an overall plant efficiency equal to 7.65 %.

The theoretical analysis and the experimental data demonstrated how to further improve expander technology via simple design implementations which are under development. Preliminary prototypes demonstrate the capability to reach overall power plant efficiencies of 10 %.

It is the authors' opinion that with the known advantages of the sliding vane technology, i.e. efficiency, reliability, scalability, there is a potential for the development of a range of SVRE's applicable to any compressed air installation.

REFERENCES

1. Lopes, J., Douglas, R. et al. Review of Rankine Cycle System Components for hybrid Engines Waste Heat Recovery, SAE Paper 2012-01-1942.
2. Quoilin, S., & Lemort, V. Technological and Economical Survey of Organic Rankine Cycle Systems. 5th European Conference on Economics and Management of Energy in Industry, Vilamoura, 2009.
3. Wang, E. H., Zhang, H. G., Fan, B. Y., Ouyang, M. G., Zhao, Y., & Mu, Study of working fluid selection of organic Rankine cycle (ORC) for engine waste heat recovery. Energy, 36(5), 3406–3418, 2012.
4. Lakew, A. A., & Bolland, O. 2010. Working fluids for low-temperature heat source. Applied Thermal Engineering,30(10), 1262–1268.
5. Maizza, V., & Maizza, A. 2001. Unconventional working fluids in organic Rankine-cycles for waste energy recovery systems. Applied Thermal Engineering, 21(3), 381–390.
6. Angelino, G.; Colonna, P. Multi component working fluids for Organic Rankine Cycles (ORCs), Energy 1997; 23, 449-63.
7. Quoilin, S., Declaye, S., Legros, A., Guillaume, L., Lemort, V. 2008. Working fluid selection and operating maps for Organic Rankine Cycle expansion machines International Compressor Engineering Conference at Purdue, July 16-19, Paper n. 1546, 1-12, 2012.
8. Mikielewicz, D., et al. Delft, NL, Experiences from operation of different expansion devices in domestic micro-ORC., Gdansk University of Technology, 2011.
9. Yamamoto, Takahisa, et al. s.l. Design and Testing of the Rankine Cycle, Elsevier, 2001. Energy 26 (2001)239-251.
10. Peterson, B R, Wang, H and Herron, T. Performance of a small-scale regenerative Rankine power cycle employing a scroll expander, IMechE, 2008. Proc. IMechE Vol. 222 Part A: J. Power and Energy.
11. Kim, H J, et al. Scroll expander for power generation from a low-grade steam source. IMechE, 2007. Proc. IMechE Vol. 221 Part A: J. Power and Energy.
12. Lemort, V., Quoilin, S., Cuevas, C., & Lebrun, J. 2009. Testing and modeling a scroll expander integrated into an Organic Rankine Cycle. Applied Thermal Engineering, 29(14-15), 3094–3102.
13. Mathias, James A., et al. Experimental Testing of Gerotor and Scroll Expanders Used in, and Energetic and Exergetic Modeling of, an Organic Rankine Cycle. s.l.: Journal of Energy Resources Technology, 2009. Vol. 131 / 012201-1.

14. Tahir, Musthafah b. Mohd., Yamada, Noboru and Hoshino, Tetsuya. Efficiency of Compact Organic Rankine Cycle System with Rotary-Vane-Type Expander for Low-Temperature Waste Heat Recovery, International Journal of Civil and Environmental Engineering, 2010. 2:1 2010.
15. Experimental investigation on the internal working process of a CO2 rotary vane expander. Yang, B, et al., Elsevier, 2009. Applied Thermal Engineering 29 (2009) 2289-2296.
16. J J Brasz, I K Smith and N Stosic, Development of a twin screw expressor as a throttle valve replacement for water cooled chillers, International Compressor Engineering Conference Purdue University, Purdue, USA, 2000.
17. I K Smith, N Stosic, A Kovacevic, R Langson Cost effective small scale ORC systems for power recovery from low enthalpy geothermal resources Proceedings of IMECE, 2006 ASME International Mechanical Engineering Congress and Exposition November 5-10, 2006, Chicago, Illinois, USA.
18. Cipollone R., Contaldi G., Villante C., Tufano R., *A theoretical model and experimental validation of a sliding vane rotary compressor*, 18th International Compressor Engineering Conference at Purdue, July 17-20, 2006.
19. Cipollone R., Contaldi G., Capoferri A., Valente R., *Theoretical and experimental study of the p-V diagram for a sliding vane rotary compressor*, ImechE International conference on compressors and their systems, 2009.
20. Cipollone R., A. Sciarretta, The Quasi-Propagatory Model: a New Approach for Describing Transient Phenomena in Engine Manifolds, SAE Technical paper no. 2001-01-0579.
21. Cipollone, R., Bianchi, G., G. Contaldi, Sliding vane rotary compressor Energy optimization, Proceedings of the ASME 2012 International Mechanical Engineering Congress & Exposition, November 9-15, 2012, Houston, Texas, USA, IMECE2012-85955.

SCREW COMPRESSORS

Calculation of discharge pressure pulsations of a screw compressor using the one-dimensional method of characteristics

A Linkamp, A Brümmer
Chair of Fluidics, Department of Mechanical Engineering,
TU Dortmund University, Germany

ABSTRACT

This paper presents a method to simulate the discharge process of screw compressors. The calculation is based on the method of characteristics. The main focus is the transient, compressible, dissipative flow through the compressor discharge port. The time-dependent velocity at the outlet is calculated using a circular orifice with time-dependent diameter and is initially validated by an exemplary screw compressor. The one-dimensional model for the working chamber is presented and the transient flow on the discharge side is calculated for different discharge pressures. The calculated results are compared to experimental data.

NOTATION

A	[m²]	area	a	[m/s]	speed of sound	
c	[m/s]	velocity	d	[m]	diameter	
f_s	[Hz]	sampling frequency	f_{pp}	[Hz]	pocket pass frequency	
f_x	[m/s²]	specific frictional force	h	[J/kg]	specific enthalpy	
i	[-]	index of spatial discretisation	l_{seg}	[m]	segmental acoustical length	
l_0	[m]	initial length	$\dot{m}$	[kg/s]	mass flow	
n	[-]	index of temporal discretisation	n_{mr}	[min⁻¹]	rotational speed of the male rotor	
p	[Pa]	pressure	R	[J/kg K]	specific gas constant	
T	[K]	temperature	u	[J/kg]	specific internal energy	
v_i	[-]	inner volume ratio	V	[m³]	volume	
z	[-]	number of lobes	ρ	[kg/m³]	density	
ζ	[-]	loss coefficient	Π	[-]	pressure ratio	

1. INTRODUCTION

Pressure pulsations of screw compressors are an important criterion for the evaluation of operating behaviour. High pressure pulsations are not only a sign of high power dissipation due to over-compression or under-compression in the working chamber of the machine, but are also the main cause of noise and vibrations of an attached system (1). This does not only concern the environment and industrial safety, but also the operating safety of systems in which screw compressors are used. Main factors influencing induced pressure pulsations are geometrical parameters like the rotor profile, the number of lobes, the wrap angle and the gap situation on the one hand. On the other hand the operational parameters rotational speed and the difference between chamber pressure and discharge pressure have a significant influence on discharge pressure pulsations (2-5).

In order to be able to predict the acoustic behaviour of screw compressors in the case of variations in the geometry or operating state, it makes good sense to develop reliable numerical methods, using which the induced pulsations can be simulated as a function of the geometry of a compressor and of operating parameters. Different approaches for calculating the transient operating behaviour of rotary displacement machines have been implemented so far. Mujic et al. present a zero-dimensional model. Comparisons with experimental data show a sound consistency with the simulations (4). Stosic et al. (6) and Huster (7) developed one-dimensional models for pipings connected to displacement machines. In both models, however, the working chamber is modelled zero-dimensional and thus simulated thermodynamically only. A comparison of an entirely zero-dimensional and the partly one-dimensional model is made, whereby, for transient considerations, the assumption of constantly applied system pressures in the case of zero-dimensional approaches proves doubtful (7). It stands to reason to assume that one-dimensional approaches are closer-to-reality than zero-dimensional models due in particular to the consideration of the conservation of momentum. A completely three-dimensional simulation of the flows in screw compressors is also possible today (8). A three-dimensional calculation demands very much computational effort however. Meshing the complex geometry of the working chambers of screw machines is particularly time-consuming, intensified by the constantly changing flow domain (9).

In this paper, a method for the one-dimensional modelling and simulation of screw compressors in combination with a surrounding piping is presented. The aim of the presented method is to obtain as accurate a prediction as possible regarding the excitation of pressure and flow pulsations by screw compressors, while keeping the modelling effort and the computational effort with regard to an extensive geometry variation within reasonable limits. The main part of this work is the one-dimensional modelling of the working chamber as well as the method for calculating the transient, compressible and dissipative flow through the outlet port with time-dependent area. The method of characteristics is regarded as being suitable for this problem because the resulting set of linear equations is solved exactly, thus no iterative errors occur except in the calculation of the flow through the outlet port. Generally, the newly developed method for calculating the flow through a time-dependent opening can be implemented in any one-dimensional CFD-method.

2. MODELLING

2.1 Calculation of pulsating flow with *DINO*
All calculations were carried out with the software *DINO* available at the Chair of Fluidics. This is based on the method of characteristics and has so far been used primarily for the calculation of transient-flow pipeline systems and reciprocating piston compressors. The application of the method of characteristics is based on the work of Schweinfurter (10). The resulting equations, which will be derived in the following, are extended by a transient form of the conservation of energy and applied to gas flow as opposed to liquid flow in Schweinfurter's work (10).

2.1.1 Basic equations
The conservation equations in differential form are the basis for calculating the flow and state variables velocity c, pressure p, density ρ and temperature T along segments with constant diameter d_{seg} (direction x, time t). The one-dimensional, transient form of the conservation of mass reads

$$\frac{\partial \rho}{\partial t} + c \cdot \frac{\partial \rho}{\partial x} + \rho \cdot \frac{\partial c}{\partial x} = 0. \tag{1}$$

The conservation of momentum in the compressible, one-dimensional, transient case results in

$$\frac{\partial c}{\partial t} + c \cdot \frac{\partial c}{\partial x} + \frac{1}{\rho} \cdot \frac{\partial p}{\partial x} + f_x = 0 \tag{2}$$

Equation (2) implies the common assumption for gas flows that the influence of gravity is negligible. The gravitation term is therefore not included. The specific force f_x is a term for the pipe wall friction. As regards the conservation of energy, an adiabatic system is always assumed, meaning that the conservation of energy, assuming an ideal gas with $du = c_v \, dT$, can be formulated as follows:

$$\frac{\partial T}{\partial t} + c \cdot \frac{\partial T}{\partial x} + \frac{RT}{c_v} \cdot \frac{\partial c}{\partial x} = 0 \tag{3}$$

The equation of state is used as the fourth equation. This generally reads

$$p = \rho \cdot R \cdot T \tag{4}$$

Thus four equations are available for the calculation of four state or flow parameters. Using the definition for the speed of sound a

$$\frac{dp}{d\rho} = a^2 \tag{5}$$

equation (1) can be rearranged to give

$$\frac{\partial p}{\partial t} + c \cdot \frac{\partial p}{\partial x} + \rho \cdot a^2 \frac{\partial c}{\partial x} = 0 \tag{6}$$

The equations (2) and (6) thus result in a partial differential equation system for the state variables p and c. Building the normal form of this differential equation system results in two equations, which each contain a spatial and time derivation pair of the variables sought.

$$\frac{\partial p}{\partial t} + (c + a) \cdot \frac{\partial p}{\partial x} + \rho \cdot a \cdot \left(\frac{\partial c}{\partial t} + (c + a) \cdot \frac{\partial c}{\partial x}\right) + \rho \cdot a \cdot f_x = 0 \tag{7}$$

$$\frac{\partial p}{\partial t} + (c - a) \cdot \frac{\partial p}{\partial x} - \rho \cdot a \cdot \left(\frac{\partial c}{\partial t} + (c - a) \cdot \frac{\partial c}{\partial x}\right) - \rho \cdot a \cdot f_x = 0 \tag{8}$$

Equation (7) and equation (8) show

$$\frac{dp}{dt} + \rho \cdot a \cdot \frac{dc}{dt} + \rho \cdot a \cdot f_x = 0 \quad \text{for} \quad \left.\frac{dx}{dt}\right|_r = c + a \quad \text{(right-running)} \tag{9}$$

$$\frac{dp}{dt} - \rho \cdot a \cdot \frac{dc}{dt} - \rho \cdot a \cdot f_x = 0 \quad \text{for} \quad \left.\frac{dx}{dt}\right|_l = c - a \quad \text{(left-running)} \tag{10}$$

The ordinary differential equation system (9-10) is from now on only valid along the characteristic directions of the derivations. These are called characteristics and read, for the one-dimensional flow problem dealt with here,

$$\left.\frac{dx}{dt}\right|_r = c + a \tag{11}$$

$$\left.\frac{dx}{dt}\right|_l = c - a \tag{12}$$

The physical significance of the characteristics lies in the propagation of information, i.e. of changes in the gas state, in the one-dimensional flow domain. The information propagation takes place along the characteristics, at the speed (c+a) in the direction of flow or at the speed (c-a) against the direction of flow.

2.1.2 Numerical solution of the ordinary differential equation system

For a numerical solution, the differentials are transformed into differences. To start with, the characteristics are linearised in sections on the assumption of a sufficient spatial and temporal resolution. This means that the gradient of the characteristics (Eq. (11) and (12)) within a space (Δx) or time interval (Δt) is regarded as constant. Thus the following results:

$$\left.\frac{dx}{dt}\right|_r \approx \left.\frac{\Delta x}{\Delta t}\right|_r = c + a \tag{13}$$

$$\left.\frac{dx}{dt}\right|_l \approx \left.\frac{\Delta x}{\Delta t}\right|_l = c - a \tag{14}$$

The computational domain is spatially discretised in a single dimension, meaning that a two-dimensional space-time grid results through the temporal discretisation. The variables sought are to be calculated on the discrete points of this grid (see Fig. 1). Based on the known state variables at time n, the variables at time n+1 at point i are now calculated. For this purpose, equations (9) and (10) are also linearised and the differential quotients replaced with difference quotients. In line with the grid shown in Fig. 1, these can be formulated for the right running characteristic (c+a) based on point A as follows:

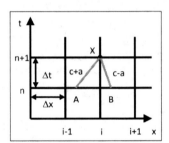

Fig. 1: Time-space-grid

$$\frac{dp}{dt} \approx \frac{\Delta p}{\Delta t} = \frac{p_X - p_A}{\Delta t} \tag{15}$$

$$\frac{dc}{dt} \approx \frac{\Delta c}{\Delta t} = \frac{c_X - c_A}{\Delta t} \tag{16}$$

The difference quotients for the left running characteristic (c-a) are analogously formed in relation to point B. By inserting in equation (9) and (10) respectively, the linear equation system (17-18) for the state variables p and c at point X results.

$$(p_X - p_A) + \rho \cdot a \cdot (c_X - c_A) + \rho \cdot a \cdot f_x \cdot \Delta t = 0 \quad \text{for} \quad \left.\frac{\Delta x}{\Delta t}\right|_r = c + a \tag{17}$$

$$(p_B - p_X) + \rho \cdot a \cdot (c_X - c_B) + \rho \cdot a \cdot f_x \cdot \Delta t = 0 \quad \text{for} \quad \left.\frac{\Delta x}{\Delta t}\right|_l = c - a \tag{18}$$

Thus the pressure p_x and the velocity c_x at point X can be calculated. The known value from the nodal point (n,i) is used for the impedance $(\rho a)_i^n$. Moreover, the intersections A and B are calculated through linear interpolation of the state variables at the nodal points (n,i-1), (n,i) and (n,i+1) (10). For calculating the specific frictional force f_x, the following functional is used:

$$f_x = \frac{\lambda}{2d_{seg}} c_i^n |c_i^n| \tag{19}$$

In order to calculate the temperature T, the energy equation is also linearised. For this purpose the differentials are in turn transformed into quotients of finite differences. Based on the time-space-grid (Fig. 1), the following results for a positive direction of flow.

$$\frac{dT}{dt} \approx \frac{\Delta T}{\Delta t} = \frac{T_i^{n+1}-T_i^n}{\Delta t} \tag{20}$$

$$\frac{dT}{dx} \approx \frac{\Delta T}{\Delta x} = \frac{T_i^n-T_{i-1}^n}{\Delta t} \tag{21}$$

With equations (20-21), the following linear relationship follows from equation (3) for the temperatures in the time-space-grid as a function of the positive flow velocity c.

$$\frac{T_i^{n+1}-T_i^n}{\Delta t} + c_i^n \cdot \frac{T_i^n-T_{i-1}^n}{\Delta x} + \frac{R \cdot T_i^n}{c_v} \cdot \frac{c_i^n-c_{i-1}^n}{\Delta x} = 0 \tag{22}$$

Subsequently, the density ρ can be determined via the equation of state (4), meaning the state at point X at time (n+1) is fully known.

2.2 Modelling of a screw compressor

The fundamental elements of a screw compressor, which are to be modelled in order to simulate the excitation of discharge pressure pulsations, are the working chamber and the outlet area. To calculate the pulsation excitation as accurate as possible, a method for calculating the dissipative, compressible and transient flow was developed. In order to take account of the flow losses in the convergent and the divergent part of the outlet flow separately, the flow and state variables in the narrowest cross-section are calculated explicitly. By doing so, an accurate prediction of the speed of sound and thus the occurance of critical or subcritical flow is obtained, which is an important phenomenon in the discharge process of screw compressors. The working chamber is modelled one-dimensional, meaning that the model is entirely one-dimensional. By doing so, transient gas-dynamic effects inside the working chamber are taken into account as well. The volume curve and outlet area curve used later on for the simulation of a screw compressor are determined by means of the geometry analysis software KaSim Pre developed at the Chair of Fluidics (11).

2.2.1 Modelling of the working chamber

Due to the one-dimensionality, the working chamber is modelled as a cylindrical element with a constant diameter and variable length (see Fig. 2). The initial length of the working chamber l_0 is determined in accordance with the difference between the crown and root circle diameter, $d_{c,m}$ and $d_{r,m}$ respectively, of the male rotor.

$$l_0 = \frac{d_{c,m}-d_{r,m}}{2} \tag{23}$$

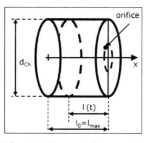

Fig. 2: Geometric model of the working chamber

The acoustic length of the working chamber thus lies in the order of magnitude of the profile depth of the male rotor. In the case of primarily radial discharge, this is regarded as a suitable abstraction of the geometrically complex working chamber of a screw machine. The diameter of the model d_{ch} is determined from the initial length l_0 and the maximum working chamber volume $V_{ch,max}$. This diameter always remains constant.

$$d_{ch} = const = \sqrt{\frac{4}{\pi}V_{ch,max}/l_0} \tag{24}$$

The time-dependent length of the working chamber $l_{ch}(t)$ results from the quotient of current volume $V_{ch}(t)$ and the constant "piston"-area A_{ch}. The working chamber is rediscretised in each time step (adaptive grid) and the state variables are then converted from the old grid to the new grid. The "piston"-velocity $c_{ch}(t)$ determined from the derivation of the volume curve $V_{ch}(t)$ and the constant "piston"-area A_{Ch} is defined as a boundary condition on the "left side" of the segment:

$$c_{ch}(t) = \frac{dV_{ch}}{dt} \cdot \frac{1}{A_{ch}} = \frac{dV_{ch}}{dt} \cdot \frac{4}{\pi d_{ch}^2} \tag{25}$$

The boundary condition on the right side of the working chamber is an orifice with a time-dependent diameter, the calculation of which is described in more detail later on.

2.2.2 Modelling of the flow through the outlet area

The flow through the outlet area is modelled as flow through a circular orifice with a time-dependent diameter. The state variables are calculated in three spatial points: the last grid point of the "left" segment (1), the point of the narrowest cross-section (O) and the first point of the computational grid in the "right" segment (2). The spatial expansion between points 1 and 2 remains unconsidered in the simulation. Moreover, a homogeneous flow state is still assumed for the derivation of the transition conditions in each cross-section. This assumption is only deviated from on the outflow side of the orifice in the case of a supercritical expansion.

For the clear determination of the states at the three points 1, O and 2, each of the state and flow parameters p, T, ρ, c are needed. Twelve equations are required for these twelve variables accordingly. The known form of the transient energy equation (22) for the realisation of the energy transfer through the orifice, is applied to point 1. Moreover, the equation for the right running characteristic (17) must be satisfied for this point. The equation for the left running characteristic (18) applies analogously for point 2. By way of simplification, the characteristic equations are formulated without pipe wall friction for the calculation of the orifice flow, meaning the frictionless equations (26) and (27) result.

$$(p_X - p_A) + \rho \cdot a \cdot (c_X - c_A) = 0 \qquad \text{for } \frac{\Delta x}{\Delta t}\bigg|_r = c + a \tag{26}$$

$$(p_B - p_X) + \rho \cdot a \cdot (c_X - c_B) = 0 \qquad \text{for } \frac{\Delta x}{\Delta t}\bigg|_l = c - a \tag{27}$$

Between the points 1, O and 2 the fluidic conservation equations must always be satisfied. These are formulated in the integral form, disregarding a storing effect of the fictitious volume between 1 and 2. The conservation of mass thus results in

$$\dot{m} = \rho \cdot A \cdot c = const \tag{28}$$

The loss coefficient of the orifice is to be explicitly given. Therefore, a loss coefficient for the convergent part of the flow up to the narrowest cross-section $\zeta_{O,in}$ is used instead of the conservation of momentum for calculating the total pressure loss between point 1 and point O $\Delta p_{t,1 \to O}$. It is related to the dynamic pressure in the orifice for the case of an isentropic change of state between points 1 and O.

$$\zeta_{O,in} = \frac{p_{t1} - p_{tO}}{(p_{tO} - p_O)_s} \tag{29}$$

If relating the loss coefficient $\zeta_{O,in}$ to the friction-affected dynamic pressure in the orifice, there is no clear solution for the mass flow in the case of compressible flow in the transonic flow range. Relating to the isentropic dynamic pressure supplies a clear solution and is thus better suited for the flow problem at hand. The conservation of energy can be formulated with the total temperature T_t.

$$T_t = \frac{h_t}{c_p} = T + \frac{c^2}{2 \cdot c_p} = const. \tag{30}$$

Since the flow is assumed as adiabatic, the total temperature T_t remains constant. Since the total variables p_t and T_t associated with the static state variables p and T are calculated from an isentropic change of state, the static temperature at point O can be determined from the total temperature using the isentropic equation (31). Moreover, the equation of state, equation (4), is in turn available at point O.

$$T_O = T_{tO}\left(\frac{p_O}{p_{tO}}\right)^{\frac{\kappa-1}{\kappa}} = T_{t1}\left(\frac{p_O}{p_{tO}}\right)^{\frac{\kappa-1}{\kappa}} \tag{31}$$

For calculating the state at point 2, the energy equation (30) can again be used. Of course, the conservation of mass (28) must also be satisfied from point O to point 2 and the equation of state (4) is also available. Furthermore, the conservation of momentum can be established in integral form for subcritical flow in the narrowest cross-section:

$$\dot{m} \cdot c_2 - \dot{m} \cdot c_O = (p_O - p_2) \cdot A_2 \tag{32}$$

Both in the orifice cross-section O and in the cross-section 2, no pressure gradient arises perpendicularly to the flow in the case of subcritical flow, since the flow has more or less straight streamlines in both cross-sections. Accordingly, both the pressure in the narrowest cross-section p_O and the pressure p_2 act over the entire area A_2 of the right-hand control volume. The pressure loss in the divergent part of the flow is implicitly set by the area ratio A_O/A_2 through applying the conservation of momentum. This becomes clear when looking at the rearranged equation (32):

$$p_2 = p_O - \frac{A_O}{A_2} \cdot \rho_O \cdot c_O \cdot (c_2 - c_O) \tag{33}$$

Thus a total of twelve equations are available for the twelve unknowns (point 1: Energy equation, right running characteristic, equation of state; from point 1 (O) to O (2): Energy equation, total pressure loss, conservation of mass, equation of state; Point 2: left running characteristic). In the case of a supercritical expansion and thus of critical flow in the narrowest cross-section, equation (33) is no longer valid, since diagonal compression shocks arise at the orifice outlet and so the assumption of a constant pressure over the cross-section is no longer given. As soon as a solution for critical flow in the orifice is found, which satisfies the above-mentioned equations up to point O, the sought variables in the smallest cross-section depend exclusively on inlet state 1. In this case, the state at point 2 is determined via the conservation of energy (30) and mass (28) as well as the equation for the left running characteristic (27) and the equation of state (4).

2.2.3 Idealisations and boundary conditions

Concerning the simulated screw compressor all gap flows are ignored in the model examined here. This means that solely the flow through the outlet area and through temporary chamber connections via the outlet port is calculated. Since only the discharge pulsations are of interest, the chamber filling process is also not simulated. The working chambers are always initialised in closed state at the time of maximum chamber volume. The gas state at the time of initialisation corresponds to the assumed suction-side state. This approach corresponds to the assumption of a complete and non-throttled chamber filling. On the high pressure side of the compressor an anechoic end is modelled, which corresponds to a theoretically infinite pipeline. This means the pure excitation of pulsations through a compressor can be calculated first, independently of the piping system. For the simulations carried out, the loss coefficient for the inflow into the orifice $\zeta_{0,in}$ is set to zero. Due to the very small minimum size of the segment of the working chamber, the spatial discretisation during the simulation of the compressor lies in the order of magnitude of $\Delta x = 10^{-5}$ m.

In order to compare the simulation results with experimental results, the pipeline system of the test stand of the examined screw compressor is also simulated. The calculated periodical velocity fluctuation at the discharge control edges of the compressor is prescribed for the simulation as a boundary condition. Thus the system response to the pulsation excitation induced by the screw compressor is calculated. This separation was carried out due to the very different geometrical orders of magnitude of the working chamber and the piping. The separation of excitation and system response leads to reactions from the pipeline system to the flow processes within the compressor not being considered. In order to achieve sufficiently accurate results with reasonable computational effort, a grid study was carried out, which led to the result of a required spatial discretisation of $\Delta x = 6 \cdot 10^{-4}$ m. The piping, which is modelled and calculated up to a pressure tank, has a cumulative length of approximately 12 m.

The application of one-dimensional approaches and theories implies that solely plane waves are calculated. Multidimensional effects such as transverse modes cannot be taken into account. The average measured values of pressure and temperature on the suction and pressure side of the compressor are always given as boundary or initial conditions during the simulations.

3. EXPERIMENTAL PROCEDURE

For the validation of the 1D simulations, experimental investigations are carried out. The examined machine is a screw compressor with an asymmetric SRM profile with four lobes on the male rotor and six lobes on the female rotor. The inner volume ratio of the original machine amounted to $v_i = 2,2$. The outlet housing was modified however in order to realise a smaller inner volume ratio. For this purpose, the contour of the axial outlet area was cut to a depth of 3 mm in circumferential direction resulting in large front gaps coming up during the compression at the point of $v_i = 1,15$.

The medium suction pressure and the medium discharge pressure are measured with capacitive relative pressure transducers. Moreover, on the high and low pressure side, the temperature is measured with thermocouples. For the measurement of the pressure pulsations, three piezoelectric dynamic pressure sensors, which are axially offset from one another, are installed, which work with a sampling frequency of $f_s = 50$ kHz. The dynamic pressure transducers are attached in the high-pressure side pipe. The measurement point closest to the compressor with a distance to outlet port of 1100 mm is referred to as measurement point 0 (mp0) in the following, and the measurement points following in the direction of flow are accordingly referred to as mp1 and mp2. The distance between each of the sensors is 210 mm.

4. RESULTS AND DISCUSSION

4.1 Calculation of the excitation
First of all, the pure excitation of the compressor is looked at. In consistency with the measurements, three different operating points were simulated, whereby the rotational speed was kept constant at $n_{mr} = 8100$ min^{-1}. This corresponds to a

discharge frequency of f_{pp} = 540 Hz with a number of lobes of the male rotor of z = 4. Table 1 provides an overview of the boundary conditions of the different operating points simulated and measured. The operating point op3 is close to the designed working conditions, thus the pressure in the working chamber and the discharge pressure correspond to each other at the start of the discharge process. In op3 the calculated pressure difference between the working chamber and high-pressure side at this point in time amounts to approx. $3 \cdot 10^4$ Pa, meaning a slight over-compression takes place. The remaining operating points deviate more strongly from the designed working conditions, meaning that an increasing over-compression takes place in the working chamber.

Table 1: Boundary conditions of the simulated operating points

Operating point	op1	op2	op3
Pressure on pressure side [Pa]	$2.25 \cdot 10^5$	$2.5 \cdot 10^5$	$2.8 \cdot 10^5$
Temperature on pressure side [°C]	29	35	52
Outer pressure difference Δp_o [Pa]	$0.5 \cdot 10^5$	$1.0 \cdot 10^5$	$1.5 \cdot 10^5$

The results of the simulations are presented in Fig. 3. The first three harmonics alone are looked at. At op1 and op2 the basic harmonic with a frequency of f_1 = 540 Hz is present as a dominant amplitude in the spectrum, while the amplitudes of the second and third harmonics steadily decrease as the frequency increases. At op3 however, the amplitudes of the first three harmonics steadily rise as the frequency increases, and the basic harmonic amplitude is weak. The dominance of low frequencies in the pressure pulsations in the case of over-compression has already been observed on several occasions (5,9). The amplitudes of the pulsations in op3 cannot be clearly interpreted due to their low strength.

4.2 Comparison of excitation and system response

In order to make the influence of the pressure-side pipeline on the resulting pulsations clear, the excitation and the resulting system response at the measurement points mp0 to mp2 are compared for the three operating points in Fig. 3. When looking at the basic harmonics, it becomes clear that the amplitudes of the system response also increase as the amplitudes in the excitation rise. No clear trend can be seen regarding the amplitudes of the second and third harmonics. The strongly increased amplitudes of the first harmonic at op1 and op2 indicate that these signals are characterised by resonances. A further indicator for this is the highly unequal spectra at the different measurement points, both at op1 and op2. Acoustic waves passing through would produce roughly the same spectra at the different measurement points.

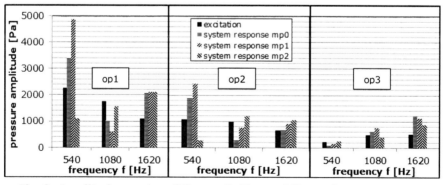

Fig. 3: Amplitude spectra of the excitation and the system response at different measurement points for op1, op2, and op3

4.3 Experimental verification

The simulation results are now evaluated based on initial experimental data. For this purpose, the amplitude spectra of the measured and calculated pulsations are compared in Fig. 4. The values at mp1 are represented for the different operating points. A comparison of simulation and measurement at op1 shows that the ratios of the amplitudes of different harmonics from the simulation correspond qualitatively to the measurement results. The first harmonic is most pronounced and the second harmonic is least pronounced. A comparable picture is shown in the results for op2. The amplitudes from the simulations as well as the measurements decrease, since op2 is closer to the designed working conditions. The amplitudes of different harmonics show comparable ratios in the measurements and the simulations at this operating point also. A comparison of the spectra at op3 does not show any such clearly comparable relationship between simulation and measurement.

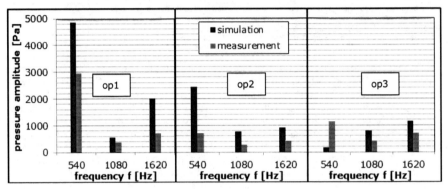

Fig. 4: Comparison of simulated and measured pressure pulsations at mp1

Although the second and third harmonics again have comparable relations with one another, the amplitude of the basic harmonic is calculated as being much smaller than it occurs in the measurements. It is to be assumed that, during the measurements at this operating point, an acoustic resonance arises, which is not calculated in the simulations. This could be down to modelling errors especially concerning the discharge piping. The compensator in particular, which is attached behind the outlet port of the compressor in the pipeline system, can constitute a difference in the acoustic characteristics of the calculated and of the real system, e.g. the damping behaviour or the speed of sound. The speed of sound, for instance, is reduced locally by the elastic wall. Until now, such an effect hasn't been considered in the modelling however.

5. CONCLUSIONS

A one-dimensional model for calculating the discharge pressure pulsations of screw compressors is presented. It is shown that, when calculating the excitation of pulsations by the newly introduced method for calculating the flow at the outlet control edges in combination with the method of characteristics for calculating the flow inside the working chamber and the piping, fundamental trends are shown. E.g., it is shown that in case of increasing deviation from the designed working conditions, amplitudes increase as simulated and measured. For the initial experimental validation, the excitation was prescribed as a boundary condition for a pipeline system in order to be able to compare the resulting pulsations with measured values. General trends from the measurements could also be recognised in the simulation results here. Future work will aim for example to also model the gaps in the working chamber of a compressor. The aim is to develop a validated tool for the 1D calculation of pressure pulsations of twin-screw machines in piping networks.

REFERENCES

(1) Nickel, A.: "Entstehung von Geräuschen an trockenlaufenden Schraubenverdichtern und Möglichkeiten zu ihrer Minderung", Fortschr.-Ber. VDI, Reihe 7, Nr. 158, VDI-Verlag, Düsseldorf, 1989
(2) Mujic, E.; Kovacevic, A.; Stosic, N.; Smith, I.K.: "Noise prediction in screw compressors", Int. Conf. on Compressors and their Systems, 2005, p.447-454
(3) Stosic, N.; Mujic, E.; Smith, I.K.; Kovacevic, A.: "Development of a rotor profile for silent screw compressor operation", Int. Conf. on Compressors and their Systems, 2007, p.133-145
(4) Mujic, E.; Kovacevic, A.; Stosic, N.; Smith, I.K.: "Noise generation and suppression in twin-screw compressors", Proc Instn. Mech. Engrs. Part E, Vol. 225 (2011) No. 2, p.127-148
(5) Wu, H.; Xing, Z.; Peng, X.; Shu, P.: "Simulation of discharge pressure pulsation within twin screw compressors", Proc. Instn. Mech. Engrs. Part A, Vol. 218 (2004) No.4, p.257-264
(6) Stosic, N., Hanjalic, K.: "Contribution towards modelling of two-stage reciprocating compressors", Int. J. mech. Sci., Vol. 19, (1977), p.439-445
(7) Huster, A.: "Untersuchung des instationären Füllvorgangs bei Schraubenmotoren", Dissertation, Universität Dortmund, 1998
(8) Kovacevic, A., Stosic, N., Smith, I.: "Screw Compressors", Springer Berlin Heidelberg New York, 2007
(9) Mujic, E.: "A numerical and experimental investigation of pulsation induced noise in screw compressors", Diss., City University London, 2008
(10) Schweinfurter, F.: "Beitrag zur rechnerischen Bestimmung von Druckschwingungen in Rohrleitungssystemen bei Erregung durch ein- und mehrzylindrige oszillierende Verdrängerpumpen", Diss., Universität Erlangen-Nürnberg, 1988
(11) Nadler, K.; Brümmer, A.: "A method for the geometrical analysis of twin-shaft rotary displacement machines", Schraubenmaschinen 2010, VDI-Verlag Düsseldorf, 2010

Discussion of actual profile clearances' calculation method in rotary compressors in the absence of rotor timing units

R R Yakupov[1], T N Mustafin[1], V N Nalimov[2], M S Khamidullin[1], I G Khisameev[1]
[1] Kazan National Research Technology University, Russian Federation
[2] NIIturbocompressor n.a. V.B.Shnepp, Russian Federation

ABSTRACT

Rotary compressors are widely used due to their reliability, ease in operation, relative simplicity in design and high efficiency.

In the present article a method has been proposed for the analysis of the rotors' meshing conditions in the rotary machines without rotor timing gears, taking into account the actual profile clearances. Employment of the results obtained will help to analyze and improve the rotor profiles and further to optimize the algorithm of underrating thus improving the operational characteristics of the newly developed compressors.

1 INTRODUCTION

At present time rotary compressors are widely used in low- and medium- capacity applications. The main working components of given compressors are the profile surfaces of the rotors, precision make and defining the basic power characteristics of a compressor.

At present just the optimization of theoretical as well as of the actual profiles of the rotors is one of the main trends in the updating of rotary compressors (1, 2, 3). At the same time, actual working clearances in the compressor and the actual position of profile surfaces of the rotors relative to the housing and relative to one another are required for this optimization. Now there are plenty of works dedicated to this issue in the literary sources, including their variations in time (transient conditions) (1, 2, 3, 4, 5, 6), however, frequently the influence of just separate aspects is examined, instead of considering them as a group. That is why the authors of this work have pooled knowledge, available today on this subject, into a unified procedure for the calculation of actual position of profile surfaces and profile clearances. The procedure being proposed is considered with reference to the screw oil-flooded compressors, being the most commonly used, however, if required, it can be easily adapted to the other types of compressors. In particular, the authors of present work have tested it for a machine with internal gearing (Trochoid compressor).

© The author(s) and/or their employer(s), 2013

2 DEFINITION OF THE PROBLEM, CALCULATION PLAN

Generally, rotors of compressors may be thought as coarse pitch gears.

However, there is one fundamental difference between the rotors and the gearings: in the gearings the torque is transferred from one gear to another through direct contact; but in rotary compressors the torque occurs through the total moment of gas forces, affecting the rotors, and through the torque attributed to the friction of the rotors against the gas and oil mixture in the gaps. It gives more freedom to the rotor in terms of potential side clearance in the engagement. This is especially clearly seen when the torque on the rotor changes its sign during the rotation, causing loss of direct contact between the rotors and shocks when the rotors come in contact. The given side clearance in the engagement is defined by the algorithm of the rotors' underrating, compressor design features and operating conditions. The last factors are considered in detail in the works (3, 4) including a diagram shown on Figure 1 of this paper. The above mentioned factors are caused by the lack of parallelism between the rotors in the working mode of the compressor, due to the housing parts' manufacturing error, difference between the changes in center distance of the compressor housing at the suction and discharge sides caused by thermal deformations, and possible precession of the rotors in the bearings. Purpose of the present paper is to integrate in a common procedure the calculation of numerical values of the factors, mentioned above, and their influence on the rotors' engagement conditions.

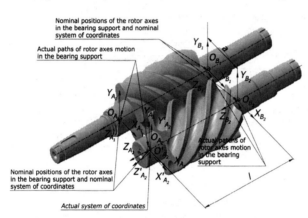

Figure 1. Diagram of the rotors' position in the bearing support

3 COORDINATES OF THE ROTOR PROFILE

Coordinates of the actual rotor profile, as a rule, can be obtained by underrating the coordinates of the theoretical one. It will be expedient to their represent nominal values in a parametrical form using radius vector:

$$\overline{r}_{Mn} = r_{Mn}\left(X_{Mn}(t_{Mn}; \theta_{Mn}); Y_{Mn}(t_{Mn}; \theta_{Mn}); p_{Mn} \cdot \theta_{Mn}\right) = r_{Mn}(t_{Mn}; \theta_{Mn}), \qquad (1)$$

$$\overline{r}_{Gn} = r_{Gn}\left(X_{Gn}(t_{Gn}; \theta_{Gn}); Y_{Gn}(t_{Gn}; \theta_{Gn}); p_{Gn} \cdot \theta_{Gn}\right) = r_{Gn}(t_{Gn}; \theta_{Gn}), \qquad (2)$$

where X,Y are the profile coordinates; t – shape parameter of the profile; θ - angle of the rotation; p – rotor lead; the index "Mn" means that the parameter refers to the driven rotor; the index "Gn" means that the parameter refers to the free rotor. Under the assumption that the temperature field is uniform for any rotor section normal to the rotation axis, change of the profile coordinates can be calculated using the following equation (7):

$$\overline{r}(t_n; \theta_n; T_{Ri}) = \overline{r}_n(t_n; \theta_n) \cdot \left(1 + \alpha \cdot (T_{Ri} - T_0)\right), \qquad (3)$$

where T_{RI} is the mean rotor temperature in "i"-section; T_0 is the rotor temperature in "cold" condition during the clearances measurement, α is a linear thermal expansion coefficient of the rotor material.

The assumption made, according to the results of (8, 9), is almost invalid only for the sections close to the discharge face of the rotor and during the change of the operation mode. Temperature of the rotor face at the discharge side will be determined mainly by the gas discharge temperature; rotor face temperature at the suction side will be governed by the temperature of the gas being sucked in and of the oil being injected. Experimental investigation of temperature patterns of the drive rotor (7) helped to determine the types of correlative functions describing these relationships. According to them, mean temperatures for the discharge and suction faces of the male rotor can be determined as follows:

$$T_{RD} = 0,8T_D + 5, \qquad (4)$$

$$T_{RS} = T_S + 0,34T_{OIL} - 4, \qquad (5)$$

where T_S, T_D, T_{OIL} are accordingly the temperatures of the suction, discharge and of the oil being injected. All the temperatures have the dimensions in °C. According to the same source, change of the temperature along the length of the rotor for any "i"- th section will be the following:

$$T_{Ri} = T_{RS} + (T_{RD} - T_{RS}) \cdot \left(\frac{l_i}{l_R}\right)^{1,56}, \qquad (6)$$

where l_i is the distance from the suction end to the "i"- th section of the rotor, l_R is the length of the rotor profile section.

The authors suggest to extend the received relationships to include the analysis of temperature patterns of the driven rotor. Validity of this assumption is confirmed indirectly by the analysis of results, obtained in the work (7).

Deviation of the bearings' axes from their nominal positions is due to the manufacturing error and to the radial clearance in the bearing (for rolling bearings), due to design features (for sliding bearings), an error of mutual position of basic support points of the case under the bearings, and also due to thermal deformations of the casing in the course of work.

These factors generally lead to crossing of axes of rotors (to rotation of the axis relative to the basic axes $O_A X_A$ and $O_A Y_A$ for the angles ζ and Σ accordingly). The analysis of their influence is in detail discussed in the works (3, 4) and consequently is not presented in the given article. Presence of the given angles will lead to the condition in which the normal planes of rotors will cease to coincide. For simplification of mathematical calculations it is expedient to use basic normal planes with the subsequent correction of position of the sections' centers and the systems of coordinates. On the basis of the scheme (Figure 2) it is possible to determine the positions of a projection of the section center on a normal plane:

$$\begin{cases} X_{Ci} = \delta X_1 + l_i \cdot \sin\Sigma \\ Y_{Ci} = \delta Y_1 + l_i \cdot \sin\zeta \end{cases}. \qquad (7)$$

The profile equation in a basic normal plane relative to the displaced center of coordinates will become:

$$\begin{cases} X' = X \cdot \cos\Sigma \\ Y' = Y \cdot \cos\zeta \\ Z' = Z \cdot \cos\zeta \cdot \cos\Sigma = l_i \cdot \cos\zeta \cdot \cos\Sigma = p \cdot \theta \cdot \cos\zeta \cdot \cos\Sigma \end{cases} \qquad (8)$$

where $Z = l_i = p \cdot \theta$, $\Sigma = \arctan\left(\frac{\delta X_2 - \delta X_1}{l}\right)$, $\zeta = \arctan\left(\frac{\delta Y_2 - \delta Y_1}{l}\right)$, thus it is necessary to notice, that use of systems (7) and (8) is fair for both rotors at assumption acceptance about their absolute rigidity (absence of deflections of rotors which have higher order of infinitesimals in comparison with sizes δX and δY).

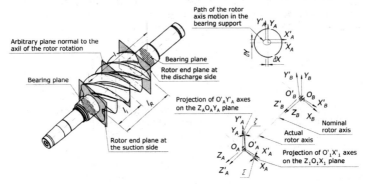

Figure 2. Diagram of the rotor position in the bearing support

4 DETERMINATION OF THE ANGULAR CLEARANCE BETWEEN THE ROTORS

Positions of the free rotor are defined by the extreme positions in contact of the rotors by one of the sides, or by some intermediate positions at transition to one or another extreme position. For their definition it is required to determine the frameworks in which the free rotor has the right to rotate before the contact with the drive rotor is achieved. The given problem was studied in the work (5). We will apply the similar approach on an example of definition of a correction angle required for the

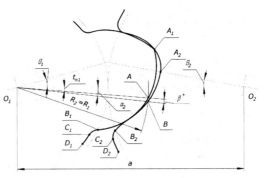

Figure 3. Diagram for the determination of the male rotor correction angle

driven rotor to come in contact with the free rotor. For this purpose we will consider position of the rotors when the male rotor is rotated for an angle Θ_1 (Figure 3), however, the female rotor, accordingly, should be nominally rotated for an angle:

$$\theta_2 = \theta_1 \frac{z_1}{z_2}, \qquad (9)$$

where z_1, z_2 – is the number of teeth of male and female rotors. The scheme presented on Figure 3, is given for nominal positions of the centers of the rotors and does not require basic replacement to take an account of their displacement in the method being presented.

212

Let's choose any point A in the section A_1D_1 with angular coordinate a_1. Let's find a point B, radial coordinate R_2 of which is equal to R_1, on a front part of a profile of the female rotor (section A_2D_2), rotated for a Θ_2 angle. Difference between the angular coordinates β of the points A and B will give us the size of correction angle of the male rotor to contact the profile of the female rotor with the point A:

$$\beta = \alpha_2 - t_{n1}.$$ (10)

Let's notice, that:

$$R_2 - R_1 = 0.$$ (11)

Thus the radial coordinate of the point A is defined under the equation:

$$R_1 = S_1(t_{n1}, \theta_1),$$ (12)

where S_1 is the function describing a profile of the male rotor. And the relationship for determination of coordinates of the point B is the following:

$$R_2 = f\left[S_2(\alpha_2), \theta_2\right],$$ (13)

where S_2 is the function describing a profile of the female rotor. Solving the equations (10) - (13) together we will receive the following relationship:

$$\beta^+ = f(t_{n1}, \theta_1).$$ (14)

Given equations are solved using numerical methods.

Having found a minimum of the function (14) for the variable a_2, we will receive the correction angle size for the male rotor to its contact to the profile of the female rotor for the position being considered:

$$\beta^+_{min\alpha} = f(\theta_1).$$ (15)

Let's determine the minimum correction angle for the male rotor $\beta^+{}_{min}$ from the relationship (15), for all the range of the rotation angle θ_1, within the existence of the contact line on the front side of the rotor tooth profile.

Similarly the advance angle $\beta^-{}_{min}$ can be calculated. However it is necessary to notice the fact that at a time in theoretical contact there can be several pairs of teeth. This is characteristic in particular for screw compressors as well. Due to the lack of parallelism between the axes of the rotors and to the presence of a gradient of temperatures along the rotor length, and hence to the different thermal deformations on the opposite ends, advance angles and correction angles will be different for each of the pairs. True values of the given angles will be their minimum values among all the couples of teeth being concurrently in theoretical gearing. The valid angular clearance between the rotors will be equal to the sum of correction and advance angles.

5 DYNAMICS OF A FREE ROTOR AND POSSIBILITY OF ITS EXIT FROM DIRECT CONTACT WITH A MALE ROTOR

Change of the free rotor correction angle leads to that it is compelled to rotate with some certain acceleration. The acceleration may be caused both by direct contact between the rotors, as well as by the torque due to gas forces. In a screw compressor working in a nominal mode, usually there is the last: when drive is implemented by a male rotor, female rotor is turned through a correction angle in the direction of the rotors' motion up to the contact with the male rotor. The size of the given correction angle can be calculated by the method stated above. Hence, for continuation of the rotors' being in engagement, possible acceleration of the rotor due to gas forces should exceed the necessary acceleration related to the change of the correction angle.

Believing as a first approximation that the motion of the male rotor is uniform due to its considerably large moment of inertia thanks to the mounted parts and to the partial compensation by the torque of the drive, then, taking into account all the above, the condition for continuation of the rotors' being in engagement can be written down as follows:

$$d\theta_2 \geq \omega_1 \cdot \frac{z_1}{z_2} \cdot d\tau + d\beta_2 , \qquad (16)$$

or:

$$\frac{d\theta_2}{d\tau} \geq \omega_1 \cdot \frac{z_1}{z_2} + \frac{d\beta_2}{d\tau} , \qquad (16a)$$

where β_2 - a correction angle of the female rotor before the contact with the male rotor, ω_1 - angular speed of the drive rotor. The sign «>» in inequalities (16) and (16a) will testify that the free rotor may «try to tighten up» the drive rotor. Having differentiated the expression (16a) with respect to $d\tau$ finally we receive:

$$\frac{d^2\theta_2}{d\tau^2} \geq \frac{d^2\beta_2}{d\tau^2} . \qquad (17)$$

Taking into consideration that $\omega_1 = \frac{d\theta_1}{d\tau}$ we receive:

$$\frac{d^2\theta_2}{d\tau^2} \geq \frac{d^2\beta_2}{d\theta_1^2} \cdot \omega_1^2 , \qquad (17a)$$

The approach to the definition of $\frac{d^2\theta_2}{d\tau^2}$ has been considered in the work (10), further some evolution of the given approach has been presented. From dynamics equation it is possible to write down:

$$\frac{d^2\theta_2}{d\tau^2} = \frac{M_{GF} + M_{AF} - M_R - M_{F1} - M_{F2} - M_{MEC}}{J_2} , \qquad (18)$$

where J_2 is the moment of inertia of the free rotor; M_{GF} is the moment created by gas forces, defined on the basis of (P-V) diagram using a method of "subtenses" (3, 7); M_{MEC} - the moment of resistance of the compressor mechanical assemblies such as the seals, balance pistons and bearings (its size is defined in many respects by design of mechanical assemblies of the rotor); M_R, $M_{F1,2}$ - the moments of friction forces against the fluid being compressed in radial and face clearances accordingly; M_{AF} - the moment created by forces of adhesion. It is not difficult to show, that for the given case the concept "adhesion" can be replaced by concept "cohesion" being related to it. Then the given moment will be proportional to the surface stress factor, the area of the contact spot and the wetting corner. However, it is necessary to notice, that the given moment will be much less than the other moments of forces, thus in turn giving us a chance to neglect it in the further calculations: $M_{AF} \approx 0$.

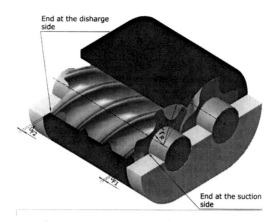

Figure 4. Diagram of clearances for determining the moments of resistance

The moments of the forces of friction against the fluid being compressed in the end clearances, according to the scheme (Figure 4) depend on the operating mode (7):

$$M_{F1,2} = \begin{cases} \dfrac{\pi \cdot \omega_2 \cdot \eta_{MIX} \cdot R_{eq}^4}{2 \cdot \delta_{F1,2}} \cdot \left(1 - \dfrac{r_R^4}{R_{eq}^4}\right), & \text{if } Re_{F1,2} < 10^4 \\ \dfrac{\rho_{MIX} \cdot \omega_2^3}{4} \cdot \left(c'_{M1,2} \cdot R_{eq}^5 - c''_{M1,2} \cdot r_R^5\right), & \text{if } 10^6 \geq Re_{F1,2} \geq 10^4 \end{cases}, \quad (19)$$

where $c_{M1,2} = 0{,}0277 \cdot Re_F^{-0,2} \cdot \left(\dfrac{\delta_{F1,2}}{R}\right)^{-0,2}$ - factor of the moment of friction; for c''_M definition r_R ($R = r_R$) is used as the defining size, in all other cases - R_{eq} ($R = R_{eq}$);

$Re_F = \dfrac{R^2 \cdot \omega_2 \cdot \rho_{MIX}}{\eta_{MIX}}$ - Reynolds's number; $\omega_2 = \omega_1 \cdot \dfrac{z_1}{z_2}$ - nominal angular speed of the free rotor; ρ_{MIX}, η_{MIX} - accordingly the density and kinematic viscosity of the gas and oil mixture in a working cavity of the compressor; δ_{F1}, δ_{F2} - end clearances between the rotor and the housing at the suction and discharge end accordingly.

The equivalent radius is defined as $R_{eq} = \sqrt[4]{\dfrac{1}{2 \cdot \pi} \int_0^{2 \cdot \pi} \left(r(t;0)\right)^4 \cdot dt}$ where $r(t;0)$ - is defined using the equations (1) or (2) for some fixed rotation angle θ in which as the initial position it is expedient to take: $\theta = 0$.

In analogy, the moments of forces of friction against the fluid being compressed in the radial clearance can be written down as (7):

$$M_R = \begin{cases} \dfrac{2 \cdot \pi \cdot \xi \cdot \omega_2 \cdot \eta_{MIX} \cdot R_{2*}^3 \cdot z_2 \cdot b_R}{(1+\xi) \cdot \delta_R}, & \text{if } Re_R < 2500 \\ \pi \cdot C_R \cdot \rho_{MIX} \cdot \omega_2^2 \cdot z_2 \cdot b_R \cdot R_{2*}^4, & \text{if } 10^5 \geq Re_R \geq 2500 \end{cases}, \quad (20)$$

where $C_R \approx 0{,}0076 \cdot Re_R^{-0,25}$ - factor of the friction (11), $Re_R = \dfrac{\omega_2 \cdot R_{2*} \cdot \delta_R \cdot \rho_{MIX}}{\eta_{MIX}}$ - Reynolds's number, b_R - thickness of the rotor tooth in the direction of the axis Z,

215

R_{2^*} - outside diameter of a rotor, δ_R - radial clearance between the rotor and the compressor case, $\xi = \dfrac{m_{OIL}}{m_{GAS}}$ - gas and oil ratio.

Thermodynamic parameters of the mix are defined using the following dependences (7):

$$\eta_{MIX} = \left(\dfrac{1}{1+\xi}\right)\cdot \eta_{GAS} + \left(\dfrac{\xi}{1+\xi}\right)\cdot \eta_{OIL}, \qquad (21)$$

$$\rho_{MIX} = (1+\xi)\cdot \rho_{GAS}. \qquad (22)$$

For the equation (19) the suction temperature for the clearance at the suction side and the discharge temperature for the clearance at the discharge side may be taken to be the defining temperatures in preliminary calculations. For the equation (20) the arithmetic-mean temperature between the suction and discharge may be taken as the defining temperature. For simplification of calculations it is possible to take the value ξ as constant for all sections of the rotor.

6 RESULTS

As an object of research the standard screw compressor is chosen developed in Joint-Stock Company «NIIturbocompressor n.a. V.B.Shnepp» (the assignee of «SKBK») with diameter of rotors 200 mm, length of a profile part of rotors 270 mm, pitch of 480 mm and the ratio of 4/6. For a profile surface the advanced SKBK profile is chosen with the standard underrating described in the work (12), the essence of which consists of the following: the profile of the male rotor remains theoretical while the profile of the female rotor is underrated equidistantly in relation to the theoretical one. The value of this equidistant underrating of the female rotor is variable on length and changes from the maximum radial value of underrating (0,13 mm) for the points with minimum radial coordinate, with smooth (linear) reduction down to the circumferential (end) value of the underrating (0,07 mm). For the face rotor-to-housing clearances the following values are accepted: from the suction side - 0,6 mm, from the discharge side - 0,07 mm. As a radial clearance the value 0,2 mm is accepted.

The diagram of correction angles for the female rotor to come into contact with the male rotor is presented on Figure 5a. It is possible to allocate two characteristic sections on it: section AB – the section of the entry into the contact with subsequent reduction of the value of correction angle to some minimum size in the point B; section BC - increase of the correction angle in conditions of the contact along the rest of the profile surface. However, it is necessary to notice, that the change of correction angle as it is shown on Figure 5a corresponds only to one section or to spur rotors. In a helical profile there will be at least one section with contact in a point with the minimum correction angle. Therefore it is useless to analyze only sizes of underrating of profiles in the statement offered above; it will be expedient to analyze them in a context with other factors, such as thermal deformations of rotors and housing, as it is made further. As a mode the compressor operating mode on air with compression from 1 bar to 7 bar, suction temperature of 25°C, discharge temperature 90°C, temperature of the oil being injected of 40°C, gas-to-oil ratio $\xi = 3,88$ and built-in volume ratio equal 4,5 is chosen.

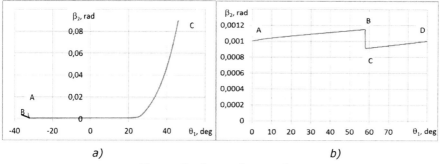

Figure 5. Correction angles

On Figure 5a correction angle change for the end section at the suction side is shown, for the other sections, the nature of change of the correction angle remains the same and is proportionally displaced towards greater values as the contact point moves to the discharge end. Sizes of true correction angles are presented on Figure 5b. Sections AB and CD on it show changes of a minimum of a correction angle (positions of point B on Figure 5a for various sections) while section BC belongs to section AB (Figure 5a).

It can be shown that rotor's normal deviation from theoretically position (which calculated by gear ratio) is proportional to correctional angle and moving of rotor's contact point along Z axis is proportional to rotation angle. With this it can be concluded that results shown on Figure 5b have good correlations with research results of Dr. C.S. Holmes (13, 14).

Comparison of the results received according to the formula (17a) with the accelerations due the torques are presented on Figure 6a and 6b, plotted for cases when the rotation speed of the male rotor corresponds to 3000 and 4500 rpm accordingly. From the graphs received it is visible, that at the speed of the male rotor of 3000 rpm the exit from gearing occurs only at the moment of the contact rerun between the teeth. At the same time at the speed of the male rotor of 4500 rpm the contact is absent for the long enough period and can be the reason of occurrence of additional vibrations. The similar picture is observed during the analysis of the other operating modes of the compressor with the smaller pressure ratio.

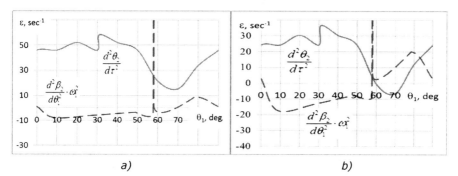

Figure 6. Acceleration of the female rotor

7 CONCLUSIONS

The developed model allows to analyze the gearing of rotors in the screw compressors, taking into account the complex of parameters influencing the gearing, such as thermal and power deformations of the rotors and housings, clearances in the bearings, errors in manufacturing of the rotors and housings, fluctuations of the moment from gas forces. The given technique is the base to the further research of the gearing of rotors and vibration, and can become the criterion for the estimation of efficiency of profiles and underrating techniques.

REFERENCE LIST

(1) Burckney, D., Kovacevic, A., Stosic, N., 2011, Consideration of clearances in the design of screw compressor rotors, 7th International Conference on Compressors and their Systems, City University London.
(2) Sakun, I.A., 1970: Vintovyi kompresorii (Screw Compressors (in Russian)), Mashinostroenie Leningrad.
(3) Stosic, N., Smith, I. K., Kovacevic, A., 2005: Screw Compressors Mathematical Modelling and Performance Calculation, ISBN-10 3-540-24275-9, Springer Berlin Heidelberg, New York.
(4) Stosic N., Smith I.K. and Kovacevic A., 2001: Calculation of Rotor Interference in Screw Compressors, International Compressor Technique Conference, Wuxi, China.
(5) Xiao, D. Z., Gao, Y., Wang, Z. Q., Liu, D. M., 1999, Mathematical basis for clearance analysis in twin screw compressor, International Conference on Compressors and their Systems, City University London.
(6) Holmes, C. S., Williamson, T., 2001, The manufacture of hardened screw compressor rotors, International Conference on Compressors and their Systems, City University London.
(7) Khisameev,I.G., Maksimov,V.A.,2000, Dvukhrotornye vintovye I pryamozubye kompressory. Teoriy, raschet i proyektirovanie (Twin rotors screw and spur compressors. Theory, calculation and design (in Russian)) ISBN 5-7544-0153-1, Kazan Fen.
(8) Weathers, B., Sauls, J., Powell, G., 2006, Transient thermal analysis of screw compressors, part II: Transient thermal analysis of a screw compressor to determine rotor-to-housing clearances, International Compressor Engineering Conference, Purdue University.
(9) Hsieh, S.H., Hsieh, W.H., Huang, C.S., Huang, Y.H., 2012, Numerical analysis of performance, rotor temperature distributions and rotor thermal deformation of an R134a screw compressor, International Compressor Engineering Conference, Purdue University.
(10) Adams, G. P., Soedel, W., 1994, Dynamic Simulation of Rotor Contact Forces in Twin Screw Compressors, International Compressor Engineering Conference, Purdue University.
(11) Koshmarov,Yu. A., 1962, Gidrodinamika I teploobmen turbulentnogo potoka neszhimayemoy zhidkosti v zazore mezhdu vrashchayushchimisya koaksialnymi tsilindrami (Hydrodynamic and heat exchange of incompressible liquid's turbulent flow in the gap between rotating coaxial cylinders (in Russian)), Journal of Engineering Physics and Thermophysics № 5.
(12) Amosov P.E. et al., 1977: Vintovyi kompresornyi mashinii – Spravochnik (Screw Compression Machines-Handbook), Mashinstroienie, Leningrad.
(13) Holmes C.S., 2005, Transmission error in screw compressors, and methods of compensation during rotor manufacture, IMechE Compressor Conference, London.
(14) Holmes C.S., 2006, Noise reduction in screw compressors by the control of Rotor transmission error, International Compressor Engineering Conference, Purdue University.

Theoretical analysis of loads on the gate rotor bearings in the single screw compressor

T Li, Z L Wang, R Huang, W F Wu, Q K Feng
School of Energy and Power Engineering, Xi'an Jiaotong University, China

ABSTRACT

Single screw compressor is one of the positive displacement compressors, which has excellent mechanical properties. Gas forces applied on screw are well balanced. However, gas forces applied on gate rotor cannot be balanced without bearings' support. In order to provide the basis for designing bearing structure and selecting bearing model, loads on the gate rotor bearings are analyzed in this paper. Gas force applied on gate rotor, radial force and axial force applied by gate rotor bearings are respectively analyzed and calculated in two different compressors with air displacement of 6 m^3/min and $17m^3/min$.

1. INTRODUCTION

The single screw compressor was invented by B. Zimmern in 1960s. It has always been regarded as a good compressor because advantages of simple structure, good dynamic balance, low noise and high reliability[1-2]. The typical structure of the single screw compressor is CP-type. It is mainly composed of a cylindrical screw, a pair of gate rotors and a casing. A groove of the screw, tooth of the gate rotor and the compressor casing form a closed volume. This closed volume is rotated along the rotating of the screw, and it decreases gradually with the rotation, thus the gas is compressed.

There are two gate rotors, which are centrosymmetric in a compressor. Thus the closed volumes appear in pairs and are also centrosymmetric, which results in very small radial loads of gas pressure on the screw rotor. On the other hand, axial loads of gas pressure on the screw rotor are inner force. For these reasons, gas pressure loads on the bearing of the screw rotor are very small.

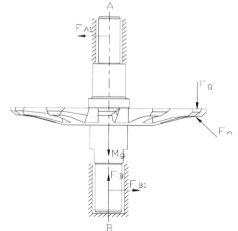

Figure 1. Force Analysis of Gate Rotor Shaft

Compared to screw, force state of gate rotor are shown in Figure 1. The compressed gas applied an axial force F_g and a torque M is generated. Both of them can not be balanced by the gate rotor itself but the bearings of the gate rotor shaft. The gate rotor carries the engaging force F_n applied by the screw as well. In the quasi-static state, F_n is far less than gas force F_g[3]. Therefore, F_n can be negligible in calculating the force acting on the bearings of the gate rotor shaft. To design the bearings structure of the gate rotor and select the bearings reasonably, it is necessary to work out the gas force F_g and the torque M.

In this paper, compression chamber volume under different gate rotor rotating angle is modeled and calculated firstly, then pressure in the closed chamber under different rotating angle can be worked out. According to the area of the gate rotor tooth which engages in the screw and the pressure distribution along the gate rotor rotating angle, the gas force F_g and the torque M can be obtained. The axial force, radial force of the gate rotor bearings are derived and calculated at the end of this paper.

2. GEOMETRIC MODEL

To calculate compression chamber volume, a geometric model is developed as follows. According to the geometric feature, compression chamber volume is separated into two parts. Part I is the volume under the condition that the gate rotor rotating angle in the interval of Expression (1) and part II is the volume under the condition that the gate rotor rotating angle in the Interval of Expression (2):

$$\varphi_{swin} \leq \varphi_{sw} \leq \varphi_{swmid} \tag{1}$$

$$\varphi_{swmid} \leq \varphi_{sw} \leq \varphi_{swout} \tag{2}$$

Where
φ_{sw} is the rotating angle of gate rotor
φ_{swin} is the rotating angle under which the screw groove, gate rotor tooth, casing just form the closed volume
φ_{swmid} is the rotating angle under which the top end of front-side of the gate rotor tooth exactly right breaks away from the screw groove
φ_{swout} is the rotating angle under which the whole gate rotor tooth exactly right breaks away from the screw groove

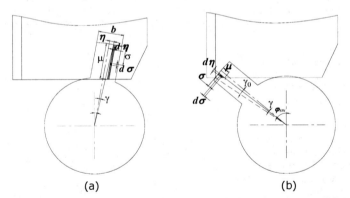

Figure 2. Computation of the Closed Volume: (a) Part I (b) Part II

2.1 Calculation of part I
The compression chamber volume can be calculated by double integration. Firstly, an area element dS is selected as shown by region shadowed in Figure 2. Secondly, the volume element can be obtained by sweeping the area element in the screw. Finally, the volume element is double integrated to work out the compression chamber volume.

$$dS = \mu d\eta \tag{3}$$

$$\mu = \sqrt{R_{sw}^2 - \eta^2} - \eta \cdot \tan \varphi_{sw} - \frac{a - R_{sr}}{\cos \varphi_{sw}} \tag{4}$$

Where μ is the length of element area, R_{sw} is the radius of gate rotor, R_{sr} is the radius of screw, a is the center distance between screw and gate rotor.

For Part I shown in Figure 2(a), the closed volume can be obtained by double integration with respect to φ_{sw} and η:

$$V_1 = \frac{P}{2} \int_{\varphi_{swin}}^{\varphi_{swmid}} \int_{-\frac{b}{2}}^{\frac{b}{2}} \frac{R_{sr}^2 - \left(a - \sqrt{R_{sw}^2 - \eta^2} \cos \varphi_{sw} + \eta \sin \varphi_{sw}\right)^2}{\cos \varphi_{sw}} d\varphi_{sw} d\eta \tag{5}$$

2.2 Calculation of part II
Calculation of part II is similar to calculation of part I except different integral interval:

$$V_2 = \frac{P}{2} \int_{\varphi_{swmid}}^{\varphi_{swout}} \int_{-\frac{b}{2}}^{R_{sw} \sin \gamma_0} \frac{R_{sr}^2 - \left(a - \sqrt{R_{sw}^2 - \eta^2} \cos \varphi_{sw} + \eta \sin \varphi_{sw}\right)^2}{\cos \varphi_{sw}} d\varphi_{sw} d\eta \tag{6}$$

Where γ_0 is shown in Figure 2(b).

2.3 Compression chamber volume $V(\varphi_{sw})$
Basic-volume is expressed in Equation (7):

$$V_0 = V_1 + V_2 \tag{7}$$

Therefore, the function of compression chamber volume under different rotating angle can be obtained by subtracting sweeping volume from basic volume:

$$V(\varphi_{sw}) = \begin{cases} V_0 - \frac{P}{2} \int_{\varphi_{swmid}}^{\varphi_{swind}} \int_{-\frac{b}{2}}^{\frac{b}{2}} f(\varphi_{sw}, \eta) d\varphi_{sw} d\eta & \varphi_{swin} \leq \varphi_{sw} \leq \varphi_{swmid} \\ V_0 - \frac{P}{2} \int_{\varphi_{swmid}}^{\varphi_{swind}} \int_{-\frac{b}{2}}^{R_{sw} \sin \gamma_0} f(\varphi_{sw}, \eta) d\varphi_{sw} d\eta & \varphi_{swmid} \leq \varphi_{sw} \leq \varphi_{swout} \end{cases} \tag{8}$$

Where

$$f(\varphi_{sw}, \eta) = \frac{R_{sr}^2 - \left(a - \sqrt{R_{sw}^2 - \eta^2} \cos \varphi_{sw} + \eta \sin \varphi_{sw}\right)^2}{\cos \varphi_{sw}} \tag{9}$$

3. FORCE ANALYSIS OF THE GATE ROTOR SHAFT BEARINGS

In this paper, analyses are based on quasi-statics. To obtain gas pressure in compression chamber under different rotating angle, several hypotheses are made as follows:
(1) Pressure loss and preheat of suction are neglected. Therefore, when the closed chamber just forms, pressure and temperature in screw groove are respectively environment pressure P_0 and environment temperature T_0.
(2) In view of oil-injected, compression process is assumed to be a polytropic process, and polytropic index is taken as 1.28[4-5].

(3) It is assumed that only lubricating oil leak out through the leakage paths without compressed gas, the gas mass in compression chamber maintains a constant value. Leakage has no impact on gas pressure.

Thus, gas pressure in compression chamber under different rotating angle is expressed in Equation (10):

$$p(\varphi_{sw}) = \begin{cases} p_0 & \varphi_{sw} < \varphi_{swin} \\ \dfrac{V_0^n}{V(\varphi_{sw})^n} P_0 & \varphi_{swin} \leq \varphi_{sw} \leq \varphi_{swd} \\ p_d & \varphi_{sw} \geq \varphi_{swd} \end{cases} \quad (10)$$

3.1 Calculation of axial force

As shown in Figure 1, in axial direction, the gas force F_g applied on gate rotor is balanced by reaction force of the bearing B which is at the reverse side of the gate rotor working surface. The direction of F_g is changeless. Therefore, only bearing B affords axial support F_{B2}. There are three gate rotor teeth engaged with screw at a same time. Gas pressure applied on one tooth under different rotating angle can be obtained through multiplying gas pressure by the action area. As shown in Figure 2, the element action area is dS, action area S is expressed in Equation (11):

$$S = \begin{cases} \int_{-\frac{b}{2}}^{\frac{b}{2}} \left(\sqrt{R_{sw}^2 - \eta^2} - \eta \cdot \tan\varphi_{sw} - \dfrac{a - R_{sr}}{\cos\varphi_{sw}} \right) d\eta & \varphi_{swin} \leq \varphi_{sw} \leq \varphi_{swmid} \\ \int_{-\frac{b}{2}}^{R_{sw}\cdot\sin\gamma_0} \left(\sqrt{R_{sw}^2 - \eta^2} - \eta \cdot \tan\varphi_{sw} - \dfrac{a - R_{sr}}{\cos\varphi_{sw}} \right) d\eta & \varphi_{swmid} \leq \varphi_{sw} \leq \varphi_{swout} \end{cases} \quad (11)$$

The gas force applied on the i-th tooth is expressed in Equation (12):

$$F_{g,i} = \begin{cases} \int_{-\frac{b}{2}}^{\frac{b}{2}} \left(p(\varphi_{sw,i}) - p_0 \right) \cdot dS_i & \varphi_{swin} \leq \varphi_{sw} \leq \varphi_{swmid} \\ \int_{-\frac{b}{2}}^{R_{sw}\cdot\sin\gamma_0} \left(p(\varphi_{sw,i}) - p_0 \right) \cdot dS_i & \varphi_{swmid} \leq \varphi_{sw} \leq \varphi_{swout} \end{cases} \quad (12)$$

Where $\varphi_{sw,i}$ is rotating angle of i-th tooth

In the three teeth engaged in screw, the last tooth engaged in screw is selected as the basic tooth to calculate resultant forces $F_g(\varphi_{sw})$ which is expressed in Equation (13):

$$F_g(\varphi_{sw}) = F_{g,1}(\varphi_{sw,1}) + F_{g,2}(\varphi_{sw,2}) + F_{g,3}(\varphi_{sw,3})$$

$$= F_{g,1}(\varphi_{sw}) + F_{g,1}(\varphi_{sw} + \lambda) + F_{g,1}(\varphi_{sw} + 2\lambda) \quad (13)$$

Axial force of bearing B F_{B2} is expressed in Equation (14):

$$F_{B2} = F_g(\varphi_{sw}) \quad (14)$$

3.2 Calculation of radial force

Because the position of gas force F_g applied on gate rotor has a distance with the center of gate rotor, a torque $M(\varphi_{sw})$ is generated by F_g. This torque can be balanced only by the torque generated by radial forces of the two bearings A and B.

An element area *dA* is selected as shown by overlapping shadow region in Figure 2. The element torque *dM* can be obtained through multiplying the gas force applied on *dA* by the force arm r. Decomposing *dM* into two parts, one is dM_x which is about the x-axis and the other is dM_y, which is about the y-axis. $M_{x,i}$, $M_{y,i}$ are expressed in Equation(15),(16)

$$M_{x,i}(\varphi_{sw,i}) = \begin{cases} \int_{-\frac{b}{2}}^{\frac{b}{2}} \int_0^\mu (p(\varphi_{sw,i}) - p_0) \cdot r_x d\sigma d\eta & \varphi_{swin} \leq \varphi_{sw} \leq \varphi_{swmid} \\ \int_{-\frac{b}{2}}^{R_{sw} \cdot \sin\gamma_0} \int_0^\mu (p(\varphi_{sw,i}) - p_0) \cdot r_x d\sigma d\eta & \varphi_{swmid} \leq \varphi_{sw} \leq \varphi_{swout} \\ 0 & else \end{cases} \quad (15)$$

$$M_{y,i}(\varphi_{sw,i}) = \begin{cases} \int_{-\frac{b}{2}}^{\frac{b}{2}} \int_0^\mu (p(\varphi_{sw,i}) - p_0) \cdot r_y d\sigma d\eta & \varphi_{swin} \leq \varphi_{sw} \leq \varphi_{swmid} \\ \int_{-\frac{b}{2}}^{R_{sw} \cdot \sin\gamma_0} \int_0^\mu (p(\varphi_{sw,i}) - p_0) \cdot r_y d\sigma d\eta & \varphi_{swmid} \leq \varphi_{sw} \leq \varphi_{swout} \\ 0 & else \end{cases} \quad (16)$$

Where r_x, r_y are the projection of the distance between area element and the center of the gate rotor to x-axis, y-axis:

$$r_x = \sqrt{\left(\sqrt{R_{sw}^2 - \eta^2} - \sigma\right)^2 + \eta^2} \cdot \cos\left(\varphi_{sw,i} + a\tan\left(\frac{\eta}{\sqrt{R_{sw}^2 - \eta^2} - \sigma}\right)\right) \quad (17)$$

$$r_y = \sqrt{\left(\sqrt{R_{sw}^2 - \eta^2} - \sigma\right)^2 + \eta^2} \cdot \sin\left(\varphi_{sw,i} + a\tan\left(\frac{\eta}{\sqrt{R_{sw}^2 - \eta^2} - \sigma}\right)\right) \quad (18)$$

Where M_x, M_y are expressed in Equation (19)

$$M_x(\varphi_{sw}) = M_{x,1}(\varphi_{sw}) + M_{x,1}(\varphi_{sw} + \lambda) + M_{x,1}(\varphi_{sw} + 2\lambda)$$
$$M_y(\varphi_{sw}) = M_{y,1}(\varphi_{sw}) + M_{y,1}(\varphi_{sw} + \lambda) + M_{y,1}(\varphi_{sw} + 2\lambda) \quad (19)$$

The torque $M(\varphi_{sw})$ generated by F_g is expressed in Equation (20)

$$M(\varphi_{sw}) = \sqrt{M_x^2(\varphi_{sw}) + M_y^2(\varphi_{sw})} \quad (20)$$

$M(\varphi_{sw})$ is balanced by the torque generated by radial force $F_{A,1}$ and $F_{B,1}$ of bearing A and bearing B. $F_{A,1}$ and $F_{B,1}$ are expressed in Equation (21)

$$F_r(\varphi_{sw}) = F_{A,1}(\varphi_{sw}) = F_{B,1}(\varphi_{sw}) = \frac{M(\varphi_{sw})}{L_{AB}} \quad (21)$$

Where L_{AB} is the distance between bearing A and bearing B.

4. RESULT

In this paper, a 37 KW compressor driven by motor and a 160 KW compressor driven by diesel engine are respectively calculated and compared. The main parameters of the two compressors are listed in Table 1.

Table 1 Main parameters of the calculation model

	37KW	160KW
Air displacement (m³/min)	6	17
Screw diameter (mm)	180	285
Gate rotor diameter (mm)	194	305
Center distance (mm)	144	228
Width of gate rotor tooth (mm)	28	44
Exhaust pressure (G, bar)	9	9
Rotation speed (r/min)	2970	2200

Gas force applied on one tooth at the interval of φ_{swin} and φ_{swout} is shown in Figure 3. When the closing chamber forms, there is no gas force applied on the tooth. After that, gas force increases with pressure increment in compression chamber and appears a wave crest which is located near the angle where compressed gas began to exhaust. Then, in exhaust process, gas force decreases because of constant exhaust pressure and decrement of action area.

Axial force of bearing B F_{B2} is equal and opposite to gas force applied on whole gate rotor, which is change with the period of included angle between two adjacent gate rotor teeth. Gas forces applied on whole gate rotor and every tooth are shown in Figure 4. The time closing volume just forms is set to be the origin of the cycle of axial force. With increment of rotating angle, axial force reaches its minimum 538N at the basic tooth rotating angle of -9° and reaches its maximum 876N at 2.5° in 37KW compressor. In 160Kw compressor, axial force reaches its minimum 1420N at -10° and reaches its maximum 2210N at 1°.When F_g reaches its maximum value, the tooth before the basis one is about to connect exhaust orifice.

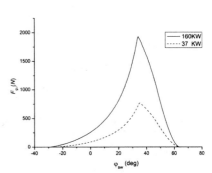

Figure 3. Gas Force Applied on One Tooth

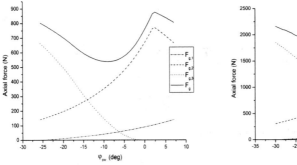
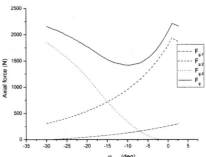

Figure 4. Gas Force(Axial Force of Bearing B)

Variation trend of radial force is similar to axial force's. The calculation result is shown in Figure 5.

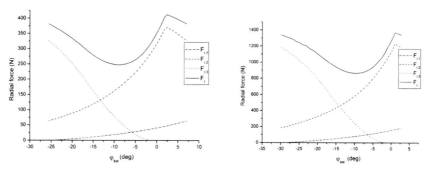

Figure 5. Radial Force of Bearing B

5. CONCLUSION

In this paper, loads of the Gate Rotor bearings in Single Screw Compressor is carried out. A quasi-static model is developed. Based on this study, the following conclusions can be drawn:

1. Only the bearing at the reverse side of gate rotor working surface affords axial force support. Axial force of bearing changes with the period of included angle between two adjacent gate rotor teeth and reaches its maximum at the moment that one of the teeth is about to connect exhaust orifice.

2. Bearings at both side of gate rotor afford radial support. Radial forces of each bearing are equal and opposite in direction.

Further studies should to be carried out in analysis of mechanics characteristics of gate rotor under dynamic load applied. The computing model proposed in this paper can be used as the base of further studies.

REFERENCES

[1] Zimmern B. Worm rotary compressors with liquid joints: USA, 3,180,565[P], 1965.
[2] Fasano B, Davis R. Rotary screw or reciprocating air compressor: Which one is right? [J]. Plan Engineering, 1998: p122-126.
[3] Zhou L., Jin GX. Dynamic Characteristics of Gate Rotor for Single Screw Compressor[J] Xi'an Jiaotong University Press, 1998: p58-62.
[4] Sangfors B. Computer simulation of oil injected twin-screw compressor[C]. Purdue, 1984: 528-535.
[5] Singh PJ, Patel GC. A generized performance computer program for oil flooded twin-screw compressors[C]. Purdue, 1984: 544-553.

Investigation of heat exchange in the working chamber of rotary compressors

I I Sharapov, A G Saifetdinov, A M Ibraev, M S Khamidullin, I G Khisameev
Kazan National Research Technological University, Russian Federation

ABSTRACT

This work is devoted to the heat exchange study in rotary compressors with the external (Roots type) and internal compression which are in demand in the industry.

The paper describes the implemented methods of measuring time-varying temperatures of gas and walls of the working chamber of the rotary compressors being investigated in conditions of the rotors motion. Results of the experiments conducted are presented. Influence of heat exchange on characteristics of rotary compressors by means of a method of mathematical simulation is established.

1 INTRODUCTION

In the analysis and study of the operating cycle of a rotary compressor using the method of mathematical simulation it is necessary to consider, whenever possible, a maximum quantity of physical processes taking place in it and influencing the reliability of calculations received. One of such processes is the heat exchange between the gas and the walls of the working chamber. Gas heating in the course of suction and subsequent compression negatively affects the productivity of compressor and increases its compression work.

In the literature there is rather limited quantity of works devoted to research of heat exchange in rotary compressors. This is, apparently, a design feature of rotary compressors, which consists in the fact that throughout the process the volume of the work space is "swept" by the rotors, and installation of sensors to measure the temperature gas is difficult.

At present the process of heat transfer among the volumetric machines is studied deeply enough only in reciprocating compressors (1, 2, 3). There are papers devoted to scroll compressors (4, 5). Therefore the definition of the heat transfer coefficient values in rotary machines by calculations and experimentally, is now an urgent task.

The objects of the study were two types of rotary compressors: with internal and external compression. Compressors have a similar design, but a different character of the gas compression. Experimental definition of gas temperature, temperatures of internal and external surfaces of compressor housing in the course of working process was the objective of research. On the basis of these data values of heat transfer coefficient between the gas and the walls are recalculated and summarized.

2 ROTARY COMPRESSOR WITH EXTERNAL COMPRESSION (ROOTS TYPE)

Let's consider a rotary compressor with external compression, known as the Roots type gas blower (Figure 1).

For the beginning of the working process of the machine position of rotors is accepted at $\theta_r = 22$ deg. It is possible to present the working process of compressor consisting of the following periods:

1. Period of gas suction into the pair cavity with increasing volume and formation of a portable working space ($\theta_r = 22 \div 68$ deg).
2. Transfer of the isolated working space to the discharge side ($\theta_r = 68 \div 112$ deg).
3. Disclosure of the working cavity at the discharge ($\theta_r = 112 \div 116$ deg).
4. Discharge of gas from the decreasing pair cavity ($\theta_r = 116 \div 218$ deg).

It was decided to use L-type thermocouples for measuring the gas and the wall temperatures. The problem of measurement of the gas temperature in the working cavity in conditions of rotors motion was solved as follows. For the ability to set the thermocouple junction into the working cavity a groove was made at the rotor top 1 mm wide and a 5 mm deep in the plane perpendicular to the axis of the rotor. Thermocouple sensors were set strictly in the same plane, remaining in the gap formed by the groove (6), and when the rotary compressor is in operation, the thermocouple junctions were not damaged by the rotor blades.

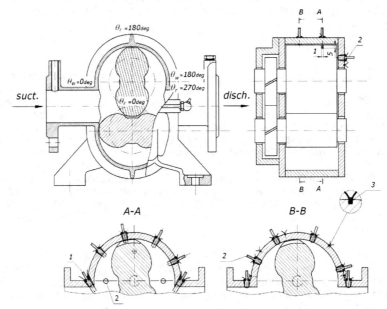

Figure 1. Temperature measurement in the compressor with external compression
1 - gas temperature sensor; 2 - inner wall temperature sensor;
3 - external casing temperature thermocouple

For registration of gas and internal wall temperatures quick-response thermocouple sensors were made. Design of the sensor for registration of the instant gas temperature 1 (Figure 1) represents a steel conic pin 6 mm in diameter and 8 mm in length equal to the thickness of the wall. Diameter of the Chromel and Copel wire of the sensitive element is d = 0,02 mm. Hot junction is welded to Chromel and Copel wire with diameter d = 0,5 mm, extending for 4 mm towards the working cavity. Terminal wires are fed out through the longitudinal holes in the sensor. Terminal wires electrically are isolated from the sensor body by a varnish coat and fixed using special glue.

The sensor measuring the wall 2 internal surface temperature was similar in design to the gas temperature sensor with the exception that the working thermocouple junction is welded to its end. The temperature of the external surface of the housing was accepted stationary, therefore it was measured by L-type thermocouples 3 with the wire diameter d = 0,5 mm. Thermocouples' working junctions were buried in the wall of the body to a depth equal to the diameter of the working junction. The sensors were arranged to cover the whole working process.

As it was shown by the results of the experiment, the given way to measure the gas temperatures requires some additional adjustment by calculation of the values obtained. This is due to the influence of the high-temperature gas stream flowing from the discharge cavity into the suction cavity through the duct in the rotor (7) on the readings of the thermocouple junction. Therefore, in the subsequent study of the heat transfer in the rotary compressor with internal compression another method of placing the gas temperature sensor in the working cavity was carried out.

3 ROTARY COMPRESSOR WITH INTERNAL COMPRESSION

Rotary compressor with internal compression (Figure 2), based on the Roots type machine, thanks to two gas distribution spools installed in the discharge cavity, has a more efficient compression process taking place in the working chamber (8).

We will divide the working process into the following stages:
1. Period of gas suction into the pair cavity. ($0 \deg \leq \theta_r < 171 \deg$);
2. Period of transfer and internal gas compression ($171 \deg \leq \theta_r \leq 229 \deg$);
3. Gas discharge ($229 \deg < \theta_r < 360 \deg$).

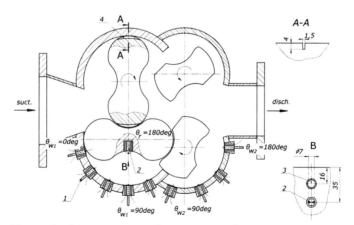

Figure 2. Arrangement of sensors on the casing and rotor of compressor
1 - sensor for measuring the temperature of internal walls;
2 - sensor for measuring the gas temperature in the working chamber;
3 - pressure sensor; 4 - bore

Use of the data on heat exchange, received for the compressor with external compression, in mathematical simulation of the working process of the compressor with internal compression requires further study. Since the nature of the gas compression process and machine designs are different, it should be reflected in heat exchange parameters. Therefore for the given type of machines similar experimental research has been carried out.

To measure the temperature of the inner wall surface, thermocouple sensors 1 (Figure 2) were installed from the suction pipe to the discharge pipe. L-type thermocouples wire diameter of these sensors is the same: 0,02 mm.

Gas temperature was measured by the same thermocouple probe 2, located on the rotor of the compressor. This helped to avoid the influence of high-temperature gas stream through the groove in the rotor. Since the junction of the gas temperature sensor 2 extends for 3 mm into the working cavity, a 4 mm deep and 1.5 mm wide boring 4 was made on the external surface of the response rotor to avoid the junction damage during the compressor operation. On the same rotor, for measuring pressure in the machine cavity, the pressure sensitive element 3 was installed. The signal from the rotating temperature sensor and the pressure sensor was picked up by a mercury current collector.

Temperature of the external surface of the wall was measured by analogy to the compressor with external compression by means of thermocouples with 0,5 mm wire diameter. Rotor position sensors were additionally used in the experiments for both compressors. Signals from the sensors were recorded simultaneously with the signals of the temperature sensors, allowing to synchronize the changes of temperature in the compressor with the working process. Measuring signals from all thermocouple sensors, pressure sensor and rotor position sensor were produced as output on a specialized measuring system designed to study fast processes. Sampling frequency of the sensors was established equal to 20000 Hz that at a frequency of the rotors rotation n = 2900 rpm amounts to 414 measurements for one turn of the rotor.

Presence of filtering circuit in the measuring complex and additional measures to reduce noise at the hardware level helped to receive a signal with relatively low noise level from the sensors (Figure 3).

The signals 2 received from the sensors located on the internal wall of the compressor casing, have shown, that its temperature at the established operating mode has insignificant fluctuations in limits 0,3°C. Therefore further they were accepted as the stationary. The signal 3 received from pressure sensor logically reflected the character of the processes taking place in the compressor. Peaks of a signal 4 from the angular marker correspond to position of compressor rotors on figure 2.

Gas temperature detected by sensor 1 in the nature of its variations did not agree with variations of pressure in the working cavity. Errors of methodical character due to the contact method of temperature measurement could be the reason for this. These are errors from heat exchange by radiation, the heat sink on wires and a lag effect of sensors. The analysis and the account of these errors have been carried out and the true temperature of the gas was determined which in the nature of its variations will be in agreement with variations of pressure in the working cavity of the compressor.

The maximum uncertainty of measurement of temperatures of walls of the compressor made 0,65%, temperatures and gas pressure in the working cavity - 0,6%.

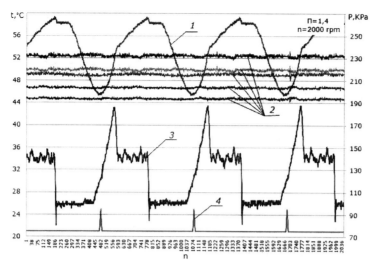

**Figure 3. Graphs of experimental data
(compressor with the internal compression)**
1 - temperature of the thermocouple junction, located in the working space on the rotating rotor; 2 - temperatures of the internal walls of the compressor; 3 - pressure in the machine chamber; 4 - signal from the angular rotor position marker; n - number of measurement points

4 RESULTS OF THE THERMOMETRY IN THE WORKING CAVITY OF ROTARY COMPRESSORS

For rotary machine with external compression the distribution of the wall temperature along the heat-exchange surface is presented in the form of a dependence on the stator angular coordinate $\theta_w = 12 \div 192$ deg. The origin of the rotor angular coordinate θ_r is displaced from the origin of the stator angular coordinate (walls) θ_w for 90 deg (Figure 1).

On Figure 4 the variations are presented of the internal wall $t_{in.w}$ and external wall $t_{ext.w}$ temperatures along the stator angular coordinate θ_w in one of the modes. On figure 5 the variations are shown of the gas temperature t_g averaged along the working cavity with respect to the rotor angular coordinate θ_r for the compressor with external compression.

On figure 6 distributions are shown of the internal $t_{in.w}$ (1) and the external $t_{ext.w}$ (2) wall temperatures of the rotary compressor with internal compression along the angular coordinate θ_{w1} of the basic cavity stator and θ_{w2} of the discharge cavity stator. Operational parameters are characterized by the rotor speed n and the pressure ratio П.

On Figure 7 dependences are shown of gas temperature on the rotor angular coordinate θ_r.

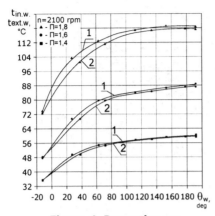
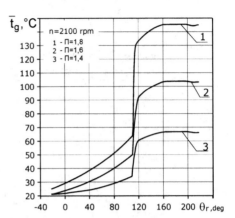

Figure 4. Dependences
1- $t_{in.w}=f(\theta_w)$ and 2- $t_{ext.w}=f(\theta_w)$
at n=2100 rpm and П=1,4÷1,8

Figure 5. Dependence $t_g=f(\theta_r)$ at
n=2100 rpm and П=1,4÷1,8

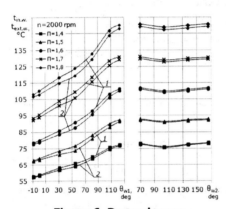
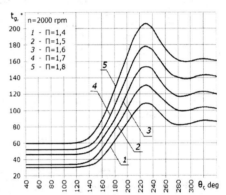

Figure 6. Dependences
1 - $t_{in.w}=f(\theta_w)$ and 2 - $t_{ext.w}=f(\theta_w)$

Figure 7. Dependence $t_g=f(\theta_r)$

5 METHOD OF DETERMINING HEAT TRANSFER COEFFICIENT

Design and experimental techniques for determining heat transfer coefficients are based on a gradient method of finding heat fluxes. Distribution of temperature in the wall, differing from the initial one, necessary for calculation of the heat fluxes values, was found using a method of elementary thermal balances (9).

Local values of the heat flux density and convective heat transfer coefficient at the τ_{k+1} moment of time were defined inside the computational grid on the basis of the temperature field obtained. To use the results of mathematical simulation of the working process values of the heat transfer coefficient, averaged over the working cavity, were calculated from the joint solution of Fourier and Newton-Richman equations with certain increments in θ_r :

$$\bar{h}(\theta_r) = \frac{\bar{q}(\theta_r)}{\bar{t}_g(\theta_r) - \bar{t}_{int.w,k+1}(\theta_r)}, \quad (1)$$

where $\bar{q}(\theta_r)$, $\bar{t}_g(\theta_r)$ and $\bar{t}_{int.w,k+1}(\theta_r)$ - are the values of heat flux, gas temperature and inner wall surface temperature at the τ_{k+1} moment of time, averaged over the working cavity. The maximum uncertainty of determination of heat transfer coefficient makes 8,6%.

Values of heat transfer coefficients, averaged over the volume of the working cavity, plotted against the rotor angular coordinate for compressor with external compression are shown on Figure 8, and for compressor with internal compression- on Figure 9.

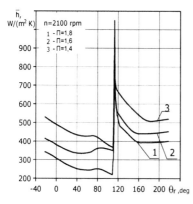
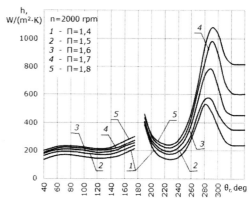

Figure 8. Dependence h=f(θ$_r$) at Π = 1,4÷1,8 and n=2100 rpm

Figure 9. Dependence h=f(θ$_r$) at Π = 1,4÷1,8 and n=2000 rpm

With the beginning of compression process the gas temperature in the working chamber starts to grow, and at the certain moment (at $\theta_r \approx 110 \div 115$ deg for compressor with external compression and $\theta_r \approx 180 \div 185$ deg - for compressor with internal compression) it becomes equal to the temperature of the internal wall. In this place there is a rupture in the function $h = f(\theta_r)$ as the denominator in the equation 1 becomes equal to zero, and the values of heat transfer coefficient increase in absolute size to a maximum and tend to infinity. In this connection values of h were not calculated in the given intervals of θ_r.

6 GENERALIZATION OF THE HEAT TRANSFER COEFFICIENT IN THE DIMENSIONLESS FORM

Results of the heat exchange study were summarized in dimensionless form. Heat transfer coefficient values in the working cavity were presented in the form of $Nu = f(Re)$.

The gas temperature and velocity having a major influence on the heat transfer rate and determining the value of dimensionless complexes have their own levels of values for each of the periods of the working process. The volume of the working chamber is also a variable in the course of the working process, so the defining characteristic size in numbers $Re(\theta_r)$ and $Nu(\theta_r)$ will change accordingly. As the characteristic size at calculation of number Re equivalent diameter is accepted. It is defined as diameter of a sphere equal to volume of a working cavity.

Thus, each period of the working process was considered separately and the equations were received for each of them. Dependences Nu = f(Re) received for all the periods considered were described by the linear equation and for the compressor with external compression the following equations were received:

$$Nu(\theta_r) = B Re(\theta_r) + A_1 \cdot Pr + A_2 ; \qquad (2)$$

- for the period of suction and transfer:

$$B = 0{,}0322 - 0{,}0442 \cdot \Pi + 0{,}0206 \cdot \Pi^2 ; \qquad (3)$$

- for the discharge period:

$$B = 0{,}0256 \text{ (at } \Pi = 1{,}4 \div 1{,}6);$$
$$B = 0{,}0341 \text{ (at } \Pi = 1{,}6 \div 2). \qquad (4)$$

Factors A_1 and A_2 for the period of suction and transfer are resulted in table 2, for the discharge period - in table 3.

Table 1

П	A_1	A_2
1,4 ÷ 1,6	$-1{,}40878 \cdot 10^6 \cdot \Pi + 2{,}38895 \cdot 10^6$	$990432 \cdot \Pi - 1{,}67893 \cdot 10^6$
1,6 ÷ 1,8	$-38670 \cdot \Pi + 196768$	$27910 \cdot \Pi - 138895$
1,8 ÷ 2	$540205 \cdot \Pi - 845207$	$-376417 \cdot \Pi + 588894$

Table 2

П	A_1	A_2
1,4 ÷ 1,6	$-373621 \cdot \Pi + 602073$	$259524 \cdot \Pi - 418204$
1,6 ÷ 2	$-322190 + 329323 \cdot \Pi - 78299{,}7 \cdot \Pi^2$	$223279 - 228612 \cdot \Pi + 54505{,}6 \cdot \Pi^2$

At the section of disclosure of the working cavity at the discharge due to the complexity of process, the heat exchange has been described by the empirical equation of a following kind:

$$\overline{h} = 836{,}451\Pi + 0{,}518n - 1734{,}75 \qquad (5)$$

Relationships received for compressor with internal compression are presented in table 3.

Table 3

Period	Relationship
Suction	$Nu(\theta_r) = 0{,}015 \cdot Re(\theta_r) + 434{,}8 \cdot \Pi - 669{,}7$
Compression	$Nu(\theta_r) = 0{,}025 \cdot Re(\theta_r) + 665{,}1 \cdot \Pi - 1090{,}8$
Discharge	$Nu(\theta_r) = 0{,}025 \cdot Re(\theta_r) + 819{,}4 \cdot \Pi - 1319{,}9$

7 INFLUENCE OF HEAT EXCHANGE ON THE RESULTS OF MATHEMATICAL SIMULATION OF THE WORKING PROCESSES IN ROTARY COMPRESSORS

The basic equations describing the course of working process in rotary compressors:

$$\begin{cases} \dfrac{dP}{d\theta} = \dfrac{k-1}{\omega \cdot V}\left(\omega \cdot \dfrac{dQ}{d\theta} + M_{inf} \cdot i_{inf} - M_{lks} \cdot i_{lks} - \dfrac{k}{k-1} \cdot \omega \cdot P \cdot \dfrac{dV}{d\theta}\right), \\ \dfrac{dT}{d\theta} = \dfrac{(k-1) \cdot T}{P \cdot \omega \cdot V}\cdot \left(\omega \cdot \dfrac{dQ}{d\theta} + \dfrac{k}{k-1} \cdot i \cdot (M_{inf} - M_{lks}) + M_{inf} \cdot i_{inf} - M_{inf} \cdot i - \omega \cdot P \cdot \dfrac{dV}{d\theta}\right). \end{cases} \quad (6)$$

where M_{inf} - gas inflow into the considered cavity; M_{lks} - leaks of gas from the considered cavity; i_{inf} - enthalpy of the inflowing gas; i - enthalpy of the gas in the working chamber; $V(\theta)$, $\dfrac{dV}{d\theta}$ - volume of the working chamber and speed of its variation according to the angle of the rotor rotation. $\omega \cdot \dfrac{dQ}{d\theta}$ is the member accounting for the heat transfer in the working chamber, which can be determined by solution of the Newton-Rihman equation. To evaluate the influence of heat exchange between the gas being compressed and the walls on compressor characteristics, calculations have been carried out ignoring the heat exchange in the mathematical model (MM), and also with due account of it. Results of calculations have shown, that heat exchange between gas and walls makes an essential impact on the results of simulation only in the modes where the gas discharge temperature is high enough, i.e. at high pressure ratios П. Ignoring the process of heat exchange in rotary compressor with external compression gives the maximum divergence between the experimental and calculated values of the volume efficiency η_v in the order of 3 % (Figure 10,a). The maximum divergence for the adiabatic temperature efficiency $\eta_{ad.t}$ ($\eta_{ad.t} = \dfrac{\Delta T_{ad.}}{\Delta T}$, where $\Delta T_{ad.}$ is an increase of temperature in adiabatic process; ΔT is an increase of temperature in real process) is more essential, up to 10 % for the mode with П = 2.

On figure 10,b results are presented of η_v and $\eta_{ad.t}$ calculations for compressor with internal compression. Calculations ignoring the heat exchange on this n show the maximum disagreement (of an order of 8%) between the experimental and calculated η_v values. The maximum disagreement for $\eta_{ad.t}$ makes 4 % for the mode with П = 1,8.

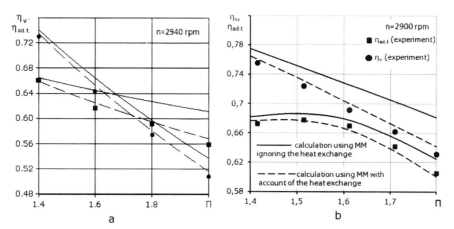

Figure 10. Influence of the account of heat exchange on the volume efficiency η_v and adiabatic temperature efficiency $\eta_{ad.t}$ calculation in the mathematical model of: a - compressor with external compression, b - compressor with internal compression

8 CONCLUSIONS

Thus, values of heat exchange factor for compressors with internal and external compression have been received using calculations and experimental method. Data on heat exchange have been generalised in a dimensionless form and the equations have been received for calculation of the factor of heat exchange between the gas and the walls in various operating modes for each period of the working process. The calculation of the parameters of both compressors using the equations obtained in this research for the heat transfer coefficient in a working chamber gives good convergence of calculated and experimental values of η_v and $\eta_{ad.t}$. These equations can be applied in mathematical model of compressors with a similar design and nature of course of working process.

9 REFERENCE LIST

(1) Fernanda P. DISCONZI, Evandro L.L. PEREIRA, Cesar J. DESCHAMPS «Development of an In-Cylinder Heat Transfer Correlation for Reciprocating Compressors», Proc. International Compressor Engineering Conference at Purdue, p.10, West Lafayette, USA, 2012.

(2) Evandro L. L. PEREIRA, Cesar J. DESCHAMPS, Fernando A. RIBAS Jr. «Numerical Analysis of Heat Transfer inside the Cylinder of Reciprocating Compressors in the Presence of Suction and Discharge Processes», Proc. International Compressor Engineering Conference at Purdue, p.9, West Lafayette, USA, 2010.

(3) P. Grolier «Performance evaluation of reciprocating compressor using an engineering toolbox», Proc. International Conference on Compressors and their Systems, pp. 165-174, London, UK, 2005.

(4) G H LEE and G W KIM «Performance simulation of scroll compressors», Proc. International Conference on Compressors and their Systems, pp. 123-132, London, UK, 2001.

(5) Marco C. DINIZ1, Evandro L. L. PEREIRA, Cesar J. DESCHAMPS «A Lumped Thermal Parameter Model for Scroll Compressors Including the Solution of the Temperature Distribution along the Scroll Wraps», Proc. International Compressor Engineering Conference at Purdue, p.10, West Lafayette, USA, 2012.

(6) A.M. Ibraev, A.G. Saifetdinov, M.S. Khamidullin, I.G. Khisameev, I.I. Sharapov. 2013. Sposob opredeleniya temperatury gaza v rabochey polosti rotornoy mashiny [Process for determining the gas temperature in the working cavity of rotary machines (in Russian)]. Patent number 2474797 (RF).

(7) I.I. Sharapov. 2009. Razrabotka metodiki izmereniya i rascheta parametrov protsessa teploobmena v shesterenchatom kompressore s tselyu povysheniya tochnosti rascheta rabochego protsessa [Development of the method of measurement and calculation of parameters of the heat transfer process in the gear-compressor to increase the accuracy of the calculation of the workflow (in Russian)]. Dissertation, Kazan national research technological university, 146 p.

(8) G.N. Chekushkin, M.S. Khamidullin. 1983. Rotornyi compressor [Rotary compressor (in Russian)]. Copyright certificate number 1044824 (USSR).

(9) E.I. Guigo. 1976. Teoreticheskie osnovy teplo- i khladotekhniki [Theoretical basis of heat and refrigeration (in Russian)]. Leningrad. 224 p.

Experimental validation of a geometry model for twin screw machines

D Buckney[1], A Kovacevic[2], N Stosic[2]
[1] Howden Compressors Ltd, Research and Development, UK
[2] City University London, Centre for Positive Displacement Compressor Technology, UK

ABSTRACT

Aspects of screw compressor geometry are often idealised in mathematical models used for design and optimisation of twin screw machines which can lead to errors in performance predictions. A model that accurately represents geometrical elements such as the volume and flow area history, variable interlobe clearances and geometry of actual compressor port shape has been previously presented by the authors. In this paper the geometry values from that model were used with an existing thermodynamic model to predict the compressor performance with consideration of different rotor profiles, actual discharge port and thermal effects on clearances for an oil free compressor. Results were verified using experimental test data.

NOMENCLATURE

A	rotor centre distance / area	F	interpolation factor
G	clearance gap	h	enthalpy
m	mass	p	pressure
$\dot{Q}$	heat flow rate	R	rotor outer radius
T	temperature	U	internal energy
V	volume	Δ	delta, small change
μ	coefficient of thermal expansion	θ	cycle angle
ω	rotational speed	Vi	volume index

Subscripts

A	axial	a	ambient
c	casing	g	gas
I	interlobe	i	flow into system
o	flow out of system	R	radial
r	rotor	x	component in x-axis

1 INTRODUCTION

Twin screw compressors are extensively used in refrigeration, gas processing and energy industries. Within these industries, compressor applications can greatly vary in terms of flow-rates, pressures, temperatures and working fluids. For each specific application, performance can be improved by use of a uniquely optimised rotor profile (1). This is possible with modern rotor profiling methods that deliver general improvements in efficiency and displacement whilst providing a high degree of versatility in the shape of the profile (2). Stosic (1) emphasises the need to simultaneously consider all design parameters such as the type of bearings and the quantity of oil injection during this rotor optimisation. Such a complex multi-variable design process is possible with the use of numerical computer simulations. Quasi one-dimensional numerical chamber models are now well established (3), (4), (5) for predicting the performance of twin screw compressors. Their ability to predict overall performance with accuracy and speed makes them indispensible and the continuing tool of choice for the preliminary design and optimisation of positive displacement machines. The thermodynamic model presented in (3) calculates the thermodynamic properties of the working fluid at successive angular positions in the compression cycle by considering conservation of energy and continuity within a control volume. It is assumed that the fluid properties are homogeneous over the control volume and kinetic energy is neglected on the basis that it is insignificant compared to the internal energy of the fluid (3).

$$\omega \left(\frac{dU}{d\theta}\right) = \dot{m}_i h_i - \dot{m}_o h_o + \dot{Q} - \omega p \frac{dV}{d\theta} \qquad (1)$$

$$\omega \left(\frac{dm}{d\theta}\right) = \dot{m}_i - \dot{m}_o \qquad (2)$$

Knowing the rate of mass flow in and out of the compressor chamber is critical for solution of the internal energy (equation 1) and the net mass (equation 2) within the control volume. Mass flows account for suction and discharge processes, leakage flows, recirculation and fluid injection. Estimating each of these flows relies on accurate geometry representation and suitable flow models.

Precise calculation of actual port area history is required to fully evaluate flow fluctuations in ports and give useful insight into particular aspects of compressor design. Mujic (6) developed a procedure to accurately represent arbitrary port shape area for quasi-one dimensional and integrated models. He then used the actual discharge port to assess the impact of the port shape on pressure pulsations and noise generation. Chen (7) modelled the part load performance of a refrigeration cycle by including accurate area curves for the slide valve recirculation and radial discharge ports into an established model (8).

Leakage paths occur due to radial (rotor tip to casing bore), axial (rotor end face to casing end face) and interlobe (rotor to rotor) clearances. The area of each leakage path is proportional to the length of its sealing line which is calculated as a function of the compression cycle. Assuming a constant, uniform clearance gap the area of the leakage path is known. In reality it is common for the interlobe sealing line to have a non-uniform clearance distribution which can have a notable effect on the overall leakage area and the resulting thermodynamic performance (9). Accurate representation of the change in clearance gaps due to thermal and other effects is therefore critical for reliable performance predictions.

More often than not, thermodynamic models rely on some empirical feedback to compensate errors due to unknowns such as the complex effect of operating deflections or reduction of the leakage flow areas due to presence of oil. For example, Hsieh (10) presented a procedure to determine empirical constants in a 'training process' which optimises multiple constants using a range of test cases. Hsieh (10) and Wu (8) presented models which once calibrated, or trained, have good agreement between simulated and measured results. Caution should be exercised when using such models, which rely on empirical constants or coefficients, to predict the performance of compressors different than the machine used to derive the coefficients. Even two equivalent compressors from the same manufacturing batch can perform differently on test due to the influence of manufacturing and assembly tolerances which are significant relative to the small clearances in screw machines. The widely utilised model developed by Hanjalic and Stosic (3) uses standard, verified flow equations and coefficients. Since operational clearances are unique to each compressor and the operating parameters (11), accurate absolute performance prediction can be achieved only if the input of the operational clearances is accurately defined. These are normally based on nominal design clearances corrected on the basis of measured correction factors or thermal deformations accounted for by complex 3D FEM or CFD calculations (12).

In this paper, the procedure for calculation of geometry characteristics (13) provided the input to the widely used thermodynamic model for performance prediction of screw compressors of Stosic et al. (2). This is used to assess the impact of specific geometry features on the overall thermodynamic performance of the oil free compressor. Performance predictions are validated using a range of test data obtained by measurements on in-house test facilities.

2　GEOMETRY MODEL

2.1　Summary of the geometry model

The development of this geometry model and its validation based on comparison with 3D CAD model is presented by authors in (13). The main objective was to accurately calculate required geometric characteristics of a screw compressor in advance of thermodynamic calculations. These values are represented as a function of the main rotor angle, θ and comprise of the set of volume and flow area curves stored in a single matrix. The flow areas include axial and radial inlet and outlet ports; radial, axial, interlobe and blow hole leakage flow areas; economiser, liquid injection, oil injection, recirculation ports etc. The aim of the model is to accept rotor profile of any type and from any source either calculated or measured, allowing flexibility and independent rotor comparisons and to support the use of arbitrary ports. The model is based on the direct numerical integration of the actual chamber volume and flow areas.

2.2　Different profile types

Two distinctly different rotor profiles have been evaluated using this geometry model presented in (13). The objective is to assess how accurate the thermodynamic simulation is for different profile shapes with considerably different blow-hole and sealing line characteristics. The performance was calculated based on the coordinates of SRM symmetric profile and the retrofitted N-profile which has improved displacement and efficiency. The verification was performed by comparing calculated performance with measurements of a Howden oil free compressor with original symmetric circular profile and retrofitted "N" profile, as shown in Figure 1.

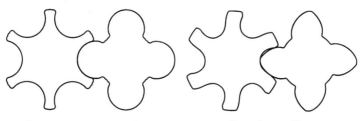

Symmetric – rotor profile　　　　　N – rotor profile

Figure 1 – Profiles of rotors used in this research

2.3 Representation of actual ports

For maximum model accuracy and validity it is necessary to represent real geometry of suction ports, discharge ports, part load re-circulation passages, liquid injection ports and economizer ports. Commonly, theoretical ports are generated from the rotor profiles based on the required volume index. The geometry model utilised for this research calculates port areas using actual point coordinates from drawings or measurements which may differ from the theoretically ideal port shape. In this case the actual volume index is derived from the real port geometry.

The left hand side of Figure 2 shows the actual port geometry of the compressor under investigation. The retrofit N-rotor profile is overlaid to show the mismatch in the rotor and port shape. By assessing area A_2, it is clear that the shape of the port does not closely follow the shape of the rotors which can potentially cause difference in the actual timing of port closing and opening and consequently different Vi compared to the original profile. Also, due to the constraints in casing design, the minimum radius on the actual port does not coincide with the rotor root so that the maximum theoretical port area cannot be achieved.

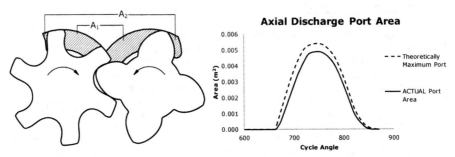

Figure 2 – Actual axial discharge port and resulting actual area curve

The right side of Figure 2 shows the history of the theoretical and actual axial discharge port flow areas. The dashed line represents the theoretically maximum port area curve that could be achieved for the current N-rotor profile with set volume index. The continuous line represents actual port area history with this combination of rotor profile and port shape. The opening timing for both ports is within 1 degree of male rotor rotation but due to the greater width on the retrofitted gate rotor lobe there is a slight delay of opening at the female side. The actual port area increases more gradually and doesn't achieve the same magnitude.

2.4 Operational clearances

The thermodynamic model calculates port and leakage flows based on flow equations which utilise a flow coefficient and a friction coefficient (3). These coefficients are well established and are not the subject of this investigation which instead focuses on the area used for flow calculation. The clearance gaps, used to calculate leakage area, are parameters that have a significant bearing on the performance predictions. The following procedure represents an attempt to approximate the operational effects of the rotor and casing temperature deflections using empirical factors that would apply for this compressor.

In oil free compressors, the clearance gaps change predominantly due to thermal expansion of the rotors and casing (11). The local change in the radial gap, ΔG_R, depends on the local relative expansion between the rotors and the casing. In practice the actual rotor and casing temperatures are difficult to determine and are not uniform. Thermal displacements are typically calculated using finite element analysis, for example in (12). This is not used in the current, generic modelling approach. Instead, the following method which assumes an 'effective temperature increase', ΔT_{rc} was applied (Equation 3). This represents the uniform increase in rotor temperature from suction to discharge that is equivalent to the average net effect of rotor (subscript 'r') and casing (subscript 'c') expansion. It is calculated here as a function of the operational gas discharge temperature and the suction temperature:

$$\Delta T_{rc} = F_{rc}(T_{2g} - T_a) \qquad (3)$$

F_{rc} is an interpolation factor which is specific for this compressor and derived experimentally. This procedure allows the gas discharge temperature, T_{g2}, to be used for estimating changes in the radial clearance gap due to thermal deformation at specific duties as per Equation 4.

$$\Delta G_R = \mu R \Delta T_{rc} \qquad (4)$$

Similarly, equations 5 and 6 apply to the change of the interlobe gap size, ΔG_I. In this case the effective temperature increase, ΔT_{rr}, describes the uniform rotor temperature increase that is equivalent to the average net effect of rotor expansion and any change in the rotor centre distance due to casing expansion. This uses a second interpolation factor, F_{rr}:

$$\Delta T_{rr} = F_{rr}(T_{2g} - T_a) \qquad (5)$$

In equation 6 the calculated value ΔG_{Ix} is the component of the change in the interlobe gap that occurs in the x-direction along the line which connects rotor centres, A:

$$\Delta G_{Ix} = \mu A \Delta T_{rr} \qquad (6)$$

ΔG_{Ix} is used to calculate the local change in the interlobe gap normal to the rotor surface. Figure 3 shows the resulting operational clearance distribution of the interlobe gap normal to the rotor surface due to a change ΔG_{Ix}. The original clearance distribution on the left is uniform and the new distribution on the right is predominantly reduced at the rotor roots:

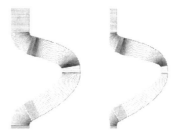

Figure 3 – Clearance distribution change due to rotor thermal expansion

The factors F_{rc} and F_{rr} must be established empirically in the first instance if they are to represent a reasonable approximation of thermal effects on operational clearances. Once the factors are known and verified, the simulation procedure is as follows: The interlobe and radial leakage areas are initially calculated using the 'cold clearances' specified in the compressor design. These initial area curves are used to estimate the gas discharge temperature which is then fed back to the geometry program which re-calculates the leakages areas using the 'operational clearances' estimated using equations 4 and 6. The procedure is then iterated as necessary.

3 MODEL VERIFICATION

3.1 Compressor test details

Test results used in this paper are from a Howden HS/204/165/26 oil free compressor – Figure 4a. The rotors have a 4/6 lobe combination and equal diameters of 204mm. The ratio of the length to the diameter is 1.65, the built in volume ratio is 1.98 and the wrap angle of the main rotor is 300 degrees. The main rotor is directly driven from the suction end where timing gears are fitted to drive the gate rotor. This design includes a number of mechanical seals to prevent oil entering the compression chamber and also features a cooling water jacket on the compressor casing. All testing was performed with atmospheric air at suction. Tests were performed over a range of pressure ratios and speeds. Test parameters were monitored and recorded continuously using a data logging system. Averaged steady state test points were later extracted from the real time test data. Raw results were corrected to p1 = 1 Bara and T1 = 20 degrees C and to the targeted speed. The flow measured on test for each of the profile types is shown in Figure 4b. All other compressor design parameters, including the design clearances have been held constant.

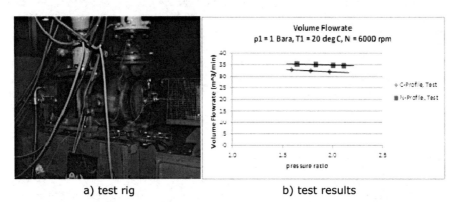

a) test rig   b) test results

Figure 4 – Oil free compressor HS204/165/260

3.2 Different Profile Types Results

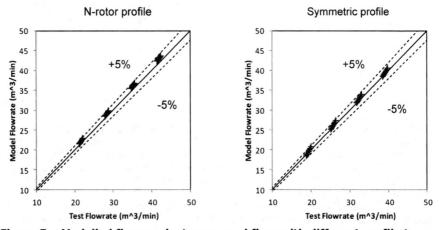

Figure 5 – Modelled flow against measured flow with different profile types

Figure 5 shows predicted volume flow against measured volume flow for speeds of 4000, 5000, 6000 and 7000 rpm at a range of pressure ratios from 1.6 to 2.2. The clearances used for modelling were adjusted to match for the N-rotor profile volume flow rate at 6000 rpm. In the case the same clearance input parameters were then used for all test points and both profile types. The accuracy of the predictions, shown in Figure 5, is similar between profiles that have significantly different geometry characteristics.

3.3 Actual Port Geometry results

Figure 6 shows the history of modelled internal pressure with the angle of rotation for theoretical and real ports which are presented in Figure 2. The slower opening of the actual discharge port has resulted in a slight lag in the time taken for the internal pressure to stabilise to the discharge pressure but the peak pressure is unaffected. The effect of this extended period of over-compression on the performance is shown in Table 1 over a range of pressure ratios. The main effect on performance is a small increase of power less than 1% at a pressure ratio of 1.6. As the pressure ratio is increased (reducing the over-compression) this power increase is reduced.

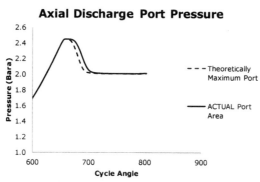

Figure 6 – Pressure history during the discharge process

Table 1 – Change in results with actual port area

Pressure ratio	Change in modelled result with actual port		
	Power	Flow	Temperature
1.6	0.88%	-0.28%	-0.23%
1.8	0.81%	-0.17%	-0.30%
2.0	0.76%	-0.12%	-0.31%
2.2	0.17%	-0.15%	0.56%

The ability to perform relative comparisons of different ports in the model offers useful insight when considering compressor design and optimisation. In this case the results suggest that investing resources to re-designing and manufacturing an updated casing will offer limited benefits.

3.4 Sensitivity to operational clearance corrections

Using the test results for the N-rotor profile a number of simulations were run to investigate the suitability of the equations 3 to 6 which are used to adjust the leakage area through the radial and interlobe clearance gaps to approximate the thermal distortion during operation. The model was initially run with the actual design clearances to provide a baseline; this was achieved by setting both factors, F_{rc} and F_{rr}, to 0. The baseline results are shown by the dashed line in Figure 7 and Figure 8. With cold clearances the model under-predicts flow and over-predicts temperature which suggests that the modelled net leakage flow-rate is too high.

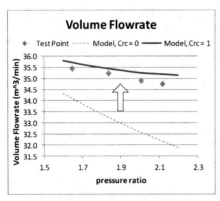 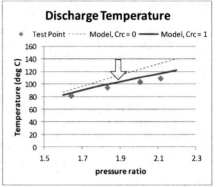

Figure 7 – Effect of modifying rotor to casing clearance gap

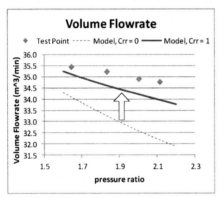 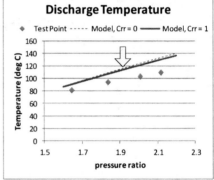

Figure 8 – Effect of modifying rotor to rotor clearance gap

In Figure 7 only the rotor to casing radial clearance gap was reduced by arbitrarily setting the factor F_{rc} to a value of 1. Modification of the radial operational gap alone is capable of bringing the modelled flow very close to the flow measured on test for all pressure ratios. The gradient of the modelled discharge temperature against the pressure ratio is improved when the radial gap is reduced however the modelled discharge temperature is still slightly higher than the measured values. In Figure 8 only the rotor to rotor interlobe clearance gap was reduced by arbitrarily setting the factor F_{rr} to a value of 1. This clearance reduction again shifts the predicted flow closer to the test flow however the effect on the discharge temperature is small in this case.

3.5 Results with operational clearance corrections

For the oil free compressor investigated in this research which has jacket cooling, the rotor to casing thermal reduction factor was set to $F_{rc} = 0.8$. This value closely reproduces the reduction in the radial gap measured on test for this type of compressor (14). The rotor to rotor thermal reduction factor was then modified to target the flow measured on test resulting in a value $F_{rr} = 0.2$.

The results from the computer model are compared against the test results at 6000rpm in Table 2. The flow calculation is consistently good across all pressure ratios with a maximum error of 1.1%. The temperature prediction was reasonably good across all the pressure ratios at this speed with error ranging from -1.2% to 5.1%. As this overall compressor power is dependent on assumed mechanical losses the main parameters of interest for verification of the current compressor model with updated geometry calculation are the delivered flow and the discharge temperature.

Table 2 – Difference between model and test results

4	Difference between model and test		
Pressure ratio	Power	Flow	Temperature
1.6	-3.57%	0.64%	-1.19%
1.8	0.89%	0.56%	2.68%
2.0	3.35%	0.76%	4.41%
2.2	4.52%	1.07%	5.07%

Figure 9 shows predicted volume flow against measured volume flow over pressure ratios of 1.6 to 2.2 bar and speeds of 4000rpm to 7000rpm. The maximum error for the modelled flow was 2.4% at a pressure ratio of 2.2 and a speed of 4000rpm.

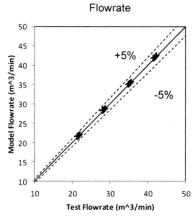

Figure 9 –"N" rotor profile – Comparison of modeled and test flow rate

4 CONCLUSIONS AND FURTHER WORK

The geometry model which calculates volume and flow area history was validated against measurements on oil free compressor with two different rotor profiles. The predicted performance is in good agreement over the full range of test results. This provides further verification of the established compressor model (3) and validates the application of the updated geometry calculation procedures. In section 3.2 the model shows a consistent level of accuracy for different rotor profiles.

The results in section 3.3 suggest that in the case of this particular "N" rotor retrofit the benefit of investing resources to re-design and manufacture a casing port modification is limited. The updated model is a useful tool when strategically prioritising development work.

In section 3.5 the model accounts for operational effects and provides improved correlation with test data over the full range of test pressures and speeds. There is some debate at this stage as to how applicable the empirical correction factors derived from this test are for different compressor builds that will have different assembly clearances.

Further verification and model development are underway – specifically to assess the use of operational clearance adjustments with oil injected compressors in which the leakage flow area and component temperatures are significantly influenced by the presence of oil. Additional cases with elevated suction pressures and different fluids would also be beneficial. Measuring PV diagrams on test will allow more detailed assessment of the results. In future the developed geometry program will be used to add more novel modelling capabilities such as compressors with variable lead rotors.

5   ACKNOWLEDGEMENT

The authors would like to thank Howden Compressors Ltd. for supporting this research.

6   REFERENCES

(1) Stosic N, Smith IK, Kovacevic A. *Optimisation of screw compressor design*, VIII European Congress on Fluid Machinery for the Oil, Gas and Petrochemical Industry, October 2002.
(2) Stosic N, Smith I, Kovacevic A. *Screw Compressors, Mathematical Modelling and Performance Calculation*: Springer; 2005.
(3) Hanjalic K, Stosic N. *Development and optimization of screw machines with a simulation model - Part II: Thermodynamic performance simulation and design optimization*. Journal of fluids engineering. 1997 September; 119: p. 664 - 670.
(4) Fujiwara M, Kasuya K, Matsunaga T, Watanabe M. Computer *modelling for performance analysis of rotary screw compressor*. In Purdue; 1984.
(5) Fleming JS, Tang Y, Cook G. *The twin helical screw compressor part 2: a mathematical model of the working process*. Proceedings of the Institute of Mechanical Engineers. 1998; 212.
(6) Mujic E, Kovacevic A, Stosic N, Smith I. *The influence of port shape on gas pulsations in a screw compressor discharge chamber*. Proc. IMechE. Part E. 2008; Vol. 222(no. 4): p. 211-223.
(7) Chen W, Xing Z, Tang T, Wu H. *Theoretical and experimental investigation on the performance of screw refrigeration compressor under part-load conditions*. International Journal of Refrigeration. 2011; 34: p. 1141 - 1150.
(8) Wu H, Xing Z, Shu P. *Theoretical and experimental study on indicator diagram of twin screw refrigeration compressor*. International Journal of Refrigeration. 2004 January; 27: p. 331–338
(9) HSIAO HH, WU YR, HSIEH HC. *Non-Uniform Clearance between Rotor Surfaces and Its Effect on Machine Performance in Twin-Screw Compressors*. In International Compressor Engineering Conference at Purdue; 2012; Purdue.
(10) Hsieh SH, Shih YC, Hsieh WH, Lin FY, Tsai MJ. *Performance analysis of screw compressors – numerical simulation and experimental verification*. Proc. IMechE Part C: J. Mechanical Engineering Science. 2012 March; 226: p. 968 - 980.
(11) Kovacevic A, Stosic N, Smith I. *The influence or rotor deflection upon screw compressor performance*. VDI BERICHTE. 2002.
(12) Sauls J, Powell G, Weathers B. *Transient thermal analysis of screw compressors part I - development of thermal model*. VDI Berichte. 2006(1932):19-29.
(13) Buckney D, Kovacevic A, Mujic E, Stosic N. *Some Aspects of Estimating Geometric Characteristics of Screw Compressors*. In International Compressor Engineering Conference at Purdue; 2012; Purdue.
(14) Buckney D, Kovacevic A, Stosic N. *Consideration of clearances in the design of screw compressor rotors*. IMechE International Conference on Compressors and their Systems, 2011, London.

Thermodynamic simulation of multi-stage screw compressors using chamber-based screw model

J Hauser, M Beinert, S Herlemann
GHH Rand Schraubenkompressoren GmbH, Germany

ABSTRACT

Compressed air is one of the most versatile and convenient power sources available in industry today. More than two-thirds of the overall cost of a compressed air package goes towards energy cost. Therefore, energy efficiency in compressed air systems is very important and must be continuously optimized.

Based on these facts, the paper will give an introduction in thermodynamic simulation of screws using a chamber-based model and its functionality. First, the performance characteristics e.g. volumetric efficiency and specific power consumption of two stage compressor will be calculated and compared with test results. Here the internal gap flows in screws as well as inner volume ratio of the compressor become key parts. Both of them will determine the operation behaviour especially the overall efficiency of the compressor unit. Next to that the intermediate pressure is directly linked to both stages and customer operating parameters e.g. drive speed and discharge pressure of the compressed air system. Finally the simulation tool will present the interaction of all different parameters to analyse the effect on compressor performance.

1 INTRODUCTION

Through progressing development of production-technology in the recent decades the production of screw compressors has steadily grown. These new technologies enable minimal tolerances and an extreme precise assembly of the compressors. Both of these attributes are critical for the operating characteristics and enabled the industrial relevance of today's screw compressors. Screw compressors are used wherever compressed, oil-free air is needed. The reliable prediction of the machine characteristics enables the minimizing of costs and time-consuming validation and examination of prototypes. For these cases software programs which supply comprehensive predictions of the operating performance are used.

The abstraction with simplified models, which describe the behaviour of the real machine, is essential for the development of screws. The continuous improvement of the product in case of screw compressors demand enormous calculating cycles for the predictions of the performance and contribute essentially to the development (1). This paper describes the investigation of the thermodynamic operational characteristics of oil free two stage screw compressors. Especially the resulting pressure between the compressor-stages and its impact has to be considered for the calculation of the power consumption. This investigation with validation by reference to real machines results in calculation-software for screw compressors which could be used in future designing and development processes.

2 MODELING OF SCREW COMPRESSORS

The modelling of positive displacement machines should especially shorten the phase of prototyping and in consequence thereof the development cycle should decrease the total effort of the development process in general. Before the development of a suitable model can start analyse of the system by incremental abstraction has to be done(2).

2.1 Chamber-based screw compressor modeling

One approach to model positive displacement machines represents the so called chamber-based model. It has been invented by Woschni (3) for the use of machines with internal combustion and has later transferred to screw compressors by Naujoks (4). One advantage of the modeling using the chamber-based model is the knowledge of the change of state of the fluid throughout the working cycle. The foundation for the simulation of the fluid's change of state is an advanced chamber-based model. This model serves as description of the fluid's change of state throughout the whole working cycle of an oil free single as well as two stage screw compressors.

For a chamber-based model the assumption is set down that the entire working chamber can consist of one or multiple chambers which contain homogenous fluid properties for each point in time. The knowledge of the screw's properties of geometry is hereby essential for the calculation of the fluid's change of state. Figure 1 shows the standardized volume curve as well as the inlet and outlet area of a typical screw compressor.

Other substantial characteristics of a screw compressor are the gaps which exist between the rotors themselves and between the rotors and the chasing. These gaps connect different chambers and therefor enable the exchange of fluid. Figure 2 shows the standardized volume curve as well as the gap areas of a typical screw compressor.

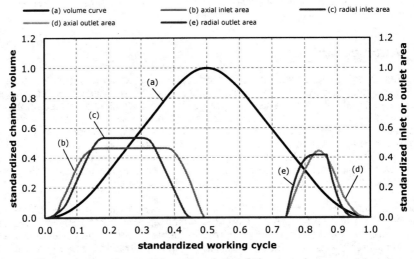

Figure 1: Necessary geometry data of the screw compressor

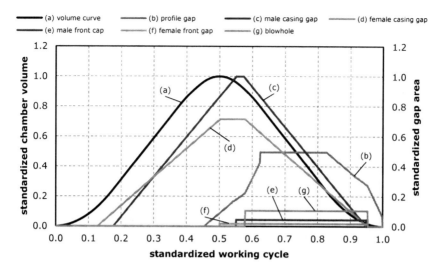

Figure 2: Necessary gap data of the screw compressor

All areas are used to describe the size of the connections which exist between chambers themselves and between chambers and inlet or outlet. The resulting gap flows directly affect the screw compressor's efficiency and are therefore the key element of the simulation. To realize the calculation of the thermodynamic behaviour inside a working chamber, the exchange of mass and energy through the gaps at each point of the working cycle has to be calculated. Depending on the geometric parameters of the machine the properties of the chamber connections has to be previously determined.

Geometric properties and defined boundary conditions serve as input for the prediction tool; results are absolute and specific fluid properties for each chamber to every point in time through the working cycle. Depending on the phase of the working cycle and the position of the rotors different connections exist between chambers, inlet and outlet. This chamber-model includes the simplification of an adiabatic thermal behaviour of the chambers to decrease the calculation time. This simplification has little impact for fast spinning screw compressors. The number of modelled chambers affects the quality of the results and the necessary calculation time significantly.

2.2 Multi chamber model

The multi-chamber model is an advancement of the initially mentioned one-chamber model and has been introduced by Naujoks (4) and Peveling (5). In comparison to the one-chamber model the multi-chamber model approximate the simulation results towards the results of the real machine and therefore increases quality of the simulation results. This is caused by the lack of inflowing mass flow through gaps for the one-chamber model which causes significant deviations in the temperature of the fluid.

Figure 3 shows the simulated results for the pressure and temperature inside the chamber over the working cycle for different tip speeds. The pressure curve indicates the indicated power consumption of the compressor while the temperature curve serves as an indicator for the quality of the simulation results. The differences in the pressure increase at the beginning of the discharging are caused by the influence of the speed; throttling effects during end phase of compression at discharge port of compressor. The differences in the chamber temperature during the compression are the result of the influence of the speed towards the exchange of heat energy.

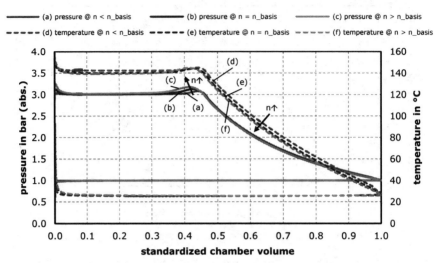

Figure 3: Results of multi chamber model

The simulation model of compressor is build up with chambers, connections between different chambers and of course linked with the suction and discharge side of the compressor. These connections allow the exchange of mass and energy in various ways. Dependent on the phase of the working cycle and the position of the rotors specific connections will or will not exist.

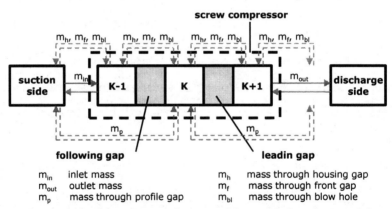

Figure 4: Existing connections used in multi chamber model

Figure 4 shows the different types of gaps and how they are connected between the chambers. The inlet creates a connection between the chamber and the suction side while the outlet creates a connection between the chamber and the discharge side. The gaps between the chambers are defined after their position and the machine parts they are between. The casing gaps, the front gaps and the blow hole create connections between adjacent chambers while the profile gap connects a chamber in the suction phase with a compressing or discharging chamber. Each of these gaps may exist to both pre- and after chambers. The amount and direction of mass and energy exchanged through these connections is determined by fluid's properties and the properties of the connection.

2.3 Modeling of a two stage screw compressor

The validation of the simulation model is realised by comparison towards the existing screw compressor CD8D which can be seen in Figure 5. The comparison supports the development of the simulation tools to ensure realistic results. The compressor CD8D is a oil free, two stage rotary screw compressor with a driving gearbox.

description	unit	LP-stage	HP-stage
Profile type	[-]	asym.	asym.
Male rotor teeth no.	[-]	4	4
Female rotor teeth no.	[-]	6	6
Center distance ratio	[-]	1	0.675
Swept volume ratio	[-]	1	0.25
Var. Discharge pressure	[bar]	3.0 – 3.6	7 – 11.5

Figure 5: Two stage screw compressor

The compressor block consists of the two compressors stages, the gearing box as well as supporting machine parts like the oil pump etcetera. The rotors of both stages are designed with an asymmetric profile type; four male rotor teeth and six female rotor teeth. The compressor block is applicable for discharge pressures between 7 and 11.5 bar abs.

3 THERMODYNAMIC CALCULATIONS

The calculation consists of the chambers passing the working cycle while the fluid's properties change and mass and energy are exchanged through the connections. The determination of the fluid's properties is realised within an iterative procedure which is based on Peveling's perceptions (5). This iterative procedure allows the determination of the behaviour of the fluid's properties pressure, temperature, mass inside the chamber while requiring the properties of the chamber volume and of the connections throughout the working cycle. For this procedure the working cycle is separated into differential time steps and the calculation is accomplished for each time step considering the change of the fluid's properties with changing chamber volume and exchange of mass and energy through the connections. The results of each iteration step are used as the starting point of the next iteration until convergence is reached while the ideal compression serves as the starting point of the first iteration step.

The principle of differential changes allows the separation of the fluid's change of state into several independent steps which could be calculated separately (5). This results the equation for the temperature inside the chamber for the next step as

$$T_{\varphi+\Delta\varphi} = \frac{m_\varphi}{m_{\varphi+\Delta\varphi}} T_\varphi \left(\frac{V_\varphi}{V_{\varphi+\Delta\varphi}}\right)^{\kappa-1} + \kappa \frac{\sum m_{in} T_{in} - \sum m_{out} T_{out}}{m_{\varphi+\Delta\varphi}} \tag{1}.$$

While φ stands for the actual time step and $\varphi+\Delta\varphi$ stands for the new time step. The exchanged mass is calculated as done by Peveling (5) with his assumptions applied. The rest of the fluid's properties can be determined through the equation of state when the change of mass is known through the mass balance. For following chambers the fluid's properties of the actual iteration step are used while for the fluid's properties of the leading chambers the last iteration step is used, because they have not yet been calculated in the actual step.

4   SIMULATION TOOL

The investigation of the operating characteristics of single and two stage screw compressors is realized with the help of specific simulation-software. Therefore the thermodynamic process is modelled through the usage of the described chamber-based model with iterative time step based calculation procedure, Figure 6. Geometrical parameters of the compressor are needed to use such a calculation program. The input will be generated with a special pretool automatically.

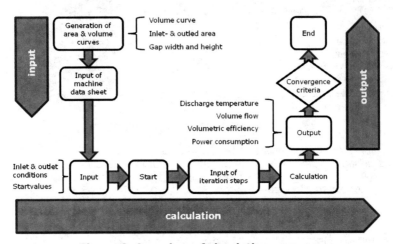

Figure 6: Overview of simulation process

4.1   Simulation of a two stage screw compressor

The simulation of a single stage compressor has been extended to contain the possibility to calculate two stage compressors with the purpose of the investigation of these kinds of machines. Especially the pressure between both stages has an important influence for the performance and is influenced by the operating condition and behaviour of both stages itself. Therefore the interaction of each component has to be considered.

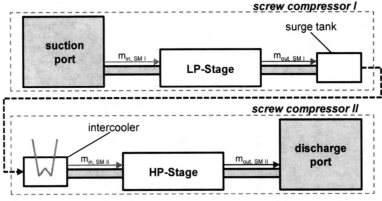

Figure 7: Overview of two stage simulation process

For the implementation of the thermodynamic calculation the two stage compressor block has to be simplified to reduce the number of components. Figure 7 shows the schematic of the two stage compressor model including the way of the mass flow.

The compressor system consists of a suction port which is connected to the low pressure stage as well as a connection between stage and surge tank. This model is similar to the model of the single stage compressor. The extra part for a two stage compressor is the surge tank, linked with an intercooler and this system represents the suction port of the high pressure stage of the airend. Therefore the calculation for each component has to be integrated into a single calculation process to ensure the modelling with correct interaction of all components.

5 SIMULATION RESULTS

The simulation of two stage screw compressors matches a series connection of two single stage compressors. The fluid's mass inside the surge tank is a result of the fluid streams out of the low pressure stage and into the high pressure stage.

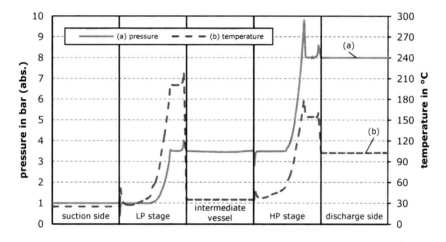

Figure 8: Result of two stage simulation iteration process

Figure 8 shows the fluid's pressure and temperature over the combined standardized working cycle off all five components, defined as suction side, low pressure stage, intermediate vessel, high pressure stage and discharge vessel. This figure enables the engineer to achieve comprehensive knowledge about the function of the two stage compressor and the interdependency between its components.

To validate the prediction tool the following figures show the intermediate pressure, the power consumption and the volume efficiency for both stages. The validation is accomplished by comparison with measurements on real machines.

Figure 9 shows the intermediate pressure as function of the low pressure stage's male rotor tip speed with fixed gearing to the high pressure stage. The expected deviation of the measurements is assumed within 5% and shows a really good correlation to reality. With higher speeds the intermediate pressure decreases because of better gap flow situation in the high pressure stage of the compressor.

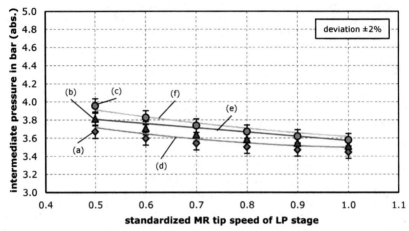

Figure 9: Intermediate pressure of two stage compressor

Figure 10 shows the power consumption as a function of the low pressure stage's male rotor tip speed. In case of the power consumption a good correlation between the simulated results and the measurements can also be achieved. Figure 11 shows the individual volumetric efficiency of both stages as a function of the low pressure stage's male rotor tip speed. And even in this case a very good correlation between the simulated results and the measurements can be achieved. With higher speeds higher volumetric efficiencies can be realised.

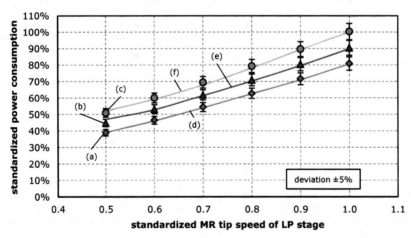

Figure 10: Power consumption of two stage compressor

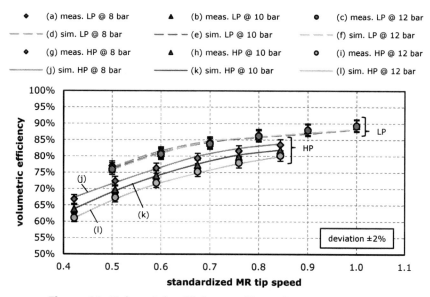

Figure 11: Volumetric efficiency of two stage compressor

5.1 Influence of the speed control

Two stage compressor units consist mainly of the two compressor stages and a gearbox which connects both compressor stages by a fixed gear ratio. For synchronous machine the gear ratio defines the rotors speeds and the volume flow. Therefore gearboxes with different gear ratios exist. For asynchronous machines the volume flow can be regulated via variable speed. The intermediate pressure is a function of the speeds of each compressor stage as shown in figure 9. Therefore the potential of separate drives for both stages has been investigated. This aims towards a reduction of the power consumption by optimising the intermediate pressure for the actual requested volume flow instead of the fixed optimisation for one volume flow. Figure 12 shows the combined power consumption of both compressor stages as a function of the high pressure stage's male rotor tip speed with a constant low pressure stage's male rotor tip speed. The reference tip speed of the fixed gearbox is marked in black and is not optimised for this volume flow.

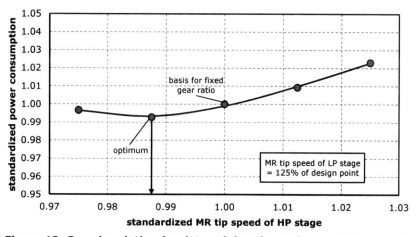

Figure 12: Speed variation for determining the optimum MR tip speed

It could be recognised, that a small change of the gear ratio, respectively a reduction of the tip speed would result in lower power consumption of approximately 1% for this operation point. This shows the optimisation potential of such a multi stage chamber model.

6 CONCLUSION

The implemented analysis of the thermodynamic operating characteristics of oil free screw compressors and the used chamber-based simulation model could be successfully verified. The introduced simulation tool empowers the engineer to investigate the comprehensive influence of various parameters with the purpose for optimisation of existing and the advanced development of new screw compressors. In addition the simulation tool enables advanced comprehension of influencing parameters and the interaction of rotor geometry and operating parameters.

REFERENCE LIST

(1) Hanjalic, K., Stosic, N.: Development and Optimization of Screw Machines with a Simulation Model, Part I: Profile Generation, Part II: Thermodynamic Performance Simulation and Design Optimisation ASME Transactions, Journal of Power and Engineering, Volume 119, 1997
(2) Rau, B.: Ein Beitrag zur Auslegung trockenlaufender Schraubenkompressoren, Dissertation Universität Dortmund, 1994
(3) Woschni, G.: Elektronische Berechnung von Verbrennungsmotor-Kreisprozessen. In: MTZ, Jahrgang 26, Heft 11, S. 439-446, 1965
(4) Naujoks, R.: Zustandsänderungen in trockenlaufenden Schraubenmaschinen – Ein Vergleich von Rechnung und Experiment, Dissertation Universität Dortmund, 1982
(5) Peveling, F.-J.: Ein Beitrag zur Optimierung adiabater Schraubenmaschinen in Simulationsrechnungen, Dissertation Universität Dortmund, 1987

Effect of variable volume index on performance of single screw compressor

M Masuda, H Ueno, T Inoue, K Hori, M A Hossain
Daikin Industries Ltd, Japan

ABSTRACT

The single screw compressor is widely used in air and water cooled central air conditioning systems for various unique characteristics compared to other compressors. It is possible to control the cooling or heating capacity by adjusting the mechanical slides built inside this type of compressor. However, the efficiency of the compressor is drastically reduced at part load conditions if the capacity is controlled by mechanical slides. We have studied the effect of volume index on performance when the capacity is controlled by varying the speed of the compressor. The volume index is varied by changing the position of the mechanical slides and the optimum volume index is obtained for rated and part load conditions. This paper will present the results of this study and propose the optimum volume index for obtaining best annual performance of the compressor.

1. INTRODUCTION

The single screw compressor first developed by Zimmern in the 1960's, is now widely used in refrigeration and air conditioning, air compression and chemical engineering systems. There are four different types of single screw compressors: Cylindrical-Plate (CP), Plate-Plate (PP), Plate-Cylindrical (PC), and Cylindrical-Cylindrical (CC) type, depending on the shape of the screw rotor and gate rotor (Fig.1). The most widely used single screw compressor is the CP type single screw compressor, which consists of one cylindrical screw rotor and two gate rotors. A screw rotor meshes symmetrically with two gate rotors to double the swept volume and balance thrust. The gate rotor is usually made of plastics. Therefore it is possible to maintain very small clearances between the contacts of the screw rotor and gate rotors, which results in not only achieving higher volumetric efficiency, but also lowering noise and vibration compared to twin screw compressors [1-5].

Compressors are selected for full-load design operation, which is commonly defined as the condition of maximum required capacity and pressure rise. Most of the time, however, compressors will run at operating conditions corresponding to a lower flow rate and a smaller pressure ratio than the original full-loaded design point. Recently a new approach to evaluate the efficiency of air conditioning system efficiency at full and part loads are introduced, which are called Integrated Part load Value (IPLV) [6].

The major loss mechanism of the single screw compressor are: leakage loss, oil drag loss, exit port flow loss, and pressure drop loss in the flow path including the oil separator. The first two mechanisms are independent of capacity and become relatively more dominant at lower flow rates. The last two losses are proportional to the square of velocity and become less pronounced at lower flow conditions. The built-in volume ratio of the screw compressor results in over-or-under compression losses at lower and higher pressure ratios than design, respectively [7].

Single screw compressors have two slide valves which are positioned on the screw rotor and are typically used to control the flow rate of the compressor. Similar types of valves are sometimes used to adjust the volume ratio depending on the operating condition of the air conditioning system. The structure is very complex if the flow rate and volume ratio are controlled by using only slide valves. Furthermore, there is more leakage loss; and the reliability is also a concern as there are more moving parts inside the compressor.

Therefore, we proposed a new type compressor where the flow rate is controlled by a variable frequency drive (VFD) system and slide valves are used to adjust the volume index (Vi), depending on the operating conditions. The purpose of this paper is to introduce the effect of variable volume index on the full load and part load performance; more specifically on the compressor's Integrated Part Load Value (IPLV) when it is run by a variable frequency drive system.

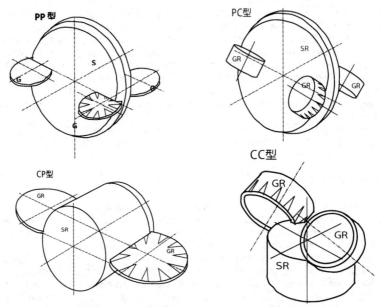

Fig.1 Single screw compressor types

2. INTEGRATED PART LOAD VALUE (IPLV)

The Integrated Part Load Value (IPLV) is a performance characteristic of modular, screw, and centrifugal chillers, which are developed by the Air-Conditioning and Refrigeration Institute (ARI) [6]. IPLV is most commonly used to describe the performance of a chiller capable of capacity modulation. Unlike an EER (Energy Efficiency Ratio) or COP (Coefficient of Performance), which describe the efficiency at full load conditions, the IPLV is derived from the chiller efficiency while operating at full load as well as various part load conditions. Since a chiller does not always run at 100% capacity, the EER and/or COP are not ideal representations of the typical chiller performance. The IPLV is a very important value to consider since it can affect energy usage and operating costs throughout the lifetime of the chiller.

The IPLV is calculated using the efficiency of the chiller system while operating at capacities of 100%, 75%, 50%, and 25%. For the purpose of chiller, the operational conditions are described in the ARI Standard 550/590-2003, which is shown in Table 1. IPLV rating can be calculated using the following equation:

IPLV = 0.01A+0.42B+0.45C+0.12D

Where:
A=COP or EER@100%Load
B=COP or EER@75%Load
C=COP or EER@50%Load
D=COP or EER@25%Load

Table 1: IPLV conditions described in standard 550/590-2003 declared by Air-conditioning and Refrigeration Institute (ARI)

	100% Load	75% Load	50% Load	25% Load
Air cooled condenser [EDB]	35.0°C	26.7°C	18.3°C	12.8°C
Water cooled condenser [EWT]	29.4°C	23.9°C	18.3°C	18.3°C
Evaporator (all types) [LWT]	6.7°C	6.7°C	6.7°C	6.7°C

EDB: Entering air dry-bulb temperature
EWT: Entering water (liquid) temperature
LWT: Leaving water (liquid) temperature

3. VARIABLE VOLUME INDEX (VI) SYSTEM

Screw compressor achieve its objective of raising the pressure of a gas by trapping a fixed volume of gas on the suction side and progressively decreasing that volume of gas which results raising its pressure at the discharge side. The ratio of the volume of the gas trapped in the thread of the screw at the start of the compression process to the volume of trapped gas in the thread when it begins to open into the discharge port is known as the compressor's "volume ratio" or "volume index", Vi. Screw compressor have slide valve mechanism which is typically used to adjust the flow rate of fixed frequency drive (50Hz or 60Hz) compressor. The inlet gas is bypassed by moving the slide valve to discharge side of the screw rotor which are shown in Fig.2. In order to obtain an improved IPLV of the compressor, the flow rate should be controlled by VFD and the slide valve should be used to adjust the volume index, depending on operating pressure ratio of the system. This is because when the flow rate is controlled by slide valves, it is not possible to adjust the volume index with the ideal volume index obtained from the operating conditions of the compressor.

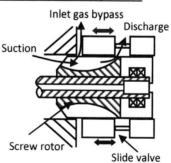

Fig.2 Traditional capacity control system which operates through bypassing the inlet gas by sliding the slide valve to the discharge direction

Fig.3 shows how the volume index is adjusted depending on the operating conditions of the system. In this study, the design has one slide valve per compression side of the gas end which moves along the axis of the screw rotor. For example when the compressor operates at a low condensing pressure condition, a "volume index slide valve" within the compressor moves in a direction towards the suction side of the screw rotor. This relieves the trapped refrigerant vapour from the compressor earlier in the compression process before it has the opportunity to rise in pressure through a decrease in volume to a point that would result in over-compressing the gas. During high condensing pressure conditions, a "volume index slide valve" moves away from the suction side of the compressor to delay the discharge of compressed gas until it reaches a higher pressure. The suction timing is fixed but the discharge timing is varied by moving the slide in a backward or forward direction, resulting in a change in volume index.

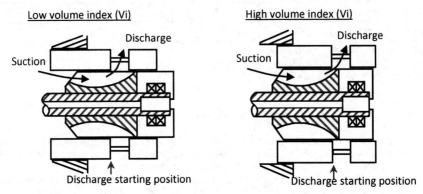

Fig.3 Variable volume index adjustment system which operates by positioning the slide valve so that the discharge timing can be changed

Fig. 4 describes an example of how the volume index varies with the variation of the slide valve position. As shown, it is possible to decrease the volume index from 3.4 down to nearly 1.0 by moving the slide valve to the suction side from its maximum position.

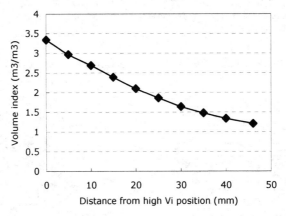

Fig.4 Change of volume index with positioning of the slide valve

4. RESULTS

Performance data of a single screw compressor of size 200RT at various operating conditions have been tested on a test rig, which is shown in Fig.5. This test data is important not only to obtain an integral impression of performance characteristics of variable frequency drive (VFD) and variable volume index (Vi) compressors, but also to get a clear comprehension of the influence of VFD and Vi on performance. Moreover, experimental data obtained from a wide range of operating conditions is very useful for validating the simulation results. All the results are at R134a.

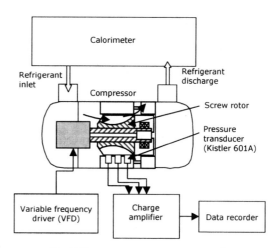

Fig.5 Screw compressor test rig

The pressure-volume diagram of a single screw compressor for different volume indexes measured experimentally is shown in Fig.6. A set of stationary transducers was positioned in compressor housing so that the pressure of a groove could be measured from the start of compression to the end. It is possible to estimate different losses like inlet pressure drop loss, outlet pressure drop loss, and the volume index loss from the measured pressure volume diagram. The volume index loss is the loss caused by the pressure ratio (ratio of the compressor outlet and inlet pressure) not matching the actual volume index. As shown in Fig.6, for a constant compressor inlet and outlet pressure (for example pressure ratio of IPLV 25% load which is 1.5), there are two pressure-volume diagrams for two different built-in volume indexes. The volume index is varied by changing the outlet timing of the gas which was done by changing the position of the slide valve. As shown in Fig 6a, a large overpressure is measured inside the compressor when the volume index is 3.3. The pressure-volume diagram obtained by experiment is very close to the theoretical pressure-volume diagram when the compressor is operated at a volume index of 1.2 (as shown in Fig 6b). These results indicate that there is a large indicated efficiency loss when the compressor is operated at different volume indexes instead of actual (ideal) volume indexes.

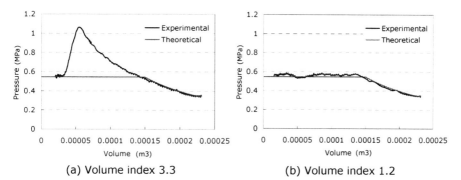

(a) Volume index 3.3   (b) Volume index 1.2

Fig.6 Experimental pressure-volume diagram for different volume indexes at a constant speed (25% of the full load) and constant suction (0.35 MPa) as well as discharge pressure (0.54 MPa)

To explore the influence of volume index on compressor efficiency, the full load and part load compressor efficiency for different volume indexes are compared at four different condenser and evaporator temperatures. The compressor load is adjusted by using a variable frequency drive (VFD) system. Fig.7 shows results of the compressor efficiency for volume index 1.2 to 3.4. For a constant condenser and evaporator temperature, the pressure ratio is fixed. For example, at full load conditions, for a condenser temperature of 49.5°C and evaporator temperature of 2.3°C, the compressor needs to develop a pressure ratio of 4.1 with an estimated ideal volume ratio of 3.0. As shown in Fig.7, the performance of the compressor is maximum for a volume index of approximately 2.9, which is close to the ideal volume index. The same phenomenon is observed when the compressor operates at 75%, 50% and 25% of IPLV load conditions. It is clearly observed that at every IPLV load conditions there is a particular volume index for which optimum performance of the compressor is achieved.

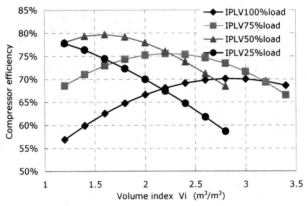

Fig.7 Compressor isentropic efficiency with regards to volume index at four IPLV load conditions

A comparison of full load and part load performance of a single screw compressor for slide valve capacity control and VFD driven capacity control while it is running at fixed as well as variable volume index is shown in Fig.8.

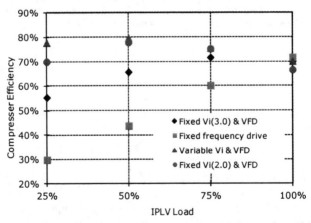

Fig.8 Comparison of compressor isentropic efficiency for slide valve capacity control and variable frequency driven capacity control (fixed and variable volume index)

Results are shown for IPLV load of 100%, 75%, 50% and 25%. The pressure ratio of each IPLV load condition is 4.1, 2.9, 2.0 and 1.5 respectively. As shown in Fig.8, when the load of a fixed frequency drive compressor is controlled by slide valve, isentropic efficiency of the compressor drastically falls with regards to decrease in its load. Isentropic efficiency of the compressor at full load operation is 71.5% while the isentropic efficiency at 25% load operation falls to 30%. When the compressor with a fixed volume index of 3 and the load are adjusted by the VFD, it is possible to keep the full load efficiency at nearly 70%; however the isentropic efficiency at 25% load operation falls to 55%. It is clearly observed that although the part load efficiency of a variable frequency drive compressor with a fixed volume index of 2 is relatively higher than with a fixed volume index of 3, the full load efficiency falls to 66.5%. The graph shows that it is possible to obtain higher efficiency for full load as well as part load when the compressor is designed with variable volume index.

A comparison of numerical results with experimental results is shown in Fig.9. The numerical results were predicted by using STAR-CD, the most commonly used CFD software. The k-ε high Reynolds number turbulence model was used for numerical simulation. Similar to experimentally evaluated results, the numerical simulation was carried out for different volume indexes when the compressor was run at different speeds. The results are presented for a suction pressure of 0.34 MPa and a discharge pressure of 0.69 MPa with a volume index of 2.4 and 50% of full load speed. A large over pressure was observed in the numerically predicted indicator diagram, which is same as the experimental results when the volume index did not match with the ideal volume index. The numerical results agree well with the experimental results.

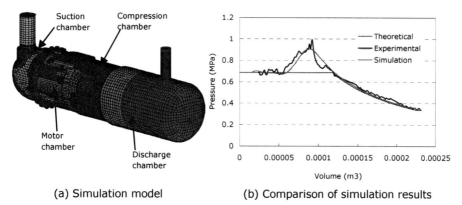

(a) Simulation model   (b) Comparison of simulation results

Fig.9 Pressure volume diagram obtained by numerical simulation and comparison of numerical results with experimental results

5. CONCLUSION

This paper deals with the effect of volume index on the performance of a variable frequency drive compressor when it is operated at full load as well as part load (IPLV) conditions. A large over pressure loss was observed when the volume index did not match with actual volume index predicted from the compressor operating conditions. An optimum volume index was obtained at each IPLV operating condition for which the efficiency of the compressor was at maximum. By using the combination of a variable frequency drive capacity control system and variable volume index adjustment system using slide valves, the maximum efficiency of the compressor was obtained at full load as well as part load conditions - compared to a fixed volume index and slide valve capacity control system.

In comparison with the experimental results the simulation achieved a moderate degree of accuracy.

REFERENCES

[1] Zimmern, B. and Patel, G. C., "Design and operating characteristics of the Zimmern single screw compressor", International Compressor Engineering Conference, Purdue (1972)
[2] Zimmern, B. "From water to refrigerant: twenty years to develop the oil injection-free single screw compressor", International Compressor Engineering Conference, Purdue (1984)
[3] Clark J. R., Hodge J. M, Hundy G. F. and Zimmern B., "A new generation of screw compressors for refrigeration", Institute of Refrigeration (1975)
[4] Sun G., "The investigation of basic geometric problems of the single screw compressor" International Compressor Engineering Conference, Purdue (1988)
[5] Murono T., Ueno H., Ohtsuka K., Takahashi T. and Susa T., "Development of single screw compressor using new tooth profile", International Conference on Compressors and their Systems, London (2007)
[6] ARI Standard 550/590, "Performance rating of water-chilling packages using the vapour compression cycle", Air-conditioning and Refrigeration Institute, (2003)
[7] Brasz, J. J., "Comparison of part load efficiency characteristics of screw and centrifugal compressor", International Compressor Engineering Conference, Purdue (2006)

Investigations of deformation failure of the slide valve in single screw refrigeration compressors

F Liu, Z L Wang, W F Wu, Q K Feng
School of Energy and Power Engineering, Xi'an Jiaotong University, China

ABSTRACT

The slide valve is used to regulate the capacity of a single screw refrigeration compressor. However, many cases have shown that it always stops doing the regulation a short period of time after the boot-start of the compressor. This failure might be attributed to the thermal and gas force deformation of the slide valve or the cylinder. In this paper, simulation of thermal and gas force deformation of a single screw refrigeration compressor with finite element method was carried out. Results showed that size of the deformation was far bigger than the clearance. It means a hard contact between the slide valve and its groove. It was believed that the deformation resulted in the failure of the slide valve.

Keywords: single screw refrigeration compressor, slide valve, deformation, failure

1. INTRODUCTION

The single screw compressor was invented by Zimmern B. in 1962 with a prototype at the power of 10 HP [1]. Then, in the middle 1970s, the application of the single screw compressor was extended to the refrigeration field, meanwhile, the lubrication system of oil injection was also developed to water injection (air compressor) and refrigerant injection (refrigeration compressor)[2,3]. The single screw compressor is now commonly used in the field of air compression, air condition, heat pump, petrochemical industry and etc.

For refrigeration compressors, slide valves are always used to regulate the capacity. However, many cases have shown that, the slide valve stops work a short period of time after the boot-start of the compressor. This might be attributed to the thermal and gas force deformation of the slide valve or the cylinder.

This paper built the three dimensional model of a single screw refrigerating compressor. Finite element method was used to analyze the thermal and gas force deformation of the cylinder and the slide valve with ANSYS. According to the obtained results, deformations caused by the thermal and gas force factors were the main reason for the failure of the slide valve.

2. MODELING

The models were built by using the Solidworks and then were input into the Workbench to do the finite element analysis. The models of the slide valve and the cylinder were all individual models without other assembly parts and they were built on the basis of the compressor's schedule drawing.

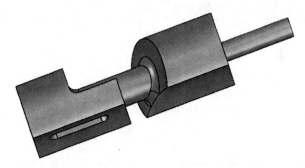

Fig.1: Simplified model of the slide valve

Some characteristics, such as the small chamfers near the suction and exhaust side and some small holes of the slide valve were ignored to simplify the model of the slide valve (Fig.1).

Fig.2: Simplified model of the cylinder

In order to reduce the computing time, the model of the cylinder was also simplified (Fig.2) as follows, under the premise of ensuring the practicability of the computing results [4].

1. Ignore the small characteristics of the compressor, including most of the small lug bosses, chamfers, round corners and etc.;
2. Ignore the characteristics which have little influence on the deformation, for instance, all the bolt holes in the motor side, the power distribution box side, the star wheel side and the exhaust side;
3. Try to maintain the cylinder structure design's consistency which was provided by the manufacturer, thus to reflect the structure characteristics of the components accurately.

3. FINITE ELEMENT ANALYSIS

Thermal analysis includes steady-state analysis and transit analysis. The temperature and pressure inside the compression chamber change constantly during the operating of the compressor. However, in terms of the slide valve and the cylinder, the temperature's rapid change has little influence on the temperature field. After the compressor's stable operation for a period of time, the temperature field would remain constant. Therefore, in this paper, steady-state analysis was used as the thermal analysis for the slide valve and the cylinder. Theoretically, the spatial discretization mode of the elasticity mechanics and the three-dimensional temperature field are the same, as do the unit and the displacement mode. The main different is that the displacement is a vector, while the temperature field is a scalar. Consequently, the temperature field, acquired by the thermal analysis, was used as the boundary condition of the model and then was input into the static structure analysis to do the thermal-structure coupling analysis.

3.1. Mesh generation

In this section, the input module of the ANSY Workbench was used to input the model of the cylinder and the slide valve. Then, both of the cylinder and the slide valve were meshed after the parameters of the material properties were set in the Workbench. The material parameters of the single screw refrigeration compressor is given in Table 1.

Table 1 Material parameters of the single screw refrigeration compressor

Components	Material	Poisson ratio	Elastic modulus GPa	Density kg/m^3	Linear expansion coefficient 10^{-6}/K	Thermal conductivity W/(m^2·K)
Cylinder	HT250	0.25	120	7300	11.8	50
Slide valve	QT500-7	0.27	173	7200	12	35.5

Directly affecting the computing accuracy and speed, mesh generation is a key part of the finite element analysis. Five mesh generation methods are provided by the Workbench:

1. Automatic method. If the geometry is irregular, the program will generate the tetrahedral mesh automatically. In the opposite, hexahedral mesh will be generated.
2. Tetrahedrons method, including Patch Conforming method and Patch Independent method.
3. Sweep Meshing method.
4. Multi Zone method.
5. Hex Dominant method.

In this paper, the model was meshed by using Patch conforming method (Fig.3). Local grid refinement was used for some of the key surfaces according to the existing computer resources. Here, the Workbench used the Solid 187 Geometry Unit to mesh the model. Having secondary displacement, plasticity, hyperelasticity, stress stiffening and etc., this unit is suitable for simulating the irregular meshes.

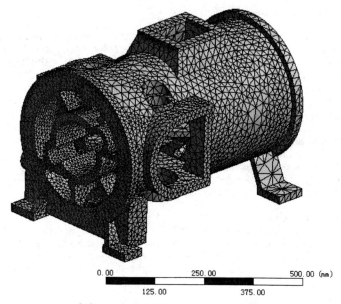

(a) Mesh generation of the cylinder

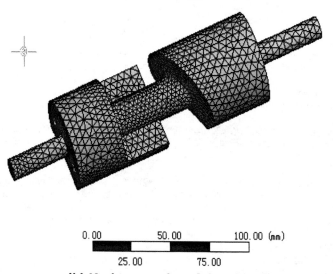

(b) Mesh generation of the slide valve

Fig.3: Mesh generation

The mesh quality is shown in Fig.4. The number of the grid units and nodes for the cylinder were 55561 and 97749, respectively. The slide valve had 28557 grid units and 44202 nodes. The average mesh quality was 0.72 for the cylinder and 0.83 for the slide valve. After times of local grid refinement and regulation, the deformation error was within 2%. And the mesh quality could satisfy the requirements for the thermal-structure coupling analysis.

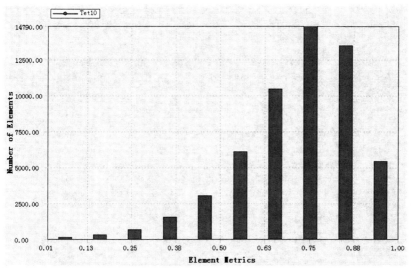

(a) Mesh quality of the cylinder

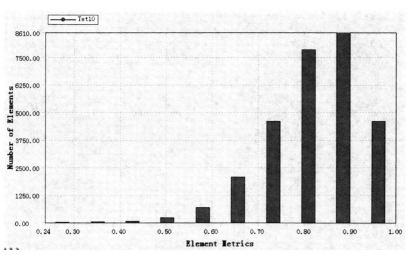

(b) Mesh quality of the slide valve

Fig.4: Mesh quality

3.2. The ANSYS finite element analysis

The boundary condition was exerted to the finite element model on the basis of the single screw refrigeration compressor's operating conditions. The gas pressure and the temperature load inside the basic-volume consisted by the screw channel, the star wheel and the cylinder could be calculated. The details would not be expatiated here. The temperature inside the screw channel rose from the suction end to the exhaust end. The temperature for the suction end was given (the suction temperature) and the temperature for the exhaust end was related to the refrigerant's physical property, the compressor's rotational speed, the oil injection quantity and the exhaust pressure.

In this paper, a full-load condition (a gas flow rate of 100%) was taken as an example to do the ANSYS finite element analysis. The refrigerant was R134a. Noticed the preheating of the inlet gas, the suction temperature was assumed to be 25°C and the exhaust temperature was 80°C. The suction, exhaust and condensing pressure were 0.3496MPa, 3.260MPa and 1.681MPa, respectively.

The three-dimensional steady-state heat conduction differential equation without inner heat source but constant property was taken as the numerical model for the temperature distribution, seen in Equation (1).

$$\frac{\partial^2 t}{\partial x^2} + \frac{\partial^2 t}{\partial y^2} + \frac{\partial^2 t}{\partial z^2} = 0 \tag{1}$$

In the steady-state heat conduction problem, the boundary condition is the only definite condition and the three general boundary conditions are as follows [5]:
1. Dirichlet condition: the boundary temperature of the subject is given, t_w=constant.
2. Neumann condition: the boundary heat flux is given, q_w=constant.
3. Robin condition: $-\lambda(\frac{\partial t}{\partial n})_w = h(t_w - t_f)$

Where t_w is the boundary temperature, t_f is the fluid's temperature and h is the heat transfer coefficient.

In this paper, the Dirichlet condition was used for the finite element model. The temperature distribution achieved from the steady-state thermal analysis was then input into the static structural analysis through the Workbench to realize the thermal-structural coupling analysis. The analysis process is shown in Fig.5.

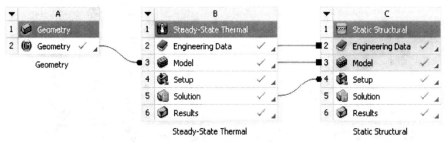

Fig.5: Process of the thermal-structural coupling analysis

3.3. Results and discussion

The two slide valves and their grooves were distinguished by left and right for convenient descriptions, shown in Fig.6.

Fig.7 shows the cylinder's deformation computing results through the direction of X and Y, under the full-load condition with a gas flow rate of 100% and an exhaust temperature of 80°C. It can be seen that a concentrated deformation appeared in the exhaust end of the compressor through the X direction with a maximum

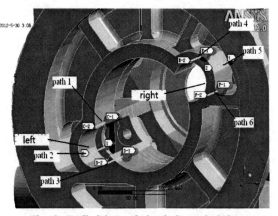

Fig.6: Definition of the left and right

deformation of 0.118mm and a minimum deformation of -0.111mm. In the Y direction, the largest deformation took place at the top of the exhaust end which was 0.256mm and the minimum was -0.259mm.

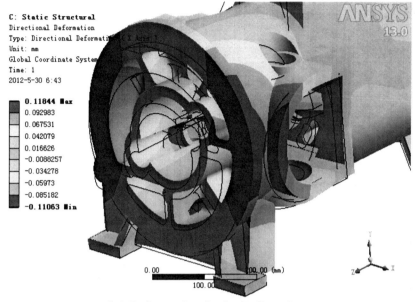

(a) Deformation in the X direction

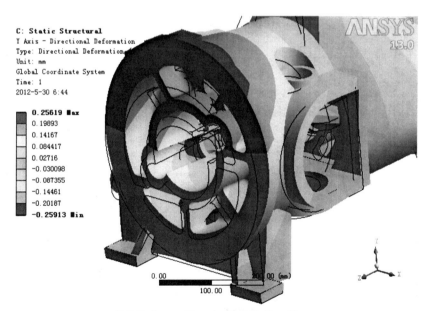

(b) Deformation in the Y direction

Fig.7: Deformation of the cylinder

Fig.8 describes the deformation computing results of the two grooves (left groove and right groove). Obviously, the deformation increased from the suction end to the exhaust end through the axial direction of the screw rotor for both of the grooves. Specifically, they all had the largest deformation in the exhaust end (0.121mm for the left and 0.157mm for the right) and the minimum deformation in the suction end (0.0208mm for the left and 0.0913 for the right).

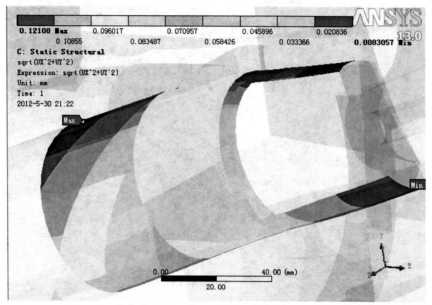

(a) Deformation of the left groove

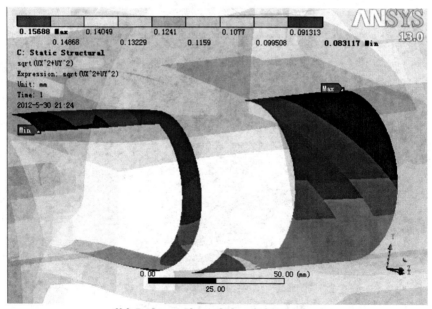

(b) Deformation of the right groove

Fig.8: Deformation of the groove

The computing results of the two slide valves are shown in Fig.9. It can be seen clearly that, in contrast to the cylinder, the maximum deformation for the slide valve happened in the suction end of the compressor (0.216 mm for the left and 0.260mm for the right), while the minimum deformation appeared in the exhaust end (0.0905mm for the left and 0.103mm for the right).

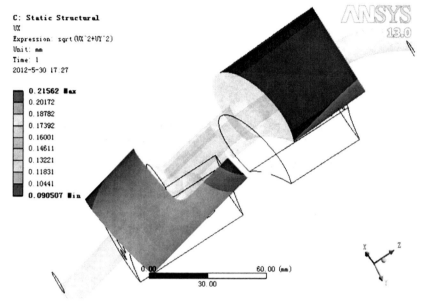

(a) Deformation of the left slide valve

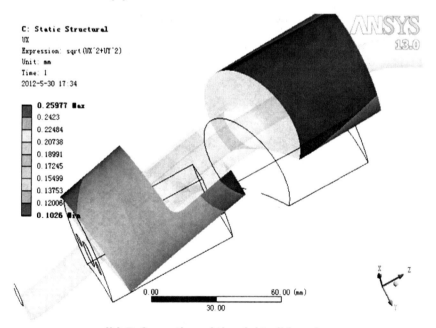

(b) Deformation of the right slide valve

Fig.9: Deformation of the slide valve

Hence, the results show that the largest deformation for the left slide valve and its groove were 0.216mm and 0.121mm with a difference of 0.095mm and for those in the right side, these were 0.260mm, 0.157mm and 0.103mm. However, the designed clearance was just 0.03mm which was much smaller than the deformation difference, therefore a hard contact was certainly occurred between the slide valve and its groove and finally led to the failure of the slide valve.

4. CONCLUSIONS

(1) A twist deformation could certainly happen to the cylinder and the slide valve for the single screw refrigeration compressor due to the thermal and gas force factors.

(2) The largest deformation of the slide valve appears in the suction end, while that of its groove happens in the exhaust end. Besides, the deformation of the slide valve is much larger compared to its groove in the suction end.

(3) The deformation difference between the slide valve and its groove is far bigger than the designed clearance. As a result, a hard contact will turn up and finally lead to the failure of the slide valve.

(4) It is suggested to do the thermal and gas force deformation analysis for the cylinder and the slide valve before the determination of the designed clearance to avoid the deformation failure.

REFERENCES

[1] Zimmern B. Worm rotary compressors with liquid joints: USA, 3,180,565[P], 1965.
[2] Zimmern B., Patel GC. Design and operating characters of the Zimmern single screw compressors [C]. Purdue, 1972: 96-99.
[3] Zimmern B. From water to refrigerant: twenty years to develop the oil injection-free single screw compressor [C]. Purdue, 1984: 513-518.
[4] G.Y. Pu. Basic Course and Example Explanation for ANSYS Workbench 12, China WaterPower Press, Beijing, 2010. p.367.
[5] J.H.Wu, G.X. Jing. Investigation on Suction Process of Oil-Flooded Rotary Positive Displacement Compressors [J].Fluid Machinery, 1995, (5): 7~10.

SCROLL COMPRESSORS

Assessment of reed valve dynamic behavior in a scroll compressor through visualization

A J P Zimmermann[1], P S Hrnjak[1,2]
[1] ACRC, MechSE Department, University of Illinois at Urbana-Champaign, USA
[2] CTS - Creative Thermal Solutions Inc., USA

ABSTRACT

Measurement of reed valve dynamics and movement in compressors is a well-known technique and can be performed by commercial proximity sensors, strain gauges or in house developed sensors embedded into the valve plate. Although timing and lift information is available through these methods, other modes of fluttering such as torsional movement a more difficult to be obtained. This work tries to address these unknowns by providing real operation information about valve dynamics through high speed imaging and processing of valve lift, valve speed and the presence of torsional movement. A scroll compressor was used and ran in a full system set up, controlling condensing and evaporating temperatures, as well as the speed of the compressor so the results can be used for variable speed systems. It was found that the information obtained from videos fits well with what has been reported in the literature and that the torsional movement was indeed observed and estimated. However, due to low impact velocities, even with torsional movement, valve reliability is not at risk under the conditions in which the visualization was performed.

1. INTRODUCTION

Valve reliability is one of the most crucial parameters in compressor design, as is efficiency. In scroll compressors only discharge valves are present since there is no need to suction valves due to design characteristics.

In order to properly design a compressor valve system, the engineer has to get access to data about valve dynamic behavior, that means, valve displacement, speed and modes of vibration as a function of time, to calculate valve stresses and reduce failures by fatigue or impact [1,2]. Traditionally, capacitive or inductive transducers have been used to gather such information from compressors being tested under real operating conditions. In addition, strain gauges have been used at the base of the valve so that numerical models can be validated and used to determine dynamic behavior [3]. Optical methods have also been used [4,5], however, very little has been made available about actual images of real operating valves under real running conditions. It is quite easy to justify the lack of such information since the majority of compressing machines have complex, intricate geometries that make accessing the locus of the valve very difficult. However, visualization of valve motion has been done before [4] in small hermetic reciprocating compressors for household refrigerators. In some cases, access to these locations is quite simple and easy to realize, as in low pressure shell scroll compressors. There, the only barrier between the discharge valve site and the naked eye is a steel shell cover. Figure 1 shows typical shapes for curves of displacement for reed valves measure in small hermetic reciprocating compressors [4].

With the objective of start filling this information gap, this work focuses on the high speed visualization of the discharge valve movement in a scroll compressor and the extraction of displacement, speed, impact velocity and indication of torsional movement (wobbling) through image processing.

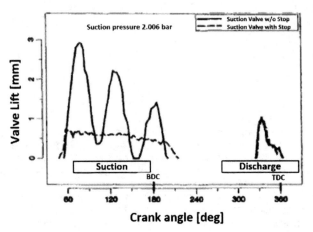

Figure 1. Typical shapes of reed valve displacement[4].

2. METHODS

2.1.   System configuration and valve visualization arrangement

In order to visualize the flow inside the discharge plenum of the scroll compressor, a total of 5 sight glasses were installed around the discharge cover. The sight glasses provided access for lighting and also to capture the images. There was also a need to modify the internal structure of the discharge plenum in order to remove physical barriers to the visualization of the valves by machining out a large amount of material, however it is not believed that it will interfere with the valve dynamic characteristics.

A high speed CMOS camera with maximum resolution of 512x512 pixels at 2200 frames per second (fps) and 256 shades of gray was used to record the videos. Two lenses were used in the visualizations: a macro lens for a broader area view so there could be an understanding of how the valves opening sequence and timing work; and a microscope lens assembly used for better spatial resolution in determining valve displacement. Videos were recorded at 5,000 frames per second (fps), since the compressor ran at 38Hz.

Figure 2a shows the internal machining and Figure 2b shows the way the compressor is assembled in the facility with the lighting fixtures and camera positioned. A sample image of each of the lenses is shown on Figures 2c for macro, and 2d for microscope.

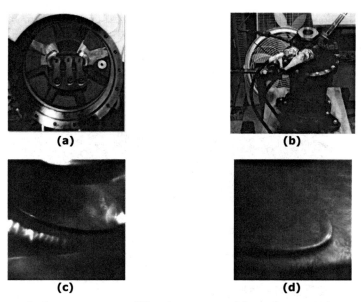

Figure 2. Compressor modifications, assembly and sample images.

The compressor used is an R410A scroll compressor, using PVE oil (FVC32), with low pressure shell, designed to operate in residential air conditioning. It is installed in a full residential air conditioning system facility comprised of two environmental chambers (indoor and outdoor). Inside each of the chambers, the respective heat exchangers are inserted into wind tunnels so air flow rate and temperatures can be controlled to set the desired suction and discharge pressure conditions to the compressor. The compressor speed is controlled with a variable frequency drive and was kept at 38Hz, which is lower than the nominal speed of 50Hz. This had to be done to avoid dense mist inside the plenum which made visualization impossible at some conditions.

The system is instrumented with a coriolis type mass flow meter with uncertainty of ±0.5% of the actual measurement, suction and discharge pressures are measured with pressure transducers with uncertainty of ±1%. Temperatures were measured at the discharge pipe using a T-type immersion thermocouple with uncertainty of ±0.25°C.

2.2. Image processing

Information about valve timing and displacement were obtained using Vision Research Phantom 675.2 software built-in tools. The valve speed was obtained by differentiating the valve displacement with respect to time. An edge Sobel horizontal filter was used to enhance the valve edges so displacement measurement could be easily performed. The displacement and speed were determined by tracking the position of three points in the valve, two on the edge and one on the top surface of the valve. The pixel values on each of these points were recorded such as for each frame of the video, the position of the points could be determined and therefore displacement could be calculated in relation to the closed valve position. The resultant of the vertical and horizontal displacement is reported in the charts. Figure 3 illustrates an example calculation. A ten frames difference was used on purpose to exaggerate the displacement between frames.

The three points chosen are shown on Figures 5, and 6 and 7 inside the charts. The scaling factor was obtained by using the valve thickness (0.3mm) as a reference dimension. In this case one pixel corresponds to 15µm. The estimated measurement uncertainty was ±1 pixel due to the manual nature of the measurement and therefore results in ±15µm for displacement, this leads to a speed uncertainty of ±0.075m/s.

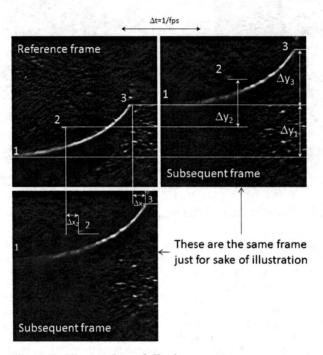

Figure 3. Illustration of displacement measurement.

3. RESULTS AND DISCUSSION

All results presented in this section are based on one video recorded at 5,000 fps, of one of the auxiliary valves, with compressor running at 38 Hz, discharge pressure was 2500 kPa, discharge temperature was 125 °C, the mass flow rate of refrigerant was 40g/s, OCR was 0.4%. The first cycle of valve operation of the video is shown as a series of six images on Figure 4.

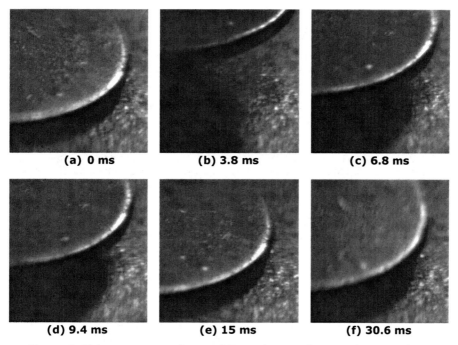

Figure 4. Time sequence of one of the valves under steady operation;
P_{disch}=2.5 MPa, T_{disch}=125 °C, $\dot{m}$=40 g/s, f=38 Hz, OCR=0.4 %.

3.1. Valve timing

Observing videos from several cycles one can determine the consistency of valve timing. Table 1 shows timing information for several cycles including cycle time period based on the operating frequency of the compressor, total cycle time, open time, percent open time and cycle to cycle variations for the variables of interest.

Table 1. Valve timing.

Cycle #	1/f [ms]	Total [ms]	Open [ms]	% Open	Var. total [ms]	Var. open [ms]
1		29.8	15.0	50.33	0	0
2		29.8	14.6	48.99	0	-0.4
3		29.8	14.8	49.66	0	0.2
4		29.8	14.6	48.99	0	-0.2
5	26.3	29.8	14.6	48.99	0	0
6		29.8	14.8	49.66	0	0.2
7		29.8	14.8	49.66	0	0
8		30.0	14.8	49.33	0.2	0
9		29.6	14.6	49.32	-0.4	-0.2
10		30.0	14.8	49.33	0.4	-0.2

Results shown in the table do not indicate significant variations in valve timing from cycle to cycle for the sample number of cycles taken for this analysis. The small variations present are on the order of the frame spacing and therefore should not be taken into account as having any physical meaning.

3.2. Valve velocity and displacement

Three curves in Figures 5 and 6 represent three different locations that were tracked for displacement. The shapes of the three curves have a good agreement but their values differ a little. It is important to note that the speed values are absolute, that means after about 4 ms, the valve is actually moving down towards the seat.

The most interesting observation comes from the displacement difference between points 3 and 1, which indicates presence of a torsional movement. This is very important, especially when it manifests itself close to the moment when the valve is approaching the seat for closing. Figure 7 shows a plot of the difference between point 3 and point 1 displacements. There is some significant torsional movement when the valve reaches its highest lift and also some minor movement of this type when the valve is approaching the closing moment. However, it is also noteworthy that the impact velocity in this case is very small in between 0.2 m/s and 0.3 m/s, posing no great risks for reliability of the valve.

As far as the authors know, this is the first time that such data has been extracted from visualizations in real operation of scroll compressors, which not only brings the physical quantities usually obtained with proximity transducers but also provides better understanding by actually watching the phenomena take place.

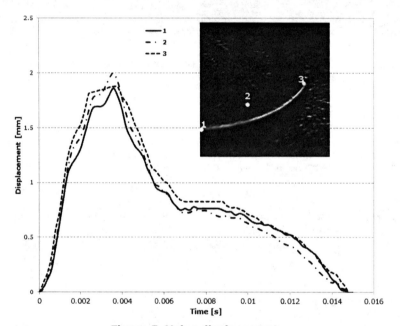

Figure 5. Valve displacement.

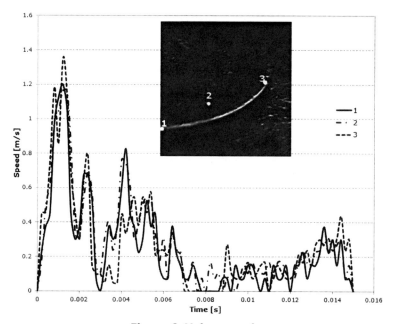

Figure 6. Valve speed.

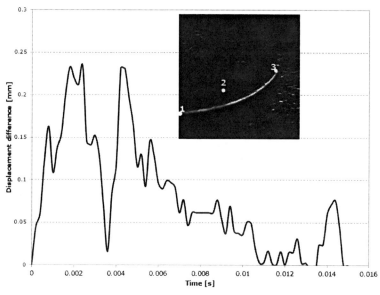

Figure 7. Estimative of displacement difference between points 1 and 3.

4. CONCLUSIONS

The visualizations performed and results extracted from the videos support to the following conclusions:
- Valve displacement and speed information can be obtained by the use of high speed visualization techniques.
- The shapes of the curves obtained agree reasonably well with what has been presented in the literature.
- Torsional movement manifests itself more significantly when the valve is wide open, and is more subtle at the closing moment.
- Impact velocity was determined not to be a concern since the values were low and valve reliability should not be an issue even with indication of a small degree of torsional movement at closing[1].
- More experiments are needed in order to expand the data base in terms of operating speeds, pressures and mass flow rates and the activity continues.

5. ACKNOWLEDGEMENTS

The authors would like to acknowledge the company sponsors of the Air Conditioning and Refrigeration Center for their technical and financial support to this work. Special thanks go to Nishant Mathure and Philippe Dewitte of Danfoss for supplying the compressor prototype. The authors are also thankful for the collaboration of Randy Lee, undergraduate assistant in the realization of this work.

6. REFERENCE LIST

1. Soedel, W., *On Dynamic Stresses in Compressor Valve Reeds or Plates during Collinear Impact on Valve Seats*, Proceedings of the Purdue Compressor Technology Conference, West Lafayette, IN, 1974.
2. Cohen, R., *Valve Stress Analysis – For Fatigue Problems,* Proceedings of the Purdue Compressor Technology Conference, West Lafayette, IN, 1972.
3. Nagata, S., Nozaki, T., Akizawa, T., *Analysis of Dynamic Behavior of Suction Valve Using Strain Gauge in Reciprocating Compressor*, International Compressor Engineering Conference at Purdue, West Lafayette, IN, 2010.
4. Ludu, A., Betto, A., Hegner, G., *Endoscope Video of Compressor Valve Motion and Pressure Measurement Assist Simulations for Design Improvements*, International Compressor Engineering Conference at Purdue, West Lafayette, IN, 2000.
5. Nagy, D., Almbauer, R. A., Lang, W., Burgstaller, A., *Valve Lift Measurement for the Validation of a Compressor Simulation Model*, Proceedings of the International Compressor Engineering Conference, West Lafayette, USA, 2008.

Fatigue design and safety factor for scroll compressor wraps

D Gross
DANFOSS Commercial Compressors, Simulation & Test Department, France

ABSTRACT

Understanding the mechanical behaviour of scroll compressor components under cyclic load (as the compressor runs) is one of the most important considerations when designing a compressor. The Fatigue design approach has been improved in numerous ways: improvement of numerical models, improvement of the Dang Van Fatigue analysis and experimental data acquisition methods. Due to these technical advances, results are more relevant and more reliable.

A major component of the mechanical behaviour is the wraps fatigue behaviour. This component has a critical impact on the reliability of a scroll compressor and, when using numerical methods, it is necessary that wraps fatigue is correctly modelled in order to determine a compressor safety coefficient for the entire operating envelope.

1. INTRODUCTION

1.1 The scroll compressor
A scroll compressor (Fig 3) consists of a fixed (Fig 1) and an orbiting scroll (Fig 2) arranged within a hermetic shell. The eccentric motor shaft and the Oldham coupling perform the orbiting motion. The orbiting scroll is in orbit around the main compressor axis, but the axis of this orbiting scroll maintains the same direction in relation to the fixed system coordinates. Pairs of compression pockets are formed between the scrolls and the pocket volume decreases in time as it moves towards the center of the scrolls as a result of the orbital motion.

Figure 1: Fixed scroll

Figure 2: Orbiting scroll

Figure 3: Scroll compressor

The orbiting scroll is driven using a controlled orbit design, which means that the motor shaft directly drives the orbiting scroll without any sliding intermediate parts such as swing links or slider blocks. A precise machining of the components enables the use of this driving principle of controlled orbit. If the tightness needs to be improved because of the machining precision (or clearances and tolerances), a sliding part can allow a radial movement of the orbiting scroll to correct the relative position of the scrolls, using either the force of inertia or a spring.

Due to its controlled orbit and precise scroll profile, the orbiting scroll slides on an oil film with no friction or wear, securing absolute radial tightness.

1.2 Scroll compression process

Internal pressure loads (Fig 4)
The duties of the involutes come from pressure pockets. Indeed, during the compression cycle, pressure pockets are created inside the involutes. See Fig 4.

Suction - 1st orbit
As the bottom scroll orbits, two refrigerant gas pockets are formed and enclosed
Compression - 2nd orbit
The refrigerant gas is compressed as the volume is reduced due to its movement to the center of the scroll
Discharge - 3rd orbit
The gas is compressed further and discharged through a small port in the center of the fixed scroll

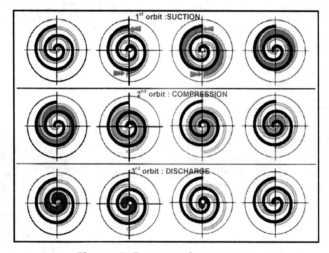

Figure 4: Compression process

Variations of the internal forces
The result of the compressive forces breaks down into a thrust load and two horizontal forces, these forces vary during one turn of the crankshaft (as it is shown in Figure 6).

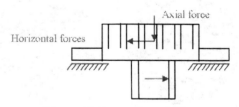

Figure 5: Force description

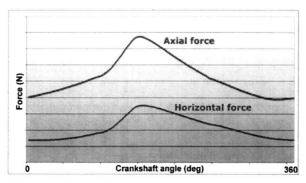

Figure 6: Variations of the internal forces

1.3 Operating conditions

Danfoss compressors serve the refrigeration, heat pump and air conditioning industry. Depending on the application, the compressor operating condition varies according to a specific operating envelope.

An example of operating envelope is shown in figure 7, where the condensing and evaporating temperatures represent the range for steady-state operation. The operating limits serve to define the envelope within which reliable operations of compressor are guaranteed.

The Scroll compressor is a volumetric machine that performs best when the system's pressure ratio conditions are similar to, or the same as the built in pressure ratio.

During the scroll set design, the first suction pocket volume defines the mass flow and the last trapped volume before the discharge process defines the Natural Pressure Ratio (NPR).

This Natural Pressure Ratio is selected in order to have the best efficiency compromise for dedicated application.

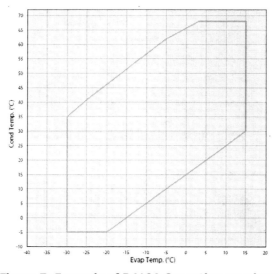

Figure 7: Example of R410A Operating envelope

As the scroll compressor is a fixed compression machine, the output pressure is always equal to the input pressure multiplied by the Natural Pressure Ratio (NPR), which depends on the length of the spiral wrap. The longer the wrap, the higher the NPR is. The Applied Pressure Ratio (APR) depends on the installation and mainly on the weather conditions, so the APR cannot be controlled. The APR is defined by the ratio of the condensing pressure divided by the evaporating pressure.

If the APR is higher than the NPR, when the last trapped gas pocket volume opens onto the discharge port, there is a "back flow" phenomenon at the centre of the scroll.

If the APR is lower than the NPR, when the last trapped gas pocket volume opens onto the discharge port there is an "overshoot" phenomenon at the centre of the wrap.

The pressure conditions in the pockets and the consideration of interaction between the final pocket volume and discharge port during compressor operating determine a cyclic load (duty) inside the wraps.

A first approach that can be used to assess the stresses at the wrap inner end is the calculation of the variation of the pressure loads on the inner and outer side of this inner end just prior to discharge. This ensures that the highest pressure loadings are accounted for in the calculation of the stresses. On the outer side of the wrap inner end the pressure in the pocket is equal to "NPR * Evaporation Pressure" whilst on the inner side of the wrap inner end the pressure is equal to condensation pressure. Figure 8 illustrates these different pressures on a typical compressor operating envelope.

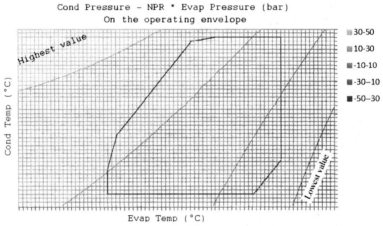

Figure 8: Pressure on wrap inner end

2. FATIGUE ANALYSIS

2.1 Choice of the fatigue criterion
All products have a minimum expected lifespan which is compatible with a normal use of the product and the customer's needs. For Danfoss compressors, which are hermetic and need no maintenance, the expected lifespan is:

10 years @66% running time	60,000 - 65,000 hrs
15 years @50% running time	

It appears that the demand in lifespan is relatively long for compressors. If one considers that there is one duty cycle per motor turn, the cycle number is higher than $1e^{10}$ cycles during the compressor life. Consequently, machines with a cyclic load are subject to fatigue and this has to be taken into account in order to achieve a suitable lifespan.

In most cases, mechanical components experience multi-axial solicitations. For scroll wraps, the main problem is high cycle fatigue for which the solicitations are multi-axial and periodic but not harmonic. For these types of solicitations, two families of fatigue criteria can be used:

- Criteria with a global approach where the stress is averaged in a considered volume
- Criteria with a critical plane approach where the stress is averaged over a critical face

Moreover, the principal direction of stress matrix could be considered fixed, so it is preferable to choose a criterion with a critical plan approach. The Dang Van criterion was chosen.

2.2 Dang Van criterion

The Dang Van criterion is a three-dimensional multi-axial fatigue limit criterion dealing with high cycle fatigue conditions where crack initiation occurs at a microscopic level. The method assumes that around the fatigue limit, cyclic plasticity will occur in critically oriented individual grains and therefore local stress tensors need to be considered. These microscopic stresses differ from the macroscopic ones by the presence of a microscopic residual stress field and the fatigue limit tends towards a pseudo "shakedown state".

The Dang Van method states that, at the stabilized state (shakedown state), crack initiation will happen whenever a function of the microscopic stress exceeds a threshold at any time in a stabilized cycle. The formulation uses a linear combination of the current local shear stress $\tau(t)$, creating plasticity, and hydrostatic pressure $p(t)$, acting on opening of micro-cracks, to calculate an "equivalent" stress and compare it to a limit.

This criterion is based on the determination of critical plane where $|\tau(t)| + \alpha \cdot p(t)$ is at its worst i.e. highest. The tensor of macroscopic stresses is simplified from 3D to 2D by using this critical plane. So the Dang Van criterion is a Macro-Micro approach in High-Cycle Multi-axial Fatigue.

2.3 Dang Van analysis

The Dang Van method is designed to provide a safety factor calculation for multi-axially loaded components in the endurance regime (large number of cycles to failure). The theoretical basis can be summarized as follows:

- Fatigue crack initiation normally occurs due to repeated plasticity on shear planes in individual grains.
- The most important factor as to whether a crack will propagate past the first grain and go on to cause fatigue failure is the microscopic shear stress in critical grains.
- The ability of a shear crack to propagate is modified by the hydrostatic stress (which can increase damage by opening existing cracks).
- A process of shakedown will affect the state of stress in a grain where repeated plasticity is occurring.

2.4 Safety factor
One of the Dang Van fatigue analysis results is the Safety Factor, which is defined by:

$$SF = \frac{b}{Max(\tau(t) + a \times p(t))}$$

The Dang Van parameters a and b define a fatigue threshold condition which is described by the equation $\tau + a.Ph = b$ where τ is the microscopic shear stress and Ph is the hydrostatic pressure.

3. MATERIAL: DUCTILE AND GREY CAST IRON

3.1 Generality
The studied part is made with either grey cast iron or ductile cast iron.

Grey cast iron is a ferrous alloy which has been heated until it liquefies and is then poured into a mould to solidify. Grey cast iron, or grey iron, has graphitic flakes which deflect a passing crack and initiate countless new cracks as the material breaks.

Carbon (C) and silicon (Si) are the main alloying elements, with the amount ranging from 2.5 to 4 wt% and 1 to 3 wt%, respectively. This material is a ternary Fe-C-Si alloy.

Grey cast iron is characterized by its graphitic microstructure. Most cast irons have a chemical composition of 2.5 to 4.0% carbon, 1 to 3% silicon, and the remainder is iron.

Nodular or ductile cast iron is a more recent development. Tiny amounts of magnesium or cerium added to these alloys slow down the growth of graphite precipitates by bonding to the edges of the graphite planes. Along with careful control of other elements and timing, this allows the carbon to separate as spheroid particles as the material solidifies.

The common defining characteristic of this group of materials is the morphological structure of the graphite. In ductile irons, the graphite is in the form of spherical nodules rather than flakes (as in grey iron), thus inhibiting the creation of cracks and providing the enhanced ductility that gives the alloy its name.

While most varieties of cast iron are brittle, ductile iron is much more flexible and elastic due to its nodular graphite inclusions.

3.2 Micrographic views + 3D views
Figure 9 presents images of the different micrographic structures of grey cast iron and ductile cast iron.

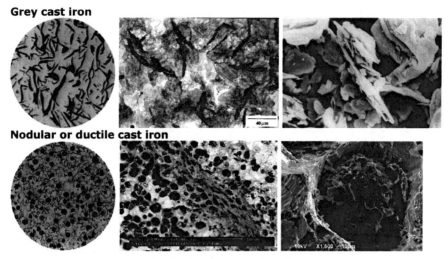

Figure 9: Micrographic views

3.3 Consideration of the material imperfections
Due to the inclusion of graphite, these materials are non homogeneous. It could considered that each graphite precipitates (spherical or flakes shape) could be a "default" or a non-discontinuity of the material. Although the size, the shape and the concentration of the graphite is mastered by engineering standards, this non-discontinuity can effectively be located anywhere inside the part. Thus it is necessary to consider the worst case scenario when there is a non-discontinuity located at the point which has the greatest fatigue stresses.

3.4 Fracture toughness characterization
Figure 10 from [4], presents the crack propagation speed da/dN vs the stress intensity factor variation for a nodular graphite cast iron. From this figure, the value of ΔK_{th} could be estimated as:

$\Delta K_{th} \approx 10$ MPa $\sqrt{m}$ = 316 MPa $\sqrt{mm}$.

To confirm this ΔK_{th} estimation, [6] gives
$\Delta K_{th} \approx 15$ MPa $\sqrt{m}$ = 474 MPa $\sqrt{mm}$.

4. FRACTURE MECHANICS APPROACH

4.1 Generality
Since there are non-discontinuities in the material due to graphite flakes or nodules, a notch could be considered to be present in the material. This assumption leads to a Fracture mechanics approach for the numerical calculation.

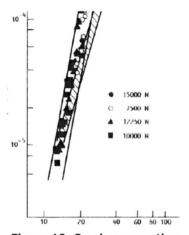

Figure 10: Crack propagation speed da/dN (mm/cycle) vs ΔK (MPa$\sqrt{m}$)

Fracture mechanics is the field of mechanics concerned with the study of the propagation of cracks in materials. It uses methods of analytical solid mechanics to calculate the driving force on a crack and those of experimental solid mechanics to characterize the material's resistance to fracture.

In modern materials science, fracture mechanics is an important tool for improving the mechanical performance of materials and components. It applies the physics of stress and strain, in particular the theories of elasticity and plasticity, to the microscopic crystallographic defects found in real materials in order to predict the macroscopic mechanical failure of bodies. The prediction of crack growth is at the heart of the damage tolerance discipline.

4.2 Stress Intensity Factor K

A modification of Griffith's solids theory emerged from this fracture machancis work; a term called stress intensity replaced the strain energy release rate and a term called fracture toughness replaced the surface weakness energy. Both of these terms are simply related to the energy terms that Griffith used:

$$K_I = \sigma\sqrt{\pi a}$$

where K_I is the stress intensity, K_c the fracture toughness, and (a) is flaw length. It is important to recognize the fact that the fracture parameter K_c has different values when measured under plane stress and plane strain. Fracture occurs when $K_I \geq K_c$. For the special case of plane strain deformation, K_c becomes K_{Ic} and is considered a material property. The subscript I arises because of the different ways of loading a material to enable a crack to propagate. It refers to the so-called "mode I" loading as opposed to mode II or III.

There are three ways of applying a force to enable a crack to propagate:

- Mode I Opening mode (a tensile stress normal to the plane of the crack)
- Mode II Sliding mode (a shear stress acting parallel to the crack plane and perpendicular to the crack front)
- Mode III Tearing mode (a shear stress acting parallel to the plane of the crack and parallel to the crack front)

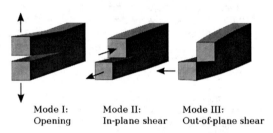

Mode I: Opening   Mode II: In-plane shear   Mode III: Out-of-plane shear

Figure 11: The three fracture modes

In terms of fatigue with high-cycle fatigue, a "Stress Intensity Factor variation" has to be introduced.

Stress Intensity Factor variation is $\Delta K = K_{max} - K_{min}$,

$$\Delta K = K_{max} - K_{min} = (\sigma_{max} - \sigma_{min})\alpha\sqrt{\pi a}$$

where:
$\sigma_{max}/\sigma_{min}$ is the minimal/ maximal stress applied on the part
α is a geometrical parameter

In figure 12, the crack growth rate is split into 3 stages.
        Stage 1 - Crack nucleation;
        Stage 2 - Crack-growth;
        Stage 3 - Ultimate ductile failure.

In the stage 1, there is a critical value of ΔK called K_{th} which defines a threshold of visibility of the crack. For $\Delta K < \Delta K_{th}$ the crack is considering frozen.

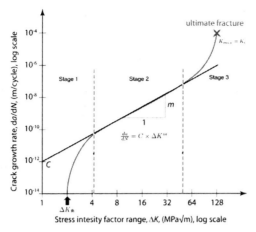

Figure 12: Typical fatigue crack growth rate graph

5. SIMULATION MODELS

5.1 Introduction
The numerical models impact the Fatigue analysis results. In order to obtain relevant and reliable results, the numerical models have to simulate as close as possible the mechanical behaviour of the parts while the compressor operates. As well as the pocket pressure loading, the inertia forces and contact forces between parts also have to be considered. The objective of the model is to describe as finely as possible the cycle duty or loading while the compressor operates. Experience at Danfoss has shown that the numerical model has to simulate the compressor loading or duty every 3 degrees of crankshaft rotation in order to obtain relevant results.

The modelling process has been broken down into three steps. The first step is a global modelling approach in order to simulate the orbiting scroll mechanical behaviour as realistically as is possible. The second step consists of simulating local hot spots with all geometrical details and a finer mesh to obtain precise stress matrix. The third step is more microscopic in nature: it is to consider a crack corresponding to a graphite flake or nodule in the material at the worst location a default or imperfection.

5.2 Global model: static analysis

5.2.1 Overview
All calculations were carried out with ANSYS and nCode Softwares.

A first numerical approach is to consider the orbiting scroll lying on the crankcase. The parts are flexible in the calculation, thus the flexibility of both the thrust bearing and of the orbiting scroll are taken into account. By experience, the wrap solicitations could be impacted by this axial contact behaviour. The inertia forces on the orbiting scroll are introduced in the model. Every

Figure 13: FEA model

3° of crankshaft rotation, the orbiting scroll moves on its orbiting location and the pocket pressure loading is updated. The calculation is non linear due to the contacts and the large displacement.

5.2.2 Boundary conditions: pockets pressures

To simulate precisely one crankshaft revolution, calculations are carried out for every 3° of crankshaft rotation. Danfoss's thermodynamic simulation software provides the pocket position and pocket pressure for each crankshaft angle. One issue here is how to apply the boundaries conditions of the pocket pressure for each angle of crankshaft. In fact the meshing of the orbiting scroll remains the same for all calculations, whereas the pocket position and pocket pressure change during a turn of crankshaft. A computer program (coded in APDL) was developed to automatically apply, independently of the structural grid, the pressure inside the involutes on the ANSYS model boundaries. The main difficulty in this process is that the element boundaries did not necessarily correspond to the pocket pressures boundaries.

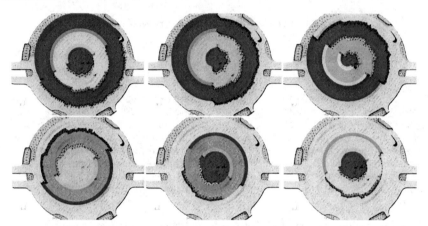

Figure 14: Pocket pressure view at different crankshaft angle during compression process on orbiting scroll

5.3 Local model: Employing submodeling for a specific hotspot

One of the sensitive locations is the connection between wrap inner end and the scroll base plate. To obtain a precise stress matrix, submodeling was used. This method consists of generating a local model with a finer meshing in order to obtain a more precise stress matrix. Moreover all tiny radii or geometrical features can be taken into account in this submodel.

The predicted stress at the connection between the wrap and the base plate depends on the meshing and scroll geometrical features. Within the context of the global model, it is difficult or numerically impossible to obtain a good stress matrix representation. From the global model the stresses are discontinuous while approaching the base plate due to FEA numerical approximation.

The Fatigue analysis on the global model localized a stress hot spot at the connection between wrap inner end and the base plate. Consequently, the sub model is created by cutting a piece of sphere centred on the hot spot. The simulation software imports the displacements every 3° of crankshaft rotation from the global model and applied these displacement on the boundary areas. Due to this procedure, a cyclic loading can be calculated on the sub model. The results extracted are very fine of which one result is the orientation of the first principal stress.

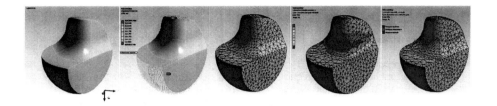

Machining tool radius   BC from global model   Fine meshing   Precise stress matrix   Stress orientations

Figure 15: Submodelling process flow

5.4 Fracture mechanics model

Fracture mechanics accounts for the cracks or flaws in a structure. The fracture mechanics approach to the design of structures includes flaw size as one important variable and fracture toughness replaces strength of material as a relevant material parameter.

Fracture analysis is typically carried out either using the energy criterion or the stress-intensity-factor criterion. When the energy criterion is used, the energy required for a unit extension of the crack (the energy-release rate) characterizes the fracture toughness. When the stress-intensity-factor criterion is used, the critical value of the amplitude of the stress and deformation fields characterizes the fracture toughness. Under certain circumstances, the two criteria are equivalent.

In the case presented here, the stress-intensity-factor criterion has been chosen. At the weak node given by the fatigue analysis, a local specific coordinate system should be created with the orientations that correspond to the principal stresses orientations. Due to this, the crack would have the worst orientation i.e. the first principal stress tends to open the flaw in Fracture Mode I. The size of the flaw corresponds to the flake or nodal graphite shape size in the real material (several microns in size). This method has been inspired from [7].

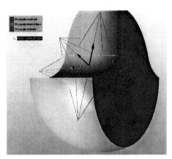

Figure 16: Orientation of the principal stresses and specific coordinate system

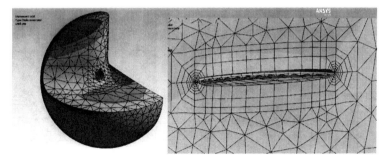

Figure 17: Crack meshing

6. RESULTS

6.1 Global model

6.1.1 Static results
The global numerical model calculates for each operating condition the evolution of the maximum stress which is close to the wrap floor. The fillet radius is not modelled in the global model resulting in sharp corner at the connection between the wrap wall and the wrap floor. Due to this, the stresses calculated by FEA in this corner are incorrect hence the maximum stress extracted from these results is at given distance from this corner.

The variation of the stress at the highest solicited node is not harmonic and always positive. The major change observed for a crank angle of 174 deg in Figure 18 corresponds to the opening of the final gas pocket onto the discharge chamber. At this point in time, the number of pressurised pockets inside the wrap switches instantly from 6 to 4.

For this example of a scroll compressor, it can be observed that the maximum stress is at low crankshaft angles (between 15° and 60°) before the opening of the highest pressurised pocket onto the discharge chamber.

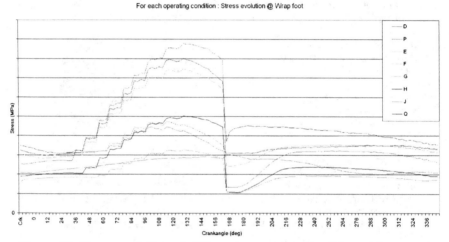

Figure 18: Stress variation at different operation conditions (Figure 20)

Eight operating conditions (D,P,E,F,G,H,J,Q) have been calculated. Some operating points are inside the compressor operating envelope, some others are outside (see Fig 20).

6.1.2 Safety factor
From the results of the FEA load cases and for each node in the FEA model, the results analysis leads to a 3D view of the safety factor in the part. Predictably, the weakest location is at the wrap foot. In addition to identifying this weakest / critical location, the results can also be used to assess a safety factor.

Figure 19: Safety factor view

Figure 20 presents a comparison of the calculated safety factor for different operating conditions. It can be noted that there is a correlation between the safety factor on the operating envelope and the maximum difference of pressure applied on the wrap inner end. For reasons of confidentiality, the safety coefficients indicated on figure 20 are <u>normalized</u> to obtain an arbitrary value of 1 for the operating point "P".

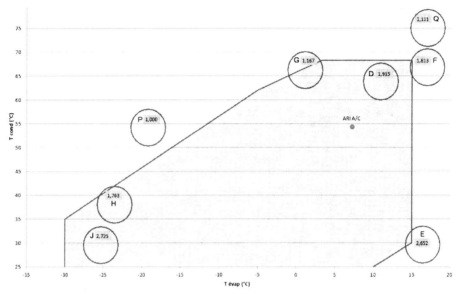

Figure 20: Normalized safety factor on operating envelope

6.1.3 Dang Van loading path

Another result of this analysis is the loading path at the weak node where the safety factor is at a minimum. During one revolution of the crankshaft, all FEA load cases results are analysed in order to define, for this node, the couple P(t), τ (t) which defines a loading path in the Dang Van diagram. Damage will occur only if the loading path crosses the Dang Van threshold line.

The more the loading path is far from the Dang Van threshold line, the more the normalized safety factor is high. On figure 21 the Dang Van threshold line is adjusted to illustrate normalized safety coefficient values.

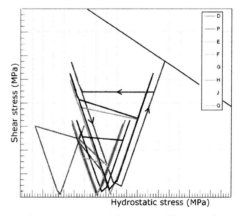

Figure 21: Dang Van Loading path

6.2 Local model

6.2.1 Geometries
The numerical model of the second step or local model gives a fine stress assessment at the weak node identified by the global model. The simulations at this finer level are also completed for every 3° of crankshaft rotation. The boundaries conditions are imported from global model. The goal of this local model is to be able to compare in terms of safety factor different machining shapes. Here, 5 geometries have been simulated, as shown in Figure 22.

Figure 22: Views of 5 different machining shapes at the wrap foot

6.2.2 Fatigue results and Safety factor
After solving FEA models and processing the fatigue analysis, the results are post processed and analysed. An analysis of the safety factor confirms that whatever the local shape, the weak node is at the same location, as illustrated in Figures 23 and 24. On figure 23 the Dang Van threshold line is adjusted to illustrate normalized safety coefficient values.

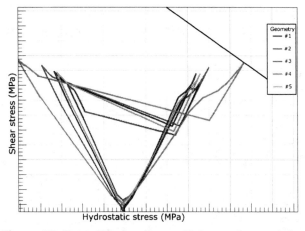

Figure 23: Dang Van loading path for each geometry

Figure 24: Views of the safety factor with the same colour scale

For reasons of confidentiality, the safety coefficients indicated below are normalized to obtain an arbitrary value of 1 for the geometry #4.

For the 5 shapes used in this example, the safety factor could vary by up to 25%.

Geometry	#1	#2	#3	#4	#5
Node	2397	2203	2326	2312	3309
Normalized Safety Factor	1,255	1,199	1,119	1,000	1,187

6.3 Fracture mechanics results

From the results around the crack as illustrated in Figure 25, the stress-intensity-factor can be calculated for every 3° of crankshaft rotation.

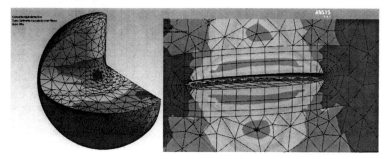

Figure 25: Views of results around the crack

In the crack itself, which corresponds to a flake shape graphite at the worst orientation and location at the wrap foot inner end, the maximum variation of stress intensity factor is about $\Delta K_1 = 170 \text{MPa}.\sqrt{\text{mm}}$ during on duty cycle for the most solicited operating condition, as seen in Figure 26.

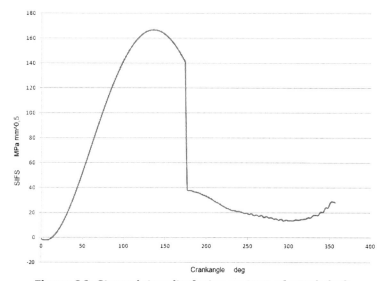

Figure 26: Stress intensity factor on turn of crankshaft

As $\Delta K_1 < \Delta K_{th}$ (cf §3.4 and §3.6) it can be concluded that even if there is a flake shape graphite, the crack will not grow.

7. FURTHER WORK

A more complex numerical approach which is closer to the reality in a scroll compressor could be implemented and which would consider all the parts of the moto-compressor. This approach would avoid some of the assumptions concerning boundary conditions. All contact would be simulated between parts and, as per the "simple" model described above, the calculation could be completed for every 3° of crankshaft rotation. This calculation would be non linear due to the contacts and the large displacements.

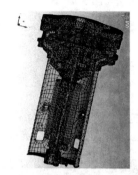

Figure 27: Moto-compressor model

8. CONCLUSIONS

The Fatigue design approach used at Danfoss has been improved. The numerical modelling is now undertaken in three steps. In the first step, the modelling has been improved by introducing the flexibility of the thrust bearing. In order to refine the Fatigue analysis, a second step has been introduced with sub modelling in order to calculate the safety factor at the weakest (most stressed) location with finer geometrical detail.

Lab tests on samples have been completed in order to determine the best material to be used for the various parts of a scroll compressor. Data provided by these tests the Dang Van threshold line could have been adjusted to determine the safety factor.

Finally, the assessment of a safety coefficient, irrespective of the compressor operating point, is a key-result to ensure the level of reliability of Danfoss compressors.

A third step has been introduced into the modelling procedure in order to consider a material imperfection such as a graphite flake or nodule which is located at the worst orientation at the wrap foot inner end. It has been concluded that for the most solicited operating condition, a crack will not grow.

Due to these latest advances in the modelling process, the results thus obtained are more relevant and more reliable.

9. REFERENCES

[1] Ansys 14.5 documentation
[2] nCode 9 documentation
[3] Ecole centrale de Lyon, LTDS UMR 5513, Ecully
[4] Manuel des fontes moulées, 1983
[5] Purdue 2010, Paper ID1490, Fatigue design for scroll compressor wraps, D GROSS, C ANCEL, L GUGLIELMI
[6] Fatigue Assessment of Notches and Cracks in Ductile Cast Iron, H ZAMBRANO
[7] Presentation "news release v14.5" by Ansys France support team 2012-11-28

Simulation model to predict temperature distribution along scroll wraps

M C Diniz[1], E L L Pereira[2], C J Deschamps[1]
[1] Federal University of Santa Catarina, Brazil
[2] Embraco, Brazil

ABSTRACT

The volumetric and isentropic efficiencies of scroll compressors are affected by the heat transfer that takes place inside the pockets during suction and compression processes. The present paper details a numerical model developed to predict the conduction heat transfer and temperature distribution of scroll wraps. The model was developed via the finite volume method and coupled to a thermodynamic model of the compression cycle. The results showed that the discharge temperature predicted with the solution of heat conduction through the scroll wraps was slightly lower than that obtained when a linear temperature profile was prescribed. It was also found that the heat transfer taking place in the metallic contact between the scrolls wraps acts to produce a linear temperature variation along their length.

NOMENCLATURE

Roman
A – Area (m^2)
h – Heat transfer coefficient (W/m^2K)
H – Height (m)
k – Conductivity (W/mK)
R – Thermal resistivity (m^2K/W)
t – Scroll thickness (m)
T – Temperature (K)

Greek
φ – involute angle

Subscripts
gas – Compressed gas
scr - Scroll
dsp – Discharge plenum
con – Contact thermal resistance
bas – scroll base

1 INTRODUCTION

Scroll compressors are positive displacement compressors usually employed in residential and commercial air conditioning, refrigeration, and heat pump applications (1). In these compressors, gas is admitted in the periphery and discharged in the central region of a mechanical compressing element. This element is formed by two identical spiral-shaped metallic pieces (scrolls). One of the scrolls remains stationary, while the other moves in an orbital pattern that initiates gas migration from the periphery to the central region of the scrolls. During this migration, the volume of the gas chambers is continuously reduced. As a consequence, there is an increase in both the pressure and temperature of the gas.

Usually, leakage between the chambers with varying pressures is considered to be the main irreversibility in scroll compressors (2). The heat transfer inside the compressor also has a significant impact on the thermodynamic efficiency. Fundamental to heat transfer characterization are the gas temperature inside the suction pockets (suction temperature), which is higher than the inlet temperature as a result of refrigerant interaction with compressor hot parts, and the temperature profile along the scroll wraps (3). Many authors have developed models to predict suction temperature in scroll compressors (4-6), usually in a coupled manner with the numerical simulation of the compression process. As far as the heat transfer during compression process is concerned, attention can be focused on the gas temperature rise (7) and scroll temperature profile (8-10). Most studies conclude that a linear temperature profile in relation to the scroll involute angle is an acceptable assumption.

When differential models are adopted to simulate the compression process, suction temperature is an important initial condition with the scroll temperature profile being the required boundary condition. In this paper, a steady state one-dimensional model was applied to predict heat conduction and temperature distribution in the scrolls. The conduction model was coupled to a thermodynamic model of the compression process developed by Pereira (2) and a simplified thermal model described by Diniz et al. (6). Therefore, the simulation was performed iteratively, allowing complete thermodynamic characterization of the compressor.

2 MATHEMATICAL MODEL

As a consequence of the compression process, there exists a temperature gradient between the center and periphery of the scrolls, which gives rise to heat conduction along the scroll wraps. Physically, heat conduction occurs in the longitudinal direction along the scroll, in the radial direction across the thickness of the scroll, and in the transversal direction from the tip to the base of the scroll (Figure 1).

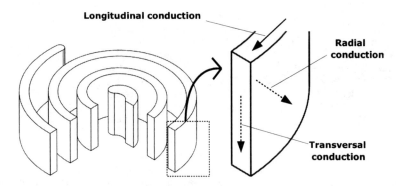

Figure 1 - Heat fluxes along the scroll wrap.

In this study, it was assumed that the radial temperature gradient is negligible. Moreover, transversal heat conduction was also considered to be negligible as small temperature differences are usually found between the center of the scroll and its base (10). Therefore, only the longitudinal heat conduction was modeled and, as such, a one-dimensional formulation was adopted.

In this one-dimensional formulation, three modes were considered in thermal energy balance. The first of them was the convective heat transfer between the gas and the scroll walls, which occurred from scrolls to gas in the suction region of the compressing element and vice-versa in the discharge region. A convective heat transfer coefficient developed by Pereira (2) through CFD was adopted in this work. This heat transfer coefficient consists in a correction of the Dittus-Boelter equation for turbulent flows (12). Another important phenomenon taken into account is the kissing heat transfer, described by Sunder (11) as the most important contribution to the resulting temperature profile. This heat flux occurs in the metallic contact between the scrolls. The convective heat transfer between gas and the contact heat transfer are depicted in Figure 2(a). This figure shows that at any given shaft angle, different regions of the scroll are in contact with the gas in the suction, compression, and discharge chambers. The third mode is related to the gas high temperature in the discharge plenum, which was represented as a source term in the scroll conduction model. This interaction is illustrated in Figure 2(b). Due to the high thermal inertia of the scrolls, the time scales regarding the compression process and the convective and kissing heat transfer mechanisms are considered to be much smaller than those related to heat conduction along the scrolls. Therefore, the time variation of the scroll temperature was neglected.

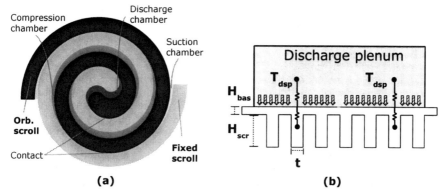

Figure 2 - Heat transfer along the scroll: (a) Convective and kissing heat transfer; (b) Thermal interaction with discharge plenum.

With regard to a small portion of the fixed scroll (Figure 3(a)), the following heat fluxes were taken into account in the energy balance: heat from the discharge plenum (A), convective heat transfer between the gas and the scroll (B), kissing heat transfer (C), and conduction along the scroll (D). The energy balance for this small volume can be described with the help of Equation 1. The thermal resistance terms adopted to estimate the kissing heat transfer and the heat transfer from the discharge plenum are given in Equations 2 and 3, respectively.

$$kA\frac{d^2T}{dx^2} + h_{cha}(T - T_{gas})H_{scr} + H_{scr}\frac{\Delta T_{kis}}{\sum R_{kis}} + t\frac{\Delta T_{dsp}}{\sum R_{dsp}} = 0 \quad (1)$$

$$\sum R_{kis} = \frac{t}{k} + R_{con} \quad (2)$$

$$\sum R_{dsp} = \frac{H_{bas} + H_{scr}/2}{k} + \frac{1}{h_{dsp}} \quad (3)$$

An analytical solution of Equation 1 is not an easy task due to the difficulties of finding a suitable expression to describe both the average convective heat transfer coefficients and gas temperature facing the internal and external faces of the scroll. For this reason, a numerical simulation approach was chosen.

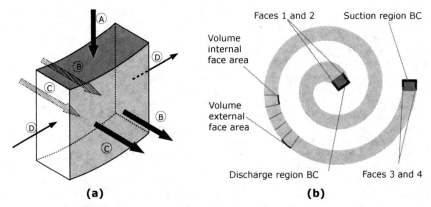

Figure 3 - Heat transfer along the scroll: (a) Energy balance in a control volume of the scroll; (b) Boundary conditions.

3 SOLUTION PROCEDURE

3.1 Boundary conditions

The heat conduction problem was solved by using the finite volume method. A model developed by (2) for the compression process provided the instantaneous convective heat transfer coefficients and the gas temperature along an entire cycle of compression. Since the conduction heat transfer in the scroll is solved for steady state, the convective heat transfer coefficients and the gas temperature on the internal and external faces of every control volume are averaged over an entire shaft revolution. The convective heat transfer between the external faces of the volumes located in the periphery of the scroll and the gas inside the compressor internal ambient were also taken into account. To this extent, a convective heat transfer coefficient of 6.5 W/m^2K was estimated considering the hypothesis of laminar flow in a rectangular duct (12), which is indicated for the geometrical characteristics of the compressor adopted in this work.

Kissing heat transfer was modeled via a simplified procedure. Following such an approach, the volumes on the opposite scroll influencing a control volume for each revolution angle are identified through the extent of the contact line between the scrolls. This contact line is estimated through hertzian stress calculations (11). The temperatures of the volumes on the opposite scroll that are limited by the contact line are then averaged, providing a representative temperature for the kissing heat transfer modeling. The temperature of this control volume is related to the representative temperature of the opposite scroll using an association of thermal resistances (Equation 2). Heat generation due to friction between the scrolls was not considered. The heat transfer contribution from the discharge plenum was implemented in the model via a thermal resistance (Equation 3) between the plenum gas temperature, T_{dsp}, calculated with a lumped thermal model (6) and the temperature of the scroll wrap.

Boundary conditions were applied to the volumes at scroll extremities. For the volume at the outer extremity, convective heat transfer was assumed with the gas at the suction temperature. In a similar way, convective heat transfer was also considered between the gas in the discharge pocket and the surfaces of the volumes in the discharge region. The heat transfer coefficients were defined as the average values between the inner and outer faces of the border volumes, as illustrated in Figure 3(b). Accordingly, faces 1 and 2 were used to calculate the heat transfer coefficient for the discharge region and faces 3 and 4 to determine the heat transfer coefficient for the suction region. Figure 3(b) also illustrates a typical control volume. The area of external surfaces is greater than that of internal surfaces due to the scroll curvature. Therefore, the volume of the control volume is calculated using the medium scroll length. A central difference scheme was adopted to interpolate the quantities needed at the control volume faces. The resulting system of linear equations was solved with the tridiagonal matrix algorithm (TDMA).

3.2 Coupled solution

A schematic view of the compressor adopted in this work is presented in figure 4. It represents a hermetic R410A compressor operating between 6000 and 10000 rpm. The compressor swept volume is 6.1 cm^3 and the pressure ratio 3.4. A lumped thermal model as described by Diniz et al. (6) is used to characterize the heat interactions inside the compressor, allowing suction temperature calculation, which is essential to both scroll conduction and compression process models.

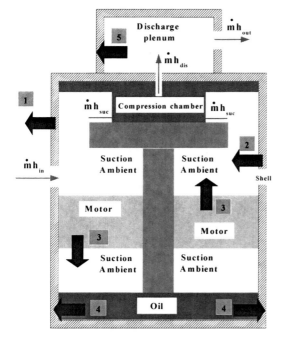

Heat flow rates:

1: Shell to external ambient;
2: Shell to suction ambient;
3: Motor to suction ambient;
4: Oil to shell;
5: Discharge plenum to shell.

Figure 4 - Compressor thermal network scheme.

The compression process model adopted in this work employs a novel methodology to determine pressure and temperature during gas compression. Instead of simulation all chambers simultaneously (4), it considers the same amount of gas from the beginning to the end of the compression cycle. As a result of this solution procedure, consecutive cycles must be simulated to achieve convergence. The thermal models require input from the compression process, which in turn, uses results delivered by the models (Figure 5(a)). Therefore, the models were interconnected and the simulation is carried out in an iterative manner. At the beginning of each cycle, the suction temperature and the scroll temperature profile are updated using the respective models. The solution procedure is represented by Figure 5(b).

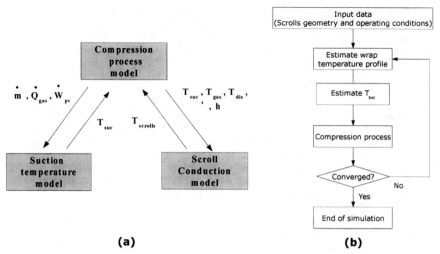

(a) (b)

Figure 5 - (a) Coupled solution scheme; (b) Solution flowchart.

4 RESULTS

4.1 Thermal profiles and heat flux

Preliminary tests have shown that using 4000 control volumes in the discretization of the solution domain was sufficient to guarantee negligible truncation error in temperature distribution along the scroll.

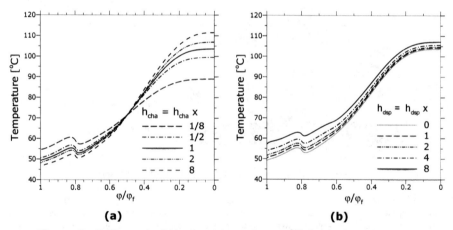

(a) (b)

Figure 6 - Influence of the heat transfer coefficients (HTC) on the temperature profile: (a) compression chamber; (b) discharge plenum.

Figure 6(a) presents numerical predictions for temperature profiles in relation to the normalized involute angle that was obtained by multiplying different factors to the average heat transfer coefficient provided by (2). As can be seen, the correct characterization of this coefficient is important, since it affects the temperature profile. Figure 6(b) shows the results of a similar test, which was carried out by adopting a multiplying factor to the heat transfer coefficient of the discharge plenum, whose reference value was h_{dsp} = 100 W/m²K. It can be noted that this heat transfer coefficient modified the levels of the profile, while not altering its shape significantly.

Figure 7 presents temperature profile as a function of different values of contact thermal resistivity, which were obtained with the reference value ($R_{con} = 10^{-4}$ m²K/W) multiplied by different factors. When the contact thermal resistance is very high, kissing heat transfer becomes negligible and convective phenomena determine the temperature profile.

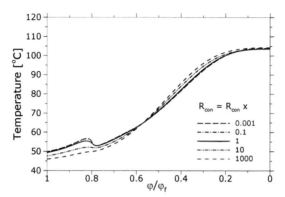

Figure 7 - Effect of contact thermal resistivity on temperature profile.

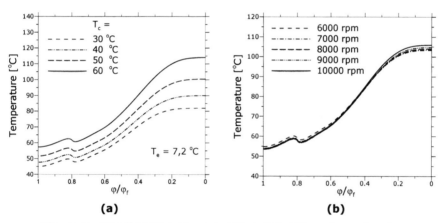

(a) (b)

Figure 8 - Scroll temperature profiles: (a) different condensing temperatures and 10000 rpm; (b) different speeds and 7.2°C/54.4°C.

Figure 8 shows the temperature profiles for different operating conditions. Figure 8(a) indicates that the shape of temperature profiles is almost insensitive to variations in condensing temperature. However, higher condensing temperatures tend to increase the level of temperature profile due to the associated higher suction temperature. These results make evident the importance of adopting the compressor thermal model to predict suction temperature required in the heat conduction model. Figure 8(b) shows that the temperature profile is very weakly affected by the compressor speed.

Figure 9 presents the heat fluxes magnitudes. Following the phenomena modeled in this work, kissing heat transfer is the decisive heat flux concerning the conduction problem along the scrolls. In the outer region of the scroll, kissing heat transfer does not occur in the external faces of the volumes. This explains the negative heat flux (heat flux from periphery to center) obtained near normalized involute angle of 0.6. This phenomenon was experimentally observed by Jang and Jeong (9).

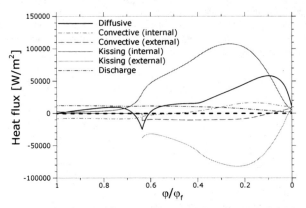

Figure 9 - Heat fluxes in the scroll; 7.2°C/54.4°C and 10000 rpm.

4.2 Discharge temperature and efficiencies

Figures 10–12 present the discharge temperature and thermodynamic efficiencies predicted by using three different thermal conditions for the scrolls: (i) a linear temperature profile, (ii) a temperature profile obtained with the heat conduction model described herein (SHC), and (iii) adiabatic walls. Figure 10(a) shows that the simulation with condition (ii) returns a lower discharge temperature than that obtained with condition (i). The discharge temperatures predicted with conditions (ii) and (iii) are virtually the same. Similarly, Figure 10(b) indicates a decrease in the discharge temperature when the heat conduction model is adopted, when compared to the result obtained with a linear temperature profile, regardless of compressor speed. A minimum discharge temperature is obtained in intermediate speeds, probably because the increase of the discharge temperature caused by leakage at low speeds and heat transfer at high speeds are minimized (2).

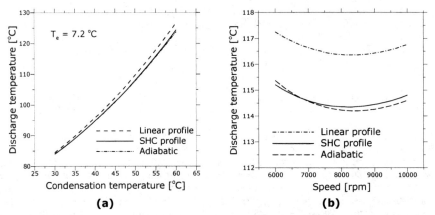

Figure 10 - Discharge temperature: (a) different condensing temperatures at 10000 rpm; (b) different speeds at 7.2°C/54.4°C condition.

The simulation model predicts the maximum isentropic efficiency near the condensing temperature of 55°C (Figure 12a). When the compressor operates far from this condition, the isentropic efficiency is decreased as a result of under- and over-compression. The reduction of volumetric efficiency with condensing temperature (Figure 12b) is mostly due to leakage. Similar to the discharge temperature, the maximum isentropic efficiency is also obtained at intermediate speeds (Figure 12a). The volumetric efficiency increases with compressor speed as more gas gets trapped in the suction pocket.

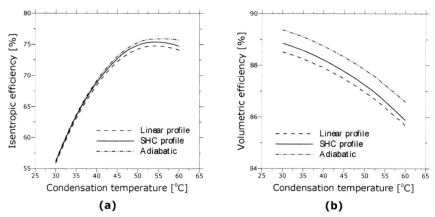

Figure 11 - Compressor efficiency at different condensing temperatures.

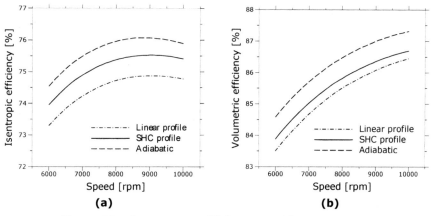

Figure 12 - Compressor efficiency at different speeds.

5 CONCLUSIONS

This paper presented a numerical model developed to determine the temperature profile along the scroll wraps of a scroll compressor by using a steady state one-dimensional formulation. The model was developed via the finite volume method and coupled to a second model devised for the compression process and a to a lumped thermal model. It was found that the kissing heat transfer that takes place in the metallic contact between the scroll wraps is the main aspect that explains the linear profile usually assumed by some studies in the literature. The coupled solution procedure presented in this paper allows a complete thermal characterization of scroll compressors. A worthwhile improvement of the simulation model could be achieved by introducing a more elaborated mathematical description of kissing heat transfer and by considering the frictional losses in the energy balance. In addition, an experimental analysis is also programmed to validate the model.

ACKNOWLEDGEMENTS

The present study was developed as part of a technical-scientific cooperation program between the Federal University of Santa Catarina and EMBRACO. The authors also acknowledge the support provided by EMBRACO and CNPq (Brazilian Research Council) through Grant No. 573581/2008-8 (National Institute of Science and Technology in Refrigeration and Thermophysics) and CAPES (Coordination for the Improvement of High Level Personnel).

REFERENCE LIST

(1) ASHRAE. Handbook – HVAC Systems and Equipment. ASHRAE, 2004.
(2) Pereira, E.L.L., 2012, "Modeling and analysis of leakage and heat transfer in scroll compressors", Ph.D. Thesis, Federal University of Santa Catarina, (in Portuguese).
(3) Wagner, T.C., Marchese, A.J., McFarlin, D.J., 1992, "Characterization of thermal processes in scroll compressors", Proc. Int. Compressor Engineering Conference at Purdue University, pp. 97-106.
(4) Chen, Y., Halm, N.P., Groll, E.A., Braun, J.E., 2002, "Mathematical modeling of scroll compressors – part II: overall scroll compressor modeling", *Int. J. Refrig.*, vol. 25, p. 751-764.
(5) Lee, G.H., "Performance simulation of scroll compressors", Proc. Institution of Mechanical Engineers, Part A: Journal of Power and Energy, vol. 216, p.169-179, 2002.
(6) Diniz, M. C., Pereira, E. L. L., Deschamps, C. J., 2012, "A Lumped Thermal Parameter Model for Scroll Compressors Including the Solution of the Temperature Distribution along the Scroll Wraps, Proc. Int. Compressor Engineering Conference at Purdue University, pp. 97-106.
(7) Ooi, K. T., Zhu, J. Convective heat transfer in a scroll compressor chamber: a 2-d simulation. Int. J. of Thermal Sciences, vol. 43, p 677-688, 2004.
(8) Jang, K., Jeong, S., 2006, "Experimental investigation on convective heat transfer mechanism in a scroll compressor", *Int. J. Refrig.*, vol. 29, p. 744-753.
(9) Jang, K., Jeong, S., 1999, "Temperature and heat flux measurement inside variable-speed scroll compressor", Proc. Int. Congress of Refrigeration, pp. 97-106.
(10) Lin, C., Chang, Y., Liang, K., Hung, C., 2005, "Temperature and thermal deformation analysis on scrolls of a scroll compressor", *Applied Thermal Engineering*, vol. 25, p. 1724-1739.
(11) Sunder, S., 1997, "Thermodynamic and heat transfer modeling of a scroll compressor", Ph.D. Thesis, Massachusetts Institute of Technology.
(12) Bejan, A., Kraus, A. D., 2003, "Heat transfer handbook", Wiley-Interscience.

Oil flow at discharge valve in a scroll compressor

A J P Zimmermann[1], P S Hrnjak[1,2]
[1] ACRC, MechSE Department, University of Illinois at Urbana-Champaign, USA
[2] CTS - Creative Thermal Solutions Inc., USA

ABSTRACT

This work focuses on high speed visualization of the three reed valves in a scroll compressor. The visualization was performed to obtain qualitative description of the oil flow in terms of morphology and breakup and to quantify the droplet size and velocity distribution functions both for start-up and cyclic operation conditions. It was observed that at startup, part of the oil film between the valve and its seat is first pushed out before the valve can start moving, after that the valve starts opening and the film is stretched so that liquid columns in organized spacing appear and are blown out by the gas originating the droplets that form a mist. During steady operation of the compressor, no oil film breakup was clearly observed at the valve-seat interface, suggesting that the oil droplets might be generated from vapor shear through the orifice since there is a recess that can hold up oil inside the orifice.

1. INTRODUCTION

Oil in circulation in an vapor compression refrigeration system is known to cause a reduction in heat transfer coefficient and increase in pressure drop in heat exchangers and connecting lines (1,2,3), as well as reducing the oil level inside the compressor crankcase. It is reported in the literature (4) that OCR reduction plays a pivotal role in scroll compressors design applied in residential air conditioning. It is not uncommon to see OCR values in the 2-4% range, especially in automotive systems.

In order to reduce the OCR of such systems by keeping the oil inside the compressor, separation strategies need to be ideally integrated into the discharge plenum of the compressor. Obtaining a good liquid separation requires knowledge of the liquid flow characteristics.

One way of obtaining the two-phase flow characteristics is to investigate the developing flow at the discharge tube after it leaves the compressor (5). Previous studies show that the predominant flow regime is of the mist-annular type due to low liquid loading and high vapor velocities. Separation in a horizontal tube by letting the droplets deposit into the film has shown an asymptotic behavior when it comes to the reduction of the liquid flow in the form of droplets as the flow progresses down the tube. This happens due to the eventual balancing of the entrainment and deposition rates for the fully developed flow (5).

External oil separators have been used for very long time in the refrigeration and air conditioning industry with quite success. However, in order to effectively separate the liquid from the vapor one should consider the following aspects: (i) volume required; (ii) allowable pressure drop; (iii) amount of refrigerant dissolved in the oil.

The first aspect is related to the most simple separation mechanism one can think of: gravity. If gravitational settling is to be used as a separation mechanism, one should look into the terminal velocity of the droplets to determine which maximum vapor velocity is allowed so that the droplets do not get carried away with the vapor. Such velocity is directly proportional to the droplet diameter and therefore the smaller the droplet the lower the velocity and the higher the volume. The volume that a gravitational separator can take is often prohibitive.

The second aspect refers to compact separators that might use obstacles or centrifugal forces to drive the droplets towards a wall and/or metal mesh, so they can be collected at a certain location and returned to the compressor. Such devices can partially solve the volume problem but they introduce pressure drop at the discharge line and that can, sometimes, be detrimental to the performance of the system. Pressure drop required is also a function of the droplet size, since smaller droplets present smaller Stokes number and have a tendency to follow the gaseous flow, they might not have enough inertia to be driven to the wall.

The last aspect has to be taken into consideration whenever an oil-refrigerant mixture is miscible at the compressor discharge conditions. For all practical purposes, the refrigerant mass fraction that is dissolved into the oil is the lowest right after it flows through the discharge orifice, since it is at its highest apparent superheat. As the mixture flows away from the compressor, it is cooled and more refrigerant is allowed to be absorbed into the oil, resulting in a loss in cooling capacity when that oil is directed back to the compressor crankcase, in this case a low pressure sump.

All of what was exposed until now leads to the conclusion that the best place to separate the oil is at the discharge plenum of the compressor. Oil can then be returned with a minimum amount of refrigerant dissolved, and no external volume or imposed pressure drop are necessary to improve system performance, and compressor reliability by managing the amount of oil that is circulating.

Figure 1 shows a schematic (a) of the preliminary experiment and the pictures taken for two distinct compressor types: (b) automotive swash-plate compressor; (c) small hermetic reciprocating compressor. The gas used was nitrogen and the oil was deposited in the gap between the valve and the valve seat in differing amounts (10μl, 20μl, 30μl) with the use of a syringe. Experiments were run in open atmosphere and the discharge orifice average velocities were kept in line with real values for the compressor types selected, 13 m/s for automotive and 9 m/s for small hermetic reciprocating. Figures 2a through 2c show the time sequence of the valve opening and ligament and droplet breakup from the automotive type compressor and figures 2d through 2e shows the same time sequence and the interesting film breakup pattern observed in the hermetic reciprocating compressor. Such an organized breakup pattern is suggestive of an onset of instabilities at the film interface and questions the validity of a set film thickness value for rupture in valve stiction models.

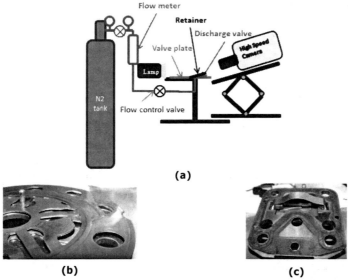

Figure 1. Open air experiments of two discharge valves:
(a) schematics; (b) automotive; (c) hermetic reciprocating.

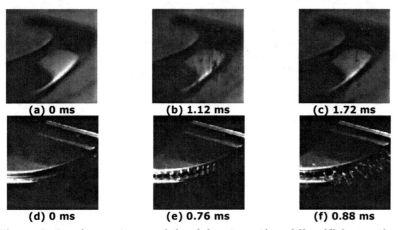

Figure 2. Breakup patterns: (a) – (c) automotive; (d) – (f) hermetic.

In this paper, the focus was on the visualization of the three discharge valves present in a scroll compressor that provides relatively easy access to the discharge plenum. The visualization was concentrated on the interface between the valves and the valve seat since earlier preliminary experiments outside the compressor suggested that the breakup of the oil film that is responsible for the stiction effect would be the greater source of droplets that get entrained into the vapor flow and leave the compressor.

2. METHODS

2.1. System configuration and valve visualization arrangement

In order to visualize the flow inside the discharge plenum of the scroll compressor, a total of 5 sight glasses were installed around the discharge cover. The sight glasses provided access for lighting and also to capture the images. There was also a need to modify the internal structure of the discharge plenum in order to remove physical barriers to the visualization of the valves by machining out a large amount of material, however it is not believed that it will interfere with the droplet breakup and size distribution.

A high speed CMOS camera with maximum resolution of 512x512 pixels at 2200 frames per second (fps) and 256 shades of gray was used to record the videos. Two lenses were used in the visualizations: a macro lens for a broader area view so there could be an understanding of how the valves opening sequence and timing interfere with the droplet generation; and a microscope lens assembly used for better spatial resolution and to attempt to capture the size of the smallest droplets. Videos were recorded at speeds ranging from 4,000 frames per second (fps) to 20,000 fps. Higher framing rates require lower resolution, therefore those were used to determine valve timing, speed and displacement and not for droplet size determination.

Figure 3a shows the internal machining and Figure 3b shows the way the compressor is assembled in the facility with the lighting fixtures and camera positioned. A sample image of each of the lenses is shown on Figures 3c for macro, and 3d for microscope.

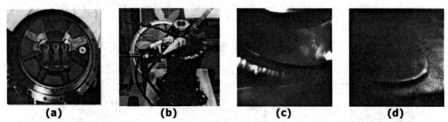

(a)   (b)   (c)   (d)

Figure 3. Compressor modifications, assembly and sample images.

The compressor used is an R410A scroll compressor, using PVE oil (FVC32), with low pressure shell, designed to operate in residential air conditioning. It is installed in a full residential air conditioning system facility comprised of two environmental chambers (indoor and outdoor). Inside each of the chambers, the respective heat exchangers are inserted into wind tunnels so air flow rate and temperatures can be controlled to set the desired suction and discharge pressure conditions to the compressor. The compressor speed is controlled with a variable frequency drive and was kept ranging from 30Hz to 38Hz, which is lower than the nominal speed of 50Hz. This had to be done to avoid dense mist inside the plenum which made visualization impossible at some conditions.

The system is instrumented with a coriolis type mass flow meter with uncertainty of ±0.5% of the actual measurement, suction and discharge pressures are measured with pressure transducers with uncertainty of ±1%. Temperatures were measured at the discharge pipe using a T-type immersion thermocouple with uncertainty of ±0.25°C.

2.2. Image processing

Information about droplet size, speed, valve displacement, valve speed, and distances were obtained using Vision Research Phantom 675.2 software functionalities. For each of the videos, a scaling image was taken that included the valve stopper in focus so a known dimension is present and a pixel to millimeter conversion factor was accurately determined. Due to the lighting conditions and relative position of the camera, the droplets in the videos show as light areas and not dark ones as can be observed in previous studies of this kind (5). This imposes some difficulties and can lead to a low number of droplets visible since the background is not completely dark.

In order to determine droplet sizes with reasonable accuracy, the diameter is measured in four directions and an average is then taken. For droplet speed, the center of each droplet is taken as the reference point to determine the displacement from frame to frame and with the framing rate it is possible to calculate the speed.

3. RESULTS AND DISCUSSION

3.1. Qualitative aspects of the flow at start-up

It is widely accepted that start up conditions for reed valves impose boundary conditions that go beyond the design for steady operation in terms of mass flow rate and consequently valve lift. Therefore it was decided to investigate what happens in terms of oil breakup during these conditions since this situation would be very similar to that seen in the open air experiments. It is important to note that experimental conditions in this case are unimportant and also that OCR cannot be determined. Figure 3a and 4a show the relative placement of the valves from a top view with the compressor open and also from the visualization view.

Right before the compressor is turned on, the situation inside the discharge plenum can be seen in Figure 4a. A small amount of oil is present as a film between the valve and the seat. Figures 4b to 4f show a time evolution of such film from the moment the compressor is turned on forward. These images were captured at full resolution and 2,200fps. It can be observed that after the compressor starts to run the oil film is actually pushed out as a pure liquid flow, without actually breaking and the valve movement is almost imperceptible (Fig.4b). In the next image (Fig.4c), one can see that oil droplets are already flowing inside the discharge plenum, coming from the film breakup. It was observed that during the startup experiments, after a few milliseconds, a significant amount of liquid comes to the compressor and actually enters the compression mechanism. That is reflected on Figure 4d. It can also be noted that the visualization becomes increasingly difficult due to this high liquid loading as it is shown on Figures 4e and 4f.

The same startup condition was also recorded at a lower resolution of 128x128 pixels to allow the use of higher framing rate at 19,000 fps. Figures 5a to 5f show the time evolution of the startup procedure and show interesting features that were previously observed in the open air experiments with the small reciprocating compressor. Figure 5b shows that at the left side close to the tip of the valve, there is film breakup present and it also indicates that the flow around the valve is not symmetrical. Later on Figures 5c and 5d, it can be observed the kind of instable structures that were present on the open air experiment. However the length scale in the real case is different than the open air case. The length scale between the instabilities was determined using the Phantom 675.2 software and the valve thickness was used as the scaling dimension in this case. Indicated dimensions a, b and c on Figure 6 are equal to 274 µm, 307 µm and 285µm respectively, as opposed to about 400 µm to 540 µm observed in the open air experiment. The information of the length scale is very important when validating models for the film breakup and resulting liquid structures like ligaments or droplets. Figures 5e and 5f show the later stages when visualization becomes blurry.

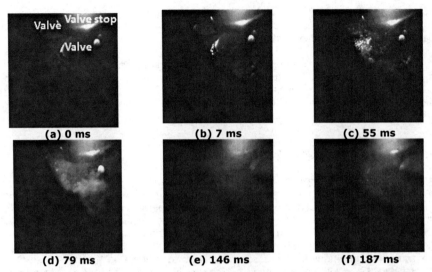

Figure 4. Startup image sequence for main discharge valve at 2,200fps.

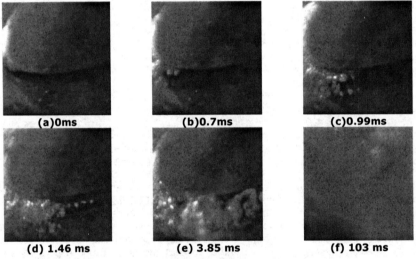

Figure 5. Startup image sequence for main discharge valve at 19,000 fps.

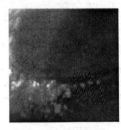

Figure 6. Length scales of the instabilities observed in the oil film.

All the images discussed so far were obtained by starting the compressor after a long period of down time, usually overnight. Another interesting situation emerges when the compressor is running at steady state and is stopped and then restarted after about thirty minutes of down time. This is somewhat different as a puddle of oil is formed and the valves are actually submerged during the next restart. Figures 7a to 7f show the time evolution of the phenomenon. In this sequence of images it is interesting to note the greater volume of oil that is displaced

It can then be said that during startup conditions, although the very first droplets are generated by the oil film breakup, liquid admission to the compression chambers is responsible for the vast majority of the liquid droplets that follow suit. Therefore, attention is then turned to the visualizations under steady state operation.

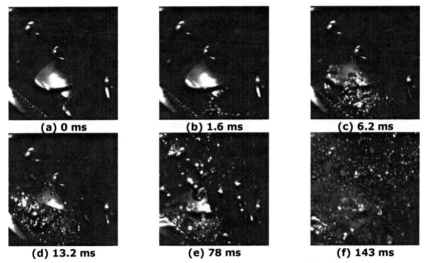

Figure 7. Startup image sequence for submerged discharge valve.

3.2. Flow visualization under steady state conditions

As previously mentioned the compressor had to be set to run at a lower speed then the nominal rated speed of 50Hz to reduce cloudiness inside the discharge plenum. Therefore, the speed during the visualizations were held constant at values that range from 30Hz to 38Hz. What we are after in steady state conditions is to establish firstly the mechanism of droplet generation and secondly to quantify the droplet size and velocity distribution of these droplets and the relation between what is generated and what leaves the compressor to the rest of the system. In this section we will focus on the first objective.

Although in startup conditions, it could be observed that the film breakup and instabilities are responsible for the formation of the very first droplets, in steady state operation, the same cannot be said with such certainty. Figures 8a to 8f show a time evolution of one of the auxiliary valves opening and closing during steady state operation. Figures 8c and 8d show a few droplets but overall the droplet population is not very high. Here it is not possible to identify the expected oil film that should be present between the valve and seat, this may be due to the fact that the volume of oil in that space is very small and it is not possible to be visualized. The other possibility is that due to the valve seat and discharge orifice geometry, the oil is not present at that interface and is probably deposited in a volume inside the discharge orifice, see Figure 9.

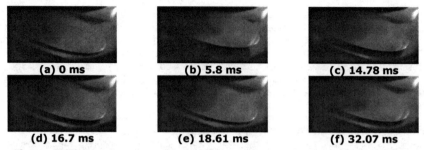

Figure 8. Time sequence of one of the valves under steady operation; P_{disch}=2400 kPa, T_{disch}=130 °C, $\dot{m}$=36 g/s, f=35 Hz, OCR=1.4 %.

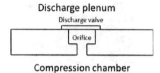

Figure 9. Sketch of the discharge orifice geometry and valve arrangement.

Figure 10a to 10f shows the same visualization but at a higher operating speed of 38Hz and with higher magnification. In this case practically no oil was observed leaving the valve since OCR was very low. The time sequence corresponds to a full open-close-open cycle of the valve.

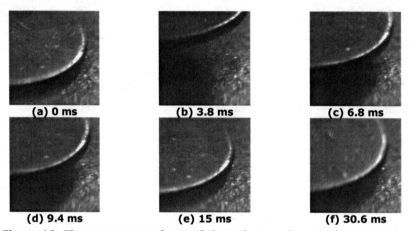

Figure 10. Time sequence of one of the valves under steady operation; P_{disch}=2500 kPa, T_{disch}=125 °C, $\dot{m}$=40 g/s, f=38 Hz, OCR=0.4 %.

It is worth saying that one may wonder that many other information can be extracted from such visualizations as it relates to the valve displacement, velocity, timing, time of the cycle that the valve is open, impact velocity, presence or not of torsional movement, fluttering, etc. This aspect of the visualization is treated in a different paper in this same conference.

3.3. Droplet size and speed under steady state conditions

As mentioned in previous sections, the actual identification of the droplets close to the valve is very difficult given the fact that the droplets show as light regions and not dark, with a somewhat clear background. A different approach was then taken to enhance droplet detection probability. A video was shot against a dark region inside the compressor plenum. Although light regions are clearly identifiable, only a few of them are in focus, making the droplet count be low. This happens because the volume of the discharge plenum is immensely larger than the volume of the region that is in focus, or the viewing area times the depth of field. Even with all these shortcomings, it was decided to report the droplet sizes that were able to be determined. In order to enhance the image for better size determination, a built in high-pass 3x3 filter was applied to the image to enhance the edges. Figures 11a and 11b show the raw and the filtered image. The authors realize the limited statistical significance of these results and are working towards a solution for this problem.

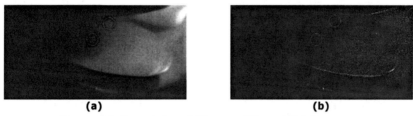

(a) (b)
Figure 11. Raw (a) and Hi-pass filtered (b) images.

Figure 12 shows an attempt to develop a histogram of droplet sizes for one of the videos where droplets were seen. The video length corresponds to three opening and closing cycles of the valves or 97ms. The values for OCR and refrigerant mass flow rate were 1.4% and 36 g/s respectively. It can be seen that for this case the droplet size occurrence is all under 150µm and it concentrates more towards 50 µm which indicates actually the resolution. It is worth noting that there might be smaller droplets that were not captured due to this spatial resolution limitation.

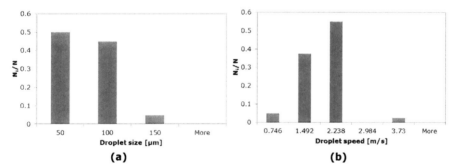

(a) (b)
Figure 12. Histogram of (a) droplet size and (b) droplet speed.

For droplet speed, it is more forgiving to consider out of focus droplets size their centers can be identified and used to calculate the displacement and speed. One drawback of this visualization is that it is only possible to account for a two-dimensional velocity and not in the direction that parallel to the lens axis. Figure 13 shows the droplet speed distribution and it can be observed that the most occurring value was centered at around 2.238 m/s.

4. CONCLUSIONS

The images and results presented so far give support to the following conclusions:
- The two major mechanisms of droplet generation during startup are oil film breakup at the very beginning and excess liquid admission to the compression chamber.

- During start up the oil film presents instabilities that were also observed in the open air experiment however with a shorter length scale of about 289µm.
- Under steady state operating conditions the oil film breakup is expected to play a minimal role in droplet generation.
- Vapor shear interaction with accumulated oil inside the discharge orifice recess is believed to be the main source of droplet generation in steady state.
- Although limited in resolution, the droplet size determination showed that droplets coming from the discharge valve are smaller than 150µm and possibly much smaller than 50µm. This is consistent with previously reported results from (4,5) for compressors that had reed valves at the discharge.
- Droplet speed showed a peak frequency of occurrence at 2.24m/s.

5. FUTURE WORK

The authors acknowledge that this paper reports a work in progress in which promising results are shown from the visualization perspective. However, more effort needs to be exercised into exploring the data that will be available from further experiments and better techniques to analyze such data. For example, the influence of pressure and temperature at the discharge on oil droplet size needs to be studied, as well as the operating frequency of the compressor.

6. ACKNOWLEDGEMENTS

The authors would like to acknowledge the company sponsors of the Air Conditioning and Refrigeration Center for their technical and financial support to this work. Special thanks go to Nishant Mathure and Philippe Dewitte of Danfoss for supplying the compressor prototype. The authors are also thankful for the collaboration of Randy Lee, undergraduate assistant in the realization of this work.

7. REFERENCE LIST

1. DeAngelis, J.M., Hrnjak, P.S., *Experimental Study of System Performance Improvements in Transcritical R744 Systems with Applications to Bottle Coolers*, ACRC CR-57, University of Illinois at Urbana-Champaign, 2005.
2. Kim, S., Pehlivanoglu, N., Hrnjak, P.S., *R744 flow boiling heat transfer with and without oil at low temperatures in 11.2mm horizontal smooth tube*, 13th International Refrigeration and Air Conditioning Conference at Purdue, July 12-15, 2010, West Lafayette, IN, 2010.
3. Pehlivanoglu, N., Kim, S., Hrnjak, P.S., *Effect of oil on heat transfer and pressure drop of R744 in 6.1mm horizontal smooth tube*, 13th International Refrigeration and Air Conditioning Conference at Purdue, July 12- 15, 2010, West Lafayette, IN, 2010.
4. Toyama, T., Matsuura, H., Yoshida, Y., *Visual Techniques to Quantify Behavior of Oil Droplets in a Scroll Compressor*, International Compressor Engineering Conference at Purdue, West Lafayette, IN, 2006.
5. Wujek, S.S., P.S. Hrnjak, *Mist to annular flow development quantified by novel video analysis*, ACRC Report TR285, University of Illinois at Urbana-Champaign, Urbana, IL, 2011.

NOVEL COMPRESSORS AND MANUFACTURING TECHNOLOGIES

Analysis and development of a new compressor device based on the new finned piston

M Heidari, A Rufer
LEI, École Polytechnique Fédérale de Lausanne (EPFL), Switzerland

ABSTRACT

In the general frame of Compressed Air Energy Storage system (CAES), the LEI Laboratory of EPFL has introduced the concept of dry finned piston. The main goal is to achieve energy storage by means of compressed air thanks to high isothermal efficiency compression/expansion processes.

For achieving this goal, a new compression and expansion machine has been defined, using a new piston-cylinder assembly consisting of a series of concentric annuli's that fit together during compression/expansion. This will increase the heat transfer surface, allowing the heat to be removed from the system during compression and to be absorbed by the system during expansion. This feature together with low speed movement makes the process close to isothermal.

First the concept of annular compression chambers is explained, with the goal of increasing the heat exchange surface. Then the new compression system with imbricated annulis is described.

A test bench has been developed in order to propose an experimental validation of such a compression process concept. It is also of a great interest to compare the dry finned piston to the classic pistons available in the market. The experimental results show a higher energetic efficiency for the finned piston due to its close to isothermal behavior thanks to increased heat transfer surface.

1. INTRODUCTION

The compressed air storage system is simply composed of a compression cycle, which delivers the gas to high-pressure storage, and an expansion cycle that expands the compressed air to extract its energy. The air compression system is driven by a hydraulic pump and cylinder, which is fed by an electric motor. On the other side the expander runs a hydraulic motor that is connected to an electric generator (Figure 1). Such typical systems are described in (1), (2) and (3).

This system uses an efficient hydraulic drivetrain to convert electrical energy into potential energy stored as compressed air. The use of hydraulics, among other advantages, allows to precisely control gas expansion and compression, thus maximizing thermal efficiency, allowing for high-efficiency conversion of mechanical to electrical energy. The system works similarly in reverse mode to convert the potential energy stored in the compressed air back into electrical energy. The proposed concept belongs to a family of storage devices based on the compression and expansion of air, but where only electrical energy is used as changing resource and also only electric energy is produced. Such systems have their own advantages in the sense of sustainable development and emission free generation as alternatives to other CAES systems using combustion machines as described in (4), (5), (6) and (7). Compressed air energy storage is also intended to replace electrochemical batteries and to serve as low aging systems.

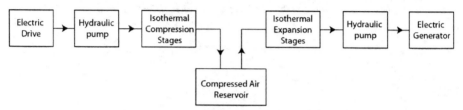

Fig 1. Schematic compressed air storage system

2. CONCEPTUAL DESIGN

2.1. Getting closer to Isothermal line

The main goal of the new finned piston is solving the problem of non-isothermal behavior of the compression system. In this regard two ways were introduced:
- Inter-stage heat exchangers
- Directly integrated exchanger

The first approach is quiet classic and conventional, but it is not enough to reach the level of isothermal efficiency that is needed. Thus, a second approach is proposed, namely directly integrated exchanger, to reach higher isothermal efficiency.

Figure 2 shows the P-V diagram of compression and expansion round trip cycle. Visibly in compression phase since the gas gets hot, we need to cool down the gas using coolers. In contrast, during expansion since the gas gets cool, we need to implement heating system.

In figures 2 (a) to (d), the compression process is represented and in figure 2 (e) to (h) the corresponding expansion is also illustrated. Four different cases are illustrated namely first the ideal isothermal process (curves (a) and (e)), and second the one stage polytropic process (curves (b) and (f)), where the shaded surfaces represent the spent respectively recovered energy. The compression and expansion surfaces correspond to the done or to be extracted work from the system for a state change from point 1 to 2 or in the inverse sequence.

The third case (curves (c) and (g)), represents a 3 stage compression expansion machine with interstage heat exchangers.

Finally, the fourth case represented in figure 2 ((d) and (h)) corresponds to a compression/expansion machine with increased thermodynamic performance due to the use of directly integrated exchangers (DIE).

To get a better illustration of the benefits provided by the directly integrated versus the benefits of the inter-stage exchangers the curves of figure 2 is superposed in figure 3. By the classical approach of a multistage compression machine, the process gets close to isothermal line, as shown by large arrows. Furthermore, with the help of the second approach, using DIE, it is possible to go further close to isothermal line, which is demonstrated using small arrows. This explains the concept of the dry piston system design.

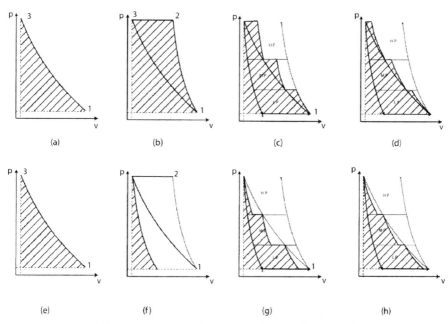

Fig 2. P-V diagram of compression and expansion

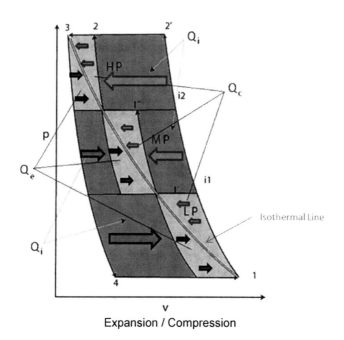

Expansion / Compression

Fig 3. P-V diagram of compression expansion roundtrip cycle

The details of this system including cooling system have been demonstrated in figure 4. As it is shown air enters the LP, MP and HP stages respectively and stored in a compressed air reservoir. The hot air is cooled down using cooling water, which gets cooled down itself in turn, using an air cooling fan. Each stage of double-acting compressor is driven by a hydraulic cylinder fed by an electrically driven motor-pump system.

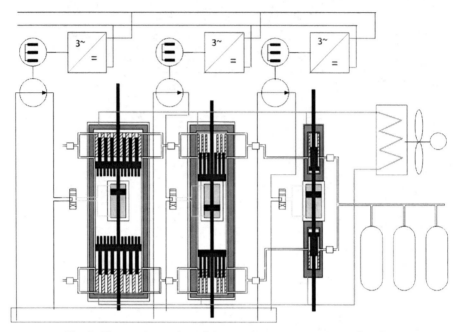

Fig 4. Three-stage dry piston compressor system circuit

2.2. Multilayered annular piston-cylinder assembly

2.2.1. Increasing the heat transfer from and to the compression chamber

In order to increase the energetic performances of a compression/expansion chamber, the heat exchange between the active volume of the gas inside of the chamber and the surrounding should be increased. In addition to the principle of the interstage coolers, the trial here is to reduce the mechanical work during the compression / expansion process. This better heat transfer should allow « moving » the PV curve of the polytropic compression/expansion in the direction of the curve of the isothermal behavior. For this purpose, the heat exchange surface of the cylinder is increased.

2.2.2. From the normal cylinder to the annular concept

The representation of figs 5(a) to (l) show the active surface of one piston, realized in three different executions. The figs 5(a) to 5(d) show the active surface of a normal piston that corresponds to a simple circular surface. The piston surface is drawn for a situation where the pressure is increased, namely from 10 bar to successively 50 bar, 150 bar and finally 250 bar. This force-active surfaces, A_{press} are calculated so as the forces developed by the pistons keep constant, independently from the pressure level. In this example, the exterior diameter for 10 bars is chosen equal to 5.5 cm. For the calculation of the active volume and for the maximum heat exchange surface, a stroke-length of 5.5 cm is also defined.

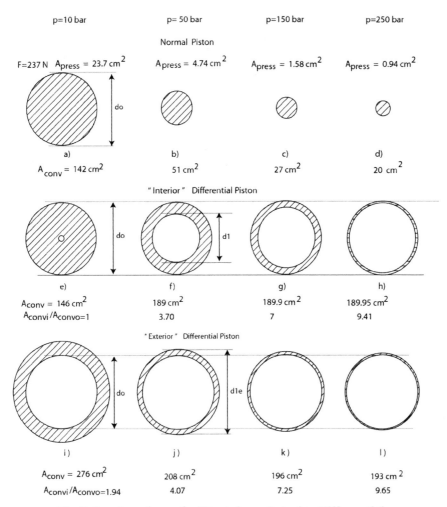

Fig 5. Front surface of «Normal», «Interior-Differential» and «Exterior-Differential» pistons

In the first column, the force-active surfaces are chosen in order to get the forces developed by all pistons are equal for all pressure levels. The force-active surface, which is indicated by A_{press}, varies from 23.7 Cm² for 10 bar to 0.94 Cm² for the 250 bar pressure level.

For such cylinders, the heat exchange surfaces are then calculated. For the normal cylinder, the heat exchange surface is equal to the external surface of the cylinder. These values are noted A_{Conv0}. It is evident that for a piston suited for a higher pressure its diameter is reduced. Then the heat exchange surface is evaluated, under the assumption that the stroke of all pistons is identical. This leads to the fact that the mechanical works produced by different pistons of a multistage compression machine are also identical. In the figure, the heat exchange surface of the normal piston varies from 142 Cm² for a 10 bar piston to 20 Cm² for a 250 bar one.

Figure 5 (e) to (h) show a different concept, called here the « Interior Differential » Annular Concept. In this case, a differential piston is considered, that means that its active surface corresponds to an annular surface. The « Interior Differential Piston» (IDP) concept is based on the fact that the external diameter is identical to the diameter of a normal piston, and that an internal non-compressible cylindrical piece is used as a piston. The reduction of the force-active surface for an increased operating pressure is obtained by increasing the interior diameter of the non-compressible piece.

In the middle part of the figure 5 (e) to (h), the piston front-sides are represented for the interior differential pistons. In this case, the interior diameter varies from nearly zero for the lower pressure case (10 bar) to a diameter near to the exterior diameter for the high pressure storage (250 bar). The force-active surface is equal to the value of the corresponding normal piston for all indicated pressure levels. The interesting results in these cases are that the heat exchange surface A_{Convi} is increased when the force active surface A_{press} is decreased. The values indicated in the figure goes from 146 Cm² for a 10 bar piston to 189.95 Cm² for the 250 bar piston.

It has to be noted that in the calculation of the interior diameter, the outer piston diameter is kept constant, and that the force-active surface in « inside » of the external (primitive) diameter.

The possibility exists also to design a so called « Exterior Differential Piston » (EDP), by placing the force-active surface « outside » the primitive diameter. Four different cases of this design are represented in Fig 5 (i) to (l). The first case considers a piston designed for the same case as in fig 5(a). The same value of force is developed by the front-side of the piston. The same procedure is followed by the other pressure levels.

Fig 6 (a) shows the variation of the diameter for the normal piston, the variation of the interior piece for the IDP and the diameter of the outer cylinder for the EDP in dependency of gas pressure. The force is kept constant independently of the pressure. Fig 6 (b) shows the heat exchange surfaces for the normal, interior differential and exterior differential pistons respectively. A comparison between these pistons shows that for the same force developed by increasing the pressure level, differential pistons can achieve around 9 times more heat exchange surface comparing to normal pistons.

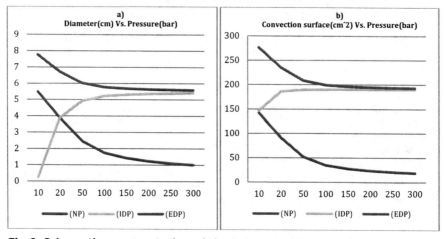

Fig 6. Schematic representation of the Integrated Heat Exchanger Concept

3. COMPRESSOR DESCRIPTION

The principle of increasing the heat exchange surface through the use of differential pistons can be extended to the situation where the full volume occupied by a cylinder/piston unit is filled with imbricated fins. In such a case, multiple layers of annular differential pistons are arranged within the same compression/expansion chamber, and correspond effectively to a concept of the integrated heat exchanger. Such geometry is represented in its principle in figure 7. Only the upper half of the fins is represented. In such a case, one mobile assembly of differential layers is sliding inside of a fixed assembly of fins. In the center of the cylinder, there is a shaft for guiding the mobile equipment inside the fixed one, and that allows designing the system with a small distance between the fins. Considerations related to the inter-fin space will be done in relation with the calculation of the effect of a « dead » volume.

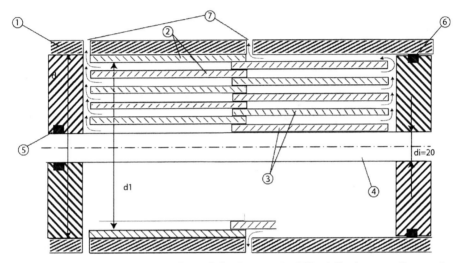

Fig 7. Schematic representation of the Integrated Heat Exchanger Concept

The parts numbered on the figure are:
1) External cylinder
2) Stator fins
3) Sliding fins
4) Guiding rod
5) Stator seal (rod-seal)
6) Piston-seal
7) Gas Inlet/outlet

The inlet/outlet of the compressed/expanded gas in such a concept is realized through radial paths, allowing the access to the internal layers. The radial paths can be arranged on both sides of the cylinder, eventually at the whole periphery.

On the left side of figure 7, the radial in/outlet is shown, and corresponds to the access to the annular compression/expansion annular chambers of the fixed (stator) part. In the middle of the figure, the radial in/outlet path is represented for the mobile (piston-side) annular chambers. It is evident that the two paths must be connected to the same in/outlet circuit or valve, using a manifold collector.

The force-active surface of such an Integrated Heat Exchanger Piston is equal to the full surface of a normal piston with identical external diameter, from which the occupied surface of the guiding shaft in the center must be subtracted.

4. EVALUATION OF PERFORMANCE

4.1. Modeling and simulation

The modeling and simulation of a classical and of a finned piston system as represented in Fig 7 has been realized at LEI. While the first estimation of performance considering heat exchange between fines and the compressed air has been done using finite element method (9). A detailed analytic estimation of isothermal performance using thermoelectric analogy method accounting for all phenomena's like conduction, convection and heat capacity effect of fins, is still under way; following the method based on graphical representation tools (8).

4.2. Experimental results

The present paper shows experimental estimation of the performance of the new piston system with integrated heat exchanger.

To evaluate and characterize the performance of developed finned piston, a test bench has been developed. First the finned piston was fabricated from aluminum with a high tolerance to avoid friction and leaks. The test bench includes an electrical motor to provide needed work, a ball-screw driver for transforming rotational to linear movement, the finned cylinder and apparatus needed for controlling speed and measuring force, temperature and pressure. (Figs 8 and 9.)

The experimentations on the compressor setup were done under two conditions: No-load and filling mode. In the no-load mode, in order to measure losses due to friction and pressure drop caused by anti-return valve and sensors the cylinder is operated without being connected to the reservoir.

In the filling mode the compressor outlet is connected to reservoir. The power consumption in each case is shown in Fig 10 for the last 11 cycles of the filling operation of the reservoir at 7bar.

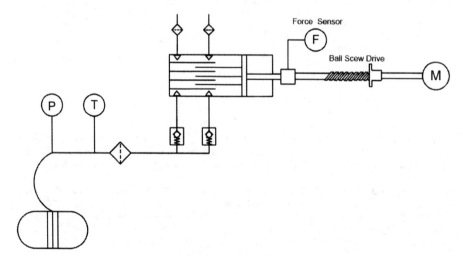

Fig 8. Setup schematic circuit in filling mode

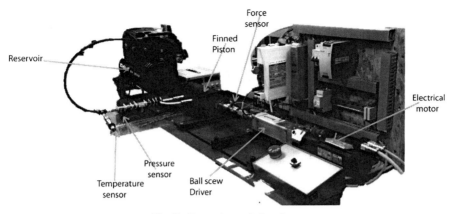

Fig 9. Experimental setup

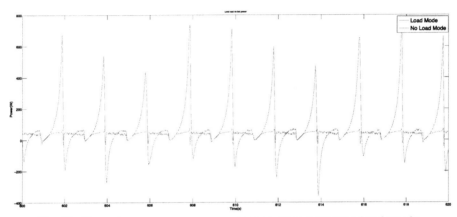

Fig 10. Finned compressor power comparison between load mode and no load mode. (Last 10 cycles)

The same experiment was done using a classic piston on the same test bench. The classic piston is a non-finned SMC-CP96 piston. Modeling and analysis of this conventional piston is available in reference (9).

5. COMPARISON AND CONCLUSION

In this section, the efficiency of both the classic and finned piston are evaluated. Both pistons have the same surface area and a same stroke so the volume is the same.

Figure 11 shows the evolution in compressors exhaust pressure during the filling of the reservoir. It is evident that the finned piston fills the reservoir in shorter time. (691s instead of 781s). The most important result recorded from the test bench is related to the exhaust temperature at the output of both the non-finned (classical) and the finned compressor stage. The temperature profile of the output of the stages is represented in Figure 12. The temperature rise is much less in finned piston, which shows a higher isothermal efficiency.

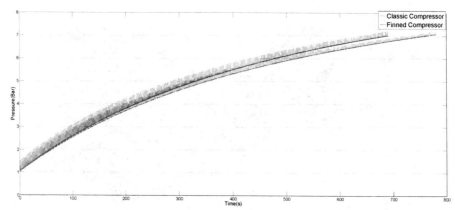

Fig 11. Pressure evolution in during filling the reservoir up to 7 Bar

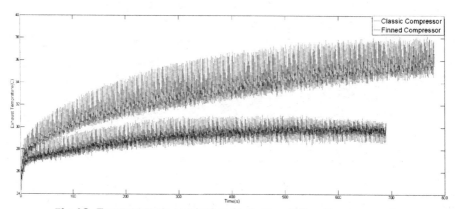

Fig 12. Temperature evolution in during filling the reservoir

Figure 13 shows the power needed to run each compressor in no load mode. Here, the finned piston shows a higher friction amount. The energy needed to run the compressor in one cycle was calculated to be 60.2J for Standard piston and 74.1J for Finned Piston.

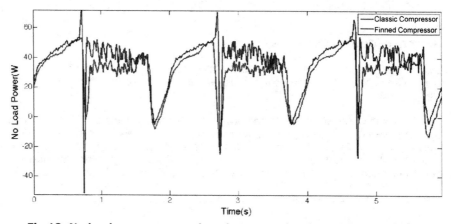

Fig 13. No load power comparison between classic and finned piston

Figures 14 and 15 show the power consumption curve in no load and load mode for classic and finned piston respectively. The total work consumed by the classic and the finned compressor is calculated by integrating numerically the power curves that have been recorded during the rise of pressure in the reservoir for 1 Bar to 7 Bar. These values were found to be 51.5 kJ for the finned piston and 59.45 kJ by classic SMC piston.

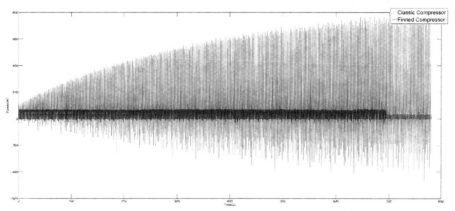

Fig 14. Classic and Finned compressor power comparison

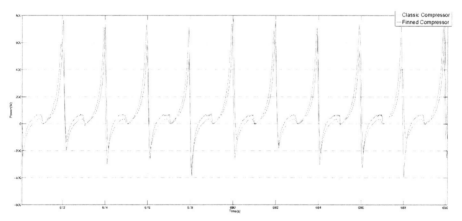

Fig 15. Classic and Finned compressor power comparison (zoomed on the last 10 cycles of filling up to 7 Bar)

In order to find the efficiency of the process, the maximum extractable energy content of reservoir should be calculated. The work potential (exergy) of the compressed air with pressure p_2 can be found from the relation:

$$E = p_2 V \left(\frac{1}{p_2} + \ln\left(\frac{p_2}{1}\right) - 1\right) \quad (kJ)$$

Since $p_2 = 7$ Bar, the energy content of the reservoir is equal to 19.05 kJ.

The power consumptions and exergetic efficiency of each compressor are summarized in table 1.

Table 1. Exergetic efficiency comparison between the pistons based on effective energy spent

	No Load Energy (kJ)	Load mode Energy (kJ)	Effective Energy (kJ)	Exergetic Efficiency(%)
Classic SMC	23.7	72.5	48.8	39.0
Finned Piston	25.8	63.4	37.5	50.7

The results show that however the finned piston has a higher friction power, it consumes less effective energy, showing a higher thermal efficiency since the compression process is closer to isothermal conditions.

Applying the existing compressor/expanders for isothermal CAES systems have a few drawbacks, but these can be addressed by employing some innovative concepts like DIE. Since the same design will be used in expansion mode, the interest in approaching isothermal condition is double. The same concept can be applied to multistage compression/expansion for reaching high compression ratios. By doing so, interstage intercoolers may be avoided, resulting in reduced pressure drop and equipment cost.

ACKNOWLEDGEMENTS

The authors are grateful to the European Commission and Solution consortium for their financial support.

REFERENCE LIST

(1) S. Lemofouet, A. Rufer. A hybrid energy storage system based on compressed air and supercapacitors with Maximum Efficiency Point Tracking. IEEE Transaction on Industrial Electronics, Vol.53, No 4,08.2006.
(2) A. Reller, A.Rufer, Cyphelly I. Usage of compressed air Storage System. Swiss Federal Office of Energy Report, May 2004.
(3) T. Kenschke, Eurosolar/Energieagentur NRW meeting: Dispersed energy storage - raising the productivity of renewable energy sources with cutting-edge energy storage solutions Proceedings, 2003, Germany.
(4) Kepshire, Dr. *Dax Demonstration of Isothermal* Compressed Air Energy Storage to Support *Renewable Energy* Production, November 3rd 2010. Update conference US DOE ESS, Washington D.C.
(5) *Lehman J, Air storage gas turbine power plants*: a major distribution for energy *storage*. In "Proc. Int. Conf. Energy *Storage*" MIT Press, Cambridge, MA.
(6) *van der Linden S. CAES for Today's Market*. Electrical Energy Storage Applications & Technology (EESAT) Conference, San Francisco, CA. April 2002, pp.15-16.
(7) Hoffeins H., Huntorf, air storage gas turbine power plant, Energy Supply, Brown Boveri Publication DGK 90 202 E, 1994.
(8) Heidari M., Rufer A., Modeling of a Reciprocating Linear Compressor using Energetic Macroscopic Representation, International Compressor Engineering Conference at Purdue, July 16-19, 2012.
(9) M. Heidari, P. Barrade and Rufer A. Modeling and Simulation of a Three-stage Air Compressor Based on Dry Piston Technology, COMSOL Conference, Stuttgart, Germany, October 2011.

Oil-flooded screw vacuum pumps with a novel flexible discharge port design

Y Tang
Kaishan Compressor, China

ABSTRACT

Industrial oil flooded screw vacuum pumps have two major functions: pulling down and maintaining the required vacuum level. As a result of the two major functions, the screw vacuum pumps have to handle a great range of pressure ratios. A screw vacuum pump with a fixed volume ratio is going to have either the over compression during the pulling down process or the under compression during the required vacuum level maintaining. This paper presents oil flooded screw vacuum pumps with a patented flexible discharge port. Such a screw vacuum pump has its volume ratio or discharge port matched with the real pressure ratio. The most under-compression and over-compression are eliminated, and the vacuum pump noise level is reduced. The design and optimization techniques of the flexible discharge port of a screw vacuum pump are discussed in this paper, and the measurement results of a screw vacuum pump with the flexible discharge port are presented. At typical industrial vacuum levels, a crew vacuum pump with the flexible discharge port design can reduce the power consumption by about 20%.

1. INTRODUCTION

Industrial oil flooded screw vacuum pumps have two typical applications: pulling down and maintaining the required vacuum level. The pulling down process is to vacuum a given space or volume from the atmosphere pressure to a required vacuum. The vacuum maintaining process is to keep a given space or volume at the required vacuum level and the vacuum pump keeps running to maintain the required system vacuum level against any air leakage to the system or any air sucked into the system. Most applications involve both the pulling down and then maintaining the required vacuum level. There are also applications only involved the pulling down. When the demanded vacuum level is achieved, the vacuum pump is shut down. Although properly designed industrial oil flooded screw vacuum pumps can reach the vacuum level as high as 29.9 inHg, they are typically used to obtain the vacuum level of 15 to 28 inHg.

Obviously, the screw vacuum pumps have a very significant pressure ratio variation. During a pulling down process, the vacuum level in the system keeps increasing, and thus the pressure ratio keeps increasing too. In the example of screw vacuum pump SKY220V presented in this paper, the pressure ratio can increase from 1.7, when the system pressure is at atmosphere pressure, to 16.7, when the system vacuum level reaches 28 inHg. Typical screw compressors used for the vacuum pumps have a fixed low volume ratio, say 1.9 to 2.1. The reason to use a low volume ratio is to avoid the serious over compression during the pulling down process when the pressure ratio is low, and thus to avoid using a huge main motor to drive the screw vacuum pump. However such a low volume ratio is going to result in a serious under compression, when the system vacuum level reaches 15 to 28 inHg, which is correspondent to the pressure ratios of 3.0 to 16.7 as per the example in this paper. The under compression shall cause the extra energy consumption when the vacuum pump maintains the system vacuum level. The ideal screw compressors used for the vacuum pumps are the ones with the variable volume ratio. The discharge port defined by the volume ratio always matches the working condition, that is, the actual pressure ratio.

The author presented a novel flexible discharge port design for screw compressors in [1], and the design has received the patent in some countries and is patent-pending in some other countries. The design does not depend on the pressure sensors, the electrical and mechanical mechanism to adjust the discharge port location to match the real pressure ratio like a traditional volume ratio valve. The performance characteristics of the flexible discharge port design are more like a discharge valve used in a reciprocating compressor. The volume ratio or the discharge port defined by the volume ratio always matches the pressure ratio automatically. In the other words, when the pressure ratio increases, the volume ratio increases automatically to match the internal pressure ratio with the external pressure ratio. The screw compressor with the flexible discharge port shall eliminate any potential over-compression or under-compression when the external pressure ratio varies. Compared with the traditional volume ratio valve, the flexible discharge port design is an automatic valve, and it is significantly simplified and more reliable by eliminating the pressure sensors, the controls and the executive mechanism. The flexible discharge port is ideal for the oil flooded screw vacuum pumps, and this paper presents the design and measurement results of an oil flooded screw vacuum pump.

2. THE DESIGN OF SCREW VACUUM PUMP WITH FLEXIBLE DISCHARGE PORT

Fig. 1 shows a cross-sectional drawing of a screw vacuum pump with the flexible discharge port. Slide valve 1 is connected to two pistons 2 and 3, and each piston has its own cylinder 4 and 5 respectively; one piston is at the inlet end of the pump, and another one is at the discharge end. The pistons may slide inside the cylinders together with the slide valve. Cylinder 4 at the inlet end is connected to the discharge pressure through hole 7 drilled in the slide valve, and thus inside cylinder 4 the pressure is the discharge pressure of the vacuum pump, typically slightly higher than the atmosphere pressure. Cylinder 5 at the discharge end is connected to the last closed lobe before the discharge process through hole 8 drilled in the slide valve, and it has the average pressure of the last closed lobe.

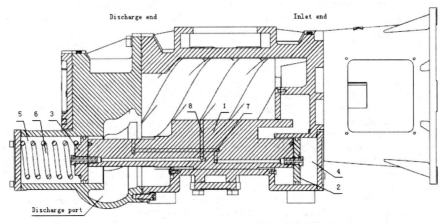

Fig. 1: A screw compressor with the flexible discharge port

It is assumed that the two cylinders and the slide valve have the same cross-sectional area, which of course can be different depending on the actual design. Before the vacuum pump starts up, the inlet pressure and discharge pressure of the vacuum pump are the same, that is, at the atmosphere pressure. The slide valve is located against the inlet end due to the loading of spring 6, so that the vacuum pump has the lowest volume ratio. When the vacuum pump starts up, the spring loading and internal pressure inside the last closed lobe shall keep the slide valve toward the inlet end, and thus the lowest volume ratio. The lowest volume ratio results in the lowest over-compression during the start-up process, and thus the lowest starting torque. As the inlet vacuum increases during the pulling down process, the average pressure inside the last closed lobe and inside cylinder 5 decreases gradually, and the slide valve moves to the discharge end until the axial forces of the two pistons and spring balance, and the discharge port defined by the slide valve is at its optimum position. If the inlet vacuum has any variation, the slide valve is going to be relocated to its new optimum position. If the inlet vacuum decreases, the pressure ratio across the vacuum pump decreases, the slide valve moves to the inlet end, and the volume ratio drops accordingly. If the inlet vacuum increases, the pressure ratio across the vacuum pump increases, the slide valve moves to the discharge end, and the volume ratio increases too.

3. THE AVERAGE PRESSURE OF THE LAST CLOSED LOBE

It is critical to define the spring loading and the spring stiffness. In order to design the spring, the following parameters are required: the slide valve travel, the spring loadings at the maximum travel, the spring loading at zero travel. The spring loadings can be calculated based on the discharge pressure, which is almost constant for an oil flooded screw vacuum pump and is slightly higher than the atmosphere pressure, and the average pressure of the last closed lobe. According to the working conditions, the optimum volume ratio can be decided according to the working process simulation. The mathematical model used in the working process simulation is presented in [2], and the correspondent software package is described in [3]. The average pressure of the last closed lobe is calculated according to the simulation results of the working process.

In this paper, Kaishan screw vacuum pump model SKY220V is used as an example. The working conditions and the simulation results of SKY220V are presented as follows. Table 1 describes the geometrical parameters and working conditions, and Table 2 presents the simulation results.

Table 1: The geometrical parameters and working conditions of SKY220V

Rotor profile	Y-1
Lobe configuration	5+6
Male rotor diameter (mm)	301.200
Female rotor diameter (mm)	254.300
Rotor length (mm)	500.000
Male rotor wrap angle (°)	336.000
Male rotor rotational speed (rpm)	2950
Nominal discharge pressure (barA)	1.35
The lowest inlet pressure (barA)	0.25
The maximum inlet pressure (barA)	0.50

Table 2: The simulation results of SKY220V

Volume ratio for axial discharge port	4.0
Minimum volume ratio for radial discharge port	2.0
Maximum volume ratio for radial discharge port	4.0
At the inlet pressure of 0.50 barA	
The optimum volume ratio for radial discharge port	2.0
The average pressure of the last closed lobe (barA)	0.990
At the discharge pressure of 0.25 barA	
The optimum volume ratio for radial discharge port	3.7
The average pressure of the last closed lobe (barA)	0.809

From the volume ratios for radial discharge port the slide valve travel can be calculated. The spring loading can be calculated from the discharge pressure and its correspondent average pressure of the last closed lobe. The spring can be designed according to the slide valve travels and the correspondent spring loadings at the different slide valve travels.

4. MEASUREMENT RESULTS OF A SCREW VACUUM PUMP WITH THE FLEXIBLE DISCHARGE PORT AND DISCUSSIONS

SKY220V with the flexible volume ratio of 2.0 to 4.0 is installed in a Kaishan screw vacuum unit VP90-59, and the brake horsepower is measured at the different inlet vacuum levels. The measurement results are presented in Fig. 2. Table 3 indicates the actual pressure ratios across SKY220V at the different inlet vacuum levels.

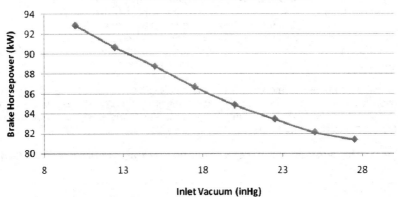

Fig. 2: The brake horsepower of SKY220V vs. the inlet vacuum

Table 3: The pressure ratios of SKY220V at different vacuum level

Inlet Vacuum (inHg)	Pressure Ratio
10.0	2.42
12.5	2.70
15.0	3.00
17.5	3.60
20.0	4.21
22.5	5.62
25.0	8.00
27.5	15.78

SKY220V has a flexible discharge port with the volume ratio range of 2.0 to 4.0. The measured results indicate that the brake horsepower has a linear relationship with the inlet vacuum from 10 to 22.5 inHg, which is correspondent to the pressure ratio across SKY220V of 2.42 to 5.62. The higher the vacuum level is, the lower the brake horsepower is. This linear performance characteristic from 10 to 22.5 inHg is different from the screw vacuum pump with a fixed discharge port. In the range of 10 to 22.5 inHg, the flexible discharge port design is always automatically adjust its discharge port to the optimum position, which matches with the actual operation pressure ratio, and the over-compression and under-compression are eliminated. The performance characteristic is similar to a performance characteristic of a reciprocating compressor, and the flexible discharge port design has the same behavior of the discharge valve in a reciprocating compressor.

When the inlet vacuum level of SKY220V is above 22.5 inHg, the actual pressure ratio is above 5.62 and the volume ratio of SKY220V reaches its maximum, that is, 4.0. The brake horsepower does not have a linear relationship any more with the inlet vacuum level. The performance characteristic of SKY220V above the vacuum level of 22.5 inHg is just like a screw vacuum pump with a fixed volume ratio, and in this case the fixed volume ratio is 4.0.

As mentioned above, typical screw vacuum pumps have a fixed low volume ratio, say 1.9 to 2.1, in order to avoid the serious over compression during the pulling down process and thus to avoid using a huge main motor to drive the screw vacuum pump. Fig. 3 presented two simulated p-V diagrams of SKY220V at the inlet vacuum of 22.5 inHg: the p-V diagram with the fixed volume ratio of 2.0 and the p-V diagram with the flexible discharge port. At 22.5 inHg inlet vacuum, the indicated power of the vacuum pump with the fixed volume ration of 2.0 is 22.4% higher than the vacuum pump with the flexible discharge power. For a continuous operated industrial screw vacuum pump, the energy saving of the flexible discharge port is very significant.

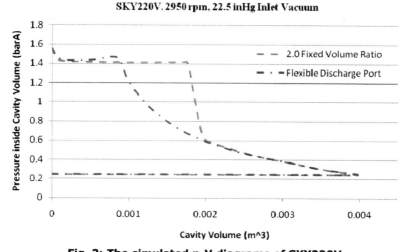

Fig. 3: The simulated p-V diagrams of SKY220V

Another advantage of the flexible discharge port is that it eliminates the discharge noise due to the unmatched discharge port and pressure ratio. The screw vacuum pumps with the novel flexible discharge port design presented in this paper has been in production for two years, and the customers are satisfied with their energy saving and their low noise level.

The further research work can be done to improve the design of the screw vacuum pumps with the flexible discharge port. The experimental study shall be carried on to optimize the design of the spring. The slide valve travel shall be increased to reduce the minimum volume ratio and to increase the maximum volume ratio. The reduction of the minimum volume ratio will reduce the starting torque of the screw vacuum pump and make its start-up easier, especially at lower ambient temperature. The maximization of the volume ratio will make the flexible discharge port to cover higher vacuum level.

5. CONCLUSIONS

1. The screw vacuum pumps with a novel flexible discharge port design have been presented. When the inlet vacuum level varies, the flexible discharge port is adjusted automatically to the actual pressure ratio and keeps the discharge port location optimum.
2. In order to properly design the spring, the optimum volume ratio and the average pressure of the last closed lobe are required, and they are obtained through a mathematical model of working process and the simulation software developed according to the mathematical model. The simulation results of SKY220V are presented as an example.
3. The measured results of brake horsepower of SKY220V indicates that the brake horsepower has a linear relationship with the inlet vacuum level up to 22.5 inHg. The brake horsepower has a direct proportional reduction as the inlet vacuum level increases. When the vacuum level is above 22.5 inHg, the volume ratio of SKY220V reaches its maximum value of 4.0, and its performance behavior is just like the screw vacuum pump with fixed volume ratio of 4.0.
4. Oil flooded industrial screw vacuum pumps are used for the vacuum level of 15 to 18 inHg typically. At the inlet vacuum level of 22.5 inHg, a screw vacuum pump with a fixed volume ratio of 2.0 consumes over 20% more power than a screw vacuum pump with a flexible discharge port design.
5. Although the screw vacuum pumps with the novel flexible discharge port have been in production for two years, and customers are satisfied with the energy saving and low noise level, the further research work shall focus on the experimental study to optimize the design of the spring and on the further increase of the slide valve travel.

ACKNOWLEDGEMENT

The author gratefully acknowledges the management of Kaishan Compressor for permission to publish the paper.

REFERENCES

1. Tang, Y. Screw compressors with a novel flexible discharge port design. International Compressor Engineering Conference at Purdue, July 16 – 19, 2012, 1445.
2. Fleming, J.S., Tang, Y. and Cook, G. The twin screw compressor Part 2: a mathematical model of the working process. Proc. Instn Mech. Engrs, Vol 212 Part C, 1998, 369 – 380.
3. Tang, Y. Computer aided design of twin screw compressors. PhD thesis, University of Strathclyde, Scotland, 1995.

Spool compressor tip seal design considerations

C R Bradshaw
Torad Engineering LLC, USA

ABSTRACT

The rotary spool compressor is a novel compressor type which combines various aspects of rotary and reciprocating devices. The tip seal of this compressor is a dynamic sealing element which should be well understood for maximum performance. An analysis which combines the hydrodynamic film theory balanced with the tip seal dynamics is presented. This model is explored over a variety of tip seal radii, widths, and mechanical spring rates. The experimental volumetric efficiency of a prototype compressor with various seal radii and mechanical spring rates is compared against the model predicted results and shows similar trends.

1 INTRODUCTION

The rotating spool compressor is a novel rotary compressor mechanism most similar to the sliding vane compressor. Primary differences are described by Kemp et al. (1, 2) and include three key differences from a sliding vane compressor.

- The vane is constrained by means of an eccentric cam allowing its distal end to be held in very close proximity to the housing bore (typically less than 0.30mm) while never contacting the bore.
- The rotor has affixed endplates that rotate with the central hub and vane forming a rotating spool.
- The practical use of dynamic sealing elements to minimize leakage between the suction and compression pockets as well as between the process pockets and the compressor containment.

These differences are shown in Figure 1 which presents a cutaway view of a rotating spool compressor with the key geometric features highlighted.

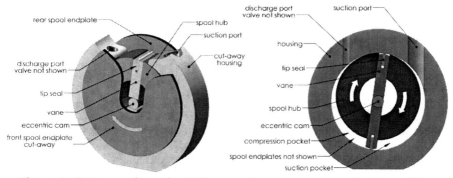

Figure 1: Cutaway view of rotating spool compressor mechanism with key components highlighted.

More demand for higher efficiency components has resulted in a renewed interest in detailed compressor modeling to predict the performance of novel compressors. A recent approach to compressor modeling, called the comprehensive approach, has provided a complete analysis of positive displacement compressors. Recently, a comprehensive model for a spool compressor was developed and presented by Bradshaw and Groll (2,3).

However, this approach relies on sub-models for things such as leakage, heat transfer, and friction. The unique behaviour of the tip seal warrants the need for additional investigation to develop a deeper understanding of the device and improve the sub-model which represents it.

The spool compressor tip seal shares similarities to an Apex seal in a Wankel compressor (5). The kinematics of the Wankel mechanism have been studied in the past (6) as well as the kinematics and forces associated with the Apex seal (7,8). In addition, it was hypothesized that the Wankel apex seal derived part of its ability to seal by the oil in the compressor chamber that it slid on. For this reason, a typical approach to model the Wankel Apex seal included a combination of hydrodynamic lubrication theory and the dynamics of the mechanism itself (9).

This work uses the approach of a combination of hydrodynamic film theory and the dynamics and kinematics of the spool mechanism to solve for the potential leakage gap in a spool compressor tip seal. This is compared with the experimentally measured volumetric efficiency of a prototype spool compressor to obtain a reasonable understanding if the model trends predicted are realistic.

2 TIP SEAL BEHAVIOR

The tip seal in a spool compressor is a compliant sealing element at the distal end of the rotating vane. A schematic view of the tip seal mechanism is shown in Figure 2.

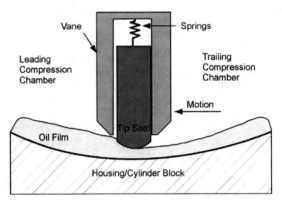

Figure 2: Schematic of tip seal assembly and proposed behaviour in spool compressor.

The spool compressor vane is constrained near the axis of rotation by means of a round cam set eccentrically on the shaft. By controlling the vane using this cam it is no longer necessary for the vane and housing to come into contact as this interface is not needed to position the vane. In addition, the geometry of the eccentric cam generates a non-uniform gap between the vane and housing. This geometry is further explored in Bradshaw and Groll (2013). An example gap width between the vane distal end and the housing as a function of the rotation angle of the compressor rotor is given in Figure 3.

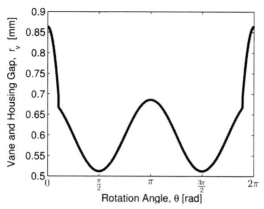

Figure 3: Profile of gap between vane and housing caused by the circular cam, generating the need for an additional compliant seal.

The example gap shown in Figure 3 is from a prototype spool compressor with a displacement of 39.3 cm³ (2.4 in³) and housing diameter of 5.46 cm (2.15 in) and a nominal air conditioning capacity of 17.6 kW (5 tonsR). Figure 3 shows that the sealing element must account for a fixed gap of roughly 0.53 mm as well as move dynamically to account for an additional 0.35 mm.

The compressor mechanism sits in a case filled with lubricant oil at discharge pressure. This oil will tend to leak across the spool seals, shown in Figure 1, and into the compression process chambers. Therefore, it is assumed that this constitutes enough oil to allow the tip seal to ride on a hydrodynamic film. This will generate a pressure change which will tend to push the tip seal away from the housing, much like a hydrodynamic bearing. An analysis modeling this behaviour will be described in the next section.

To counteract the hydrodynamic bearing behaviour the tip seal must be biased against the housing. Figure 2 shows that the designed mechanism to achieve this actuation is generated by a combination of force generated by mechanical springs, leading edge pressure which travels behind the seal to push the seal toward the housing and the weight of the seal which generates centripetal acceleration and accelerate the seal toward the housing. The modelling approach used to account for this is detailed in Section 4.

3 HYDRODYNAMIC ANALYSIS

Starting with the general Reynold's equation and assuming that the side-leakage (axial flow) is negligible and that the fluid properties will not change substantially when in contact with the tip the Reynold's equation reduces to:

$$\frac{dp}{dx} = 6u\mu \frac{h - h_m}{h^3} \qquad (1)$$

where u is the linear speed that the tip seal travels past the housing, μ is the viscosity and h is the gap between the tip seal and housing and h_m is the gap height where the pressure gradient is zero. This expression can be integrated to obtain an expression for the pressure distribution under the tip seal:

$$P(x) = 6u\mu \int \frac{h-h_m}{h^3} dx + C \tag{2}$$

where C is the integration constant. The gap between the housing and tip seal, h, is a function of the linear position, x, as shown in Figure 4.

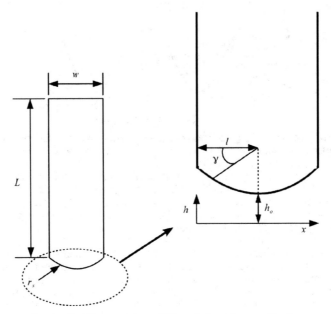

Figure 4: Dimensions of tip seal and coordinates used in hydrodynamic analysis.

The tip seal shape is circular and placed on the center of the tip seal. Therefore, the expression to describe the tip seal profile can be expressed as a function x, or parametrically:

$$\begin{aligned} x &= -r_s \cos\gamma + l \\ h &= r_s \sin\gamma + r_s + h_o \end{aligned} \qquad \gamma_{min} \le \gamma \le \pi/2 \tag{3}$$

where,

$$\gamma_{min} = \cos^{-1}\left(\frac{l}{r_s}\right) \tag{4}$$

This now leaves three unknowns in Equation (1), h_m, h_o, and the integration constant C. The values of h_m and C can be solved for by applying the appropriate boundary conditions:

$$\begin{aligned} P(x=0) &= P_l \\ P(x=L) &= P_t \end{aligned} \tag{5}$$

C can be solved directly but h_m requires iteration, which is outlined in Section 5. The minimum gap height, h_o, must also be solved iteratively. This variable is a strong function of the load applied by the combination of tip seal forces, which will be solved in the next section.

4 DYNAMIC ANALYSIS

The tip seal in a spool compressor must react dynamically during the compression process. The tip seal design shown in Figure 2 achieves this by utilizing the higher leading edge pressure, mechanical biasing springs, and the weight of the seal to load the tip seal against the housing. As detailed in the previous section this is reacted by the hydrodynamic load carrying capability of the tip seal as it rides across a film of oil. The summation of these forces is shown in the free-body diagram in Figure 5.

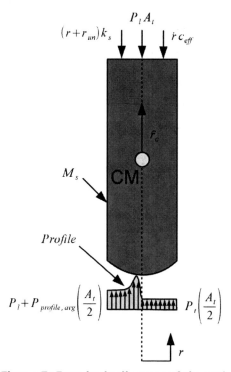

Figure 5: Free-body diagram of tip seal.

These forces are used to generate an equation of motion for the tip seal:

$$M_s(\ddot{r}+\ddot{r}_c)+c_{eff}\dot{r}+k_s r = \frac{A_t}{2}(P_{profile,avg}(h_o)+P(\theta)_t - P(\theta)_l) \tag{6}$$

where the centripetal acceleration, $\ddot{r}_c$, is defined as $\dot{\theta}^2 R_g$. In addition, the effective damping terms are generated from the viscous shearing of the oil between the tip seal and vane as the tip seal moves inside the vane. The effective damping from the leading and trailing interface is calculated independently as the gap width between each is likely to vary greatly due to the higher pressure on the leading edge of the tip seal.

$$c_{eff} = c_l + c_t \frac{\mu L z_{stator}}{h_l} + \frac{\mu L z_{stator}}{h_t} \tag{7}$$

where h_l and h_t are assumed constant and 12.7 and 2.5 μm, respectively, and z_{stator} is the axial depth of the compression chamber. Equation (6) is a nonlinear ODE which requires a simultaneous solution with Equation (2). In addition, Equation (6) is constrained because the tip seal cannot move beyond the housing boundary. This constraint is shown in Figure 3 which represents the maximum extension of the tip seal (i.e. when the tip seal is in contact with the bore).

5 MODEL SOLUTION

The models presented in Sections 3 and 4 are coupled by the pressure distribution in the oil film under the tip seal. This provides a challenging mathematical coupling to solve. The ultimate goal is to understand the impact of the tip seal on global compressor performance metrics (e.g. volumetric efficiency). The mathematical coupling will yield the most accurate dynamics of tip seal motion at a high computational cost. An alternative to this is to recast Equation (6) such that an accurate estimate of the net force applied to the bore can be estimated which can be used as an input to the hydrodynamic model from Section 3. This results in the following expression:

$$M_s(\ddot{r} + \ddot{r}_c) + c_{\mathit{eff}}\dot{r} + k_s r = \frac{A_t}{2}(+P(\theta)_t - P(\theta)_l) + F_{\mathit{input}} \qquad (8)$$

where F_{input} is the force required to constrain the motion of the tip seal to operate within the bounds of the compressor housing. This is now an independent nonlinear ODE which can be solved numerically to obtain the net force that needs to be reacted by the oil film over the course of one rotation. An example of this force profile is shown in Figure 6.

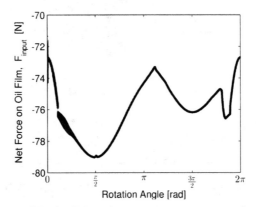

Figure 6: (Typical) Net force onto compressor housing due to tip seal from dynamic analysis.

The variation of this force is due to the combination of the changes in geometry and pressure during the compressor rotation. It is acknowledged that this will impact the instantaneous minimum oil film thickness (h_o) the goal is to obtain the average oil film thickness. Therefore, the force profiles obtained similar to Figure 6 are averaged to obtain a net input force for the hydrodynamic model.

This workflow is outlined in Figure 7 which describes the complete solution process including the iteration of the three unknown variables.

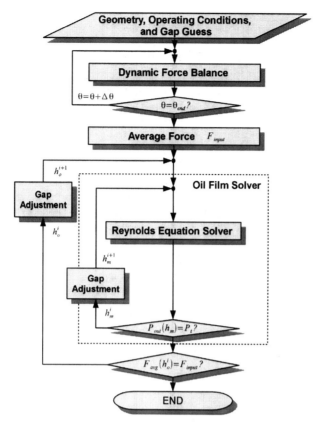

Figure 7: Solution flowchart of tip seal analysis.

Using the input force on the oil film a guess for the minimum oil film thickness is made. Then the film height that the pressure gradient is zero (h_m) is iterated on until the outlet boundary condition is satisfied. Once this loop is completed the pressure profile is integrated to obtain the average reaction force from the oil film. This is compared against F_{input}. If the values do not match h_o is adjusted and the process repeats.

6 MODEL RESULTS

This model was used to explore the impact of design parameters on the predicted minimum oil thickness. The minimum oil thickness represents a gap which leakage between chambers can occur. A larger minimum oil thickness would suggest a higher potential for leakage to occur.

Two geometric parameters were chosen to study the impact on leakage the tip seal radius and width of the tip seal. In addition, the spring rate of the mechanical spring was varied to explore the influence of the mechanical springs. The tip seal radius was varied between 1.78mm (0.070 in) and 17.8mm (0.700 in) at three widths 0.762, 1.27, and 1.78 mm (0.030, 0.050, and 0.07 in). The spring rates are studied between 20 and 80 kNm^{-1}.

Figure 8 shows the predicted minimum oil film thickness (leakage gap) as a function of tip seal radius for three tip seal widths.

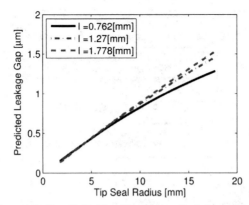

Figure 8: Predicted leakage gap (h_o) for various tip seal radii at three different seal widths.

The leakage gap increases with increasing radius regardless of the width of the tip seal. At radii less than roughly 7mm the predicted leakage gap becomes similar for all widths of seal. However, at larger radii the width of the seal begins to play a larger role. At the largest radii the smallest width tends to produce a smaller leakage gap.

To explain this behaviour it is necessary to look at the net increase in pressure under the tip seal due to the hydrodynamic effect. Figure 9 shows the pressure profile of the tip seal from the leading edge of the seal to the centreline for three seal radii.

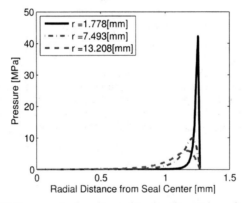

Figure 9: Pressure change from tip seal leading edge to under the middle of the seal for three different seal radii.

The profile of each seal looks similar in shape by entering at near leading edge pressure (pressure difference of zero) and building pressure as the oil film is pushed under the seal and minimizing at trailing edge pressure at the seal centerline. For larger radii the peak pressure increase is the lowest but stretches over a wider area. This effect results in a higher average force compared with the smallest radius which have a high peak pressure over a small linear distance. The higher net force of the largest radius tips result in the oil film pushing the tip seal further off the compressor housing and generating a larger potential leakage path compared with the smallest radii. This result of this study suggests that the thinnest seals with a small tip radius would likely seal the compression pockets the best.

In addition to geometric parameters the stiffness of the mechanical spring is also studied. Figure 10 shows the model predicted leakage gap (h_o) as a function of various spring stiffness's.

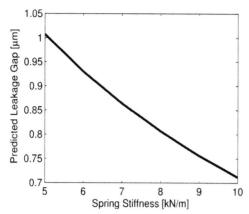

Figure 10: Model predicted leakage gap as a function of various actuating spring stiffness's.

This figure shows that the model predicted leakage gap decreases with increasing spring stiffness. This is a result of the increase in spring stiffness increasing the average force applied by the spring. This suggests that adding stiffer springs would prove useful in providing additional sealing the compressor.

7 EXPERIMENTAL RESULTS

Using a prototype compressor and experimental procedure outlined by Orosz et al. (10) an array of tip seals and spring combinations were tested. Tip seals with radii of 2.06, 7.62, and 15.2 mm (0.081, 0.3, and 0.6 in) were tested with a spring stiffness of 60 kNm^{-1}. In addition a tip seal with a tip radius of 2.06 mm (0.081 in) was tested with spring stiffness ranging from roughly 18 to 70 kNm^{-1}. Figure 11 shows the experimentally obtained volumetric efficiency at the various tip seal radii and spring rates.

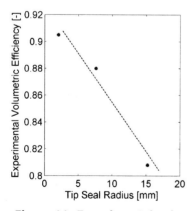 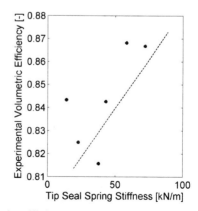

Figure 11: Experimental volumetric efficiency of prototype spool compressor with variable tip seal radius (left) and variable mechanical spring rates (right).

349

In the left plot in Figure 11 the trend as the tip seal radius increases the volumetric efficiency decreases. This corresponds to the prediction of the tip seal model as the predicted leakage gap (h_o) and the volumetric efficiency will generally be inversely proportional. Meaning as the leakage gap increases the volumetric efficiency will tend to go down.

In the right plot of Figure 11 the trend shows that as the spring stiffness increases the volumetric efficiency also increases. This data displays more disparity which is likely a result of increased uncertainty from inconsistent mechanical springs. However, this too correlates with the trend found in the previous study that as the spring stiffness increases the leakage gap will tend to decrease and the volumetric efficiency will tend to increase.

8 CONCLUSION

A detailed model of the spool compressor tip seal is presented. This model includes the balance of the hydrodynamic force generated by the oil film under the tip seal and the dynamic loading applied by the pressure differential surrounding the seal, mechanical springs, and the weight of the seal itself.

This model was used to explore the influence of the tip radius, tip width, and the spring rate of the mechanical springs. It was found that the potential leakage gap would tend to increase as the tip radius increases and that smaller widths would tend to decrease the sensitivity to this effect. This came as a result of an increased reaction load from the oil film. In addition, an increase in mechanical spring stiffness would tend to reduce the potential leakage gap by adding additional load for the oil film to overcome.

These trends were compared to a series of experimental studies which looked at various tip seal radii and spring rates. A direct comparison to the predicted leakage gap and experimental could not be made but the trends seen in the compressor volumetric efficiency gives an indication of the influence. It was found that the trends shown by the model are seen experimentally as the smallest radii tip seal showed the highest volumetric efficiency. In addition, the highest spring rate also showed the highest volumetric efficiency which also correlates with the findings of the study.

While this works focuses on the sealing component to tip seal design additional considerations must be taken to account for the frictional losses associated with the reduction in leakage gap. Future work will include extending this work to account for frictional losses of the tip seal.

9 ACKNOWLEDGEMENTS

The author acknowledges the assistance of Greg Kemp, President of Torad Engineering, and Joe Orosz, COO of Torad Engineering in the design and testing of prototype compressor hardware.

10 REFERENCES

(1) Kemp, G., Garrett, N., Groll, E., 2008. Novel Rotary Spool Compressor Design and Preliminary Prototype Performance. In: Proceedings of the International Compressor Engineering Conference. Purdue University, West Lafayette, IN USA. No. 1328.

(2) Kemp, G., Elwood, L., Groll, E., 2010. Evaluation of a Prototype Rotating Spool Compressor in Liquid Flooded Operation. In: Proceedings of the International Compressor Engineering Conference. Purdue University, West Lafayette, IN USA. No. 1389.

(3) Bradshaw, C., Groll, E., 2013. A comprehensive model of a novel rotating spool compressor . Int. J. Refrigeration. (in review).

(4) Bradshaw, C., Orosz, J., Kemp, G., Groll, E., 2012. A comprehensive model of a novel rotating spool compressor. In: Proceedings of the International Compressor Engineering Conference. Purdue University, West Lafayette, IN USA. No. 1142.

(5) Froede, W. G., 1961. The NSU-Wankel Rotating Combustion Engine. Society of Automotive Engineers International (SAE), SAE 610017.

(6) Leemhuis, R. S., Soedel, W., 1976. Kinematics of Wankel Compressors (or Engines)by Way of Vector Loops. In: Proceedings of the International Compressor Engineering Conference. Purdue University, West Lafayette, IN USA. No. 228.

(7) Beard, J., Pennock, G., 2000. Acceleration of the Apex Seals in a Wankel Rotary Compressor Including Manufacturing Process Variation in Location of the Seals. In: Proceedings of the International Compressor Engineering Conference. Purdue University, West Lafayette, IN USA. No 1457.

(8) Pennock, G. R. & Beard, J. E. 1997. Force analysis of the apex seals in the wankel rotary compressor including the influence of fluctuations in the crankshaft speed. Mechanism and Machine Theory 32(3), 349 - 361.

(9) Iskra, A. & Babiak, M. 2009. Problems With Representation Of The Oil Film Generating Conditions On The Wankel Engine Cylinder Sliding Surface. POLISH CIMAC, 47.

(10) Orosz, J., Kemp, G., Bradshaw, C., Groll, E., 2012. Performance and Operating Characteristics of a Novel Rotating Spool Compressor. In: Proceedings of the International Compressor Engineering Conference. Purdue University, West Lafayette, IN USA. No. 1257.

Generating grinding in rotor production – KAPP rotor grinding machine RX 120

A Köhler, J Heyder, F Wölfel
KAPP Werkzeugmaschinen GmbH, Germany

ABSTRACT

There has been an increasing trend in recent years from piston and scroll compressors towards screw-type compressors in the manufacturing of small compressors. These changes are accompanied by the need of a cost-efficient production of rotors used in screw-type compressors. KAPP developed the generating grinding process for rotor manufacturing in order to meet this requirement. This new solution enables a grinding time reduction of up to 40% compared to conventional production methods.

1 MANUFACTURING CHAIN OF SMALL ROTORS - STATE OF THE ART

Screw-type compressors offer various advantages when compared to piston compressors [1], [2], however, using screw-type compressors is only reasonable when their manufacturing costs are at least comparable to those of piston compressors. Successful competing against piston compressors requires significantly more efficient production methods. The rotor manufacturing costs mainly depend on the process of profile grinding.

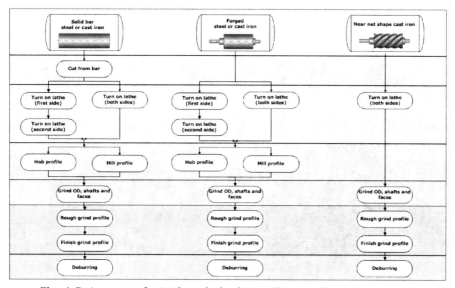

Fig. 1 Rotor manufacturing chain depending on the raw material

Conventional rotor production requires pre-machining by profile or generating milling and finishing by profile grinding. Rotors are often produced from cast blanks in order to shorten the process chain of smaller rotors and to reduce costs. Neither generating milling nor profile milling is part of the production chain anymore (Fig. 1).

Cast rotor blanks typically have a profile stock of up to 3mm. Removing this stock with a high degree of reliability and profitability is a key requirement for the grinding technology applied. Thus, a two-stage profile grinding process (roughing and finishing) using non-dressable CBN grinding wheels is often preferred in volume production today.

Fig. 2 Profile grinding of a rotor using non-dressable CBN profile grinding wheels

Using this technology the achievable stock removal rate has been steadily increased over the last few years by raising the cutting speed up to 80 m/s as an example. However, distinctive improvements of profile grinding as sole technology do not seem possible from today's point of view.

2   FUNDAMENTALS OF GENERATING GRINDING

2.1   Generating grinding kinematics of involute gears and rotors

Generating grinding is a well-known process for efficient involute gear production. The worm-shaped tool having a rack profile in axial section performs the grinding process by means of a workpiece with an external gear (Fig. 3). The form of the involute is created in a continuous generating cut process by the simultaneous grinding movement of the workpiece and the worm [3]. Generating cut process means that the tooth profile is created by profiling cuts in gear grinding processes similar to generating grinding [4] (Fig. 4). The generated cut deviations known from hobbing can be avoided using the closed worm-shape.

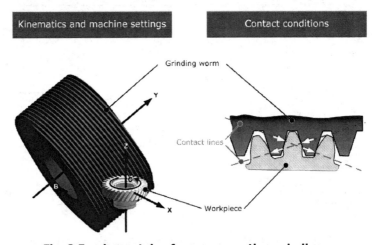

Fig. 3 Fundamentals of gear generating grinding

As every screw-type compressor is basically a profile form that can be generated by hobbing [1], the generating process can also be applied in the grinding of rotor profiles by means of a respective modification of the tool profile.

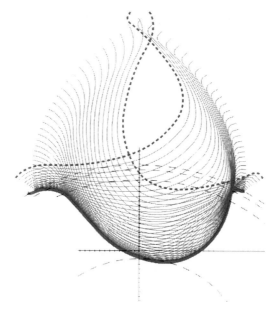

Fig. 4 Rotor profile created in a generating cut process

Generating grinding offers a significantly higher cutting capacity than profile grinding resulting from the simultaneous, point-shaped contact of several tool areas with the workpiece. Moreover, the generating movement assures a uniform material removal within all gaps of the workpiece during the axial grinding travel. This leads to a lower and more consistent heat input into the workpiece, which is a basic prerequisite for the production of high-quality rotors.

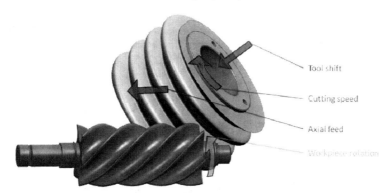

Fig. 5 Generating grinding kinematics of screw-type compressor rotors

Generating grinding also offers the advantage that all gaps of the workpiece are ground with one axial stroke. Thus, indexing into the next gaps including the repetition of the axial stroke, as required for profile grinding, can be avoided. A significant machining time reduction is achieved in this way.

2.2 Shifting

The grinding worm is shifted in longitudinal direction so that it can be used as long as possible. Therefore, the width of the grinding worm required in order to fully grind out the rotor profile (= profile form zone) is smaller than the total width. This enables several working positions depending on the respective rotor profile and possible worm width, which significantly increases the number of grindable workpieces per grinding worm coating (Fig. 6).

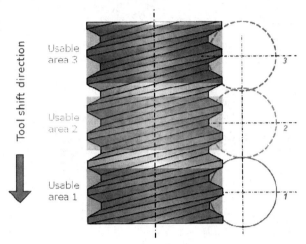

Fig. 6 Shifting areas of the CBN grinding worm

3   GENERATING GRINDING OF SCREW-TYPE COMPRESSOR ROTORS

Continuous generating grinding of small module gears has gained more importance and increased its market share in recent years. In comparison to profile grinding, generating grinding provides higher productivity since its idle times are considerably lower based on the total machining time of a part.

The characteristics of the gear grinding process can be transferred to the grinding processes of small rotors as well. However, there are differences between gears and rotors regarding the profile geometry to be ground: The dimensions of gears are, without doubt, significantly smaller than those of rotors.

The combination of the two grinding processes was the most obvious solution to meet the requirement of high productivity on the one hand and high quality rotors on the other hand. This idea is also described, claimed and granted in the German Patent DE 10 2008 035 525 B3.

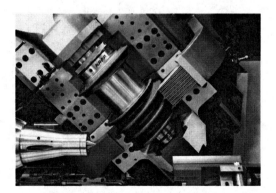

Fig. 7 Tool concept RX 120

4 PROCESS PERFORMANCE

Basic technological comparison:

- Profile grinding with **KAPP** CBN wheels (roughing)
  Cutting speed 60 to 80 m/s
  Feedrate 1500 to 2000 mm/min
- Profile grinding with **KAPP** CBN wheels (finishing)
  Cutting speed 50 to 70 m/s
  Feedrate 1800 to 2500 mm/min
- Generating grinding with **KAPP** CBN worms (roughing)
  Cutting speed 50 to 63 m/s
  Feedrate 250 to 700 mm/min

Based on these technological data and the advanced machine concept the grinding times for different rotors are significantly reduced on the RX 120 (Fig. 8).

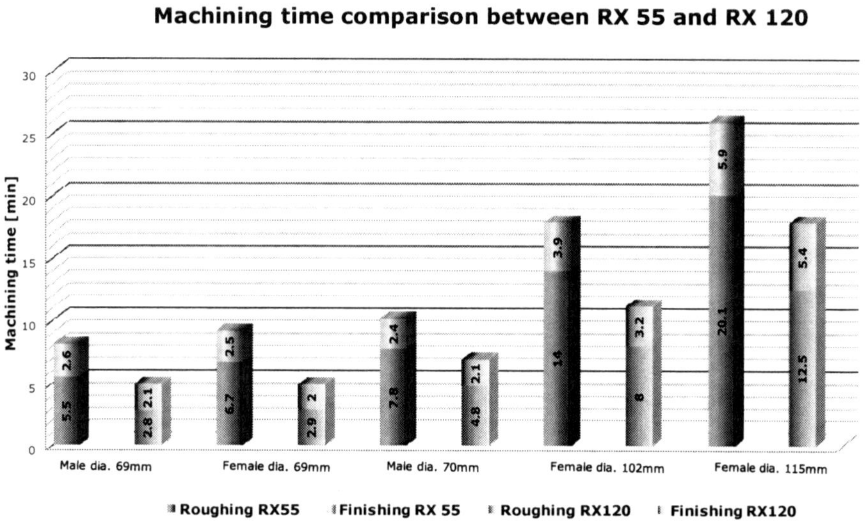

Machining time comparison between RX 55 and RX 120

Fig. 8 Machining time comparison

Depending on the material volume to be removed machining time reductions between 30 and 40% are achieved in comparison to pure profile grinding.

5 MACHINE CONCEPT RX 120

5.1 Machine design

The established bed of a **KAPP** generating gear grinding machine serves as basis for the RX 120. This machine is designed in the form of a rigid and thermally stable cast construction and was optimised for minimum space requirements (without automation and coolant system) (Fig. 9). All machine movements are realised by NC axes (Fig. 10).

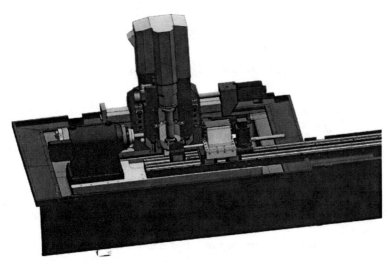

Fig. 9 Basic machine

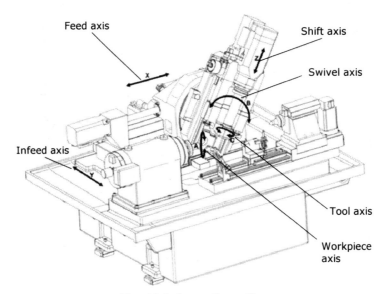

Fig. 10 Axis configuration

5.2 FEM-optimised components

All process-influencing components were examined using the finite element method and optimised in several stages during the construction of the RX 120.

The result of these comprehensive examinations and improvements is a machine that is optimised for the dynamic force progressions of rotor processing.

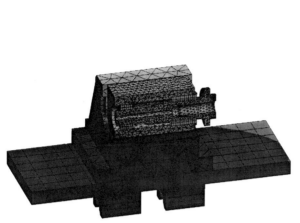 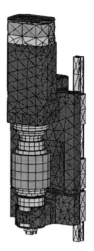

Fig. 11 FEM analysis of tailstock   Fig. 12 FEM analysis
                                       of tool spindle

5.3 Directly driven workpiece and tool axis

The workpiece axis is designed in the form of a directly driven and forcibly cooled spindle with a maximum speed of 2,000 rpm. The interface of the clamping chuck is carried out via a self-centring short taper.

 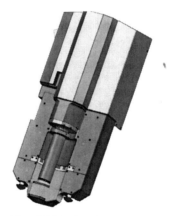

Fig. 13 Workpiece axis   Fig. 14 Tool spindle with
                           counter-bearing

The workpiece axis is also designed in the form of a directly driven and forcibly cooled spindle with a maximum speed of 7,500 rpm. The spindle is equipped with a counter-bearing for optimum stiffness and maximum precision. The workpiece spindle is connected to the drive motor via an HSK interface. This tool holding-arbour is replaced by another tool holding-arbour to be prepared outside the machine during set-up. This design guarantees very short set-up times.

5.4 NC tailstock

The tailstock is designed in the form of an NC axis. The tailstock spindle is forcibly cooled for higher precision. The NC tailstock design enables very short idle times during workpiece changes. The tailstock travels into a workpiece change position (Fig. 15) outside the machine working area for loading and unloading.

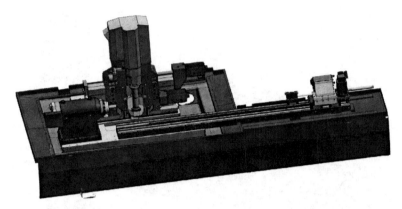

Fig. 15 Tailstock in workpiece change position

5.5 Optimised coolant supply

The process supply is implemented via a novel high-pressure needle nozzle for rotor grinding at the inlet side. The tools are rinsed by means of a special low-pressure nozzle at outlet side (Fig. 16). This enables the optimum coolant supply of the tools as well as a low oil flow rate. The high-pressure and low-pressure nozzles can be mounted at the inlet or outlet side of the tools and always receive the correct oil quantity with the respectively required oil pressure.

5.6 Outside diameter grinding

The high-precision grinding of the outside workpiece diameter by means of the profile grinding machine is today's state of the art. The RX 120 is the first rotor grinding machine to enable a technologically optimised OD grinding process due to the design of its workpiece axis in the form of a direct drive. The OD grinding process can be performed by means of the finishing tool or a special CBN OD grinding wheel for this purpose.

Fig. 16 High-pressure needle nozzle

5.7 Integrated measuring probe

The established measuring probe used in the larger machines RX 59 and RX 55 was also integrated in the machine concept of the RX 120. Profile and OD grinding operations can be performed in a measurement-controlled way by means of this probe. Furthermore, this probe can also be used as set-up aid to measure the rotor profile.

5.8 Energy efficiency

KAPP attached great importance to the compliance with the EU Directive on Energy-Using Products during the development process of the RX 120. A significant amount of energy is, for example, saved by using efficient and controlled synchronous motors, controlled feed and feedback modules, low-watt valves and the automatic switching-off function of the machine when processing has been finished. KAPP Group is an Alliance Member of "Blue Competence", a partner of the sustainability initiative in mechanical engineering and plant construction in Germany.

6 INTEGRATED LOADING AND DEBURRING

The RX 120 was specially designed for the simple connection of an automation system according to customer requests. The complete moving-out of the tailstock including the workpiece from the machine housing facilitates pick-up by automation.

Fig. 17 Front view of RX 120 rotor finishing cell

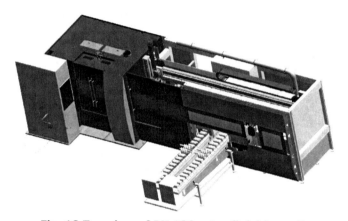

Fig. 18 Top view of RX 120 rotor finishing cell

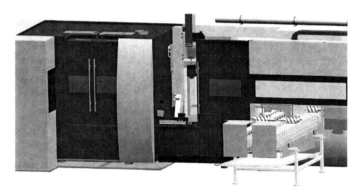

Fig. 19 Loading station

The optional automation cell available for the machine was specifically developed to achieve a design with minimum set-up requirements. It is optionally equipped with a deburring station for the ground workpieces. The burrs occurring at the profile face sides during the grinding process are removed in a functional manner by means of special brushing. A cylindrical brush can optionally be used to remove the burr occurring at the outside rotor diameter of some profile forms.

7 SUMMARY AND CONCLUSIONS

The RX 120 enables its user to meet the market requirements for the production of small screw-type compressor rotors by means of latest technologies and machine tool solutions. Significant cost-savings can be achieved in connection with shorter set-up times and its energy-efficient design when compared to conventional production solutions. This new machine demonstrates the translation of market requirements into new technologies and advanced technological concepts in the development of machine tools.

REFERENCE LIST

[1] L. Rinder. Schraubenverdichter, Springer-Verlag Wien New York 1979
[2] R. Müller. Kolben- oder Schrauben-Kompressor?, Industriebedarf 9/2004, Verlag W. Sachon
[3] T. Bausch. Moderne Zahnradfertigung. Verfahren und Maschinen zur kostengünstigen Herstellung von Stirn- und Kegelrädern mit hoher Qualität, 3. Auflage, Expert Verlag, Renningen-Malmsheim, 2006
[4] G. Sulzer. Werkzeugmaschinen zum Feinbearbeiten der Zahnradflanken von vorverzahnten Zahnrädern, Europäische Patentschrift EP0282046, Liebherr Verzahntechnik, Kempten, 1993

Experimental study of noise and vibration reduction in a medium-size oil-flooded twin-screw compressor by the application of helix relief

K Matsuo
Mayekawa Mfg. Co., Ltd, Japan
C S Holmes
PTG Advanced Developments, Holroyd, UK

ABSTRACT

MYCOM and HOLROYD have studied noise and vibration reduction in medium-size oil-flooded twin-screw compressor in collaboration. The oil-flooded twin-screw compressor works with lobe contact between male and female rotors. Rotor misalignments in operation lead to transmission errors which can excite noise and vibration. In this study, rotor misalignments of a target machine were estimated and helix modifications which absorb the bad influence of these misalignments were designed by using transmission error simulation. As a result of experiment of prototype rotors, a reduction in the noise and vibration was confirmed under target conditions and part load conditions.

1   INTRODUCTION

There are two kinds of noise in an oil-flooded twin-screw compressor. One is caused by fluid flow such as pressure pulsations of the discharge gas, and the other is related to mechanical motion, for instance the separation and collision of rotor lobes.

Computational fluid dynamics has made possible prediction of noise related fluid flow, as described in Refs. [1] to [5], and the reduction of this type of noise in the early design phase. On the other hand, it was difficult to predict mechanical noise theoretically, especially if related to lobe contact, because it has many factors such as the machining tolerance of parts, accuracy of assembly, gas forces, and various external forces in operation, and these factors are both variable, and hard to predict theoretically.

Under these circumstances, Holmes proposed a method of noise and vibration prediction which is based on the transmission error simulation of rotor motion, and a noise reduction method by helical compensation of the rotor lobes. The method is described in [6] and builds on the gears work of Gregory and Harris [7] and Munro [8]. MYCOM was interested in this new method as a key technology for reducing noise, and have conducted an experimental study of noise and vibration reduction in medium-sized oil-flooded twin-screw compressors in cooperation with Holroyd.

2 TRANSMISSION ERROR SIMULATION

Transmission error occurs in the driven component of a pair when its angular position differs from the theoretical angular position determined by the gear ratio. By convention, transmission error is positive at any instant when the driven component is ahead of its theoretical position. The error is expressed in microns at the pitch circle radius.

The displacements and accelerations can be transformed by Fourier analysis, and the acceleration spectrum has been found to be especially relevant. Sources of transmission error are:

1. Lead mismatch
2. Lead non-linearity
3. Housing bore locations
4. Bearing deflections, including thermal effects
5. Bending under gas forces
6. Divide (pitch) errors
7. Run-out
8. Profile shape at drive band
9. Unwanted profile contact, e.g. at root.

For a realistic simulation, estimates must therefore be made of the above. In the present study items 6 to 9 were disregarded, either because they are considered negligible compared to helical effects (6,7,8), or because they should not occur at all (9).

Magnitudes depend on design, manufacture, and operating conditions. It is convenient to assume that deviations which cause advance of idling rotor (e.g. closure of axes) are positive.

2.1 Estimation of rotor misalignment

Rotor misalignment was estimated from the machining tolerance of the casings, thermal expansion of materials, measured production data of rotors and oil film thickness of bearings in operation.

The centre distance and height of the bearings have some machining tolerance. Thermal expansion of the casings occurs at operating temperatures. These are required inputs for the simulation. Helix mismatch of the rotors from production data is also entered. The oil film thicknesses of the radial bearings under target operating conditions are calculated and centre positions of journals in operation are estimated. These calculated results are also used as inputs of this transmission error simulation.

2.2 Transmission Error Simulation

Table 1 is used to collect the estimated values for the geometrical model. Items marked * are bilateral (can go in either direction). Based on the values estimated, the transmission error was plotted assuming 3 scenarios, namely when the bilateral deviations were all positive, all negative, or all zero. Fig. 1 shows the worst of the three cases. It is referred to as a Harris map after its originator, the late Prof. Stephen Harris. A full explanation of this is given in [6].

2.3 Noise & Vibration Simulation

These displacements can be transformed into the frequency domain to produce Fig. 2, and the acceleration spectrum can be derived as in Fig. 3.

Table 1 Inputs required for transmission error simulation

Rotor misalignments & distortions in operation		
Enter combined misalignment of end 2 relative to end 1. Enter transverse plane data except where stated. Inputs with bilateral tolerances are marked *. Closing gap is positive (+ material) Relief (metal off) is negative. Positive Transmission Error = advance of idler. Z is measured along the rotor axis.	Inputs (green)	Calculations
Bilateral state (1, -1, or 0):		Combined gas & belt effects and bearing deformation
Max. closure from housing tolerances, mm	*	
Closure from gas & belt forces (linear part only), mm		
Closure - other (linear part only), mm		
Thermal closure, mm		
Total closure, mm		
Skew from housing tolerances, mm	*	Combined gas & belt effects and bearing deformation
Skew from gas & belt forces (linear part only), mm		
Skew - other (linear part only), mm		
Total skew		
Normal Mismatch, mm		
Helix mismatch from manufacture, mm	*	
Total normal mismatch (linear only), mm		Maximum combined relative deflection, negative when separating.
Slope of mismatch		
Helix Form errors (gas forces)	Dmax=	

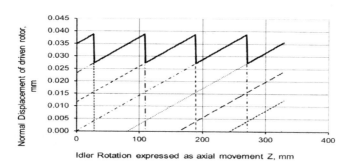

Fig. 1 Harris map of normal displacements of the driven rotor without relief for bilateral deviations positive

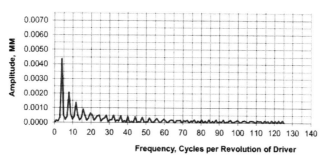

Fig. 2 Fourier spectrum of displacement for positive bilateral deviations and zero relief

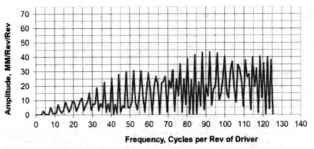

Fig. 3 Fourier spectrum of accelerations for positive bilateral deviations and zero relief

3 DESIGN OF RELIEF

Fig. 4 shows an unmodified helix (heavy line), with reliefs at each end RA1 and RA2. These can take any value and can differ. Relief can be specified as linear (as shown) or parabolic. The lengths RL1 and RL2 can also take any values up to half the body length. In the case of the deviations chosen, the type, amount, and length of relief was optimized to minimize the displacements for the expected mismatch. The combined effect of the assumed deviations and the relief is shown in Fig. 5. The net transmission error has been reduced to about 3 microns. Most importantly, the sudden accelerations have been smoothed out. As before, these can be transformed to give displacement and acceleration spectra. See Fig. 6 and Fig. 7.

Machine tool control software was developed to apply relief to the helix. In addition to zero relief and optimum relief, additional pairs were produced to allow testing on either side of the predicted value. Four pairs of rotors were made for this noise test. #01 is zero relief, #03 is the calculated target amount, and #02 and #04 are amounts below and above the amount.

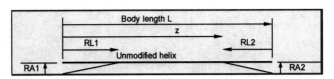

Fig. 4 Helix relief locations

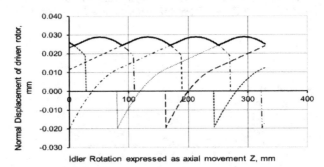

Fig. 5 Harris map of normal displacements of the driven rotor for positive bilateral deviations and parabolic relief

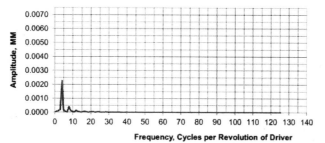

Fig. 6 Fourier displacement spectrum for positive bilateral deviations and parabolic relief

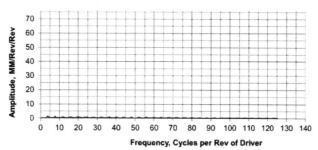

Fig. 7 Fourier acceleration spectrum for positive bilateral deviations and parabolic relief

4 NOISE TEST

4.1 Method and conditions

4.1.1 Model machine
Specifications of the test compressor is shown in Table 2.

Table 2 Specifications of the test compressor

Compressor type	Oil-flooded twin-screw
Rotor diameter	163.2 mm (Male and Female)
Rotor length	270 mm
Lobe combination	Male 4 / Female 6
Bearing type	Radial: Plain sleeve type
	Axial: Angular Contact Ball Bearings
Displacement	622 m³/hr @ 3000 rpm
Main application	Refrigeration system

4.1.2 Test conditions
Table 3 shows the operating conditions and range of control parameters.

Table 3 Operating conditions and range of control parameters

Refrigerant	NH_3 (ammonia)
Suction Pressure (P_s)	0.018 MPaG to 0.33 MPaG
Discharge Pressure (P_d)	1.25 MPaG
Condensing temperature (T_c)	35 deg. C
Evaporating temperature (T_e)	-30 deg. C to 0 deg. C
Shaft speed (N)	2400 rpm to 3600 rpm
Internal volume ratio	2.63 to 5.80
Load conditions	100 % to 50 %

4.2 Results and discussion

4.2.1 Effect of Helix modification on noise and vibration under full load conditions

Measurement results of sound pressure level at 1.0 meter from the machine side are shown in Fig. 8. The abscissa is evaporating temperature T_e that can be converted to suction pressure P_s. The ordinate is sound pressure level (SPL) expressed by 0 - 20 kHz overall level. #01 is the reference rotor pair. #02, #03 and #04 are rotor pairs with helix relief. A suitable volume index V_i was selected for each evaporating temperature. The sound pressure level of the rotor pairs with helix modification was lower than that of the reference pair in all the tested T_e range. There is nearly 3dB noise reduction when T_e=-15 deg. C to -25 deg. C and we can confirm the effect of helix modification. At the targeted condition T_e=-15 deg. C, pair #02 and #03 has the maximum effect. Pair #04 has different tendency. The sound pressure level of this pair has a maximum value at T_e=-15 deg. C. So far, we are not able to explain this apparent anomaly.

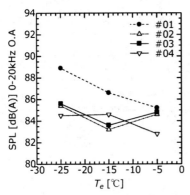

Fig. 8 Sound pressure level under full load conditions (T_c=35 deg. C, N=3000 rpm, 1.0m from the machine side)

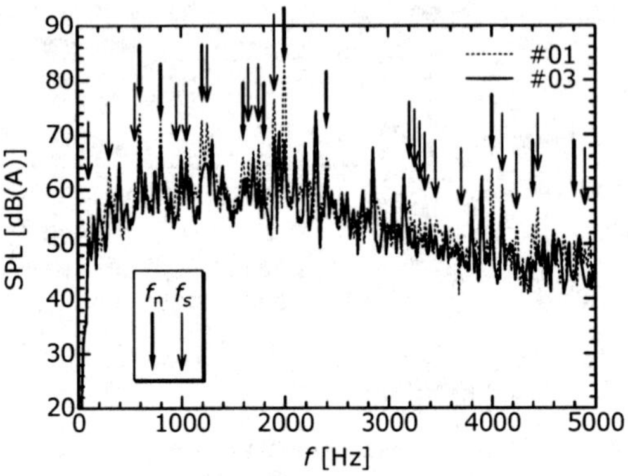

Fig. 9 Frequency domain view of sound pressure level (T_c=35 deg. C, N=3000 rpm, 1.0 m from the machine side)

Fig. 9 is a comparison between sound pressure level of pair #01 i.e. the reference pair, and pair #03 in the frequency domain under the condition T_e=-15 deg. C. By studying Fig. 9 we can find that harmonic components of lobe meshing (f_n Hz) of pair #03 indicated by heavy line arrows are smaller than those of the reference pair. Furthermore, sub-harmonic components observed in the data of the reference pair (f_s Hz) indicated by fine line arrows are kept to a low level in the data of pair #03. These tendencies suggest that the intensity of lobe meshing impacts is reduced and the rotor motion is stabilized by helix modification. The sound pressure level of pair #03 fell about 3dBA as a result of these effects.

Here,
$f_n = n \cdot Z_1 \cdot f_0$ (n=1, 2, 3, …)
n: positive integer
Z_1: lobe number of the drive rotor (4)
f_0: rotational frequency of drive rotor (50 when N=3000 rpm)
f_s: sub-harmonics ($1/m$ harmonics of lobe meshing, m=2, 4, 6, 12)

Fig. 10 shows the effect of helix modification on vibration under full load conditions. The ordinate is vibration acceleration in the horizontal direction a expressed in 0 to 5 kHz overall level. The abscissa is evaporating temperature T_e. Under the condition

T_e=-15 deg. C. the difference between reference pair #01 and the helix modification pairs is small. However the vibration acceleration of the helix modification pairs are smaller than those of the reference pair when T_e=-25 deg. C. This effect has a close connection with the relief amount. In the data shown in Fig. 10, pair #03 has the smallest vibration acceleration.

Fig. 11 shows a comparison between vibration acceleration of pair #01 and pair #03 in the frequency domain under the condition T_e=-15 deg. C. Several sub-harmonic components (f_s Hz: fine line arrow) are observed in the data of #01 pair. Large component of harmonic of lobe meshing frequency (f_n Hz: heavy line arrow) are also observed in pair #01. These sub-harmonic and harmonic components are controlled in that of pair #03.

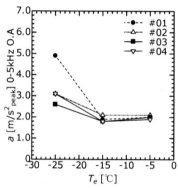

Fig. 10 Vibration acceleration under full load conditions (T_c=35 deg. C, N=3000 rpm, Horizontal direction)

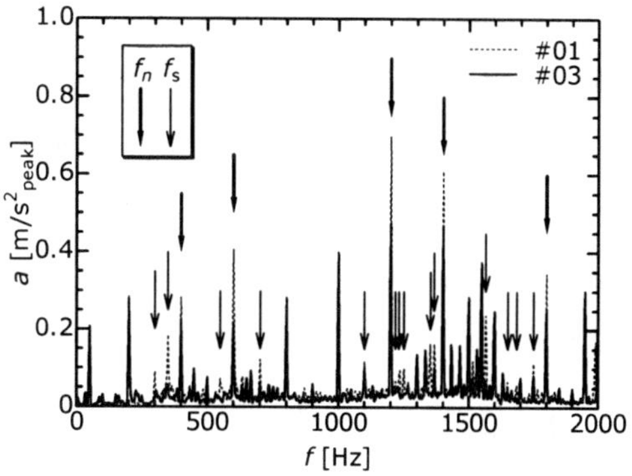

Fig. 11 Frequency domain view of vibration acceleration (T_c/T_e=35/-15 deg. C, N=3000 rpm, Horizontal direction)

4.2.2 Effect of Helix modification on noise and vibration under part-load conditions

The effects of helix modification under part-load conditions are indicated in Fig. 12 to Fig. 15. Fig. 12 shows measured results under the condition that N=3000rpm, T_c/T_e=35/-15 deg. C. The ordinate is sound pressure level SPL expressed by 0 - 20 kHz overall level and the abscissa is load conditions expressed in percentages.

At 100% load, the sound pressure level of helix modification pairs #02, #03 and #04 is lower by about 2 dB than that of reference pair #01. Pair #02 has the smallest value but the differences between the helix modification pairs are small. The sound pressure level of pair #01 gradually increases with a decrease in part-load conditions from 100% to 80 % and then slightly decreases to 50%. On the other hand, the sound pressure level of pair #02 increases with a decrease in load and the effect of helix modification disappears at 50% load. Pair #03 indicates a slight increase and decrease when the load changes from 100% to 50% in the same way as pair #01. Pair #04 does not depend on load condition. The difference between reference pair #01 and helix modification pairs #03 and #04 is slightly increased with a decrease in load. The results show that helix modifications have a noise reduction effect under part-load conditions and this effect becomes significant with an increase in the amount of modification.

Fig. 13 is a comparison between reference pair #01 and pair #04 in the frequency domain under a 60% load condition. We can confirm that fundamental and its harmonic components of lobe meshing of pair #04 f_n are smaller than those of the reference pair. Large half-component and sub-harmonic components f_s are observed in the data of reference pair, while these components are reduced drastically in the data of pair #04. This result suggests that transmission error of male to female rotors under the part-load conditions was managed by helix modification.

Fig. 14 shows vibration acceleration under the condition that N=3000rpm, T_c/T_e=35/-15 deg. C. The ordinate is vibration acceleration and the abscissa is load conditions expressed as percentages. At 100% load, there is no large difference between reference pair #01 and helix relief pairs #02, #03 and #04. There is a marked increase in vibration acceleration of the reference pair from 100% to 90% load and it becomes almost steady from 90% to 50%. On the other hand, there is no rapid increase in that of the helix relief pairs during 100% to 90% and vibration acceleration of these pairs increase with a decrease in load linearly and gradually. Comparing pairs #01 and #04, vibration acceleration is halved from 90% to 60% load where compressors are used regularly. The vibration acceleration of other helix modification pairs is also smaller than that of reference pair #01 in all part-load conditions under this operating condition. From these results, we confirmed that helix modification has a large effect in part load conditions.

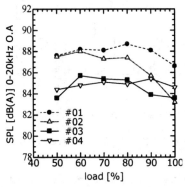

Fig. 12 Sound pressure level under part load conditions (T_c/T_e=35/-15 deg. C, N=3000 rpm, 1.0m from machine side)

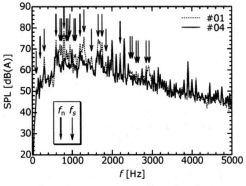

Fig. 13 Frequency domain view of sound pressure level (T_c/T_e=35/-15 deg. C, N=3000 rpm, 60% load 1.0m from machine side)

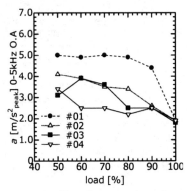

Fig. 14 Vibration acceleration under part load conditions (T_c/T_e=35/-15 deg. C, N=3000 rpm, Horizontal direction)

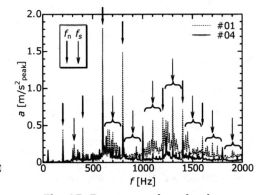

Fig. 15 Frequency domain view of the vibration acceleration (T_c/T_e=35/-15 deg. C, N=3000 rpm, 60% load, Horizontal direction)

Fig. 15 is a comparison between reference pair #01 and pair #04 in the frequency domain under 60% load condition. We can find the same tendency as in Fig. 13. The fundamental and its harmonic components of lobe meshing of pair #04 f_n are smaller than the reference pair. The half-component and sub-harmonic components f_s observed in pair #01 are controlled in pair #04.

4.2.3 Influence of Helix modification on performance of compressor

Volumetric efficiency decreases with an increase in relief amount in this nose test. The loss of efficiency for the target relief amount was approximately 1.2%. However, there are accuracy variations in rotor production and this can cause efficiency variations of a similar magnitude. It is therefore not possible to conclude that the difference measured is a result of the relief. To finally answer this will need statistical averaging over a batch, which at the time of writing had not been carried out.

5 CONCLUSIONS

- The effect of helix modification on noise was confirmed experimentally. Noise reduction by 3dB was achieved under full load conditions.
- Helix modification has a clear effect in part load conditions. In the most effective case, vibration acceleration is cut in half.
- The loss of efficiency for the target relief amount was approximately 1.2% in this noise test.

REFERENCES

[1] Koai K. & Soedel W., 1990: *Gas pulsation in twin screw compressors - Part I: Determination of port flow and interpretation of periodic volume source*, The 1990 International Compressor Engineering Conference at Purdue, pp.369-377.

[2] Koai K. & Soedel W., 1990: *Gas pulsation in twin screw compressors - Part II: Dynamics of discherge system and its interaction with port flow*, The 1990 International Compressor Engineering Conference at Purdue, pp.378-387.

[3] Sangfors B, Modelling, *Measurement and Analysis of Gas-Flow generated Noise from Twin-Screw Compressors*, The 2000 International Compressor Engineering Conference at Purdue, pp.971-978.

[4] 4. Mujic, E., Kovacevic A., Stosic, N., Smith I. K., 2005: *Noise prediction in screw compressors*, International Conference on Compressor and Their Systems, London, pp.447-454.

[5] 5. Mujic E., Kovacevic, A. Stosic, N. Smith I. K., 2009: *Numerical modelling of gas pulsations in a screw compressor*, International Conference on Compressor and Their Systems, London, pp.33-40.

[6] 6. Holmes C. S. 2006: *Noise reduction in screw compressors by the control of rotor transmission error*, International Compressor Engineering Conference at Purdue, C145.

[7] Gregory R.W, Harris S.L., Munro R.G. *Dynamic behaviour of spur gears*. Proc. Inst. Mech Eng Vol 178, 1963-4, Part 1 No.8.

[8] Munro R. G. 1979. *A Review of the Single Flank Method for Testing Gears*. Annals of the CIRP 28/1/1979: 325-9.

RECIPROCATING COMPRESSORS

Survey of factors influencing reciprocating compressor efficiencies and discharge gas temperatures

E H Machu
Consulting Mechanical Engineer, Austria

ABSTRACT

The present paper tries to give a summary of mathematical models dealing with compressor efficiencies, volumetric as well as energetic, as well as temperature rises, considering ventilation work losses in valves and pockets, heat exchange, gas leakages and gas inertia, on the basis of research work published in the past. Due to lack of space, many details have to be taken from the literature cited.

NOMENCLATURE

Symb.	unit	comment	Symb.	unit	Comment
c	m/s	velocity of sound	η	1	efficiency
C	1	rel. clearance volume	λ	1	eff. delivery
D	m	cylinder bore	ω	rad/s	angular velocity
h	J/kg	spec. enthalpy	$\chi \cdot R \cdot T$	J/kg	$\Delta u_{real\,gas}$
m	1	isentropic exponent	θ	rad	crank angle
M	kg	mass	Φ	m²	effect. flow area
Ma	1	MACH number	**Subscripts, superscripts**		
n	r.p.m.	rotational speed	*		ideal
p	Pa	pressure	1,2		nominal suction, discharge
q	1	valve adequacy	c		end of compression
Q	J	heat transferred	dpl		discharge plenum
r	m	crank radius	DV		discharge valve
R	J/kg.K	gas constant	e		end of discharge event
s	M	bumping clearance	f		final, in discharge flange
s	J/kg.K	specific entropy	i		index, chamber upstream
t	s	time	ind		indicated
T	K	abs. temperature	is		isentropic
u	J/kg	specific intern. energy	k		index, chamber in calcult.
v	M³/kg	spec. volume	leak		leaking
v	m/s	velocity	m		mean
V	m³	volume	p		at constant pressure
W	J	work	s		end of suction event
Z	1	real gas compr.	spl		suction plenum
α	W/m²K	heat trans. c.	st		stroke
α	1	coeff. flow cont.	SV		Suction valve
γ	1	duct parameter	thr		throttling
ρ	kg/m³	density	T		temperature isentrope
π	1	=3.141593 ...	v		volume isentrope
ξ	1	duct parameter	vol		Volumetric

© The author(s) and/or their employer(s), 2013

Disclaimer: This paper was prepared according to the best knowledge available to the author when writing this paper. The author accepts no liability for any kind of damage, direct or consequential, if any, suffered by any third party following actions taken or decisions made based on this paper.

1. INTRODUCTION

The performance of a reciprocating compressor working under given operating conditions, i.e. capacity, energy requirement and discharge temperature, must be known in the design state with the best possible precision to optimise compressor performance versus cost of investment, comply with safety regulations such as upper limits for discharge temperatures and avoid warranty problems during commissioning. However the formulas published in many compressor text books give only approximate values. They often neglect the fact that, cylinder temperatures may be higher than what is estimated by simple models, overestimating capacity and underestimating indicated power, they neglect leaking sealing elements and efficiency degradations due to heavily throttling valves.

2. ENERGY BALANCE, SPECIFIC ENTHALPIES AND TEMPERATURES

Applying the 1st law of thermodynamics (conservation of energy) to a compressor cylinder operating at steady state, with M as the mass of gas taken in at p_1, T_1, the energy input and output have to be equal.

- Input is equal to the mechanical work supplied by the piston $W = -\int p \cdot dV$ plus the heat transferred Q plus the energy transported by the gas, that is the flow work $M \cdot p_1 \cdot v_1$ plus internal energy $M \cdot u_1$.
- Output is the energy leaving with the discharged gas at p_2 and final temperature $T_{2,f}$, being equal to flow work $M \cdot p_2 \cdot v_{2,f}$ plus internal energy $M \cdot u_{2,f}$.

$$W + Q + M \cdot (u_1 + p_1 \cdot v_1) = M \cdot (u_{2,f} + p_2 \cdot v_{2,f}) \qquad (1)$$

Neglecting heat transfer, i.e. with $Q=0$ and using the definition of specific enthalpy $h = u + p \cdot v$, the energy balance simplifies to

$$\frac{W}{M} = u_{2,f} + p_2 \cdot v_{2,f} - (u_1 + p_1 \cdot v_1) = h_{2,f} - h_1 = \Delta h_{real} \quad (2), \qquad \eta_{is} = \frac{\Delta h_{is}^*}{\Delta h_{real}} \qquad (3)$$

Equation 2 is an interesting result: Any mechanical work supplied by the piston increases the enthalpy of the gas and hence its temperature. Applied to the suction event: The bigger the work loss in the suction valves plus valve pockets, the hotter will be the gas inside the cylinder and the smaller its density before compression starts. Since reciprocating compressors are volumetric machines, where power requirement depends on the volume, not on the mass of the gas to be compressed, this heating of the gas reduces compressor capacity. It also may increase specific power requirement in [J/kg] as can be seen from the examples of Tables 1 and 2.

In general, with heat transfer and valve leakage, total energy required to transport unit mass of gas from the suction to the discharge line, hence total specific enthalpy increase Δh_{real}, Fig. 1, will be composed out of the following increments:

- $h_{1,spl} - h_1$ = change of specific enthalpy in the suction plenum, due to heat exchange and mixing with hot gas flowing back through leaking suction valves.

As stated above, a hotter gas reduces compressor capacity. Inversely cooling of the gas involves the danger of condensate formation and liquid contamination, to be avoided according to API618 ([7], point 2.6.3.3).
- $h_{1,s} - h_{1,spl}$ = work loss in the suction valve during intake, plus heat transferred from the walls of the suction valve;
- $h_{2,c} - h_{1,s}$ = work for compression, for heat transfer and the enthalpy change due to gas flows through leaking discharge valves and piston rings;
- $h_{2,e} - h_{2,c}$ = work to cover work losses in the discharge valve during the discharge event plus heat transferred from the walls of the discharge valve;
- $h_{2,f} - h_{2,e}$ = change of specific enthalpy in the discharge plenum, due to heat exchange and mixing of gas flows from head end and crank end working chambers arriving with different temperatures $T_{2,e}$. For safety reasons, there are recommended temperature limits, see API618 ([7], point 2.3.1).
- The ideal loss-free compressor only needs $\Delta h^*_{is} = h^*_{2,is} - h_1$, Figure 2.

Subscripts of gas temperatures are those of the corresponding specific enthalpies.

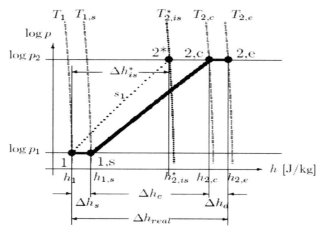

Figure 1: *MOLLIER*– or h-log(p)-diagram of a compression cycle

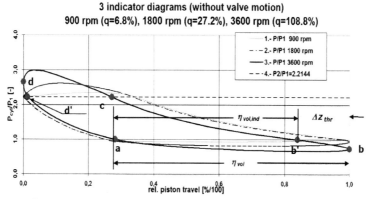

Figure 2: Three computed indicator diagrams, curves ...
1.- $q_{SV}=0.068$, 2.- $q_{SV}=0.272$, 3.- $q_{SV}=1.088$,
(no valve motion, valves open and close immediately in points 1 and 2' for the SV, 3 and 4' for the DV)

377

3. MATHEMATICAL MODELS TO SIMULATE COMPRESSOR OPERATION

From the above it becomes clear that a good model should try to quantify all these factors. The following two types of models have proven to be useful in practice:

1. Model 1 uses approximate formulas. It is therefore less precise but very fast in yielding results ([13], [14]). Many of the approximate formulas used in Model 1 have been confirmed by regression analysis of results obtained with Model 2 mentioned hereafter, such as work loss in suction valves, as is the case with the suction throttling loss $\Delta z_{thr,sv}$, equation 8.

2. Better precision can be obtained by using a system of differential equations for pressure and temperature evolutions, called Model 2. It can consider simultaneous changes of pressure and temperature in every chamber of interest, simultaneous gas flow through many orifices adjacent to each chamber, heat exchange, valve leakage, special effects like gas inertia inside the working chamber and real gas behaviour. It can be coupled with valve dynamics equations. In the papers [12] and [17] such a model has been derived from equation 1, a result (here without valve dynamics) is shown in Figures 2 through 4.

4. IMPORTANCE OF CONSIDERING REAL GAS BEHAVIOUR

Any change of real gas compressibility Z along the compression isentrope, i.e. with $Z_{1,s} \neq Z_{2,c}$, implies that the exponents of temperature isentrope m_T and volume isentrope m_v, equations 4 and 5, local or mean, are different in general.

$$\frac{T_2}{T_1} = \left(\frac{p_2}{p_1}\right)^{\frac{m_{T,m}-1}{m_{T,m}}} \quad (4) \qquad \frac{v_2}{v_1} = \left(\frac{p_1}{p_2}\right)^{\frac{1}{m_{v,m}}} \quad (5)$$

Taking the example of an isentropic compression of carbon dioxide CO_2 from 30 to 90 [barabs], $T_{1,s} = 40$ [°C], Table 1 shows some data, all found from the REDLICH – KWONG equation of state [18], [19]. $T_{2,c}$ was obtained by integrating the proper thermodynamic equations along the isentrope from states $1,s$ to p_2, the other values are local ones. The change in specific enthalpy during compression $\Delta h_c = h_{2,c} - h_{1,s}$ can also be taken from an h-log(p) diagram for CO_2, or found approximately by

$$\Delta h_c \approx p_1 \cdot v_{1,s} \cdot \frac{Z_m}{Z_{1,s}} \cdot \frac{m_{T,m}}{m_{T,m}-1} \cdot \left[\left(\frac{p_2}{p_1}\right)^{\frac{m_{T,m}-1}{m_{T,m}}} - 1\right] \quad (6) \qquad Z_{m,arith} = \frac{Z_{1,s} + Z_{2,c}}{2}, \quad Z_{m,geo} = \sqrt{Z_{1,s} \cdot Z_{2,c}} \quad (7)$$

Table 1: Isentropic compression of CO_2 from p_1, $T_{1,s}$ to p_2.

	P	T	Z	m_T	m_v	$\kappa^* =$	Δu_c	Δh_c
	[bar abs]	[K]	[1]	[1]	[1]	$=c^*_p/c^*_v$	[J/kg]	[J/kg]
1,s	30.0	313.15	0.86085	1.3253	1.2525	1.2822	0	0
2,c	90.0	406.81	0.82753	1.3005	1.2786	1.2504	49629	62301
	Mean values, equations 4, 5:			1.3126	12535	1.2663		
and after having increased $T_{1,s}$ by 5 [K], pressures remaining unchanged:								
1,s	30.0	318.15	0.86851	1.3211	1.2523	1.2801	0	0
2,c	90.0	412.30	0.83619	1.2973	1.2762	1.2489	50885	63838
	Mean values, equations 4, 5:			1.3088	1.2529	1.2644		

As $T_{1,s}$ changes from 313.15 by 1.6% to 318.15 [K], compressor capacity and gas density decrease proportionally to $1/(T_{1,s} \cdot Z_{1,s})$ or by 2.5%, while specific enthalpy rises by 2.5% from 62301 to 63838 [J/kg]. Another danger of neglecting real gas behaviour: When using ideal gas $\kappa^* = c^*_p/c^*_v = 1.2822$ at $T_{1,s}$ instead of the temperature exponent $m_{T,m} = 1.3126$, a $T_{2,c}$ too small by some 8 [K] is obtained from equation 4. A still smaller estimate is obtained when taking the mean κ^*_m as recommended in [6], page 79: $\kappa^*_m = (1.2822 + 1.2504)/2 = 1.2663$ only.

Table 2: Comparison of Δh_c [J/kg] (REDLICH-KWONG versus equation 6)

	$T_{1,s} = 313.15$ [K]		$T_{1,s} = 318.15$ [K]	
	Δh_c		Δh_c	
from Redlich-Kwong	62301	=reference	63838	=reference
from equation 6:		error		error
with $m_{T,m}$, geom.. $Z_{m,geo}$	62704	+0.648 %	64226	+0.608 %
with $m_{T,m}$, arithm. $Z_{m,arith.}$	62717	+0.667 %	64237	+0.626 %
with $m_{v,m}$, no Z	62655	+0.568 %	64178	+0.533 %
with κ^*, geom. $Z_{m,geo}$	61711	-0.948 %	63249	-0.923 %

5. FACTORS INFLUENCING THE GAS FLOW RATE

5.1 Volumetric efficiency η_{vol} and efficiency of delivery λ:

Volumetric efficiency η_{vol} given by equation 8 is defined as the useful part of stroke volume. The indicator diagram, Figure 2, drawn in full black lines (heavy suction valve throttling) shows that efficiency η_{vol} has to be corrected by the throttling loss $\Delta z_{thr,sv}$ (see also [22] Fig. 20.14) to give indicated volumetric efficiency $\eta_{vol,ind}$ that can be taken from a measured indicator diagram. $\Delta z_{thr,sv} > 0$ means that cylinder pressure, Fig. 3, is so far down during intake that it cannot recover and rise again up to nominal p_1 in due time, before the suction event ends.

The efficiency of delivery λ is defined as the mass of gas actually delivered per compression cycle divided by the mass of gas with nominal suction density ρ_1 filling the stroke volume. It will be smaller than $\eta_{vol,ind}$ even with tight sealing elements ($\Delta\lambda_{leakage} = 0$) since temperature $T_{1,s} > T_1$ in point b', as can be seen from Figures 3 and 4, and as outlined in Section 2, paragraph following equation 3.

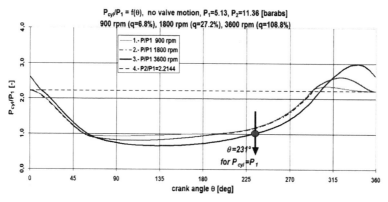

Figure 3: Evolutions of $p_{cyl}/p_1 = f(\theta)$ With $q_{sv} = 1.08$, $p_{cyl}/p_1 = 1$ is reached at crank angle $\theta = 231°$, and $\Delta z_{thr,sv} = 0.16$ of relative piston travel.
In other words, η_{vol} is reduced by some 16% due to suction valve throttling.

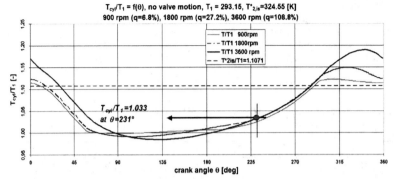

Figure 4: Evolutions of $T_{cyl}/T_1 = f(\theta)$
With $q_{SV}=1.088$, $T_{cyl}/T_1 =1.033$ when $p_{cyl}/p_1 =1$
In other words, the mass of gas inside the cylinder ready
for compression is reduced by some 3.3%.

$$\eta_{vol} = 1 - C \cdot \left[\frac{Z_{1,s}}{Z_{2,c}} \cdot \left(\frac{p_2}{p_1} \right)^{1/m_{T,m}} - 1 \right] \quad (8) \qquad \Delta z_{thr,SV} \approx \max\left[0, \frac{q_{SV} - 0.15}{6} \right] \quad (9)$$

$$\lambda = \underbrace{\left(\eta_{vol} - \Delta z_{thr,SV} \right) \cdot \frac{T_1 \cdot Z_1}{T_{1,s} \cdot Z_{1,s}}}_{=\eta_{vol,ind}} - \Delta\lambda_{leakage} \quad (10)$$

$$Ma_{SV} = \frac{v_{m,\Phi,SV}}{c_1} = \frac{v_{m,\Phi,SV}}{\sqrt{m_{v,1} \cdot R \cdot T_1 \cdot Z_1}} \quad (11) \qquad v_{m,\Phi,SV} = \frac{V_{st}}{\Phi_{SV}} \cdot \frac{n}{30} \quad (12)$$

$$q_{SV} = \left(v_{m,\Phi,SV} \right)^2 \cdot \frac{\pi^2}{8} \cdot \frac{\rho_1}{p_1} = \frac{\pi^2}{8} \cdot m_{v,1} \cdot \left(Ma_{SV} \right)^2 \quad (13)$$

From Model 2 which has been derived from equation 1 it can be shown that all relevant machine and operating data like stroke volume V_{stroke}, rotational speed n, suction valve equivalent flow area Φ_{SV} and gas properties (ρ_1, p_1, $m_{v,1}$) lump together in a single term called suction valve adequacy q_{SV} or MACH number Ma_{SV}, equations 11 or 13 (in [12], equation 40, or [22], Section 20.4.5) This suction valve adequacy q_{SV} is therefore the most important parameter to characterise the quality of valve layout. It determines pressure evolution inside the cylinder during the suction event, suction work loss and throttling loss $\Delta z_{thr,SV}$.

On the discharge side throttling loss $\Delta z_{thr,DV}$ is much smaller if not negligible. The reason can be seen from [12], equation 20: The higher the pressure (occurring in the numerator) and the smaller then cylinder volume (occurring in the denominator), the steeper the pressure gradients, and the faster any pressure recovery.

As to the temperature rise in the suction plenum at partial load, when one cylinder side is unloaded by holding the suction valves open, please refer to [13], paper II. For the leakage loss $\Delta\lambda_{leakage}$ used in equation 12 please refer to [14], Section 4.1.

5.2 Gas inertia, its influence on efficiency of delivery λ
In some very rare cases inertia of the gas inside the cylinder may reduce the pressure difference across the discharge valve, thus reducing the gas flow rate

therein. Figure 6 shows a measured indicator diagram of such a case, Figure 7 the trials to simulate it (efficiency of delivery λ = 35.9% for curve 4 "with gas dynamics"). Figure 8 shows the same simulations after some cylinder modifications. The result is encouraging: λ = 45.8% for curve 4 "with gas dynamics".

Unfortunately the modifications simulated in Figure 8 could not be checked any more in practice for the following reason: The author of the present paper came across this solution years later by chance, when investigating uneven pressure distribution inside a cylinder trying to explain piston rod failures due to bending [15]. At this time, this propane compressor had already been scrapped, being considered as a "hopeless case", since all the classical modifications did not improve anything. The Operating Company decided to buy another compressor.

This phenomenon seems to occur with high speed, short stroke compressors having a big cylinder bore D and a small bumping clearance s, this is the safety distance between piston and cover in dead centre position. All this causes high gas velocity changes inside the cylinder near the inner dead centre position during operation. When working with heavy gases the resulting dynamic pressure changes Δp_{dyn} are rather big. If $\Delta p_{orifice} = \Delta p_{undisturbed}$ (Fig. 5) is normally available for pushing the gas through the discharge valve, this will reduce to $\Delta p_{orifice} = \Delta p_{undisturbed} - \Delta p_{dyn}$ and gas flow in the discharge valve will stall or almost. Dynamic pressure changes in the discharge plenum can usually be neglected.

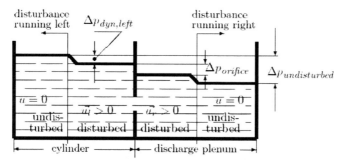

Figure 5: Duct of Figure 9 shortly after having opened a slot (valve) in the separation. Two pressure waves (disturbances) travel out, one to the left, one to the right

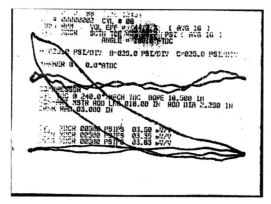

Figure 6: Measured indicator diagram of a propane C_3H_8 compressor: Due to gas inertia in the cylinder, discharge is hampered.
η_{vol}=44.9% P=71.57 kW

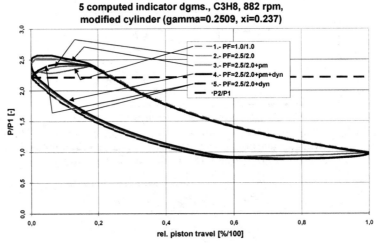

Figure 7: Simulations of the situation of Figure 10 with several options.
1.- valves alone; 2.- valves + pocket factors; 3.- valves + pocket factors + piston masking; 4.- valves + pocket factors + piston masking + dynamic pressure losses,
$\xi=1.097$, $\eta_{vol}=35.9\%$, $\lambda=35.90\%$, $P=61.44\ kW$

Figure 8: Situation improved by increasing the piston bumping clearance s,
$\xi=0.237$, $\eta_{vol}=48.2\%$, $\lambda=45.84\%$, $P=77.26\ kW$

The well known Method of Characteristics ([26], [27]) offers a simple way to model this case: With the symbols of the nomenclature and ΔA_{duct} as area of an additional duct in the cylinder cover to facilitate gas flow towards the discharge valve, parameter ξ, equation 16, gives a rough estimate for this effect. With $\zeta > 0.5$ there may be efficiency loss due to gas inertia.

$$\Delta p_{dyn} = -\rho \cdot c \cdot \Delta v \quad (14) \qquad \gamma = \frac{s}{r} + 2 \cdot \frac{\Delta A_{duct} \cdot D}{V_{stroke}} \quad (15) \qquad \xi \approx \frac{\Delta p_{dyn}}{p_2} \approx \frac{p_2 \cdot c_2 \cdot D \cdot \varpi}{p_2 \cdot \sqrt{2 \cdot \gamma + \gamma^2}} \quad (16)$$

Modelling the problem by CFD (Computational Fluid Dynamics) methods has been done since ([24], [25]), having the advantage of being more flexible, and the disadvantage of considerably increased computing time. Gunther Machu in [24] has shown that reflected cross waves may be harmful to valve reliability.

5.3 Work loss in valves, leaking sealing elements, heat transfer

Due to lack of space only a brief summary can be given here. Work loss and leakage loss as found by a Model 2 approach have been given in [12] and [17].

Heat transfer coefficients α can be modelled following the instructions given in [3], diagram XXVIII, or [20] etc. However, it seems that cylinder heat transfer can be neglected altogether. According to RETORET and MOREAU [8], cooling water does nothing else but equalise temperatures in the cylinder body thus avoiding thermal stresses and deformations; BROK, TOUBER and VAN DER MEER have published a paper [9] with the title "Modelling of cylinder heat transfer, much effort, little effect"; K.LEE and J.K.SMITH assure in [10] that "Time resolved mass flow measurements of a reciprocating compressor" that there is practically no cylinder heat transfer. The same result was found by the author in [12]: Even when running the Model with heat transfer coefficients ten times as big as suggested in [20], the effect on compressor capacity is next to nothing. In fact, inside the cylinder temperature differences and the time for substantial heat transfer are small. It seems that 90% or more of the heat carried away by cylinder cooling water originates from the discharge plenum, where wall areas, temperature differences as well as the time available for heat transfer are the biggest in a cylinder.

6. CONCLUSION

It was the intention of this paper to show that calculation of reciprocating compressor performance may be quite a demanding job and that many simplifications, although rather popular, may lead to misjudgements, concerning compressor capacity, power requirement and discharge temperatures.

REFERENCES

[1] Adolf Hinz, *Thermodynamische Grundlagen für Kolben.- und Turbokompressoren (Thermodynamic Principles of Reciprocating and Rotating Compressors)*, Springer Berlin 1914.
[2] David Allan Low, *Heat Engines,* Longmans, Green and Co, London 1952.
[3] Franz Fröhlich, *Kolbenverdichter (Reciprocating Compressors)*, Springer Berlin / Göttingen / Heidelberg, 1961.
[4] M. I. Frenkel, *Kolbenverdichter (Reciprocating Compressors)*, German translation Verlag Technik, Berlin, 1969, original textbook in Russian language 1960.
[5] Engineering Data Book by GPSA (Gas Processors Suppliers Association), Tulsa, Oklahoma, USA, 8th edition 1972, edition in SI units 1980.
[6] CT2 Compressor Tech2, Compression Technology Sourcing Supplements 2008 and 2012, Diesel & Gas Turbine Publications, Waukesha, Wisconsin, USA.
[7] API Standard 618, Reciprocating Compressors for General Refinery Services, 4th edition 1995, American Petroleum Institute, Washington, USA.
[8] J.P. Retoret, M. Moreau, *Les Compresseurs Alternatifs (Reciprocating Compressors)*, periodical L'Industrie du Petrole en Europe, Paris, July and Sept. 1972.
[9] W. Brok, S. Touber, J.F.T. Mac Laren, *Modelling of Cylinder Heat Transfer, much effort, little effect*, International Compressor Engineering Conference at Purdue, July 1980, proceedings.

[10] K. Lee, J.L. Smith, *Time resolved mass flow measurement of a reciprocating compressor,* International Compressor Engineering Conference at Purdue, July 1980, proceedings.

[11] F. Bauer, *Flow resistance of valves installed in reciprocating compressors,* International Compressor Engineering Conference at Purdue, July 1988, proceedings.

[12] E.H. Machu, *How Valve leakage can influence volumetric and isentropic efficiencies of reciprocating compressors,* International Compressor Engineering Conference at Purdue, July 1990, proceedings.

[13] E.H. Machu, *Valve throttling, its influence on compressor efficiency and gas temperatures,* Part I: *Full load operation,* Part II: *Half load and zero load operation,* International Compressor Engineering Conference at Purdue, July 1992, proceedings.

[14] E.H. Machu, *Compressor valve selection, some points to be considered for optimising life time and compressor efficiency,* International Conference on Compressors and their Systems, IMechE, London September 1994, proceedings ref. C477/020/94.

[15] E.H. Machu, *Problems with high speed short stroke reciprocating compressors,* Gas Machinery Conference, Denver, USA, 1998, proceedings.

[16] E. Huttar, V. Kacani, *Simulation of leakage in piston rings of reciproating compressor,* 7th workshop on Reciprocating Compressors, Kötter, Rheine, Germany, 2003, proceedings.

[17] E.H. Machu, *Einfluss der Ventile auf Kolbenverdichter-Wirkungsgrade (Impact of valves on reciprocating compressors efficiencies)* 12th workshop on reciprocating compressors, Kötter, Rheine, Germany, 2008, proceedings.

[18] Edmister, *Applied Hydrocarbon Thermodynamics,* Vol. 2, Gulf Publishing, Houston Texas, 1974.

[19] Baehr, *Thermodynamik,* Springer Berlin, Heidelberg, New York, 9th editn. 1996

[20] VDI Wärmeatlas, (Techn. Data on Heat Transfer) VDI-Verlag Düsseldorf, 1984.

[21] Kleinert, *Taschenbuch Maschinenbau,* Bd. 5, (pocket book *Mechanical Engineering,* Berlin VEB-Verlag Technik 1969, also published in Dubbel, Taschenbuch f.d. Maschinenbau, Springer Verlag, 19. Auflage 1997, W.Beitz und K.-H.Grote ISBN 3-540-62467-8, page P38, equ. (35).

[22] P.C. Hanlon, *Compressor Handbook,* McGraw-Hill 2001.

[23] E.H.Machu, PhD-Thesis, Vienna University of Technology, 1989.

[24] G. Machu, *Pulsationen im Verdichtungsraum – eine potenzielle Schadensursache? (Pulsations inside the working cylinder – a potential cause for troubles?) 8th Workshop on Reciprocating Compressors,* Kötter Consulting Engineers Rheine, Germany, 2004.

[25] R. Aigner, H. Steinrück, (Vienna University of Technology), *Modelling Fluid Dynamics, Heat Transfer and Valve Dynamics in a Reciprocating Compressor,* 5th Conference of the European Forum for Reciprocating Compressors (EFRC), Prague, 2007.

[26] J. Kestin, J.S.Glass, *Application of the Method of Characteristics to the Transient Flow of Gases,* IMechE Council of Publishing, January 1949.

[27] D. Woollatt, *A Pulsation calculation algorithm well suited for use with valve dynamics calculations,* International Compressor Engineering Conference at Purdue, July 1990, proceedings, pp. 354.

An investigation of the heat transfer phenomena between the hermetic reciprocating compressor components

A R Ozdemir[1], E Oguz[1], S Onbasioglu[2]
[1] Arcelik Research and Development Center, Turkey
[2] ITU Mechanical Faculty, Mechanical Engineering Department, Turkey

ABSTRACT

A hermetic reciprocating compressor is the most effective component (90%) with respect to the energy consumption of the refrigerators. Detailed investigation studies on the compressor play an important role for contributing the overall efficiency increment of the compressors.

In this study, heat transfer processes between the components of a hermetic reciprocating compressor were studied theoretically and experimentally. Compressor is divided into different control volumes (Shell, suction /discharge mufflers/pipes/plenums, cylinder, lubricant, electrical motor, body, inner gas etc) for determining the heat transfer characteristics between these control volumes theoretically. In the experimental part of the study, conceptual designs on the components of the hermetic compressor were made to examine the effects on the heat transfer characteristics between the components and also compressor performance.

The results of the theoretical and experimental studies are used for investigating the network including conduction, convection and radiation forms of heat transfer inside the compressor. Generated heat transfer network helps to characterize the thermal functions of the main components which leads to the new and better compressor designs.

Keywords: Reciprocating compressor, Heat Transfer, Thermal Map, Coefficient of performance (COP)

NOMENCLATURE

δ	: Fresh gas ratio	cool	: Cooling capacity
ε	: Emissivity	H	: Total Enthalpy
ρ	: Density	h_c	: Heat convection coefficient
cir	: Circulation inside the compressor shell	muf	: Suction muffler inlet
		P	: Pressure
cond	: Condensation	sub	: Subcool
cs	: Compressor suction inlet	sup	: Superheat
dif	: Distribution inside the shell	T	: Temperature
dir	: Directly absorbed	ΔT	: Temperature difference
eva	: Evaporation	tot	: Total
h	: Refrigerant enthalpy	ote	: Oil impelling tube eliminated
$\dot{m}$	: Flow rate	dme	: Discharge muffler eliminated
$\dot{Q}$	: Heat transfer rate		
$\dot{q}$	: Heat Flux		
$\dot{W}$	: Work done per unit of time		

1. INTRODUCTION

Development of new systems having maximum performance with minimum energy consumption is the main issue for the research and development studies. According to this state, improving the household type refrigerators energy efficiency becomes an important subject for the researchers. A hermetic reciprocating compressor is the most critical component of a household refrigerator. It consumes approximately 90% energy of overall electrical input power of the refrigerator. Due to the increasing importance of energy efficiency, performance improvement studies of the compressor play an important role to reduce overall energy consumption of the refrigerators.

In order to decrease the power consumption of the compressor, The mechanical, electric motor and thermodynamic efficiencies are three main focus points. Relatively mechanical and electrical motor losses are more effective on compressor overall efficiency, however they are very close to their limits under current conditions. For this reason in the near future thermodynamic efficiency is thought to be more critical issue for compressors. Consequently characterizing the thermal behavior of the compressor becomes very important phenomenon for improving the thermodynamic efficiency.

Several studies concerning the thermodynamics, heat transfer and fluid mechanics phenomena of the compressor have been proposed in the literature. Fagotti et al (1994) heat transfer inside the cylinder is evaluated with different correlations and experimental studies. Perez-Segerra et al (2005) applied different numerical methods and experimental techniques to solve heat transfer mechanism inside the compressor. Compressor is divided into different geometrically simplified control volumes for investigating the conduction and convection forms of heat transfer phenomena by Ooi, K. T. According to Dutra T. *et al* (2010) lubricating oil has great effect on the heat transfer process at the upper region of the compressor. Almbauer R. A. *et al* (2006) investigated 3 dimensional simulation with thermal network calculation which includes lumped conductance method. Kara S. *et al (2010)* investigated the temperature distribution of crankcase both experimentally and numerically. Pereira E. *et al* (2010) numerically analyzed the heat transfer inside the cylinder of a simplified geometry under actual operating conditions.

In this study, heat transfer processes between the components of a hermetic reciprocating compressor were studied theoretically and experimentally. Compressor is divided into different control volumes (Shell, suction /discharge mufflers/pipes/plenums, cylinder, lubricant, electrical motor, body, inner gas etc) for determining the heat transfer characteristics between these control volumes theoretically. In the experimental part of the study, conceptual designs on the components of the hermetic compressor were made to examine the effects on the heat transfer characteristics between the components and also compressor performance. The measurements on the compressor, includes; temperature measurements of the hermetic compressor components by thermocouples and analyzing the thermal map of the compressor and also realizing the heat flux measurements, which were conducted on the specified components of the hermetic compressor. In addition to this, conceptual designs, which have affected the heat transfer on the components and performance of the compressor, were also applied during the study.

The results of the theoretical and experimental studies are used for investigating the network including conduction, convection and radiation forms of heat transfer inside the compressor. Generated heat transfer network helps to characterize the thermal functions of the main components which leads to the new and better compressor designs.

2. EXPERIMENTAL STUDIES

The experimental studies were performed on the compressor including temperature measurements of the hermetic compressor components by using the thermocouples and analyzing the thermal map of the compressor and also realizing the heat flux measurements, which were conducted on the specified components of the hermetic compressor. Tested compressors are designed for isobutane (R600a) **refrigerant** and the cooling capacity of the compressor at ASHRAE conditions is 100 W. Determined measurement points and number of thermocouples are given in the Table 1.

Table 1. Temperature measurement points

No	Measurement Point	Number	No	Measurement Point	Number
1	Lubrication Oil	4	10	Cylinder Head	2
2	Suction Muffler Backside	4	11	Discharge Muffler	1
3	Suction Muffler Front side	4	12	Discharge Tube Inlet	1
4	Inner Gas	6	13	Discharge Tube Outlet	1
5	Cylinder Surface	4	14	Electrical Motor	3
6	Body	4	15	Shell	8
7	Discharge Plenum	2	16	Compressor Outlet.	2
8	Compressor Inlet	2	17	Suction Muffler Inlet	1
9	Suction Plenum	1	18	Suction Muffler Outlet	1
				Total:51	

Figure 1- General view of the compressor with thermocouples

Compressor performance tests were performed while the compressors were running on a fully automated calorimeter system that can activate determined operating conditions during the temperature measurements. Determined ASHRAE working conditions are given in the below table. Data acquisition system was used for collecting temperature data after the thermal stabilization of the compressor was provided.

Table 2. Operating Conditions

Operating Condition	P$_{eva.}$ (bar)	P$_{cond.}$ (bar)	T$_{eva.}$ (°C)	T$_{cond.}$ (°C)	T$_{sub.}$ (°C)	T$_{sup.}$ (°C)
ASHRAE	0.624	7.61	-23.3	54.4	32.2	32.2

The figure below explains the measurement methodology of the cooling capacity generated by the compressor: In this systeme subcooling & superheating temperatures and the condensation & evaporation pressures are measured. With these values the enthalpies can be determined. Also the average refrigerant massflow is being measured. By multiplying the enthalpy difference with the refrigerant massflow the cooling capacity can be determined:

$$\dot{Q}_{cool} = (h_{after\ evaporator} - h_{before\ expansion}) \cdot \dot{m}_{refrigerant} \qquad (1)$$

The evaporation process takes place inside a insulated vessel and it is assumed here that there is no heat transfer between the vessel and the ambient. Inside this vessel an electrical heating element is positioned. To keep the vessel at an equal temperature the heating element has to generate an amount of heat equal to the cooling capacity of the evaporator.

$$\dot{Q}_{cool} = \dot{Q}_{heater} \qquad (2)$$

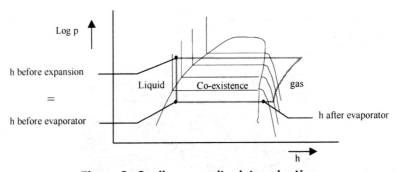

Figure 2- Cooling capacity determination

Second major experimental subject of this study was conceptual designs. These designs were used for determining effects of the heat transfer on the components thermal behaviors and the performance of the compressor.

Firstly, an air duct with a fan application was designed to improve the heat transfer coefficient of the compressor shell. Then, temperature and performance measurements were conducted with varying voltage rates to investigate the effects of the design.

Table 3. Temperature measurement points an air duct with a fan application

Temperature (°C)	Original	Fan: 4V	Fan: 7V	Fan: 10V	Fan: 14V
Suction Muffler Backside	61.9	58.3	55.7	53.8	52.6
Suction Muffler Front side	57.9	54.4	51.6	50.0	48.9
Compressor Inlet	44.6	41.7	39.9	38.4	37.7
Suction Muffler Inlet	49.5	47.6	45.9	44.7	44.1
Suction Muffler Outlet	59.1	55.9	54.0	52.5	51.4
Suction Plenum	67.9	64.9	62.7	61.0	59.8
Inner Gas	70.2	65.9	62.6	60.6	58.6
Cylinder Head	85.9	83.7	79.7	77.8	76.8
Discharge Plenum	97.2	94.5	91.4	89.2	88.4
Discharge Muffler	82.8	79.7	76.5	74.2	73.2
Discharge Tube Inlet	80.1	79.7	73.9	71.5	70.5
Discharge Tube Outlet	70.2	66.4	63.0	60.8	59.6
Cylinder Surface	79.6	76.5	73.7	71.2	70.3
Body	72.1	68.3	65.2	62.9	61.7
Electrical Motor	69.3	65.2	62.0	59.7	58.3
Shell inside	64.7	59.8	56.4	54.0	52.5
Shell outside	61.9	57.0	53.0	50.2	48.2
Lubrication oil	61.7	56.6	53.8	51.7	50.2

Average massflow rate and constant pressure values taken from the compressor calorimeter system were used with compressor inlet /outlet temperature measurement results for calculating the enthalpy difference of the system. Steady regime work was calculated from the flow rate and enthalpy difference. The difference between the compressor inlet power and the steady regime work assumed as the heat transfered from the whole shell of the compressor.

Average temperatures of the components decrease between 3K /4 K at 4 V Fan application. Increasing the voltage of the fan decreased overall temperatures of the components however it wasn't proportional to the voltage level. Heat transfer coefficient for the baseline of the compressor 1.98 W/K becomes %24.7 higher (2.47 W/K) for 4V operation voltage of the fan. Due to the increased total heat transfer coefficient, heat transfer rate increases and average shell temperature decreases. According to this situation thermal energy transferred from shell to ambient didn't change at the same rate with the heat transfer coefficient variation. As a result cooling the shell didn't provide significant benefits on the coefficient of performance (COP) value of the system.

Table 4. Effects of cooling the shell

	Original	Fan: 4V	Fan: 7V	Fan: 10V	Fan: 14V
Flow Rate (g/s)	0.284	0.295	0.293	0.290	0.293
Enthalpy Difference (kJ/kg)	25.60	23.03	19.13	17.16	16.07
Compressor Inlet Power (W)	66.2	68.6	68.6	68.6	69.4
Heat Transfer from shell (W)	58.9	61.8	63.0	63.6	64.7
Shell UA (W/K)	1.98	2.47	3.00	3.49	3.99
Steady Regime Work (W)	7.27	6.79	5.61	4.98	4.71
Δ COP (%)	0	0	-1.4	-3.5	-6.3

Secondly, the amount of the oil in the crankcase was reduced for researching the impression of the oil volume on the performance and the temperature distribution of the hermetic compressor. Lubrication oil level of the recent compressor reduced (%20) and temperature distribution results were examined. Critical effects haven't defined with respect to the results. Oil level doesn't affect the temperature distribution ($\Delta T = \pm 1$ K) inside the compressor.

In the further studies, path of the lubrication oil was reviewed. Oil impelling tube in the compressor was eliminated to obtain the cooling effect of the oil in the compressor. Eliminating the oil impelling tube in the compressor influenced the lubrication sealing between the cylinder and the piston. Leakage due to the inadequate lubrication had negative effects on cooling capacity. In addition to this general average temperatures inside the compressor increased and this also negatively effects the cooling capacity and performance of the compressor. After this study, discharge tube's and discharge muffler's heating effects in the compressor were examined. In this study, existing discharge muffler volume cancelled and a new discharge line connected to the cylinder head of the compressor Detailed temperature measurement results are shared in Table 5. Eliminating the discharge muffler and reduction of the discharge tube prevented the cooling of the compressed high pressurized and heated refrigerant inside the compressor. According to this issue general average temperature of the components decreased. This change in temperature definitely effects the temperature distribution inside the compressor. As a result eliminating discharge muffler of the compressor provides benefits on the cooling capacity of the compressor however it doesn't affect the coefficient of performance (COP) value due to the increased inlet power of the system.

Table 5. Temperature measurements for oil impelling tube eliminated compressor and discharge muffler eliminated compressor

Temperature (°C)	Original	T_{ote}	ΔT (K)	T_{dme}	ΔT (K)
Compressor Inlet	44.6	44.4	-0.2	40.7	-3.9
Suction Muffler Inlet	49.5	50.4	+0.9	45.7	-3.8
Suction Muffler Outlet	59.1	60.1	+1.0	58.3	-0.8
Suction Plenum	67.8	71.0	+3.2	75.1	+7.3
Suction Muffler Backside	61.9	62.7	+0.6	52.9	-9.0
Suction Muffler Front side	57.9	58.3	+0.4	54.7	-3.2
Discharge Plenum	97.2	99.7	+2.5	n/a	n/a
Discharge Muffler	82.8	84.3	+1.5	64.7	-18.1
Discharge Tube Inlet	80.1	80.0	-0.1	87.6	7.5
Discharge Tube Outlet	70.1	72.1	+2.0	78.0	7.9
Cylinder Head	85.9	89.4	+3.5	n/a	n/a
Inner Gas	70.2	69.9	-0.3	63.0	-7.2
Cylinder Surface	79.5	81.4	+1.9	72.0	-7.5
Electrical Motor	69.3	71.1	+1.8	62.8	-6.5
Body	71.1	73.6	+2.5	63.0	-8.1
Shell	61.8	60.4	-1.4	51.6	-10.2
Lubrication oil	61.7	63.5	+1.8	56.6	-5.1

3. ANALYTICAL MODELLING

There are several heat transfer mechanisms inside the compressor with respect to the various temperature gradients between the components. First law of thermodynamics conservation of energy was used for developing the heat transfer network. Control volumes with energy and mass transfer available control surfaces were described. Inlet and outlet energy balance without energy storage inside the control volumes was stated during the heat transfer solutions. Compressor was separated to different control volumes for determining the heat transfer characteristics between these control volumes. Creating the heat transfer network and modeling this network were the main scopes of this section.

Results of the conceptual experimental studies were evaluated during the semi empiric modeling studies. According to these studies compressor is divided into twelve different control volumes. Determined control volumes inside the compressor given in Table 6.

Table 6. Control volumes determined inside the compressor

	CONTROL VOLUMES		
1	Shell	7	Discharge Plenum
2	Inner Gas	8	Discharge Muffler
3	Suction Muffler	9	Discharge Tube
4	Suction Plenum	10	Body
5	Cylinder	11	Lubrication Oil
6	Cylinder Head	12	Electrical Motor

Control volumes and assumptions with respect to the correlations in literature and experimental studies summarized in this section:

Upper and lower part of the compressor shell was combined and named as shell. Homogeneous temperature distribution used, the movement of the lubrication oil in the sump assumed as rotational and impingement movement of the oil is neglected inside the shell for simplifying the model. Fresh gas ratio approach given in the below was used for refrigerant inlet conditions:

$$\delta = \frac{\text{Suction Muffler Inlet Refrigerant\_Mass}}{\text{Compressor Suction Tube Inlet Refrigerant\_Mass}} \tag{3}$$

Ideal gas and steady specific heat approach was used. Oil and refrigerant assumed independently without mixture and first law of thermodynamics was used for modeling.

$$\dot{Q} = \dot{W} + \dot{m}_{tot} h_{cs} + \dot{m}_{dif} h_{cir} - \dot{m}_{tot} h_{muf} - \dot{m}_{dif} h_{cs} \tag{4}$$

$\dot{m}_{tot}$: Total flow rate of the refrigerant coming from evaporator,
$\dot{m}_{dir}$: Flow rate of the refrigerant directly absorbed by the suction muffler,
$\dot{m}_{dif}$: Flow rate of the refrigerant distributed inside the shell,
h_{cs} : Enthalpy of the refrigerant at the compressor inlet,
h_{cir} : Enthalpy of the refrigerant circulated inside the compressor shell,
h_{muf} : Enthalpy of the refrigerant at the suction muffler inlet,
T_{cs} : Temperature of the refrigerant at the compressor inlet,
T_{cir} : Temperature of the refrigerant circulated inside the compressor shell,
T_{muf} : Temperature of the refrigerant at the suction muffler inlet.

In this system work haven't occurred with respect to this situation we assumed the work term $\dot{W} = 0$. In addition to this heat transfer process between refrigerant coming from evaporator and distributed gas inside the shell without any interaction with the ambient was assumed.

$$\dot{m}_{tot}\left(h_{cs}-h_{muf}\right)=\dot{m}_{dis}\left(h_{cs}-h_{cir}\right) \tag{5}$$

$$\frac{\dot{m}_{dis}}{\dot{m}_{tot}}=\frac{h_{cs}-h_{muf}}{h_{cs}-h_{cir}} \tag{6}$$

$$\delta=\frac{\dot{m}_{dir}}{\dot{m}_{tot}}=1-\frac{\dot{m}_{dis}}{\dot{m}_{tot}} \tag{7}$$

$$\delta=\frac{h_{muf}-h_{cir}}{h_{cs}-h_{cir}}=\frac{T_{muf}-T_{cir}}{T_{cs}-T_{cir}} \tag{8}$$

Fresh gas ratio was calculated from detailed temperature measurements and it was used during the simulations:

$$\delta=\frac{T_{sus}-T_{sir}}{T_{kg}-T_{sir}}=\frac{51-69.5}{45.6-69.5}=\frac{-18.5}{-23.9}=0.77$$

Average component temperatures used, impingement movement and mixture of the oil with inner gas weren't directly included inside the model. These effects evaluated inside the ampiric expressions by tuning the heat transfer coefficients. Several heat transfer relations have been studied and implemented in the model. Determined heat transfer coefficients compared with the typical values for the natural, forced heat convection for the gases and liquids.

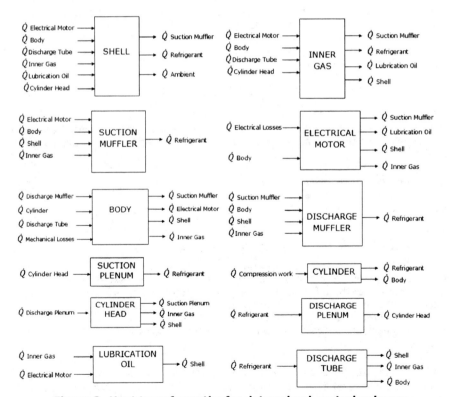

Figure 3- Heat transfer paths for determined control volumes

Heat transfer coefficients taken from theoretical, numerical and experimental studies for determined control volumes given in the Table 7.

Table 7. Heat transfer Coefficients

Heat Transfer Coefficient	(W/m²K)	Emissivity Coefficient	
Internal heat transfer coefficient shell-wall	60	Shell	0.98
External heat transfer coefficient shell-wall	15	Cylinder head	0.79
Suction muffler-inner gas	30	Discharge tube	0.65
Discharge tube -inner gas	4	Motor	0.90
Oil-shell	35	Suction muffler	0.85
Body-gas	50	Body	0.70
Discharge muffler-Body	155	**Heat Transfer Rate**	(W/K)
Discharge plenum-Cylinder head	60	Oil-inner gas	30
Thermal conductivity coefficient	(W/mK)	Oil-body	3
Cylinder head-suction plenum	175	Discharge suction plenum	0.4

4. RESULTS

Heat transfer paths for determined control volumes combined for the development of the whole heat transfer network of the compressor given in the Figure 4.

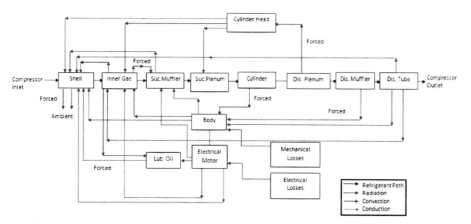

Figure 4- General view of heat transfer network inside the compressor

The results of general heat transfer network was used for developing the semi empirical model for thermal characterization of the compressor. Detailed geometrical parameters used for simulations of the model. Original situation and simulation results were given in Table 8. Temperature difference between the average experimental and numerical results were lower than ~15 K. Critical control volumes mainly oil and inner gas should be more detailed modelled by increasing the number of sub control volumes and modifying the semi empiric models. This modification will improve the temperature distribution convergency of the numerical results by including the local temperature variations.

Table 8. Temperature simulation results

Temperature (°C)	Original	Simulation	ΔT (K)
Cylinder*	79.5	79.5	
Lubrication oil	61.7	50.5	-11.2
Suction Muffler	49.5	54.6	-3.3
Inner Gas	70.2	68.1	-2.1
Cylinder Head	85.9	71.1	-14.2
Body	70.5	67.5	-3.0
Discharge Tube Inlet	80.1	79.8	-0.3
Discharge Tube Outlet	70.2	69.8	-0.4
Electrical Motor	69.3	57.7	-11.6
Suction Plenum	67.8	71.5	+3.7
Discharge Muffler	82.8	74.7	-8.1
Shell	61.8	48.8	-13.0

*In all heat transfer calculation the cylinder wall temperature is taken to be constant and needs to be determined by either experiment.

Based on the results of this study it can be concluded that:

- Average temperature increase magnitude on suction line was determined as 13.5K. Nearly 1/3 of this temperature increase occurred before the compressor shell inlet.
- Mixture of the refrigerant coming from the inlet of the compressor and inner gas inside shell affects the %13 of the total heating on the suction path.
- Cooling gradient at the discharge path determined *(discharge plenum-discharge muffler 13.1K; discharge muffler-discharge pipe inlet 4K; along the discharge line 10K).*
- Average temperatures of the components decrease between 3K /4K at 4 V Fan application. Increasing the voltage of the fan decreased overall temperatures of the components however it wasn't proportional to the voltage level.
- Eliminating the oil impelling tube in the compressor had negative effects on cooling capacity. In addition to this general average temperatures inside the compressor increased and this also effects the cooling capacity too.
- Reduction of the discharge tube and excluding the discharge muffler and prevented the cooling of the compressed high pressurized and heated refrigerant. According to this issue general average temperature of the components decreased.
- Temperature difference between the average experimental data and simulation results were lower than ~15K. Temperature difference will be decrease by modifying the semi empiric model.

5. CONCLUSION

The results of the theoretical and experimental studies are used for investigating the network including conduction, convection and radiation forms of heat transfer inside the compressor. Generated heat transfer network helps to characterize the basic thermal functions of the main components which leads to the new and better compressor designs.

Based on the results of this study it will be extended that:

- The investigation of the semi empirical model of the heat transfer network will be extended by numerical analysis. Simulation of the model coupled with the computational fluid dynamics analyzes can be used for performance calculations of the compressor.
- Increasing the number of control volumes for lubricating oil will improve the characterization of heat transfer network.
- Heat flux measurements can be detailed at the high local temperature distribution points.
- Generalizing the model for all kind of parametric analyses by validation of experimental studies.

6. ACKNOWLEDGEMENT

The authors would like to express their sincere appreciation to Arçelik A.Ş.

7. REFERENCE LIST

[1] **Dutra, T.et al** 2010. Experimental Investigation of Heat Transfer in Components of a Hermetic Reciprocaiting Compressor, International Compressor Engineering Conference at Purdue, Purdue University, USA, 1346.
[2] **Almbauer R. A. et al** 2006. 3-Di,mensional Simulation for Obtaining the Heat Transfer Correlations of a thermal Network Calculation for a Hermetic Reciprocating Compressor, International Compressor Engineering Conference at Purdue, Purdue University, USA, 079.
[3] **Kara S. et al** 2010. Thermal Analysis of a Small Hermetic Reciprocating Compressor, International Compressor Engineering Conference at Purdue, Purdue University, USA, 1307.
[4] **Pereira E. et al** 2010. Numerical Analysis of Heat Transfer inside the Cylinder of Reciprocating Compressors in the Presence of Suction and Discharge Processes, International Compressor Engineering Conference at Purdue, Purdue University, USA, 1310.
[5] **Ooi, K. T.,** 2003. Heat Transfer Study of a Hermetic Refrigeration Compressor, *Applied Thermal Engineering*, 23,1931-1945.
[6] **Rigola, J., et al** 2005. Parametric Studies on Hermetic Reciprocating Compressors *Internal Journal of Refrigeration,* **28**, 253-266.
[7] **Fagotti, F.,** 1994. Heat Transfer Modelling in Reciprocating Compressor, *International Compressor Technology Engineering Conference at Purdue*, Purdue University, USA, 320-327.
[8] **Özdemir, A.R.,** 2007. An Investigation of the Heat Transfer Characteristics Among the Hermetic Reciprocating Compressor Components, M.Sc. Thesis, Istanbul Technical University, Istanbul, Turkey.

Study of aerodynamic noise in hermetic refrigerator compressor

S Lin, Z He, J Guo
Xi'an Jiaotong University, School of Energy and Power Engineering, People's Republic of China

ABSTRACT

This paper is to study the aerodynamic noise in hermetic refrigerator compressor. First, a compressor model is built and its working process is numerically simulated to obtain data of velocity and pressure in suction. Next, use the above data to calculate aerodynamic noise of two compressors, one with and the other without a suction muffler. Then improve the suction muffler and calculate it aerodynamic noise to show that it has gained better performance. Finally experimental results obtained from noise experiments verify the above conclusion.

1 INTRODUCTION OF COMPRESSOR NOISE

In a refrigerator, the main noise source is the internal refrigerant compressor. Currently, the majority of compressors used in refrigerators are hermetic compressors. For hermetic compressor noise, the mechanical noise and the acoustic performance of mufflers have been studied widely. While the aerodynamic noise is the most dominant noise source, little research has been conducted on it. When the compressor works, the intermittent suction and exhaust actions at valves cause pressure fluctuations that generate aerodynamic noise. Meanwhile, the hole injection phenomenon, the resonating noise, the impact noise[1] and the regenerative noise are all easy to be induced.

For piston compressor, the exhaust noise is weaker than the suction noise after multiple buffers, so the aerodynamic noise is mainly the suction noise. Suction mufflers are commonly used in practical applications to reduce the aerodynamic noise of compressors. The design and optimization of a muffler is an ideal noise reduction method with the lowest cost[2]. So this article is of great theoretical and practical research value.

2 SIMULATION OF WORKING PROCESS OF COMPRESSOR

2.1 Models and grids
Use Compressor QDH51G as the model. Compressor cylinder diameter=0.0206m, piston stroke=0.0154m, speed=2900r/min, suction pressure P_1=0.133MPa, suction temperature T_1=305K, exhaust pressure P_2=1.32MPa. The Working fluid is R134a. The exhaust gas temperature T_2=393K by calculation. The material for valves is valve sheet steel, and the modulus of elasticity is 2.07GPa and density is 7800kg/m^3.

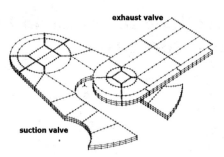

Figure 1 Suction and exhaust valves models

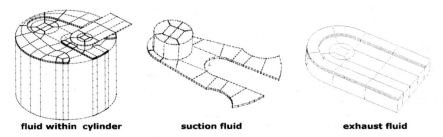

Figure 2 Compressor fluid models

The compressor fluid model can be divided into three parts as shown in Figure 2, they are the fluid within cylinder, the suction fluid and the exhaust fluid. The underside of the fluid within cylinder model is set as a moving wall, acting as the piston. Displacement of the moving wall is calculated by the reciprocating compressor piston displacement formula.

In the fluid model, surfaces in contact with valve plates are set as fluid-structure interaction boundaries. The suction fluid and the exhaust fluid are divided from the fluid within the cylinder by two GAP boundary conditions respectively. The rest surfaces of fluid are set as adiabatic. GAP boundary lines are initially turned off until the suction and exhaust valves are deformed to a certain extent. Leader-Follower is set on the fluid-structure interaction boundary to prevent unit deformity or grid superimposition around the boundary. Also, in suction and exhaust valves solid models as shown in Figure 1, surfaces in contact with the fluid are set as the fluid-structure interaction boundaries.

The fluid model is calculated with non-dimensional calculation, and the time step is set as 1×10^{-5}s, and together there are 2150 steps. The solid model is calculated using the automatic time step. The compressor working process is calculated within a cycle.

2.2 Results and analysis

When the suction valve opens, a high-speed jet stream is formed within the compressor together with a large number of vortexes with irregular movements. This causes sharp flow and pressure disturbances generating injection noise. Therefore, the suction flow velocity and pressure pulsation must be controlled to reduce the aerodynamic noise.

In this chapter we gained the pressure and velocity values within the suction channel as shown in Figure 3, which could be used as boundary conditions to study the impact of different pressures and flow rates on aerodynamic noise.

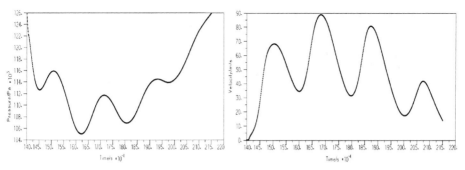

Figure 3 Fluid pressure and velocity within suction channel

3 CALCULATIONS AND ANALYSIS OF THE SUCTION NOISE

3.1 Models and grids

In this section we study two compressor models, one without suction muffler (model 1) and the other equipped with an expansion chamber suction muffler (model 2) as shown in Figure 4. The suction muffler includes a resonant cavity, an inlet intubation and an expansion chamber. Points indicated in figures are sound pressure monitoring points.

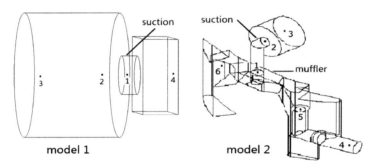

Figure 4 Structures of the two models

3.2 Process of calculation

In this paper, the aerodynamic noise is calculated by the CAA method. For model 1, we discrete the suction pressure and velocity fluctuation values from the last chapter, and choose three groups of values as shown in table 1 to calculate the aerodynamic noise.

Table 1 Suction Noise Boundary conditions of model 1

No.	Inlet pressure MPa	Outlet pressure MPa	Inlet velocity m/s
1	0.133000	0.104355	9.68
2	0.133000	0.114559	7.27
3	0.133000	0.119550	4.94

For model 2, as the installation of mufflers will affect the distribution of flow field during suction, the outlet suction pressure is initially set as 0.114559MPa to calculate the corresponding velocity, which turns out to be 14.48m/s. Then use these two values as boundary conditions to calculate the sound pressure level (SPL) in suction.

3.3 Calculation results analysis

In Figure 5, all the points are on the same axis of flow channel. Monitoring point 1 at the entrance of the suction valve has the maximum SPL, mainly due to the sudden formation of the high-speed jet flow. The two monitoring points 2 and 3 within the cylinder have larger SPLs than monitoring point 4 mainly due to the jet noise directivity, which means the SPL in the ejecting direction is higher than that in the opposite direction.

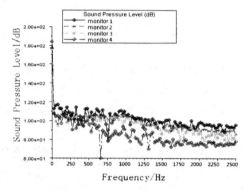

Figure 5 Comparison of SPLs of monitoring points 1,2,3,4 of model 1 under boundary condition 2

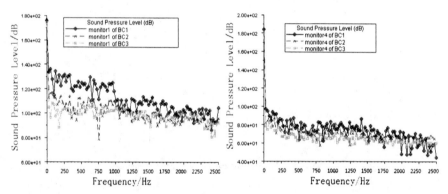

Figure 6 Comparison of SPLs of monitoring points 1 and 4 of model 1 under different boundary conditions

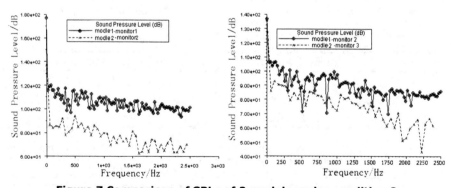

Figure 7 Comparison of SPLs of 2 models under condition 2

400

SPLs of monitoring point 1 and 4 under the 3 different boundary conditions are shown in Figure 6. It can be seen that for both points, SPLs increase with the increase in inlet velocity and the reduction in outlet pressure. For the ejection speed of the jet flow and the velocity gradient grow larger when the inlet flow velocity elevates. And also the pressure pulsations and the noise increase. Therefore, it can be concluded that it's necessary to take measures to reduce the flow velocity and the pressure loss to decrease the jet noise.

In Figure 7, monitor point 1 of model 1 and monitor point 2 of model 2 are compared, for both points are at the entrance of suction channel. And monitor point 2 of model 1 and monitor point 3 of model 2 are compared, for both points are set within the cylinder. We can see that for both cases, SPL of model 1 is lager than that of model 2 by around 10 dB, due to the cushioning effect of the muffler that reduces the flow rate and noise in model 2. So it can be seen that the silencing effect of muffler is quiet outstanding and installing mufflers is a preferable choice to reduce the compressor noise. Average SPLs of monitoring points 1-6 of model 2 are: (1)112.8dB, (2)99.5dB, (3)98.3dB, (4)107.9dB, (5)111dB, (6)95.8dB. SPLs of point 1 and 5 are greater, because they are close to the bend of intubation. SPLs of the two points within the cylinder (2 and 3) are lower because of the cushioning effect of the muffler. Monitoring point 6 is in the resonant cavity, within which there is only a small amount of flow, so the velocity and pressure pulsation is lower.

3.4 Chapter Summary
The high-speed flow though the suction inspires highly directional jet noise in the flow field; reducing the pressure loss in suction will help to reduce the jet noise. The muffler has a dampening effect on the flow velocity to slow down pressure pulsation in the suction process, but structural mutations are places easy to produce large aerodynamic noise.

4 ANALYSIS OF MODIFIED MUFFLER

4.1 Model and grid
Considering the impact of muffler structure on the aerodynamic noise and the transmission loss, we improved the original muffler by the following steps:
(1) Increase the diameter of intubation, and use the dual intubation;
(2) Increase the width and thickness of the expansion chamber, and improve the expansion ratio of the muffler;
(3) Change the original the resonant cavity into expansion chamber, and simplify the shape of expansion chamber.

The model of the modified muffler is shown in Figure 8. And the aerodynamic noise of the modified muffler is calculated in accordance with methods described in the previous chapter.

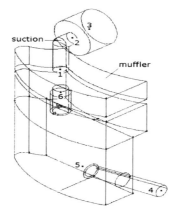

Figure 8 Model of modified muffler

4.2 Calculation results analysis
Just as the original muffler, on size mutations there are high-speed flows in the modified muffler. But as the modified muffler has an increased intubation diameter, the flow velocity into the cylinder reduces. The increased expansion ratio and volume allow more space to buffer the flows to reduce the local flow velocity. But in the mean time, it is easy to cause vortex in the export intubation. Generally, the modified

muffler allows lower flow velocity, lower vortex magnitude and lower pressure fluctuating values.

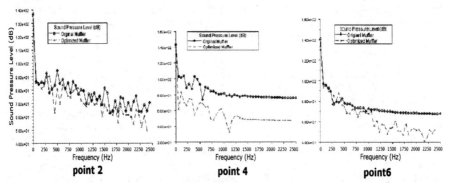

Figure 9 Comparison of SPLs of monitoring point 2, 4 and 6 of original model and modified model

Comparison of SPLs of the original muffler model and the modified muffler model on monitoring points 2, 4 and 6 is as shown in Figure 9. On monitoring points 2 and 4, SPLs of the modified muffler are lower than that of the original muffler. While on point 6, SPLs of the two models have little difference, primarily because of the second expansion chamber, within which the velocity gradient and the pressure pulsation become greater. Generally speaking, the noise level reduces within the modified muffler model.

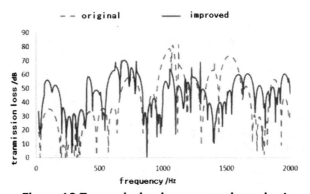

Figure 10 Transmission loss comparison chart

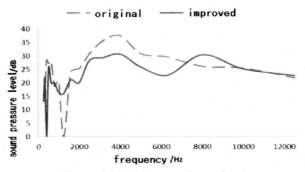

Figure 11 SPL comparison chart

Use the Sysnoise software to compare the transmission loss of the original muffler model and the modified model. The original muffler model peaks in the transmission loss due to the presence of the cavity resonator. The modified muffler is added with an expansion chamber to widen the muffling frequency. The performance of the modified muffler improves within frequency below 1000Hz, while on the high-frequency range, the performance drops. Overall, the modified model changes for the better, but the gap between the two is not large.

Main indicators to evaluate muffler performance are noise elimination and power loss. In the design and selection of mufflers, it's important to obtain the minimum pressure and power loss while at the same time, the maximum noise elimination.

The above two different mufflers are measured with noise experiments. The average sound power and the A SPL of compressor installed with the original muffler are 42.3dB and 42dB. As to the compressor installed with the modified muffler, the average sound power and the A SPL are 38.4 dB and 38.5dB respectively. The noise spectrum comparison is shown in Figure 11, which indicates that after installing the modified muffler, the sound pressure level is reduced within the low frequency spectrum. The experimental data confirm that the calculation of compressor aerodynamic noise is reliable and the muffler structure optimization is of great reference value in this article.

5 CONCLUSION

It is important to reduce the flow velocity and pressure loss in suction to eliminate the compressor aerodynamic noise and improving the suction muffler is an ideal method to achieve this target. We found out that increasing the volume of intubation and expansion chamber and the expansion ratio of the muffler can reduce the aerodynamic noise of the compressor. While changing the original cavity into expansion chamber or adding additional cavity is likely to inspire new aerodynamic noise. In the design of muffler structure, both the characteristic of the aerodynamic noise and the performance of the muffler have to be taken into account to guarantee the muffler performance while minimize the aerodynamic noise.

REFERENCE LIST

(1) Qiu Ying, Li Hong-qi, Lv Ya-dong. Analysis of Suction and Discharge Valve Noise in a Hermetic Reciprocating Compressor by Using Jet Noise Theory[J]. Fluid Machinery, 2007, 35 (1): 25-28.
(2) Chiu Min-Chie. Optimal Design of Multichamber Mufflers Hybridized With Perforated Intruding Inlets and Resonating Tubes Using Simulated Annealing[J]. Journal of Vibration and Acoustics, 2010, 132 (5): 54503.
(3) Esteve S. J., Johnson M. E. Reduction of sound transmission into a circular cylindrical shell using distributed vibration absorbers and Helmholtz resonators[J]. Journal of the Acoustical Society of America, 2002, 112 (6): 2840-2848.

EXPANDERS

Physics of a dry running unsynchronized twin screw expander

J Hütker, A Brümmer
Chair of Fluidics, Department of Mechanical Engineering, TU Dortmund, Germany

ABSTRACT

The concerned dry running gearless screw machine (GL 51.2) has already been investigated in detail as a super charger and vacuum blower. Aim of the present paper is the physical explanation of the machine behaviour as twin screw expander. This will be done by extensive experimental results including pressure-angle diagrams and indicator diagrams together with numerical calculations using a multi-chamber model of the machine. It is shown that the primary loss mechanisms of a screw expander are the inlet throttling during chamber filling as well as the gap mass flows in the expansion start range. As the rotational speed increases, the inlet throttling becomes stronger, while the gap losses become smaller.

In addition, the same machine has been evaluated numerically by use of full 3D CFD at the City University London. Corresponding results for 3D CFD will be presented in a parallel paper at the same conference.

NOTATION

a	[m]	shaft distance
$\dot{m}$	[kg/s]	mass flow
n	[min^{-1}]	rotational speed of the male rotor
p_C	[bar]	chamber pressure
P_{eff}	[W]	effective power
P_i	[W]	internal power
p_i	[bar]	inlet pressure
v_i	[-]	inner volume ratio
α_{HR}	[°]	rotational angle
ϑ_i	[°C]	temperature at inlet side
LP		low pressure
HP		high pressure
MR		male rotor
FR		female rotor
PM		profile intermesh

1. INTRODUCTION

The screw machine is primarily used in applications of compressed air supply and process gas technology. In these applications, about half of all used compressors are now screw machines. Thanks to the robust construction, the favourable operating characteristics and the high energy conversion quality, the range of application of the screw machines has been constantly extended. Alongside applications in vacuum technology and in automotive applications as a mechanical charging unit, the use of screw machines as screw expanders for the conversion of thermal into mechanically usable shaft work in different power ranges is being researched. Although the screw machine functioning as an expander equates to a reversal of the operating principle of the screw compressors in theory, the constructional and energy-related design principles cannot be transferred without restrictions (1), (2).

In this article, the energy-related processes in the conversion of thermal into mechanical energy with a screw machine modified to an expander are discussed. The basis for the evaluation of dissipative procedures during an operating cycle is supplied in particular by experimentally determined chamber pressure curves as a function of rotor rotation. The experimental data is supplemented by results from thermodynamic simulation calculations, in order to assess the influence of throttling losses during the high-pressure side charge change and of the gap flows during the entire operating cycle as quantitatively as possible. Investigating the gap mass flows depending on the rotation angle also allows the dominance of the individual gap types to be weighted in terms of their dissipative effect and their influence on the "refilling" of the working chamber in the expansion phase.

2. THE SCREW EXPANDER GL51.2-M

The screw machine used as an experimental machine is a screw charger modified for expander operation. This was originally developed at the TU Dortmund Chair of Fluidics for the mechanical charging of combustion engines and as a module in fuel cell concepts. The type designation of the machine GL51.2-M is a reference to the unsynchronised design (GL = gear less), the distance between shafts (a = 51 mm), the 2nd machine generation and to the modifications to a screw expander (M=motor). The latter essentially involve constructional redesigns in the area of high-pressure side flow control and constructional changes for introducing pressure sensors to dimension the indicator diagrams. The used rotor profile is a slightly modified asymmetrical SRM profile. A machine-specific feature is that the profiled part of the rotors has an on average 3 µm thick tungsten carbide/carbon wear-protection coating. This means that, even if going without a synchronizing pair of gears, the machine can be used as a "dry runner", without the need for a water or oil injection (3). Owing to its original range of application, the screw expander GL51.2-M has an internal volume ratio v_i = 1.47, which is small for expanders, and allows a broad rotational speed range of up to n = 20,000 min^{-1} to be investigated. The geometrical machine parameters and a picture of the machine in the experimental area can be seen in Fig. 1 and Table 1.

Fig. 1: Experimental machine GL51.2-M and machine environment

Table 1: Geometrical data of the screw expander GL51.2-M

	Male rotor	Female rotor
Number of lobes	3	5
Wrap angle	200°	120°
Head diameter	72 mm	67.5 mm
Length	101 mm	
Internal volume ratio	1.47	
Displaced volume per male rotor revolution	285 cm³	
Rotor profile	mod. asym. SRM-profile	

The volume curve of the screw expander GL51.2-M (Fig. 2) begins in rotor neutral position $\alpha_{HR} = 0°$, in which, by definition, a low-pressure-side lobe head of the male rotor is in the connecting line of both rotor shafts. The rotors rotate out of the lobe meshing and the chamber volume rises progressively at first. The area of the inlet opening also increases and reaches its maximum with a rotation angle of $\alpha_{HR} = 104°$. From roughly this rotation angle position, the chamber volume shows a linear dependence on the rotation angle. The expansion phase starts after the subsequent lobe head of the male rotor has moved over the inlet area with a rotation angle of $\alpha_{HR} = 207°$ and ends at $\alpha_{HR} = 342°$ when the low-pressure-side control edges have been reached. As the rotor rotation continues, the outlet area (not shown) forms and the discharge procedure starts.

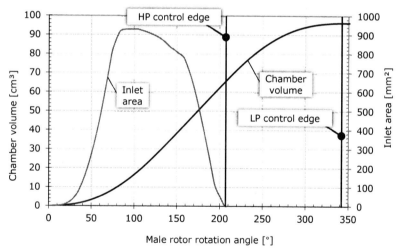

Fig. 2: Inlet area and volume curve of the screw expander GL51.2-M

3. BASIS OF THE EXPERIMENTAL AND THEORETICAL INVESTIGATIONS

The basis for the experimental investigations is an open air circuit with technically dry compressed air as a working medium. The air is sucked in by two oil-injected compressors and compressed. Following the direction of flow, the working fluid is mechanically filtered in several stages and divided into a main and bypass mass flow, before it is led to an electrically-powered flow heater. A heat supply can thus be provided at the inlet of the screw expander thanks to this construction, which is freely selectable in terms of its parameters (consisting of pressure, temperature and mass flow) within the operational limits of the test facility.

The simulation of rotary displacement machines on the basis of a chamber model represents a recognized method for their analysis and further development. The chamber model method is primarily based on the common characteristic of all displacement machines, on one or more cyclically changing working chambers. For the purpose of simplicity, it is assumed that spatial gradients in the intensive state variables are negligibly small and the fluid condition within a working chamber can thus be regarded as homogeneous. Based on this principle, the fluid condition is described by the extensive state variables of energy and mass and the related working chamber volume. A thermodynamic change of the fluid state can occur e.g. by means of volume change, heat convections or mass and energy flows. The calculation by the time-stepping method is made under consideration of the law of conservation of mass and energy. The simulation program "KaSim" represents an implementation of a multi-chamber model method. A fundamental distinction is made between capacities and connections. Capacities represent a storage for different physical manifestations, whereas connections enable an exchange between the capacities and, in their state, they are at any time defined by the related capacities. The working chambers of the rotary displacement machine are an example for a finite fluid capacity with time-variant volume, internal energy and mass; the gaps are an example of a connection for the exchange of mass and energy (4).

4. INVESTIGATION RESULTS

The investigation results are first discussed based on a comparison of the experimentally determined integral indices and those determined using simulation calculations. The physical processes during the operating cycle are then dealt with using the indicator diagrams and the distribution of gap mass flow dependant on the rotation angle.

4.1 Performance map of GL51.2-M

As an example of experimentally and theoretically determined integral indices, the power and mass flow dependence on the rotational speed shall be addressed in the following (Fig. 3).

The effective power of the screw expander is experimentally determined by recording the torque and the rotational speed. The effective power increases over the total rotational speed range and reaches its maximum at $P_{eff} \approx 4600$ W. The experimental internal power results from the measured effective power, taking the mechanical efficiency into account, which is determined from the experimentally established indicator diagrams. The comparison between experimentally determined internal power and that determined by means of thermodynamic simulation shows a good consistency of results over the total rotational speed range. The measured and simulated internal power values show a maximum difference of 4.6% (Fig. 4). The high modelling quality of the simulation system also becomes clear when looking at the mass flow rate, where the results show a very good consistency. For this parameter, the measured and simulated values differ only in the range of low rotational speeds by up to 4.1%. Both the deformation of components as a consequence of thermal expansions and the leak mass flow through labyrinth seals were considered in the calculation. Compared to the "cold" machine, thermal expansion leads to a housing gap heights increasing and thus to an increase of the mass flow rate by around 5-10%. Also shown in Fig.

4 is the measured effective, isentropic efficiency as a function of rotational speed. The efficiency reaches values up to 70%. Even if these values can be achieved by turbo-expanders, the screw expander is distinguished by the good part load behavior. Additionally, the investigated screw expander has the ability to handle multi-phase fluids with relatively high liquid content.

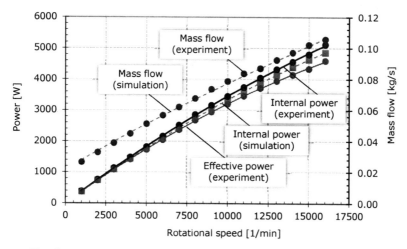

Fig. 3: Measured and calculated powers and mass flows as a function of the rotational speed [p_i = 2 bar and ϑ_i = 75 °C]

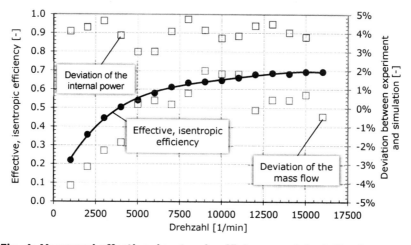

Fig. 4: Measured effective, isentropic efficiency and deviation between experiment and simulation as a function of the rotational speed [p_i = 2 bar and ϑ_i = 75 °C]

4.2 Physical mechanisms during the operating cycle

The physical processes during the operating cycle shall be discussed in the following, starting with an operating point in the middle parameter range. For this purpose, the rotational speed is set at n = 10,000 min^{-1}, the inlet pressure at p_i = 2 bar and the inlet temperature at ϑ_i = 75 °C. The effects of a variation in the inlet pressure and rotational speed are then looked at. Alongside the experimentally established indicator diagram in Fig. 5, the simulated, outflowing gap mass flows are represented in Fig. 6 as a function of the rotation angle of the male rotor.

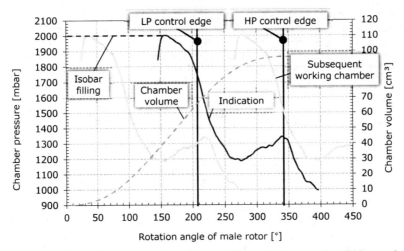

Fig. 5: Measured pressure curves over the rotation angle of the male rotor [n = 10,000 min⁻¹, p_i = 2 bar and ϑ_i = 75 °C]

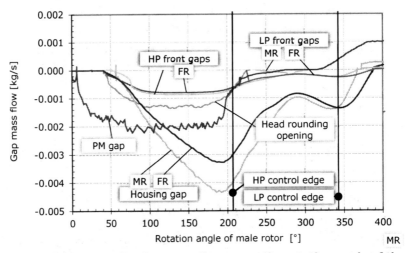

Fig. 6: Simulated outflowing mass flows over the rotation angle of the male rotor [n = 10,000 min⁻¹, p_i = 2 bar and ϑ_i = 75 °C]

As well as the representative working chamber, Fig. 4 shows the previous and subsequent working chamber offset by the 120° male rotor rotation angle. Since the chamber pressure can only be metrologically recorded from a rotation angle of α_{HR} = 158° due to geometrical restrictions, an isobaric filling is assumed at the machine high-pressure level for calculating the internal work for the range of smaller rotation angles.

Up to a rotation angle of $\alpha_{HR} \approx 80°$, the gap mass flows dominate over the profile mesh gap at the beginning of the operating cycle (fig. 6). From a rotation angle position of $\alpha_{HR} \approx 50°$, the area of the head rounding opening first forms completely. The areas of the high-pressure-side front gaps and of the housing gaps at the male and female rotor grow at the same time. As the rotor continues to rotate, the gap masses outflowing via the housing gap become more and more significant accordingly. For rotation angle $\alpha_{HR} > 110°$ (so roughly from the middle of the filling procedure) they dominate the masses outflowing from the working chamber.

The real expansion already starts at $\alpha_{HR} \approx 160°$ and so roughly 50° before reaching the high-pressure-side control edges. The premature start of expansion can be explained by the decreasing inlet area at a chamber volume now increasing linearly with the rotation angle (increasing throttling losses during chamber filling). Integrally speaking, the outflowing gap mass flows assume their maximum value towards the end of the filling procedure, which also contributes to the pressure drop in the working chamber. At the theoretical start of expansion, the chamber pressure is just $p_C \approx 1750$ mbar (Fig. 5).

After the control edges at the high-pressure side have been crossed, the operational behaviour of the screw expander is considerably affected by the gap mass flows. As well as the outflowing mass flows represented in Fig. 6, the inflowing mass flows now become more significant. Fig. 7 shows the inflowing and outflowing gap mass flows via the housing gaps as well as the added gap mass flows via the head rounding openings and the housing gaps. The mass flows entering via the housing gap starting from $\alpha_{HR} \approx 175°$ (so before reaching the high-pressure-side control edge) is explained by the geometry of the inlet area, which first causes a separation of the working chamber at the female rotor, before the male rotor lobe triggers the theoretical start of expansion through moving over the control edge.

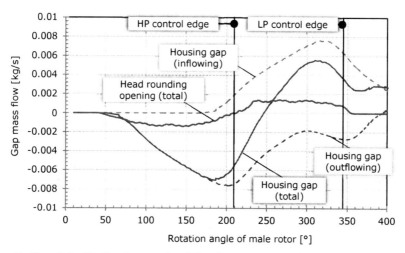

Fig. 7: Simulated inflowing and outflowing gap mass flows of the housing gaps and integral gap mass flows of the housing gaps and of the head rounding openings [n = 10,000 min^{-1}, p$_i$ = 2 bar and ϑ_i = 75 °C]

During the expansion, the chamber pressure steadily decreases up to a rotor rotation angle of $\alpha_{HR} \approx 275°$ and reaches a local minimum at $p_C \approx 1190$ mbar (Fig. 5). At this point in the operating cycle, the inflowing gap mass flows already dominate over the outflowing gap mass flows at the head rounding opening and at the housing gaps, which is reflected by an integrally speaking positive gap mass flow balance via the respective gap type. This refilling of the working chamber leads to the chamber pressure rising in the second expansion phase, although the chamber volume still increases.

After the leading lobe flanks have moved over the low-pressure-side control edges, the load is discharged from the working chamber starting from a rotation angle position of $\alpha_{HR} = 342°$. The pressure compensation in the working chamber to the level of the machine back-pressure is made relatively slowly over a rotation angle of about 50°.

4.3 Influence of inlet pressure and rotational speed

The indicator diagram having, for the operating point discussed, allowed an insight into the technical and physical mechanisms in the middle parameter range, the influence of the operating parameters shall be dealt with in the following. This is done using the example of a variation in the inlet pressure and in the rotational speed.

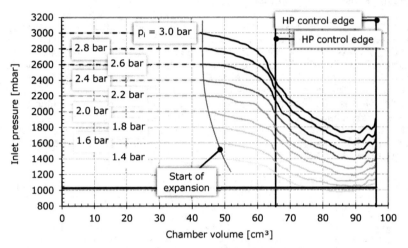

Fig. 8: Measured indicator diagrams for various inlet pressures
[n = 10,000 min^{-1} and ϑ_i = 75 °C]

Fig. 8 shows the pressure curves as a function of the chamber volume in the case of an inlet pressure variation between p_i = 1.4 and 3.0 bar. At a constant rotational speed and in good approximation, raising the inlet pressure generally leads to a linear increase of the internal work and thus of the internal power. In their detail however, pressure curves are different in terms of the start of expansion and progression. In the case of low inlet pressures, the chamber pressure remains at the high pressure level over a larger rotation angle. The throttling losses rising with the square of the flow velocity and linearly with the inlet pressure (density) explain these different times of expansion beginning. Over the progression, the pressure curves with higher inlet pressures show a steeper dropping of pressure in the range of the high-pressure-side control edge. The increasing gap mass flows out of the working chamber are responsible for this. The local pressure minimum in the middle part of the expansion phase is already reached at smaller chamber volumes in the case of low inlet pressure. This is on the one hand a consequence of smaller refilling due to the lower pressure ratios to the subsequent chamber and on the other hand a consequence of the smaller gap mass flows moving out of the chamber. Due to the refilling, the chamber pressures rise in the final expansion phase for all investigated inlet pressures until the low-pressure-side control edges are reached, before dropping to the level of the back pressure during the discharge procedure.

The variation of the rotational speed also shows a significant influence on the chamber pressure curve (Fig. 9). Under ideal conditions, the output power of the expander rises linearly with the rotational speed. Due to the shorter duration of the operating cycle as the rotational speed rises, and the throttling effect of the inlet area that increases as a result, an expansion starting increasingly earlier (i.e. before reaching the high-pressure-side control edges) occurs. The measured output power of the expander therefore declines over the rotational speed (Fig. 4). This behaviour is also explained by the decreasing gap mass losses as the rotational speed increases. This causes a shift of the local pressure minimum in the middle range of the expansion towards larger chamber volumes and thus causes a reduction of the refilling. The working area in the indicator diagram accordingly becomes smaller as the rotational speed increases. The chamber pressure at the end of the expansion procedure, right before the low-pressure-side control edges are reached, makes the significant influence of the refilling on the energy conversion quality clear.

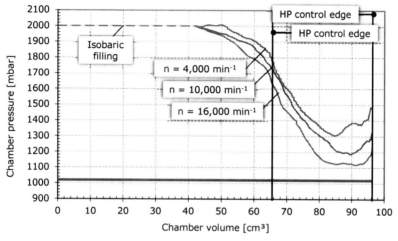

Fig. 9: Measured indicator diagrams for various rotational speeds
[p_i = 2 bar and ϑ_i = 75 °C]

5. SUMMARY

Using extensive experiments (e.g. pressure indication) as well as thermodynamic simulations on the basis of a multi-chamber model, the technical and physical processes within a screw expander during the conversion of thermal into mechanical energy are described in detail. It is shown that, in contrast to the ideal process (isobaric filling, isentropic expansion, isochoric-isobaric discharge), a significant dropping of the chamber pressure takes place in the real process before the high-pressure-side control edge is even reached, particularly due to throttling losses during inflows into the working chamber. This premature starting of the expansion becomes more premature as the rotational speed rises and the inlet pressure increases. In terms of gap priority, the outflowing gap mass flows via the profile mesh gap dominate first at the beginning of chamber filling. However, the housing gap becomes continually more significant, meaning that the mass percentages outflowing via the housing gap essentially contribute to a pressure drop in the working chamber towards the end of chamber filling and upon the real expansion beginning. The leading working chamber is refilled in this way however, meaning that the chamber pressure in the second expansion phase increases in spite of the increasing volume and therefore makes the indicated work greater. The positive effect of refilling becomes smaller as the rotational speed increases however, since the fluid mass exchanged between the chambers decreases as a result of the shorter operating cycle times.

REFERENCES

(1) Hütker, J.; Brümmer, A.: "A comparative examination of steam-powered screw motors for specific installation conditions". 8. VDI-Fachtagung Schraubenmaschinen 2010, VDI Bericht 2101, S. 109-123
(2) Brümmer, A.; Hütker J.: "Influence of geometric parameters on inlet losses during the filling process of screw-type motors". Developments in mechanical engineering, vol. 4, pp. 105-121
(3) Temming, J. "Stationärer und instationärer Betrieb eines unsynchronisierten Schraubenladers". Dissertation, TU – Dortmund, 2005
(4) Janicki, M.: "Modellierung und Simulation von Rotationsverdrängermaschinen". Dissertation, TU – Dortmund, 2005

3D CFD analysis of a twin screw expander

A Kovacevic, S Rane
City University London,
Centre for Positive Displacement Compressor Technology, UK

ABSTRACT

Twin screw machines can be used as expanders for variety of applications. This paper describes how the performance of an oil free twin screw air expander of 3/5 lobe configuration was estimated by use of full 3D Computational Fluid Dynamics (CFD) applying a procedure similar to that used for screw compressors. The grid generator SCORG© was employed for pre-processing of the moving domains between the rotors while the stationary grids for the ports were derived from a commercial grid generator. Flow calculations were carried out using the ANSYS CFX® solver. Pressure-angle diagrams, mass flow rates and expansion power at different operating conditions were estimated and compared with experimental test results. The overall performance predictions obtained by simulation agreed very well with measured data. It was concluded that correct design of the high pressure port is vital in order to obtain optimum power output. Leakage flows have a significant effect on screw expander efficiency and have the greatest influence at low speeds and high filling pressures.

ABBREVIATIONS

CFD – Computational Fluid Dynamics
FVM – Finite Volume Method
FEM – Finite Element Method
GGI – Generalized Grid Interface
SCORG© – Screw Compressor Rotor Grid Generator
ORC – Organic Rankine Cycle

1 INTRODUCTION

Twin screw expanders are positive displacement machines that can be used to recover mechanical power from fluids. Their expansion efficiency depends on several factors that include the leakage flows, the size and shape of the high pressure port and the volume ratio of the machine, as reported by *Smith et al.* (1). The analysis of such machines has been reported in several papers by *Stosic, Smith and Kovacevic* (2), (3) using a quasi 1D thermodynamic model. A detailed review of the methods used in the mathematical modelling of screw machines has been presented by *Stosic et al.* (4). *Bruemmer* and *Hutker,* (5), (6), more recently used a zero dimensional chamber model, based on the conservation of mass and energy, to predict the performance of such machines, in order to evaluate the influence of internal volume ratio V_i, length to diameter (L/D) ratio, rotor wrap angle $\varPhi_w$, size of leakage path and some other parameters. Similarly, *Nikolov et al.* (7), estimated the influence of rotor and casing thermal deformation on the performance of a screw expander in an ORC system, based on the iterative coupling of a thermodynamic chamber model and FEM thermal analysis.

© The author(s) and/or their employer(s), 2013

The use of three dimensional transient CFD analyses of twin screw compressors is based on the conservation of energy, momentum and space for deforming domains by use of the finite volume method, the principles of which are extensively elaborated in (8). The application of this procedure in positive displacement screw machines has been pioneered by Kovacevic and has been presented in (9), (10) and (11). The main challenge in the application of such methods is the construction of the grids required for the deforming rotor domains. The authors are only aware of SCORG©, the software developed in house, as a tool available for this purpose and for this reason it has been utilized in the present study. *Kovacevic et al., (10)* presented a numerical simulation of a combined screw compressor–expander machine for use in high pressure refrigeration systems. *Kethidi et al., (12),* conducted further studies on the influence of turbulence modelling on the CFD predictions of local velocity fields in twin screw compressors. Optimization of the discharge port area based on flow behaviour in the discharge chamber has been the main subject of research performed by *Mujic et al., (13)* and *Pascu et al., (14)*.

GL-51.2 is a dry running, unsynchronized twin screw machine designed at TU Dortmund, (6). It has been tested for use as an oil free expander of high pressure air at various operating conditions. The experimental data, obtained from this machine, has been used in the present study to validate the use of 3D transient CFD calculation procedures to predict expander performance.

2 CFD ANALYSIS OF A TWIN SCREW EXPANDER

In a compressor, the rotation of rotors displaces the gas from the low pressure port to the high pressure port with gradual decrease in the contained volume. Expander screw rotors rotate in the opposite direction to compressor rotors in such a way that the gas is displaced from the high pressure port to the low pressure port with gradual increase in volume. Since these two processes more or less mirror each other, the same CFD analysis calculation procedures used for screw compressors, can be applied to expanders. The process of obtaining performance predictions starts with the extraction of the fluid flow domains from a 3D CAD model of the expander components, namely the male and female rotors, casings, the high pressure and low pressure connection flanges and the pipes. The expander fluid domain is decomposed into four main regions namely, the male rotor flow domain, the female rotor flow domain, the suction flow domain and the discharge flow domain. The male and female rotor flow domains are separated by a unique plane defined by a rack as explained in (11). The rotor flow domains are mapped in SCORG© and integrated in the same software, with the numerical grids of the port domains generated by commercial software directly from CAD. This is followed by the simulation setup which comprises importing the integral numerical grid in the CFD solver and definition of the operating parameters and solution schemes. The post processing is usually performed within the CFD solver.

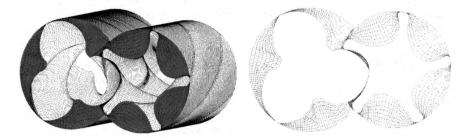

Figure 1. Numerical grid of the GL-51.2 screw expander rotors

GL 51.2 is the twin screw expander designed by TU Dortmund with 3/5 lobe rotor combination. The outer diameters of the male and female rotors are 71.847 mm and 67.494 mm respectively. The L/D ratio is 1.406 and the centre distance between the rotors is 51.222 mm. The male rotor wrap angle is $\varPhi_w = 200°$.

2.1 Grid Generation

SCORG© (11) was used to generate hexahedral grid in the rotor domain. The grid of the rotors used in this research is comprised of 40x8x40 divisions in the circumferential, radial and axial direction per interlobe respectively in the male and female fluid zones. The complete rotor grid for one of the cross section is shown in Figure 1.

The mesh in the high and low pressure ports is tetrahedral and generated by use of the ANSYS grid generator. These three domains were assembled together in the solver through non-conformal GGI interfaces. The connecting pipes were meshed by a hexahedral grid. All the domains were integrated to form a single working domain which represented all the flow domains, including the leakage paths through the interlobe and radial leakage gaps. Figure 2 shows the complete model of the analysed expander with all flow sub domains.

Initially, the end clearance leakage area was not included in the calculation but after preliminary evaluation it was noticed that this leakage area plays important role in the performance of the machine. Therefore a hexahedral numerical grid with 5 cells across its thickness was introduced in the model to account for that leakage.

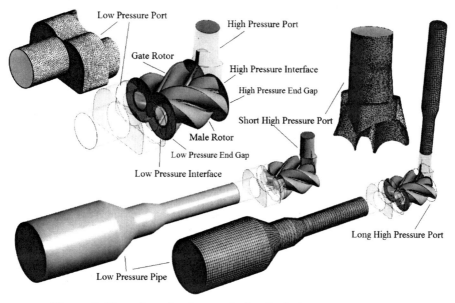

Figure 2. Flow domains for analysis of a twin screw expander

The numerical grid for screw compressor rotors was comprised of 213328 cells. The numerical grid for short ports accounted for 574813 cells while for the long ports it was 585761 cells. The overall number of computational numerical cells with short ports was 788141 while with the long ports it was 799089.

2.2 Simulation

A detailed description of the principles of the CFD simulation of flow in twin screw machines, based on grid generation with SCORG©, can be found in (11). Numerical grids for all time steps are generated in advance of the CFD calculation and are passed to the solver initially in the model setup. At each time step the mesh for the appropriate position of the rotors (Figure 1) is updated in the solver by use of an external subroutine. The solver is set with a higher order advection scheme and the second order backward Euler temporal discretization. The working fluid was air following an ideal gas law with a molar mass of 28.96 kg kmol^{-1}, Specific Heat Capacity 1004.4 J kg^{-1} K^{-1}, Dynamic Viscosity 1.831x10^{-5} kg m^{-1} s^{-1} and Thermal Conductivity 2.61x10^{-2} W m^{-1} K^{-1}.

The convergence criteria for all equations were set to 1.0x10^{-3} applying 10 coefficient loops for every time step. During solution, r.m.s residuals for all time steps achieved values between 1.0x10^{-3} and 5.0x10^{-3} for the momentum equation, while the values for the continuity and energy equations were always less than 1.0x10^{-3}. The calculations were allowed to run for a sufficient number of time steps in order to ensure that cyclic repetition of the flow and pressure values was achieved at the boundaries.

The analysis was carried out in two stages. The objective of the first stage was to identify the modelling parameters that influence the prediction of results from the CFD calculations and select those parameters that provide best results in comparison with experimental data. The main objective of the second stage was to calculate the machine performance, using a selected set of modelling parameters, obtained from the first stage, which best coincided with measured test results.

2.2.1 Stage I: Selection of modelling parameters

The accuracy of performance predictions with CFD model depends on many factors which are dependent on the level of assumptions introduced in the setup. The first phase in analysing this screw expander was to determine the influence of modelling parameters such as:

i) Size of clearance gaps,

Clearance gaps have a significant effect on the performance of screw machines. It was noticed in previous studies (11) that they affect the fluid flow much more significantly than the pressure distribution in the working domain. Three different clearance gaps were evaluated namely the interlobe, radial and end face. The design clearances of this machine were defined as 50-80 μm in the interlobes, 80 μm radially, 100 μm at the high pressure end and 250 μm at the low pressure end. Due to thermal deformation and bearing clearances, these gaps change during the operation of the machine. In order to evaluate the effects of individual gaps on the performance, a number of different values for each of these gaps were compared.

ii) Location of the domain boundary,

Figure 2 shows a model of the screw expander with short and long domains on the suction and discharge ends of the machine. It was expected that the length of the ports and location of boundaries might affect the solution.

iii) Type of boundary condition,

Due to the cyclic variation of pressure, temperature and other values of the flow in the compressor ports, it is difficult to define fixed conditions at the boundaries of the flow domain. Usually it is possible to define an "opening" boundary with a specified non-reflecting pressure head at the boundary location. This type of boundary permits the flow to enter and leave the computational domain but affects the solution, depending on the location of the boundary relative to the source of the pulsations. An alternative approach, proposed in (11), is to define a computational domain at the end of the machine flow domains which will account for a mass and energy source or sink to simulate constant conditions in the machine reservoirs. Such a source domain adds to or removes mass and energy from the gas so that pressure oscillations in the ports are damped and maintained at a given level.

iv) Flow regime, Laminar or Turbulent simulation,
Flow calculations assuming both laminar flow and the Spalart-Allmaras turbulence model have been checked in various test cases to evaluate their influence on the results. The Spalart-Allmaras turbulence model (12) is a one-equation model that solves a modeled transport equation for the kinematic eddy (turbulent) viscosity. It will be shown that the turbulence modelling influences the mass flow through the expander.

The designed operating speed range of the expander was from 2000 to 16000 rpm. Hence two points at 4000 and 10000 rpm were selected for *Stage I* calculations. Similarly the high end port pressure can range from 1.5 to 3.0 bar, hence a Pressure of 2.0 bar was selected. The results obtained were compared with the measured pressure variation inside the working chamber presented in a Pressure – Alfa diagram. Details of the cases analysed are given in Table 1.

Table 1. Cases evaluated for the analysis in *Stage I*

Case No	High Pressure Port	Male Rotor Speed		High Pressure Boundary	Turbulence	Clearance (um)			
		rpm	rpm			Interlobe	Radial	HP End	LP End
1	Short	4000	10000	Opening Pressure	Spalart-Allmaras	50	80	400	0
2	Short	4000	10000	Opening Pressure	Spalart-Allmaras	50	80	100	0
3	Short	4000	10000	Opening Pressure	Spalart-Allmaras	50	80	100	250
4	Short	4000	10000	Opening Pressure	Spalart-Allmaras	10	10	0	0
5	Long	4000	10000	Opening Pressure	Spalart-Allmaras	50	80	100	0
6	Short	4000	10000	Opening Pressure	Laminar	50	80	100	0
7	Long	4000	10000	Source Domain	Spalart-Allmaras	50	80	100	0

2.2.2 Stage II: Performance evaluation
In the second stage, the CFD models, with setup conditions selected from *Stage I*, were analysed to evaluate the performance of the expander over a wide range of operating conditions.

Experimental values of flow and power, together with the corresponding pressure history inside the machine, as obtained from the TU Dortmund expander test rig are given in (15). The relevant test results, thus obtained are presented in Table 2.

Table 2. Experimental test results for Expander GL 51.2

Speed	2000 rpm		5000 rpm		10000 rpm	
Pressure	Flow rate [kg/s]	Power [W]	Flow rate [kg/s]	Power [W]	Flow rate [kg/s]	Power [W]
1.6 bar	0.0292	511.93	0.0451	1318.84	0.0615	1748.06
2.0 bar	0.0327	758.50	0.0511	1817.65	0.0786	3444.82
3.0 bar	0.0409	1492.35	0.0737	3589.64	0.1179	7148.29

All measured points were taken at a high pressure air temperature of 350° K.

3 RESULTS AND DISCUSSION

3.1 Pressure – Alfa Diagram

A specimen Pressure – Alfa diagram, presented in Figure 3, shows the phases in the expansion process. The working cycle starts with filling, which is characterised by an increase in the volume formed between the rotors exposed to the high pressure port. This part of the process is characterised by pressure fluctuations resulting from the moving rotor and gas interaction. The inlet port of the machine has a relatively small area which results in throttling losses and a pressure drop in the expander inlet. The inlet area initially increases but it then decreases sharply, causing a further pressure reduction in the suction process which is defined as pre-expansion in the diagram.

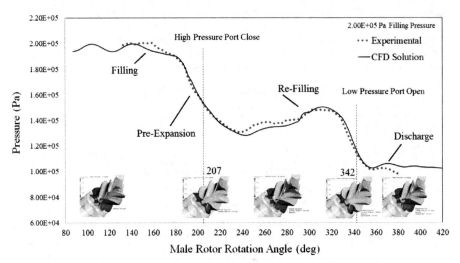

Figure 3. Diagram of indicated pressure in a Twin Screw Expander

Pre-expansion reduces the pressure and hence the generated power. Useful work is produced during the expansion of the gas trapped in the enlarging working chamber. As expansion proceeds, the mass in the working chamber increases due to leakage flow from adjacent chambers which are at higher pressure. If the leakage paths are sufficiently large, this gain in mass can induce a pressure rise in the working chamber, despite its increasing volume. This process is referred to as re-filling and affects the expander power output. The final phase of the process is discharge. Ideally, this should start with the pressure in the working domain near the pressure downstream of the machine.

3.1.1 Influence of the clearance size on the performance of the expander

Figure 4 shows the P-Alfa diagram for an inlet pressure of 2.0 bar and an expander speed of 4000 rpm, with various sizes of clearance gaps. It is shown that the size of clearance gaps does not significantly affect pressure during the filling, pre-expansion and discharge phases. However, it has a significant effect on the re-filling process. The nominal clearances specified in the design of this machine are: Interlobe 50 μm, Radial 80 μm and Axial 100 μm. To evaluate their influence on the performance, the following cases were calculated. In Case 1, both the interlobe and the radial clearances were assumed to have their design value, but the axial clearance on the high pressure side was assumed to be 400μm. This case, with deliberately over-specified clearances, produced a P-Alfa diagram close to that of the experiment, but resulted in a very high mass flow rate and a very low indicated specific power. The small clearances assumed in Case 4 inhibited the leakage and thus reduced the mass flow through the expander, thereby increasing the specific power output. The design values were assumed for interlobe, radial and high pressure axial clearances in Case 2 but zero axial clearance was taken for the low pressure end. This predicted the mass flow rate and specific power well but under predicted the pressure in the re-filling phase. Case 3 with the same interlobe and radial gaps as in Case 2, but with a 250 μm axial low pressure clearance gap also predicted a similar re-filling pressure, with a nearly equal mass flow rate. These cases and several others, not presented in the paper in order to maintain clarity, confirmed that the axial clearance on the high pressure end makes a significant contribution to the direct leakage which increases the mass flow rate through the machine. The axial clearance on the low pressure end, although contributing to the internal leakage, does not affect the mass flow rate through the machine.

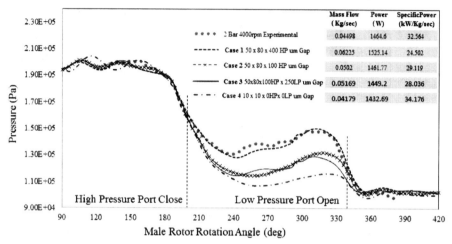

Figure 4. Influence of clearance variation on P-Alfa at 4000rpm

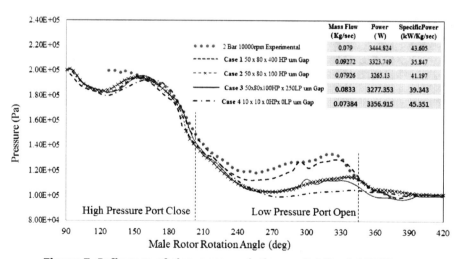

Figure 5. Influence of clearance variation on P-Alfa at 10000rpm

The diagram for 2.0 bar filling pressure and 10000rpm is shown in Figure 5. It is clear from this that the influence of the clearance gaps is larger at lower speeds but it was also noticed that the increase in the interlobe and radial leakage gaps significantly increases mass flow rate through the machine. The design clearance gaps in Case 2 were selected for further analysis of the machine performance in stage II since they gave the best agreement with the test results in terms of predicting mass flow rate, indicated power and the P-Alfa diagram at both speeds.

As usual, the area under the pressure-volume diagram gives the indicated power, which has been used for comparison in this paper. Although the measured and estimated mass flow rates and indicated powers in Case 2 agree well, the measured refilling pressure is higher at both speeds. This can be explained by the dominant influence of the high pressure region on the indicated power due to the steep change in trapped volume and high pressure level. Therefore it appears to be vital to select operating clearances which will result in an accurately predicted mass flow rate rather than to closely match the pressure curve in the refilling process. It appears that these clearances coincide with the design clearances. However, during the operation of the machine, these clearances will certainly change and will result in different machine performance. The authors are of the opinion that some of the clearance gaps change in such a way that the refilling process does not affect the main mass flow as it occurs after the high pressure port is closed. Further investigation by use of Fluid-Solid Interaction studies may be useful to validate this theory.

3.1.2 Influence of the length of high pressure ports on performance

The P-Alfa diagrams shown in Figure 6 are obtained with different lengths of high pressure ports for the 4000 and 10000rpm studies. The arrangements for the long and the short ports are shown in Figure 2. The long ports place the boundary condition far upstream of the inlet to the rotor filling. This produces pressure fluctuations in the filling process. These fluctuations are not noticeable in the experimental results. On the other hand, the short ports in which the boundary is placed much closer to the expansion chamber provide better predictions of the filling and pre-expansion phase and this results in a much closer match with the test results. Hence the short high pressure ports were selected for *Stage II* calculations.

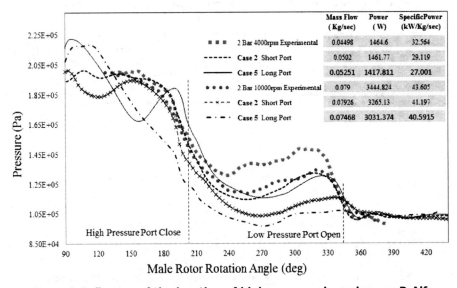

Figure 6. Influence of the location of high pressure boundary on P-Alfa

3.1.3 Setting boundary conditions for the high pressure inlet

Previous authors (11) described the use of a boundary domain for providing quick and stable calculation of the performance of screw compressors. A similar approach was applied to evaluate the expander performance and was compared with non-reflecting pressure boundary conditions at the high pressure opening. Figure 7 shows a comparison between the estimated P-Alfa diagrams and the test results, with two different types of boundary conditions at the high pressure inlet side. The introduction of a source domain caused damping of the pressure fluctuations in the filling stage. Since non-reflecting pressure boundary conditions at the short port openings gave better predictions of both power and mass flow, these were selected for all further performance calculations.

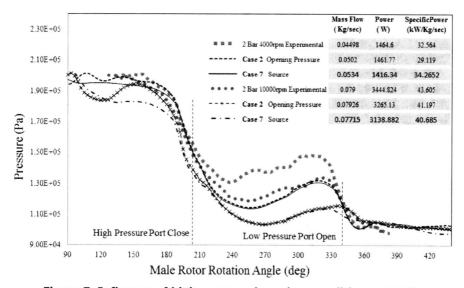

Figure 7. Influence of high pressure boundary condition on P-Alfa

3.1.4 Influence of the Turbulence model

Figure 8 shows P-Alfa diagrams of predicted performance at 4000 and 10000 rpm for both laminar flow and turbulent flow, based on the Spalart-Allmaras model, compared with test results. The inclusion of turbulence reduces the leakage flows and increases the power, thus indicating that it plays a role in clearance flows. Based on the values of power and flow, also given in Figure 8, the Spalart-Allmaras turbulence model was used for all further calculations. Comparing the results in Figure 8 with those in Figure 4, it can be seen that leakage flow and power output are far more sensitive to the size of the clearance gaps than to turbulence effects since the Pressure-Alfa diagram is significantly affected by clearance gap changes but only slightly by the inclusion of turbulence.

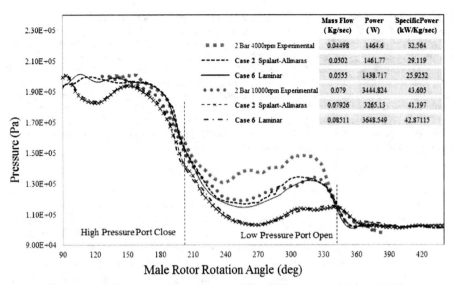

Figure 8. Influence of Laminar and Turbulence model on P-Alfa

3.2 Mass flow rate and Indicated Power

Estimated mass flow rates and indicated power outputs for the cases studied, compared with measured values, are shown in Figure 9 and 10 respectively. These were obtained with the same modelling conditions i.e. short high pressure port, opening pressure boundary condition, the Spalart-Allmaras turbulence model and a clearance distribution of 50μm interlobe, 80μm radial and 100μm axial clearance on the high pressure side and no axial gap on the low pressure side. These show good agreement for Case 2 studies at 4000rpm and 10000rpm. These selected modelling parameters were used for further performance evaluation carried out at operating conditions listed in Table 2.

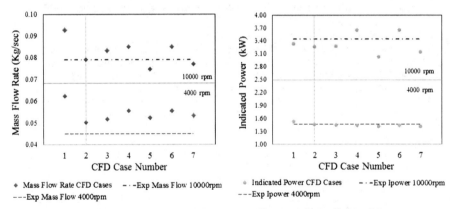

Figure 9. Mass Flow Rate vs CFD Case variants

Figure 10. Indicated Power vs CFD Case variants

3.3 Evaluation of overall Performance characteristics

The performance of the expander can be evaluated by comparing the indicated power and mass flow rate with test results at the operating speeds and operating pressures given in Table 2. The results obtained from *Stage II* CFD calculations are presented in Table 3 for the respective operating conditions.

Table 3. CFD model performance prediction of GL 51.2

Speed	2000 rpm		5000 rpm		10000 rpm	
Pressure	Flow rate [kg/s]	Power [W]	Flow rate [kg/s]	Power [W]	Flow rate [kg/s]	Power [W]
1.6 bar	0.0297	435.531	0.0425	993.606	0.0621	1690.72
2.0 bar	0.0395	759.192	0.0551	1793.303	0.0794	3265.13
3.0 bar	0.0622	1594.74	0.0852	3877.391	0.1215	7361.67

All calculated points assume the high pressure air temperature to be 350° K.

The estimated performance characteristics of the expander in terms of indicated power and mass flow rate versus male rotor speed at different filling pressures are shown in Figures 11 and 12 respectively. The trends obtained by CFD modelling vary progressively and follow those of the experimental results.

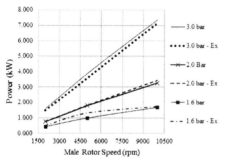

Figure 11. Power vs Speed at different filling pressures   **Figure 12. Mass Flow Rate vs Speed at different filling pressures**

These characteristics represent the following:
- The indicated Power increases with both higher rotor speeds and filling pressures. Also, the Specific Power increases at higher pressures. CFD Model predictions of power are relatively over estimated at higher pressure.
- The Mass flow rate increases with both higher speeds and filling pressures, due to the higher gas density in the filling port. CFD Model mass flow rate predictions are very accurate at low pressure but are over estimated at high pressures and Low Speed conditions, where the influence of leakage is higher. This region is highlighted in Figure 12 and further investigation of the CFD models is required to get better results in this regime.

4 CONCLUSION

Transient 3D CFD analysis of a twin screw expander has been successfully carried out using the SCORG© grid generator and ANSYS CFX Solver. In the model setup stage, experimental Pressure-Alfa diagrams were used as a guideline for the selection of appropriate modelling parameters for CFD calculations. In the performance study, a set of different pressure ratios, with different operating speeds at fixed inlet gas temperature were used to validate the established CFD model.

- It is possible to apply the modelling procedures, developed for the analysis of twin screw compressors, to twin screw expanders.
- It is important to define the appropriate geometry and position of the pressure boundary of the high pressure port in order to enable adequate pressure fluctuations and throttling in the port. Further investigation is necessary to generalise this procedure.
- The inclusion of turbulence modelling in CFD calculations does not significantly affect the pressure history within the machine, but it does influence the gap flows and hence the predicted mass flow rate and specific power. The Spalart-Allmaras turbulence model was used in this study but further investigation will need to be performed in order to determine the best turbulence model for screw expanders.
- Clearance distribution in interlobe, radial and axial gaps has a strong influence on the performance of the expander. It is difficult to predict their exact operating values. These were obtained by iterative trials in this case. A comprehensive Fluid-Solid interaction study would help to better understand their dynamic behaviour.

The procedure for the analysis of screw expander developed in this research can be used in future to optimise all aspects of screw expander performance and improve its design.

REFERENCE LIST

(1) Smith I.K, Stosic N, Aldis C A. (1996), *Development of the trilateral flash cycle system, Part3: design of high-efficiency two phase screw expanders*, Proceedings of IMechE, Part A: Journal of Power and Energy, Volume 210.

(2) Smith I.K, Stosic N, Kovacevic A. (2004), *An Improved System for Power Recovery from High Enthalpy Liquid Dominated Field*, GRC Annual Meeting, Indian Wells, California.

(3) Stosic N., Smith I. K., Kovacevic A. (2002) *A Twin Screw Combined Compressor and Expander for CO_2 Refrigeration Systems*. Proc. Int. Compressor Conf. at Purdue, pp. C21-2.

(4) Stosic N., Smith I.K., Kovacevic A. (2005) *Screw Compressors: Mathematical Modeling and Performance Calculation*, Springer Verlag, Berlin.

(5) Brummer A., Hutker J. (2009) *Influence of geometric parameters on inlet-losses during the filling process of screw-type motors*. Developments in mechanical engineering, vol. 4, pp. 105-121.

(6) Hutker J., Brummer A. (2012) *Thermodynamic Design of Screw Motors for Constant Heat Flow at Medium Temperature Level*. Proc. Int. Compressor Conf. at Purdue, pp. 1478.

(7) Nikolov A., Huck C., Brummer A. (2012) *Influence of Thermal Deformation on the Characteristics Diagram of a Screw Expander in Automotive Application of Exhaust Heat Recovery*. Proc. Int. Compressor Conf. at Purdue, pp. 1447.

(8) Ferziger J. H., Peric M. (1996) *Computational Methods for Fluid Dynamics*, Springer, Berlin, Germany, Berlin.

(9) Kovacevic A., Stosic N., Smith I. K. (2003) *3-D Numerical Analysis of Screw Compressor Performance*. Journal of Computational Methods in Sciences and Engineering 3: 259-284.

(10) Kovacevic A., Stosic N., Smith I. K. (2006) *Numerical simulation of combined screw compressor–expander machines for use in high pressure refrigeration systems*. Simulation Modeling Practice and Theory 14: 1143–1150.

(11) Kovacevic A., Stosic N., Smith I. K. (2007) *Screw compressors - Three dimensional computational fluid dynamics and solid fluid interaction* Springer-Verlag Berlin Heidelberg New York.

(12) Kethidi M., Kovacevic A., Stosic N., Smith I. K. (2011) *Evaluation of various turbulence models in predicting screw compressor flow processes by CFD*. 7th

International Conference on Compressors and their Systems, City University London.
(13) Mujic E., Kovacevic A., Stosic N., and Smith I. K., (2008) *The influence of port shape on gas pulsations in a screw compressor discharge chamber*, Proceedings of IMechE, Vol. 222 Part E: Journal of Process Mechanical Engineering JPME205.
(14) Pascu M., Kovacevic A., Udo N. (2012) *Performance Optimization of Screw Compressors Based on Numerical Investigation of the Flow Behaviour in the Discharge Chamber*. Proc. Int. Compressor Conf. at Purdue, pp. 1145.
(15) Hütker J., Brümmer A., (2013) *Physics of a dry running unsynchronized twin screw expander*, International Conference on Compressors and their Systems, City University London.

Sizing models and performance analysis of volumetric expansion machines for waste heat recovery through organic Rankine cycles on passenger cars

L Guillaume, A Legros, S Quoilin, S Declaye, V Lemort
Thermodynamics Laboratory, University of Liège,
Aerospace and Mechanical Engineering Department, Belgium

ABSTRACT

This paper aims at helping designers of waste heat recovery organic (or non-organic) Rankine cycles on internal combustion engines to best select the expander among the piston, scroll and screw machines, and the working fluids among R245fa, ethanol and water. The first part of the paper presents the technical constraints inherent to each machine through a state of the art of the three technologies. The second part of the paper deals with the modeling of such expanders. Finally, in the last part of the paper, performances of the various Rankine systems are compared and a decision array is built to select the most appropriate couple of fluid and expander.

NOTATIONS

Symbols

A	Area	[m²]
cf	Dead volume factor	[-]
f	factor	[-]
h	Specific enthalpy	[J/kg]
$\dot{M}$	Mass flow rate	[kg/s]
N	Rotational speed	[RPM]
P	Pressure	[Pa]
$\dot{Q}$	Heat flow rate	[W]
r_v	Volume ratio	[-]
s	Specific entropy	[J/kg.K]
t	Temperature	[°C]
T	Torque	[N.m]
U	Heat transfer coefficient	[W/m².K]
v	Specific volume	[m³/kg]
V	Volume	[m³]
V_s	Displacement	[m³]
w	Specific work	[J/kg]
$\dot{W}$	Power	[W]

Subscripts

0	Recirculated
a	Aspired
adm	Admission
amb	Ambiant
c	Compression
calc	Calculated
cp	Compressor
ex	Exhaust
exp	Expander
in	Internal
leak	Leakage
loss	Loss
meas	Measured
r	Residual
ref	Reference
sh	Shaft
su	Supply
thr	throttle

© The author(s) and/or their employer(s), 2013

Greek letters
- α   Proportional losses factor   [-]
- ε   Efficiency   [-]
- ΔP Pressure difference   [-]
- ρ   Density   [kg/m³]

INTRODUCTION

The interest in organic Rankine cycles for waste heat recovery on internal combustion engines has grown significantly for the past few years. Indeed, in such engines, only about one third of the energy available is actually converted into effective power, what remains being dissipated into heat. Therefore, since it becomes really challenging to increase the engine efficiency itself, solutions that focus on the recovery of this waste heat are increasingly investigated to improve the energetic efficiency of vehicles. Among these solutions, Organic Rankine Cycle systems are particularly appropriate.

The adoption of such technology in the automotive domain requires a specific R&D activity to select and develop the components and identify the most appropriate system architecture. Particularly, the selection of the working fluid and of the expansion machine technology constitutes an important part of this research. Once identified the technical constraints inherent to each machine (rotational speed, pressure ratios, maximum temperatures, volumetric expansion ratios, etc.) and the performance reached in the scientific literature, a way to achieve the selection is to develop simulation models of the expanders. The performance of the Rankine system integrating these expander models could then be compared for various working fluids.

This paper attempts to address the problematic of selecting the expander and the working fluid for a waste heat recovery organic (or non-organic) Rankine cycles on a particular internal combustion engine. It focuses especially on three technologies: the scroll, the piston and the screw expanders. Displacement expanders are generally preferred because of their lower rotational speeds compared to turbines and their ability to operate under large pressure ratios. Three working fluids: R245fa, ethanol and water are also investigated.

1   STATE OF THE ART

This state of the art aims at highlighting the limitations in terms of rotational speed, pressure and temperature. It is based on information available in open access scientific literature and data coming from manufacturers as well as on the experience gained at the Laboratory of Thermodynamics of Liège in the field of expansion machines.

More detailed information can be found in [1].

1.1   Displacement and rotational speed

In theory, the displacement and rotational speed of piston expanders can reach values similar to those obtained for internal combustion engines. In Exoes [2], for example, the rotational speed of these machines usually varies between 500 and 6000 RPM. The upper limit of the speed may be due to mechanical constraints on the mechanisms of valve lifts. The importance of friction losses can also result in a limitation on the speed. The lower bound of the speed is relatively low, which can be explained by low internal leakage in this type of machine.

In practice, the range of scroll machine sizes is very broad and extends from mini-scroll of a few cubic centimeters to very large machine. Currently, the trend in the compressor industry is to increase their size. The compressor Emerson ZP725K [3] is an example since its cooling capacity is about 158kW. According to the values observed in the literature, the displacement of the scroll expander generally varies from 1.1 to 49 l /s. Regarding the rotational speed, it often varies within a range of values between 1000 and 4000 RPM. However, speeds around 10,000 RPM are reached with scroll compressors in vehicle air-conditioning applications.

Currently, screw expanders are often characterized by high capacities. Thus expanders whose capacity varies between 25 and 1100 l / s for produced powers of around 20kWe to 1 MWe can be found in the literature. There are very few studies on mini-screw expanders. Regarding the speed of these machines, it can reach relatively high values, up to 25,000 RPM.

1.2 Inlet temperature

Piston expanders can tolerate high inlet temperatures beyond 500 °C [4].

On scroll compressors, limitation relates to the discharge temperature. The maximum discharge temperature is usually between 154 and 160°C for hermetic compressor to avoid excessive thermal expansion of the machine and a potential degradation of the lubricant. In expander mode, an inlet temperature of 165 °C was achieved with air [5] and a temperature of 215 °C was reached with water vapor [6].

Screw expanders can be exposed to high inlet temperatures. For example, Hutker considered a value of 490 °C with water vapor [7].

1.3 Built-in volume ratio

The built-in volume ratio of the piston expander can also reach high values similar to those obtained with internal combustion engines. However, for a given displacement, an increase in the built-in volume ratio yields a decrease in specific work produced by the machine and a relative increase in friction losses. This is why, in practice, the value of built-in volume of piston machines usually varies between 6 and 14.

In refrigeration, the built-in volume ratio of scroll machines generally varies in a range between 1.5 and 3.5. Beyond a built-in volume ratio of 3.5, it becomes difficult to ensure adequate performance of the machine for a reasonable manufacturing cost. It is indeed difficult (and not economically viable) to maintain a good lateral seal between the spirals while increasing the number of pairs of contact points. However, in the literature, machines with built-in volume ratio greater than 4 exist (e.g., a scroll with a built-in volume ratio of 5.25 was proposed by Air Squared). They are often air compressors adapted to be tested as expanders.

In literature the built-in volume ratio of the screw machines is often limited to values of 4 or 5. Nevertheless, some studies show values of up to 8 [8].

1.4 Pressure ratio

The pressure ratio of the piston expanders could reach, by analogy to the displacement, similar values to those obtained by internal combustion engines.

The limitation on the pressure ratio of scroll machines is usually a consequence of the maximum allowable outlet or inlet temperature depending on whether the machine is operating in compressor or expander mode respectively. Thus, in refrigeration, the pressure ratio can reach a value of 11. In expander operation, a pressure ratio of 15 was achieved with water vapor [6]. The Eneftech company is working on the development of an expander capable of operating under a pressure ratio of 25 [5].

Screw expanders can be subjected to high pressure ratios. For example, Nikolov considers a value of 50 with ethanol [9].

2 EXPANDER MODELS

Semi-empirical simulation models of 3 different expansion machines were built into the Engineering Equation Solver (EES) software [10].

The proposed simulation models retain the most important physical phenomena inherent to the expansion machine and involve a limited number of parameters (~ 10) which are identified on the basis of performance points. Studies in laboratory showed that these models are able to predict the performance of the expansion machines with good accuracy. Moreover, their semi-empirical nature enables to extrapolate the performance of the machine for different operating conditions and design characteristics (displacement, sections of the inlet and exhaust ports, etc...).

2.1 Scroll and screw expanders model

The model used is the one proposed by Lemort and al. [11] which has been validated for the scroll expander (Figure 1). The different processes and losses occurring during the passage of fluid through the machine are described below.

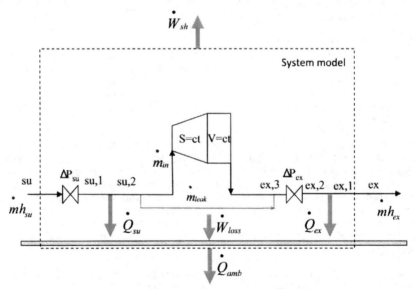

Figure 1: Volumetric expander model

2.1.1 Description of the model

During the admission phase, the fluid undergoes a pressure drop as it enters through the inlet port. This pressure drop is modeled through an equivalent nozzle.

This supply pressure drop is followed by heat transfer losses when the fluid comes in contact with the wall of the expander. This heat exchange is assumed to occur between the fluid and fictitious isothermal wall.

After that, a part of the fluid flow rate is assumed to pass through the expansion machine directly from the inlet to the outlet without interacting with the mobile part and therefore without producing any mechanical power. The leakage path is modeled by means of a fictitious nozzle whose cross section is A_{leak}. The internal flow rate is thus defined as follow.

$$\dot{M}_{in} = \dot{M} - \dot{M}_{leak}$$

The fluid is then subjected to expansion itself. The first part of this expansion is assumed to be an isentropic expansion from the inlet pressure to the pressure adapted to the built-in volume ratio of the machine. The second part of the expansion is assumed to be a constant machine volume expansion from the adapted pressure to the exhaust pressure (plus the pressure drop). This is actually a lamination of a part of the fluid caused by the contact of the old expansion chambers with the exhaust line.

The leakage flow is then mixed with the internal flow from the constant machine volume expansion. The enthalpy of the mixing is calculated using an energy balance, the process being assumed isobaric.

The fluid is finally discharged and it undergoes a pressure drop. The model used for the pressure drop at the outlet is similar to that used for the pressure drop at the inlet. A section of the throat of a fictitious nozzle $A_{thr,ex}$ is identified using measurement points.

During the discharge phase are also expected losses by heat transfer between the fictitious isothermal wall and the working fluid. The model used is analogous to that used for the supply heat transfer losses. A nominal transfer coefficient $AU_{ex,n}$ is identified by means of experimental data.

In the model developed by Lemort, the shaft power produced by the expander consists of three terms:
- The internal power produced by the expansion: $\dot{W}_{in} = \dot{M}_{in} * (w_{exp,1} + w_{exp,2})$
- The constant mechanical losses: $\dot{W}_{loss,0} = 2 * \pi * N * T_{loss}$
- The losses proportional to the internal power: $\alpha * \dot{W}_{in}$

Both parameters T_{loss} (mechanical torque) and α are determined experimentally, and N is the rotational speed of the expander.

Finally, the temperature of the fictitious isothermal wall is determined by means of a heat balance.

The coefficient of heat transfer between the wall and the atmosphere AU_{amb} is determined through experimental measurements.

In conclusion, there are finally 8 parameters to identify.

2.2 Piston expander model
A piston expander model is also proposed. It is based on the model developed in [12]. The same phenomena as those described above are expected to occur within this machine. The model is therefore composed of the same eight basic parameters.

However, in piston expanders, part of the flow going through the machine is re-compressed during the ascension of the piston in the cylinder. The description of this phenomenon increases the total number of parameters of the model.

As for the expansion, the recompression is assumed to occur in two stages, one isentropic compression until the adapted pressure followed by a constant machine volume compression.

A part of the available power is lost during the recompression and, similarly to the expansion, assumed to result from two contributions corresponding to the two compression types.

The internal flow rate is here defined taking into account the re-compressed mass flow rate as follow.

$$\dot{M}_{in} = \dot{M} + \dot{M}_0 - \dot{M}_{leak}$$

In the case of piston engines, durations for intake and discharge may be defined by means of two parameters, f_a and f_c respectively, expressed as function of the displacement of the engine and shown in the P-V diagram below.

In this figure, c_f is a factor representative of the dead volume, f_{adm} determines the volume of fluid in the cylinder at the moment when the intake valve closes. These parameters are linked by the following relations:

$$f_{adm} = c_f + f_a$$
$$f_r = 1 - f_{adm}$$

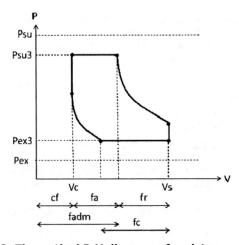

Figure 2: Theoretical P-V diagram of a piston expander

From these parameters can be defined the volume ratios of the machine in expander and compressor modes.

$$r_{v,in,exp} = \frac{1}{f_{adm}}$$

$$r_{v,in,cp} = \frac{c_f}{1 - f_c}$$

2.3 Models calibration
As mentioned above, the different parameters involved in the models are identified using measurements performed on real machines.

This identification is realized by minimizing the error between the predicted and measured values of the shaft power of the expander, of the outlet temperature of the expander and of the fluid mass flow rate.

In general, measurements are only available for a limited number of operating points. The model can therefore only be calibrated and validated in the range of the corresponding operating conditions. Outside this range, the bound on the error between the predicted and measured values is a priori unknown.

However, in this study, the assumption is made that the model can be extrapolated outside the range in which it has been validated, and the same set of parameters is used anyway. This approach is justified by the physical meaning of the parameters.

2.3.1 Screw expander

The screw expander model was calibrated on a machine running with Solkatherm. Figure 3 shows the predicted shaft power as a function of the power measured during tests on the real machine after calibration. The obtained parameters are as follows.

Table 1: Parameters of the screw expander model

$A_{thr,su}$ [m²]	AU_{su} [W/K]	A_{leak} [m²]	$A_{thr,ex}$ [m²]	AU_{ex} [W/K]	α [-]	T_{loss} [Nm]	AU_{amb} [W/K]
0.000128	10	0.000006151	0.001099	25	0.3	2.155	9.4

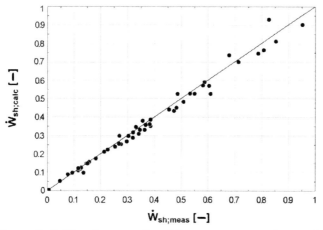

Figure 3: Calibration results of the screw expander model

2.3.2 Scroll expander

In this case, the measurements were performed in the laboratory of Liège with air and water for different operating conditions on an expander 38 cm³ and internal volume ratio of 4.1.

The various parameters of the model have been published [11].

2.3.3 Piston expander

The expander piston model has been calibrated on a machine using water. A set of parameters is available in the literature [12].

2.4 Scaling of the parameters

Since the first task is to size the expansion machines to compare the performance of different Rankine systems that would suit for the application, it is obviously not possible to use directly the previous models of expanders.

One solution is to calibrate the models on the existing machines, as done in the previous section, and then use scaling relations for the parameters to adapt the models to the machines being currently sized.

Regarding the coefficients of convective heat transfer between the working fluid and the wall at the admission and discharge, they can be expected to vary as follows.

$$\frac{AU_i}{AU_{ref,i}} = \left(\frac{\dot{M}}{\dot{M}_{ref}}\right)^{0.8} \text{(Assumption of turbulent flow)}$$

Similarly, the throat sections of the fictitious nozzles can be assumed to vary with the square of the characteristic dimension of the expanders.

$$\frac{A_i}{A_{ref,i}} = \left(\frac{V_s}{V_{s,ref}}\right)^{\frac{2}{3}}$$

For the parameters related to mechanical losses, the proportional loss coefficient is retained while the mechanical loss torque is determined with the assumption that the ratio of the constant losses to the internal power is kept.

Finally, regarding the built-in volume ratio, the parameters of the existing machines were kept.

3 PERFORMANCE ANALYSIS

The expander models can then be integrated into complete cycle models. In fact, there are several possible architectures of Rankine systems depending on the heat sources available on the vehicle. In the frame of this paper, only the basic architecture, consisting in recovering the waste heat of the exhaust gases, is considered. Thus the Rankine system is constituted of a single evaporator, a condenser, a pump and an expander as depicted in figure 4.

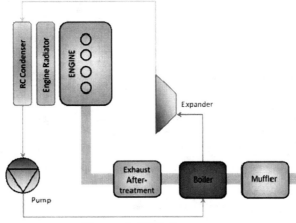

Figure 4: ORC diagram

Thus, the performance of the systems can be evaluated and compared for the three expanders and the three working fluids. This comparison is achieved in the following sections for nominal conditions of a car engine and then for a less loaded point of the engine in order to compare the performance of the different system at part load.

3.1 Nominal operating conditions

The Rankine cycle and the expanders are sized on the nominal engine operating conditions (table 2). This sizing is obtained by mean of an optimization calculation whose goal is to get for each expander technology and each working fluid the Rankine system that produces the best performance in terms of shaft power. This optimization calculation is out of the scope of this study. Basically, the evaporating pressure of the cycle and the rotational speed of the expanders are optimized, taking into account the limitations presented in the state of the art, to maximize the shaft power of the expander. These two optimal values, combined with the assumptions made on the other parameters of the cycle (fixed superheating, subcooling and condensing pressure), enable to size the various components of the Rankine system.

Table 2: Operating conditions used for the design of the systems

Exhaust gases temperature [°C]	550
Exhaust gases mass flow rate [kg/s]	0.025
Superheating [°C]	5
Subcooling [°C]	5

3.1.1 Results

The simulation results for nominal operating conditions are presented below. In the two first columns of table 2, for each couple of expander and working fluid, the values of the two optimization variables (the supply pressure and the rotational speed of the expanders) are shown. As it can be seen, the supply pressure is generally higher in the case of R245fa. Indeed, in the case of this fluid, the condensing pressure (or exhaust pressure of the expander) is relatively high compared to the one of the other fluids. Consequently, the evaporating pressure or supply pressure of the expander resulting from the optimization calculation is also higher in order to maintain a sufficient pressure ratio on the expander. Regarding the rotational speed, as it could be deduced from the state of the art, the piston expander rotates at a relatively low speed while the speed of the screw expander is much higher. The opposite behavior is observed for the displacements of the machine since, apart from the leakage flow rate, the flow rate, the displacement and the rotational speed of volumetric expanders can be linked by means of a linear relation. The last column of table 2 presents the obtained values of the isentropic efficiencies of the expanders.

Table 3: Simulation results in nominal regime

Exp. Type	Fluid	P_{su} [bar]	N [RPM]	P_{ex} [bar]	V_s [cm³]	ϵ_s [%]
Piston	R245fa	30	1600	3.43	12.9	60.57
	Water	21	2350	1	12.83	58.50
	Ethanol	18	1800	1	16.07	57.31
Scroll	R245fa	21.3	3600	3.43	6.26	65.14
	Water	15	7200	1	4.18	54.32
	Ethanol	15	7500	1	3.69	54.24
Screw	R245fa	30	13000	3.43	1.11	54.01
	Water	30	25000	1	0.73	45.61
	Ethanol	24	14500	1	1.2	48.1

The performance obtained for the different couples of expander and working fluid can be compared in terms of shaft power. In figure 5, it is observed that even if the piston expander coupled to R245fa seems to be the most appropriate couple for the application, the results are too close to take a decision. Therefore, simulations were performed again but for a less loaded engine operating point.

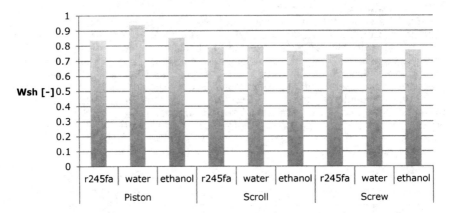

Figure 5: Predicted shaft power of the expanders in nominal regime

3.2 Part load conditions

A less loaded engine operating point yields a lower mass flow rate and a lower temperature of the exhaust gases of the engine. Simulations were performed under these conditions to compare the performance of the systems at part load. The simulation results are presented below.

3.2.1 Results

As for nominal results, table 3 presents the obtained values of the two optimization variables and of the isentropic efficiencies of the expanders. These values are obviously reduced but the same tendency is observed.

Table 4: Simulation results in part load regime

Exp. Type	Fluid	P_{su} [bar]	N [RPM]	P_{ex} [bar]	ϵ_s [%]
Piston	R245fa	17.2	525	3.43	22.93
	Water	10	575	1	23.55
	Ethanol	9.25	550	1	30.46
Scroll	R245fa	12.8	1000	3.43	51.83
	Water	6.2	1900	1	32.51
	Ethanol	8	2000	1	38.49
Screw	R245fa	23.5	2500	3.43	39.69
	Water	11	7000	1	24.15
	Ethanol	10.75	4700	1	35.74

Regarding the shaft power, results are compared in figure 6. As opposed to nominal results, best values are now obtained for the scroll and the screw expander when R245fa is used as working fluid. Therefore, part load results do not really help to achieve the selection and a new way has to be investigated.

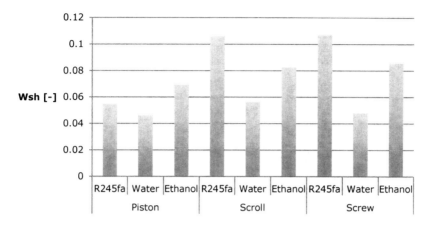

Figure 6: Predicted shaft power of the expanders in part load regime

4 QUALITY OF THE ASSUMPTIONS AND IMPROVEMENTS

Several assumptions were made to build the models of the three expander technologies and particularly for scaling the machines in order to adapt them to the considered waste heat recovery application. The strongest assumption is the variation of the throat sections of the fictitious nozzles with the square of the characteristic dimension of the expanders. Indeed, even if this assumption can be accepted for small variations of the size of the machines, it is not the case for larger variations.

The leakage area, which is one of the parameters that contributes the most to the losses, is submitted to this assumption. Therefore, this parameter can be misevaluated when the scaled machine is much larger than the initial one on which the parameters of the model had been calibrated. As a result, losses occurring in the machine can also be misevaluated and so can be the performance.

Therefore, further investigation should be achieved in order to identify how exactly varied this parameter with the size of the machine and thus predict more accurately the performance of the machine.

5 DECISION ARRAY

A way to make the selection feasible is to compare the machines and the fluids on the basic of more qualitative criteria such as the freezing point of the fluid or the compactness of the machine. In the following array, several criteria are investigated. A weight from 1 to 5 is attributed to each criterion and a rating from 1 to 5 is attributed to each couple of machine and working fluid for each of these criteria. The weighted sum taking into account all the criteria should then enable to select a fluid and a machine. Obviously, the list of criteria is non-exhaustive and the rating attributed for each criterion to each couple remains subjective. But, when achieved in a proper way by a sufficient number of experts, this solution should give good results.

In the array, it can be seen that the couple formed by the scroll expander and water, according to the selected criteria and the ratings and weights attributed, is the most appropriate solution.

Nonetheless, when criteria on the fluid are irrelevant, they can be set aside and the couple constituted of the scroll expander and R245fa is then the most appropriate.

Parameter	Weighting Factor (0 neglegtable - 5 important)	Technology Option							
		R245fa			Water			Ethanol		
		Scroll	Piston	Screw	Scroll	Piston	Screw	Scroll	Piston	Screw
Fluid	**3**	**104**	**104**	**104**	**116**	**116**	**116**	**92**	**92**	**92**
Thermal stability at operating temperatures	5	2	2	2	5	5	5	4	4	4
Low freezing point	5	5	5	5	1	1	1	5	5	5
Flamable risk	5	4	4	4	5	5	5	1	1	1
Environmental Point	5	3	3	3	5	5	5	3	3	3
Serviceability	2	1	1	1	5	5	5	3	3	3
Ability to operate open	2	0	0	0	5	5	5	0	0	0
Heat recovery potential	5	4	4	4	2	2	2	3	3	3
Corrosion (Aggresivness)	3	4	4	4	2	2	2	2	2	2
System (Expander)	**5**	**95**	**91**	**85**	**91**	**87**	**74**	**86**	**83**	**78**
Nominal Performance	5	3	4	3	3	4	3	3	3	3
Nominal Efficiency	3	4	4	3	3	3	2	3	3	2
Compacity	4	2	1	3	2	1	3	2	1	3
Weight	4	3	3	2	3	3	2	3	3	2
Rotating speed	3	5	5	4	4	5	2	4	5	3
System Efficiency	3	3	3	3	4	3	3	3	3	3
Nominal Heat rejection	2	2	2	2	3	3	4	2	2	3
Part load behaviour	3	4	3	4	3	2	2	3	3	3
Maturity	2	4	3	2	4	3	2	4	3	2
Weighted Score		**986**	**962**	**926**	**1010**	**986**	**908**	**884**	**866**	**836**

Figure 7: Decision array

6 CONCLUSION

Nine simulation models of Rankine systems were built for two operating points of the engine. The goal was to compare the performance of each system taking into account the limitations on the expander technologies. Unfortunately, the results did not enable to take a decision on the technology and the fluid to select. A decision array was then built to take criteria other than performance into account. According to the selected criteria and the ratings and weights attributed, the couple formed by the scroll expander and water appeared to be the most appropriate solution for the studied application.

7 REFERENCES

[1] Lemort V., Guillaume L., Legros A., Declaye S., Quoilin S., A comparison of piston, screw and scroll expanders for small-scale rankine cycle systems, 2013.
[2] http://www.exoes.com
[3] R410A Copeland Scroll for Systems Up to 1 MW, http://www.emersonclimate.com/europe/en-eu/About_Us/News/Press_Releases (March 27, 2012)
[4] S.E. Eckard, R.D Brooks, Design of reciprocating single cylinder expanders for steam, Final report, Prepared for U.S. Environmental Protection Agency, Office of Air Pollution Control, Alternative Automotive Power Systems Division, Ann Arbor, Michigan (1973).
[5] Kane, M., D. Cretegny, D. Favrat, J. Maquet, Projet HTScroll, Nouveau système de cogénération à turbine spirale haute température, Rapport final, Département fédéral de l'environnement, des transports, de l'énergie et de la communication DETEC, Office fédéral de l'énergie OFEN, 29 octobre 2009.
[6] Lemort V., I.V. Teodorese, and J. Lebrun, Experimental Study of the Integration of a Scroll Expander into a Heat Recovery Rankine Cycle. 2006. 18th International Compressor Engineering Conference, Purdue, USA.
[7] Hütker, J., and A. Brümmer. 2012. Thermodynamic Design of Screw Motors for Constant Waste Heat Flow at Medium Temperature Level. In: Proc. of the In the InternaWonal Compressor Engineering Conference at Purdue. July 16-19 2012. Paper 1478.
[8] Brummer, Energy efficiency – waste heat utilization with screw expanders, Pumps, Compressors and Process Components, 2012.
[9] Nikolov A., Huck C., Brïmmer A., Influence of Thermal Deformation on the Characteristic Diagram of a Screw Expander in Automotive Application of Exhaust Heat Recovery, International Compressor Engineering Conference at Purdue, July 16-19, 2012.
[10] http://www.fchart.com/ees/
[11] Lemort V., Quoilin S., Cuevas C., Lebrun J., Testing and modeling a scroll expander integrated into an Organic Rankine Cycle, Applied Thermal Engineering 29, 3094-3102, 2009.
[12] Glavatskaya Y., Podevin P., Lemort V., Shonda O., Descombes G. 2012. Reciprocating Expander for an Exhaust Heat Recovery Rankine Cycle for a Passenger Car Application. Energies 5(6):1751-1765.

Simulation of expansion process in positive displacement expander

K M Ignatiev, M M Perevozchikov
Emerson Climate Technologies Inc., Research Department, USA

ABSTRACT

The CO2 is gaining popularity as natural refrigerant for multiple applications. However, low critical temperature of CO2 has negative impact on cycle efficiency during operation at transcritical conditions.

Expander significantly improves efficiency of transcritical refrigeration cycle utilizing CO2 as a refrigerant. In order to develop an expander, it is very beneficial to have simulation tool in place, which is capable of modeling expansion process. During expansion process, fluid state can transition from transcritical to single phase or two-phase one. Therefore, simulation tool should be capable of functioning in each of these states and achieve seamless transition between them. A new thermodynamic parameter has being introduced for better description of the fluid state.

NOMENCLATURE

sf	Superfraction	-
X	Vapor fraction	-
T	Temperature	deg C
U	Internal Energy	J
u	Specific Internal Energy	J/kg
P	Pressure	Pa
V	Volume	m3
h	Enthalpy	J/kg
s	Entropy	J/kg/C
m	Mass flow	kg/s
M	Mass	kg
t	Time	sec
ρ	Density	kg/m3
v	Specific Volume	m3/kg
v	Velocity	m/s
c	Speed of sound	m/s
λ	Flow Coefficient	-

1. INTRODUCTION

Expansion machines and their application in refrigeration cycle are well known and there has been a lot of publications describing those machines. Though expansion machines in general are described as a machine where the working fluid is expanding, thus reducing its internal energy which is captured in a form of mechanical work, there are primary two applications of expanders. One application is the expander as a part of heat engine, the best example of which is a reciprocating steam engine; the other application is the application of an expander in a refrigeration system, where the expander is used instead of an expansion valve to increase capacity and improve efficiency of the refrigeration system. The difference between those two applications from thermodynamic standpoint is the fact that the heat engine expander is operating primarily in the superheated vapor region, while the specifics of the refrigeration application of the expander requires operation primarily in supercritical and two-phase region.

While coming from the compressor simulation background, we would like apply classification of traditional levels of analysis to the modeling scrutiny depth of expander machines. Level 1 would be a bulk parameter description, using average values of thermodynamic parameters and thermodynamic conditions before and after the machine. Level 2 refers to an assumption of thermodynamic uniformity of parameters within selected control volumes, connected with flow paths, so that the model consists of system of ordinary differential equations; Level 3 utilizes approach to apply computational fluid dynamics methods, therefore representing working process as a set of Navier-Stokes equations solving numerically utilizing CFD software package.

If we apply this classification to publications dedicated to description of the expander machines, then we can see an interesting picture: There is a vast group of publications, like Basz , et.al (1), Stosic et.al (2), describing specific designs and basic working process of the expanders, classified as Level 1, there are also publications like Kovacevic et.al (3), describing working process in a very detailed manner using CFD package, which can be classified al Level 3; However, we could not find a publication on 2 phase expander describing the Level 2 model.

In the current publications we would like to explain specifics of thermodynamics for numeric description of the expansion process, which enables to build a level 2 model for the expansion process in a positive displacement machine for refrigeration application, operating in supercritical, single phase and two phase regions with a seamless transitions from region to region.

2. INTRODUCTION TO SUPERFRACTION

While describing the thermodynamic processes in a single phase, or supercritical region (referring to Fig.1, those are zones A, C and D respectively), it is traditional and very convenient to use pressure P and Temperature T.

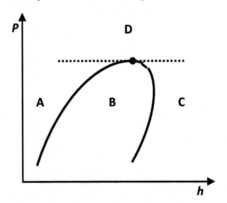

Figure 1. Pressure-enthalpy diagram and zones of different states of the fluid

However, in the 2-phase region (zone B), temperature is no longer an independent variable from pressure P. Usually, instead of temperature, a Vapor fraction x in the 2-phase mixture is used. Of course, as the second independent parameter, it is possible to use Internal Energy, or Entropy, or Enthalpy. That might make sense for internal variables in the analysis software, however, it is not as convenient for the interface purposes as using temperature or vapor fraction.

Therefore, a new parameter, which we decided to call superfraction (sf), can be introduced:

In the 2-phase subcritical zone B: $\quad sf = x$ \hfill (1)

In the subcritical liquid zone A: $\quad sf = \frac{T - T_{sat}}{T_{scale}}$ \hfill (2)

In the subcritical vapor zone C: $\quad sf = \frac{T - T_{sat}}{T_{scale}} + 1$ \hfill (3)

In the supercritical zone D: $\quad sf = \frac{T - T_{critical}}{T_{scale}} + 0.5$ \hfill (4)

Superfraction is the extension of the vapor fraction all over the operating zone, therefore any state of the fluid can be defined by two parameters: Pressure P and Superfraction sf. From those, all other thermodynamic parameters can be calculated.

It can easily be observed, that in the subcritical liquid zone, the superfraction value is equivalent to scaled value of subcooling, while in subcritical vapor zone it is equal to scaled superheat minus 1.

3. THERMODYNAMIC MODEL SPECIFICS OF THE EXPANSION PROCESS

Traditionally, differential equations describing energy and mass in a control volume are presented in the following form (at this point we decided not to include heat transfer):

$$dU = PdV + \sum_{j=1}^{N_{in}} h_j m_j dt - h \sum_{i=1}^{N_{OUT}} m_i dt$$

$$dM = \sum_{j=1}^{N_{in}} m_j dt - \sum_{i=1}^{N_{OUT}} m_i dt$$

\hfill (5)

By slight transformation, they can be converted to a specific internal energy and density differentials:

$$du = \frac{V\rho u + PdV + \sum_{j=1}^{N_{in}} h_j m_j dt - h \sum_{i=1}^{N_{OUT}} m_i dt}{M + \sum_{j=1}^{N_{in}} m_j dt - \sum_{i=1}^{N_{OUT}} m_i dt} - u$$

$$d\rho = \frac{V\rho + \sum_{j=1}^{N_{in}} m_j dt - \sum_{i=1}^{N_{OUT}} m_i dt}{M + \sum_{j=1}^{N_{in}} m_j dt - \sum_{i=1}^{N_{OUT}} m_i dt} - \rho$$

\hfill (6)

This form of equations would lead to selection of internal energy u and density ρ as the primary arguments, which can be used to re-calculate all other dependent thermodynamic parameters.

These differential equation are intended to be solved numerically using Refprop 9.0 DLLs. However, with the 9.0 version of Refprop, we were not able to calculate all refrigerant properties from internal energy u and density ρ if the fluid state lands into 2-phase zone. Nevertheless, Refprop 9.0 was able to calculate pressure as a function of internal energy and density:

$$P = f(u, \rho) \tag{7}$$

From the pair of values P and u, the actual state of the fluid can be easily determined and all the remaining thermodynamic parameters can be calculated.

4. PORT FLOW MODEL SPECIFICS

To calculate a mass flow through an orifice, a form of Bernoulli equation was used to define velocity and mass flow:

$$v = \sqrt{\frac{2(P_{high} - P_{low})}{\rho(P_{low}, S_{high})}} \tag{8}$$

$$m = \lambda S v \rho(P_{low}, S_{high}) \tag{9}$$

In case of choke condition, the fluid velocity is equal to local speed of sound and the pressure and therefore density at the orifice are different:

$$v = c(P_{choke}, S_{high}) = \sqrt{\frac{2(P_{high} - P_{choke})}{\rho(P_{choke}, S_{high})}} \tag{10}$$

In case of ideal gas with a known gas constant, there is a well known formula for the choke flow, however, it is not very straightforward in case of using numerical computations with Refprop functions.

There is a Refprop function c=SpeedOfSound(P,s) which can be applied to calculate the local speed of sound. Unfortunately, 9.0 version of Refprop, unable to calculate the speed of sound in the 2-phase zone.

To overcome this problem, a very general formula for the local speed of sound was used:

$$c(P, s) = \sqrt{-v^2 \frac{\partial P}{\partial v}}\bigg|_{S=const} ; \tag{11}$$

In the computation procedure, numerical procedure was used to calculate the speed of sound according to this formula. Iterative procedure was implemented to calculate the critical flow.

In order to evaluate the orifice flow model, a flow process through an orifice was simulated, using CO_2 as a fluid, with the incoming pressure of 8.275 MPa and temperature of 32.2 C, which represents a typical supercritical condition of the outlet of the gas cooler of a CO_2 refrigeration system.

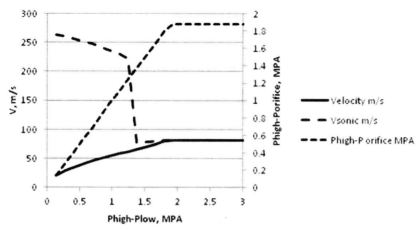

Figure 2. Velocity, speed of sound and orifice pressure drop vs. overall pressure drop

Fig.2 represents the results of this simulation. It is interesting to observe, that in the supercritical and single state, the sonic velocity is in the range of 230 to 260 m/s, while in the 2-phase zone it is substantially lower being around 80 m/s. Therefore, the choke condition takes place much earlier compared to a single phase expansion because the speed of sound is substantially reduced. The iterative procedure needs to be stable enough to handle the significant non-linearity of the speed of sound. From the mathematical perspective, there is a possibility of non-convergence, when the choke takes place right around the transition from single phase to two phase zone, where the speed of sound experiences abrupt change. In this case, it makes sense to select an orifice pressure value just above the entry point into 2-phase zone B and calculate the velocity, density and mass flow at its value, though the flow would not yet be critical.

In the future, there are plans to extend this approach to flow through leak paths, similar to the one described by Bell et. al. (4), but including the effects of the real fluid with phase changes.

5. EXPANSION RPOCESS ANALYSIS EXAMPLE

To verify the approach, a very simple model of an expander was developed. It is a generic rotary expander machine with a realistic inlet porting. The geometry modeling analysis of this machine has been performed using the approach described in greater detail by Ignatiev (5). The results of the geometry analysis (Inlet port cross-sectional area, working chamber volume as a function of crank angle) are illustrated on Fig.3.

In this particular example, we did not consider an outlet port at this point in time, because we were focused on expansion process transitioning from zone D to zone B most likely through zone A.

Inlet pressure and temperature was selected at the values of 8.275 MPa and 32.2 C respectively. The working fluid was CO_2. Initial conditions inside the cavity were also selected at the same conditions. The scale temperature for Superfraction was selected as 5/9 deg C (1 deg F).

Fig. 4 represents the results of the computational example.

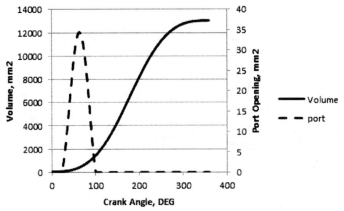

Figure 3. Expander volume and port opening as the function of crank angle

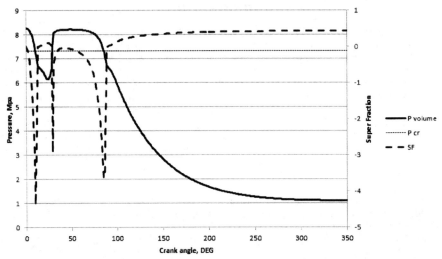

Figure 4. Pressure and Superfraction in the expander as a function of crank angle

The selected porting and initial conditions caused the initial charge of the fluid to expand before opening the inlet port. This is the explanation of a drop in pressure within first 35 DEG of crank angle. Then the inlet port opens and pressure in the expander equalizes with the incoming fluid pressure. Then, around 90 DEG of crank angle, port starts to close and expansion process takes place. A dotted line representing critical pressure value is provided for reference. It is interesting to observe, that expansion from supercritical zone B places the fluid into liquid zone A, which corresponds to crossing of the Critical pressure line, and also significant reduction in Superfraction value, referring to a subcooled liquid condition. During expansion in those zones, there is a very steep expansion slope illustrating the reverse of almost "non-compressibility" of the liquid refrigerant.

Sudden jumps in Superfraction coincide with pressure curve value crossing critical pressure threshold.

In this expansion process example, fluid state transitions from zone D to zone A and then into zone B. This transition can clearly be seen by the value of superfraction becoming negative, while the pressure is below critical.

It is also interesting to observe, that during expansion in zone B (2-phase), after initial transition, the value of superfraction stays almost constant.

6. CONCLUSIONS

1. Mathematical description enabling simulation of dynamic process of expansion in supercritical, liquid and 2-phase zones, including seamless transitions between those zones, has been presented. This approach has been programmed to numerically solve those equations in conjunction with Refrop-sourced DLL for refrigerant properties.
2. A new dimensionless thermodynamic parameter- superfraction- was introduced.
3. An example of expansion process simulation was provided demonstrated to prove the functionality of the algorithm based on mathematical description introduced. Expansion process was illustrated using the Superfraction parameter and its advantage for better interpretation of the process was demonstrated.

ACKNOWLEDGEMENT

The authors would like to acknowledge the support of the management of the Research Department, Emerson Climate Technologies.

REFERENCE LIST

1. Brasz, J.J., Smith, I.K., Stosic N., 2000, Development of a Twin Screw Expressor as a Throttle Valve Replacement for Water-Cooled Chillers, *Proc. 2000 Int.Compressor Eng. Conf. at Purdue,* West Lafayette, IN, USA, July 25-28: p. 979-986
2. Stosic N., Smith, I.K., Kovacevic A., 2002, A Twin Screw Combined Compressor and Expander For CO2 Refrigeration Systems, *Proc. 2002 Int. Compressor Eng. Conf. at Purdue,* West Lafayette, IN, USA: # C21-2
3. Kovacevic, A., Stosic N., Smith, I.K., 2006,Numerical simulation of Combined screw Compressor-Expander Machines For Use in High Pressure Refrigeration Systems, Simulation, modelling practice and theory 14 (2006) 1143-1154, International Journal of the Federation of European Simulation Societies ELSEVIER
4. Bell I., Groll E., Braun J., Horton W.T., 2012, A Computationally Efficient Hybrid Leakage Model for Modeling Leakage in Positive Displacement Compressors, *Proc. 2012 Int. Compressor Eng. Conf. at Purdue,* West Lafayette, IN, USA, #1103
5. Ignatiev, K.M. 2012, Approach To The Numeric Geometry Analysis of Positive Displacement Compressors, Its Application To A Single Screw Compressor Simulation And Verification By Experiment, *Proc. 2012 Int. Compressor Eng. Conf. at Purdue,* West Lafayette, IN, USA, #1171

Investigation on a scroll expander for waste heat recovery on internal combustion engines

A Legros[1,2], L Guillaume[2], V Lemort[2], M Diny[1], I Bell[2], S Quoilin[2]
[1] DRD/DRIA/APWE, PSA Peugeot Citroen, France
[2] Thermodynamics Laboratory, University of Liège, Belgium

ABSTRACT

In the present article, a model of scroll expander will be introduced. This model is able to evaluate the performance of a given machine with influence of the geometry. Several losses are also included by the model such as internal leakages, heat transfers or mechanical losses. The forces generated by the gas pressure on the involutes can also be calculated.

That expander model is used in order to demonstrate its potential and provide some guidelines to the design of a new expander suitable for the application.

NOTATIONS

Mathematical symbols

A	Area	m^2
D_h	Hydraulic diameter	m
f	Friction coefficient	–
h	Specific enthalpy	J/kg
H	Height	m
K_1, K_2	Coefficients	–
L	Length	m
$\dot{M}$	Mass flow rate	kg/s
n	Normal direction	m
P	Pressure	Pa
Pr	Prandtl number	–
$\dot{Q}$	Heat transfer rate	W
Re	Reynolds number	–
r_b	Base radius	m
r_v	Volume ratio	-
t	Wall thickness	m
U	Internal energy	J
V	Volume	m^3
x, y	x and y coordinates	m
$\dot{W}$	Power	W

Greek letters

γ	Heat capacity ratio	–
ϵ	Efficiency	–
ΔP	Pressure difference	Pa
θ	Crank angle	rad
λ	Thermal conductivity	$W/(m\,K)$
ρ	Density	kg/m^3
φ	Construction angle	rad
φ_0	Initial angle	rad
ω	Angular velocity	rad/s

Subscripts

ax	Axial
h	High pressure side
i	Inner involute
in	Coming in the chamber
l	Low pressure side
$loss$	Losses
o	Outer involute
out	Coming out of the chamber
s	Isentropic
rad	Radial
x, y	According to the x or y axis

© The author(s) and/or their employer(s), 2013

1. INTRODUCTION

Waste heat recovery Rankine cycle on mobile application is not a new idea. A first concept on a train has already been commercialized in the 1920s, taking advantage of the price difference between diesel and coal [1]. Unfortunately, this system quickly became not competitive because that difference was not profitable anymore. The research on WHR systems rose when oil crisis occurred, in the 1970s. Several systems were developed, mostly for trucks or marine applications. After that period, the interest disappeared until the 2000s, when automotive manufacturers started being interested in that technology again. Actually, due to the restrictive standards on CO_2 emissions placed by Europe, it is necessary to find new ways to reduce car consumption and, thus, improve overall engine efficiency. Waste heat recovery technologies are among the possible solutions. This work focuses on Rankine cycles, mainly because of its potential in efficiency and produced power, demonstrated in previous simulation works [2, 3].

The present article studies a waste heat recovery Rankine cycle for conventional gasoline engines. The first part will present a model of a scroll expander developed to evaluate the performances of a scroll expander considering various parameters such as geometry, internal leakages, heat transfers or mechanical losses. The second part will present some simulation results and provide some guidelines to the design of an expander.

The experimental investigations are currently conducted and the results of the calibration will be written soon.

2. MODELIZATION OF THE SCROLL EXPANDER

Several complex models for scroll compressors have already been developed and are available in literature [4, 5] whereas only one work has been for scroll expanders [6]. The work presented in this paper is based mainly on those three thesis. The goal of this model is to be able to size an expander and design one that is adapted to waste heat recovery in a light vehicle.

The model presented here is split into two consecutive sub-models. The first one describes the geometry based on several input parameters and evaluates different geometrical characteristics. The last one is about the thermodynamic evolution of the fluid inside the machine and is able to provide the fluid state in each chamber at every moment of a revolution. This sub-model also evaluates the performances of the machine, the efficiencies and the produced work by calculating the frictional losses and other kinds of losses. Lastly, an additional part of that sub-model consists in evaluating the resultant forces on the moving involutes. Those forces can be used in order to size the shaft and the bearings for example.

The scroll model has been developed in Matlab and the thermodynamic properties are evaluated with Refprop library. A huge effort has been placed on the numerical side. In order to decrease the computation time, lookup tables are generated before the first simulation and interpolations are used inside those tables to evaluate the properties. This can decrease the computation time up to 20%. Additional work has been carried out to decrease the CPU time but it's not presented here.

2.1 Geometric sub-model
The profile of a scroll spiral is usually a circle involute. The Cartesian equation system for the circle involute is given by:

$$\begin{cases} x = r_b(\cos(\varphi) + (\varphi - \varphi_0)\sin(\varphi)) \\ y = r_b(\sin(\varphi) - (\varphi - \varphi_0)\cos(\varphi)) \end{cases}$$

Each spiral consists of two circle involutes whose only initial angles φ_0 are different and those two curves are named inner and outer involutes. Those two involutes produce a spiral whose thickness is constant and equal to

$$t = r_b(\varphi_{i,0} - \varphi_{o,0})$$

In scroll compressors, the central tip geometry is not a circle involute because this would lead to design that would be difficult to machine. The linking curves between the inner and outer involutes could consist of two circle arcs [7]. Those arcs are tangent to each other and to its neighboring involute. Those conditions provide an equations system with infinity of solutions. One parameter can be set in order to adjust the design and needs to be imposed for the uniqueness of the solution.

Once the geometry is defined, the geometry sub-model evaluates the geometric characteristics over one revolution that will be useful to the other sub-model, such as chamber volumes, heat transfer areas, leakage areas, etc.

Almost every characteristic is based on the determination of the limits of each chamber. The limits are defined by the "contact" points between the two spirals. Those points are called conjugacy points and can be represented by a conjugacy angle, written φ_{conj} with i- or o- subscript depending on which involute is considered (inner or outer). Based on those angles every other geometric characteristic is numerically evaluated by special algorithms. Note that most of the characteristics can also be derived from analytical expressions but the peculiar tip geometry requires some numerical tools because analytical expressions do not exist.

The interference between the moving spiral and the supply port can be evaluated using a specific Matlab function. However those functions are time consumers and it is indeed better to avoid using them. In order to decrease the calculation time, a detection algorithm has been implemented in order to guess when the interference will happen and then start using the interference Matlab functions.

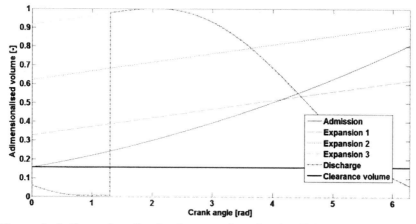

Figure 1: A dimensionalized volume – given by the division of the actual volume by a constant one – of the different chambers. The clearance volume is also displayed.

The volume of the different chambers are present in Figure 1. In this Figure, it can be seen the evolution of the different volumes of each pocket of fluid. The clearance volume is displayed and it corresponds to the minimal volume of the suction chamber. A saturation lower limit has been imposed of the discharge volume in order to avoid too small crank angle step.

2.2 Thermo-mechanical sub-model

2.2.1 Global considerations

This sub-model is an iterative model on the exhaust enthalpy. The exhaust enthalpy is evaluated after every complete revolution of the scroll expander. Then a convergence criteria based on that enthalpy is checked and the algorithm decides whether or not another revolution is necessary.

One revolution is divided in several constant crank angle steps. The algorithm is sub divided in three parts which are consecutively run at every step of the complete revolution. Those calculations are done simultaneously in every chamber of the expander in order to reduce the calculation time. The three parts are the update of the thermodynamic variables, the evaluation of the different mass flows and the calculations of the heat transfers. They will be described in the next sub-sections.

2.2.2 Thermodynamic variables

In the first part, every useful thermodynamic variable, i.e. pressure, temperature, enthalpy, entropy, specific volume and quality, are evaluated based on the determination of the internal energy and the mass of fluid in the chambers. In order to evaluate the internal energy and the mass of gas, some analytical expressions of the variation of internal energy and mass must be derived in function of time, or in function of the crank angle since rotation speed is known. Those expressions can be derived from the conservations of energy and of mass. The differential equations, solved at each crank angle step, are presented hereunder.

$$\frac{dU}{d\theta} = \frac{\dot{Q}}{\omega} - \frac{PdV}{d\theta} + \frac{1}{\omega}\sum(\dot{M}_{in}h_{in} - \dot{M}_{out}h_{out})$$

$$\frac{dM}{d\theta} = \frac{1}{\omega}\sum(\dot{M}_{in} - \dot{M}_{out})$$

Those derivatives are function of the variables at the previous crank angle step and can be used in an Euler-Forward solver.

Euler Forward solver has firstly been chosen for its simplicity of implementation. However it is known that it does not show the largest accuracy. The latter is actually of order of the squared value of the crank angle step. Moreover Euler Forward is not an adequate solver for stiff problems such as the small volume encountered in a scroll expander in the discharge chamber. The choice of the solver is very important and should result from a compromise between the precision and the computational time required.

A literature review has been conducted about the different explicit solvers that can be used in stiff problems. Fourth order Runge Kutta (RK4) is the second most famous solver. Its precision is of the order of the crank angle step exponent 4. The inconvenient of that solver is that it creates intermediate points between the points n and $n + 1$. To overcome this issue that leads to an increase of the computational time, a fourth and fifth order Runge-Kutta-Fehlberg – or simply RKF 45 - solver could be used. It uses the RK developments of fourth and fifth order and allows modifying the crank angle step to decrease the computational time. All those solvers are one step solver which are functions of the previous crank angle step only. The last solver is a multistep solver named Adams-Bashforth-Moulton solver – or ABM. This solver is based on a predictor-corrector scheme. The predictor is an explicit step that will define a new fictitious point. That point is used by an implicit step, called the corrector, and defines the point $n+1$. The main advantage is that this solver only creates one intermediate point because it is function of the previous values already known. However the first steps need to be solved by a different solver since it requires several previous steps.

Those solvers have been compared on a simple stiff problem given by

$$\frac{dY}{dt} = -aY(t)$$

where a is a positive constant. The solution of this differential equation is rather simple and is equal to the exponential of the product of $-a$ times t. The error between the solver solution and the analytical solution is presented in Figure 2. RK4 and ABM solvers are pretty close in terms of precision and in computational time. The precision of RKF45 is very good and can be modified by specifying the crank angle step limits which define the modification of the crank angle step size. From that simple example, it can clearly be seen that Euler Forward is not the ideal solution but provides a good compromise between precision and computational time. In order to improve the simulation accuracy, RKF45 or ABM solver should be implemented.

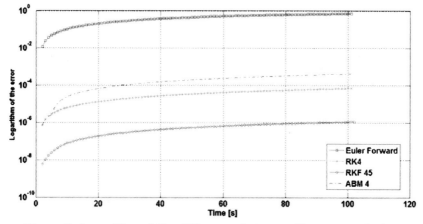

Figure 2: **Logarithm of the difference between the exact solution and the solver solution for four solvers.**

2.2.3 Mass flow modeling
The second part consists in the evaluation of the different mass flow rates. This includes the supply, the exhaust and the leakage mass flows. The modeling of the mass flows in a scroll compressor is a subject widely treated in literature, from simple 0D models to highly complex 2D model [5, 8]. In order to choose correctly how to model those flows, a literature review has been conducted. Firstly, some rather simple 0D models have been chosen. This is mostly to avoid too much calculation time of the algorithm but Yong et al. confirm our choice. Indeed, they developed in [8] a 2D model and conclude that a simple 0D model can be accurate enough depending on whether the ratio of the height of the leakage gap over the length of the leakage path is small or not. If it is small, a 0D model is enough, otherwise the complexity should be increased.

The following equations define the most common mass flow model, which is the isentropic flow model. It has been studied by several authors and compared with experimental results [9]. The biggest disadvantages of this model is that it does not take into account friction. So, if the flow path is narrow and long, friction will have a major importance. However, its efficiency for leakage with a short flow path has been demonstrated with experimental results.

$$\dot{M} = \begin{cases} A\sqrt{2P_h \rho_h} \sqrt{\left(\dfrac{\gamma}{\gamma-1}\right)\left(\left(\dfrac{P_l}{P_h}\right)^{\frac{2}{\gamma}} - \left(\dfrac{P_l}{P_h}\right)^{\frac{\gamma+1}{\gamma}}\right)} & if \dfrac{P_l}{P_h} > \left(1 + \dfrac{\gamma-1}{2}\right)^{\frac{\gamma}{\gamma-1}} \\ A\sqrt{2P_h\rho_h \dfrac{\gamma}{\gamma+1}\left(\dfrac{2}{\gamma+1}\right)^{\frac{1}{\gamma-1}}} & otherwise \end{cases}$$

To overcome the friction issue of the isentropic flow model, different models have been proposed. Bell fixes the problem by introducing a correction coefficient on the mass flow [10] but it has to be calibrated with experiments. Ishii et al. also proposed a solution [9] where they did some tests and compared different models of flow and come to the conclusion that a model of incompressible, viscous and fully turbulent is accurate enough. In such a model, the mass flow rate is computed by

$$\dot{M} = A\sqrt{\dfrac{2D_h \rho_h \Delta P}{fL}}$$

So, for the model of the supply and the exhaust mass flows, we choose to represent them by an isentropic flow. Indeed, the path flow is rather short and should not be submitted to friction too much.

For the internal leakages, literature distinguishes two types of leakages. The first one is the one called flank leakage and is happening between two consecutive chambers along the spirals. This one can also be modelled as an isentropic flow for the same reasons as above.

Finally, the radial leakages are the second type of leakage inside a scroll machine. It goes from one chamber to the next one radially and over the spiral. This leakage is more likely to have friction involved because the leakage path is quite long comparing to its height. So, we choose a frictional model as [6, 9] mentioned it.

2.2.4 Heat transfers
Heat transfers appear between the fluid and the scroll at the inlet and the exhaust of the machine, between the fluid and the scroll wraps and between the scroll wraps and the ambiance.

The heat transfer coefficients are modeled by the Dittus-Boelter relation with some corrections. The heat transfer coefficient equation for the Dittus-Boelter relation is given hereafter.

$$h_{DB} = 0.023 \dfrac{\lambda}{D_h} Re^{0.8} Pr^{0.4}$$

The heat transfer coefficient is then corrected by multiplying the Dittus-Boelter coefficient by two correction factors. The first one - K_1 - has been investigated for the spiral heat exchanger [11] and it is used to take into account the spiral geometry. The moving involute introduce a forced convection heat transfer and it is the purpose of the second coefficient - K_2 - to include that phenomenon [12]. The relations for the coefficients are given by

$$K_1 = 1 + 1.77 \frac{D_h}{r_c}$$
$$K_2 = 1 + 8.48(1 - \exp(-5.35\, St))$$

The heat transfers are computed at each crank angle step between each chamber and its walls and between the ambiance and the scroll expander.

2.2.5 Forces evaluation
The axial force is easy to compute and is given by the following equation

$$F_{ax} = P \frac{V}{H}$$

The radial force is decomposed into its x and y components. The problem resumes to find the normal direction to the involute because the area on which the force is applied can be numerically computed knowing the conjugacy angles of the chamber. The components of the radial force are then given by

$$F_x = PA \frac{n_x}{\sqrt{n_x^2 + n_y^2}}$$
$$F_y = PA \frac{n_y}{\sqrt{n_x^2 + n_y^2}}$$

Where the normal components are defined by

$$n_x = -r_b(\theta - \varphi_0)\sin(\theta)$$
$$n_y = r_b(\theta - \varphi_0)\cos(\theta)$$

Finally, the radial force can be obtained by

$$F_{rad} = \sqrt{F_x^2 + F_y^2}$$

2.2.6 Mechanical losses
The most important source of mechanical losses in a scroll compressor is often the friction between the moving involute and the thrust bearing [13]. In order to compute this loss, a simple force balance on the moving involute can provide the effective force acting on that involute. The effective force is the difference between the axial force resulting from pressure inside the chamber and the counter force that is applied in order to maintain the two spirals together. With the proper friction coefficient and a dry friction model, the effective force can be transformed into a friction force and a friction torque with the radius where the force applies. Finally, including the rotation speed, the friction power loss is obtained.

3. NUMERICAL SIMULATIONS

Simulation results are presented hereunder in graphs with normalized axes. These are the first results and in order to be trustworthy the model still need to be calibrated. However those results can provide trends in design. In the following paragraphs, every geometric characteristic, such as leak gaps, geometric angles... is kept constant. Otherwise it is specified. The simulations are run with water as working fluid.

Figure 3 presents a P-V diagram for two different supply pressures. As we can see the model can take into account under- and over-expansion. The pressure loss at the supply port of the expander is also modeled and visible in the P-V diagram. The pressure loss at the admission is however better illustrated in the Figure 4. In that Figure, the pressure during a full revolution is plotted with respect to the crank angle. The different curves correspond to cases where the radius of the admission port or the rotation speed vary. Figure 4 illustrates the pressure loss that occurs in the suction chamber.

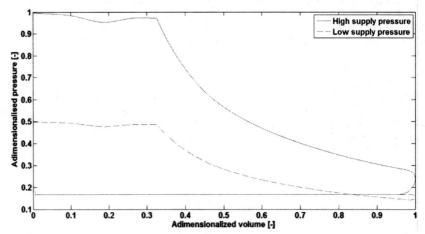

Figure 3: Pressure-volume diagram for different rotational speeds. Both axes have been adimensionalized by dividing the variable by a constant.

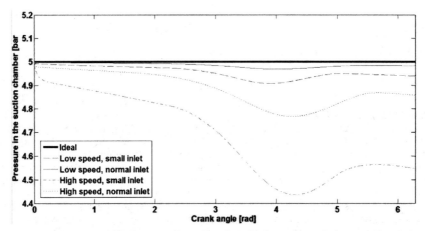

Figure 4: Pressure in the suction chamber in function of the crank angle. Results are presented with different cases where rotational speed and inlet radius vary.

The ideal case would be to have a constant pressure along the revolution but two phenomenon reduce the efficiency, especially at high speed. The first one is that the flow can be choked and this should not happen normally since the pressure ratio is very low but it is more likely to happen at high speeds. The major issue comes from the fact that the moving spiral can pass over the admission port. Tip geometry can be modified to avoid that, however, this is not implemented in the algorithm. From Figure 4, it is clearly obvious that operating at high speed could lead to significant pressure drop and thus, to efficiency and produced power reductions. This leads to a major conclusion: since the clearance volume is not a source of loss in scroll expander, the tip geometry should be designed in such a way that the suction port has the most important area.

The isentropic efficiency is defined by the following equation and is presented in the Figure 5 with respect to pressure ratio and rotation speed.

$$\epsilon_s = \frac{\dot{W} - \dot{W}_{loss}}{\dot{W}_s}$$

From this graph, we can clearly see that the model is able to represent correctly the efficiency curve of the scroll machines. It is also known that the speed has a positive impact on leakage and since the frictional losses are low in this case the efficiency tends to increase if the speed is increased. It is also important to note that the gain obtained is a degressive function of the rotational speed. Working with a scroll expander using high rotational speed seems to be a good solution if friction can be maintained low. Moreover, leakage is a difficult parameter to control while the control of friction is easier. Indeed, the choice of adequate materials or coatings and the anti-rotation mechanism could greatly help reducing friction. On the other side, internal leakage, particularly flank leakage, can only be reduced by introducing a high viscosity fluid. The choice of that fluid is particularly difficult since the operating environment is very harsh (high temperature, high pressure …). In Figure 5, the maximum of the efficiency curve usually corresponds to the built-in volume ratio, which can be related to a pressure ratio regarding the operating conditions. However we can see in that Figure that the maximum is drifting with an increase of rotational speed. This is due to the fact that the leakage increase a little bit the pressure at the end of the expansion. Increasing the speed reduces the leakage and show a maximum closer to the built-in volume ratio.

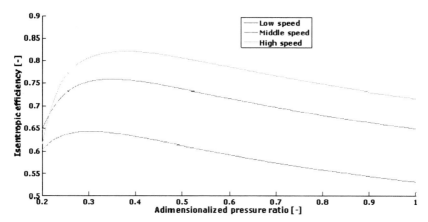

Figure 5: Isentropic efficiency in function of a normalized pressure ratio for different rotational speeds. The pressure ratio has been normalized regarding to the pressure ratio corresponding to the built-in volume ratio.

Figure 6 illustrates the different losses taken into account by the model and their impact on the scroll performance. For this graph, we have taken parameter close from what we could expect in an actual scroll compressor. The flank leakages represent the biggest source of losses and the other are much smaller. Every loss seems also not to vary significantly with the pressure ratio, except the loss due to fixed volume ratio of the expander, which are called "Inadapted r_v loss". That loss induces over- or under- expansion.

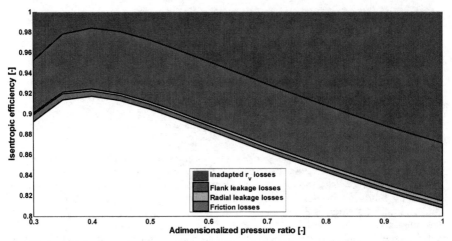

Figure 6: Losses in a scroll expander plotted in an efficiency graph.

4. CONCLUSION

A model of scroll expander has been developed and we are able to see the influence of various parameters such as the geometry, the leakage gaps or the working conditions. This model will be used to design a scroll expander suited for waste heat recovery in a light vehicle. Some important conclusions can already be drawn. Indeed, the design of the inlet port should be carried out correctly since a low molar mass fluid will require higher rotational speeds.

The scroll expander model could also benefit from a few improvements. The calculation time needs to be optimized and some numerical solvers with adaptive crank angle step could be a solution. Other types of solvers can also be implemented. Last improvement should be the calibration of the model on some experiments.

In conclusion, several improvements can still be made but the model is ready and properly working. It provides results that are for now more qualitative but still useful to provide guidelines for the design of an expander.

5. BIBLIOGRAPHY

[1] W. J. Still, "Internal combustion engine". England Patent 1230617, 19 June 1917.
[2] PSA Peugeot Citroën, "Thermal recuperation: Rankine Cycle", in *Highly Integrated Combustion Electric Propulsion System: Final event*, Florence, 2011.
[3] PSA Peugeot Citroën, "Internal Work", 2011.
[4] Y. Chen, Mathematical modeling of scroll compressors, 2000.
[5] I. Bell, Theoretical and experimental analysis of liquid flooded compression in scroll compressors, 2011.
[6] V. LEMORT, Contribution to the Characterization of Scroll Machines in Compressor and Expander Modes, 2008.
[7] B. R. SCHAFFER and E. A. GROLL, "Parametric representation of scroll geometry with variable wall thickness", in *International Compressor Engineering Conference*, Purdue, 2012.

[8] H. YONG, "Leakage calculation through clearances", in *Internation Compressor Engineering Conference*, Purdue, 1994.

[9] N. Ishii, K. Bird, K. Sano, M. Oono, S. Iwamura and T. Otokura, "Refrigerant leakage flow evaluation for scroll compressor", in *International Compressor Engineering Conference*, Purdue, 1996.

[10] I. H. Bell, E. A. Groll, J. E. Braun and T. Horton, "A computationally efficient hybrid leakage model for modeling leakage in positive displacement compressors", in *International Compressor Engineering Conference*, Purdue, 2012.

[11] N. Tagri and R. Jayaraman, "Heat transfer studies on a spiral plate heat exchanger", *Trans. Inst. Chem. Eng.*, pp. 161-168, 1962.

[12] K. Jang and S. Jeong, "Experimental investigation on convective heat transfer mechanism in a scroll compressor", *International Journal of Refrigeration*, pp. 744-753, 2006.

[13] S. HAJIME, I. TAKAHIDE and K. HIROYUKI, "Frictional characteristics of thrust bearing in scroll compressor", in *International Compressor Engineering Conference*, Purdue, 2004.

TURBO MACHINES

Oil-free centrifugal refrigeration compressors: from HFC134a to HFO1234ze(E)

J J Brasz
Danfoss Turbocor Compressors Inc., USA

ABSTRACT

The high global warming potential of HFC134a (GWP=1300) has led to the development of a new family of man-made refrigerants with much lower GWP values. HFO1234ze(E) is one of those fluids. It has a GWP value of 6 and has been in commercial production as a blowing agent for several years. Based on cost and availability this new fluid has been selected a potential candidate to replace HFC134a in commercial chillers.

A new family of oil-free direct-drive centrifugal compressors with HFO1234ze(E) as working fluid is being introduced commercially, covering a cooling capacity range from 200 – 300 $kW_{thermal}$. These new compressors are a spin-off of an existing platform of oil-free HFC134a products. Being oil-free eliminated oil-refrigerant compatibility issues, which were a major stumbling block during the transition from CFC12/HCFC22 towards HFC134a in the early 1990's.

Due to its somewhat lower pressure and vapor density HFO1234ze(E) requires a slightly larger fluid module to achieve the same refrigeration capacity as HFC134a. Impeller tip speed is reduced at equal temperature lift (=difference between condenser and evaporator saturation temperatures) as a result of the lower sonic velocity of HFO1234ze(E). Overall compressor efficiency improves as a consequence of these two changes.

1. INTRODUCTION

During the nineties, the air-conditioning and refrigeration industry saw a transition from CFC's towards HCFC's and HFC's. The main driver for this transition was the discovery of stratospheric ozone layer depletion by the chlorine atoms found in CFC's and HCFC's.

After the resolution of the ozone layer depletion problem by the introduction of chlorine-free HFC refrigerants, the environmental concern started to focus on the global warming impact of these man-made refrigerants. Legislation intended to limit the future use of refrigerants in systems with a high direct effect on global warming has led to the development of new systems using natural refrigerants such as CO_2 as well as the development of new man-made refrigerants with much lower global warming potential than the currently used HFC refrigerants.

Table 1 shows a number of medium pressure refrigerants used in water-cooled chillers. The refrigerants have transitioned from fluids with high ozone depletion potential (ODP) such as CFC12 and HCFC22 towards fluids with zero ODP but still a large global warming potential (GWP) such as HFC134a. Today there are candidate refrigerants with both zero ODP and very low GWP. HFO1234yf is the result of a joint effort by DuPont and Honeywell to develop a drop-in replacement fluid for HFC134a for mobile air conditioning (MAC) applications [1]. HFO1234ze(E) is a low GWP fluid developed by Honeywell. It is currently used as a foam blowing agent [2].

Table 1. Ozone layer depletion (ODP) and global warming potential (GWP) of various fluids.

Refrigerant	ODP	GWP
CFC12	1	8500
HCFC22	0.05	1700
HFC134a	0	1300
HFO1234yf	0	4
HFO1234ze	0	6

2. CENTRIFUGAL CHILLERS AND THEIR OPERATING CONDITIONS

In the refrigeration industry centrifugal compressors are predominantly used in water-cooled chillers although smaller capacity direct-drive centrifugal compressors are now also applied to air-cooled chillers. Figure 1 shows a simplified equipment diagram of a water-cooled chiller.

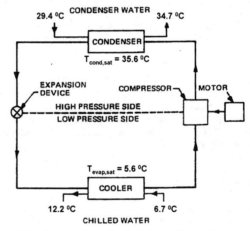

Figure 1. Simplified equipment diagram of a water-cooled chiller.

Typical full-load entering and leaving evaporator water temperatures as specified by the ARI rating code [3] are 12.2°C (54°F) and 6.7°C (44°F), respectively. Evaporator and condenser water flow rates are specified by the code as 0.043 l/s per kW of refrigeration (2.4 gpm/ton) and 0.054 l/s per kW of refrigeration (3.0 gpm/ton). The heat absorbed by the evaporating refrigerant is rejected in the condenser together with the heat of compression. The rejection of this heat by the condensing refrigerant increases the temperature of the water returning from the cooling tower entering the condenser at 29.4°C (85°F) to a leaving water temperature of around 34.7°C (94.5°F) – the exact value of the condenser leaving water temperature varies slightly depending on the ideal cycle efficiency of the refrigerant and the compressor efficiency. The evaporator saturation temperature will be just below the leaving chilled water temperature of 6.7°C (44°F), say 5.6°C

(42°F) and, similarly the saturation temperature of the refrigerant in the condenser will be just above the leaving condenser water temperature of 34.7°C (94.5°F), say 35.6°C (96°F).

Overall chiller efficiency is defined using of a coefficient of performance (COP) number which is defined as follows:

$$COP = \frac{Cooling\ Capacity\ (kW_{thermal})}{Compressor\ Input\ Power\ (kW_{electric})}$$

For refrigeration systems this number is always larger than 1 (or 100%), which is the reason that the word efficiency has been replaced with the term coefficient of performance. In the US it is customary to specify chiller performance in terms of a kW/ton number defined as:

$$kW/ton = \frac{Compressor\ Input\ Power\ (kW_{electric})}{Cooling\ Capacity\ (tons\ of\ refrigeration)}$$

Since a ton of refrigeration (defined as the cooling capacity obtained from the melting of a so-called short ton (=2000 pounds) of ice in 24 hours) is equivalent to a cooling capacity of 3.52 kW, the relationship between these two quantities is as follows:

$$COP = \frac{3.52}{\frac{kW}{ton}}$$

Actual state of the art COP numbers for water-cooled centrifugal chillers at ARI full-load conditions vary from 6.18 (0.57 kW/ton) to 6.77 (0.52 kW/ton). This variation is caused by the choice of refrigerant, compressor efficiency and refrigeration cycle details such as sub-cooling and economizing. Advertised centrifugal chiller performance figures can be up to 7.5% percent better as a result of the measurement tolerances allowed by the ARI-550 test code, including a 5% error in power and water flow rates and a 0.28°C (0.5°F) deviation in temperature.

3. NEW REFRIGERANTS IN OIL-FREE COMPRESSORS

3.1 Oil-free compressors allow easy transition to new refrigerants
Oil-free centrifugal compressors can relatively easy be converted to alternative refrigerants. The oil-free compressor operation eliminates the identification or development and subsequent qualification of lubricants that are compatible with the new refrigerant. The new refrigerant compatibility studies are limited to the elastomer (O-rings) and motor insulation materials.

3.2 Chiller global warming impact of HFO1234yf and HFO1234ze(E)
The global warming impact of a refrigeration system consists of at least two effects:
- a direct effect (representing the global warming due to the leakage of refrigerant molecules into the atmosphere) and
- an indirect effect (representing the amount of global warming as a result of energy utilization of the refrigeration system which is related to its efficiency).

The industry has accepted the TEWI concept (Total Equivalent Warming Impact) [4] to account for both effects. For systems notorious for their high refrigerant leak rates such as commercial refrigeration and automotive air conditioning systems the direct effect is the predominant cause of global warming. For residential and commercial air conditioning equipment with its much lower leak rates the indirect effect is the main contributor of global warming, with the exact amount of global warming being determined by the percentage of power being generated by coal, oil and natural gas versus renewable and nuclear power generation. The cycle efficiency of a low GWP refrigerant relative to the one it replaces is the dominant characteristic for refrigeration systems with low leak rates. It can be categorically stated that for commercial and residential HVAC equipment the cycle efficiency of a low GWP alternative refrigerant should not be inferior to the existing refrigerant it is intended to replace. Table 2 compares the ideal (= assuming 100% compressor efficiency) coefficient of performance (COP) for typical water-cooled chiller conditions, using the latest Refprop 9.0 refrigerant properties [5]. As can be seen from that table HFO1234yf has a 3.2% lower COP than HFC134a and should therefore to be eliminated as a low GWP alternative for commercial HVAC applications where the energy consumption is the major contributor to global warming.

HFO1234ze(E) has an ideal COP that is equal to that of HFC134a. A transition from HFC134a to HFO1234ze does not increase the indirect global warming – like a transition towards HFO1234yf would – and dramatically reduces the direct global warming effect.

Table 2. Ideal cycle COP comparison for HFC134a, HFO1234yf and HFO1234ze showing an almost equal COP for HFC134a and HFO1234ze and a 3.2% lower COP for HFO1234yf.

INPUT VALUES					
Evaporature saturation		$T_{evap,sat}$	5.56 [°C]		
Evaporator superheat		$\Delta T_{evap,sup}$	0.00 [°C]		
Condenser saturation		$T_{cond.sat}$	36.11 [°C]		
Condenser subcooling		$\Delta T_{cond,sub}$	3.89 [°C]		
Fluid			R134a	R1234yf	R1234ze
OUTPUT VALUES					
$\Delta h_{evaporator}$	[kJ/kg]		156.89	122.70	144.13
$\Delta h_{s,compressor}$	[kJ/kg]		19.55	15.80	17.98
Pressure ratio	-		2.57	2.43	2.60
COP_{cycle}	-		8.02	7.77	8.02
COP relative to R134a	-		100.0%	96.8%	99.9%

3.3 Predicted centrifugal compressor capacity change when replacing HFC134a with HFO1234ze(E)

The refrigeration capacity of a compressor is the product of the evaporator enthalpy rise and the compressor mass flow rate. Table 3 shows the inlet volumetric flow rates required to achieve 350 ton of cooling. As can be seen HFO1234ze(E) requires a 33% (.601/.450) larger volumetric flow rate. Since compressors are essentially constant volumetric flow machines, drop-in compressor behavior will result in a capacity shortfall of 25.2% for HFC1234ze(E). However, drop-in of alternative refrigerants in centrifugal compressors requires a speed adjustment to insure surge-free operation without over-compressor (choke) in order to maintain compressor peak efficiency and turn-down capability [6]. As can be seen from Table 3, the isentropic enthalpy rise or head needed to compress the refrigerant from the evaporator saturation temperature of 5.56 °C to the condenser saturation temperature of 36.11 °C is quite different for these refrigerants. A compressor designed for HFC134a (19.55 kJ/kg) delivers 9% more head than is required for HFO1234ze(E) (17.98 kJ/kg). Compressor speed has to be reduced and capacity will further diminish. For pressure ratios around 2.5 the centrifugal compressor flow varies with compressor speed to the power 1.6 and head varies with compressor speed to the power 1.9 [7]. An additional capacity reduction of 7% with thus occur a result of the speed reduction needed for optimum compressor performance. The refrigeration capacity of an existing R134a compressor will be reduced by 32.2 % (25.2% + 7%) for HFO1234ze(E).

The platform of the four existing HFC134a centrifugal compressors come in frame sizes that have upward capacity jumps of about 50% (= downward capacity jumps of about 33%). Existing R134a compressor applications can be retrofitted for HFO1234ze(E) by just selecting the next available compressor frame size.

Table 3. Calculation of thermodynamic properties used to predict the compressor performance change in terms of capacity and speed when switching from HFC134a to HFO1234ze(E).

INPUT VALUES				
	Evaporature saturation	$T_{evap,sat}$	5.56 [°C]	
	Evaporator superheat	$\Delta T_{evap,sup}$	0.00 [°C]	
	Condenser saturation	$T_{cond,sat}$	36.11 [°C]	
	Condenser subcooling	$\Delta T_{cond,sub}$	3.89 [°C]	
	Refrigeration capacity		264 [kW]	
	Fluid		R134a	R1234ze
OUTPUT VALUES				
	Speed of sound [m/s]		146.71	139.07
	$\Delta h_{evaporator}$ [kJ/kg]		156.89	144.13
	$\Delta h_{s,compressor}$ [kJ/kg]		19.55	17.98
	mdot [kg/s]		1.68	1.83
	density [kg/m^3]		17.45	14.20
	Vdot [m^3/s]		0.096	0.129
	Pr(Pressure ratio)	-	2.57	2.60

3.4 Maintaining refrigeration capacity by moving up a frame size when replacing HFC134a with HFO1234ze

Moving up a compressor frame size has an efficiency advantage. Compressor efficiency increases with impeller size - both due to Reynolds number effect and reduced relative surface roughness. For the relatively small, high-speed centrifugal compressors in the 200 to 700 kW$_{th}$ capacity range, peak efficiency increases about 2% for each jump in frame size. Figure 2 shows the increase in efficiency when going from one frame size (TT300) to the next larger frame size (TT350).

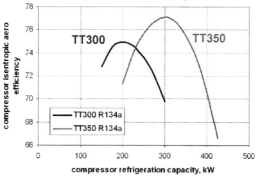

Figure 2. Comparison of aerodynamic efficiency for two adjacent compressor frame sizes.

Running the larger TT350 compressor with HFO1234ze(E) and comparing its performance against the performance of the TT300 compressor with HFC134a assuming identical aero efficiency results in equal refrigeration capacity with a 2% boost in compressor efficiency for HFO1234ze as shown in Figure 3.

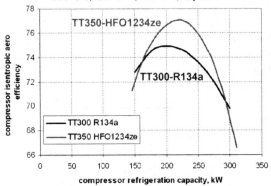

Figure 3. Comparison of aerodynamic efficiency for a TT300 DTC compressor with HFC134a against a next frame size TT350 compressor with HFO1234ze(E) assuming identical aero efficiency for HFO1234ze(E) and HFC134a.

A remaining question is whether compressor efficiency is affected by the change from HFC134a to HFO1234ze(E). Compressor performance is controlled by many factors. At identical impeller tip Mach number (u_2/a_0) we should expect identical performance only to be corrected for differences in frictional losses. The 4.5% drop in actual impeller speed - required for head and flow factor similarity - means that all fluid velocities will be 4.5% lower when the compressor is running with HFO1234ze(E) compared to HFC134a. Given the identical vapor kinematic viscosities of HFC134a and HFO1234ze(E), it is to be expected that the frictional and mixing losses which are proportional to the square of the fluid velocities will be reduced by almost 10.0%. Assuming an 80% fluid efficiency this would mean a reduction of the 20% fluid loss by 10.0% resulting in a 2 point increase in aero efficiency.

That anticipated compressor aero efficiency improvement was confirmed during back-to-back testing of TT300 and TT350 compressors with HFC134a and HFO1234ze(E). The test results for the TT300 compressor are summarized in Figures 4 and 5. The 2 point higher aero efficiencies at equivalent tip Mach numbers shown in Figure 4 will result in a correspondingly higher head factor for equal tip Mach number lines on the compressor map, as shown in Figure 5.

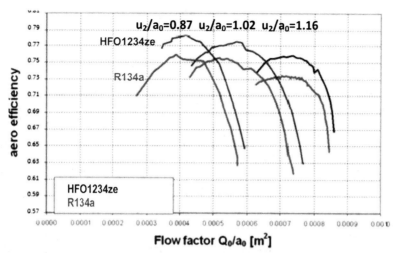

Figure 4. Back-to-back test results of a TT300 compressor with HFC134a and HFO1234ze(E) at three identical tip Mach numbers showing a 2-2.5 point aero efficiency improvement for HFO1234ze(E).

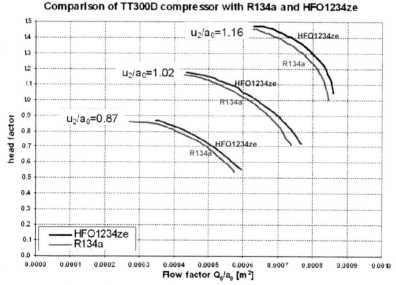

Figure 5. Back-to-back test results of the DTC TT300D compressor with HFC134a and HFO1234ze(E) at three identical tip Mach numbers showing a head factor increase corresponding to the measured aero efficiency improvement for HFO1234ze(E).

4. THE FIRST AIR-COOLED CHILLER WITH HFO1234ze(E)

An initial batch of twenty HFO1234ze(E) compressors was released in 2012 for research and development work and system integration by original equipment manufacturers. The higher volumetric flow rate of HFO1234ze(E) for a given chiller capacity will cause higher pressure drop in air-cooled condensers optimized for HFC134a duty, could result in undesirable liquid carry-over in the flow leaving the evaporator and might require a larger expansion valve.

A summary of the capacity and efficiency test results of an air-cooled chiller developed by STAR Refrigeration for the HFO1234ze(E) compressor is presented in Figure 6 [8]. This figure shows measured HFO1234ze(E) chiller capacities (diamonds) and coefficient of performance data (squares) as a function of ambient dry bulb temperature. The coefficient of performance falls as the ambient increases, as expected. The interesting part is that in all cases the coefficient of performance exceeds the COP value of same size HFC-134a chiller, shown by the solid red line. This chiller performance is agreement with the compressor test results shown earlier (Figures 3 and 4) that indicated improved aero efficiency with HFO1234ze(E).

After completion of the factory tests this first centrifugal chiller with a synthetic low-GWP refrigerant was shipped to its customer in Buffalo, NY, where it was installed and commissioned in the fall of 2012.

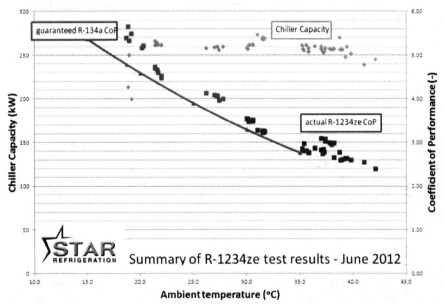

Figure 6. Test results of the STAR Refrigeration air-cooled chiller with the DTC TG310 HFO1234ze(E) compressor as a function of ambient temperature.

CONCLUSIONS

- DTC compressor capacity can be maintained when replacing HFC134a with HFO1234ze(E) by switching the existing fluid module to the next frame size fluid module.

- Testing at DTC has indicated the potential of a 4.0 to 4.5 % efficiency improvement for commercial chillers when switching from HFC134a to HFO1234ze(E). This improvement consists of a 2% benefit obtained by selecting the next frame size compressor combined with a 2-2.5% benefit as a result of the apparent lower viscous losses of HFO1234ze(E) versus HFC134a thanks to its lower impeller speed.
- This improved compressor efficiency has a bigger impact on reducing the carbon footprint of the chiller than the extremely low GWP value of HFO1234ze(E) of 6 versus 1300 for HFC134a.
- For air-cooled chiller systems using HFO1234ze(E) a condenser redesign might be is required to prevent excessive pressure drop as a result of the 50% higher condenser volumetric flow rate at equal capacity that could negate the potential performance benefit of HFO1234ze(E).

REFERENCES

1. Minor, B, Spatz, M, *HFO-1234yf low GWP refrigerant update*, Paper 2349 of the International Refrigeration and Air Conditioning Conference at Purdue, July 14-17, 2008.
2. Yana Motta, S.F., Vera Becerra, E.D., Spatz, M.W, *Analysis of LGWP Alternatives for Small Refrigeration (Plugin) Applications*, Paper 2499 of the International Refrigeration and Air Conditioning Conference at Purdue, July 12-15, 2010.
3. ARI 550/590 (I-P)-2011, *Performance Rating of Water Chilling Packages Using the Vapor Compression Cycle*, Air-Conditioning, Heating, and Refrigeration Institute (formerly ARI), 2011.
4. Sand, J.R., Fisher, S.K., Baxter, V.D., *TEWI Analysis: Its Utility, Its Shortcomings, and Its Results*, International Conference on Atmospheric Protection, Taipei, Taiwan, September 13-14, 1999.
5. NIST Reference Fluid Thermodynamic and Transport Properties Database (REFPROP): Version 9.0 http://www.nist.gov/srd/nist23.cfm
6. Brasz, J.J., *Centrifugal Compressor Behavior with Alternate Refrigerants* paper 96-WA/PID-2 presented at the 1996 ASME International Mechanical Engineering Congress and Exhibition, Atlanta, Ga. November 17-22, 1996.
7. Brasz, J.J., *Variable-Speed Centrifugal Compressor Behavior with Low GWP Refrigerants*, 2009 IMechE conference on Compressors and their Systems, September 7-9, 2009.
8. Pearson, A.B., *R-1234ze for variable speed centrifugal chillers*, Proceedings of the Institute of Refrigeration, London, UK, April 11, 2013.

The application of the Universal Modeling Method to development of centrifugal compressor model stages

Y B Galerkin, K V Soldatova
Compressor Dept., TU Saint-Petersburg, Russia

ABSTRACT

The test data of 16 compressors (2 – 8 stages, power 4,5 - 25 MW, delivery pressure up to 12,5 MPa) was reduced by the advanced version of Universal modeling computer programs (basic information on the Universal modeling was presented at the Conferences 1999 and 2001).

The stages of the compressors can be considered as 99 model stages with the flow rate coefficients 0,025 – 0,064, Euler work coefficients 0,40 – 0,85, relative hub 0,258 – 0,483, outer relative diameter of a diffuser 1,316 -1,720. Stages polytropic efficiency is 0,765 – 0,885 and surge limit ratio is 0,30 – 0,93 depending on a stages specific speed.

NOTATION

a_{s1} - sonic velocity at a blade suction side, m/s;
b_1 - blade height at an impeller inlet diameter, m;
c - absolute velocity, m/s;
c_r - radial velocity, m/s;
c_u - tangential velocity, m/s;
c_w - drag force coefficient;
D_2 - impeller diameter, m;
D_1 - impeller inlet diameter, m;
$\overline{F}$ - symbol of an exact flow path normalized dimensions;
h_{wmix} - head loss due to mixing at an impeller exit;
k - isentropic coefficient;
M - Mach number;
$M_u = \dfrac{u_2}{\sqrt{kRT_{inl\,tot}}}$;
p - pressure, Pa;
R_{bl} - radius of blade curvature, m;
Re - Reynolds number;

$\mathrm{Re}_u \dfrac{u_2 D_2}{\mu_{inl}} \dfrac{p_{inl\,tot}}{RT_{inl\,tot}}$;

T - temperature, K;
u_2 - impeller periphery speed, m/c;
$\overline{V}_{inl}$ - inlet flow rate, m³/min;
w - flow relative velocity, m/c;
w_1'' - flow relative velocity in a blade row throat, m/c;
X_i - empirical coefficient;
$\Phi = \dfrac{\overline{V}_{inl}}{0,785 D_2^2 u_2}$ - flow rate coefficient;

$\eta = \dfrac{\psi_p}{\psi_i}$ - polytropic efficiency;

$\pi = \dfrac{p_{exit}}{p_{inl}}$ - pressure ratio;

ρ - gas density, kg/m³;
τ - blade blockage coefficient;
$\psi = \dfrac{h_p}{u_2^2}$ - polytropic head coefficient;

$\psi_T = c_{u2}/u_2$ - Euler work coefficient;
ς - loss coefficient;

© The author(s) and/or their employer(s), 2013

Subscripts
1   - blade row inlet;
crit- - critical;
des - design regime;
exit - exit;
h   - hub;
inl - stage inlet;

jet - jet;
m   - mean value;
p   - blade pressure side;
s   - blade suction side, shroud;
sh - shroud.

INTRODUCTION

Industrial centrifugal compressors are used in wide range of flow rates, delivery pressures and pressure ratio. Constructive limitations add diversity to compressor designs. In most cases compressor manufacturers develop new compressor flow paths as a sum of previously designed and tested model stages. Problem arouses if existed model stages do not correspond to necessary gas dynamic and constructive parameters, or to modern efficiency level. New model stage development is long and costly process (1). Russian and some foreign manufacturers invite TU Saint-Petersburg Compressor Department to new compressor gas dynamic design when similarity theory cannot be applied. The Universal modeling computer programs (2, 3, 4) include primary design methodology (5) and the head loss model. Therefore the programs are able to design any necessary flow path and calculate proper performance curves. Variants' comparison leads to flow path optimization.

Two-decade long design practice demonstrates consistence of Universal modeling. Several dozens of designs are realized, guaranteeing necessary delivery pressure at given flow rate without preliminary tests. More than 350 compressors with total power above 4 megawatts operate in pipeline and other industries – a sample in Fig. 1.

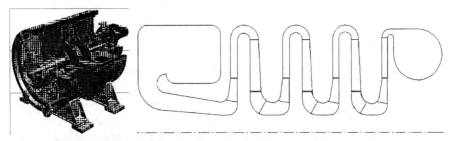

Fig. 1. Left – view of internal parts of two-stage pipeline compressor with delivery pressure 7,5 MPa, right – scheme of typical flow path of a pipeline booster compressor (5)

The typical sample of project and measured performance curves is presented in Fig. 2.

There is complete correspondence of parameters at $\overline{V}_{inl\,des}$ =270 m³/min and visible deviation at off-design flow rates. The Authors analyzed algorithms and models of 4th generation computer programs that were applied for design purposes. Some simplifications connected with low ability of old PC and incomplete modeling were responsible for modest results at off-design flow rates. The algorithms and models were improved and realized in the computer programs of 5th generation. The plant test data of 16 compressors were used to find new set of model empirical coefficients. The satisfactory modeling of performance curves in all performance range demonstrated validity of the new modeling instrument.

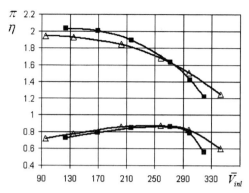

Fig. 2. Four-stage centrifugal compressor performance curves (6).
■ – project, Δ – plant test

The Authors suppose that good modeling of complete compressor performance curves means also that performance curve of its stages are predicted with acceptable accuracy. Altogether with known flow path shape and calculation method to predict similarity criteria influence the stages can be used in new designs the same way as traditional model stages are used.

MODERNIZATION OF MODELING TECHNOLOGY

Non dimensional stage performance curves $\psi, \eta = f(\Phi)$ depend on a flow path shape $\overline{F}$ and similarity criteria M, Re, k. I.e.:

$$\psi, \eta = f(\overline{F}, M, \mathrm{Re}, k, \Phi). \tag{1}$$

The general idea of the Universal modeling is to substitute the principally correct but non-applicable to calculation equation (1) by the equation (2):

$$\psi, \eta = f(\overline{W}, \overline{F}', M, \mathrm{Re}, k, \Phi). \tag{2}$$

Here $\overline{W}$ is a symbol of non viscid velocity diagram. Symbol $\overline{F}'$ represents several simple geometry parameters - aspect ratio, blade row density, etc.

This equation is deployed as sum of algebraic equations. The equations are for one-dimensional thermo - gas dynamic calculation of flow parameters in control planes and for head loss evaluation based on proper loss models. For instance, friction loss coefficient for shroud and hub surfaces of an impeller is equal (5):

$$\zeta_o = c_{wo} \frac{\left(1-(D_1/D_2)^2\right)\tau_m}{\Phi} \frac{w_m}{u_2} \frac{\rho_{inl}}{\rho_m}\left(\frac{w_m}{w_1}\right)^2, \tag{3}$$

where D_1 is mean inlet diameter.

The equation (3) includes a drag force coefficient that is a subject of modeling:

$$c_{w0} = c_f \left(1 + X_i \left(1 - \frac{w_2}{w_1}\right)^{X_{i-1}}\right)\left(1 + X_{i+2}\left(4 \pm \frac{w_m/u_2}{R_{bl}/D_2}\right)^{X_{i+3}}\right). \tag{4}$$

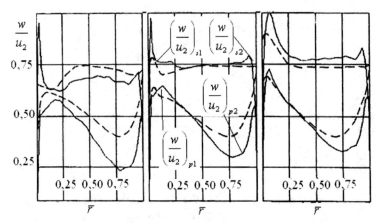

Fig. 3. Measured (stroke) and calculated (solid) non-viscid velocity diagrams at an industrial compressor impeller blade. Design flow rate. Hub, mean and shroud surfaces (6) $\bar{r} = (r - r_1)/(r_2 - r_1)$

Here c_f is a drag force coefficient for thin plate that is calculated by well-known formulae of Prandtl–Schlichting that take into account Reynolds number and roughness influence. The member in the first parentheses reflects influence of flow deceleration, in the second parentheses – of a normalized velocity gradient (analog of a Rossby number). Empirical coefficients X_i are objects of loss model identification, i.e. of mathematical reduction of numerous model stages test data. Unlike it the 4-th generation model has not taken into account surface roughness. It created uncertainties and inconveniencies. In fact the 4-th generation model treats surface as hydraulically smooth. It leads to incorrect results for high – pressure compressors. The problem is eliminated in 5-th generation model.

The loss model for an impeller blade row takes into account the difference of flow velocities at pressure and suction sides of blades. In equations equivalent to (3), (4) participate mean velocities along suction and pressure sides w_{ms}, w_{mp}. The definition of w_{ms}, w_{mp} is based on four parameters of non-viscid flow velocity diagrams at blades as shown in Fig. 3. Non viscid calculated velocity diagrams are compared with measured ones there.

The measurement technique is shortly presented in [6]. The principle of a multi-point pressure transducer operation is presented in fig. 4. An impeller with drainage pipes prepared for pressure measurements is shown in fig. 5. The real device used in experiments is presented in fig. 6. This is the combination of the single point pressure transducer and of 34 points selection mechanism to pass measured pressures at different points to a manometer.

Surface velocities are calculated on a base of measured static pressures and total pressures at inviscid core. The last are easily calculated by the Bernoulli equation for rotating coordinate system.

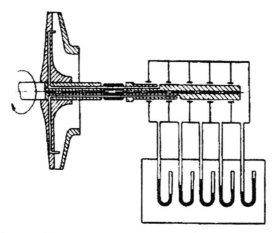

Fig. 4. Scheme of pressure measurements in rotating impeller [6]

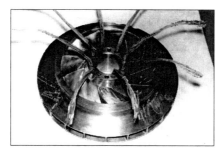

Fig. 5. Impeller with drainage pipes prepared for pressure measurements [6]

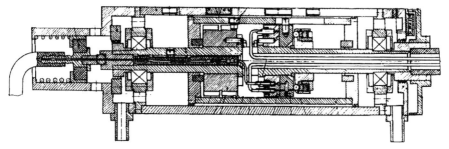

Fig. 6. Single-point pressure transducer with 34 points selection mechanism for pressure measurements [6]

The comparison demonstrates alikeness at mean and shroud surfaces - everywhere but the end of suctions side where wake zone appears. In accordance with applied loss model it is necessary to define velocities near leading and trailing edges on both sides of a blade as shown in Fig. 7. Kinetic energy $0,5(w_1 + w_2)^2$ controls surface friction on any of blade sides. Flow deceleration at a suction side w_{s2}/w_{s1} controls wake formation and mixing losses. Maximum Mach number of a relative flow w_{s1}/a_{s1} controls negative effects of high compressibility.

Fast – operating engineering level programs do not include non-viscid calculations. At the loss models of 4th and earlier generations the necessary four velocities were calculated on a base of very approximate schematization. Solid straight lines shown in Fig. 7, left diagram, demonstrate this method.

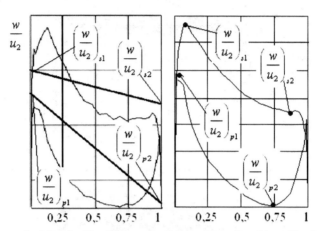

Fig. 7. Non-viscid velocity diagrams along normalized length at a shroud blade-to-blade surface and two ways of its schematization

The Authors applied results of numerical experiment presented at (7, 8) to improve the velocity diagram description. About one hundred of impellers in range of flow rate coefficients 0,020 – 0,090, Euler work coefficients 0,45 – 0,80, relative hub 0,20 – 0,45 were designed and their velocity diagrams were calculated and mathematically reduced. The approximating equations for all four velocities were developed. One of equations was proposed by MSc Eng. A. Drozdov, who also has made programming of new models and to whom the Authors are grateful for co-operation. To calculate maximum local velocity on a blade MSc Eng. A. Drozdov proposed the equation that use geometry and gas dynamic parameters:

$$\frac{w_{s1}}{u_2} = \frac{w_1''}{u_2} + 0,5\left(2,06\frac{b_1}{1-D_1} - 0,261\psi_{T\,des} + 1,69\left(\frac{D_1}{D_2}\right)^{2,8}\right)\frac{(w_s - w_p)_{mean}}{u_2}. \quad (5)$$

The sample of calculation by the approximation formulae is presented in Fig. 7 (right) as black points.

More realistic velocity diagram description increases precision of friction losses calculation and that of mixing losses calculation as well. Elevated level of a flow velocity at impeller exit in comparison with viscid flow is visible in Fig. 3.

The principle of mixing losses evaluation is illustrated by Fig. 8. Two simplifications are accepted: zero velocity at a wake zone, instant mixing after blade ends. Sudden expansion equation is applicable in this case:

$$h_{w\,mix} = 0,5(c_{r2\,jet} - c_{r2})^2 \cdot X_i. \quad (6)$$

Unlike the earlier models, the empirical coefficient X_i in the Eq. (6) is not a figure but the function of velocity deceleration in a blade row and of a flow deflection from an inertial direction.

These and a dozen of other improvements have lead to sufficient improvement of computer programs of 5th generation. As it is shown below the test performance of 16 high – efficiency modern compressors were accurately modeled with minimum variation of empirical coefficients X_i.

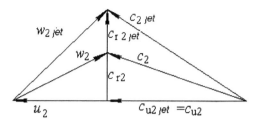

Fig. 8. Jet – wake and uniform flow velocity triangles at an impeller exit

PLANT TEST PERFORMANCE CURVE MODELING

Samples of test results for several compressors are presented in Fig. 9.

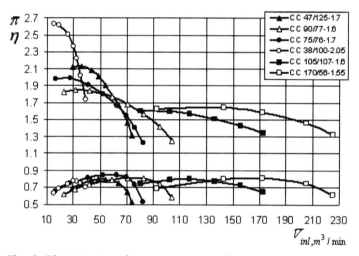

Fig. 9. Plant test performance curves for compressors with
$\overline{V}_{inl}$ = 44 – 170 m³/min

The highest efficiency registered at a design point in a course of 16 different compressor plant test (air) is 87,6%, the lowest – 76%. All compressors were designed according to the same principles (5). The difference in efficiency is due to stages' specific speed, Euler work coefficient, hub ratio, vaneless diffuser radial length.

Typical results of performance modeling are demonstrated in Fig. 10. Presented samples relate to compressors whose names contain information on their main parameters. The name "CC246/76-1,71" means that a centrifugal compressor with $\overline{V}_{inl}$ = 246 m³/min has delivery pressure 76 bar and pressure ratio 1,71.

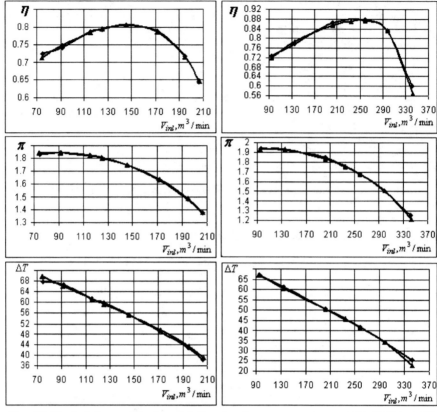

Fig. 10. Test data and their modeling for two compressors CC146/71-1,75 (left) and CC249/76-1,71(right).
Above – efficiency, middle – pressure ratio, below – temperature rise. ♦ – test, ▲ - calculation

MODEL STAGES' PERFORMANCE CURVES DERIVED FROM TEST DATA

The samples of accurate modeling of 16 compressors test data are presented in fig. 10. The compressor parameters: 2 – 8 stages, power 4,5 - 25 MW, delivery pressure up to 12,5 MPa. There were calculated performance curves of 99 model stages with the flow rate coefficients 0,025 – 0,064, Euler work coefficients 0,40 – 0,85, hub ratio 0,258 – 0,483, outer relative diameter of a diffuser 1,316 -1,720. Stages polytropic efficiency is 0,765 – 0,885 and surge limit ratio as $\overline{V}_{inl\,crit}/\overline{V}_{inl\,des}$ = 0,30 – 0,93 depend on a stages specific speed.

It is important to model properly temperature rise because compressor efficiency is calculated as:

$$\eta = \frac{\ln\left(\dfrac{P_{exit}}{P_{inl}}\right)}{\dfrac{k}{k-1}\ln\left(\dfrac{T_{exit}}{T_{inl}}\right)}. \tag{7}$$

All modeling results seem to be satisfactory, therefore calculated performance curves of their stages are reliable enough to be used in design practice.

The stages operated at certain values of Mach M_u, Reunolds Re_u and isentropic coefficient $k = 1,4$ at test conditions. While applied as model stages the other values of these criteria would be necessary. Transformation of performance curves under similarity criteria variation is estimated by 5th generation computer programs.

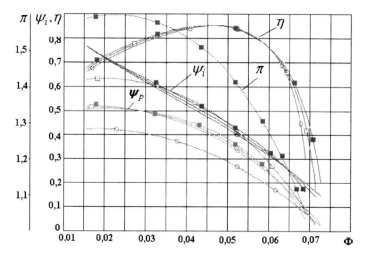

Fig. 11. Performance curves of the stage "impeller + vaneless diffuser + exit nozzle" with proportional change of Mach and Reunolds criteria.
○ - M_u =0,6, Re_u =6E6, □ - M_u =0,725, Re_u =7,25E6, ■ - M_u =0,85, Re_u =8,5E6

Samples of typical calculations are presented below. Fig. 11 demonstrates performance curves variation under RPM change in dimensionless form – as influence of proportional change of M_u and Re_u.

Variation of both parameters in Fig. 11 does not influence efficiency at design flow rate. Deterioration of efficiency at off-design flow rates is result of negative compressibility (M_u) influence. Work and polytropic coefficients are slightly bigger at higher Mach numbers due to increased compressibility.

Another situation takes place when a stage with slightly different Φ_{des} from existing stage is necessary for a designed flow path. Then existing stage flow path channel widths can be changed proportionally. The sample of possible transformation is presented in Fig. 12.

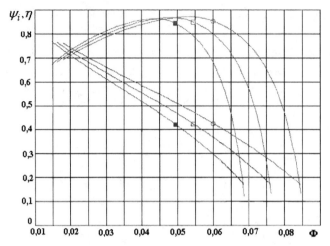

Fig. 12. Performance curves of the existing stage with $\Phi_{des} = 0{,}060$ and of derived stages with channel widths reduced by 10 and 19%

Proposed model stages are easy to use in design of single-shaft non cooled compressors. Special computer program contain information on geometry of stages and proper empirical coefficients. Standard program of the Universal modeling proposes variants with different RPM, different number of stages and different Euler coefficients of impellers. All main information on necessary stages ($\Phi_{des}, \psi_{t\,des}, M_u, Re_u$) is presented and stages can be chosen from the database. The computers program CCPM-G5E calculates performance curves in a chosen range of inlet parameters and RPM.

The new booster pipeline compressor was designed by one of compressor producers with the use of model stages of an existing generation when stages with increased Euler coefficients were not effective. Test performances of this compressor CC 105/107-1,60 are presented in Fig. 13. The alternative design was made by the Authors with use of new model stages. It promises increased efficiency up to 4%.

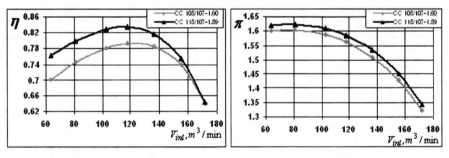

Fig. 13. Test performance curves of a booster compressor based on model stages of old generation (CC 105/107-1,60) and of its analog designed on a base of proposed stages (CC 115/107-1,59) (5)

CONCLUSION

Several improvements introduced to the loss and work input models made the 5-th generation modeling program more powerful instrument for centrifugal compressor design and calculations. Design point efficiency and pressure ratio of different stages and compressors is modeled well by a common set of empirical coefficients. The previous generation programs need individual groups of coefficients for stages with different flow rate coefficients and Euler work coefficients. Modeling of off-design flow rates is still a problem. Individual sets of coefficients for modeling incidence losses were applied for 16 compressors whose stage performances are recommended for design practice.

REFERENCES

1. Japikse, D., and Oliphant, K. N. "Turbomachinery Modeling: Explicit and Implicit Knowledge Capturing (2005A)"; presented at the ASME Turbo Expos 2005: Power for Land, Sea and Air. – ASME Paper No. GT2005-68104. – Reno-Tahoe. – Nevada. – 2005.
2. Galerkin Y.B., Danilov K.A., Popova E.Y. Universal Modeling for Centrifugal Compressors-Gas Dynamic Design and Optimization Concepts and Application. Yokohama International Gas Turbine Congress, Yokohama. – 1995.
3. Galerkin Y., Danilov K., Popova E. Design philosophy for industrial centrifugal compressor. International Conference on Compressors and their systems. – London: City University, UK. – 1999.
4. Galerkin Y., Mitrofanov V., Geller M., Toews F. Experimental and numerical investigation of flow in industrial centrifugal impeller. International Conference on Compressors and their systems. – London: City University, UK. – 2001.
5. Galerkin, Y. B. Turbo compressors. // LTD information and publishing center. – Moscow. – 2010 (in Russian).
6. Galerkin, Y. B., Soldatova, K.V. Operational process modeling of industrial centrifugal compressors. Scientific bases, development stages, current state. Monograph. // Saint-Petersburg. - SPbTU. – 2011 (in Russian).
7. Galerkin, Y.B., Lysiakova A. Analysis and generalization of charts for surface velocities for centrifugal compressor's impellers. Part 1. // Compressors & Pneumatics. – Moscow. – 2010. – No. 6. – Page 4-11 (in Russian).
8. Galerkin, Y.B., Lysiakova A. Analysis and generalization of charts for surface velocities for centrifugal compressor's impellers. Part 2. // Compressors & Pneumatics. – Moscow. – 2010. – № 7. – Page 26-33 (in Russian).

Flow and performance investigation of the specially designed channel diffuser of centrifugal compressor

M Kalinkevych, A Skoryk
Technical Thermophysics Department, Sumy State University, Ukraine

ABSTRACT

The results of flow pattern and performance investigation of the specially designed channel diffuser of centrifugal compressor are presented in the paper. The design concept is to provide high pressure recovery of the diffuser by assuming the preseparation condition of the boundary layer along one of the channel surfaces.

Experimental model of the designed channel diffuser has been constructed to verify the flow structure assumed while designing. The numerical simulation of the diffuser was implemented by means of CFD-software. Obtained gas dynamic characteristics of the designed diffuser were compared to the base diffuser of the compressor stage.

Keywords: centrifugal compressor, channel diffuser, design method

1 INTRODUCTION

The operating conditions and purpose of centrifugal compressor should be taking into account to choose the type of its diffuser. Channel diffusers (CD) could be more preferable as compared to other types in the following cases: 1) at small flow angles at the diffuser inlet; 2) when the gas passes from the diffuser channels to the separated cameras; 3) if the diffuser channels turn to the channels of the return element of compressor.

Traditional geometry of channel diffusers includes the initial section shaped as a logarithmic spiral along the one of the vane surfaces and the main section with straight walls (Figure 1a). The channel diffusers with wedge vanes are also widely used (Figure 1b).

Some authors recommend to use the Runstadler et al. [1] and Reneu et al. [2] database for the flat plane diffusers to design wedge-shaped channel diffusers. Kano et al. [3] and Clements [4] indicated that such data cannot be used to design the high-performance diffuser of centrifugal compressor because of distorted three-dimensional swirled flow at the inlet. In addition, such database does not cover the wide range of geometries and inlet flow conditions.

There is no clear indication of CD being superior to the vaned diffuser (VD) with airfoil vanes. Also it is doubtful that straight surfaces of the wedge vane provide the aerodynamically favorable flow.

Generally, the most effective vane geometry may be obtained by solving the inverse problem of gas dynamics. The principles of design method for high-performance channel diffusers, in which vane thickness increases along the radius, are presented in the paper.

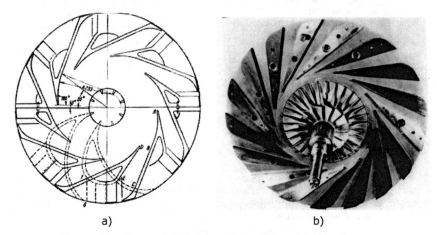

a)   b)

Figure 1 – Channel diffusers of traditional geometry:
a) channel diffuser with initial section shaped as a logarithmic spiral
b) channel diffuser with wedge vanes

2 DESIGN METHOD

According to the presented method the diffuser design is based on the preseparation condition of the boundary layer along one of the vane surfaces. Stratford [5] obtained experimentally the close to zero skin friction flow along one of the surfaces of the two-dimensional diffuser. In that case the maximum pressure rise which is possible for given diffuser dimensions should be achieved.

Mathematical model for solving the inverse problem of gas dynamics is developed for the steady adiabatic gas flow without separations.

The angular momentum change about axis z for the annular element of gas with width Δb and mass flow rate $\Delta \bar{m}$ from the diffuser inlet r_{in} to the current section r:

$$\Delta M = \Delta \bar{m} \cdot (r_{in} \cdot c_{in} \cdot \cos\alpha_{in} - r \cdot c \cdot \cos\alpha), \quad (1)$$

where r, c, α are the values of radius, velocity and flow angle at the current section respectively (Figure 2).

Moment of forces acting on the z_v vanes of diffuser from r_{in} to r:

$$\Delta M = \Delta b \cdot z_v \cdot \int_{r_{in}}^{r} \Delta p \cdot r \cdot dr, \qquad (2)$$

where $\Delta p = (p_{ps} - p_{ss})$ is the pressure difference between the pressure and suction surfaces of the vane.

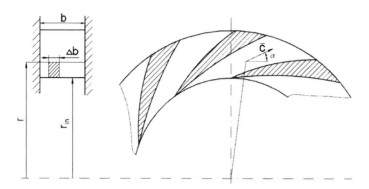

Figure 2 – Channel diffuser scheme

Continuity equation is of the form

$$\Delta \bar{m} = c_r \cdot \rho \cdot 2\pi \cdot r \cdot \Delta b \cdot \tau, \qquad (3)$$

where c_r is the radial velocity, ρ – gas density, τ – blockage factor.

The equations (1), (2) and (3) may be represented using gas dynamics relations for isentropic flow:

$$\begin{cases} r_{in} \cdot \lambda_{in} \cdot \cos\alpha_{in} - r \cdot \lambda \cdot \cos\alpha = \\ = \dfrac{b \cdot z_v \cdot p_{in}^*}{\bar{m} \cdot a_{cr}} \cdot \int_{r_{in}}^{r} \left[\pi(\lambda_{ps}) - \pi(\lambda_{ss})\right] \cdot r \cdot dr, \\ \bar{m} = \lambda \cdot \varepsilon(\lambda) \cdot a_{cr} \cdot \rho^* \cdot 2\pi \cdot r \cdot b \cdot \tau \cdot \sin\alpha, \end{cases} \qquad (4)$$

where $\lambda, \lambda_{ps}, \lambda_{ss}$ ($\lambda = c/a_{cr}$, $a_{cr} = \sqrt{\dfrac{2\gamma}{\gamma+1} RT^*}$ is the critical velocity, γ is the ratio of specific heats) are the mean flow velocity within the diffuser channel, velocity along the pressure surface and velocity along the suction surface of the vane respectively; α is the mean flow angle within the diffuser channel; stagnation parameters are designated with "*".

Pressure and density relations are determined as a function of velocity:

$$\pi(\lambda) = \frac{p}{p^*} = \left(1 - \frac{\gamma-1}{\gamma+1} \cdot \lambda^2\right)^{\frac{\gamma}{\gamma-1}}, \quad \varepsilon(\lambda) = \frac{\rho}{\rho^*} = \left(1 - \frac{\gamma-1}{\gamma+1} \cdot \lambda^2\right)^{\frac{1}{\gamma-1}}. \qquad (5)$$

The blockage factor is given by the following equation

$$\tau = 1 - \frac{\delta' \cdot z_v}{2\pi \cdot r \cdot \sin\alpha},\qquad(6)$$

where $\delta' = \delta_v + \Sigma\delta^*$ - modified vane thickness; δ_v – vane thickness, $\Sigma\delta^*$ - total displacement thickness of boundary layers in the vane channel estimated using the Loycianskiy method [6].

Velocity distribution, which provides the preseparation condition of the boundary layer along the pressure surface of the vane, is defined by the formula [7]:

$$\lambda_{ps} = \lambda_1 \cdot \left[1 + \frac{(\bar{l} - \bar{l}_1)\cdot(2+H_s)\cdot(-f_s)}{\bar{\delta}_1^{**}}\right]^{-\frac{1}{2+H_s}}.\qquad(7)$$

where H_s and f_s are the shape factors for the preseparation condition of boundary layer; $\bar{l}$ is the length coordinate along the vane surface related to the vane center line length, $\bar{\delta}^{**}$ is the momentum thickness related to the vane center line length. Parameters marked with subscript "1" are the coefficients, which affect the given velocity distribution quantitatively.

The set of equations (4) includes unknowns λ, δ_v, λ_{ss}. The dependence $\alpha = f(r)$ may be given as linear.

For linear pressure distribution along the vane pitch the relation between the velocities may be defined as

$$\pi(\lambda_{ps}) + \pi(\lambda_{ss}) = 2\cdot\pi(\lambda).\qquad(8)$$

Equation (8) allows to eliminate one unknown, so set of equations (4) can be solved using numerical methods.

The initial data for the design are:
- gas properties (gas constant R, the ratio of specific heats γ, kinematic viscousity ν);
- static pressure p_{in} and static temperature T_{in} at diffuser inlet;
- inlet and outlet flow angle (α_{in}, α_{out});
- mass flow rate;
- geometrics of the meridional contour;
- quantity of vanes z_v;
- mean flow angle distribution along the diffuser channel ($\alpha = f(r)$).

As a result of calculation the vane thickness distribution along the radius $\delta_v = f(r)$ is estimated, so the geometry of the vane is totally defined.

The compressor design point flow parameters at the impeller exit are used as mentioned above initial data for the diffuser design. Presented design method is valid for the subsonic flow along the entire diffuser.

3 APPLICATION OF THE DESIGN METHOD FOR THE CENTRIFUGAL COMPRESSOR MODEL STAGE

Using presented method, the CD for the model centrifugal compressor stage of JSC "Sumy Frunze NPO" was designed. The design was implemented for the parameters at the diffuser inlet at the design point of the stage. The parameters of the flow were obtained by numerical simulation of the base model compressor stage with VD.

The values of design parameters are shown in Table 1. The meridional contour geometry, inlet and outlet angles of the vanes are the same as for the base VD.

Table 1 – Initial data for channel diffuser design

Pressure at the diffuser inlet, Pa	118000
Temperature at the diffuser inlet, K	319
Mass flow rate, kg/s	1,775
Quantity of vanes	17
Inlet radius, r_3, m	0,2622
Outlet radius, r_4, m	0,3092
Width of diffuser, b_3, m	0,0155
Vane leading edge thickness, m	0,003
Vane leading edge centerline angle, α_{3v}, °	22
Vane trailing edge centerline angle, α_{4v}, °	37

The given velocity distribution for the CD design is shown in Figure 3. The geometry parameters of the vane are shown in Figure 4. Vane thickness distribution was obtained as a result of design calculation.

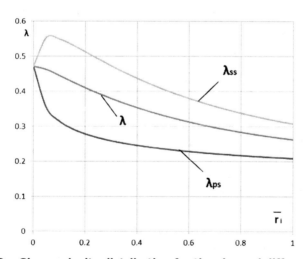

Figure 3 – Given velocity distribution for the channel diffuser design

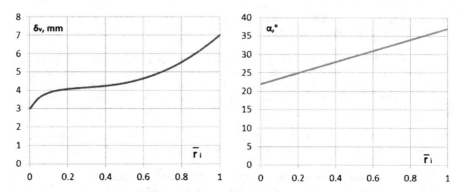

Figure 4 – Vane thickness and angle distribution for the designed channel diffuser

The relative radius in Figures 3 and 4 is defined by formula

$$\bar{r}_i = \frac{r_i - r_{in}}{r_{out} - r_{in}}. \tag{9}$$

4 EXPERIMENTAL FACILITY AND MEASUREMENTS

The experiments were carried out for the compressor stage with base VD. The sectional view of the test rig of investigated centrifugal compressor model stage is shown in Figure 5. The centrifugal compressor is driven by 400 kW electric motor. The parameters of the tested stage are shown in Table 2.

Table 2 – The parameters of the tested compressor stage

Impeller exit diameter D_2, m	0,460
Impeller exit width, m	0,0147
Quantity of impeller blades	25
Exit blade angle (from tangential axis positive in direction of rotation), deg	76
Rotation speed of impeller, rev/min	8500
Diffuser vane leading edge relative diameter, D_3/D_2	1,140
Diffuser vane trailing edge relative diameter, D_4/D_2	1,344
Diffuser width, m	0,0155
Quantity of diffuser vanes	21
Vane leading edge centerline angle, α_{3v}, deg	22
Vane trailing edge centerline angle, α_{4v}, deg	37

For the base VD following measurement scheme was used:
- four static pressure taps at diameter $D/D_2 = 1,076$;
- four static pressure taps at diameter $D/D_2 = 1,391$;
- four total pressure probes at diameter $D/D_2 = 1,076$, traversed flow in five width points each;
- four total pressure probes at diameter $D/D_2 = 1,391$, traversed flow in five width points each.

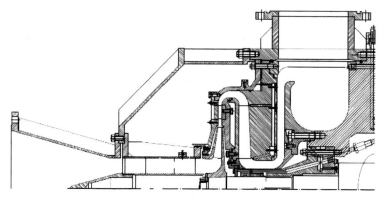

Figure 5 – Sectional view of the investigated centrifugal compressor model stage

Such measurement scheme allows to calculate area averaged static pressures and mass averaged total pressures at mentioned control sections. Mass flow was measured using an orifice meter at compressor inlet. The compressor inlet air was taken from a fresh air duct.

To evaluate the diffuser aerodynamic performance the pressure recovery coefficient

$$C_p = \frac{p_{out} - p_{in}}{p_{in}^* - p_{in}} \qquad (10)$$

and total pressure loss coefficient

$$\zeta = \frac{p_{in}^* - p_{out}^*}{p_{in}^* - p_{in}} \qquad (11)$$

were used.

Measured characteristics of VD are shown in Figure 7 for the ease of comparison with numerical simulation.

5 NUMERICAL SIMULATION

Numerical simulation was performed by use of commercial CFD-software ANSYS CFX v.14 for two different compressor stages. The first one is the model compressor stage of JSC "Sumy Frunze NPO" with base VD. The second stage has the same impeller as the first. The only difference is the diffuser, which was designed using presented method (see paragraph 3).

5.1 Grid quality and preprocessor set-up

The structured hexahedral grids for impeller and diffusers were created in ANSYS Turbogrid. SST – turbulence model is the most acceptable model for the centrifugal compressor flow simulations [8], which in the case of sufficient grid refinement shows appropriate results for the near-wall boundary layers and flow core.

While creating the near-wall prismatic layers, it was checked that value of y^+ is less than 2. The more coarse grid was created for the flow core, which is acceptable for the SST – turbulence model. The impeller grid consists of 690 690 elements (Figure

6a). The CD grid consists of 637 296 elements (Figure 6b) and VD grid (Figure 6c) consists of 618 618 elements. Due to the differences in geometry of these diffusers, it is not possible to create totally topologically identical grids. The parameters which define the topology of the near-wall prismatic layers and the quantity of grid elements in meridional plane were identical. Therefore this topological difference is acceptable for the comparison. The difference is only in the flow core elements quantity.

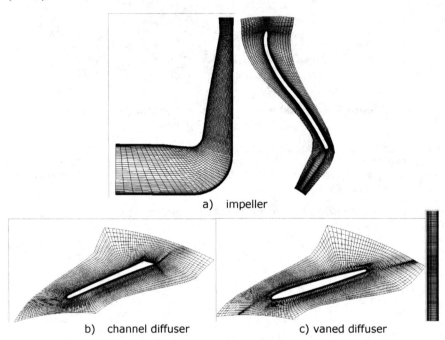

a) impeller

b) channel diffuser   c) vaned diffuser

Figure 6 – Grids of the stage elements

The main parameters specified in ANSYS CFX preprocessor are shown in Table 3. To connect the respective surfaces of the impeller and diffuser the interface "stage" was used. This type of interface is usually used for the steady state calculations; the parameters at the interface surfaces are averaged circumferentially. Therefore interface "stage" is oriented for estimating of the integral characteristics of the compressor stage.

Table 3 – Boundary conditions and models, specified in ANSYS CFX preprocessor

Impeller inlet	Total pressure, total temperature
Diffuser outlet	Mass flow rate
Interface type between impeller and diffuser	Stage
Turbulence model	SST
Heat transfer model	Total energy
Fluid model	Air Ideal Gas
Wall heat transfer model	Adiabatic

As the convergence criteria the discrepancy in static pressure recovery coefficient equal to 0,01 and in total pressure loss coefficient equal to 0,005 have been used.

The grid independence study showed that for the design point the diffusers' non-dimensional characteristics change is distinct for the diffusers node quantity less than 400 000. For off-design conditions the results of simulation are more grid-sensitive.

The main purpose of simulations was to establish the more effective diffuser by comparing their non-dimensional characteristics. Generated grids were indicated as sufficient for such calculations.

5.2 Simulation results

Obtained non-dimensional diffusers' characteristics are shown in Figure 7 as a function of incidence angle

$$i_3 = \alpha_{3v} - \alpha_3, \qquad (12)$$

where α_{3v} is vane centerline angle at the diffuser inlet, α_3 is the flow angle at the diffuser inlet.

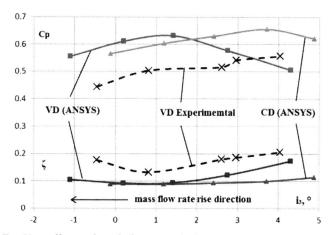

Figure 7 – Non-dimensional characteristics of designed CD and base VD

It can be seen that characteristics of CD are better than of VD at the range of positive incidence angles. Moreover, maximal value of pressure recovery of CD is slightly higher than of VD. The shape of characteristics of CD is more flat, especially for total pressure loss coefficient. In order to understand these results the flow pattern within the diffusers was examined.

Figure 8 presents the velocity vectors within both diffusers for $i_3 \approx 4°$. Predicted high loss level and low pressure recovery of VD occurs due to the wide flow separation region. Due to the high pressure gradients the flow separation occurs on the suction side surface of the vane and covers the area from the hub side to midspan. Reducing of the effective area resulted in low pressure recovery.

The flow pattern within the CD is more favorable due to the controlled flow deceleration provided by the designed vane geometry. For the operating conditions range of $i_3 \approx 0° \div 4°$ the total losses includes only the friction loss and wake mixing loss at diffuser outlet (Figure 8 and Figure 9). Therefore the shape of the loss characteristic of CD is almost straight. The flow separation within the CD was detected at $i_3 \approx 5°$.

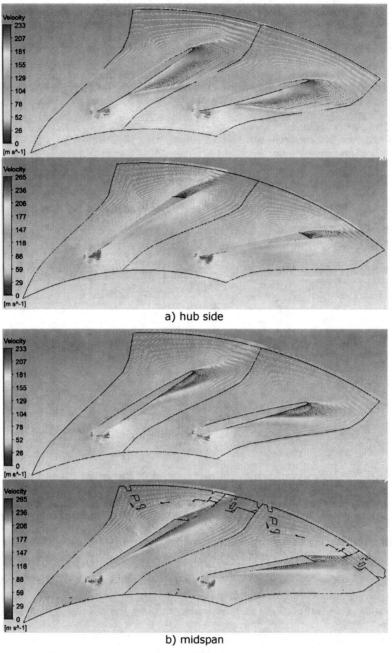

Figure 8 – Velocity vectors for different flow surfaces of VD (top) and CD (bottom) at $i_3 \approx 4°$

As it can be seen from Figure 9 for the top pressure recovery operating condition of VD at $i_3 \approx 1°$ the pressure recovery of CD is slightly lower, whereas the losses are practically the same.

The shape of experimental loss characteristic of VD is close to the obtained numerically. The experimental and numerical C_p characteristics differ more substantially, especially at low mass flow rates. Probably, such qualitative difference is due to the total pressure probe measurement uncertainty at the regions of the strong flow non-uniformity and separation.

However, qualitatively the results of the numerical simulation are seemed to be plausible. The good agreement between VD experimental loss characteristic and prediction is one of the reasons to expect that experimental characteristics of CD would be close to predicted.

Further step of this research will include the detailed experimental investigation of the designed CD to estimate the real advantages of such diffuser type and verify the validity of presented design method.

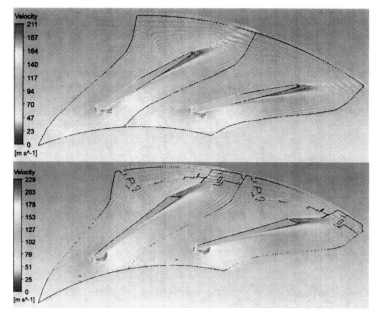

Figure 9 – Velocity vectors for the midspan plane of VD (top) and CD (bottom) at $i_3 \approx 1°$

6 CONCLUSIONS

According to the numerical simulation results, the replacement of the vaned diffuser by channel diffuser designed using presented method allows to increase the efficiency of compressor stage at the range of low mass flow rates because of the flow separation prevention and friction area decrease in diffuser.

The channel diffusers' characteristics are more flat and stable for the positive incidence angles as compared to the vaned diffuser. Therefore application of channel diffusers may be more preferable in the case of variable operating conditions of the compressor stage.

REFERENCE LIST

1. Reneau, L.R., Johnston, J.P., Kline, S.J. Performance and design of straight, two-dimensional diffusers. J. Basic Engg. 89, 1967, 141 – 150.
2. Runstadler, Jr., P. W., and Dean, Jr., R. C. Straight Channel Diffuser Performance at High Inlet Mach Numbers", 1969, ASME J. Basic Eng., 91, pp. 397–412.
3. Kano F., Tazawa N., Fukao Y. Aerodynamic Performance of Large Centrifugal Compressors.// J. Eng. Power - October 1982 - Volume 104, Issue 4, 796 (9 pages).
4. Clements, W.W. and Artt, D.W. The Influence of Diffuser Channel Length-Width Ratio on the Efficiency of a Centrifugal Compressor. Proceedings of the Institution of Mechanical Engineers, Part A: Journal of Power and Energy August 1988 202: 163-169.
5. Stratford, B.S. 1959. An experimental flow with zero skin friction throughout its region of pressure rise, *ASME J. Fluid Mech.*, Vol. 5: pp. 17-35.
6. Loycyanskij, L.G. 2003. Mekhanika zhidkosti i gaza [Fluid Mechanics (in Russian)]. Moscow: Drofa. 840 p.
7. Kalinkevych, M., Obukhov, O., Smirnov, A., Skoryk, A., 2011, The design of vaned diffusers of centrifugal compressors based on the given velocity distribution, 7th International Conference on Compressors and their Systems, Woodhead Publishing: p. 61-69.
8. Menter, F., Kunitz, M., Langtry, R., 2003, Ten Years of Industrial Experience with the SST Turbulence Model, J. Turbulence, Heat and Mass Transfer, vol. 4: p. 625-632.

Investigation of gas flow with injection in vaneless diffuser of centrifugal compressor

M Kalinkevych, O Shcherbakov, V Ihnatenko
Sumy State University, Ukraine

ABSTRACT

Efficiency of vaneless diffusers at low flow rates decreases because of flow separation and rotating stall inception. To prevent these phenomena injection into boundary layer can be used.

Mathematical model for gas flow in vaneless diffuser with mass, momentum and energy addition has been developed. Parametric variations of injector location, injection flow rate and injection direction were investigated numerically using developed model. According to the simulation results the test injection system for experimental investigation of injection in vaneless diffuser has been designed.

Keywords: vaneless diffuser, flow separation, injection

NOTATION

C	averaged absolute velocity, m/s;
C_r, C_u	averaged radial, averaged circumferential component of velocity, m/s;
r	radius, m;
ρ	averaged density, kg/m^3;
b	diffuser width, m;
$\dot{m}$	mass flow, kg/s;
a	averaged flow angle, ^{0};
β	injection angle, ^{0};
A	area, m^2;
C_f	skin friction coefficient;
p^*	averaged stagnation (or "total") pressure, Pa;
p	averaged static pressure, Pa;
T^*	stagnation (or "total") temperature, K;
T	static temperature, K;
c_p	specific heat at constant pressure, $J/(kg \cdot K)$;
q	heat addition per unit mass, J/kg;
k	constant from Eq. (14);
Re	Reynolds number;
M	Mach number;
δ_r^{**}	momentum thickness, m;
Γ	the Buri shape-factor;

v	kinematic viscosity, m^2/s;
D	diameter, m;
$\bar{b} = b/D_{2-2}$	width ratio;
$\bar{D} = D/D_{2-2}$	dimensionless diameter;
$\bar{c} = c/U_{2-2}$	relative velocity;
U	tip speed, m/s;
C_p	static pressure recovery coefficient;
ζ	total pressure loss coefficient;

Subscripts

inj	properties of injected flow;
1	properties of flow injected from the hub side;
2	properties of flow injected from the shroud side;
2-2	impeller outlet;
3	diffuser inlet;
4	diffuser outlet;
r, u	radial and circumferential components;
des	parameters at the design point.

1 INTRODUCTION

The behavior of vaneless diffusers of centrifugal compressors has been widely studied theoretically, experimentally and numerically. It was shown that the efficiency of the vaneless diffusers at low flow rates extremely decreases because of the flow separation and rotating stall inception (1, 2). The last one results in dramatically loss of compressor performance and instability and even can cause damage of the machine.

Many researchers have contributed to determine control mechanisms for boundary separation, stall and surge in both axial and centrifugal compressors over the past several years. Nelson et al. (3) reported the stabilizing effect of air injection into the diffuser channels through slots in the suction side of the vanes applied to the axial compressor of a turbo-shaft engine. Stein et al. (4) performed numerical investigation of a high-speed centrifugal impeller at surge conditions. Air injection at the impeller inlet was added to the model and several parametric variations of injection angle were analyzed. It was shown, that air injection eliminates a local separation which causes flow reversal and improves impeller stability. Spakovszky et al. (5) and Skoch (6) showed that an improvement in stable range of the centrifugal compressors could be obtained by injecting air into the vaneless region between the impeller and vaned diffuser. However both authors (5, 6) noticed that an injection produces losses of total pressure and results in poor compressor performance. So the purpose of our investigation is to improve the diffuser stable range using injection with minimal pressure losses.

This paper presents the model of compressible flow in vaneless diffuser with mass, momentum, angular momentum and energy addition. Results from a series of simulations to investigate influence of injector location, injection flow rate and injection direction are reported. According to the simulation results the prototype injection system for experimental investigation of injection in the vaneless diffuser has been designed.

2 MODEL DESCRIPTION AND ASSUMPTIONS

A sketch of a vaneless diffuser and used nomenclature are given in Figure 1.

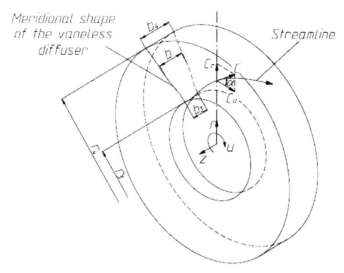

Figure 1 – Vaneless diffuser view

In the development here both mainstream and injected flow were considered to be perfect gases with constant specific heat. The flow is axisymmetric and steady. To deal with friction forces, it is assumed that any friction forces act directly opposite to the local stream direction. The developed model is applicable for diffusers with any meridional profile $b=f(r)$ symmetrical with respect to r-u plane (Figure 1). According to the developed model, the calculation of the flow consists of two stages:
1. One-dimensional analysis of compressible swirling flow with mass, momentum, angular momentum and energy addition;
2. Calculation of the mixing region and boundary layer parameters.

One-dimensional analysis of the flow is similar in concept to that presented by Anderson et al (7). According to the above assumptions and applying the conservation laws for an elementary control volume shown in Figure 2 the set of equations can be written as follows:

- conservation of mass:

$$\frac{1}{C_r}\frac{dC_r}{dr} + \frac{1}{\rho}\frac{d\rho}{dr} + \frac{1}{b}\frac{db}{dr} + \frac{1}{r} - \frac{1}{2\pi rb \cdot \rho C_r}\left(\frac{d\dot{m}_{inj1}}{dr} + \frac{d\dot{m}_{inj2}}{dr}\right) = 0; \qquad (1)$$

- conservation of momentum:

$$C_r\frac{dC_r}{dr} - \frac{C_u^2}{r} + \frac{1}{\rho}\frac{dp}{dr} + \frac{C_f \cdot C^2 \sin\alpha}{b}\sqrt{1+\left(\frac{db}{dr}\right)^2} + $$
$$+\frac{1}{\rho \cdot 2\pi rb}\left[(C_r - C_{inj1r})\frac{d\dot{m}_{inj1}}{dr} + (C_r - C_{inj2r})\frac{d\dot{m}_{inj2}}{dr}\right] = 0; \qquad (2)$$

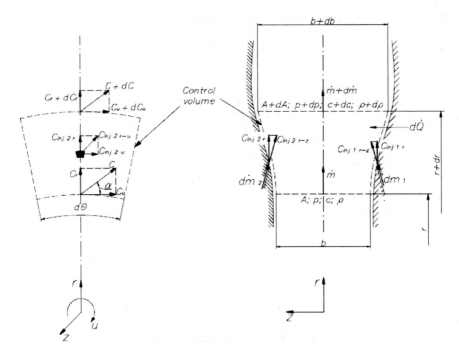

Figure 2 – Control volume in a vaneless diffuser

- conservation of angular momentum:

$$C_r \frac{dC_u}{dr} + \frac{C_r C_u}{r} + \frac{C_f \cdot C^2 \cos\alpha}{b}\sqrt{1+\left(\frac{db}{dr}\right)^2} +$$
$$+ \frac{1}{\rho \cdot 2\pi r b}\left[(C_u - C_{inj1u})\frac{d\dot{m}_{inj1}}{dr} + (C_u - C_{inj2u})\frac{d\dot{m}_{inj2}}{dr}\right] = 0; \quad (3)$$

- conservation of energy:

$$\frac{dT^*}{dr} - \frac{1}{\dot{m}}\left[(T^*_{inj1} - T^*)\frac{d\dot{m}_{inj1}}{dr} + (T^*_{inj2} - T^*)\frac{d\dot{m}_{inj2}}{dr}\right] - \frac{1}{c_p}\cdot\frac{dq}{dr} = 0. \quad (4)$$

The conservation equations must be supplemented by the equation of state:

$$\frac{dp}{p} - \frac{d\rho}{\rho} - \frac{dT}{T} = 0; \quad (5)$$

the definition of stagnation temperature:

$$T^* = T + \frac{c^2}{2c_p}; \quad (6)$$

and the relation between the velocity magnitude, c, and the two components:

$$c^2 = c_r^2 + c_u^2. \quad (7)$$

Differentiating (6) and (7):

$$\frac{dT^*}{dr} - \frac{dT}{dr} - \frac{C}{c_p}\frac{dC}{dr} = 0; \tag{8}$$

$$\frac{dC}{dr} - \cos a \frac{dC_u}{dr} - \sin a \frac{dC_r}{dr} = 0. \tag{9}$$

The resulting differential equations (1-5, 8, 9) can be numerically integrated using Runge-Kutta method. However, it is important to specify an appropriate friction coefficient C_f. Values of skin friction coefficient C_f can be used either as an average for the entire diffuser or as a function of local geometry and flow condition. The equation:

$$C_f = k\left(\frac{1,8 \cdot 10^5}{Re}\right)^{0,2}, \tag{10}$$

where k is a constant has been used for current research. It was taken from (8). According to (8) a value of $k=0,010$ is frequently used based on a review of different vaneless diffusers in industrial machines; however a half or nearly twice k value may be found in some applications.

To predict the flow separation the described model has to be supplemented by appropriate boundary layer model.

For the flow without injection a simplified two-zone model has been used. According to this model, boundary layer may be divided into two regions: 1) the so called viscous sub-layer near the walls in which the flow is laminar; 2) region of turbulent flow in which both molecular and molar friction forces exist. To calculate the flow fields, velocity profiles for both regions of the boundary layer described in (9) have been used. The Buri shape-factor has been used as a separation criterion. Results of simulations performed using the described boundary layer model have been reported in (10). Comparison with the experimental data showed a significant qualitative and quantitative correspondence between calculated and experimental data at the design point, as well as the separation point was estimated rather well.

Method developed by Ginevskiy (11) has been used for the calculation of the flow fields with injection. It is based on integration of the equations of motions for the jet using polynomial approximation of the velocity profiles and semiempirical model for turbulent friction. Geometry parameters of the jet were estimated using semiempirical equations obtained by Abramovych (12). Described method showed good correspondence to experimental data for plane jets, but its suitability for swirling flows has to be checked experimentally. Until that it can be used with care.

The developed model is appropriate for calculation of subsonic turbulent flows ($Re_{C3} > 10^5$; $M_{C3} < 0,8$).

3 RESULTS OF CALCULATIONS

The test vaneless diffuser is an element of the compressor end stage. Cross section of the diffuser is shown in Figure 3. Diffuser width ratio is $b_3/b_2 = 1,07$.

First of all an experimental investigation of the baseline vaneless diffuser has been done. The detailed description of the test system, procedure and results is presented in (13).

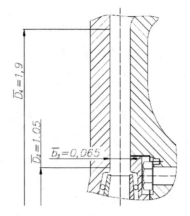

Figure 3 – Cross section of the test diffuser

The overall performance of the diffusers was evaluated using the static pressure recovery and the total pressure loss coefficients. The static pressure recovery coefficient was defined as:

$$C_p = \frac{p_4 - p_3}{p_3^* - p_3}, \qquad (11)$$

and the total pressure loss coefficient as:

$$\zeta = \frac{p_3^* - p_4^*}{p_3^* - p_3}. \qquad (12)$$

The calculated and measured total pressure loss and static pressure recovery coefficients, presented as a function of the inlet flow angle a_3, are shown in Figure 4. The diffuser inlet flow angle was defined as:

$$\alpha_3 = arctg\left(\frac{C_{r3}}{C_{u3}}\right). \qquad (13)$$

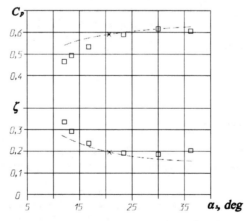

Figure 4 – The baseline diffuser performance:
▫▫▫ **Measured data;** – – – **Calculated (x – the predicted separation point)**

It can be seen from the diagram that there is a point with the minimum measured total pressure loss coefficient at $a_3=30,0^0$, while the calculated ones monotonically decrease as the mass flow rate increases. The average discrepancy between the measured and predicted total pressure loss coefficients is 14,1%. The calculated static pressure recovery coefficients are in good agreement with the experimental results. The average discrepancy between the measured and predicted static pressure recovery coefficients is 3,7%. The surge flow rate for the test compressor with the baseline diffuser corresponded approximately to $a_3=12,1^0$.

To identify the sources of losses velocity diagrams were considered. The measured flow fields are shown in Figure 5. As the diagram shows there is a flow separation zone at the diffuser hub at $a_3=22,4^0$. Flow separation results in total pressure decrease. As the mass flow rate decreases this zone becomes more prominent (at $a_3=16,8^0$), which causes higher total pressure loss. In addition, flow separation reduces the effective flow passage. As a result the flow velocity at the shroud is accelerated, which also produces an increase in total pressure loss coefficient. The developed model has also predicted that at $a_3=21,1^0$ the separation initiates (the location where separation first time appears is marked in Figure 4 as "x"). It can be thus concluded, that separation point was predicted satisfactory.

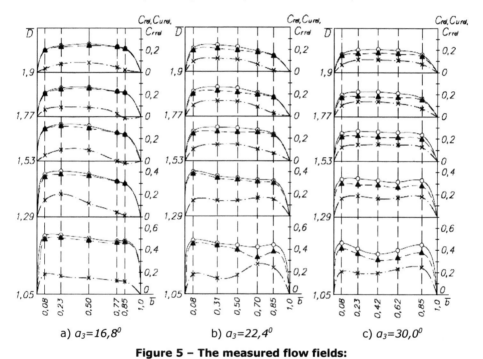

a) $a_3=16,8^0$ b) $a_3=22,4^0$ c) $a_3=30,0^0$

Figure 5 – The measured flow fields:

$C_{rel} = C_u/U_2$; $C_{urel} = C_u/U_2$; $C_{rrel} = C_r/U_2$.

Parametric variations of injector location, injection flow rate and injection direction were investigated numerically using developed model. Configuration of the calculated injection system is shown in Figure 6. During the simulations followed parameters were varied:
- injection flow rates between 1 and 3 percent of compressor design flow;
- injection angles: $\beta=5^0$, $\beta=a$, $\beta=90^0$, $\beta=175^0$;
- dimensionless diameters of the injector location: $\overline{D_{inj}} = 1,125$ and $\overline{D_{inj}} = 1,25$.

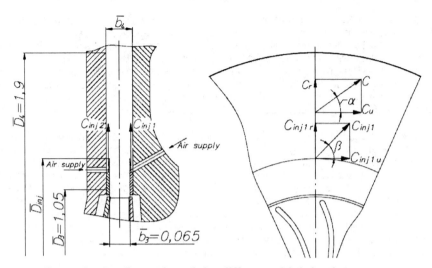

Figure 6 – Configuration of the diffuser with injection system

For all calculations it was assumed that parameters of the air injected from the hub and shroud sides are equal, the heat exchange is neglected and $\overline{b_4} = 0,075$.

The calculated total pressure loss and static pressure recovery coefficients are shown in Figure 7. As the diagrams show, at the same mass flow of injected air the total pressure loss coefficient is lower for the minor values of injection angle β. This fact goes with the experimental results presented by Skoch (6). The lowest values of total pressure loss coefficient were obtained at $\beta=5^0$ (at lower injection angles the injector was choked) and the highest values of total pressure loss coefficient were obtained at $\beta=175^0$. Increase in total pressure loss at higher injection angles can be explained by interaction between the injected and the main flows. At low injection angles, the tangential velocity component of the injected flow is high. It produces acceleration of the main flow and results in higher total pressure level. As the injection angle is increased injected jets reduce the tangential component of the main flow velocity, resulting in dynamic head decrease. This feature is more evident at low compressor flow rates, when the circumferential component of the main flow velocity is higher. Due to the same reason, there is a critical value of the injection angle β_{cr}, such as for $\beta<\beta_{cr}$ the injection mass flow rate increasing causes the decreasing of the value of loss coefficient; in the case of $\beta>\beta_{cr}$ as the mass flow rate increases the total pressure loss coefficient increases as well. Thereby, forward-tangent injection is of great interest. However, it can be realized only using discontinuous nozzles. Such configuration results in circumferential non-uniformity of the flow, so the developed model cannot be used to evaluate such injection. With regard to the injector position, as the diameter of the injector location increases and other parameters remain the same the loss coefficient decreases. This is due to the lower sudden expansion losses and more slight interaction between the main and injected flows. However, installation of the injectors close to the diffuser outlet limits the region of active flow control.

Increase in total pressure loss revealed in the diffuser can make continues use of injection inexpedient. However, injection can be used directly when the onset of instability is detected to take advantages of the significant stability improvement. Moreover the developed model cannot predict the level of the boundary separation losses, so some improvement of the real diffuser characteristics at low flow rates compared to the calculated ones is expected.

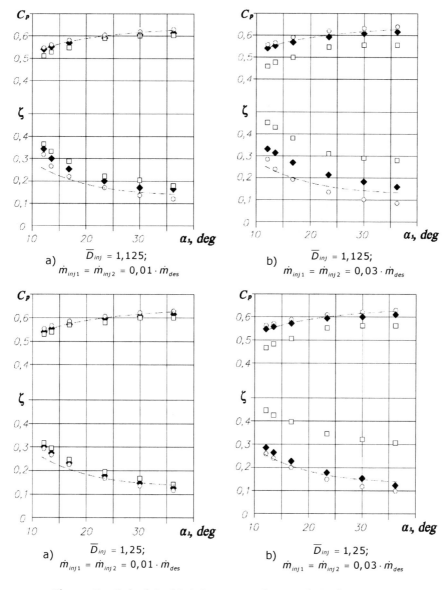

Figure 7 – Calculated total pressure loss and static pressure rise coefficients at different injection modes:

--- no injection; ○○○ $\beta=5^0$; ◆◆◆ $\beta=90^0$; □□□ $\beta=175^0$

To confirm the adequacy of the presented results and conclusions, experimental investigation of the flow with injection has to be performed. For this reason on the basis of calculation results prototype injection system has been designed. The designed model allows to vary position, direction and flow rate of the injected air. Injection system would be supplied by an external air source.

4 CONCLUSIONS

A mathematical model for gas flow in vaneless diffuser with mass, momentum and energy addition has been developed.

Parametric variations of injector location, injection flow rate and injection direction were investigated numerically using the developed model. Results of simulations have showed that an injection produces losses of total pressure across the diffuser. These losses are due to interaction between the injected jets and the main flow. The lowest values of total pressure loss coefficient were obtained when air is injected into the diffuser almost tangentially in the direction of impeller rotation. As the injection flow angle increases the total pressure loss coefficient increases as well.

Total pressure losses across the diffuser could make continuous operation using an injection disadvantageous. However, injection can be used periodically when the onset of instability is detected. Real advantages and disadvantages of injection can be evaluated experimentally. According to the simulation results the prototype injection system for experimental investigation of injection in vaneless diffuser has been designed.

REFERENCE LIST

1. Jansen, W. 1964. Rotating Stall in a Radial Vaneless Diffuser, *J. Basic. Eng.*, Vol.86: pp. 750-758.
2. Senoo, Y., Kinoshita, Y. 1977. Influence of Inlet Flow Conditions and Geometries of Centrifugal Vaneless Diffusers on Critical Flow Angle for Reverse Flow, *J. Fluids Eng.*, Vol. 99: pp. 98-103.
3. Nelson, E. B., Paduano, J. D., and Epstein, A. H. 2000. Active Stabilization of Surge in an Axial Centrifugal Turboshaft Engine, *ASME J. Turbomach.*, Vol. 122: pp. 485–493.
4. Stein, A., Niazi, S., Sankar, L. N. 2000. Computational Analysis of Stall and Separation Control in Centrifugal Compressors, *Journal of Propulsion and Power*, Vol. 16, No. 1: pp. 65-71.
5. Spakovszky, Z. S., Weigle, H. J., Paduano, J. D., et al. 1999. Rotating Stall Control in a High-Speed Stage with Inlet Distortion: Part I – Radial Distortion, *ASME J. Turbomachinery*, Vol. 121: pp. 510—516.
6. Skoch, G. J. 2003. Experimental Investigation of Centrifugal Compressor Stabilization Techniques, *ASME J. Turbomachinery*, Vol. 125: pp. 704—713.
7. Anderson, L. R., Heiser, W. H., Jackson, J. C. 1970, Axisymmetric One-Dimensional Compressible Flow – Theory and Applications, *ASME J. Applied Mechanics*, Vol. 37: pp. 917–923.
8. Japikse, D. 1996, Centrifugal Compressor Design and Performance, Concepts ETI, Inc., Wilder, VT.
9. Sherstyuk, A. N. 1974. Turbulentniy pogranichniy sloy [The Turbulent Boundary Layer (in Russian)]. Moscow: Energiya, 272 p.
10. Kalinkevych, M., Shcherbakov, O., Gusak, O., Ihnatenko, V., Investigation of the gas flow in the vaneless diffusers of the centrifugal compressors, 7th International Conference on Compressors and their Systems, City University London, 5-6 September 2011.
11. Ginevsky A. S. 1969. Teoriya turbulentnyh struy i sledov [The Theory of Turbulent Jets and Wakes (in Russian)]. Moscow; Mashinovedenie, 400 p.
12. Abramovich, G. N. 1963. The Theory of Turbulent Jets. Cambridge, Mass: MIT Press, 671 p.
13. Kalinkevych, M., Shcherbakov, O. 2011. Experimental'noe issledovanie techeniya gaza v stupeni centrobezhnogo compressora [Experimental investigation of gas flow in the vaneless diffuser of centrifugal compressor (in Russian)], *DonNTU Herald*, Vol. 22: pp. 88-100.

Experimental study of radial low specific speed turbocompressor running in reverse as turboexpander

M Arjeneh*, K R Pullen*, S Etemad**
* City University London, School of Engineering and Mathematical Science, UK
** Dynamic Boosting Systems (DBS) Ltd, UK

ABSTRACT

This paper includes small market research in the field of turboexpanders. The market research includes list of applications as well as list of some of major manufacturers of turboexpander. During the course of this research numbers of experiments were carried out to assess the performance of a small turboexpander with 65mm diameter rotor operating at relatively low speeds in comparison to standard radial turbines. The output power of the turboexpander was calculated based on the electrical output of the loading generator. The turbine rotor comprised of a TurboClaw compressor operating in reverse with two set of experiments for the following nozzles; as expander with single nozzle and a set of six nozzles. It was found that even for a non optimised geometry (rotor is a compressor not a turbine) the system efficiency for micro turboexpander is excellent.

Keywords: Turboexpander, Nozzle, Low Specific Speed, Expander Manufacturers

NOMENCLATURE

A	Cross Sectional Area	R_{es}	Electrical Resistance
CHP	Combined Heat and Power	PR	Pressure Ratio
D	Diameter	RPM	Revolutions Per Minutes
I	Current	SOFC	Solid Oxide Fuel Cell
Ma	Mach Number	T	Temperature
OTEC	Ocean-Thermal Energy Conversion	V	Voltage
P	Electrical Power	η	Efficiency
p_n	Pressure	ρ	Density
PEFC	Proton Exchange Fuel Cell	γ	Specific Heat Ration for Air is 1.4
R	Gas Constant	ω	Angular Velocity

1 INTRODUCTION

With exponential growth of the world population, the demand for energy has also increased substantially. Especially in developing countries like China, India and Brazil where the economy is booming and standard of leaving rising. The rise in standard of living usually associated with demand in power requirement per person and booming economy will require substantial increases in power consumption. Currently most countries in the world produce their energy requirement from traditional energy source such as fossil fuel. It was reported in 2004 that about

88% of our energy needs comes from fossil fuels (US Environmental Protection Agency). Fossil fuel our primary source of energy are non-renewable and they are been extracted at such speedy pace that they might be depleted in near future. Extensive use of fossil fuel has resulted substantial increase in amount of CO_2 in our planet atmosphere which is one of the main greenhouse gasses.

With increased emphasis on achieving fuel economy, engineers have been striving hard to reduce fuel consumption, and which can also comply with new emissions regulations which are getting tougher year by year. Any methods that increase the power and improve performance of a power system are hence attractive. Turboexpanders are expansion turbines. They are rotating machines which work in a similar way as a steam turbine. Turboexpanders can also be considered as rotating device that convert the pressure energy of a gas or vapour into mechanical work. The gas or vapour expands through the expander and produce work. For any system that burns fuel there will be exhaust gases. These exhaust gases contain a large percentage of fuel energy. Much cannot be recovered for thermodynamic reasons but even capturing a small amount lead to efficiency gains. Thus it is very important to extract work from this waste energy if possible and improve the overall cycle efficiency. One component for doing this is a turboexpander used as part of a bottoming cycle. If the engine power output is small then so will be the size of the turboexpander and this is where the majority of engines in terms of market size lies.

2 MARKET RESEARCH

In the next section of this paper potential applications of turboexpanders are listed and discussed in the form of a small market research. It was also interesting to see what companies already manufactured turboexpanders and the technology that is being used.

2.1 Turboexpander applications
There are numerous applications for use of turboexpander. Those that are known are given with brief description.

Solid Oxide Fuel Cell (SOFC): SOFC is different from other types of fuel cell by its high operating temperature of > 500 degree C and is mainly used in CHP. In pressurized designs where the compressor is utilised to provide adequate air to SOFC, an expander can be used to extract energy from the exhaust gas steam. In simplest recuperated SOFC/GT electrical efficiency of approximately 60 % can be achieved adding reheat the efficiency can raise to 70% [1].

Proton Exchange Fuel Cell (PEFC): PEFC are one of the best candidates for electrical vehicle prime movers to date and numerous studies being carried out to develop PEFC further. The pressurised air entering the PEFC will be expelled mainly as nitrogen and water at approximately 80-100°C. Hence the cell efficiency can be increased further if exhaust gas was used to provide some of power input to compressor.

Turbochargers: A turbocharger is mainly used in IC engines where exhaust gas leaving the exhaust manifold passes across the turbine which in turns drives the compressor that pressurises the air entering the engine intake manifolds.

Replacing the expansion valve in refrigeration and heat pump: The expansion valve can be replaced with turboexpander. Power produced by turboexpander can be used to run a generator and produce power or can be used to help drive the compressor lowering motor power demand and increasing efficiency.

Gas pressure letdown: Natural gas is transported in pipes for long distances; gas pressures in these pipes can have pressures as high as 5-7 MPa. This high pressure gas needs to be reduced to about 2.5 MPa in transmitting stations and then needs to be reduced to even lower pressure 0.3 MPa before entering household pipelines [2]. A turboexpander can be used in these stations to reduce the pressure of pipelines as well as produce power.

Binary geothermal cycle: In binary geothermal cycle the power is differ from other type of geothermal cycle that the hot fluid form geothermal never comes in contact with the turbine. Instead is hot water flows to a heat exchanger which heats up the working fluid which is in a closed cycle. The resulting vapour expands in turbine and produce work.

Ocean-Thermal Energy Conversion (OTEC): OTEC power plant uses solar heat of ocean surface to vaporize the working fluid which is Ammonia. After the fluid expands in turbine it is condensed by colder water which is pumped from the ocean depth.

2.2 List of turboexpander manufacturers
Table 1 below shows some of the main established turboexpander manufacturers in the world and their expander specification.

Table 1 List of turboexpander manufacturers and specification. Source: manufacturer website and [3]

Manufacturer	Country	Power [kw]	P_{in}[bar]	T_{in}[°C]	M_{flow} [kg/s]	Technology
Atlas Copco	Germany	to 25000	2-150	-215 to 510	0.138 to 42	Radial
Dresser Rand	USA	to 1000	3.2 to 15	700 to 760	12.5 to 178	Axial
Elliot	USA	to 37300	3.4	760	61 to 177	Axial
Mafi-Trench	USA	1000 to 15000	<125	-200 to 300	0.138 to 208	Radial
Siemens	Germany	to 45000	to 60	550		Radial
Mitsui Eng	Japan	2000 to 150000	4.5 to 20	60 to 670	278 to 2278	Axial
Man Diesel & Turbo SE	Germany	20000 to 30000	10 to 15	500	70 to 111	Axial & Radial

2.3 Organic rankine cycle (ORC)
The Organic Rankine Cycle is based on the Rankine Cycle with water and steam replaced with a hydrocarbon. It is a thermodynamic process where heat is transferred to the working fluid at constant pressure this causes the fluid to evaporate. The evaporated fluid is expanded in the turbine where it produces work. The vapour is condensed and pumped back to boiler and recycled.

The only difference between them is that in ORC the working fluid is not water (it uses organic fluid). Use of organic fluids that have lower boiling temperature compare to water. Working fluid can be Refrigerants, Ammonia, and Benzene and so on. There have been numerous research studies on effects of working fluid on overall performance of the cycle. It also has been well documented the working fluid has strong correlation with the application that is being used for.

2.4 ORC applications

ORC technology has hundreds of potential applications but only a few of them are well established. Some of the applications are waste heat recovery, combined Heat and Power (CHP), geothermal, solar Power Plant, heat recovery from IC Engines.

2.5 List of manufacturers of ORC

Table 2 shows a list of manufacturers of ORC website. It has been reported [4] that Freepower which is one of ORC manufacturers based in the UK produce expander with power output of as low as 6 KW. However, I was not able to find this expander in their product listing on the website.

**Table 2 List of ORC manufacturers.
Source: from the manufacturer's website 2012 and [4]**

Manufacturer	Application	Country	Power [KW]	Technology
Electra Therm's	WHR	US	30-65	Twin Screw Expander
Infinity Turbine	WHR	US	30-500	Radial Expander
Ormat	Various	US	200-30000	NA
Freepower	WHR	UK	85-120	NA
Cryostar	WHR, Geothermal	Multi-International	500-15000	Axial Turbine
GMK	WHR, Geothermal	Germany	35-5000	Axial Turbine
Bosch KWK	WHR	Germany	40-375	NA
Turboden	CHP	Italy	619-2833	Axial Turbine

After this phase it was clear that there are a few companies that offer a micro turboexpander solution. This is mainly due to very low efficiency and cost benefits. If Dynamic Boosting System (DBS) turbocompressor works as expander there will be a gap in market to be filled. Thus set of nozzles where designed which replaced the diffuser and performance of turboexpander was evaluated.

3 DESIGNS

Two different nozzles were designed and used during these experiments. The first design which had multi nozzle and second design had a single nozzle. Due to the lack of time and resources the nozzles were designed to be compatible with existing components, so that the motor and rotor can be fitted without any changes. Thus, once the diffuser casing has been removed it can be fitted with the nozzle casing without any other changes.

3.1 Volute geometry

The nozzles volute geometry was constructed by using method based on Bezier polynomials described by Casey [5]. Beizer curve is used in most CAD software. It is used to create complicated surfaces and geometry. Using the spline in CATIA the geometry shown in figure 1 was created. First initial point was created then final point had to be determined. The vector OA shown in the figure 3 is function of angle ψ. As angle ψ is increased from zero to 360 degree the length of Vector OA reduces at a constant rate until reaches the end point that had to be determined at earlier stage.

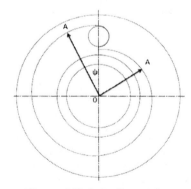

Figure 1 Volute Geometry

3.2 Nozzle area
Maximum mass flow possible through a duct for isentropic flow is given by

$$\dot{m}_{max} = \gamma^{0.5} \left(\frac{2}{\gamma+1}\right)^{\frac{\gamma+1}{2(\gamma-1)}} A\rho(RT)^{0.5} \qquad (1)$$

For $\gamma = 1.4$ then the maximum area can be simplified further and it is given by:

$$A = \frac{\dot{m}_{max}(RT)^{0.5}}{0.6847P} \qquad (2)$$

The throat area of nozzle vane has substantial effect on turbine performance so it needs to be designed so it delivers required mass flow.

$$A_n = \frac{\dot{m}(RT/\gamma)^{0.5}}{Pth} \frac{1}{Ma_n} \left(1 + \frac{\gamma-1}{2} Ma_n^2\right)^{(\gamma+1)/2(\gamma-1)} \qquad (3)$$

The subscript n refers to nozzle throat number.

3.3 Nozzle-rotor interspace
Watanabe [6] has empirically shown the relation between the nozzle-rotor interspace and the maximum efficiency of radial turbine, value of interspace geometry parameter K given by

$$K = \frac{\Delta r}{b \cos \alpha} \cong 2 \qquad (4)$$

Where Δr the radial distance between nozzle exit and the rotor tip, α is the flow angle and the b is height of nozzle.

3.4 Nozzle designs
Two types of nozzle were design during the course of this project. The figure 2a shows design of multi nozzle and figure 2b represents single nozzle. The multi nozzle was design in such way that to replicate diffuser geometry as close as possible. Since diffuser has been design in conjunction with the impeller it has the best flow angle and inlet cross sectional areas that produce the best performance. Flow angle at exit of the nozzle is the most important factor in designing the stator. Thus the nozzle throat area were designed to be the same as diffuser throat area and the inlet flow angle to be as tangential as possible.

The nozzle passages are designed in manner that flow contact area is kept to its minimum, so that the skin friction is reduced and flow entering the impeller achieves a greater velocity. While the shape of each nozzle vanes were designed to be very similar to an aerofoil in order to achieve a good aerodynamic efficiency. The aerofoil nozzle shape is a very common practice when designing stator for radial inflow turbines. The single nozzle could be considered as vaneless volute. The nozzle was design

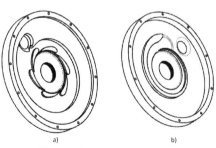

**Figure 2 Nozzle Design:
a) Multi-Nozzle, b) Single Nozzle**

in similar procedure that has been described in section Volute geometry. Advantage of this design compare to the multi nozzle is that if successful it would be cheaper to manufacture. It will reduce weight, meaning less material required and overall size will also reduce since the geometry is much simpler compare to multi nozzle.

4 EXPERIMENTAL RIG

The expander that was used during this experiment has been shown in the figure 3. The impeller is forward swept and has diameter of 65 mm. The motor that has been presented in the figure 6 is 250 watt rating and can run up to speed of 4000 KRPM.

Figure 3 Main component of DBS 65 mm diameter expander

Test rig set up with sensor position has been shown in the figure 4. The hot air was supplied with similar design compressor but slightly larger 85 mm impeller diameter. The hot air flows from the pipe was divided into two branches one flowing into expander and other was directed into the blow off valve which was vented into atmosphere. Blow off valve had to be used in order to avoid surge in the hot air supplier compressor. Temperature and pressure of air at inlet and out let of expander was measured using pressure sensors and temperature sensors. In order to measure the RPM of expander a laser tachometer was employed. The mass flow meter was used to measure the mass flow rate.

Hot air that passed through the expander caused the shaft to rotate and produce electricity. Thus load bank was used to replicate real life situation. The load bank has series of resistance within that causes reduction in current output. Five switches were mounted on the load bank and switches are turned on the resistance reduces.

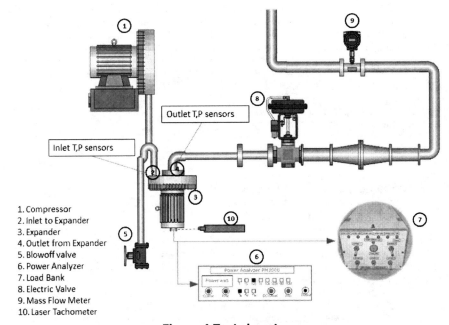

1. Compressor
2. Inlet to Expander
3. Expander
4. Outlet from Expander
5. Blowoff valve
6. Power Analyzer
7. Load Bank
8. Electric Valve
9. Mass Flow Meter
10. Laser Tachometer

Figure 4 Test rig setup

5 EXPERIMENTAL RESULTS AND DISCUSSION

Turboexpander performance was fully mapped and the data has been presented in the next section of this paper.

5.1 Turboexpander performance map

Two set of experiments were carried out to determine the performance of two different nozzle designs. Full map of performance of these two set of nozzle have been shown in the figure 5. The figure 5 a) and b) compares the PR and mass flow at various speeds and various load bank resistances (Resistance changes from highest to lowest for each speed). It is evident that the mass flow is higher across the turbine with multi_ nozzle comparing with single nozzle. The performance of these two designs in term of efficiency was quiet very similar up to speed of 30 KRPM. The expander speed with multi nozzle could reach up to 35 KRPM whereas the expander speed with single nozzle was limited to about 30 KRPM. Only in one occasion the speed of 33 KRPM was reached.

Furthermore, another interesting phenomenon which was observed during the course of these experiments was drop in PR as speed increases. It can be seen from the both set of data that at 15 to 25 KRPM expander shaft speed, the PR curve follow similar pattern, as soon as the expander shaft speed reaches speed to 30 KRPM and 35 KRPM the PR seems to be almost constant. This suggests that at high shaft speed the effect of load applied on pressure ratio across the turbine is diminished.

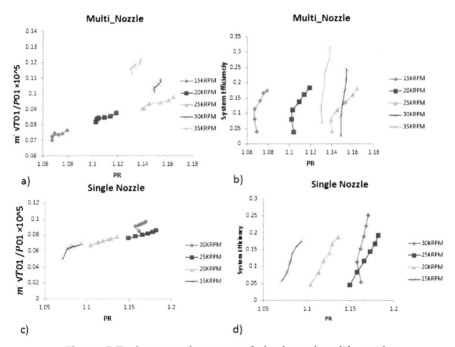

Figure 5 Turboexpander maps of single and multi-nozzle

Power output from the expander has direct relation with the overall system efficiency. The system efficiency is ratio of power output (measured by power analyser) over ($m_{flow} \times C_p \times (T_2-T_1)$). As the power output increases so does the system efficiency. The maximum power output from single nozzle was about 25 watt which almost identical to multi nozzle running at 30 KRPM. The maximum power output of multi nozzle just below 32 watt at speed of 35 KRPM. The maximum system efficiency of single nozzle was just above 25 % whereas the system efficiency of multi nozzle was just below 32%. The main reason for difference in efficiency is that the expander with multi nozzle shaft speed was higher. The system efficiencies at other speeds (15 to 30 KRPM) are virtually identical.

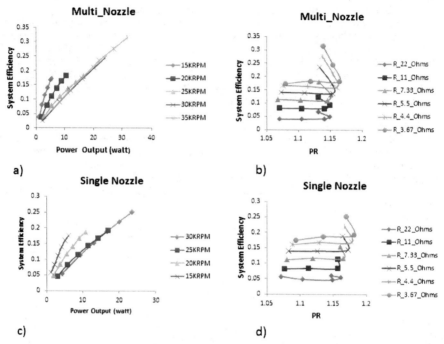

Figure 6 Comparison of two sets of nozzles

6 POWER LOSS CALCULATIONS

In this section of this paper power lost due to copper loss, windage loss and bearing loss was calculated in order to determine the efficiency of turbine rather than the system efficiency.

6.1 Copper loss

There are number of temperature sensors inside the DBS 250 watt motor. One of the sensors is responsible for measuring the motor winding temperature. The relationship between the winding temperature and current has been shown in the figure 7a. It is evident that there is a linear relation between the current and winding temperature, as current increases so does the winding temperature.

The winding resistance of one the phases was measured at room temperature and since the temperature rise of winding is known at various current, one can use the linear approximation to determine the resistance of winding at a given temperature. Since the temperature rise inside the winding was not high the resistance was slightly higher than the room temperature. Once the resistance was found the copper loss equation was used **Power$_{loss}$=3I^{2}R$_{es}$ (6)** to determine the power lost.

The graph of power lost versus I^2 has been presented in the figure 7b. Copper loss effect becomes substantial at low load this could be 11.5 % and the effect is much smaller at higher load.

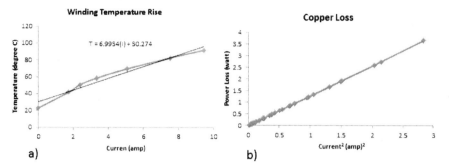

Figure 7 a) shows effect of current on winding temperature and b) shows effects of copper power loss

6.2 Windage losses

Windage is measure of fluid drag on a rotating object. Understanding the windage on an enclosed disc and drum when designing a turbomachinery is very important and this becomes even more important for low specific turboexpander, the windage usually represents huge percentage of overall power lost.

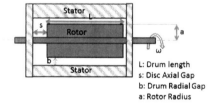

L: Drum length
s: Disc Axial Gap
b: Drum Radial Gap
a: Rotor Radius

Figure 8 Windage Geometry

Windage loss is a function of geometry of rotor and stator. In order to be able to measure the power lost as result of windage the dimensions shown in the figure 8 have to be known [7]. For these calculations superimposed mass flow of 0.5 g/s was assumed. The windage power loss increases as the RPM increases. This effect is evident from the figure 9. The windage drum power loss which includes the rotor and the magnet is substantially greater than the windage loss due to disc.

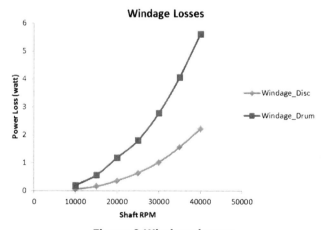

Figure 9 Windage losses

519

6.3 Bearing loss

There are two deep groove single row ball bearings inside to motor, one at the end of the shaft and another at start of the shaft. Two temperature sensors within the motor measured the temperature rise. The bearing temperature is function of shaft RPM this has been shown in figure 10a. There is very slight temperature difference when comparing the temperature rise between the two bearings.

The lubricant used for these bearings is SKF grease LH23 Lithium soap. It is very important to measure the accurate viscosity of lubricant at various temperatures. Lubricant viscosity has exponential relation with temperature this is evident from the figure 10b. Since bearing temperature is known at given speed one can measure the lubricant viscosity at given temperature in order to accurately model the bearing power loss.

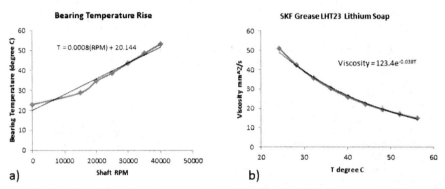

Figure 10 a) bearings temperature, b) effect of lubricant viscosity as function temperature

There are two types of loads act on the bearings one axial load and other radial loads. The axial load was assumed to be constant at 65 N throughout all speeds. The axial load is based on spring preload. For radial load unbalanced mass was used as well as mass of shaft. The unbalanced mass was 0.1 gram at radius of 6 mm, thus using this equation **F=mrω² (7)** the radial loads can be calculated. It was found that the since mass of shaft and rotor are small the effect of static load on the bearing was negligible. Power lost due to the bearings has been shown in the figure 11. Bearing power lost was calculated using the SKF website[8]. It is clear bearing power loss has linear relation with the rotational speed of the shaft as the RPM increases so the power loss. Bearing power loss seems to be substantial; more than 50% of power output is lost due to the bearings. The bearing loss is substantially greater when it is compared with windage and copper loss.

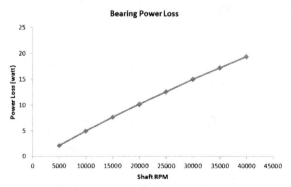

Figure 11 Bearing power losses

6.4 Expander efficiency

Since most of the power losses have now been calculated, these losses can be added to the power output from the expander and the expander efficiency can be calculated. The figure 12 only presents more accurate model of expander efficiency. The expender efficiency alone is assuredly greater than what have been presented here since not all the losses have accounted for.

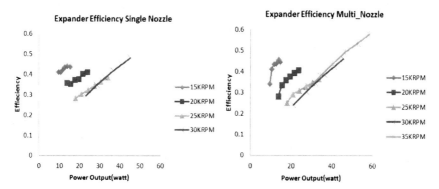

Figure 12 Expander efficiency adjusted for power losses

The expander efficiency with two sets of nozzles has been presented here. It is evident from the graphs that there has been substantial increase in efficiency of expander at lower speed, for instance at speed of 15 KRPM the system efficiency was below 10 %, whereas now the expander efficiency has risen to more than 40%. The maximum efficiency of turbine at 35 KRPM for single nozzle is just over 45% and for multi nozzle is just below 60% which is about double the system efficiency.

7 CONCLUSIONS

During the course of these project it was determined that DBS compressor works well when it is being run in reverse as expander. The system efficiency of expander using two different set of nozzles were determined. The peak system efficiency of multi nozzle was found to be just below 32 % and expander with single nozzle had system efficiency of just above 25 %. It also need to bear in mind that compressor was ran at 40 KRPM where maximum speed of expander was 35 KRPM for multi nozzle and 30 KRPM for single nozzle. It was also found that expander system efficiency is much greater at low speed (15 and 20 KRPM) that of compressor.

In order to determine the turbine efficiency not the system efficiency, the power lost due to copper loss, bearings and windage was calculated. It was discovered that bearing loss is substantial and it was greater than other losses. Copper loss was smallest due to small amount of current passing through the motor windings. The losses were added to power output from expander and the expander efficiency was calculated. The expander efficiency with multi nozzle was just short of 60 %, efficiency almost doubled. For case of single nozzle design the expander efficiency increase to over 45%.

Furthermore, there is room for improvement in test rig as well as in expander geometry which could increase efficiency of turbine even further. The purpose of the presented experiments was to determine if Turboclaw impeller could work well in principle in expanding gas. Based on these results the experiment was successful and it was determined that DBS turbocompressor works well as expander.

REFERENCES

1. Roberts, C., et al., *Demonstrations: The Bridge to Commercialization for the SOFC.* Jun, 2000.
2. Poživil, J., *Use of expansion turbines in natural gas pressure reduction stations.* Acta Montanistica Slovaca, 2004. **3**(9): p. 258-260.
3. Bloch, H.P. and C. Soares, *Turboexpanders and process applications,* 2001: Gulf Professional Publishing.
4. Quoilin, S. and V. Lemort, *Technological and economical survey of Organic Rankine Cycle systems.* 2009.
5. Casey, M., *A computational geometry for the blades and internal flow channels of centrifugal compressors.* ASME Journal of Engineering for Power, 1983. **105**(2): p. 288-295.
6. Dixon, S.L., *Fluid mechanics and thermodynamics of turbomachinery,* 2005: Butterworth-Heinemann.
7. Romero-Hernandez S, E.M., Pullen KR, *Numerical Determination of Windage Losses on High-Speed Rotating Discs.* Chapter 6 Computational Fluid Dynamics in Practice, 2000.
8. http://www.skf.com/skf/productcatalogue/calculationsFilter?lang=en&newlink=&prodid=&action=Calc5.

Shape optimization of a centrifugal compressor impeller

S Khalfallah[1], A Ghenaiet[2]
[1] Laboratory of Thermal Power Systems, Applied Mechanics, Ecole Militaire Polytechnique, Algeria
[2] Faculty of Mechanical Engineering, University of Sciences and Technology USTHB, Algeria

ABSTRACT

The multi-objective optimization algorithms become more suitable for general aerodynamic design problems, despite the number of objectives evaluated by CFD calculations. This paper presents an aerodynamic shape optimization (ASO) algorithm based on the idea of replacing some of the CFD evaluations by those provided by an approximation model known as "Radial Basis Functions" RBF. As a matter of fact that this latter requires high number of evaluations to perform an accurate approximation, herein is proposed a combination of the "Multidisciplinary Aero/Struc Shape Optimization Using Deformation" MASSOUD and the Bspline techniques based on the idea of parameterizing the shape perturbations instead of the shape itself. Ansys-CFX flow solver is used to evaluate the objectives for each element of the database. Based on MASSOUD and Bspline techniques, and the RBF used to approximate the objectives with respect to design variables, the NSGA-II algorithm is applied to produce the Pareto front. This procedure was applied for the NASA low speed centrifugal compressor LSCC to optimize its performance within a reasonable computational time.

1 INTRODUCTION

The main goal of the aerodynamic shape optimization (ASO) is to produce highly efficient turbomachinery in acceptable development cost, time and accuracy. Considerable research works addressed the challenge associated with ASO problems, using different approaches classified as direct methods (random, zero-order or stochastic) and gradient based methods (1,2). The first are usually referred as random guided methods, such as genetic algorithm (GA) and simulated annealing algorithm (3). In general, the gradient based methods converge quickly to an optimal solution requiring few evaluations. But, since only one solution is obtained, designers have no alternative options. Furthermore, these algorithms may get stuck in a local minimum (4), and it is difficult to set appropriate weights to different objectives due to lacking of information about the inner relations. Therefore, genetic algorithms are more suitable for problems with multiple objectives (2). Many types of optimization strategies may be considered in ASO dominated sorting genetic algorithm (NSGA-II) proposed by Deb et al. (5), which perform well as compared with other multi-objective GAs (2,6,7). Oyama et al. (8) optimized the NASA Rotor67 using a parallel GA and full 3D flow solver, requiring thousands of computations to get the converged optimal solution. The main drawback of the multi-objective direct algorithms is the evaluations of a high number of objectives during the optimization cycle. To circumvent this difficulty, many authors have replaced some of CFD evaluations by those provided by a response surface approximation method (RSM). This latter is constructed from a number of flow simulations and then used during optimization. Many authors assessed the potential use of surrogate method to generate a reliable response surface for multidisciplinary shape optimization. Mengistu et al. (9) used Artificial Neural Network (ANN) as a low order RSM to approximate the objective function at a relatively lower computing effort by a factor of ten. Jin et al. (10) published a description of surrogate models based on: spline, radial basis function, kernel smoothing and Kriging, with a comparison of their efficiency for mathematical functions. Peter et al. (11) performed a comparison between four types of surrogate models: least square polynomials, artificial neural networks (multi-layer perception and radial basis function) and Kriging used in an industrial context to design a stator blade optimized for the local exit pressure considering two design variables. Among all the models that have been tested, the Kriging models and radial basis function appear to give the best results in approximating the exact function. Alan Diaz-Manrfquez et al. (12) have compared between four meta-modeling techniques: polynomial approximation, kriging, radial basis functions and support vector regression from the point of view of accuracy, robustness, efficiency and scalability. Also, have identified advantages and disadvantages of each meta-modeling technique and selected the most suitable to be combined with evolutionary optimization algorithms, and showed the best approach to be used in low dimensionality problems is the Kriging. In contrast to high dimensional problems, the best technique would be RBF, which seems to give good results in terms of accuracy and time.

The objective of the present study is to implement an ASO algorithm based on a CFD solver and RBF and NSGA-II algorithms, to optimize the total pressure ratio and the isentropic efficiency for a centrifugal compressor. A large effort was devoted to improve the global optimization cycle time by combining MASSOUD (13) and Bspline techniques in the parameterization steps.

2 OPTIMIZATION PROCEDURE

The implemented ASO procedure for impeller blade shape optimization starts by generating a database using MASSOUD and Bspline techniques, which requires choosing a parametric description of the blade shape and the optimization variables with their variations space. Initially, the database contains only one baseline shape and to generate other shapes, random perturbations of the geometric parameters on the control points are added to the baseline shape. CFD simulations using the flow solver ANSYS-CFX are used to evaluate the objectives for each shape in the database, and then, RBF model is used to approximate them with respect to the optimization variables. Finally, NSGA-II is applied to produce the optimal Pareto front.

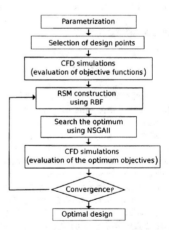

Figure 1 ASO procedure

2.1 Parameterization
To define the blade shape, some geometric parameters with a physical meaning should be introduced to allow an effective and fast search for the optimal design (14). In this study, four parameters r, θ, z respectively, radial, tangential and axial coordinates and e is blade thickness. Tangential coordinate θ can control the

curvature of the blade mean-line: $\theta = \int_0^1 \delta\theta\, ds$ with $\delta\theta = tan(\beta)dM'$, where the blade angle β is the angle between the axial direction and the tangent of the blade mean-line, ds is the fractional distance along the mean-line and $dM' = \sqrt{\delta r^2 + \delta z^2}/r$ is the fractional normalized meridional distance. Thus, (r,θ,z) are used to define the camber mean-line and the distribution of thickness e to produce a smooth profile on the suction and pressure sides, fitted with Bspline curves. θ is considered as an optimization variable since it influences the flow pattern and the aerodynamic performance. The different methods used to define the blade shape are mainly those using a set of control points fitted with Bezier, polynomial or spline curves (Fig. 2a). The shape deformation is achieved by perturbing the control points which are considered as the optimization variables (Fig. 2a). However, profiles shapes require a large number of control points to be accurately represented, resulting in a high number of optimization variables. As consequence, the MASSOUD method (13) is used in an effective way to reduce the number of optimization variables by parameterizing the shape perturbations instead of the shape itself ($\theta_{new} = \theta_{base} + \delta\theta$). In this paper, the MASSOUD and Bspline techniques are combined to reduce the time of optimization according to these steps:
- Select the number and the positions of control points as indicated in Fig. 2b.
- Impose a perturbation $\delta\theta_i$ on each control point.
- Determine the subsequent perturbation $\delta\theta$ on the intermediary positions by Bspline interpolation to ensure a smooth distribution and avoid wavy curves.
- Update the baseline geometry by adding the perturbations to the baseline shape $\theta_{new} = \theta_{base} + \delta\theta$.

There are six control points considered in this study with three points at the leading edge and three other points at the trailing edge, as shown by (Fig. 3). To reduce computational time, only three control points at the leading edge at 0%, 50% and 100% of blade span are considered. The perturbation at the trailing edge was not considered, and subsequently, only three optimization variables were considered on three control points θ_1, θ_2 and θ_3.

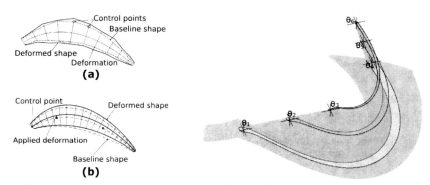

Figure 2 Blade deformation method a) MASSOUD method, b) Direct control points perturbation

Figure 3 Control points used in impeller shape optimization

2.2 Selection of design points

A set of points (training samples) is selected to be used in the RSM construction. For each training sample $x^i = (\theta_1^i, \theta_2^i, \theta_3^i)$ the blade geometry is generated, afterwards CFD simulations are applied to each blade shape to obtain its aerodynamic performance. As a result, a database is obtained which is composed of a set of variables and objective functions (x^i, PR^i, η^i). This database is used to construct approximations of the objective functions, starting by selecting the range δ_i of each optimization variable that produces an optimization space enclosing a range of perturbations ($\delta\theta_i \in [-\delta_i, \delta_i]$). The following conflicts arise when selecting the range of a value: a larger value of δ_i increases the chance to obtain a better result but increases the computational time and decreases the database accuracy. On the other hand, a reduced value of δ_i increases the database accuracy but decreases the chance of obtaining a good solution. In this study, a range of δ_i is selected to limit the time requirement and ensure an acceptable accuracy of database as presented in section 3. Therefore, 157 training samples were used to produce an accuracy of database less than 0.05%, which represents the averaged difference between the CFD results and RBF approximation. After selection of the optimization space, it has to be discretized to produce the training samples set with a sufficient size to ensure an accurate RSM approximation. In this study a database is constructed by dividing the optimization variables ranges to four segments. Thus, the perturbation of each optimization variable $\delta\theta_j$ (j=1,2,3) took one of the five values (-4°,-2°,0°,2°,4°). All the possible combinations of the three perturbations from these five values were considered, thus producing a database size of $5^3=125$ for the training simple.

2.3 CFD simulations

The optimization procedure was carried upon the tested NASA centrifugal compressor (LSCC) which is composed of a 55 degree back-sweep impeller with 20 full blades (Fig. 4) and a vaneless diffuser. The impeller has an inlet and outlet diameters of 0.87 m and 1.524 m, with a design point corresponding to a mass flow of 30 kg/s and a speed of 1862 rpm (15). First, the flow solution by using ANSYS-CFX was validated against available experimental results, to ensure the validity of the computational grids, boundary conditions and solver setting parameters. The computational domain used in these CFD simulations (Fig. 5) encompasses an intake, a blade passage and a vaneless diffuser.

Figure 4 NASA LSCC impeller (15)   **Figure 5 Computational domain**

A frozen interface was selected such that the flow in stationary parts was solved in a stationary frame, whereas that in impeller was solved in a rotating frame. The boundary conditions were a total pressure and temperature at the inlet and a mass flow rate at the diffuser outlet and the periodic boundaries were set at the lateral sides. High quality multi-block structured meshes of 585000 nodes were generated using an O-grid topology around the blades in order to guarantee boundary layers resolution, and H-grid blocks were added around the core region (Fig. 6). This non-uniform distribution of blocks allows more refinements in the vicinity of leading and trailing edges of blade and at hub and shroud. Near walls, nodes were positioned in such a way that the value of y+ varies from 1 to 120 (Fig. 7). First, the flow field was determined based on the frozen-rotor simulations, considering local time step and *SST* turbulence model with an automatic wall function. Second, the results of these simulations were used as an initial guess for the stage interface simulations to predict the global aerodynamic performance.

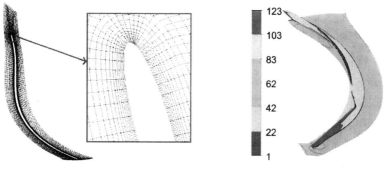

Figure 6 Impeller meshing at mid-span   **Figure 7** Y+ distribution

A comparison between predicted and measured aerodynamic performance (Fig. 8) depicts a good agreement. The relative difference between computed and measured pressure ratio is 0.2% at design point and 0.3% near stall. Similarly for the relative difference in isentropic efficiency which is only 0.31% at design point, but 2.7% near stall, due the strong transient character of the flow at stall, which must be simulated by transient simulations.

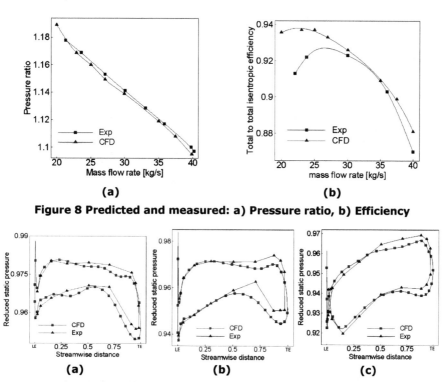

Figure 8 Predicted and measured: a) Pressure ratio, b) Efficiency

Figure 9 Static pressure around blade at different spans:
a) 5%, b) 49%, c) 93%

Figures 9 show a comparison between the distributions of computed and measured static pressure at the design mass flow rate. The overall agreement between experiments and computational data is reasonable because the local differences remain below 0.1%.

2.4 Response surface method (RSM) construction using RBF

This section presents the method used to approximate the objective functions during the optimization cycle instead of using CFD simulations. The used RSM model is the Radial Basis Functions (RBF) proposed by Hardy (16), based on the following principles: Given distinct points: $x^1,...,x^n \in R^d$ and real scalars $f^1,...,f^n$. To construct a function $s: R^d \to R$ for which $s(x^i) = f^i, i = 1,...,n$, the radial basis functions approach consists in choosing a function φ and a norm on R^d such that $s(x) = \sum_{i=1}^{n} \lambda_i \varphi(\|x - x^i\|)$. The interpolation conditions $s(x^i) = f^i$ defines a linear system $A\lambda = f$ and $A_{ij} = \varphi(\|x^i - x^j\|), 1 \le i, j \le n$. The resolution of this system produces the coefficients λ_i.

There are many types of radial functions such as the multi quadratic $(\sqrt{1+r^2})$, the inverse quadratic $1/\sqrt{1+r^2}$ and Gaussian e^{-r^2}. The use of this radial functions directly produce a piecewise approximation. In order to produce smooth approximations, the shape parameter ε is used as follows: $\varphi(r) = e^{-(\varepsilon r)^2}$ (12). The value of ε has an important effect on the approximation accuracy especially at the mid training samples positions (Fig. 10). Thus the selection of this value is done in such a way that the difference between the real and the approximate are minimum. Sobester et al (17) used a *leave-one-out* cross validation procedure to search for the optimum ε, but requiring a high computational time. In the present work, the steps used to optimize ε were as follow: First, the training samples were used to evaluate the coefficient λ_i based on an initial guess value of ε. Second, 32 points were selected at mid-distance of the training samples where the RBF error may be at maximum as shown by Fig. 10. Then the difference between CFD and RBF approximated objective functions were evaluated at these points and this difference is taken as an objective function to optimize ε. The use of a radial function depends on the type of the problem and the sampling data. In this study a Gaussian radial function is used with Euclidian norm, since they produce satisfactory results.

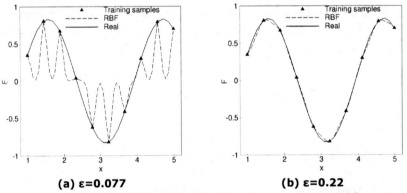

(a) ε=0.077   (b) ε=0.22

Figure 10 RBF approximation for different values of ε

2.5 Optimal design search method NSGA-II

In the recent years, a number of multi-objective GAs based on the Pareto optimal concept have been proposed, such as the widely used non-dominated sorting genetic algorithm NSGA-II proposed by Deb et al. (5), which may provide good results because it maintains a better spread of solutions on the obtained non-dominated front by using crowded comparison approach, as well as to reduce computational time, by using a fast non-dominated sorting approach and ensuring elitism. In this work, NSGA-II is used in conjunction with RBF to perform blade shape optimization. First, a test was done to get an insight about the coupling effect of NSGA-II and RBF precisions using FONT test functions (5) given by:

$$F_1(x) = 1 - \exp\left(-\sum_{i=1}^{n}\left(x_i - \frac{1}{\sqrt{n}}\right)^2\right), \quad F_2(x) = 1 - \exp\left(-\sum_{i=1}^{n}\left(x_i + \frac{1}{\sqrt{n}}\right)^2\right)$$

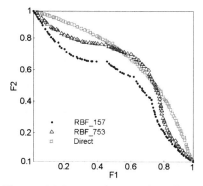

Figure 11 shows the Pareto front obtained by NSGA-II using directly the FONT functions. NSGA-II in conjunction with RBF using training samples evaluated using the FONT were used to construct the approximation model. The two sets of 157 and 753 training samples were used to highlight the effect of the number of training samples, as used in RBF construction, on the obtained Pareto front. As a result, the coupling of RBF and NSGA-II allowed the reproduction of the Pareto front which differs from that produced without approximation. This difference is shown to reduce significantly by a refinement on the RBF construction database.

Figure 11 Pareto front using direct FONT and with RBF approximation

The increase of the time needed by RBF construction is very important when increasing the number of training samples, whereas the corresponding difference between the approximated and the direct evaluated objectives as indicated by Err_{rbf} in Tab. 1 is not significant.

Table 1 Mathematical test based on FONT functions

Training sample number	Time of RBF construction (sec)	$Err_{RBF\text{-}max}$ %	$Err_{front\text{-}max}$ %
157	25	2.18	20.7
753	2707	1.23	8.9

$Err_{RBF\text{-}max}$ is the maximal relative difference between the approximated and the direct evaluated objectives.
$Err_{front\text{-}max}$ is the maximal relative difference between the front obtained from RBF approximation and that using direct FONT functions.

3 RESULTS AND DISCUSSIONS

This section presents the main results in term of the Pareto fronts obtained from the application of the present ASO procedure for the LSCC centrifugal compressor. As mentioned earlier, the quality of the Pareto front depends on the RBF approximation precision, investigated before performing optimization. A number of 155 CFD simulations carried out at different training samples were used to construct the RBF approximation using 125 points to evaluate the coefficients λ_i and 30 points to optimize the coefficient ε. As a result, the averaged difference between the CFD results and RBF approximation is about 0.05%, which is sufficient to get an accurate Pareto front. The same number of training samples used in the mathematical FONT model produced a difference of 2.18% (Tab.1), which means that the approximation quality of the RBF does not depend only on the training sample number, but also largely depends on the distribution and the spacing. The number of objective functions evaluations needed to optimize the blade shape of LSCC compressor was 40000 evaluations, whereas the RBF model has reduced this number by CFD computations to only 157, and subsequently the computations time reduced from 32000 to 125.6 hours.

The present ASO procedure was applied for two types of centrifugal compressor impellers, and the results in terms of Pareto fronts are presented in Fig. 12 and Fig. 13, depicting uniformly spacing points and a spread of solutions obtained, allowing a trade-off in the design solutions. Both objective functions are seen to be improved as compared to the baseline compressor. However, these figures show that the ranges of objectives in the Pareto fronts are small, especially for the LSCC compressor (Fig. 12). This is because of limited design space to get higher improvements in performance for the well-designed LSCC compressor, and due to the fact that a larger optimization space would require higher number of training samples to ensure an accurate RBF approximation and optimization results. Figure 13 shows the results obtained from the ASO procedure applied upon a modified blade of the Schwitzer radial compressor (18), and the ability of the present procedure to improve an arbitrary baseline shape. This latter is constructed by modifying an existed blade in such a way that the distribution of angle θ is constant from hub to shroud along the blade. As a result, when the baseline blade was deformed within the selected optimization space, it returned to its original designed blade shape, which confirms the validity of this procedure.

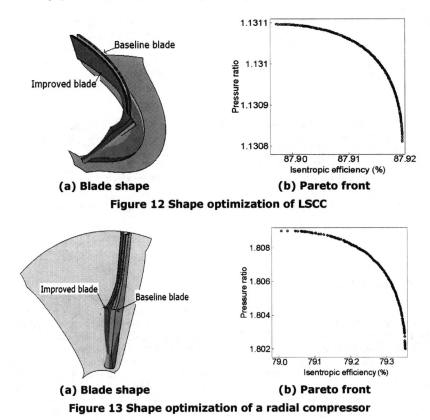

(a) Blade shape (b) Pareto front
Figure 12 Shape optimization of LSCC

(a) Blade shape (b) Pareto front
Figure 13 Shape optimization of a radial compressor

Figures 14 and 15 show a comparison between the baseline and the optimized radial compressor, depicting a clear improvement in the flow pattern after optimization. The plot of relative velocity vectors show that the baseline blade is improved by the ASO algorithm in such a way that the flow incidence decreases toward optimal values, leading to a reduction in the separations zones on the suction side and losses as revealed by the distribution of relative total pressure in Fig. 15. Furthermore, as shown, the highest deformation $\delta\theta$ is imposed near the tip where the incidence is the largest, hence confirming the validity of this ASO method.

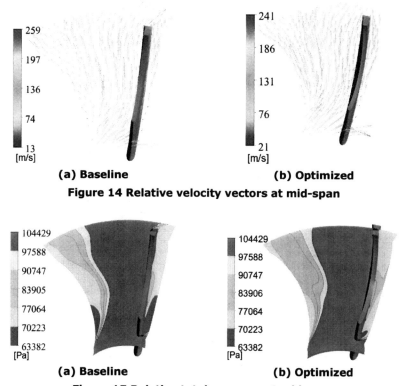

(a) Baseline   (b) Optimized

Figure 14 Relative velocity vectors at mid-span

(a) Baseline   (b) Optimized

Figure 15 Relative total pressure at mid-span

4 CONCLUSION

In the present paper, a coupled optimization procedure based on NSGA-II, RBF and MASSOUD technique was used to improve the design of a centrifugal compressor. The coupled algorithm produced the Pareto optimal fronts which enhanced the isentropic efficiency and the total pressure ratio. The number of objectives evaluations during an optimization cycle showed a subsequent reduction in CFD computations from 40000 to only 157. Also, the quality of Pareto front is shown to depend on the accuracy of RSM approximation that depends on the size of training sample, as well as spacing and distribution. Furthermore, it is shown that the improvement of the baseline shape depends on the range of the optimization space. The main disadvantage of this ASO, when considering arbitrary baseline shapes is the larger time cycle, due to the large optimization space. Further improvement for this procedure is required to allow larger optimization spaces with a lower time cycle when considering an arbitrary baseline shapes.

REFERENCES

1. Deb, K, "Optimization for engineering design, Algorithms and examples", Delphi: Prentice-Hall, 1995.
2. Wang X D, Hirsch C, Kang Sh and Lacor C, "Multi-objective optimization of turbomachinery using improved NSGA-II and approximation model", Comput. Methods Appl. Mech. Engrg., V.200, pp 883-895, 2011.
3. Tiou W T, Yiu K F C and Zangeneh M, "Application of simulated annealing to inverse design of transonic turbomachinery cascades", Journal of Power Energy, V. 216(1), pp 59-73, 2002.
4. Lohner R, "Applied CFD Techniques: An introduction based on finite element methods", Second Edition, Wiley, 2008.
5. Deb K, Pratap A, Agarwal S and Meyarivan T, "A fast and elitist multiobjective genetic algorithm: NSGA-II", IEEE Transactions on Evolutionary Computation, Vol. 6 No. 2, pp 182-197, 2002.
6. Samad A and Kim K Y, "Multi-objective optimization of an axial compressor blade", Journal of Mechanical Science and Technology, V.22, pp 999-1007, 2008.
7. Samad A, Kim K Y and Lee K S, "Multi-objective optimization of a turbomachinery blade using NSGA-II". ASME/JSME fluids engineering conference, San Diego, CA; pp: 885-91, 2007.
8. Oyama A, Liou M S and Obayashi S, "Transonic axial-flow blade shape optimization using evolutionary algorithm and three-dimensional Navier- Stoke solver, Ninth AIAA/ISSMO Symposium on Multidisciplinary Analysis and Optimization" (AIAA 2002), Atlanta, USA, September 2002.
9. Mengistu T and Ghaly W, "Aerodynamic optimization of turbomachinery blades using evolutionary methods and ANN-based surrogate models", Journal of Optimization and Engineering, Vol.9, pp 239-255, 2008.
10. Jin R, Chen W and Simpson T W, "Comparative Studies of Metamodeling Techniques Under Multiple Modeling Criteria", Journal of structural and multidisciplinary optimization", Vol 23(1), pp 1-13, 2001.
11. Peter J and Marcelet M, "Comparison of surrogate models for turbomachinery design", WSEAS Transactions on Fluid Mechanics, V.3, pp:10-17, 2008.
12. Diaz-Manriquez A, Toscano-Pulido G and Gomez-Flores W, "On the Selection of Surrogate Models in Evolutionary Optimization Algorithms", proceeding of: Evolutionary Computation (CEC), IEEE Congress 2011, pp 2155–2162.
13. Samareh J A, "Multidisciplinary Aerodynamic-Structural Shape Optimization Using Deformation (MASSOUD)", 8th AIAA/USAF/NASA/ISSMO Symposium on Multidisciplinary Analysis and Optimization, AIAA-2000-4911, Long Beach, CA, 2000.
14. Koini G N, Sarakinos S S and Nikolos I K, "A software tool for parametric design of turbomachinery blades, Advances in Engineering Software, V.40, pp 41–51, 2009.
15. Hathaway M D, Criss R M, Wood J R and Strazisar A J, "Experimental and Computational Investigation of the NASA Low-Speed Centrifugal Compressor Flow Field", ASME Journal of Turbomachinery, Vol. 115, pp 527-542, 1993.
16. Hardy R L, "Multiquadric Equations of Topography and Other Irregular Surfaces", J. Geophys. Res., vol. 76, pp 1905–1915, 1971.
17. Sobester A, Leary S J and Keane A J, "On the Design of Optimization Strategies Based on Global Response Surface Approximation Models", Journal of Global Optimization, V.33, pp 31–59, 2005.
18. Khalfallah S, Ghenaiet A, "Analyses of Impeller-Vaneless-Diffuser-Scroll Interactions in a Radial Compressor", Proceedings of the Institution of Mechanical Engineers Part A: Part A: Journal of Power and Energy, V. 224, pp 851-867, 2010.

Centrifugal compressor efficiency types and rational application

Y B Galerkin, A Drozdov, K V Soldatova
Compressor Dept., TU Saint-Petersburg, Russia

ABSTRACT

Expected efficiency of a compressor must be used in engineering calculations as it is not possible to predict head losses in a flow path. Various aspects of efficiency application (flow parameters in control planes definition, power consumption calculation) are shortly discussed. Presented samples show errors of different compressors comparison if inappropriate adiabatic efficiency is applied. The "real" efficiency based on calculation of head losses in elements of a flow path is presented. The "real" efficiency calculated by the Universal modeling method (1,7)) is compared with usual total polytropic efficiency.

NOTATION

c - absolute velocity, m/s;
f - area, m²;
H - head, j/kg;
H_p - polytropic head, j/kg;
H_i - internal head, j/kg;
H_d - dynamic head, j/kg;
H_r - head losses j/kg;
k - Isentropic coefficient;
$\dot{m}$ - mass flow rate, kg/s;
M - Mach number;
$M_u = \dfrac{u_2}{\sqrt{kRT_{inl\,tot}}}$;
n - polytropic coefficient;
N_i - power, W;
p - pressure, Pa;
R - gas constant, j/kg/K;
S - entropy, j/kg/K;
T - temperature, K;
v - volume, m³;

$\eta = \dfrac{\psi_p}{\psi_i}$ - polytropic efficiency;
η_r - real efficiency;
$\pi = \dfrac{p_{ex}}{p_{inl}}$ - pressure ratio;
ρ - gas density, kg/m³;
$\psi = \dfrac{h_p}{u_2^2}$ - polytropic head coefficient;
$\psi_T = c_{u2}/u_2$ - Euler work coefficient;
ς - loss coefficient;

Subscripts
ad – adiabatic;
des – design;
ex – stage exit;
inl – stage inlet;
ideal – ideal;
mean – mean;
meas – measured;
t – total parameter.

EFFICIENCY FOR CALCULATION OF GAS PARAMETERS IN FLOW PATH ELEMENTS

Flow parameters in control planes in the flow path are necessary to define velocity triangles (direct task) or area of the control planes in the flow path (reverse task). The compression process equation controls (for) the change of the parameters:

$$pv^n = const .\tag{1}$$

In engineering calculations the exponent n is assumed to be constant in all flow path elements (10). The graphic representation of the process is shown as a $T-S$- diagram in Fig. 1.

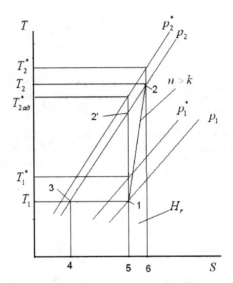

Fig. 1. $T-S$- **diagram of compression process in a stage or in a non cooled multistage compressor**

A polytropic exponent n depends on a polytropic efficiency (static parameters):

$$\frac{n}{n-1} = \eta \frac{k}{k-1} .\tag{2}$$

A polytropic efficiency depends on flow parameters measured at the inlet and exit control planes of a stage or a compressor:

$$\eta = \frac{\lg(p_2/p_1)}{\frac{k}{k-1}\lg(T_2/T_1)} .\tag{3}$$

The temperature T_{meas} is measured by the thermometer or thermocouple located in inlet/exit planes. Its value is intermediate between the static temperature T and total temperature T_t. Regulations for calculating temperatures T, T_t for thermometers and thermocouples based on T_{meas} were put forward in (2).

For calculating flow velocities the following values are needed:

$$c_{inl} = \frac{\dot{m}}{\rho_{inl} f_{inl}}, \quad c_{ex} = \frac{\dot{m}}{\rho_{ex} f_{ex}}. \quad (4)$$

Pressure and density change equations together with the compression process equation allow solving the problem of flow parameters definition in the stage control planes in principle:

$$\frac{p_2}{p_1} = \left(\frac{T_2}{T_1}\right)^{\frac{n}{n-1}}, \quad \frac{\rho_2}{\rho_1} = \left(\frac{T_2}{T_1}\right)^{\frac{1}{n-1}}. \quad (5)$$

Location of control planes is shown in Fig. 2. Static temperatures at control planes depend on total temperature and a flow velocity:

$$T = T_t - \frac{c^2}{\frac{2k}{k-1}R}. \quad (6)$$

If heat transfer can be neglected, total temperatures in all control planes previous of an impeller are equal to $T_{t\,inl}$ and after the impeller they are equal to $T_{t\,ex}$. Total temperature rise due to mechanical work input is:

$$T_{2t} = T_{1t} + \frac{H_i}{\frac{k}{k-1}R}. \quad (7)$$

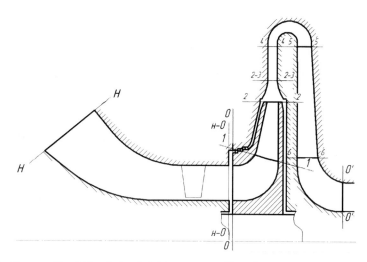

Fig. 2. Flow path of the industrial type centrifugal stage and control planes

Application of total adiabatic efficiency for calculations is also possible and is not less convenient:

$$H_{tad} = \frac{k}{k-1}RT_{tinl}\left[\left(\frac{p_{tex}}{p_{tinl}}\right)^{\frac{k}{k-1}} - 1\right], \quad H_i = \frac{k}{k-1}R(T_{tex} - T_{tinl}), \quad \eta_{tad} = \frac{H_{tad}}{H_i} = \frac{\left(\frac{p_{tex}}{p_{tinl}}\right)^{\frac{k}{k-1}} - 1}{\frac{T_{tex}}{T_{tinl}} - 1}, \quad (8)$$

the total pressure ratio in accordance with Eq. (8) is:

$$\frac{p_{tex}}{p_{tinl}} = \left[1 + \eta_{tad}\left(\frac{T_{tex}}{T_{tinl}} - 1\right)\right]^{\frac{k-1}{k}}, \quad (9)$$

the static pressure in a control plane is:

$$p = p_t \frac{p}{p_t} = p_t\left(\frac{T}{T_t}\right)^{\frac{k}{k-1}}, \quad (10)$$

As velocities in all control planes depend on density (Eq. (4)), density depends on temperatures (Eq. (5)) and temperature depends on velocity (Eq. (6)), it is necessary to apply iterative processes.

Calculation of efficiency by equations (3) or (8) requires taking into account the external heat transfer. Usually a compressor case has a higher temperature than the surrounding atmosphere, so the exit temperature is less than it would be if there were no heat transfer. The performance test code establishes measures to take heat transfer into account (12).

POWER CONSUMPTION CALCULATION

Mechanical power of the drive transferred to the compressed gas by the impellers is equal to:

$$N_i = \dot{m}H_i. \quad (11)$$

Mass flow rate $\dot{m}$ is measured in compressor tests. Some general considerations concerning an internal head are briefly presented below.

Internal head H_i transmitted by the impellers in a flow path produces polytropic head H_p, and dynamic head H_d. H_i is partly lost in a flow path due to different kinds of drag forces:

$$H_i = H_p + H_d + H_r, \quad (12)$$

$$H_p = \frac{n}{n-1}RT_{inl}\left[\left(\frac{p_{ex}}{p_{inl}}\right)^{\frac{n-1}{n}} - 1\right], \quad (13)$$

$$H_d = \frac{c_{ex}^2 - c_{inl}^2}{2}. \qquad (14)$$

It is possible to use one of the total efficiencies to estimate the head loss H_r.

Polytropic total efficiency:

$$\eta_t = \frac{H_p + H_d}{H_i}, \quad H_i = \frac{H_p + H_d}{\eta_t}. \qquad (15)$$

Adiabatic total efficiency:

$$\eta_{t\,ad} = \frac{\dfrac{k}{k-1} RT_{t\,inl}\left[\left(\dfrac{p_{t\,ex}}{p_{t\,inl}}\right)^{\frac{k-1}{k}} - 1\right]}{H_i}. \qquad (16)$$

As known the adiabatic efficiency depends not only on the head loss but also on similarity criteria M, k. It is not easy to estimate the value of $\eta_{t\,ad}$ accurately if there is no close analog for the designed compressor. Therefore, the total polytropic efficiency is used in industrial compressor analysis (12):

$$\eta_t = \frac{\lg\left(\dfrac{p_{t\,ex}}{p_{t\,inl}}\right)}{\dfrac{k}{k-1}\lg\left(\dfrac{T_{t\,ex}}{T_{t\,inl}}\right)}, \quad \frac{T_{t\,ex}}{T_{t\,inl}} = a\log\left[\dfrac{\lg\left(\dfrac{p_{t\,ex}}{p_{t\,inl}}\right)}{\dfrac{k}{k-1}\eta_{t\,pol}}\right], \quad H_i = \frac{k}{k-1}RT_{t\,inl}\left(\frac{T_{t\,ex}}{T_{t\,inl}} - 1\right). \qquad (17)$$

COMPARISON OF COMPRESSORS AERODYNAMIC PERFECTNESS

The efficiency of a better designed compressor seems to be higher. However, this may not be true, if adiabatic efficiency is used in order to compare compressors with different pressure ratios and adiabatic exponents. As dynamic head $H_d = 0,5(c_2^2 - c_1^2)$ is small in turbo compressors it will be neglected in further considerations.

In accordance with known properties of $T-S$ diagram the internal head H_i is proportional to the area 6-2-3-4 in Fig. 1. The head loss is proportional to the area under the line of a $pv^n = const$ process, i.e. to area 5-1-2-6. It means that total polytropic efficiency depends exclusively on ratio H_r/H_i (and nothing else):

$$\eta_t = \frac{H_p}{H_i} = 1 - \frac{H_r}{H_i}. \qquad (18)$$

Adiabatic head in Fig. 1 corresponds to area 5-2'-3-4. Adiabatic head ratio to the internal head (area 6-2-3-4) defines adiabatic efficiency that is always less than polytropic efficiency for the same compressor:

$$\eta_{ad} = \frac{H_{ad}}{H_i} < 1 - \frac{H_r}{H_i} = \eta. \tag{19}$$

Adiabatic efficiency is unduly treated as head loss area 1-2-2'. So, loss of efficiency $\Delta\eta_{ad} = 1 - \eta_{ad} > \Delta\eta_p$ does not reflect a share of head losses in a compression process. The relationship between adiabatic and polytropic efficiencies follows from the formulae for adiabatic and polytropic heads:

$$\eta_{ad} = \frac{\pi^{\frac{k-1}{k}} - 1}{\pi^{\frac{k-1}{k\eta}} - 1}. \tag{20}$$

Table 1 demonstrates a sample of efficiency difference $\Delta\eta = \eta - \eta_{ad}$ that depends on π, k, η in accordance with equation (20). Efficiencies are compared for compressors operated at different M_u that leads to different pressure ratio as it is shown for a single stage compressor:

$$\pi = \left(1 + (k-1)\eta\psi_i M_u^2\right)^{\frac{k}{k-1}\eta}. \tag{21}$$

The calculation of adiabatic efficiency is made for a compressor with polytropic efficiency $\eta = 0.86$ and for a compressor with $\eta = 0.82$. Polytropic efficiencies are assumed to be independent of pressure ratio, i.e. are independent of M_u in accordance with eq. (21).

Table 1. Adiabatic efficiency change with change of pressure ratio ($k = 1,4$)

η	π	1.5	2.0	4.0	8.0	16.0
0.86	η_{ad}	0.8517	0.8457	0.8308	0.8151	0.7989
0.82	η_{ad}	0.8084	0.8017	0.7827	0.7628	0.7423

The difference $\Delta\eta = \eta - \eta_{ad}$ increases from 0,5 - 1% for compressors with $\pi = 1,5$ up to 6-8% for compressors with $\pi = 16$. It demonstrates an adiabatic efficiency inability to reflect properly level of head losses in a compression process.

LOSS ESTIMATION

If polytropic efficiency is equal 0,86 for instance it means that 14% of an internal head is lost - eq. (18). However, for an efficiency calculated by eq. (3) it would be true if a polytropic exponent n would be constant.

In fact a flow path consists of compression elements with different level of compression efficiency where always $n > k$. There are also confuser elements of a flow path where $n < k$. It means that n are different in different elements. The next equation connects polytropic and isentropic exponents through a head loss and kinetic energy:

$$\frac{n}{n-1} = \frac{k}{k-1}\left(1 + \frac{h_w}{\frac{k}{k-1}R(T_2-T_1)}\right) = \frac{k}{k-1}\left(1 + \frac{0.5\zeta c_2^2}{0.5(c_1^2 - c_2^2)}\right). \qquad (22)$$

The loss model of the TU SPb Universal modeling method (3,4) is based on summarizing losses in each element of a centrifugal compressor stage. Therefore the programs provide information to present $T-S$ - diagram with $n = $ var in a stage elements. Several "real" $T-S$ - diagrams by Universal modeling method are presented below. Calculation of entropy change by well-known formula $S_2 - S_1 = \frac{R}{k-1}\left(\ln\frac{p_2}{\rho_2^k} - \ln\frac{p_1}{\rho_1^k}\right)$ (9) is not applicable for flow path elements where change of pressure and density is too small. The simplified formula based on $dS = \frac{dh_r}{T}$ was used instead:

$$\Delta S = \frac{h_w}{T_{mean}} = \frac{p_{2t\,ideal} - p_{2t}}{\rho_{mean}T_{mean}} = \frac{2R(p_{2t\,ideal} - p_{2t})}{p_2 + p_1}. \qquad (23)$$

The $T-S$ diagrams for the 1-st stage of two-stage pipeline compressor at three flow rates are presented in Fig. 3.

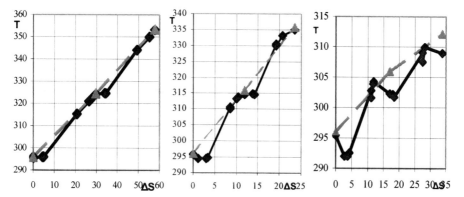

Fig. 3. $T-S$ - **diagrams of the 1st stage of two-stage pipeline compressor. Left – $\dot{m}/\dot{m}_{des}$ = 0,485 (close to surge), in the center – design flow rate, right - $\dot{m}/\dot{m}_{des}$ = 1,526.** $n = $ var - **solid lines,** $n = const$ - **stroke lines**

The area under a line of a compression process that is proportional to H_r is visibly different for $n = const$ and for $n = var$. The difference is minimal at the most effective design flow rate. As appeared, it is true also for compressors with different efficiency at design regimes too. The compressor with $T-S$ - diagrams presented at Fig. 3 has not too high efficiency. The newly designed one-stage pipeline compressor is more effective. Its diagrams at three regimes are shown at Fig. 4.

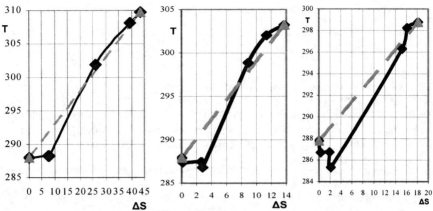

Fig. 4. $T-S$ - diagrams of one stage high effective pipeline compressor. Left – $\dot{m}/\dot{m}_{des}$ = 0,360 (close to surge), in the center – design flow rate, right - $\dot{m}/\dot{m}_{des}$ = 1.435. n = var - **solid lines**, $n = const$ - **stroke lines**

The $T-S$- diagram at a design regime for less effective multistage compressor is presented at Fig. 5.

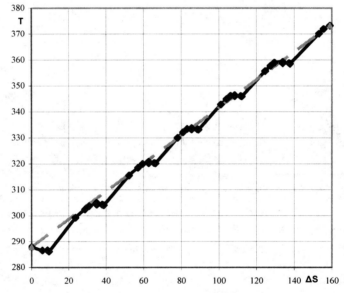

Fig. 5. $T-S$ - diagram of 6th - stage comparatively low effective pipeline compressor. Design regime. n = var - **solid lines**, $n = const$ - **stroke lines**

The computer programs of Universal modeling method calculate efficiency that we name "real" polytropic efficiency:

$$\eta_{tr} = \frac{H_i - \sum h_r}{H_i} = 1 - \frac{\sum h_r}{H_i}. \tag{24}$$

A polytropic efficiency that is used in engineering practice and that is measured in plant tests of compressors in accordance with (12) is defined by the next equation:

$$\eta_t = \frac{\lg\left(\dfrac{p_{t\,ext}}{p_{t\,inl}}\right)}{\dfrac{k}{k-1}\lg\left(\dfrac{T_{t\,ex}}{T_{t\,inl}}\right)}. \tag{25}$$

The efficiencies calculated by eq. (25) and (24) for stages and compressors at three regimes are presented Tables 2, 3, 4.

Table 2. Total, real, static polytropic efficiencies of the 1st stage medium – effective pipeline compressor

$\dot{m}/\dot{m}_{des}$	1.5258	1	0.4845
η_t	0.5467	0.8554	0.7517
η_{tr}	0.5480	0.8600	0.7600
η_{tr}/η_t	1.0023	1.0053	1.0110

Table 3. Total, real, static polytropic efficiencies of the 2nd stage medium – effective pipeline compressor

$\dot{m}/\dot{m}_{des}$	1.6939	1	0.4743
η_t	0.2518	0.8497	0.7455
η_{tr}	0.2580	0.8550	0.7560
η_{tr}/η_t	1.0246	1.0062	1.0057

Table 4. Total, real, static polytropic efficiencies of the one-stage high effective pipeline compressor

$\dot{m}/\dot{m}_{des}$	1.4336	1	0.3590
η_t	0.759	0.875	0.717
η_{tr}	0.762	0.880	0.727
η_{tr}/η_t	1.004	1.006	1.014

The "real" efficiency is higher in all cases. It is not important for engineering practice but it must be meant when results of modeling are compared with test data.

CONCLUSION

Well-known efficiency types – polytropic, adiabatic, total and static - serve well for proper to each of them kinds of application. Head losses are calculated in elements of a flow path and summarized in a process of a centrifugal compressor performance modeling. The "real" polytropic efficiency is defined by these summarized losses. This "real" efficiency appeared to be higher than a usual total polytropic efficiency in several cases presented in the paper. The difference is visible especially for off-design regimes. This must be taken into account when calculated and measured performances are compared.

REFERENCES

1. Galerkin Y. Turbo compressors. // LTD information and publishing center. - Moscow. – 2010 (In Russian).
2. Galerkin, Y, Rekstin F. Methods of research of centrifugal compressors. - Leningrad. – 1969 (In Russian).
3. Galerkin Y. Danilov K., Popova E. Design philosophy for industrial centrifugal compressor. International Conference on Compressors and their systems. – London: City University, UK. – 1999.
4. Galerkin Y.. Mitrofanov V., Geller M., Toews F. Experimental and numerical investigation of flow in industrial centrifugal impeller. International Conference on Compressors and their systems. – London: City University, UK. – 2001.
5. Galerkin, Y. B., Soldatova, K.V. Operational process modeling of industrial centrifugal compressors. Scientific bases, development stages, current state. Monograph. // Sankt-Peterburg. - SPbTU. – 2011 (In Russian).
6. Galerkin, Y.B., Soldatova, K.V., Drozdov, A.A. Specification of algorithm of calculation of parameters of a stream in the centrifugal compressor stage. [text] // Scientific and technical transactions of the TU SPb. – 2010. – No. 4. – Page – 150-157 (In Russian).
7. Galerkin Y., Drozdov A., Soldatova K. Development of computer programs of the Method of universal modeling of 1th level. Works 14 International scientific and technical conferences on compressor equipment. Volume 1. – Kazan. – 2011. – Page 276-284. -420 (In Russian).
8. Galerkin Y., Drozdov A., Soldatova K. Turbocompressor efficiency application and calculation. // Compressors & Pneumatics. – Moscow. - 2011. – № 8. – Page 18-24 (In Russian).
9. Lojtsanskij L. Mechanics of liquid and gas. – Moscow. – 1978 (In Russian).
10. Ris V. Centrifugal compressors. - Leningrad. – 1981 (In Russian).
11. Transactions of the TU SPb compressor school. –TU SPb. – 2010 (In Russian).
12. Turbocompressors - Performance test code// International Standard DD ISO 5389.1991.

VALVES

Historic review of attempts to model valve dynamics

E H Machu
Consulting Mechanical Engineer, Austria

ABSTRACT

From the design point of view, compressor valves can be subdivided into two groups: Valves with a flexible reed having no springs, typically used in small refrigeration compressors, and valves having a rigid body sealing element loaded by separate springs, such as the ported plate valve, the ring, channel or poppet valve. The present paper focuses on the last group, the ported plate valve in particular.

The present paper tries to give a summary of the important steps that have been accomplished to assess valve designs in view of compressor reliability and efficiency. Due to lack of space the paper has to focus on just a few aspects.

NOMENCLATURE

a, b, c	-	polynomial coefficients	Σ	-	summation symbol
C_D	-	drag force coefficient	ρ	kg/m³	density
C_r	N.m/rad	rotatory stiffness	φ	rad	polar co-ordinate
C_t	N/m	translatory stiffness	φ	-	coupling coefficient
d	-	differential operator	ϖ	rad/s	ang. velocity, nat. frequ.
D	m	diameter			
e	m	eccentricity	**Subscripts**		
F	N	force	0		horizontal position
h	m	valve lift, translatory displ.	1		nominal suction
I	kg.m²	moment of inertia	$1 \ldots 3$		degree of freedom
m	kg	mass	cg		centre of gravity
n	-	number	cs		crank shaft
ns	-	number of springs	D		drag force coefficient
p	Pa	pressure	g		gas drag force
R	m	radius	i		degree of freedom
t	s	time	j		number of a term
v	m/s	velocity	k		number of a term
y	-	ΔY within step	max		maximum
Y	-	relative displacement	n		beginning of nth step
α	-	coefft. of flow contraction	N		nominal
α, β	-	FOURIER coeffts.	o		outer (diameter)
γ	rad	rotatory displacement	p		(valve) plate
θ	rad	crank angle	pc		pitch circle
ϑ	rad	θ within a step	r		rotatory
ν	-	frequency ratio	s		spring
Π	-	gas drag force ratio	t		translatory
χ	-	spring force ratio	v		valve
λ	-	damping parameter			

© The author(s) and/or their employer(s), 2013

Disclaimer: This paper was prepared according to the best knowledge available to the author when writing this paper. The author accepts no liability for any kind of damage, direct or consequential, if any, suffered by any third party following actions taken or decisions made based on this paper.

1. INTRODUCTION

Valves are crucial items for compressor performance and reliability. Valve optimisation means the process of finding, in a first step, the best design features to build reliable valves, and in a second step to determine the best design parameters. Final purpose is to find the best compromise between good flow areas during suction and discharge events to reduce energy consumption and gas temperatures on one side, this requires a high valve lift, and excellent valve reliability on the other side, requiring a low lift. The choice of the words "good" and "excellent" already shows that, in general, reliability is the much more important choice.

Such a valve has to be self acting, i.e. there are no external cams or similar. It has to open and close in the right moments of a compression cycle, not earlier ad not later. When opened it has to offer the biggest possible flow area to assure minimum pressure drops and work losses. In addition, opening and closing shall happen with well controlled (impact) velocities to assure trouble free operation for years. All this requires good knowledge of its dynamic behaviour, its valve dynamics.

The purpose of the present paper is to concentrate on the mechanical part of ported plate valve modelling and assessing, assuming that coefficients of flow contraction α and of gas drag force C_d are known, either by experiment or by flow calculation (using geometric considerations or Computational Fluid Dynamics CFD). Moreover the interactions between of valve flow area and compressor performance and efficiency cannot be considered here.

2. HISTORICAL RETROSPECT

The steel plate valve was invented in an attempt to increase valve reliability. Mr. Hanns Hörbiger, working in the 1890s as chief designer for Maschinenfabrik Lang, Budapest, was charged to solve valve problems in a huge steel blower in Mezières, Northern France. It is reported that, in order to better understand how the existing valves behaved and operated, he had spent some time inside the suction plenum of this steel blower, watching, with the naked eye and under the light of a torch, the existing valves in operation. The steel plate valve he then designed was patented in Germany and Austria in 1895 [1] and is in fact the ancestor of the ported plate valve still widely used nowadays. A few years later he drew the sketch shown in Figure 1 "Sk 86, device to plot valve motion diagrams of a compressor valve" dated 1902.III.14. This modified indicator seems to be the oldest device used to investigate valve dynamics empirically.

To the knowledge of the author, the first attempt to describe the motion of a valve plate analytically can be found in a textbook by Hort [2] of 1922, pages 246 - 251. Hort models the valve of a water pump using the well known differential equation of a one degree-of-freedom oscillator. The fact that water is incompressible simplifies the problem: Instantaneous volume flow through the valve equals instantaneous piston displacement. There is no need for coupling with fluid flow equations.

Lanzendörfer [3] in 1932 seems to have been the first to publish measurements of valve plate motion in an air compressor. He also was using what he called an indicator to plot the motion of a valve plate. Due to the inertia of this measuring device he limited his investigations to compressors with a maximum speed of 600 rpm. In [3] he also published the variation of coefficients of flow contraction and of gas drag force over valve lift, both needed to evaluate the equilibrium of forces acting on the valve plate. By a simple analytical approach he specified limits for impact velocity, for allowable valve lifts and appropriate closing spring forces.

The first complete analysis of valve motion and compressor performance was presented by Costagliola in 1949 and 1950 [4]. Costagliola focused on the calculation of one valve at a time and established a system of two differential equations, one for valve motion and one for pressure evolution, to be solved simultaneously. Temperature is assumed to vary adiabatically with pressure.

Söchting [5] in 1955 used an electronic oscilloscope to investigate valve dynamics in air compressors up to 3000 rpm. One of his objectives was to minimize the mass of the part of the capacitive lift transmitter fitted to the valve plate.

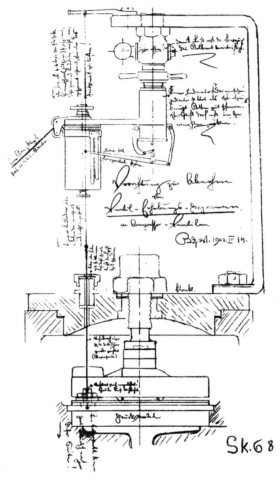

Figure 1: Valve motion indicator, hand sketch by Mr. Hanns Hörbiger (inventor of the ported plate valve, patents from 1895), dated 14.03.1902, (published with courtesy of HOERBIGER Kompressortechnik Holding GmbH, Vienna Austria)

H. Meier [6] in 1962 describes measured valve motions. Due to the extreme eccentricity of his valve pockets, shown in Figure 3 of this article, he had also mentioned having observed some tilting motion of the valve plates. However he obviously did not realize the danger for valve reliability which may be caused by this tilting motion in conjunction with late valve closure, as shown here in Figure 4.

Hackmülller [8] and Hagel [9] make use an analogue computer to model the motion of a spring loaded valve plate and describe in detail the mathematical background of this kind of modelling. It has to remembered that, back in 1968, analogue computers were much faster and much cheaper than digital ones.

MacLaren and Kerr in [10] and [12] investigate self-acting valve behaviour by means of a digital computer. Measured and computed valve motions are in good agreement as is shown in many examples in these two papers. Discrepancies are said to be caused by oil adhesion.

The paper of H. Davis [11] contains valuable information on coefficients of flow resistance and of drag force as function of valve lift based on experimental data.

Although already mentioned earlier by MacLaren and Kerr [10], an analytical approach of the phenomenon of oil sticktion of the valve plates to seat or guard has been presented by Bauer [16], Böswirth [19] and Stehr [21]. Sticktion may in fact be quite a reliability problem: By delaying the beginning of valve opening or valve closure while pressure across the valve is growing rapidly, the valve plate, once it has started moving, experiences much higher acceleration and final impact velocities on the other stop than normally. However, since details like the viscosity and thickness of the oil film as well as the "wetted" area of contact are usually not known, efforts are limited to derive design guidelines for making a valve less sensitive to oil sticktion [16].

E. Machu, having had several valve reliability problems with valve plates not moving in parallel to the valve seat (Figures 3, 4 and 5 have been taken in this compressor), presented his findings in papers [14], [18] and [20]. Here again, since the causes of valve plate tumbling (like the eccentricity of the gas drag force) can hardly be foreseen in the design stage of a compressor, the benefit of such research work is to derive guidelines for better valve designs.

Gas inertia may have some influence on valve dynamics. Gas inertia in valve ports has been examined by Trella, Soedel [13]. In addition to valve ports Böswirth in [15] and [19] also considered the gas spring effect and the unsteady work exchange between the gas flow and the moving valve plate. However, it seems that, to increase valve reliability, one has to focus above all on the valve's closing motion, where, unlike the opening motion, changes of flow velocity and hence effects of gas inertia are much less pronounced. On the other side, gas inertia inside the working chamber, i.e. uneven pressure distribution and reflected pressure waves, as examined by G. Machu [22], may indeed also influence the closing motion and hence reliability.

A very complete survey of valve dynamics has been presented by R. Habing [24], modelling the flow by CFD methods and allowing several degrees of freedom for the motion equations.

Finally a question sometimes bothering compressor users concerns the possibility of resonance between the valve plate and one of the harmonics of its forcing function. This may happen if the natural frequency ratio v_l (Table 1) of a valve happens to be in integer number. It has been shown in [17] (verbally and mathematically) that there is no danger whatsoever in resonance. Due to lack of space, no explanation can be given here, a very short discussion in given below in Section 4.

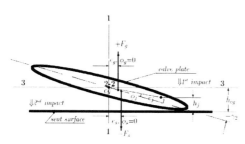

Figure 2: Tumbling valve plate

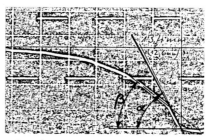

Figure 3: Due to tumbling, impact velocity is increased in the proportion $\tan(\alpha)/\tan(\beta) = 2.56$

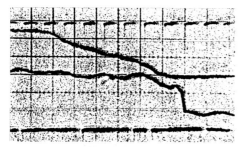

Figure 4: Late discharge valve closure, impact velocity up to 11 m/s !!!

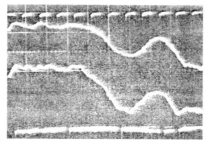

Figure 5: Early discharge valve closure, moderate impact velocities

In Figure 4, due to tumbling, impact velocities at two diametrically opposite points of a valve plate are 2.0 (upper trace) and 11.0 m/s (lower trace). Dashes mark 15° crankshaft rotation, piston dead centre position is marked by the short dash.

Figures 3, 4 and 5 are published with courtesy of HOERBIGER Kompressortechnik Holding GmbH, Vienna, Austria.

3. ANALYSIS OF VALVE PLATE MOTION

Although valve plate motions with only one degree of freedom have been analysed several times since 1949 ([4], [8], [10], [17] etc.) it seems that valuable additional conclusions for valve layout can be obtained when extending the analysis to a model handling two or three degrees of freedom.

3.1 Valve dynamics with one degree of freedom (DOF=1)

Applying Newton's second law of motion, stating that *mass × acceleration is equal to the sum of all forces acting in the direction of motion*, to a valve plate gives

$$m \cdot \frac{d^2 h}{dt^2} = \underbrace{A_p \cdot C_D \cdot (p_s - p_g)}_{\text{gas drag force = forcing function}} - \underbrace{F_n + C_t \cdot (h_n - h)}_{\text{spring force}} - \underbrace{r \cdot \frac{dh}{dt}}_{\text{damping force}} \qquad (1)$$

For a suction valve (and inversely for a discharge valve), with suction plenum pressure on the seat side given by a FOURIER series, guard side pressure in the working chamber by a polynomial, equations 3. Rearranging, writing $\lambda_1 = r/(2.m.\omega.v_l)$ for the damping coefficient, using the symbols of Table 1, changing from a time domain to a crank angle domain with coordinates $\vartheta = \theta - \theta_n$ within one step and angular crankshaft velocity $\varpi_{cs} = d\theta/dt$ finally leads to equation 4

$$p_s = p_{\text{suction plenum}} = p_1 \cdot \sum_{i=0}^{m} [\alpha_i \cdot \cos(i \cdot \theta) + \beta_i \cdot \sin(i \cdot \theta)] \qquad (2)$$

$$p_g = p_{\text{working chamber}} = p_1 \cdot \sum_{i=0}^{2} a_i \cdot \vartheta^i, \qquad \Delta p_{\text{valve}} = p_s - p_g = p_1 \cdot \sum_{i=0}^{2} b_i \cdot \vartheta^i \qquad (3)$$

$$\frac{d^2 y_1}{d\vartheta^2} + v_1 \cdot \left(2 \cdot \lambda_1 \cdot \frac{dy_1}{d\vartheta} + v_1 \cdot y_1\right) = -v_1^2 \cdot \chi_1 \pm$$
$$\pm \left[\Pi_1 \cdot \left(\sum_{k=0}^{n}(\alpha_k \cdot \cos k\vartheta + \beta_k \cdot \sin k\vartheta) - \sum_{k=0}^{2} \frac{a_k}{k+1} \cdot \vartheta^k\right)\right] =$$
$$= v_1^2 \cdot \left[-\chi_1 + \Pi_1 \cdot \sum_{k=0}^{2} \frac{b_k}{k+1} \cdot \vartheta^k\right] \qquad (4)$$

In case of ± the upper sign is for a suction valve, with the FOURIER series on the seat side, the lower sign for discharge valves. The coefficients b_k describe the evolution of valve pressure drop over crank angle within one step of integration. In principle, this is the equation of all one-degree-of-freedom models without considering gas inertia, starting with Costagliola [4], but also [10], [12] etc.

3.2 Valve dynamics with several degrees of freedom (DOF ≥ 2)

According to Figure 2, lift and velocity of any point j on a tumbling valve plate (with polar coordinates e_j, Φ_j in the plane of the valve plate), angles of inclination γ_2 and γ_3, are given by equations 5 and 6

$$h_j = h_{cg} + e_j \cdot (\gamma_2 \cdot \cos\phi_j + \gamma_3 \cdot \sin\phi_j) \qquad (5)$$

$$\frac{dh_j}{dt} = \frac{dh_{cg}}{dt} + e_j \cdot \left(\frac{d\gamma_2}{dt} \cdot \cos\phi_j + \frac{d\gamma_3}{dt} \cdot \sin\phi_j\right) \qquad (6)$$

The mere fact that equation 6 contains other terms than dh_{cg}/dt (this is the velocity of the centre of gravity h_{cg} as found by a DOF=1 model) shows that, due to tumbling, local valve plate velocity may be different. Starting deductions using the LAGRANGE set of equations (not NEWTON's as above), rearranging and using the symbols of Table 1 as before (for more details please refer to [20]), will give

$$\frac{d^2 y_i}{d\vartheta^2} + v_i^2 \cdot y_i + v_i^2 \cdot \varphi_i \cdot y_{(3-i)} = v_i^2 \cdot \left[-\chi_i + \Pi_i \cdot \sum_{k=0}^{1} \frac{b_k \cdot \vartheta^k}{k+1}\right] \qquad (7)$$

With outer valve plate diameter D_o, nominal, i.e. maximum possible rotatory displacement γ_N can be defined by as $\gamma_N \approx \sin(\gamma_N) \approx \tan(\gamma_N)$ provided γ_N is small. Subscript N stands for "nominal", subscript n for the value at the beginning of the nth step of integration of the differential equations.

It is easily seen that, for $i=1$, with vanishing coupling coefficients $\varphi_1=0$ and $\varphi_2=0$ and neglecting damping, i.e. with $\lambda_1=0$, equation 7 is identical with equation 4. These coupling coefficients φ_1 and φ_2 will vanish provided identical springs are arranged symmetrically, i.e. when for every spring location $e_j\sin(\Phi_j)$ there is an opposite $e_j\sin(\Phi_j+\pi)=-e_j\sin(\Phi_j)$. It has been shown in [20] that this is also true for an odd number of springs as long as there are at least three of them.

Table 1. The coefficients of equations 4 and 7

Comment	translation: i=1	units	rotation: i=2	units
Relative displacement	$Y_1 = h/h_N$	[1]	$Y_2 = \gamma_2/\gamma_N$	[1]
Same, within a step	$y_1 = Y_1 - Y_{1,n}$	[1]	$y_2 = Y_2 - Y_{2,n}$	[1]
total spring rate	$C_t = \sum_{j=1}^{ns} C_j$	[N/m]	$C_r = \sum_{j=1}^{ns} e_j^2 C_j$	[N.m/rad]
natural frequency	$\varpi_1 = \sqrt{C_t/m}$	[1/s]	$\varpi_2 = \sqrt{C_r/I}$	[1/s]
frequency ratio	$v_1 = \varpi_1/\varpi_{cs}$	[1]	$v_2 = \varpi_2/\varpi_{cs}$	[1]
gas drag force ratio	$\Pi_1 = \dfrac{A_p C_D p_1}{h_N C_t}$	[1]	$\Pi_2 = \dfrac{A_p C_D p_1 e_g}{\gamma_N C_r}$	[1]
Spring force ratio	$\chi_1 = \dfrac{\sum_{j=1}^{ns} F_{j,n}}{h_N C_t}$	[1]	$\chi_2 = \dfrac{\sum_{j=1}^{ns} e_j F_{j,n}}{\gamma_N C_r}$	[1]
coupling coefficient	$\varphi_1 = \dfrac{\gamma_N}{h_N} \cdot \dfrac{\sum_{j=1}^{ns} e_j C_j}{C_t}$	[1]	$\varphi_2 = \dfrac{h_N}{\gamma_N} \cdot \dfrac{\sum_{j=1}^{ns} e_j C_j}{C_r}$	[1]

3.2.1 Danger of tumbling in conjunction with late valve closure

Almost all functions of gas drag force coefficients over valve lift $C_d=f(h)$ known to the author, whether measured or computed, show that C_d is biggest when lift h is smallest and inversely. Consequently, with a tumbling valve plate in an oblique position, the resulting gas drag force will be eccentric, moving towards the point of the valve plate which is nearest to the seat. As a result, as long as flow direction in the valve is normal, from seat to guard, the gas drag force will act upwards (in Figure 2) and tend to restore the horizontal position of the valve plate. Inversely, during late valve closure and reversed flow direction from guard to seat, the gas drag force will tend to increase the oblique position. This, together with the rapidly increasing pressure differences across the valve during late closure (when the piston has passed dead centre position and either compression or expansion is on its way) explains the extremely high impact velocities at the second seat contact, observed in these cases, see the measured valve lift diagram Figure 4.

4. THERE IS NO DANGER IN RESONANCE

Due to lack of space, the reader has to refer to paper [17], only a short summary can be given here: There are two solutions of equation 4, concerning the particular integral containing the Fourier series, here equation 2, in paper [17] given by equations 2.14 (no resonance) and 2.15 (case of resonance: $i=v$, $\lambda=0$). However, since in both cases any (opening or closing) motion of the valve plate starts from rest, the solution curves for all these cases are given in Figure 6 of this paper. They show that there is virtually no significant difference between them, at least during the first 30° of crankshaft rotation. This can also be explained verbally as follows: An oscillator in or near resonance with its forcing function can only have huge amplitudes provided it can accumulate energy during several complete oscillation cycles. But this is what a valve plate cannot do. Long before it can complete the first oscillation cycle, it will bump into a stop, seat or guard, and loose all its kinetic energy. Next cycle starts again from rest with zero energy.

The real danger of a pressure pulsation seems to reside in the fact that it may be the origin of a sudden change of pressure difference across a valve. If this happens during the opening or the closing motion of the valve plate, its impact velocity may be increased, its reliability reduced in a dangerous way.

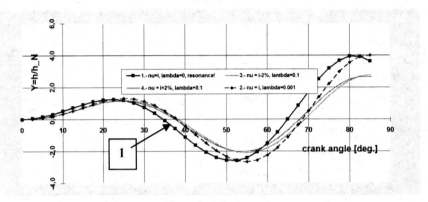

Figure 6: Valve motion starting from rest, in resonance (curve 1) and near resonance (curves 2, 3 and 4). Noted that there is no substantial difference between these curves, at least during the first 30° of crankshaft rotation

5. CONCLUSIONS

Besides a proper selection of valve lift (a recommendation based on experience can be found in [23], page P38, equ. 35), the best way to be on the save side is to assure early valve closure. This requires stronger springs having also a higher spring rate, thus tending to counteract any tilting motion of the valve plate. Furthermore by arranging the closing springs as far outside as possible, near the periphery of the valve plate, rotatory stiffness due to the closing springs is further increased.

Natural frequencies or rather frequency ratios v_1 and v_2 as defined in Table 1 shall be big, preferably not smaller than 5. This means that, if the valve plate was free to oscillate, there shall be, in each degree of freedom, at least 5 complete cycles during every complete revolution of the crank shaft.

REFERENCES

[1] Hanns Hoerbiger, "*Eine Neuerung an Ventilen und ähnlichen Verschlussorganen*" (*A Novelty concerning Valves and similar Sealing Elements*), German patent 87.267 of 7 Aug.1895, Austrian patent 47/250 of 26 Jan. 1897, priority 20 Nov. 1895.
[2] W. Hort, *Technische Schwingungslehre (Technical Vibration Analysis)*, textbook in German language, Springer, Berlin, 1922.
[3] E. Lanzendörfer, *Strömungsvorgänge und Bewegungsverhältnisse in Druckventilen schnelllaufender Kompressoren (Flow events and valve dynamics in the discharge valves of high speed compressors)*, Zeitschrift des VDI 76/14 1932, VDI-Verlag Düsseldorf.
[4] M. Costagliola, *The theory of the spring loaded Valve for Reciprocating Compressors*, Journal of Appl. Mech. Dec 1950, and *Dynamics of a reed type valve* PhD. Thesis, MIT June 1949, Journal of Applied Mechanics 17(4), 1950.

[5] F. Söchting, *Huboszillogramme von Ventilen raschlaufender Verdichter (valve lift oscillograms of valves in high speed compressors)*, ZVDI 87/1955, VDI-Verlag Düsseldorf.
[6] H. Meier, *Untersuchung der Arbeitsweise selbsttätiger Verdichterventile (Investigation of the operation of automatic compressor valves)*, Z. Maschinenbautechnik 1/1962.
[7] H. Najork, Möglichkeiten zur Vorausberechnung von Druckverlusten in den selbsttätigen Ventilen von Kältemittelkompressoren (Possibilities to forecast pressure drops in the automatic valves of refrigerating compressors), Z. Luft- und Kältetechnik 3/1967.
[8] E. Hackmüller, J.W. Hagel, *Ein elektrisches Modell des federbelasteten Plattenventils (an electric model of the spring loaded plate valve)* Z. Elektrotechnik und Maschinenbau, 12/1968.
[9] J.W. Hagel, *Untersuchung des Flatterverhaltens federbelasteter Ventile (Investigation of fluttering of spring loaded plate valves)* Z. Konstruktion, 1/1968.
[10] J.F.T. MacLaren, S.V. Kerr, *Valve behaviour in a small refrigerating compressors using a digital computer,* Journal of Refrigeration 1968/11 (nr. 6).
[11] H. Davis, *Effects of reciprocating compressor valve design on performance and reliability,* Conference on Industrial Reciprocating and Rotary Compressors: Design and Operational problems, IMechE, London, October 1970.
[12] J.F.T. MacLaren, S.V. Kerr, *An analytical and experimental study of self-acting valves in a reciprocating compressors,* Conference on Industrial Reciprocating and Rotary Compressors: Design and Operational problems, IMechE, London, October 1970.
[13] Trella, Soedel, Effect of Valve Port Gas Inertia on Valve Dynamics, International Compressor Engineering Conf. at Purdue, July 1974, proceedings.
[14] E.H. Machu, *Valve dynamics in three dimensions,* Hoerbiger internal research report, 1978.
[15] L. Böswirth, *Valve Flow taking non steady flow into account,* International Compressor Engineering Conference at Purdue, July 1984, proceedings.
[16] F. Bauer, *The influence of liquids on compressor valves*, International Compressor Engineering Conference at Purdue, July 1990, proceedings.
[17] E.H. Machu, *Valve dynamics in a pulsating environment*, PCRC conference Denver/USA, 1992.
[18] E. H. Machu, *The two-dimensional motion of the valve plate of a reciprocating compressor valve*, International Compressor Engineering Conference at Purdue, July 1994, proceedings.
[19] L. Böswirth, *Strömung und Ventilplattenbewegung in Kolbenverdichterventilen (Flow through valves and valve plate motion in reciprocating compressor valves),* text book (in German language), published by the author (Argentinierstrasse 28, A 1040 - Vienna, Austria), 1998.
[20] E. H. Machu, *Valve dynamics of reciprocating compressor valves with more than one degrees of freedom*, International Conference on Compressors and their Systems, IMechE, London, September 2001, proceedings.
[21] H. Stehr, *Oil sticktion, investigations to optimise reliability of compressor valves*, International Conference on Compressors and their Systems, IMechE, London, September 2001, proceedings.
[22] Gunther Machu, *Pulsationen im Verdichtungsraum – eine potentielle Schadensursache? (Pulsations inside the working chamber- a potential cause of troubles?)*, 8th Workshop Reciprocating Compressors, Kötter Consulting Engineers Rheine, Germany, Oct. 2004, proceedings.
[23] Kleinert, *Dubbel, Taschenbuch f.d. Maschinenbau*, (Pocket Book for Mechanical Engineering) Springer, 19th Editn. 1997, ISBN 3-540-62467-8.
[24] R. Habing, *Flow and Plate Motion in Compressor Valves,* PhD Thesis presented at the University of Twente, Enschede, Netherlands, 2005, ISBN 90-365-2179-3.

Numerical analysis of seat impact of reed type valves

F C Lajús Junior[1], **C J Deschamps**[1], **M Alves**[2]
[1] Federal University of Santa Catarina, Brazil
[2] University of São Paulo, Brazil

ABSTRACT

Reed type valves of compressors are subject to structural failure due to bending and impact fatigues. The failure by bending is a well-known and manageable problem, but the same cannot be said about the impact fatigue. Usually, the criterion to determine whether a reed will fail or not in the presence of repetitive impacts is either the impact velocity specified by the material supplier or other rule of thumb derived from experience. This paper reports the results of a numerical model developed to simulate the seat impact of reed valves. The most influential parameters of the phenomenon are identified through dimensional analysis and factorial designs. The study indicates that reed deformation and maximum stress are greatly affected by even small geometric variations.

NOMENCLATURE

Roman

c	speed of sound	V	reed tip initial velocity
C_{valv}	reed length	**Greek**	
D_{orif}	orifice diameter	δ	tip displacement
e_{valv}	reed thickness	θ	torsion angle
e_{cnt}	seat width	ρ	density
E	material Young modulus	σ_{IMP}	impact stress
L_{valv}	reed width	υ	Poisson coefficient

1. INTRODUCTION

One of the main challenges associated with the development of highly efficient compressors is the design of reed valves that do not fail due to impact fatigue. In part, this difficulty arises because the physical mechanism that originates such a fault is not fully understood (1). In general, fatigue failure is characterized by the formation of small cracks, or fractures of small chips, in a region between the edge of the seat and the free tip of the reed, i.e., at certain distance from the contact region. For this reason, it is usually assumed that impact fatigue is related with stress wave propagation. However, there is no consensus about which stress is the most relevant and how stress waves interact with each other to cause fatigue failure.

Fracture or crack in reeds occurs after several impacts, characterizing a high-cycle fatigue. Additionally, the virtual absence of a plastic deformation region suggests a fragile aspect for this type of failure. Disagreements in the literature exist regarding the origin of the cracks (2-3), whether it starts on the reed surface or internally in the material. For instance, there are reports of cracks occurring on the opposite surface of the contact region, leading to question about what would actually be the nature of the (critical) impact stress. The lack of knowledge of many issues associated with this problem also arises due to the difficulty of measuring the mechanical efforts originated during the impact. This is one of the main reasons for the use of an impact velocity as the parameter to indicate a possible reed failure rather than a value for impact stress itself.

The problem of valve reed fatigue has been primarily analyzed using accelerated life test machines to evaluate the material response of reeds under operation (2-3). Such machines are devised to evaluate the number of impacts a certain specimen can resist before the beginning of any crack, while being subjected to a controlled impact velocity. However, one occasionally finds that the limit of a critical impact velocity is not enough to ensure reed reliability in actual applications.

This paper presents the simulation results of seat impact of reeds, aimed at identifying influential aspects that may have a role in fatigue failure. A three-dimensional structural model was adopted to evaluate impact stresses for different geometries and impact velocities.

2. NUMERICAL MODEL

A numerical model was devised using the finite element method available in the commercial LS-DYNA code. Basically, we seek the solution of displacements in the linear momentum equation written in the weak form for specified initial and boundary conditions. Details regarding the mathematical formulation and discretization of the equations can be obtained from (4). Regarding the contact modeling, the penalty method was used to introduce a reactive force proportional to the nodes penetration between meshes.

A three-dimensional formulation was adopted to characterize the reed geometry and evaluate the stresses resulting from its impact against the seat. The reed was initially pre-tensioned by applying a prescribed displacement at its tip so as to represent the reed moments before contact. After this pre-tension condition is established, a linear velocity distribution is assigned to the nodes along the reed length and the simulation then is carried out via an explicit solution method. The initial condition of velocity for the simulation is that applied to the reed tip.

The seat is modeled as a rigid contact ring and therefore most of the reed initial energy is converted into deformation of the reed. This hypothesis is somewhat plausible because the seat contains much more material than the reed. Nevertheless, the model is expected to overestimate to some degree the impact effects on the reed.

It is necessary to specify the stresses to be monitored in the simulation. The adoption of one of the six components (σ_X, σ_Y, σ_Z, τ_{XY}, τ_{XZ}, τ_{YZ}) in a given coordinate system is not necessarily the right approach. For this reason, it is generally sought a relationship between the principal stresses so as to make the stress state independent of the geometry orientation in space.

The criteria of maximum energy distortion (Huber-Mises) and the maximum shear stress (Tresca) are quite common in structural mechanics. However these quantities are more related to ductile fracture of materials. Another criterion is the maximum principal stress (Rankine), which simply consists of checking the highest principal stress found in relation to a threshold stress level. This approach was created for brittle materials, which fail when subjected to tensile or compressive stresses.

The characteristics of reed fatigue failure do not reveal significant plastic deformation. Indeed, the fragile nature of the failure suggests that the maximum principal Rankine stress criterion should be suitable. However, the absence of a critical stress level for fatigue implies that this criterion and the stress magnitudes are only meaningful for a parametric analysis of the problem.

3. DIMENSIONLESS NUMBERS FOR VALVE IMPACT

A Design of Experiments (DoE) based on dimensional analysis is quite advantageous for a parametric analysis. Following the Buckingham's Π theorem (5), the impact stress is described as a function of dimensionless groups formed by combining different parameters: geometry and material of the valve system, initial velocity of the reed (V), initial displacement of the tip (δ) and reed torsion angle (θ); the latter being only relevant when the valve reed hits the seat with a torsional angle. Figure 1 shows the geometric parameters of the valve system composed by the reed and seat.

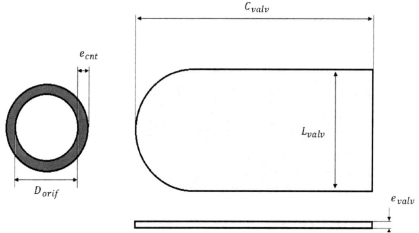

Figure 1–Geometric parameters of reed and seat

By choosing the initial velocity (V), reed density (ρ) and seat orifice diameter (D_{orif}) as repetitive parameters, one finds the following dimensionless groups:

$$\Pi_{IMP} = \frac{\sigma_{IMP}}{\rho V^2} \quad (1) \qquad \Pi_\upsilon = \upsilon \quad (2) \qquad \Pi_{\upsilon_A} = \upsilon_A \quad (3)$$

$$\Pi_{\rho_A} = \frac{\rho_A}{\rho} \quad (4) \qquad \Pi_{E_A} = \frac{E_A}{\rho V^2} \quad (5) \qquad \Pi_\delta = \frac{\delta}{D_{orif}} \quad (6)$$

$$\Pi_\theta = \theta \quad (7) \qquad \Pi_L = \frac{L_{valv}}{D_{orif}} \quad (8) \qquad \Pi_C = \frac{C_{valv}}{D_{orif}} \quad (9)$$

$$\Pi_e = \frac{e_{valv}}{D_{orif}} \quad (10) \qquad \Pi_{e_{cnt}} = \frac{e_{cnt}}{D_{orif}} \quad (11) \qquad \Pi_V = \frac{\rho V^2}{E} = \frac{V}{c} \quad (12)$$

In Eq. (12), c refers to the speed of sound in the reed represented by a one-dimensional compression wave. The dimensionless impact stress, Eq. (1), can be regarded as the ratio between the internal work produced by the impact stress (σ_{IMP}) and the initial kinetic energy of the valve (ρV^2), both per unit of volume. The geometries of the reed and seat are described by Eqs. (8-11).

The validity of the aforementioned dimensionless groups was assessed through tests of similarity. As a first test, the initial velocity had its value doubled and the density of the valve material was reduced to a fourth of its original value, in order to maintain the same value of dimensionless group represented by Eq. (1). In the second case, the initial velocity was unchanged, but the geometry was scaled down by a factor of 10:1. Dynamic similarity was observed in both cases.

In addition to reducing the number of variables for DoE and generalization of results, dimensional analysis opens the possibility of using large-scale models for measuring impact stresses.

4. RESULTS OF SENSITIVITY ANALYSIS AND DISCUSSIONS

Based on material properties (Table 1), the ranges of dimensionless parameters were defined for the reed impact (Table 2). In order to reduce the cost of simulations, some dimensionless parameters were kept fixed.

Table 1—Material properties of valve and seat.

		VALUE	UNIT
REED	E	210	GPa
REED	υ	0.3	-
REED	ρ	7.70E-06	kg/mm³
SEAT	E_A	135	GPa
SEAT	υ_A	0.27	-
SEAT	ρ_A	6.90E-06	kg/mm³

Table 2—Dimensionless numbers adopted in the analysis.

	Low	Medium	High	Variation
Π_e	0.02	0.03	0.04	± 33%
Π_L	1.3	1.45	1.6	± 10%
Π_C	3	3.3	3.6	±9%
Π_{ecnt}	0.06	0.1	0.14	± 40%
Π_V	2.00E-04*	6.00E-04*	1.00E-03*	± 67%
Π_υ	-	0.3	-	-
Π_θ	-	0	-	-
Π_δ	-	0.05	-	-

* rounded values

A total of 96 simulations of impact were carried out. The discretization meshes were generated via a blocking method to allow elements of minimal distortion, with proportions 2:2:1. The meshing process was automated by imposing five elements to discretize the valve thickness. Therefore, three mesh densities resulted by the three values of dimensionless thickness (Π_e) adopted in the analysis. The simulations computationally more expensive took an average of 18 hours of processing. A maximum error of approximately 6% was estimated based on mesh refinement tests.

Figure 2 shows results for the dimensionless impact stress (Π_{IMP}) with respect to the maximum principal stress (σ_{IMP}) for high, medium and low levels of dimensionless impact velocity (Π_V). In the same figure, the black filled symbols for each value of Π_V represent the results obtained from Eq. (13) developed for one-dimensional impact of rods and sometimes adopted to obtain estimates of reed impact stress:

$$\sigma = V_0\sqrt{E\rho} \tag{13}$$

where σ is the impact stress, V_0 is the impact velocity, E is the Young's modulus and ρ is the density of the reed.

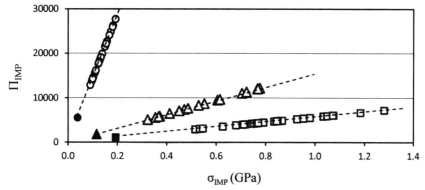

Figure 2–Relation between Π_{IMP} and σ_{IMP}; Square, triangular and circle symbols are associated with high, medium and low levels of Π_V, respectively; Black filled symbols represent estimates given by Eq. (13).

Figure 2 allows the analysis of the influence of the initial velocity and geometrical parameters on the impact stress σ_{IMP}. Each level of Π_V defines a straight line whose slope is related to the product of density and the square of the reed initial velocity (ρV^2). As the material was not varied in this study the different slopes are due solely to different velocity values. The scattering of the results along each line is originated by geometric variations, which are more pronounced in the case of a higher initial velocity. For instance, the magnitude of the principal stresses varies between 0.5 and 1.3 GPa for the highest level of Π_V.

In order to identify which geometric parameters have greater influence (effect) on Π_{IMP}, a statistical analysis was applied to quantify the main effects and interactions following the theory of factorial design (6-7). The term "main effect" provides a measure of the influence of a single variable, such as L (width), whereas the "interaction effect" stands for the influence of combined variables, such as e*L (width and thickness).

It was found that the most influential geometric parameters are different for each initial velocity level. For instance, Figure 3 shows the results for the higher velocity considered. Square and circle symbols represent positive and negative effects, i.e., effects that increase or decrease the impact stresses, respectively.

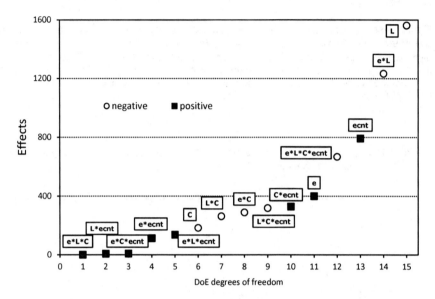

Figure 3–Effects of Π_{IMP} for different degrees of freedom in high Π_V.

As can be seen, the most significant main effects on Π_{IMP} are associated with the reed width (L) and the seat thickness (ecnt). An increase of Π_L causes an average decrease of 1560 units on the dimensionless impact stress (Π_{IMP}), which is equivalent to approximately 280 MPa for this level of Π_V. On the other hand, an increase of Π_{ecnt} brings about an increase of 790 units on Π_{IMP}, i.e., approximately 142 MPa. The influence of the combined effect (e*L) is also very important on the stress levels in this case and suggests that the width and thickness of the reed could be varied in such a way to reduce Π_{IMP}.

Numerical results for reed deformation can be used in order to physically understand what makes a given parameter or a combination of parameters to increase or decrease the impact stress. In the present study, it was found that the progression of the impact gives rise to different efforts depending on the geometric configuration. This can be better explained with reference to Figure 4, where results of deformation are shown for two reed widths (Π_L =1.3 and 1.6) subjected to the same velocity (Π_V).

As can be seen, for Π_L =1.3 there is a contact progression after the impact that makes the reed to deform in a pattern very similar to "flexural waves" on both sides of the valve (Fig. 4a). As the contact progresses, the curvatures of such deformations are increased. Eventually, both waves meet each other at the valve tip and this results the maximum stress on the surface opposite to where the contact occurs. It is also possible to notice that the innermost part of the reed is deformed into the orifice.

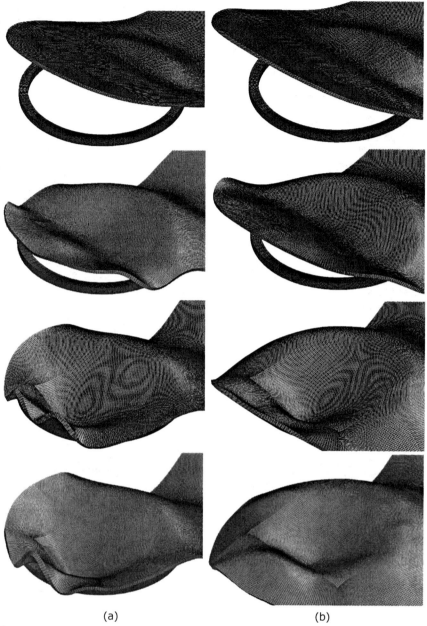

(a)   (b)

Figure 4–Results for two valve geometries subjected to an impact with the same Π_V: (a) $\Pi_L = 1.3$; (b) $\Pi_L = 1.6$.

On the other hand, Figure 4b shows that the reed has a trend to wrap up the seat in the case of larger width ($\Pi_L = 1.6$) and a smaller deformation into the seat orifice. It is also apparent that this increase of Π_L reduces the aforementioned "flexural waves". Therefore, the stress peak is lower than that verified in the first case.

In fact, the results of our study show that geometric variations may result impacts with differences of 60% in the maximum stress considering the same initial velocity. As can be seen, Eq. (13) is not able to predict such a variation and would estimate much lower stress levels, as already indicated in Fig. 2. This limitation should also be expected from two-dimensional models, since the contact progression is highly three-dimensional. Therefore, three-dimensional models are better candidates to predict efforts associated with cracks perpendicular to the reed edges as observed in the field.

Considering the results of Fig. 4, it is appropriate to discuss the relation between the impact stress and characteristics of fatigue failure. Initially, it should be noted that the maximum principal stress adopted in the present model to analyze reed impact has some limitations, since the stress peak does not always occur at the same region of the reed. However, the beneficial effect of larger reed width found herein was also verified in accelerated life test machines (3), although the authors attributed this finding only to the damping capacity of the material. Moreover, reports of crack in the region where the reed makes the last contact with the seat (2) also indicate that the contact during impact evolves in a similar manner as observed in the present study.

Some authors (8-9) state that obliquity of impact explains the occurrence of high stresses and therefore reed failure. However, the numerical results shown herein show that all impacts are oblique, but not all of them result the same stress level and reed deformation. Thus, reed failure due to impact seems to be associated with more physical aspects than just obliquity of impact (8) or even whiplash effect (9). Therefore, numerical analysis may be a useful complement to measurements in the design of reliable compressor reed valves.

5. CONCLUSIONS

Many works on impact fatigue have been conducted over the years, but the relationship between the impact and the type of cracks is not fully understood yet. The present work presents a definition for reed impact stress and investigated its sensitivity in relation to different parameters. It was found that the contact development between reed and seat lead to efforts commonly linked to impact failure. The study also indicates that impact velocity is not a sufficient parameter to explain reed failure. In fact, small geometric variations may result in quite different contact development, reed deformation and impact stress even when the impact velocity is kept fixed.

ACKNOWLEDGEMENTS

The material discussed herein forms part of a joint technical-scientific program of the Federal University of Santa Catarina and EMBRACO. The authors also acknowledge the support provided by EMBRACO and CNPq (Brazilian Research Council) through Grant No. 573581/2008-8 (National Institute of Science and Technology in Refrigeration and Thermophysics).

REFERENCES

(1) Chai, G.; Zetterholm, G.; Waldén, B., 2004: Flapper valve steels with high performance, Proc. Int. Compressor Engineering Conf. at Purdue, West Lafayette, USA, Paper C132.

(2) Svenzon, M., 1976: Impact fatigue of valve steel, Proc. Int. Compressor Engineering Conf. at Purdue, West Lafayette, USA, 65-73.
(3) Dusil, R.; Johnsson, B., 1980: Influence of seat positioning and seat design of valve fatigue performance, Proc. Int. Compressor Engineering Conf. at Purdue, West Lafayette, USA, 368-373.
(4) LIVERMORE Softwave Technology Corporation (LSTC), 2006: LS-DYNA Theory Manual.
(5) Fox, R.W., Pritchard, P.J., McDonald, A.T., 2006: Introduction to Fluid Mechanics, Wiley, 7th edition.
(6) Montgomery, D. C., 2011: Applied statistics and probability for engineers, John Wiley & Sons, 5th edition.
(7) Box, G.E.P.; Hunter, J.S.; Hunter, W.G., 2005: Statistics for experimenters: design, innovation, and discovery, John Wiley & Sons, 2nd edition.
(8) Böswirth, L., 1980: Hypothesis on the failure of spring loaded compressor valve plates, Proc. Int. Compressor Engineering Conf. at Purdue, West Lafayette, USA, 198-206.
(9) Nilsson, J.; Nilsson, L.; Oldenburg, M., 1980: Impact stresses in flapper valves - A finite element analysis, Proc. Int. Compressor Engineering Conf. at Purdue, West Lafayette, USA, 390 - 396.

The improved valve assembly of swash plate compressor for vehicle air conditioning system

G-H Lee
Doowon Technical College, Korea

T-J Lee, S-W Lee, H-J Kim
Doowon R&D Center, Korea

ABSTRACT

At the present the swash plate type compressor which is a kind of the reciprocating compressor has been used to the air conditioning system for the vehicle. This compressor consists of cylinder, piston, swash plate, shoe and valve assembly, and is operated by the rotation of swash plate which is connected to the vehicle engine drive at the high speed this compressor operates from 800rpm to 4000rpm. Also the displacement range of vehicle compressor is about from 90cc to 180cc. Therefore in the reciprocating compressor, the valve assembly is the main design parameter in order to improve the compressor performance. This paper presents the optimal design method of suction valve and discharge valve of swash plate compressor. It also shows the compressor performance analysis on the suction valve shape and the two suction port shape with the large valve flow area and the low valve lift height. Also the discharge port flow analysis for reducing the pressure drop was carried out by CFD and the two stage discharge port was designed with low over pressure and small dead volume. Finally the improved valve assembly was manufactured and the compressor performance test was carried out. As the test results, we could get the about 12.7% increase of compressor performance under general operating conditions like as suction pressure 2.8 bar,abs, discharge pressure 16 bar,abs and super heated temperature 10K.

1 INTRODUCTION

As figure 1, a compressor usded vehicle air-conditioning system jointed to enging and operates an air conditioning system. The power from the engine is transmitted to the clutch of compressor by the engine belt. Then, the compressor is operated by applying the electrical current for the magnetic clutch. While the clutch is on, clutch's rotaty motion make rotation of shaft and swahs plate and then, pistons are reciprocated by rotating swash plate and shoe. When the cylinder volume is increased according to reciprocating motion of piston inside of cylinder, the refrigerant flows into the cylinder through the suction valve owing to the pressure drop in the cylinder and when cylinder's volume is decreased, the refrigerant is compressed and discharged through discharge valve. These operation mechanism of swash plate type compressor are same with reciprotcating compressor. Since the capacity of compressor changed with the angle of swahs plate, the compressor call the fixed swash plate type compressor that plate angle is fixed and the variable swash plate type compressor that plate angle can be controlled.

As these compressors for vehicle air conditioning system use the engine power, the energy consumption from the compressor affects fuel efficiency. These are approximately 9% of total fuel consumtion yearly, and over 20% in summer due to high thermal load condition. Therefore, this paper aims to suggest to the way of efficiency improvement of compressor for rising fuel efficiency of vehicle.

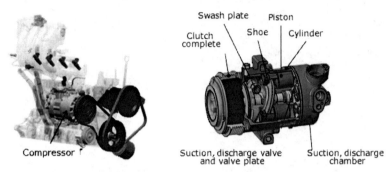

Figure 1: Engine layout and the swash plate type compressor for vehicle air conditioning system

2 THE COMPRESSOR PERFORMANCE IMPROVEMENT DESIGN

Table 1 shows test conditions and results of performance evaluations for the variable displacement swash plate compressor for this study. Operating speeds of compressor on the test condition are 800rpm for an idling condition and 3000rpm for a representative high speed driving condition of vehicle. Considering actual driving conditions, the suction and discharge pressure are high at 800rpm condition because 800rpm condition is idle state so that compressor speed is low and condenser air volume velocity is also low as a vehicle is stop, and relatively suction and discharge pressure is lower at 3000rpm on the contrary. Therefore, the suction and discharge pressure of test condition are turned by compressor's operation speed such as Table 1. In the test conditions, a cooling capacity from compressor performance is calculated by the equation (1) multiplying measured mass flow rate from mass flow meter by enthalpy difference between inlet and outlet for an evaporator. Then, it is assuming that evaporator inlet enthalpy is equal to condenser outlet enthalpy. - The enthalpy of evaporator inlet and outlet for the

Table 1. A test condition and results of compressor performance

	Unit	Cond. 1	Cond. 2
Comp. speed	[rpm]	800	3000
Suction pre.	[bar,abs]	5.0	2.8
Discharge pre.	[bar,abs]	20.0	16.0
Super heat	[K]	10	10
Sub cool	[K]	0	0
Cooling	[kW]	2,57	4.39
Power	[kW]	1.00	2.78
COP	[-]	2.57	1.58

refrigerant R134a on each condition is indicated on the Moliere diagram. In addition, required compressor power is also calculated by the equation (2) using measured operating torque of compressor.

$$\text{Cooling capacity} \quad Q[w] = \dot{m} \cdot (h_1 - h_2) \tag{1}$$

$$\text{Compressor power} \quad P[w] = 2\pi f \cdot T \tag{2}$$

2.1 Compressor performance analysis

As pressure-volume diagram of swash plate compressor, Figure 2 shows cylinder internal pressure according to volume change by reciprocating motion of the piston. The swash plate angle is determined by the control chamber pressure and the refrigerant in the suction chamber by the reciprocating motion of the piston flows into the cylinder through the suction valve and is compressed and then through the discharge valve. The compressor volumetric changes of this process can be written as the next equation:

$$V(\theta) = \frac{\pi}{4} \cdot D_p^2 \cdot \frac{PCD}{2} \cdot \tan(\beta) \cdot [1 - \cos(\theta)] + V_{cle} \tag{3}$$

where Dp = Diameter of piston, PCD = pitch circle diameter, β=swash plate angle.

The cylinder internal pressure due to volume change of cylinder is determined by the change of mass flow rate in the cylinder due to leakage at the suction valve and discharge valve and the clearance between piston and cylinder, and the change of temperature by the energe equation.

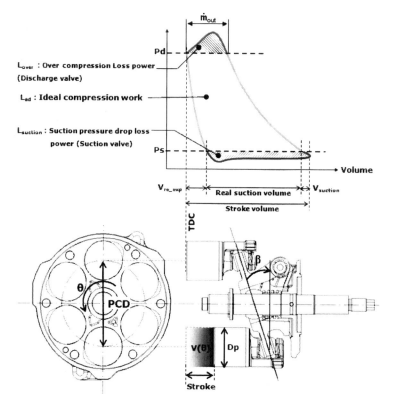

Figure 2: Pressure-volume diagram

2.1.1 Cooling capacity

As the equation (1), cooling capacity of compressor is defined as the value multiplied by the actual mass flow rate and the enthalpy difference in evaporator inlet and outlet, it determined by the mass flow rate of refrigerant because the enthalpy difference is determined by test operating condition of compressor performance. When evaluating the compressor performance, discharge refrigerant mass flow rate is measured in flowmeter of test rig while the loss factors of the refrigerant flow can be analysised at pressure-volume diagram of figure 2. As equation (4), the actual discharge mass flow rate of the compressor ($\dot{m}_{out}$) is considering the value of ideal suction refrigerant flow rate ($\dot{m}_{ideal}$), loss due to re-expansion flow ($\dot{m}_{re\_expansion}$), loss due to the suction valve flow ($\dot{m}_{suction}$) and loss due to leakage of the gap of each of the components. For loss due to re-expansion flow and loss due to the suction valve flow, it can be expressed as the value of the volume loss at pressure-volume diagram. Also, for loss due to leakage, it can be calculated by subtracting the ideal suction refrigerant flow rate measured from the test rig, re-expansion flow loss and flow loss by suction valve from ideal suction refrigerant flow rate as shown in equation (5).

$$\dot{m}_{out} = \dot{m}_{ideal} - \dot{m}_{re\_exp} - \dot{m}_{suc} - \dot{m}_{leak} \qquad (4)$$

$$\dot{m}_{out} = \dot{m}_{ideal} - V_{re\_exp} \cdot \rho_s - V_{suc} \cdot \rho_s - \dot{m}_{leak} \qquad (5)$$

2.1.2 Compressor power

The compressor power consists of the sum of the compressor work and the mechanical loss work as shown in equation (6), the compressor work is power required to compress the refrigerant and the mechanical loss work is friction losses caused by the coefficient of friction in the the sliding members in the course of these power transmission. In case that the coefficient of friction is constant, because mechanical loss work is proportional to the compression work, the compressor power including the mechanical loss can be reduced by reducing the compression work. As shown in equation (7), the refrigerant compression work is area of pressure-volume diagram of figure 2, compressor requirement power is reduced by minimizing the pressure loss in suction and discharge valves because it consists of ideal compression work (L_{ad}) to compress the refrigerant intaked at the each cylinder, over compression loss (L_{over}) due to discharge valve and suction pressure drop loss ($L_{suction}$) due to suction valve.

$$\text{Compressor power [W]}: P = L_{PV} + L_{mech} \qquad (6)$$

$$\text{Indicated Power [W]}: L_{PV} = L_{ad} + L_{over} + L_{suc} = 2\pi f \cdot \int P dv \qquad (7)$$

2.2 Cylinder pressure measurement

It was discussed that the performance reduction of compressor can be analyzed exactly in the pressure-volume diagram at chapter 2.1. With these discussions, this chapter is presented for the pressure-volume diagram on how to measure and analysis the measurement results.

2.2.1 Measurement of cylinder volume

Figure 3 shows gap sensor and measurement method in order to measure the volume of compressor. Gap sensor type is SENTEC's LS500-1 Model and can obtain the data of 200 points for one cycle as frequency is 10 kHz at 3000rpm condition.

The sensor has been installed on the valve plate of 2.3mm thickness, which is placed on the upper side of the cylinder, so that there is no impact on compressor performance like the increase of dead volume etc. As figure 4 is measurement results using gap sensor, it shows the maximum signal when the piston reaches TDC and shows a constant value when the distance of piston head and gap sensor is over the maximum measured value (2mm). After volumetric change rate following swash plate angle (θ) from eqation (3) translates function of time as equation (8), cylinder volume by gap sensor signal is able to appear as shown in figure 4 on the basis of TDC of gap sensor.

$$V(t) = \frac{\pi}{4} \cdot D_p^2 \cdot \frac{PCD}{2} \cdot \tan(\beta) \cdot \left[1 - \cos(2\pi f \cdot t)\right] + V_{cle} \tag{8}$$

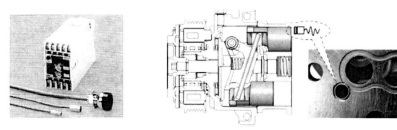

Figure 3: Test sample for volume measurement with gap sensor

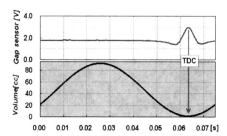

Figure 4: Gap sensor measurement data and cylinder volume

2.2.2 Cylinder pressure measurement

The static pressure sensor used to the cylinder internal pressure is Kistler's 4005B model with 100kHz response time and it was suitable to measure the pressure changes. The sensor which is installed to the upper side of valve plate is able to measure change of cylinder pressure without dead volume increase as shown in figure 5. As results of cylinder internal pressure measurement in each test condition I and II, figure 6 shows that the discharge over compression of 2.1 bar has been occurred at 800rpm and then the discharge over compression of 6.0 bar and the suction pressure drop of 1.5bar have been occurred at 3000rpm.

These pressure losses occurs in each discharge and suction valve when refrigerant is not discharged or intaked smoothly. As a result, it occurs reduction of cooling capacity of the compressor and increase of compressor power. The losses in each valve at low speed of 800rpm condition is small because mass flow through the valve is small. Whereas the losses at high speed of 3000rpm condition take place relatively increase due to increasing mass flow.

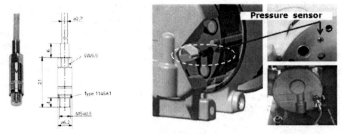

Figure 5: Test sample for pressure in cylinder with static pressure sensor

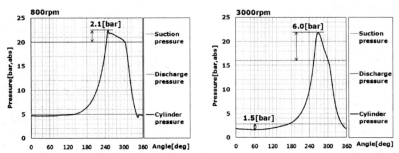

Figure 6: The measurement results of cylinder internal pressure

2.2.3 Pressure-volume diagram and analysis

This chapter discussed about the pressure-volume diagram using pressure and volume data measured as chapter 2.2.1 and 2.2.2 and the results analyzing it. Figure 7 shows measurement data of gap sensor and pressure sensor and pressure-volume diagram. The test results and analysis results of pressure and volume diameter are shown as Table 2.

(1) Cooling performance

The cooling capacity measured at compressor test rig is 2570W and 4390W at 800rpm and 3000rpm respectively. At this time, the discharge mass flow is measured by 78.98 kg/h and 128.88 kg/h respectively. Through analyzing PV diagram about these results of test rig, re-expansion loss volume is 9.4 cc and 14 cc at 800rpm and 3000rpm. When this volume multiplied by density of 23.04 kg/m3 and 13.15 kg/m3 at each suction condition and then converted to unit of time, re-expansion volume loss is 10.43 kg/h and 33.71 kg/h. In the same way as suction volume loss is 2.08 kg/h and 43.85 kg/h respectively.

The difference between compressor total discharge mass flow and loss mass flow is considered to leakage mass flow. As results of the PV analysis, it can be found that re-expansion loss is occurred at both low and high speed condition. And suction volume loss is occurred largely at high speed condition. Therefore, in order to improve the cooling capacity of the compressor, it should be provided the ways which can reduce re-expansion loss and suction volume loss by suction valve.

(2) Compressor power

The compressor power measured at compressor test rig is 1000W and 2780W at 800rpm and 3000rpm respectively, PV diagram of figure 7 shows that compression work is 820.67W and 2343W respectively. Also over compression loss by the discharge valve is 41.04W and 261W respectively, suction drop loss by the suction valve is 30.79W and 439W. These results shows these losses power by valves is fairly large at high speed, especially these losses by suction valve is larger than that by discharge valve.

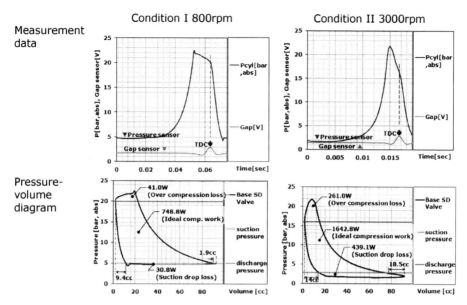

Figure 7: A measurement results of pressure-volume in the cylinder

Table 2. Performance results and pressure-volume diagram analysis results of base sample

				Unit	Cond.1 800rpm	Cond. 2 3000rpm
Q	Test rig data	Cooling capacity	Q	[W]	2570	4390
		Enthalpy difference	Δh_{eva}	[kg/h]	117.15	112.63
		Real mass flow rate	$\dot{m}_{out}$	[kJ/kg]	78.98	128.88
	PV analysis	Ideal mass flow rate	$\dot{m}_{ideal}$	[kJ/kg]	99.53	213.09
		Suction density	ρ_s	[kg/m³]	23.04	13.15
		Re-expansion loss	$\dot{m}_{re\_exp}$	[kg/h]	10.43	33.71
		Suction volume loss	$\dot{m}_{suc}$	[kg/h]	2.08	43.85
	Leakage volume loss (Test rig-PV)		$\dot{m}_{leak}$	[kg/h]	8.05	6.66
P	Test rig	Compressor power	P	[W]	1000	2780
	PV analysis	Compression work	L_{PV}	[W]	820.67	2343
		Ideal comp. work	L_{ad}	[W]	748.84	1643
		Over comp. Loss	L_{over}	[W]	41.04	261
		Suction drop Loss	L_{suc}	[W]	30.79	439
	Mechanical Loss(Test rig-PV)		L_{mech}	[W]	179.33	437
	COP			[-]	2.57	1.58

2.3 The design for performance improvement

2.3.1 The discharge valve design for reducing discharge over compression

The discharge over compression has been confirmed by measurement result of cylinder internal pressure during discharge process at chapter 2.2. And discharge over compression is occurred in the high speed especially, because discharge valve port area is same while the flow is increased. In order to solve this problem, the over compression loss during the discharge process can be improved by increasing the size of the discharge valve port. But the discharge port is included in top clearance volume as Figure 8. If the size of discharge port increase, then top clearance volume is also increased, as a result, the volumetric efficiency is decreased. Therefore, we suggests the optimized discharge valve design which is minimized the increasing of the top clearance volume. The equation (9) is the function of effective flow area about the valve geometric and the figure 9 is the result graph of equation (9). In equation (9) Ao indicating sectional area of valve has a large enough value. But Ay decided by the circumference of the valve and the lift amount of the valve has a relatively small value compared to Ao. Therefore, to increase valve effective area, Ay area need to be increased, but in order to increase Ay area the increase in Valve displacement(y) is the major cause of the compressor noise increase and valve breakage. To solve such problems, the 2-step discharge port shape is presented as shown in Figure 9 in this study. In case of Ø9mm two step ports, the rise of effective flow area was increased about 20% at the point of 1mm valve lift. As important factors in valve design, a depth of two step ports affects compressor performance. If its depth is low, the improvement is negligible. In the opposite case, it deviate an initial design purpose due to increase of clearance volume. Therefore, optimized depths of two step ports were selected by using CFD analysis in this study. Figure 10 shows calculated results of discharge mass flow on the depth of two step ports in case of 17bar for inlet pressure and 15bar for outlet pressure. While a clearance volume(V) increases in proportion to the depth(h) of two step ports, discharge flow is almost constant over 0.5mm depth. Therefore, in case of 0.5mm two step ports depth, discharge mass flow increased by 56% from existing 32.7g/sec to 51.2g/sec. Whole produced clearance volume is 146mm^3. On the other hands, clearance volume applying two step ports is 101mm^3. When comparing the two cases, clearance volume was decreased by 46%. These results show that ti can be possible to get the maximize discharge effective flow area with minimizing a clearance volume increase.

$$Ae = \frac{Ao}{\sqrt{1+(Ao/Ay)^2}} = \frac{\frac{\pi}{4} \cdot D^2}{\sqrt{1+\left(\left(\frac{\pi}{4} \cdot D^2\right)\Big/(D\pi \cdot y)\right)^2}} \quad (9)$$

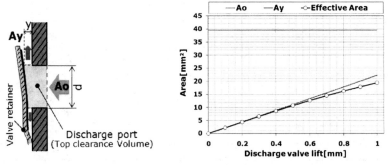

Figure 8: The discharge port and effective flow area

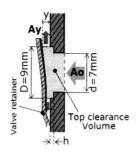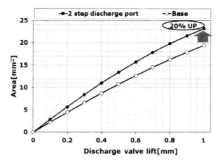

Figure 9: The 2 step discharge port and effective flow area

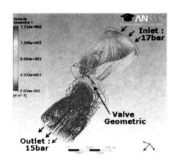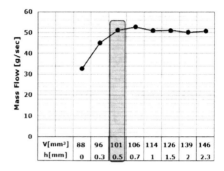

Figure 10: The CFD analysis results of 2 step discharge port

2.3.2 The suction valve design for reducing suction pressure drop

In the cylinder internal pressure measurement results at chapter 2.2.2, suction pressure drop occur up to 1.5bar and the suction pressure loss work was aout 439W at 3000 rpm. The increase of the suction pressure drop can be caused the increase of compressor power and the decrease of the cooling capacity. These losses can be improved by increasing suction port area of valve in the same way as the discharge valve. Because the discharge port is included in dead volume, it should be considered ways to minimize increase of volume. But the suction port is not included in dead volume, it is an alternative way for improvement of suction pressure drop to design as large as possible in the range of possible design. Figure 11 shows FEM analysis results for the shape of each suction valve. Fig. 11(a) is existing valve shape of compressor for this study. And Fig. 11(b) is structure that the valve area is designed as large as possible with the purpose of improvement for suction presser drop. However, the case of Fig. 11(b), Max load of valve increases greatly because of increasing of the area acted cylinder pressure according to the increase of valve area in compression process as shown in the FEM analysis results. Fig. 11(c) is structure that valve max load is markedly decreased because of dividing valve port in two, and it is structure that suction pressure drop can be improved by increasing more than doubled suction valve area compared to existing valve.

573

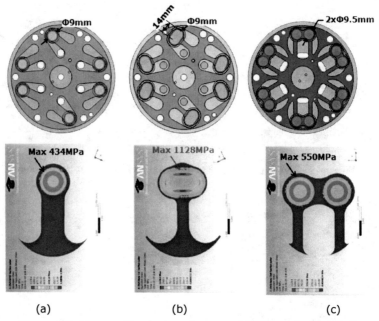

Figure 11: The suction valve shape and FEM analysis result

3 THE TEST RESULTS OF IMPROVED VALVE ASSEMBLY

The performance test has been carried out using samples which were produced by the optimized valve structure design mentioned in chapter 2. Figure 12 and Table 3 show results of tests and measured cylinder internal pressure-volume.

At the test condition 1, the cooling capacity is decreased from 2,570W to 2,460W and the compressor power is increased from 1,000W to 950W. In conclusion, COP is improved from 2.57 to 2.59 by 0.8%.

The cooling capacity is decreased by 4.3%. and the compressor power is improved by 5.0%. In this case, the over compression loss is decreased from 41.04W to 14.71W and the suction drop loss also is decreased from 30.79W to 16.04W.

At test condition 2, the cooling capacity is improved from 4,390W to 4,750W by 8.2% and the compressor power is also improved from 2,780W to 2,670W by 4%. As a result, COP is improved from 1.58 to 1.78 by 12.7%.

These performance improvement is caused by improved suction valve, the suction volume loss is decreased from 43.85 kg/h to 23.51 kg/h and the suction pressure drop loss is also decreased from 439W to 321W. And the discharge valve, over compression loss is also decreased from 43.85 kg/h to 23.51 kg/h.

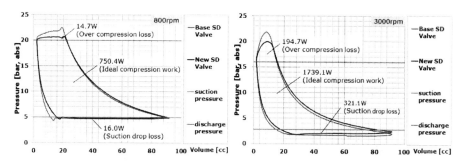

Figure 12: A pressure measurement result in the cylinder

Table 3. Performance results at test condition1 and condition 2

800rpm				Unit	Base SD valve	New SD valve	Ratio [%]
Q	Test rig data	Cooling capacity	Q	[W]	2,570	2,460	-4.3
		Enthalpy difference	Δh_{eva}	[kg/h]	117.15	117.15	-
		Real mass flow rate	$\dot{m}_{out}$	[kJ/kg]	78.98	75.60	-4.3
	PV analysis	Ideal mass flow rate	$\dot{m}_{th}$	[kJ/kg]	99.53	99.53	-
		Suction density	ρ_s	[kg/m³]	23.04	23.04	-
		Re-expansion loss	$\dot{m}_{re-exp}$	[kg/h]	10.43	15.16	45.3
		Suction volume loss	$\dot{m}_{suction}$	[kg/h]	2.08	1.02	-50.7
	Leakage volume loss (Test rig-PV)	$\dot{m}_{leak}$	$\dot{m}_{leak}$	[kg/h]	8.05	7.75	-3.7
P	Test rig	Compressor power	P	[W]	1,000	950	-5.0
	PV analysis	Compression work	L_{PV}	[W]	820.67	781.14	-4.8
		Ideal comp. work	L_{ideal}	[W]	748.84	750.39	0.2
		Over comp. Loss	L_{over}	[W]	41.04	14.71	-64.2
		Suction drop Loss	$L_{suction}$	[W]	30.79	16.04	-47.9
	Mechanical Loss(Test rig-PV)	L_{mech}	L_{mech}	[W]	179.33	168.86	-5.8
	COP			[-]	2.57	2.59	0.8
3000rpm				Unit	Base SD valve	New SD valve	Ratio [%]
Q	Test rig data	Cooling capacity	Q	[W]	4,390	4,750	8.2
		Enthalpy difference	Δh_{eva}	[kg/h]	112.63	112.63	-
		Real mass flow rate	$\dot{m}_{out}$	[kJ/kg]	128.88	139.44	8.2
	PV analysis	Ideal mass flow rate	$\dot{m}_{th}$	[kJ/kg]	213.09	213.09	-
		Suction density	ρ_s	[kg/m³]	13.15	13.15	-
		Re-expansion loss	$\dot{m}_{re-exp}$	[kg/h]	33.71	41.50	23.1
		Suction volume loss	$\dot{m}_{suction}$	[kg/h]	43.85	23.51	-46.4
	Leakage volume loss (Test rig-PV)	$\dot{m}_{leak}$		[kg/h]	6.66	8.64	29.8
P	Test rig	Compressor power	P	[W]	2,780	2,670	-4.0
	PV analysis	Compression work	L_{PV}	[W]	2,343	2,255	-3.8
		Ideal comp. work	L_{ideal}	[W]	1,643	1,739	5.9
		Over comp. Loss	L_{over}	[W]	261	195	-25.4
		Suction drop Loss	$L_{suction}$	[W]	439	321	-26.9
	Mechanical Loss(Test rig-PV)	L_{mech}		[W]	437	415	-5.0
	COP			[-]	1.58	1.78	12.7

4. CONCLUSIONS

This study has been carried out to improve the efficiency of compressor for improving fuel efficiency of the vehicle. First, cylinder internal pressure is measured by using static pressure sensor to analyze the cause of efficiency deterioration. The results show that discharge over compression and suction pressure drop occur by valves. To improve these losses, an alternative discharge valve structure optimized by using CFD is suggested with minimizing increase of the dead volume. Optimized suction valve is also designed with high performance and durability through FEM analysis. As a result, discharge over compression loss is decreased from 261W to 195W and suction pressure drop loss is decreased from 439W to 321W at high speed condition. These results lead improving performance with increasing C.O.P by 12.7% since each cooling capacity is increased by 8.2% and compressor power is reduced by 4%.

REFERENCES

1. Kazuhito Miyagawa, Hiroaki Kayukawa, 1998, Development of the swash plate type continuously variable displace compressor, SAE , 980290.
2. David M. Ebbing, 2001, Control stability and NVH improvements of the variable displacement swash plate compressor, SAE, 2001-01-3837.
3. Hiroyasu Nadamoto, Atsushi Kubota, 1999, Power saving with the use of variable displacement compressor, SAE, 1999-01-0875.
4. Lee, G. H. and Lee, T. J, 2004, A Study of variable displacement mechanism of swash plate type compressor for automotive air conditioning system, International compressor conference at Purdue, C079.
5. Tae Jin Lee, Ki Beom Kim, Seung Won Lee, and Geon Ho Lee, 2011, Development of performance analysis program and the study of substitution refrigerant R1234yf for vehicle refrigerant compressor, SAREK, Vol. 23, No.11 pp. 699-704.

Transient and dynamic numerical simulation of the fluid flow through valves based on large eddy simulation models

O Estruch[1], O Lehmkuhl[1,2], J Rigola[1], A Oliva[1], C D Pérez-Segarra[1]
[1] Centre Tecnològic de Transferència de Calor (CTTC),
   Universitat Politècnica de Catalunya (UPC), Spain
[2] TermoFluids S.L., Spain

ABSTRACT

The present paper attempts the dynamic simulation of the fluid flow through valve reed using the in-house implemented CFD and moving mesh coupled code TermoFluids. The CFD solver is based on a parallel, second-order, conservative and unstructured finite volume discretization. Large eddy simulation is performed to solve the turbulent flow, using the subgrid scale WALE model. The moving mesh technique uses RBF interpolation. As a preliminary approach, a simplified geometry of an axial hole plus a radial diffuser with a piston based inlet condition is considered. The valve dynamics is modelled by a specific law according modal analysis of valve reed.

NOMENCLATURE

RBF nomenclature

$\Omega \subset \Re^d$	Deformable domain, $2 \leq d \leq 3$
$V = \{x_k\}_{k \in Y}$	Vertices of the CFD grid covering Ω
$Y = \{1, \cdots, N_v\}$	Set of indexes of V
$|Y| = N_v$	Total number of vertices of V
$V_b = \{x_i\}_{i \in Y_b}$	Subset of pairwise vertices of the moving boundary, also called control points, where $Y_b \subset Y$ and $|Y_b| = N_{v_b}$
$s^*, * \in \{x, y, z\}$	Interpolation function
ϕ	Radial basis function
$\gamma_i^* \in \Re$	Interpolation coefficients
$g^* = \{g_i^*\}_{i \in Y_b}$	Known boundary displacements
M	Interpolation matrix

CFD nomenclature

$\mathbf{u} \in \Re^{3m}$	Velocity vector (m the total number of CVs)
$\mathbf{p} \in \Re^m$	Pressure (m the total number of CVs)
$\mathbf{u}_g$	Surface velocity vector
t	Time

Δt	Time step
ν	Kinematic viscosity
ρ	Fluid density
$\mathbf{u}_c^p$	Predicted velocity
$\tilde{p}_c$	Pseudo-pressure
$C(\mathbf{u})$	Convective operator
D	Diffusive operator
G	Gradient operator
M	Divergence operator
L_c	Discrete laplacian operator
$\bar{\mathbf{u}}$	Filtered velocity
M	Divergence operator of a tensor
T	SGS stress tensor
ν_{sgs}	SGS viscosity

SCL nomenclature

$\Omega(t)$	Moving CV
$S(t)$	Closed surface bounding $\Omega(t)$
$\mathbf{v}_g$	Velocity of the surface whose outward unit vector is $\mathbf{n}$
$\dot{m}_c$	Mass flux through a cell face c
$\delta\Omega_c$	Volume swept by the CV face c during a time step Δt

Valve dynamics nomenclature

$w(x,y,t)$	Transverse deflection of the valve reed at point (x,y) at time t
ρ	Mass density of valve material
h	Thickness of the valve reed
$p(x,y,t)$	Load per unit area at location (x,y) at time t
D	Flexural rigidity of valve reed
E	Young's modulus
v	Poisson's ratio
$\phi_m(x,y)$	Natural modes
ω_m	Natural frequencies
ξ	Damping coefficient
$\Delta p(t)$	Pressure differential across the valve
$A_F(w(x,y))$	Effective force area
ΔA_i	Area of the port hole at location (x_i, y_i)
A	Total port area
ΔA_j	Area of the geometric discretization elements of the valve reed

1 INTRODUCTION

In the majority of hermetic reciprocating compressors valves are a basic component of both the suction and discharge ports and, particularly, reed type valves are widely used for domestic or commercial reciprocating compressors. The understanding of the behaviour of the fluid flow through the valve plate and moreover the dynamic action of the valve reed is essential to improve the compressor design and to contribute in the efficiency optimization. Thus, the present paper attempts the dynamic simulation of the fluid flow through the suction valve reed including an inlet port valve condition according to valve movement due to piston displacement and modelling the valve reed by means of a specific law based on modal analysis of valve reed theory (1). Few works in the literature consider a computational model to simulate the dynamics of reed type valves of reciprocating compressors. In (2) a one-degree of freedom model for the valve motion and a finite volume methodology for incompressible flow were adopted and a moving coordinate system was employed. A software tool for simulating the dynamic behaviour of reed valves was presented in (3). Reference (4) simulated a six petals reed valves pack by the use of FLUENT coupled with a FSI tool.

In the following study, the transient simulation of the fluid and valve interaction is performed by means of the newly in-house implemented CFD&HT and moving mesh coupled code TermoFluids (5), which is detailed and validated in (6,7). This work extends previous studies, in which numerical experiments were carried out considering static geometry and constant boundary conditions (8,9,10). As a preliminary approach, a simplified geometry of an axial hole plus a rectangular valve reed is considered.

2 MATHEMACIAL FORMULATION AND NUMERICAL METHOD

2.1 The RBF interpolation method
In this section we briefly introduce the mathematical formulation of the moving grid technique based on the RBF interpolation (11, 12, 13, 14, 15, 16) and depict the mesh adaptive algorithm implemented. Thus, in this paper the following interpolation problem is considered: find a function $s^* \in T = \{s : \Re^d \to \Re\}, * \in \{x, y, z\}$ so that

$$s^*(\mathbf{x}) = \sum_{i \in Y_b} \gamma_i^* \phi(\|\mathbf{x} - \mathbf{x}_i\|) \tag{1}$$

$$s^*(\mathbf{x}) = g_i^* \quad \forall i \in Y_b \tag{2}$$

where $\|\cdot\|$ denotes the Euclidean norm. This leads to the linear system of equations

$$M\gamma^* = g^* \tag{3}$$

where M is of dimension $N_{v_b} \times N_{v_b}$ containing the evaluation of the basis function $M_{ij} = \phi(\|\mathbf{x}_i - \mathbf{x}_j\|)$. The RBF adopted in this study corresponds with the Wendland C^2 (Eq. 4), which is of compact support and strictly definite positive (a support radius r has been used to scale the compact support). Consequently, the interpolation matrix M is invertible, symmetric and strictly positive definite, and the linear system of equations can be solved by means of a conjugated gradient (CG) method (17).

$$\phi(\mathbf{x}) \equiv \Phi(\|\mathbf{x}\|/r) = \Phi(\xi) = \begin{cases} f(\xi) = (1-\xi)^4(4\xi+1) & \text{if } \xi \in [0,1] \\ 0 & \text{otherwise} \end{cases} \tag{4}$$

2.2 Navier-Stokes discretization coupled with dynamic mesh

For the present work the governing equations correspond to the incompressible Navier-Stokes and continuity equations, which can be written as

$$M(\mathbf{u} - \mathbf{u}_g) = 0 \qquad (5)$$

$$\frac{\partial \mathbf{u}}{\partial t} + C(\mathbf{u} - \mathbf{u}_g)\mathbf{u} + \nu D \mathbf{u} + \rho^{-1} G \mathbf{p} = 0 \qquad (6)$$

Convective and diffusive operators in the momentum equation for the velocity field are given by $C(\mathbf{u}) = (\mathbf{u} \cdot \nabla) \in \Re^{3m \times 3m}$, $D = -\nabla^2 \in \Re^{3m \times 3m}$ respectively. Gradient and divergence (of a vector) operators are given by $G = \nabla \in \Re^{3m \times m}$ and $M = \nabla \cdot \in \Re^{m \times 3m}$ respectively.

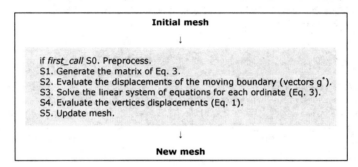

Figure 1: Move mesh tool algorithm.

The governing equations have been discretized on a collocated unstructured grid arrangement by means of second-order spectro-consistent schemes (see (18)). Such schemes are conservative, i.e. they preserve the kinetic energy equation. These conservation properties are held if, and only if the discrete convective operator is skew-symmetric ($C_c(\mathbf{u}_c) = -C_c^*(\mathbf{u}_c)$), the negative conjugate transpose of the discrete gradient operator is exactly equal to the divergence operator ($-(\Omega_c G_c)^* = M_c$) and the diffusive operator D_c is symmetric and positive-definite (the subscript c holds for the cell-centred discretization). These properties ensure both, stability and conservation of the kinetic-energy balance even at high Reynolds numbers and with coarse grids.

For the temporal discretization of the momentum equation (Eq. 6) a fully explicit second-order self-adaptive scheme (19) has been used for the convective and diffusive terms, while for the pressure gradient term an implicit first-order scheme has been used. The velocity-pressure coupling has been solved by means of a classical fractional step projection method, $\mathbf{u}_c^p = \mathbf{u}_c^{n+1} + G\tilde{\mathbf{p}}_c$, where $\tilde{\mathbf{p}}_c = p_c^{n+1} \Delta t^n / \rho$, $n+1$ is the instant where the temporal variables are calculated, and $\Delta t^n = t^{n+1} - t^n$. Taking the divergence of the predicted velocity equation and applying the incompressibility condition yields a discrete Poisson equation for $\tilde{\mathbf{p}}_c$: $L_c \tilde{\mathbf{p}}_c = M_c \mathbf{u}_c^p$. $L_c \in \Re^{m \times m}$ is, by construction, a symmetric positive definite matrix ($L_c \equiv M \Omega^{-1} M^*$). Finally the mass-conserving velocity at the faces ($M_s \mathbf{u}_s^{n+1}$) is obtained from the correction $\mathbf{u}_s^{n+1} = \mathbf{u}_s^p - G_s \tilde{\mathbf{p}}_c$, where G_s represents the discrete gradient operator at the CV faces.

2.3 Large eddy simulation model

In LES, the largest, energy-carrying scales of the flow, are computed exactly, while the effect of the smallest scales of the turbulence are modelled by means of a subgrid-scale (SGS) model. The decomposition into a large-scale component and a small SGS is done by filtering spatially the Navier-Stokes equations (Eq. 5 and 6),

$$M(\bar{\mathbf{u}} - \bar{\mathbf{u}}_g) = \mathbf{0} \tag{7}$$

$$\frac{\partial \bar{\mathbf{u}}}{\partial t} + C(\bar{\mathbf{u}} - \bar{\mathbf{u}}_g)\bar{\mathbf{u}} + \nu D\bar{\mathbf{u}} + \rho^{-1}G\bar{p} \approx MT \tag{8}$$

The right term indicates some modelling of the non-linear convective term, in which $T = -2\nu_{sgs}\bar{S} + (T:I)I/3$, where $\bar{S} = 1/2[G(\bar{\mathbf{u}}) + G^{\ast}(\bar{\mathbf{u}})]$, being $G^{\ast}$ the transpose of the gradient operator. Then, the modelling is made through a suitable expression for the SGS viscosity, ν_{sgs}.

In this paper, LES have been performed using the wall-adapting local-eddy viscosity model (WALE) (20), available in TermoFluids (21, 22). This model is based on the square of the velocity gradient tensor. In its formulation, the SGS viscosity accounts for the effects of the strain and the rotation rate of the smallest resolved turbulent fluctuations. In addition, the proportionality of the eddy viscosity near walls ($\nu_{sgs} \propto y^3$) is recovered without any dynamic procedure,

$$\nu_{sgs} = (C_w \Delta)^2 \frac{(v'_{ij} : v'_{ij})^{3/2}}{(S'_{ij} : S'_{ij})^{5/2} + (v'_{ij} : v'_{ij})^{5/4}} \tag{9}$$

$$S'_{ij} = \frac{1}{2}\left[G(\overline{u'_c}) + G \ast (\overline{u'_c})\right] \tag{10}$$

$$v'_{ij} = \frac{1}{2}\left[G(\overline{u'_c})^2 + G \ast (\overline{u'_c})^2\right] - \frac{1}{3}\left(G(\overline{u'_c})^2 I\right) \tag{11}$$

2.4 The space conservation law

When dynamic meshes are used, the computational volume must be preserved. We impose this by means of the so-called Space Conservation Law (SCL) (see Eq. 12). Hence, the governing equations correspond to the conservation principles of space, mass, momentum and energy (e.g. see (23)).

$$\frac{d}{dt}\int d\Omega + \int \mathbf{v}_g \cdot \mathbf{n} \, dS \tag{12}$$

The mass conservation is obtained by enforcing the SCL. Actually, according to (25) only the mass flux through a cell face c needs to be modified as follows:

$$\dot{m}_c^{updated} = \int_{S_c} \rho(\mathbf{v} - \mathbf{v}_g) \cdot \mathbf{n} \, dS \approx \rho_c(\mathbf{v} \cdot \mathbf{n})_c S_c - \rho_c \dot{\Omega}_c = \dot{m}_c - \rho_c \dot{\Omega}_c \tag{13}$$

where $\dot{\Omega}_c = (\mathbf{v}_g \cdot \mathbf{n})_c S_c = \frac{\delta \Omega_c}{\Delta t}$. The volume swept is evaluated by an in-house conservative method, i.e. it is calculated exactly in order to satisfy the SCL and, thus, the mass conservation.

2.5 Valve dynamics

As a first approach we consider a simplified geometry of an axial hole plus a radial diffuser with a piston based inlet condition (see Figure 4). In the context of suction compressor valves, the radial diffuser is considered like a flexible reed valve. Then, referring to (1), the dynamic action of the valve is based on a specific law according modal analysis of valve reed theory. This methodology assumes that the valve motion results from the superposition of the valve vibration modes. In the following, the equations to solve the dynamics of the plate type reed valve are derived.

The equation of motion of a plate type reed valve is, in cartesian coordinates:

$$D\nabla^4 w(x,y,t) + \rho h \ddot{w}(x,y,t) = p(x,y,t) \tag{14}$$

where $\nabla^4 = \dfrac{\partial^4}{\partial x^4} + 2\dfrac{\partial^4}{\partial x^2 \partial y^2} + \dfrac{\partial^4}{\partial y^4}$ and $D = \dfrac{Eh^3}{12(1-v^2)}$. In order to avoid needing boundary conditions at each edge of the valve reed, the deflection function is expressed as a combination of the natural modes of the reed valve,

$$w(x,y,t) = \sum_{m=1}^{\infty} q_m(t)\phi_m(x,y) \tag{15}$$

$\phi_m(x,y)$ and ω_m can be evaluated analytically for simple cases; otherwise, for complicated configurations they should be evaluated experimentally. After some effort (1), it leads to the following generalized equation, considering a valve reed of arbitrary geometry with k port holes.

$$\ddot{q}_m^{(t)} + 2\xi\omega_m \dot{q}_m^{(t)} + \omega_m^2 q_m^{(t)} = \dfrac{\Delta p(t)\sum_{i=1}^{k}\phi_m(x_i,y_i)A_F(w(x_i,y_i))\Delta A_i}{A\rho h \sum_{j=1}^{l}\phi_m^2(x_j,y_j)\Delta A_j} \tag{16}$$

In this study, only the first main natural mode is considered, what leads to a first approach of the valve dynamics. Then, for the configuration with one port hole of coordinates (x_1, y_1) Eq. 15 and Eq. 16 read, for the first main natural mode ϕ:

$$w(x,y,t) = q(t)\phi(x,y) \tag{17}$$

$$\ddot{q}^{(t)} + 2\xi\omega_0 \dot{q}^{(t)} + \omega_0^2 q^{(t)} = \dfrac{\Delta p(t)\phi(x_1,y_1)A_F(w(x_1,y_1))}{\rho h \sum_{j=1}^{l}\phi^2(x_j,y_j)\Delta A_j} \tag{18}$$

where $\xi = 0.056$, $\omega_0 = 1132.15\ rad/s$, $\phi(x_1,y_1) = 5.39 \cdot 10^3$, $\rho = 7870\ kg/m^3$, $h = 2 \cdot 10^{-4} m$ and $\sum_{j=1}^{l}\phi^2(x_j,y_j)\Delta A_j = 4.96 \cdot 10^3 m^2$. The effective force area $A_F(w(x_1,y_1))$ is extracted from the analytical method presented in (25). The shape of the first main natural mode is obtained from the finite element vibration resolution provided by a commercial software. Eq. 18 is integrated implicitly by means of a Crank Nicolson method.

3 COMPUTATIONAL DOMAIN, MESH AND BOUNDARY CONDITIONS

Figure 2 illustrates the computational mesh and domain containing the rectangular reed type valve. In the figure the inlet port hole of the suction valve is pointed out. The dimensions of the global domain are $L = 0.08m$ and $H = 0.06m$. The hole is centred at the bottom base, and its diameter is $d = 9.75 \cdot 10^{-3}m$. The valve dimensions are: length $l = 0.026m$, width $w = 0.01m$ and thickness $h = 2 \cdot 10^{-4}m$. The height of the axial hole is $e = 3.6 \cdot 10^{-3}m$. In the initial configuration the valve is completely closed, i.e. is contained in the plane $y = 3.6 \cdot 10^{-3}$, and it is translated $t = 6 \cdot 10^{-3}m$ in the x direction. The computational mesh is structured and has over 2.5 million CVs.

For the bottom inlet orifice a piston based inlet condition is assumed (see Figure 3), which is defined with a frequency of $50Hz$. A pressure based boundary condition applies for the outlet fluid exit (lateral and top walls). Non-slip boundary conditions are considered on solid walls (bottom part amb valve reed). An immersed body procedure is used to simulate solid parts inside the domain and, hence, to reproduce the inlet and bottom boundaries (see Figure 4). Therefore, the RBF method allows the simulation from null valve deformation.

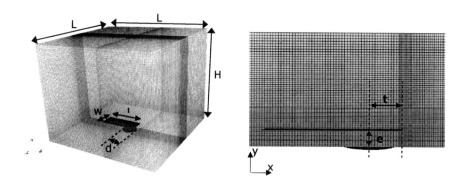

Figure 2: Computational mesh and domain. General view (left); zoom view (right).

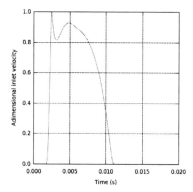

Figure 3: Piston based inlet boundary condition.

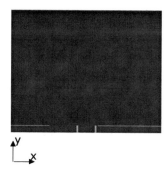

Figure 4: Solid parts using an immersed boundary procedure.

4 NUMERICAL RESULTS

The results presented in this section pretend to be a preliminary illustrative study of the transient and dynamic simulation of the fluid flow through a suction valve reed subjected to a piston based inlet boundary condition and considering a modal model for the valve dynamics. For instance, in Figure 5 the hole centre deflection as function of time during a complete suction valve opening cycle is depicted. The period of time in which the valve reed remains opened corresponds with the period of existing inlet flux (see Figure 2), although the valve reed appears to become completely closed with a little delay. During the opened stage, three peaks of oscillation can be appreciated, each of lower peak value than the one before, what seems logical if the inlet velocity tends to decrease. The first maximum peak appears in agreement with the maximum inlet velocity (Figure 2).

The pressure, velocity and vorticity profiles for different states of the valve movement cycle are shown in Figure 6, whereas Figure 7 depicts the pressure profiles along the valve reed. The flow phenomena observed is consistent with previous studies (8,9,10) and accomplishes a qualitative accurate transient simulation of the turbulent flow through the valve reed. In agreement with the sudden increment of the inlet velocity according to the piston based inlet boundary condition (Figure 3), a considerable increase of velocity is appreciated in the valve reed aperture when it starts to open. Hence, huge velocity gradients and consequently high vorticity appear in this area, where the mesh should be particularly fine to capture the smallest scales of the flow. Referring to this, the RBF method allows that the mesh quality is maintained in this region, provided that the initial mesh has sufficient number of CVs below the valve plate and the parameter radius of the RBF interpolation is chosen appropriately. Therefore, the CFD and dynamic mesh coupled code TermoFluids would be capable to carry out successfully the transient simulation of the flow through the suction reed valve, even with more complex geometries.

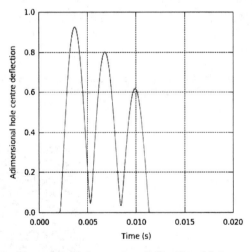

Figure 5: Hole centre deflection history during a complete valve opening cycle.

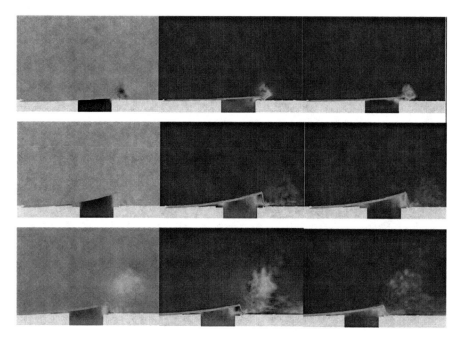

Figure 6: Pressure (left), velocity (middle) and vorticity (right) profiles for two intermediate states of the valve opening cycle.

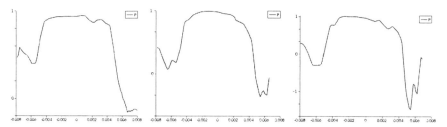

Figure 7: Adimensional pressure profile along the valve reed for the three intermediate states of the valve opening cycle in Figure 6, respectively.

5 CONCLUSIONS

The transient numerical simulation of the fluid flow through a suction valve reed has been carried out using the in-house newly implemented CFD and dynamic mesh coupled code TermoFluids. A first approach based on a simplified geometry of an axial hole plus a rectangular valve reed has been analyzed. The valve dynamic action has been modelled as a flexible reed valve by a specific law according modal analysis. The algorithm has been capable to reproduce the cycle of movement dynamically capturing the transient flow phenomena with qualitative precision, even with a mesh relatively coarse to the huge valve maximum deflection.

As future task we consider using a finer mesh and a wider radius of the RBF interpolation to obtain more accurate numerical results, particularly when capturing the smallest scales of the turbulent flow. Moreover, other geometries of the radial diffuser will be also tested, and a structural solver will be coupled with the fluid solver in order to evaluate the dynamic action of the valve reed.

ACKNOWLEDGMENTS

This work has been financially supported by the Ministerio de Educación y Ciencia, Secretaría de Estado de Universidades e Investigación, Spain (ref. ENE2010-17801) and by the Collaboration Project between Universidad Politècnica de Catalunya and TermoFluids S.L. (ref. C06650).

REFERENCE LIST

(1) W. Soedel. Mechanics, simulation and design of compressor valves, gas passages and pulsation mufflers. *Purdue University Short Courses*, IN, USA, 1992.
(2) F.F.S Matos, A.T. Prata, C.J. Deschamps. Numerical simulation of the dynamics of reed type valves. *International Compressor Engineering Conference at Purdue University*, Indiana, USA, 2002.
(3) G. Machu, M. Albrecht, O. Bielmeier, T. Daxner, P. Steinruck. A universal simulation tool for reed valve dynamics. *International Compressor Engineering Conference at Purdue University*, Indiana, USA, 2004.
(4) A. Angeletti, M.E. Biancolini, E. Costa, M. Urbinati. Optimization of reed valves dynamics by means of fluid structure interaction modelling.
(5) O. Lehmkuhl, R. Borrell, C.D. Pérez-Segarra, M. Sòria, A. Oliva. TermoFluids: A new parallel unstructured CFD code for the simulation of turbulent industrial problems on low cost PC Cluster. *Parallel Computational Fluid Dynamics 2007*, Vol. 67, pp. 275-282, 2009.
(6) O. Estruch, O. Lehmkuhl, R. Borrell, C.D. Pérez-Segarra, A. Oliva. A parallel radial basis function interpolation method for unstructured dynamics meshes. *Computer and Fluids*, http://dx.doi.org/10.1016/j.compfluid.2012.06.015, 2012.
(7) O. Estruch, O. Lehmkuhl, R. Borrell, C.D. Pérez-Segarra. Large-eddy simulation of turbulent dynamics fluid-structure interaction. *7th International Synopsium on Turbulence, Heat and Mass Transfer*, Palermo, 2012.
(8) J. Rigola, O. Lehmkuhl, A. Oliva, C.D. Pérez-Segarra. Numerical simulation of the fluid flow through valves based on Large Eddy Simulation models. *International Conference on Compressors and Their Systems*, London, pp. 137-145, 2009.
(9) J. Rigola, O. Lehmkuhl, C.D. Pérez-Segarra, A. Oliva. Numerical simulation of the fluid flow through valves reeds based on large eddy simulation models (LES). *International Compressor Engineering Conference at Purdue University*, Indiana, USA, 2008.
(10) J. Rigola, O. Lehmkuhl, J. Ventosa, C.D. Pérez-Segarra, A. Oliva. Numerical simulation of the turbulent fluid flow through valves based on low Mach models. *International Compressor Engineering Conference at Purdue University*, Indiana, USA, 2012.
(11) A. de Boer, M.S. Van der Schoot, H. Bijl. Mesh deformation based on radial basis function interpolation. *Computer & Structures*, 85:784-795, 2077.
(12) H. Wendland. Piecewise polynomial, positive definite function and compactly supported radial basis functions of minimal degree. *Adv. Comput. Math.*, 4:389-396, 1995.
(13) S. Jakobsson, O. Amoignon. Mesh deformation using radial basis functions for gradient-based aerodynamic shape optimization. *Computer & Fluids*, 36:1119-1136, 2077.
(14) J. Yoon. Approximation by conditionally positive definite functions with finitely many centres. *Trends in Approximation Theory*, 2001.
(15) A. Beckert, H. Wendland. Multivariate interpolation for fluid-structure-interaction problems using radial basis functions. *Aerospace Science Technology*, 5(2):125-134, 2001.

(16) H. Wendland. Konstruktion und Untersuchung radialer Basisfunktionen mit kompaktem Träger. *Dissertation Universität Göttingen*, 1996.
(17) Y. Saad. Iterative methods for sparse linear systems, *Society for Industrial and Applied Mathematic*. Second Edition, 2003.
(18) R.W.C.P. Verstappen, A.E.P. Veldman. Symmetry-preserving discretization of turbulent flow. *Journal of Computational Physics*, 187:343-368, 2003.
(19) F.X. Trias, O. Lehmkuhl. A self-adaptive strategy for the time integration of Navier-Stokes equations. *Numerical Heat Transfer, Part B: Fundamentals*, 60:116-134, 2011.
(20) F. Nicoud, F. Ducros. Subgrid-scale stress modelling based on the square of the velocity gradient tensor. *Flow, Turbulence and Combustion*, 62:183-200, 1999.
(21) O. Lehmkuhl, C.D. Pérez Segarra, R. Borrell, M. Sòria, A. Oliva. TERMOFLUIDS: A new parallel unstructured CFD code for the simulation of turbulent industrial problems on low cost PC cluster. *Proceedings of the Parallel CFD 2007 Conference*, 1-8, 2007.
(22) O. Lehmkuhl, J. Calafell, I. Rodríguez, A. Oliva. Large-eddy simulations of wind turbine dedicated airfoils at high Reynolds numbers. *EUROMECH Colloquium 528, Wind Energy and the impact of turbulence on the conversion process*, 2012.
(23) C. Orozco. Finite volume computation and verification of fluid flow and heat transfer phenomena in domains with moving boundaries and complex geometries. *Terrassa: ETSEIT (UPC)*, 2006.
(24) W. Shyy [et. al]. Computational fluid dynamics with moving boundaries. *Taylor and Francis*, 1996.
(25) D.D. Schwarzler, J.F. Hamilton. An analytical method for determining effective flow and force areas for refrigerating compressor valving system. *International Compressor Engineering Conference at Purdue University*, 30-36, Indiana, USA, 1972.

COMPUTATIONAL FLUID DYNAMICS

Use of computational fluid dynamics to develop improved one-dimensional thermodynamic analyses of refrigerant screw compressors

J Sauls, S Branch
Ingersoll Rand, Climate Solutions, USA

ABSTRACT

This paper illustrates the use of computational fluid dynamics (CFD) to analyze selected features of a refrigeration screw compressor. Specifically, we consider leakage flows between the housing and the tips of the male rotor and through the blowhole, a leak path formed by the two rotors and the housings together. In addition to revealing the nature of the flows in these leak paths, this study provided data for the development of simplified leakage models that can be used in an existing one-dimensional thermodynamic simulation.

Calculations were carried out for leakage flow through specific geometries representative of the radial clearance between the screw rotors and the compressor housing as well as for the blowholes of two selected rotor pairs. In the case of the radial leak path analysis, the geometry of the leak gap for a particular male rotor is used. Cases with and without relative motion between the walls were analyzed. When there is relative motion, the velocities comparable to the tip speed of the rotor and twice that level are studied. For the blowhole calculations, only models with stationary walls are considered. In all cases, calculations presented here are for refrigerant only; oil is not considered in this study.

Computed flow rates are compared to those calculated by the relatively simple model used in our one-dimensional thermodynamic simulation. The result is a set of correction factors that result in leakage flows in the simulation being better aligned with those computed using the more detailed CFD analyses.

1. INTRODUCTION

The work reported here is of the same nature as that reported in (1), where CFD calculations were used to improve the capability of models within one dimensional thermodynamic simulations for screw, scroll and reciprocating compressors. In that paper, the flow through the discharge ports of a refrigeration screw compressor was studied and an improved, albeit still simplified, model was made available to the simulation program. A second study reported was the use of CFD modeling to improve leakage flow calculations for a scroll compressor simulation.

In this paper, we address only the issue of refrigerant leakage in screw compressors. The effects of oil are not included and the geometries studied are unique to screw compressors. The general approach was to build a computational mesh for the flowpath to be examined then compute mass flow rates by systematically varying the solution domain inlet and exit pressures. Several levels of pressure and temperature were used to cover the range seen in normal operation of a refrigeration screw compressor designed for use in air cooled water chillers.

We constructed CFD models of the clearance gaps between the compressor housing and the outer diameter of the male rotor and of the blowhole, the leak path formed by the housing and rotor profile flanks near the housing cusp – the intersection of the male and female housing bores.

When looking at the flowpaths representing the gap between the rotors and the housing, cases with both stationary and moving walls were run. However, for the blowhole analysis, we only present results for the case of stationary walls.

There is no good benchmark for the flow through the leakage gaps in an actual compressor. However, an experimental study by Peveling in 1988, summarized in reference (2), provides results from tests of numerous geometries that were considered representative of leak path elements in screw compressors. We decided to run calculations for one of those geometries to establish some connection between our analyses and actual test results.

This paper continues with Section 2 providing a summary of the comparison of calculations to the data in (2). We then present results for the analysis of the radial clearance between rotors and housing in Section 3 and for the blowhole in Section 4. Finally, some concluding remarks are offered in Section 5.

2. COMPARISON TO TESTS

As far as we have been able to ascertain, there is no definitive reference as to actual flow rates through the clearances in a functioning screw compressor. Since we were setting out to run calculations to support the modeling of such flows, it nonetheless seemed appropriate that we attempt to validate our approach against a known solution that matches our problem as nearly as possible.

Peveling summarized a large body of work he carried out at the University of Dortmund in reference (2). Chapter 7 of the reference describes an experimental study of flow through 23 different clearance flowpaths made up of opposing walls of various shapes, positioned to form selected clearances varying between 0.1 mm and 0.5 mm.

Individual shapes were assembled in pairs to form a flow path with a small clearance between the parts. These were then installed into a test section connected to a pressurized plenum on one side with the other side open to atmosphere. Figure 1 shows a few of the configurations documented in (2).

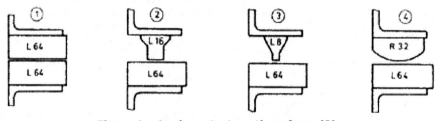

Figure 1 – Leakage test sections from (2)

The test sections were defined by flat sections and arcs; the numbers in the section elements illustrated in Figure 1 are the length or radius of the section shape, as appropriate. We selected geometry 4 to model using ANSYS-CFX as this is one for which data was available in (2) and since it is generally similar to the leakage flow path between the rotor tips and compressor housing. Tests were run in air over a range of overall pressure ratios from about 1.2:1 to 2.2:1 with the minimum clearance between the test section elements varying from 0.50 mm to 0.10 mm.

As explained in the reference, mass flow rates were measured using the experimental apparatus developed specifically for the tests. In addition, the isentropic flow through the gap at the tested upstream and downstream pressures was computed using the equation for flow through a converging nozzle. The test section outlet was open to ambient so all tests were run with downstream pressure of one atmosphere with the upstream test section plenum pressurized to generate the pressure differences.

What information that was available from the text of the reference plus what we could deduce from the various photographs of the test section allowed us to build a representative CFD model of the experimental setup. A cross-section of the three dimensional calculation grid is shown in Figure 2.

Figure 2 – Computational domain representing geometry 4 from (2)

The mesh used in all of the calculations contained 370,000 nodes. Element volume variation within the domain is less than 7:1 (< 20), minimum element face angle is greater than 32 degrees (>20) and the aspect ratio is less than 75 (400). Values in parentheses are considered representative of a "good" mesh. A mesh sensitivity study showed use of a mesh with 10,000,000 nodes produced results that were less than 1% different than when using the mesh chosen for this study.

Figure 3 shows an example of the comparison of our CFD calculations to data replicated from Figure 7.10 in (2).

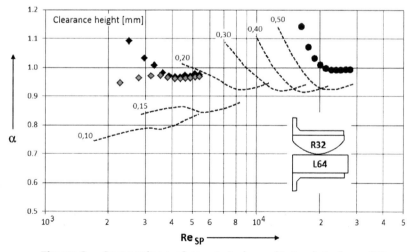

Figure 3 – Comparison of computed results to data from (2)

The data is presented in the form of a flow coefficient (α) as a function of the clearance gap Reynold's number (Re_{sp} where the representative length is twice the gap height) with gap height in mm as a parameter.

For the first round of calculations, we ran cases with 0.50 mm and 0.10 mm gaps. We assumed the walls were hydraulically smooth and ran the CFD model using the Shear Stress Transport (SST) turbulence modeling option. Results for these calculations are shown with the solid symbols in Figure 3. Circles are used for the 0.50 mm case, diamond symbols for 0.10 mm.

The flow coefficient α is defined as the ratio of the measured flow rate to the isentropic flow for the pressure ratio and flow area of the test run. This ideal flow is computed using the relationship for isentropic, compressible flow through a simple converging nozzle. The equation for this as given in (2) was used to compute the α values from the CFD results as well.

Two things stood out in this comparison: the computed flow rates (equivalent to α since the idealized reference flow rates were computed in the same fashion) are higher than the measurements and the shapes of the α vs. Re_{sp} curves for 0.10 mm gap are distinctly different.

The Reynold's numbers for the case of 0.10 mm gap cover the range from turbulent through transition to laminar. However, the turbulence modeling was not adjusted for this, so the small gap cases were re-run using a Transition-SST option. Results for this case are shown with the lightly shaded diamond symbols. This clearly improved the comparison in terms of the characteristic shape seen in the 0.10 mm data, but there was no real improvement in the comparison of flow rate levels.

Clearly there is some lack of consistency between the model and the data. However, there is little detail behind the results presented in the charts in (2). For example, actual high pressure side temperature is not reported; also, the report showed a hygrometer in the test section but there is no discussion as to humidity levels nor how that information was used in reduction and interpretation of the data. We carried out some mesh sensitivity studies, built more detailed (and larger) models of the geometry to capture smaller features and experimented with the effects of roughness. None of these studies provided any answers as to the differences. If nothing else, this study reminded us of the fact of uncertainty in simulating these flowpaths with a numerical model and the need for good practices in building the models and carrying out the analyses.

This study did provide some insight into the state of CFD calculations for small gaps; nonetheless, questions remain. With this as background, the report continues with investigation of the flow of refrigerant R134a in leak flowpaths whose models were constructed from actual compressor rotor and housing geometries. Section 3 contains results for analysis of leakage of pure refrigerant vapor through the radial clearances between rotor tips and the compressor casing. A brief review of calculations of leakage through the blowhole follows in Section 4.

3. FLOW THROUGH RADIAL GAPS BETWEEN ROTORS AND HOUSING

As the primary focus of this work, we carried out calculations of the flow through the radial clearance between the screw rotor tips and the opposing housing wall. To build the computational grid, we took a cross section of the actual screw rotor normal to the rotor helix, then extruded the resulting shape to create a three-dimensional domain. Models were built with the housing wall located a set distance away from the outermost part of the rotor profile to define the minimum clearance for a case. Figure 4 shows the calculation domain for the male rotor.

The model **Inlet** and **Outlet** are near the top of the lobes upstream and downstream of the one defining the minimum clearance area. The original plan was to have these planes in more or less the areas shown with the shaded sections labeled **I1** an **O1**. There were some issues with this and the geometry was extended and new inlet and outlet boundaries with smaller areas were created as shown in the figure. The adjustments were made in accordance with guidelines in the ANSYS-CFX help system to deal with the specific issues we saw. Gaps at the **Inlet** and **Outlet** sections, while small, are more than ten times larger than the minimum clearance. Inspection of the results showed that the stagnation pressure at the **I1** and **O1** locations in this model were virtually identical to the boundary condition values set at the **Inlet** and **Outlet** planes of the domain.

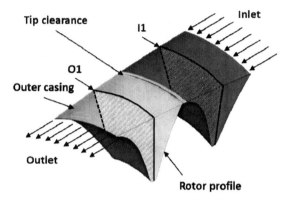

Figure 4 – Calculation domain for analysis of radial clearance leakage

Two levels of minimum clearance were included in the study: 0.127 mm and 0.0762 mm. There were also two levels of inlet stagnation pressure, each with a corresponding fixed stagnation temperature. These boundary condition pairs are (pressure - bar, temperature - °C): (6.9, 37.8) and (13.8, 51.7). Outlet pressure was varied in 6 steps providing a range of pressure ratio from 1.005 to 1.600.

The model was set up for hydraulically smooth walls with the fluid being pure refrigerant R134a vapor. The **Outer casing** wall was given a fixed velocity, moving in the direction from high pressure to low pressure (**Inlet** to **Outlet**) to simulate the relative motion between rotor and casing in an actual compressor. Wall velocities equal to 0, 1 and 2 times compressor tip speed were run. Adiabatic boundary conditions were specified for all wall surfaces.

Ideal flows were again calculated for a converging nozzle with an area equal to the area of the minimum clearance cross section (**Tip clearance** in Figure 4); the ratio of computed and ideal flows defined the flow coefficient (CD – the same as the α factor defined in (2), described in Section 2), the target result from the study.

The CD's derived are shown plotted against overall test section pressure ratio in Figure 5. All of the boundary conditions and both clearances are included in this plot. The results did seem to sort themselves out with the actual level of gap and inlet total pressure as parameters. However, considering that the deviation of results from the fit shown with the dashed line was at most on the order of 1%, we did not consider it necessary to include additional parameters.

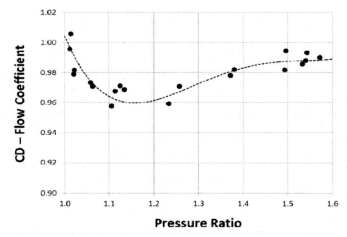

Figure 5 – Calculated flow coefficient for male rotor radial clearance

Calculations with the 0 and 2x velocities were run with the 0.127 mm gap and (13.8 bar, 51.7°C) inlet boundary conditions. We also created a coarser computational grid to simulate as closely as possible the mesh in the vicinity of the rotor tip in a model used for transient analysis of the complete screw compressor. Results of both the wall speed and mesh studies are shown in Figure 6.

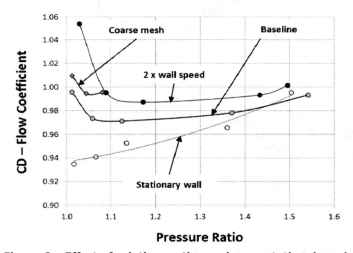

Figure 6 – Effect of relative motion and computational mesh

The **Baseline** case has higher mesh density and a wall velocity corresponding to the compressor tip speed. It contains 438,000 nodes; volume variation, face angle and aspect ratios are: 6, 23 and 165 respectively. For transient, moving/deforming mesh models, the flowpath across the rotor tips can be modeled with a coarser but generally acceptable grid. The **Coarse mesh** case is typical of what we would see in rotors meshed with a tool such as SCORG (3). We looked at this case since the deformed rotor mesh is used in computing rotor-to-rotor gap leakage. This simply shows that leakage calculations imbedded in large compressor assembly models may not be as accurate due to compromises in the mesh. When we model the leakage gaps specifically, the meshes chosen are more appropriate.

The **Coarse mesh** has 85,800 nodes; volume change, face angle and aspect ratio are 5, 24 and 139 The differences do affect the answers as shown in Figure 6, even though overall metrics for mesh quality are about the same as for the **Baseline**. However, local details are different and these give rise to differences in results. The ability of the mesh to capture boundary layer details is important in this case. Figure 7 compares the y+ metric for the **Baseline** and **Coarse** meshes. y+ is a measure of the distance of the of the computational element at the wall compared to the distance over which the boundary layer is affecting the flow; appropriate values are dependent on the choice of turbulence modeling used in the analyses.

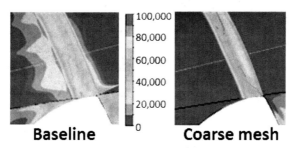

Figure 7 – Baseline and Coarse mesh - contours of y+

The SST turbulence mode was used in these calculations. It is suggested that a mesh with a y+ on the order of 100 represents a good compromise in mesh size and accuracy of results (cf. (5)). We chose this as a target for the **Baseline** mesh used in all leakage gap calculations in this report. In a very limited area at the absolute minimum gap, the y+ values for the two cases are <100. However, moving from this region, the profile surface moves away from the bore, forming slightly diverging sections – this region is highlighted by the blue-yellow-green contours where the **Coarse mesh** case shows considerably higher y+ values. The results showed a difference of about 2.5% in flow rate, giving us some insight into sources of uncertainty in models for complete compressor assemblies where a more refined mesh would be computationally expensive, if not prohibitive.

The relative velocity effect is shown by the 0, 1 and 2x tip speed cases. The effect is more pronounced at low pressure ratios where flow rates and fluid velocities are low. As pressure ratio increases and the fluid velocity in the flowpath's minimum area approaches sonic, the cases more-or-less converge to a single CD value representing choked flow. Since the relative motion of the wall is in the direction from the high pressure side of the clearance gap towards the low pressure side, the idea that the wall is "dragging" fluid from one side to the other seems confirmed.

This work lead to creation of correlations for the effect of pressure ratio and tip speed on leak path flow coefficients. A commercially available computer program (TableCurve3D) was used to fit the data in Figure 6 (baseline mesh only). The resulting relationship of CD = f(gap, wall speed, pressure ratio) can be imbedded in our screw compressor simulation to provide an automatic adjustment based on conditions at the leakage gaps within the compressor. Further work is required to build additional correlations including surface roughness and oil effects.

4. FLOW THROUGH BLOWHOLE

Analyses the leakage through the blowhole in a screw rotor pair were run in much the same way as for the radial clearances. A model of the blowhole was built using sections from male and female rotors and the housing forming the outer boundary of the domain. The geometry is difficult to illustrate - Figure 8 shows a view of the rotor and housing surfaces and the computational domain.

The triangular blowhole in the center of the figure is formed by sections of the male and female rotor flanks and the housing. The edges forming this shape are actually not in a single plane. Nonetheless, it is this 2D shape that our rotor profile design program uses to compute a flow area for leakage calculations. The process for this calculation is as described by Rinder in (4).

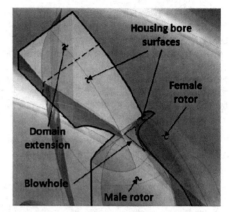

Figure 8 – View of the blowhole in the approximate direction of flow

The boundaries of the computational domain created for the flow calculations are highlighted by the solid lines. The rotor and housing bore surfaces comprise most of the boundaries. In regions where we defined the domain inlet and outlet, we added rectangular section extensions. One of these is shown in the figure, starting at the plane highlighted by the dashed line.

We built models based on two different compressors, one "small" with a nominal tip speed of about 21 m/sec and one "large," running at a bit over 40 m/sec. Flow rates and resulting CD's for both blowholes were calculated for three levels of inlet pressure and temperature: (6.9 bar, 37.8°C), (13.8, 51.7) and (24.1, 107.2). There was no relative motion built in to the analyses. Results are shown in Figure 9.

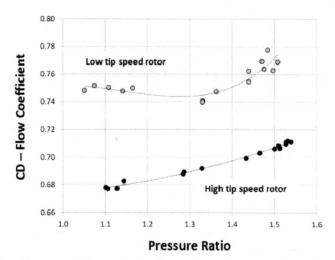

Figure 9 – Flow coefficients for leakage through blowholes of two rotors

The flow coefficients computed were fairly well described by their dependency on the blowhole geometry and pressure ratio. However, it is not clear how or whether it is even possible to use these results to derive a more a more generalized model. The different geometries have CD levels that are about 10% apart.

The rotor pairs that form the two blowhole geometries differ from each other in ways beyond overall size. They have a different number of lobes, the leads are different and the profile forms represent different design approaches. Hence, it is not obvious how to generalize the results. Past practice has been to assign blowhole CD values in the range from 0.6 to 0.8. At this point, our strategy is to provide the thermodynamic simulation with a curve that is the average of the two shown in Figure 9 and allow a user-defined adjustment in overall level. Of course, the actual values shown in Figure 8 can be used for simulations using the rotors upon which the results of this study are based.

5. CONCLUDING REMARKS

The studies reported here are the continuation of an attempt to use the power of CFD analyses to gain more insight into internal flows in screw compressors and to provide information that could be converted into improved models that are still simple enough to be readily incorporated into one-dimensional thermodynamic simulations.

As we expected, the details provided by the CFD analyses provided insights into the flows and generated data that we can use in our simpler simulation programs. These initial analyses were very much simplified compared to the environment within a running screw compressor. This was done to allow us to assess the approach at its most basic level. And even at this level, there were also few surprises.

Summarizing some our key observations:

- Using the data from reference (2) provided insights in meshing a domain which has very large changes in flow areas and showed the need to choose the proper approach to turbulence modeling. However, there is some confusion in not being able to more accurately reproduce the results reported in the reference.

- Even using a limited number of configurations, we were not able to reduce the results for the non-dimensional flow coefficient to one or two parameter models. In addition, the characteristics of the tip flow across the male rotor tip and the female rotor tip (the latter not included in this report) were significantly different enough to require different forms of representation.

- The results acquired to date and those of reference (2) might be more useful for calculations of dry running compressors. However, the effect of tip speed will likely require adjustment to any results obtained with stationary walls.

- In consideration of what we have learned so far, we have incorporated a model based on the effects of tip speed and pressure ratio as shown in Figure 6 in our thermodynamic simulation. This provides characteristics for the effects of pressure ratio and speed; users can specify a scale factor to adjust the average levels to account for oil or to otherwise calibrate specific models. We have not included gap height as a parameter since our compressors are generally set within a small range.

- The effects of oil are almost certain to require adjustment to the level of CD if not the entire approach to representing it. Addressing this question is part of our plan for continuing this study.

- An average of the two blowhole cases shown in Figure 9 has been added to the thermodynamic simulation in order to provide some measure of the pressure ratio effects. As with the radial clearance model, the user can adjust average levels as needed.

Our plan is to continue looking into the results already obtained to make sure we have a clear understanding of the quality of the CFD modeling approach and of the flow process details that such analyses provide. Introducing oil into the analyses is a priority and we will also add surface roughness as a parameter. Finally, we plan to carry out analyses similar to what is reported here for the clearance along the three dimensional sealing line between the rotors. This leak path is very important in setting compressor performance and understanding it more completely is important to improving our modeling capabilities.

ACKNOWLEDGEMENTS

The authors would like to thank our colleagues Gordon Powell, Fahu Gu and Jason Busch for their assistance and valuable insights. We also recognize the contribution of Ingersoll Rand and the Trane engineering organization in providing the environment in which work such as this is supported.

REFERENCE LIST

1. **Sauls, J**. *Use of finite element and computational fluid dynamics analyses in the development of positive displacement compressor simulations*, International Conference on Compressor and their System, IMechE Conference Transactions 2011; London; 2011.
2. **Peveling, F-J**. *Ein Beitrag zur Optmierung Adiabater Schraubenmaschinen in Simulationsrechnungen*, Fortschrittberichte VDI; Reihe 7 Nr. 135 Düsseldorf; VDI-Verlag 1988.
3. **Kovacevic, A., et. al.** *Advances in Numerical Modeling of Helical Screw Machines*; Proceedings of the 20th International Compressor Engineering Conference at Purdue; Purdue University; West Lafayette, Indiana USA; 2010.
4. **Rinder, L.** *Schraubenverdichter*, Wien, New York: Springer Verlag 1979.
5. **Menter, F.R., et. al.** *Ten Years of Industrial Experience with the SST Turbulence Model*; Turbulence, Heat and Mass Transfer IV, Antalya, Turkey; Begell House, Inc., Redding, Connecticut USA; 2003.

CFD grid generation and analysis of screw compressor with variable geometry rotors

S Rane, A Kovacevic, N Stosic, M Kethidi
City University London,
Centre for Positive Displacement Compressor Technology, UK

ABSTRACT

This paper presents development of an algebraic grid generation algorithm applicable to Finite Volume Method for Computational Fluid Dynamics (CFD) calculation of variable pitch twin screw machines. It is based on the principles developed for the uniform pitch rotors with constant cross-section profile. The same algorithm could be also used for rotors with variable profile geometry. Performance predictions are obtained by ANSYS CFX for an oil-free 4/5 lobed twin screw compressor with variable pitch rotors and uniform 'N' profile. A comparison with the performance of a compressor of the same rotor size and wrap angle, but with the uniform pitch rotors showed that the variable pitch rotors give better compression characteristics. This is achieved by reduced throttling losses, reduced length of the sealing line towards the high pressure end and a larger discharge area for the same pressure ratio.

NOMENCLATURE

L	– Rotor Length	p_e	– Ending Pitch
D	– Male Rotor Outer Diameter	z_1	– Number of lobes on the Male rotor
Φ_w	– Male Rotor Wrap Angle		
α	– Male rotor rotation angle	z_2	– Number of lobes on the Female rotor
Δα	– Increment in Male rotor angle		
Z	– Axial distance along the rotors	i	– Rotor Gear ratio = z_2/z_1
		r.p.m	– Male rotor speed
ΔZ	– Increment in Axial distance	t	– Time
p_s	– Starting Pitch	V_i	– Built in Volume Index

Abbreviations

CFD	– Computational Fluid Dynamics	PDE	– Partial Differential Equation
FVM	– Finite Volume Method	TFI	– Transfinite Interpolation
SCORG©	– Screw Compressor Rotor Grid Generator	GGI	– Generalized Grid Interface

1 INTRODUCTION

Screw compressors are usually manufactured with helical rotors of uniform lead and have profile optimized for efficient compression process. It was suggested in open literature that rotors with variable lead may provide more efficient process for high pressure applications. Figure 1 shows an example of a CAD model of an oil injected twin screw compressor. Its compression process will include leakage flows, heat transfer, oil injection and other phenomena of interest for the compressor design. Boundary conformal representation of the compressor physical space is usually obtained by grid generation techniques which start from the boundary representation and proceed to the interior. There are three main classes of mathematical techniques used for this process, a) Algebraic methods, b) Differential Methods and c) Variational methods. Authors such as (1), (9), (11), (12), (13), (14) and (15) have described different grid generation techniques in detail.

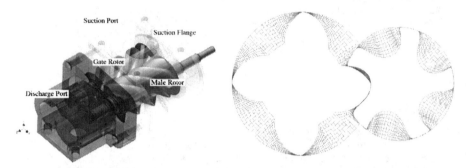

Figure 1. Oil injected Twin Screw compressor and its working chamber

Kovacevic et. al. in (5), (6), (7) and (8) have successfully used algebraic grid generation method with boundary adaptation to generate numerical mesh for twin screw machines with constant pitch rotors. This has been implemented in the custom made program for the calculation of a screw compressor numerical mesh called SCORG©. The motivation for the present work was to extend the functionality of SCORG© by including algorithms to handle twin screw rotors with variable pitch and variable section profiles. This would allow CFD simulations of these new types of machines (10). Numerical treatment of the uniform lead rotors and variable lead rotors differs mainly on grid generation. The axial distance between the grid points for constant pitch rotors is uniform. The challenge in generating a grid for variable lead rotors is that the axial distance and angular rotation of nodes change continuously. To accommodate this change, the grid needs variable axial spacing which still provides a conformal mesh. Figure 1 also shows the numerical grid in one of the cross-sections. For the variable pitch rotors this topology will remain the same over the length of the rotor but the relative position of the rotors will be different for the variable lead compared with the constant lead rotors. An additional challenge is that the grid generating procedure needs to accommodate large differences in length scales of the main domain and the clearances.

 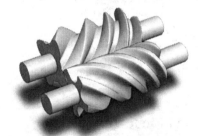

Figure 2. Meshing of Uniform Pitch Twin Screw Rotors   **Figure 3. Meshing of Variable Pitch Twin Screw Rotors**

The patent on variable pitch rotors by *Gardner* (3) dates back to 1969, but such machines are still rarely used due to the lack of efficient and economical manufacturing techniques. Figure 2 shows twin screw rotors with uniform pitch. Figure 3 shows the same size twin screw rotors with variable pitch. Gardner (3) claims that for the same rotor lengths, diameters, wrap angles and lobe profiles, variable pitch rotors can achieve higher pressure ratios, larger discharge port area and reduce throttling losses when compared to constant pitch rotors.

As confirmed by many authors, efficiency of screw compressor depends upon the rotor profile, number of lobes, length, diameter and wrap angle of rotors and rotor clearances (2), (4) and (15). Based on these and the original work of Gardner (3) it is suggested that the effects of variable lead rotor designs are as follows:

a) If all other variables are unchanged for the constant lead rotors and variable lead rotors, the variable lead rotors will have shorter sealing line towards the shorter lead side which is normally discharge pressure side. Since the leakage loss is directly proportional to the length of the sealing line and the sealing line is shorter for the variable lead rotors in the high pressure regions, the leakage loss will be reduced. This may result in higher efficiencies with variable lead rotors.

b) For variable lead rotors, the reduction of volume during the compression process will be faster than for constant lead rotors, as shown in Figure 4. Consequently the pressure will rise more rapidly for the variable lead rotors as shown in Figure 5.

c) The built-in volume index is the ratio of the suction and discharge volumes. The suction volume is the maximum volume at which the suction port is usually closed and where the compression process begins. The discharge volume is the size of the compression chamber at the moment of opening of the discharge port. As shown in Figure 4, to achieve same volume index for variable pitch rotors the discharge port should be opened earlier which allows it to be bigger than in the constant lead case. Hence it is possible to have a greater discharge area at a similar pressure ratio and this will reduce the throttling losses.

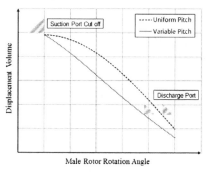 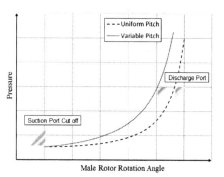

Figure 4. Volume-Angle diagram   **Figure 5. Pressure-Angle Diagram**

d) If the same size of discharge port is retained for both the variable and uniform lead rotors, the discharge pressure of the variable rotors will be higher. This indicates that variable lead rotors can achieve a higher V_i index.

The advantages and disadvantages identified on the basis of previous research have not been extensively tested on physical prototypes due to difficulties in producing such rotors with existing manufacturing methods. However, these can be investigated further in detail by use of CFD analysis if an appropriate numerical mesh can be generated.

2 GRID GENERATION FOR VARIABLE PITCH AND VARIABLE PROFILE SCREW ROTORS

For the Uniform Pitch rotors, there is a fixed relation of the axial distance between the cross sections and the unit rotation angle over the entire length of the rotor. Grid generation in these cases is convenient because if the profile is constant, the grid generated for one interlobe space can be reused in consecutive interlobes. However, for rotors of variable pitch, this relation varies along the length of the rotor. Therefore it is impossible to use the same method with a constant axial distance between sections for grid generation of such rotors. At the same time the rotors need to rotate at a constant angular speed similar to the rotors of constant pitch. In such a case the angular and axial intervals for grid definition do not relate directly to the angular rotation. Figure 6 shows the grid difference between the constant and variable pitch rotors.

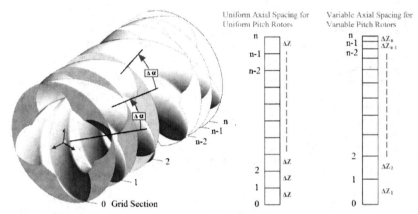

Figure 6. Axial spacing difference between uniform pitch and variable pitch rotor grids

In order to achieve this, the existing procedure used in SCORG© has been reformulated to be adaptable with the variable pitch and variable profile screw rotors. The pitch variation could be constant, linear or stepped. The former is used in screw vacuum pump technology while the latter is applied in some car superchargers. The expressions can be specified in the form of equations 1 – 4.

Constant Pitch
$$p = \text{constant} \tag{1}$$

Linear Pitch
$$p = \left(\frac{p_e - p_s}{L}\right) z + p_s \tag{2}$$

Quadratic Pitch
$$p = \left(\frac{p_e - p_s}{L^2}\right) z^2 + p_s \tag{3}$$

Constant-Quadratic Combination Pitch
$$p = 2.0 \left(\frac{p_e - p_s}{L^2}\right) z^2 + \left(\frac{p_e - p_s}{L}\right) z + p_s \tag{4}$$

Two approaches are proposed here to the solution of grid generation for the CFD analysis of such machines. Approach 1 is easier to implement by modifying the existing procedure and is suitable for variable pitch machines with a uniform rotor profile. Approach 2 is more complex in nature but is generally applicable for any cross section, including conical rotors.

2.1 Approach 1

In this approach, the pitch function is used to derive a relationship between the fixed angular increments $\Delta\alpha$, from one section to the other and the required variable axial displacements $\Delta Z_1, \Delta Z_2, \Delta Z_3.... \Delta Z_n$. By this means, the set of point coordinates generated for one angle of rotation is simply re-positioned in the axial direction with variable ΔZ.

$$Z_i \left[1 - \frac{\left(\frac{(p_e - p_s) z_1}{L}\right)(\alpha_i - \alpha_{i-1})}{2\pi} \right] = Z_{i-1} + \left[\frac{p_s \; z_1 \; (\alpha_i - \alpha_{i-1})}{2\pi} \right] \tag{5}$$

A linear variation as in Equation 5 has been used to find the axial position of each of the sections over the rotor length in an example shown in Figure 7. Additional computational effort is required compared to the uniform pitch rotors to calculate this axial position for each cross section as the axial spacing and the rotation of the rotors are not in the direct relation to each other. The assembly of a grid from 2D cross sections to a 3D structure remains the same as for constant lead rotors. However, this approach cannot be used if there are any variations in the rotor profile along the rotor length.

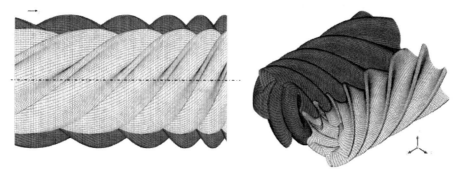

Figure 7. Example of variable pitch grid with uniform profile: 5/6 'N' rotors

2.2 Approach 2

This approach allows more generic generation of rotors with variable rotor pitch and variable cross section profile over the length of the rotor. Therefore, in addition to the variation of axial position of cross sections used in *Approach 1*, this algorithm needs to facilitate change of the rotor profile over the cross sections.

This grid generation algorithm is shown in a block diagram in Figure 8. The foundation of this approach is based on the assumption that every cross section contains a conjugate screw rotor pair. Therefore each cross section can be handled independently and the grid generation process which uses rack to divide working domain in two subdomains for the male and female rotors, can be repeated over each cross section independently.

2.2.1 Procedure
- The process starts with the division of the rotor length into n number of cross sections and proceeds with the generation of 2D vertex data (Nodal x and y coordinates) in the first cross section. This involves boundary discretization and adaptation. Interior nodes at this section are calculated using TFI and the vertex data are recorded after grid orthogonalisation and smoothing operations. This step is labelled as 'Subroutine Run – 1' in the block diagram.
- The process is repeated over the second cross section and additionally this section will receive its axial position (nodal z-coordinate) from Pitch variation function.

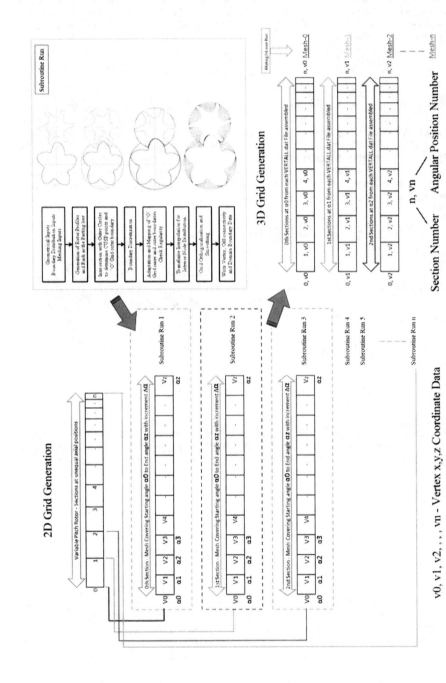

Figure 8. Variable Pitch and Variable Profile Grid Generation

- The 2D grid generation is repeated n times until all cross sections along the z axis are calculated. Calculated vertex coordinates are generated first for the initial time step and then the process is repeated for all required time steps.
- Blocks of vertices representing first rotor position in each cross section are collected to construct the 3D assembly for the first time step. The particular attention is given to the correct z coordinate position of each cross section. This initial mesh is called Mesh-0. Consequently, the second mesh called Mesh-1 is constructed from the corresponding data for the second time step and the process is repeated until meshes for all required time steps are generated.

2.2.2 Example of application of approach 2

An example of variable geometry rotor grid generated using approach 2 is shown in Figure 9. The rotor cross section profile changes continuously from the one to another end of the rotors. This set of rotors with parallel axis has a fixed centre distance so that the rotors are tapered. Namely, outer diameter of the main rotor is reducing from the suction to the discharge end while the inner diameter remains constant. On the gate rotor, the inner diameter is changing while the outer diameter is constant. Figure 10 shows the 2D cross sections grids for variable rotor "Rotor generated demonstrator" profile at the suction end, middle of the rotors and the discharge end.

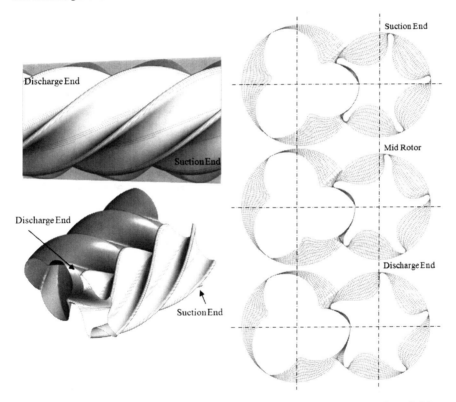

Figure 9. Example of variable geometry rotor grid with uniform pitch: 3/5 'Rotor generated demonstrator' profile

Figure 10. Grid sections of variable geometry rotor, 3/5 'Rotor generated demonstrator' profile

3. CFD ANALYSIS

The validation of the new grid generation procedures is performed by use of numerical CFD analysis. Numerical mesh for all cases is generated by use of the updated SCORG© employing Approach 1 to generate all required cases. The performance prediction of the variable lead screw compressor were obtained in order to evaluate
    a. Indicator diagrams,
    b. Increase in discharge port area for a given pressure ratio,
    c. Reduction in sealing line length towards the high pressure zones.

These characteristics are identified in the literature (3) as advantages of the variable lead rotors. Three cases were analysed, namely:

Case 1. Uniform Pitch rotors and a discharge port opening area to give a built in volume index V_i of 1.8.

Case 2. Uniform Pitch rotors and a reduced discharge port opening area to give built in volume index $V_i > 1.8$. In this case, the compression chamber is exposed to the discharge pressure relatively late in the cycle as shown in Figure 14. This corresponds to an additional male rotor rotation of about 6.75° and allows for further pressure build up in the chambers.

Case 3. Variable Pitch rotors with the discharge port opening area equivalent to that of *Case 1*.

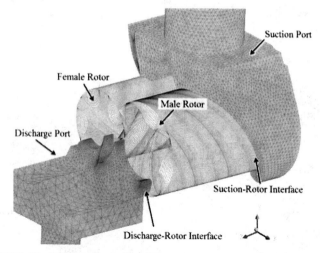

Figure 11. Different parts of the compressor numerical model

The configuration of the compressor is as follows; Rack generated 'N' profile, Male rotor with 4 lobes and a 60mm outer diameter, Gate rotor with 5 lobes, L/D ratio 1.55 and centre distance 42mm. The wrap angle on the male rotor was 306°. The compressor was run at 10000 rpm which gives a tip speed of 31.5m/s. The simulation was performed for oil free mode in order to avoid influence of oil on the process. Radial and interlobe clearances were set to 10μm. Axial end clearances were not included. The lead on the male rotor for the uniform pitch rotors was 108.8 mm. The variable lead male rotor had lead of 125.0 mm at the suction end, 108.8 mm in the middle of the rotors and 92.6mm at the discharge end. The wrap angle remains the same as for the uniform lead rotors. The rotor grids were generated with the same density in all three cases. The numerical grids for ports were generated externally using commercial grid generator. Figure 11 shows the different parts of the numerical model. The compressor domain is decomposed into four main sub-domains, namely Male rotor flow domain, Female rotor flow domain, Suction port domain and Discharge Port domain. These domains are connected through sliding interface.

Constant pressure receivers were positioned at both ends using the procedure described in (6). The suction and discharge pipes were reasonably extended to obtain good convergence of the flow and reduce the numerical discrepancies arising from the pressure pulsations in the suction and discharge flows. The pressure in the suction receiver was set to 1.0bar while the Discharge receiver was set to 3.0bar. The convergence criteria for all equations were set to 1.0×10^{-3} and coefficient loops for every time step was set at 10. During solution, r.m.s residuals for all time steps reached between 1.0×10^{-3} and 5.0×10^{-3} for momentum equation and below 1.0×10^{-3} for continuity and energy equations. The calculations were run sufficiently long such that a cyclic repetition of flow and pressure characteristics were identified at the boundaries. The working fluid was air following an ideal gas law with the molar mass of 28.96 kg kmol^{-1}, Specific Heat Capacity 1004.4 J kg^{-1} K^{-1}, Dynamic Viscosity 1.831×10^{-5} kg m^{-1} s^{-1} and Thermal Conductivity 2.61×10^{-2} W m^{-1} K^{-1}.

3.1 Results and Discussion

3.1.1 Indicator diagram

Figure 12 shows the indicator diagram for three different cases. *Case 3* with the variable rotor lead has a steeper rise in pressure than *Case 1* and *Case 2* which have constant rotor lead. At the same time *Case 1*, with the same discharge area as that in *Case 3*, builds up the least pressure before the compression chamber is exposed to the discharge pressure at about 246° male rotor rotation. Near this region, *Case 2*, with a reduced discharge port area, continues with the internal compression before it is exposed to the discharge pressure. This lag is about 6.75° of the male rotor rotation.

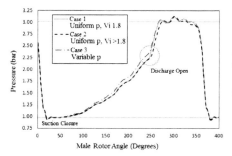

Figure 12. Absolute Pressure variation with Male Rotation

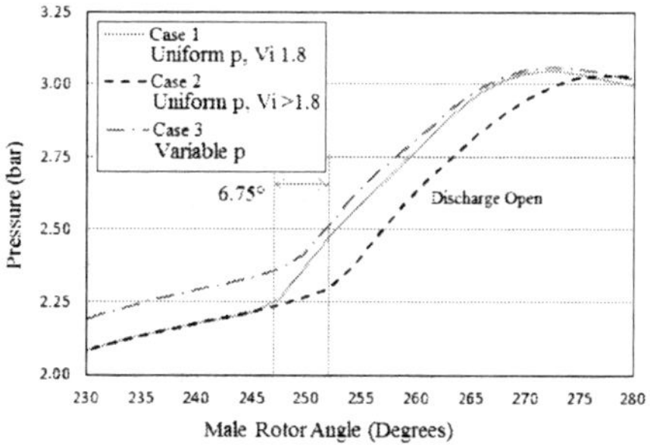

Figure 13. Absolute Pressure variation near discharge opening period of rotation

3.1.2 Discharge Port Area

Figure 14 shows an increase in the discharge port area with the variable lead rotor. In Case 2, with uniform pitch and a reduced discharge port opening area, the pressure increased to about 2.3bar before the discharge port was open. The maximum opening area of that port is 741.57 mm2. This pressure rise was close to that of the variable pitch rotor which reached around 2.4bar, for which the maximum opening area was 788.98 mm2. For the same internal pressure increase, variable lead rotors require 6% larger discharge port which is favourable for the reduction of throttling losses in the compressor.

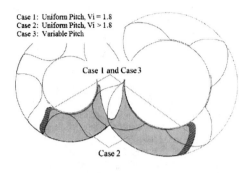
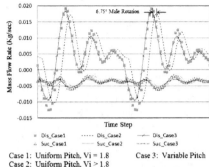

Figure 14. Discharge Port Area gain with variable lead rotor for the same delivery pressure

Figure 15. Mass Flow Rate at Suction and Discharge

Figure 15 shows the mass flow rate at the suction and discharge of the compressor. The average flow was found to be similar in all three cases. The dip in flow at the beginning of the cycle is due to under-compression in which case the discharge pressure is higher than the pressure reached in the compressor chambers just before the opening of the port. The highest dip in flow was obtained for Case 1 with uniform rotor pitch, which developed the least internal pressure (Figure 13).

3.1.3 Sealing Line Length

The interlobe sealing line is the line of closest proximity between the two rotors. The leakage of gas takes place through this gap and is proportional to the length of the sealing line and clearance normal to it (2), (4) and (15). Contours of pressure distribution on the rotors can be established from numerical calculations and the dividing line between high and low pressure levels can be considered as the sealing line. The maximum pressure gradient is present across this division and is the driving force for leakages.

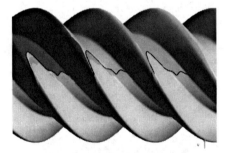

Figure 16. Sealing Line on Uniform Pitch Rotor

Figure 17. Sealing Line on Variable Pitch Rotor

Figure 16 shows the sealing line obtained on the uniform pitch rotors (*Case 1* and *Case 2*). Figure 17 shows the sealing line obtained on the variable pitch rotor (*Case 3*). The projection of the sealing line on the rotor normal plane shows the difference more clearly in Figure 18. The sealing line on the uniform pitch rotor is of the same length for each interlobe space along the rotor. However, on the variable pitch rotors the sealing line is longer at the suction end and shorter at the discharge end of the rotor.

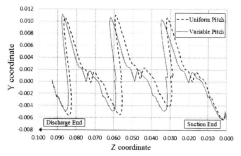

Figure 18. Comparison of Interlobe Sealing Line Length

Table 1. Comparison of Interlobe Sealing Line Length

Interlobe Loop No	Uniform (mm)	Variable (mm)	Difference (mm)
1	67.902	70.923	+3.021
2	63.194	62.652	-0.542
3	64.447	60.984	-3.463
4 (Partial)	12.784	11.778	-1.006
Total Length	208.327	206.337	

Table 1 presents the variation in the sealing line lengths between the uniform and variable pitch cases at one of the rotor positions, and also gives the magnitude of the differences from the suction to the discharge ends of the rotors. At the suction end the sealing line on variable pitch rotor is 3mm longer but at the discharge end it is 3.5mm shorter. This helps to reduce leakage as the pressure difference across the sealing line is highest at the discharge end and smallest at the suction end. Also the total length was 2mm shorter.

4 CONCLUSION

3D CFD grid generation for twin screw compressors with variable pitch rotors was formulated and implemented successfully. The framework defined in this research is suitable for rotors with variable lead and variable profiles. Examples of grids are presented in the paper and CFD analysis was performed for the variable lead, constant profile rotors.

The analysis showed that by varying the rotor lead continuously from the suction to the discharge, it is possible to improve compression characteristics with a steeper internal pressure build up. The analysis also shows that varying the rotor lead allows a larger size of the discharge port area, thereby reducing throttling losses, and provides increase in volumetric efficiency by reducing the sealing line length in the high pressure zone. These latest enhancements in the grid generation open new opportunities for further investigation of the flow behaviour and performance predictions for variable lead and variable profile screw machines by the use of CFD.

REFERENCE LIST

(1) Eiseman PR, Hauser J, Thompson JF, Weatherill NP, Numerical Grid Generation in Computational Field Simulation and Related Fields, *Proceedings of the 4th International Conference*, Pineridge Press, Swansea, Wales, UK, 1994.
(2) Fleming JS, Tang Y. The Analysis of Leakage in a Twin Screw Compressor and its Application to Performance Improvement, *Proceedings of IMechE, Part E, Journal of Process Mechanical Engineering*, 1994; 209, 125.
(3) Gardner JW. US Patent No 3,424,373 – Variable Lead Compressor. Patented 1969.
(4) Hanjalic K, Stosic N. Development and Optimization of Screw machines with a simulation Model – Part II: Thermodynamic Performance Simulation and Design Optimization. ASME Transactions. *Journal of Fluids Engineering*. 1997; 119, 664.

(5) Kovacevic A, Stosic N, Smith IK. Grid Aspects of Screw Compressor Flow Calculations, *Proceedings of the ASME Advanced Energy Systems Division*, 2000; 40, 83.
(6) Kovacevic A. Three-Dimensional Numerical Analysis for Flow Prediction in Positive Displacement Screw Machines, Ph.D. Thesis, School of Engineering and Mathematical Sciences, City University London, UK, 2002.
(7) Kovacevic A. Boundary Adaptation in Grid Generation for CFD Analysis of Screw Compressors, *Int. J. Numer. Methods Eng.*, 2005; 64, 3, 401-426.
(8) Kovacevic A, Stosic N, Smith IK. *Screw compressors - Three dimensional computational fluid dynamics and solid fluid interaction*, ISBN 3-540-36302-5, Springer-Verlag Berlin Heidelberg New York, 2007.
(9) Liseikin VD. Grid Generation Methods, ISBN 3-540-65686-3, Springer-Verlag (1999).
(10) Rane S, Kovacevic A, Stosic N, Kethidi M. Grid Deformation Strategies for CFD Analysis of Screw Compressors, *Int Journal of Refrigeration*, http://dx.doi.org /10.1016/j.ijrefrig.2013.04.008, 2013.
(11) Samareh AJ, Smith RE. A Practical Approach to Algebraic Grid Adaptation, *Computers & Mathematics with Applications*, 1992; 24, 5/6, 69-81.
(12) Shih TIP, Bailey RT, Ngoyen HL, Roelke RJ. Algebraic Grid Generation For Complex Geometries, *Int. J. Numer. Meth. Fluids*, 1991; 13, 1-31.
(13) Soni BK, Grid Generation for Internal Flow Configurations, *Computers & Mathematics with Applications*, 1992; 24, 5/6, 191-201.
(14) Steinthorsson E, Shih TIP, Roelke RJ. Enhancing Control of Grid Distribution In Algebraic Grid Generation, *Int. J. Numer. Meth. Fluids*, 1992; 15, 297-311.
(15) Stosic N, Smith IK, Kovacevic A. *Screw Compressors: Mathematical Modeling and Performance Calculation*, Springer Verlag, Berlin, ISBN: 3-540-24275-9, 2005.
(16) Thompson JF, Soni B, Weatherill NP. Handbook of Grid generation, CRC Press, 1999.

CFD modelling of scroll compressor intermediate discharge ports

B Angel
Renuda, France

P Ginies, D Gross, C Ancel
Danfoss Commercial Compressors, France

ABSTRACT

In order to improve the understanding of the impact of intermediate discharge ports (IDPs) on scroll compressor performance, a simplified CFD model for modelling scroll compressor flows was developed. Simulations were completed for a given scroll at rated and extreme conditions in order to compare the CFD results with an existing 0D model and to analyse the interaction process between the scroll gas pockets and the IDP. Results showed that the simplified CFD model is able to predict scroll compressor flows whilst providing an insight into the flow interactions between the scroll gas pocket, the IDP and the high pressure zone.

1. INTRODUCTION

The scroll compressor is a volumetric machine that has an optimal performance when the system's pressure ratio (Pr) is similar to or the same as the "built in" pressure ratio. However, when these ratios are out of sync, excessive pressure can occur inside the scroll which needs to be bled off in order to avoid excessive power consumption and possible structural damage due to excessive loading. This excess pressure is typically bled off via the use of pressure relief or an intermediate discharge port (IDP) located at a given point on the compression cycle.

Previous work analysing the impact of IDPs on scroll performance, [1], used a 0D model. This work indicated that IDPs can be used successfully to reduce the pressure overload in a scroll gas pocket and improve scroll performance under certain meteorological conditions. However, uncertainties remain with respect to this type of modelling given that it uses empirical correlations for estimating pressure and viscous losses.

In order to gain further understanding of the impact of IDPs on scroll compressor performance, a simplified CFD model was developed for this purpose. The model was validated by comparing the CFD results with the results of a 0D simulation for a scroll compressor previously developed by Danfoss.

2. SIMPLIFICATIONS AND ASSUMPTIONS

The modelling of a scroll compressor using CFD is a complex task given that the CFD model needs to account for:

- an unsteady, 3D, compressible and turbulent flow
- time varying fluid properties of a real gas during the compression process
- the orbital motion of the scroll rotor about the stator, see Figure 1
- the motion of the IDP valve
- mesh motion and deformation due to the rotor orbital motion and the IDP valve motion
- the presence of oil in the gas
- leakage flows around the flanks and tips of the rotor and stator

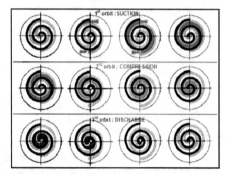

Figure 1: Scroll compression process and rotor orbital motion

This represents a significant challenge for any CFD code and whilst some commercial codes can undertake such simulations, to a greater or lesser degree of accuracy, the objective of the study presented here is to concentrate on the impact of the IDPs.

Hence, in order to simplify the CFD model and to achieve the objectives of the work presented here, certain simplifications and assumptions were applied with respect to the compression process, the compressor geometry, gas pocket motion and IDP valve motion. However, it should be noted that the simplified model is identical to the actual scroll compressor when considering the:

- compression process and the linear reduction in gas pocket volume as the gas pocket is compressed by the orbiting scroll
- distance between gas pockets, the diameter and length of the IDP canal, the IDP valve mass and diameter, the main discharge port height and the high pressure zone diameter and volume
- distance between the centres of the IDP canal and the main discharge port
- limits of the IDP valve motion
- properties of the real gas in the scroll and their variation with pressure and temperature

The simplifications applied to the CFD model are discussed below.

2.1. Compression process

The reduction in volume of a gas pocket inside the scroll is linear up until the opening of the direct gas pocket onto the main discharge port. The change in volume after this opening is non-linear but can also be approximated as linear as a first attempt given that the CFD model is to focus on the modelling of the IDP which is typically used to bleed excess pressure from the second gas pocket.

2.2. Compressor geometry

The compressor geometry can be simplified to the zone of interest, namely the scroll compression pockets, the main discharge port (DP) and the IDP and the high pressure (HP) zone. The simplifications applied to each of these geometric parts for the scroll compressor to be modelled are described in the following sub-sections.

2.2.1. Scroll gas pocket geometric form

Given that the reduction of the pocket volume is assumed to be linear from gas induction to delivery, the pocket geometry can also be simplified to account for this. As a result, the gas pockets in the CFD model are assumed to have the geometric form of a rectangular duct. The distance between gas pockets was assumed to be equal to the thickness of the tip seals used between the rotor and stator wraps and the base plates.

2.2.2. Intermediate discharge port and valve

The canal leading to the intermediate discharge valve was assumed to be straight but maintained the canal diameter and height of the compressor to be modelled. The valve was simplified to a round disc with a diameter of 15mm a thickness of 1mm. The stiffness and mass of the actual valve was used to determine the valve movement of the simplified valve (limited to a vertical motion) due to the various forces acting on it.

2.2.3. Main discharge port

The main discharge port assumed a constant diameter equal to that at the interface between the pockets and this port but maintained the same height as the discharge port of the compressor to be simulated. The distance between the centre lines of the discharge port and intermediate discharge port was also maintained.

2.2.4. High pressure zone

The high pressure zone maintained the same diameter and volume as the compressor to be simulated but the zone height was assumed to be constant instead of variable.

2.2.5. Model geometry

The simplified compressor geometry and its various components are presented in Figure 2. The geometry was built using ANSYS DesignModeler, [3].

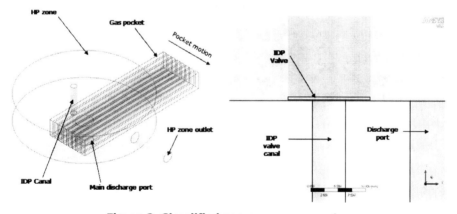

Figure 2: Simplified compressor geometry

2.3. Gas pocket motion

Given the geometric set-up presented in Figure 3, the pocket motion can also be simplified to be a linear translation motion that passes underneath the intermediate discharge and main discharge ports. This motion is indicated in Figures 2 and 4.

2.4. Intermediate discharge port valve motion

The motion of the IDP valve was assumed to be vertical in order to facilitate the coding of the valve motion and the control of the mesh deformation and remeshing process beneath and on top of the valve itself.

3. CFD MODEL

3.1. The FLUENT model

The CFD code ANSYS FLUENT, [4], was used for all simulations. The flow was assumed to be unsteady, 3D, compressible and turbulent with adiabatic walls. The gas was assumed to be a real gas, R410A, modelled using the Aungier-Redlich-Kwong model available in FLUENT. The calculation of the specific heat of R410A was taken from the DuPont database, [5], and coded as a polynomial in FLUENT. Turbulence was modelled using the k-ε RNG turbulence model with standard wall functions. The boundary conditions used were pressure-outlet, wall and interface. Second order discretisation was used to discretise the equation set on the volume mesh. The SIMPLE algorithm with default relaxation was used to progress the solution in time using a first order time stepping method. The default 20 iterations per time step was used with a constant time step size of the order of 6×10^{-5}s.

3.2. Operating conditions

Three operating conditions were used by the CFD model: low, middle and high pressure ratios. The middle pressure ratio condition corresponds to the nominal compressor operating condition where the scroll pressure ratio is matched to ambient conditions. The other two conditions represent extreme conditions where there is a mismatch between the scroll pressure ratio and ambient conditions. These latter conditions are not usual. Figure 3 illustrates these three points on a typical scroll compressor operating map.

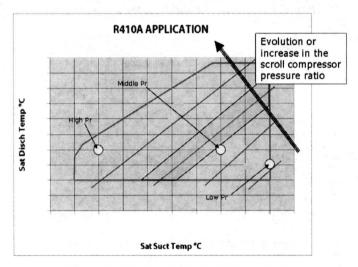

Figure 3: Compressor operating points

3.3. Volume mesh

The volume mesh was generated using ANSYS Meshing, [6], and the "sweep" method in order to generate a fully hexahedral cell mesh. The final volume mesh contained some 300000 cells and is shown in Figure 4.

 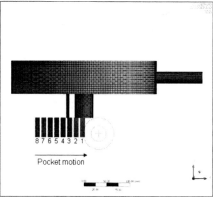

Figure 4: Views of the volume mesh

3.3.1. Gas pocket mesh motion and deformation

The mesh motion and compression process for the gas pockets was set up in the FLUENT dynamic mesh motion GUI and controlled by user coding implemented in the DEFINE_CG_MOTION and DEFINE_GRID_MOTION macros.

3.3.2. IDP valve mesh motion and deformation

The mesh motion and deformation process for the IDP valve was also set up in the FLUENT dynamic mesh motion GUI and controlled by user coding implemented in the DEFINE_CG_MOTION macro.

4. CALCULATIONS

For each operating condition, two calculations were completed, one with and one without an IDP. Each calculation took some 2 days to complete when running in parallel on 6 processors. Solution convergence per time step was achieved for all residuals using default convergence criteria. Table 1 lists these calculations where "Pr" refers to the pressure ratio.

Table 1: CFD calculations

Condition	Low Pr		Middle Pr		High Pr	
Calculation	C1	C2	C3	C4	C5	C6
IDP	No	Yes	No	Yes	No	Yes

During a calculation, the gas pockets 1 to 3 shown in Figure 4 were used to initialise the flow in the IDP, DP and HP zone and generate a periodic flow field in time. Pockets 4 to 7 were used to analyse the numerical results and pocket 8 was used to maintain the periodic flow for pocket 7.

5. RESULTS

5.1. Performance comparison of CFD and 0D results – No IDP

The results of the CFD calculations with no IDP were compared with the results from the 0D model. Figures 9 to 11 present a comparison of the pressure and temperature rise in the gas pocket 5 for calculations C1, C3 and C5 respectively. The vertical (red) dashed lines represent the end of the compression process of a given gas pocket and the start of the compression process of the subsequent gas pocket.

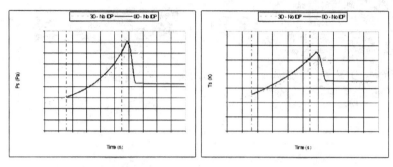

Figure 9: Comparison of C1 and 0D results – Low Pr

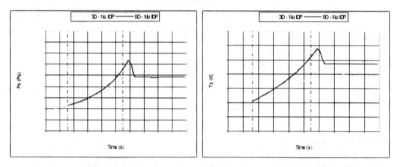

Figure 10: Comparison of C3 and 0D results – Medium Pr

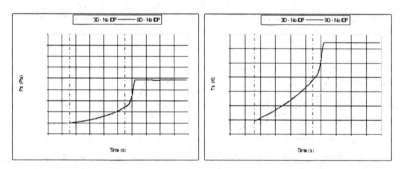

Figure 11: Comparison of C5 and 0D results – High Pr

Globally, both the CFD and 0D models predict the same trends with a steadily increasing gas pressure and temperature until the gas pocket opens onto the main discharge port. At this point, for the low and middle pressure ratios, C1 and C3 respectively, the pressure and temperature decrease rapidly as the compressed gas in the pocket mixes with that in the discharge port. For the high pressure ratio condition, C5, the gas pressure and temperature remain virtually constant.

For all CFD calculations, the pressure at the start of the second gas pocket in the scroll is identical to the 0D starting pressures but the temperature appears to be lower for calculations C3 and C5. The increase in pressure and temperature during the compression process of the second gas pocket is different for all operating conditions. The CFD results predict a lower increase in pressure and temperature resulting in lower peak values at the end of the compression of the second pocket. As the gas pocket opens onto the main discharge port, the differences between the CFD and 0D results become more pronounced. Although both sets of results predict an increase in the pressure and temperature, the peak values predicted by the CFD calculations are lower, of the order of 5 bar and 20K, than those of the 0D calculations with the CFD peak values also occurring several milliseconds later in time than the 0D results.

The differences between the two sets of results are thought to be due to loss modelling in the 0D calculations both during the compression process (perfect gas model) and the interaction between the gas pocket and the main discharge port at the point of opening (underestimated head loss coefficients). This has recently been confirmed to be the case due to a subsequent recalibration of the 0D model based on newly available data. This recalibration has resulted in better agreement between the results of the CFD and 0D models.

5.2. Performance comparison of CFD results with and without an IDP

Comparisons of the results of the CFD calculations with and without an IDP are presented in Figures 12 to 14. As before, the vertical (red) dashed lines represent the end of the compression process of a given gas pocket and the start of the compression process of the subsequent gas pocket.

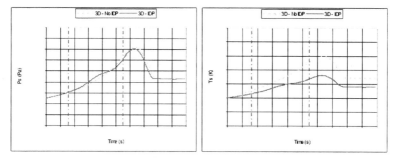

Figure 12: Comparison of C1 and C2 results – Low Pr

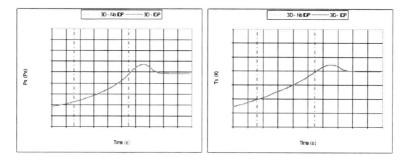

Figure 13: Comparison of C3 and C4 results – Medium Pr

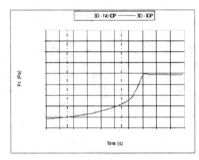 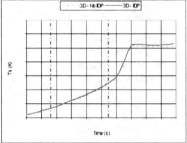

Figure 14: Comparison of C5 and C6 results – High Pr

At the low pressure ratio condition, it can be seen that the IDP has had an impact on the pressure and temperature in the gas pocket but, for the middle and high pressure ratio conditions the pocket pressure and temperature remain virtually unchanged. An analysis of the volume averaged pressure fields in the IDP canal and the high pressure zone shows that, for the low Pr condition, the canal pressure exceeds the HP zone pressure but that this is not the case for the other two conditions. Hence, for these latter two conditions, it is as if the scroll is operating without an IDP hence the virtually identical results. This is as expected.

For the low Pr condition, the reduction in peak gas pocket pressure is of the order of 1 Bar and the peak temperature of the order of 30K. This represents a potential to reduce power consumption by some 8% to 9% for this scroll compressor at this extreme condition. Whilst this is less significant than the results of [1], which showed the potential for a reduction in peak pressure of up to 10 Bar for a single IDP, the results of [1] were for a 0D code version prior to its recalibration. This means with almost no head loss calculation and no rules for the reed valve behaviour and its associated losses. Results since recalibration indicate that the predicted pressure drop in the gas pocket of the 0D model is more in line with the CFD calculation and experimental results.

5.3. Predicted low pressure ratio flow field

The pressure fields at two points in time, corresponding to a closed and a fully open IDP valve position, are shown in Figure 15 for the low Pr operating condition, calculation C2.

For the valve closed position, LHS of Figure 15, the pressure in the gas pocket and the IDP canal is less than that of the HP zone. However, as the gas is compressed due to the reduction in pocket volume, the pressure in the pocket and IDP canal starts to exceed that of the HP zone and the valve opens. When the pressure difference is high enough the IDP valve achieves its fully open position, as shown on the RHS of Figure 15.

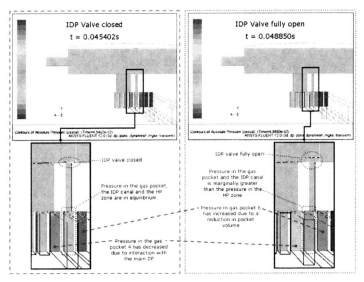

Figure 15: Scroll pressure field

The velocity fields at the same two instances in time for calculation C2 are shown in Figure 16. This figure indicates that when the pressure in the pocket and the IDP canal is in equilibrium with or less than the HP zone pressure, there is stagnant flow in the IDP canal (LHS images in Figure 16). However, as the IDP valve opens due to the pressure increase in the pocket and the canal, the gas starts to flow from the pocket into the HP zone via the IDP canal. The gas velocity in the canal increases over time to a maximum value due to the increase in pocket pressure and the reduction in contact surface area between the pocket and the IDP canal (RHS images in Figure 16). The CFD results indicate that the maximum canal velocity can be in excess of 50m/s.

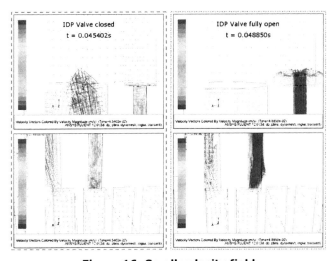

Figure 16: Scroll velocity field

Figure 16 also illustrates the interaction between the gas pocket and the main discharge port for this scroll configuration. Due to the difference in pressure in the pocket and the DP, the gas exiting the pocket at the moment of opening can achieve an initial high velocity; the results indicate in excess of 50m/s. As the pressure difference decreases, due to an increase in the contact surface area between the pocket and the DP, the gas velocity also decreases until the flow is in equilibrium. It is recognised that the results of the CFD simulation are not exact due to the simplified nature of the geometry but they are indicative of the type of interaction that can occur between the pocket and the DP.

Figure 17 presents a comparison of the pressure and velocity fields in the vicinity of the IDP valve at the same instance in time, t=0.04885s, for calculations C2 and C6, low Pr and high Pr respectively. These operating conditions are at opposite ends of the scroll operating envelope. This comparison shows that whilst the pressure in the pocket and IDP canal are greater than that of the HP zone for the low Pr condition, which results in an opening of the IDP valve, this is not the case for the high Pr condition. Here, the pocket and canal pressure is significantly lower than that of the HP zone, due to its starting point, resulting in the IDP valve remaining closed during the passing of a gas pocket underneath the IDP canal.

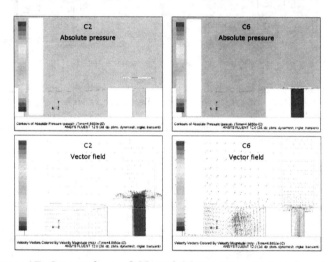

Figure 17: Comparison of C2 and C6 pressure and vector fields

Figure 18 compares the velocity field of calculations C2 and C6 (low and high Pr respectively) at t=0.04885s. Due to the lower pressures in the gas pocket for C6, the interaction between the gas pocket and the main discharge port is seemingly less intense. It can be seen that, for C6, the gas jet flowing from the pocket into the DP has a lower maximum velocity and does not penetrate as far into the main discharge port as the gas jet predicted by C2.

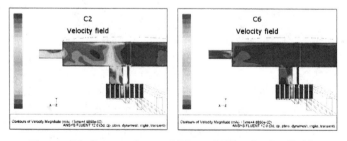

Figure 18: Comparison of C2 and C6 velocity fields

5.4. Mass transfer

The results shown in §5.3 indicate that there is a potential for a transfer of mass from the gas pocket to the IDP canal and the HP zone when the IDP valve is open. For the low Pr condition, this is indeed the case but for the other operating conditions, for which the IDP valve remains closed, there is no mass transfer.

6. CONCLUSIONS

A CFD model has been developed for modelling scroll compressor intermediate discharge ports. The model uses a geometrically simplified form of a given scroll compressor design but maintains important geometric design criteria. The model is able to simulate 3D, unsteady, compressible turbulent flow with real gases and use appropriate operating conditions for a given scroll compressor operating point.

The model has been tested using a given Danfoss scroll compressor design and validated using other numerical data at both rated and extreme operating conditions with no intermediate discharge port present in the model.

The use of an intermediate discharge port has been shown to reduce gas pocket pressure and temperature under low pressure ratio conditions. Here, initial estimations indicate that the use of an IDP could reduce the power consumed by the scroll compressor modelled by some 8 to 9%. For middle and high pressure ratio conditions, the pocket pressure has been shown to not be high enough for the IDP valve to open. This is in agreement with previous experience.

7. FUTURE WORK

The CFD model is generic enough to be used to assess the impact of different designs of intermediate discharge port for most types of scroll compressor developed by Danfoss. The next step in the development process of the CFD model is therefore to integrate and test different IDP designs.

8. REFERENCES

[1] Ginies P, Ancel C, Gross D: "Scroll compressors and intermediate valve ports" International Conference on Compressors and their Systems, City University, September 2011.
[2] Angel B, Ginies P, Gross D, Ancel C: "Development of a simplified CFD model for modelling scroll compressor intermediate discharge ports" Internal Danfoss report.
[3] ANSYS DesignModeler V12.
[4] ANSYS FLUENT V12.
[5] DuPont: "Thermodynamic properties of DuPont^{TM} Suva® 410A refrigerant" 2004.
[6] ANSYS Meshing V12.

CFD analysis of pressure pulsation in screw compressors – Combine theory with practice

J Hauser, M Beinert
GHH Rand Schraubenkompressoren GmbH, Germany

ABSTRACT

The increasing importance of environmental protection and occupational health and safety takes the noise of screw compressors in the center of attention. The reduction of these noise emissions accordingly influences the requirements in design and construction of screw compressors. A major part of the noise source can be described by the gas dynamics of the working fluid in the discharge area and pipes connected to the screw compressor.

Here, simulation of the discharged gas flow using Computational Fluid Dynamics (CFD) is a powerful tool to transfer the gas pulsations into theory. Within this paper, the development of a simulation model for the calculation of the pressure pulsation in the discharge port of a screw compressor is described. The objective is to predict the effect of operating parameters on the gas pulsation with sufficient accuracy.

To increase the accuracy of the simulation, the CFD model is compared with experimental investigations which show the pulsation at the discharge port due to various operating parameter. A variation of different computational grids will allow evaluating the ability of the tested grid types and element numbers for this application. In combination with experimental measurements, the correlation between the physical and technical characteristics is pointed out.

NOMENCLATURE

a	sonic speed	r	radius
c	velocity	Re	Reynolds number
d	diameter	rel	relative
f	frequency	t	time
i	inner	u	tip speed
i	gear ratio	x, y, z	coordinates
norm	normalized	Δ	delta
p	pressure	Π	pressure ratio

1 INTRODUCTION

Pulsation is the unsteady effect of the spreading of the fluid's change of state. The state change is a procedure where rate of spreading is limited by the speed of sound. Therefore state change of fluid can't be realized at once. This causes sonic waves within the fluid, which themselves causes noise emission and vibration. To describe the characteristics of pulsation the static pressure p_0 is defined as average value of the dynamic pressure p. The sound pressure $p_\sim$ is defined as a difference between both of them. The pulsation Δp is defined as a difference between maximum and minimum value of sound pressure (Figure 1). Similar definitions are used for density and velocity as well. Another important characteristic represents the sound intensity which is defined as the vector product of sound pressure and velocity.

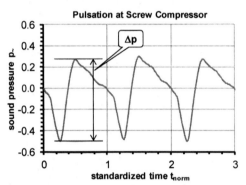

Figure 1: Sound pressure and pulsation

The wave equation is shown in eq. (1)

$$\frac{1}{a^2} \cdot \frac{\partial^2 p_\sim}{\partial t^2} - \Delta p_\sim = 0 \qquad (1)$$

including laplacian Δ and the sonic speed. The solution of the equation for the sound pressure $p_\sim$ and the velocity $v_\sim$ can be visualize in eq. (2) and eq. (3),

$$p_\sim(x,t) = p_{\sim +}(x,t) + p_{\sim -}(x,t) = \left(\hat{p}_+ \cdot e^{-ikx} + \hat{p}_- \cdot e^{ikx} \right) \cdot e^{i\omega t} \qquad (2)$$

and

$$v_\sim(x,t) = v_{\sim +}(x,t) + v_{\sim -}(x,t) = \left(\hat{p}_+ \cdot \overline{Z}^{-1} \cdot e^{-ikx} + \hat{p}_- \cdot \overline{Z}^{-1} \cdot e^{ikx} \right) \cdot e^{i\omega t} \qquad (3).$$

For the experimental work a screw compressor called CS80 was used which is an oil free single stage airend designed for silo applications. The inner pressure ratio is around 3.0 and the suction conditions are predefined as atmospheric. The discharge pressure of the compressor is normally in between 2.0 and 3.5 bar (abs.). Figure 2 gives an overview about compressor components and where the pressure transducer is located during testing phase. The pressure transducer is positioned directly in the area of the discharge port (backside of picture). For the experimental analysis a pressure probe for high temperature applications by PCB Piezotronics Inc. was used. It was chosen because of its resonant frequency of 200 kHz which is more than sufficient for the measurement of more than 100 measurement points during discharge phase of compression chamber. The probe is able to tolerate the existing fluid temperature and pressure to compensate drift effects as well.

Figure 2: Dry running screw compressor CS80

Under theoretical aspect, computational fluid dynamics (CFD) allows the simulation of fluidics by the unsteady compressible three-dimensional laws of conservation for mass, energy and momentum to describe the fluids behaviour (2). Therefore volume is divided into several individual elements using the finite element method. The used software application is ANSYS CFD. In a first step the quality of the simulation results for the flow in a pipe has to be analyzed to select the mesh which fits optimal into the area of pressure pulsation in screw compressors.

Pulsation in screw compressors has been investigated by several authors with the aim of reduction of noise emission or vibration in the total system in general. The authors mainly used experimental (3) or analytical methods (4). The use of computational fluid dynamics for the investigation of pulsation is a recently introduced approach (5).

2 EXPERIMENTAL ANALYSIS OF PRESSURE PULSATION

During experimental analysis the effect of different discharge pressures and the rotation speeds on pressure pulsation have been investigated. Figure 3 presents the pressure pulsation as a function of male rotor tip speed for different discharge pressures. In general pressure pulsation grows up with higher tip speeds for each discharge pressure because of higher gas velocity effects during discharge phase. Figure 4 shows the pulsation as a function of discharge pressure for several tip speeds. First, pressure pulsation is directly linked with compressor speed. Higher tip speed will increase the pressure pulsation delta which was already mentioned before. Second, a higher deviation between discharge pressure and the build-in pressure ratio of compressor will also affect pressure pulsation in a negative way. Because of having a pressure difference compressed air will be forced to move in each direction, at lower pressure backwards from discharge port into chamber and also the other way around. The interdependencies of internal pressure ratio, discharge pressure and pulsation allow the conclusion that adjusting the pressure ratio and therefore the volume ratio in best fit to the expected discharge pressure could bring up a reduction of pulsation for screws. The next step would be to investigate the influence of the shape of the discharge port as realized in (6).

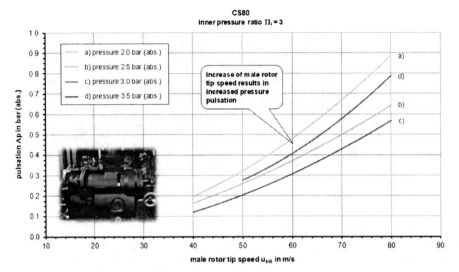

Figure 3: Influence of the tip speed on pulsation

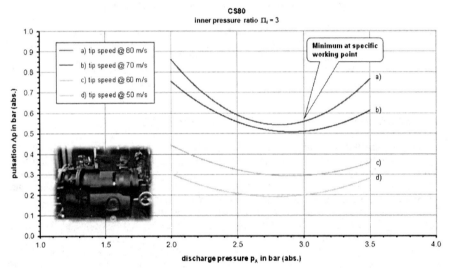

Figure 4: Influence of the discharge pressure on pulsation

3 INFLUENCE OF THE MESHES IN CFD SIMULATION

The application of computational fluid dynamics requires the modelling into separated elements which build up specific type of mesh. The sort of mesh or the number of used elements has an essential effect on the quality of the simulation results and needed simulation time. Because these are opposing effects, the right balance has to be found for each application. That is the reason why those influences are investigated using the example of a pipe to evaluate their suitability.

3.1 Stationary laminar flow in a pipe

Figure 5 represents the flow velocity over the radius of block structured and unstructured meshes as well as the analytical solution of Hagen-Poiseuille described in (7).

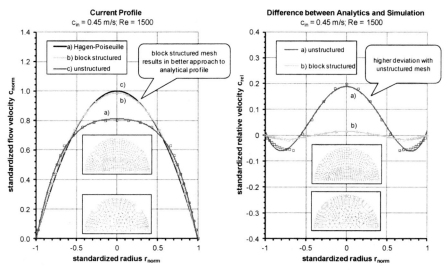

Figure 5: Simulation of unstructured and structured meshes

The block structured mesh has a significant lower deviation against the analytical solution which is basically a result of the identical cross sections over the standard pipe geometry and therefore the identical direction of the fluid's flow and connection between each mesh. Figure 6 shows the differences between the analytical solution and several block structured meshes with different numbers of elements. An increase in the amount results in better simulation correctness but therefore the calculation time takes too much time.

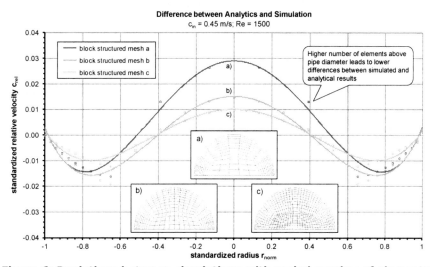

Figure 6: Deviations between simulations with varied number of elements

3.2 Unsteady laminar flow in a pipe

In the next step the stationary flow is replaced by an unsteady flow. The deviation between analytical and simulated flow velocity is shown in figure 7 as well as the inlet velocity. A very good alignment could be achieved.

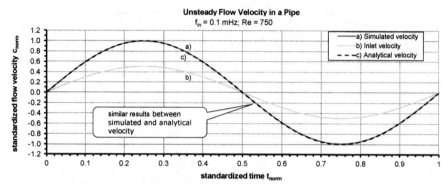

Figure 7: Deviations between Simulation and analytical theory

3.3 Two-dimensional and three-dimensional meshes

Because of the required calculation time of three-dimensional meshes we will only put our focus on rotationally symmetric two-dimensional meshes. Figure 8 shows that the deviation against the analytical solution is smaller for the rotationally symmetric two-dimensional mesh which allows more simplification to a simple volume of mesh. But rotationally symmetric meshes have the restriction that they could only applied to respective geometry as mentioned earlier. This applies to the case of a simple pipe.

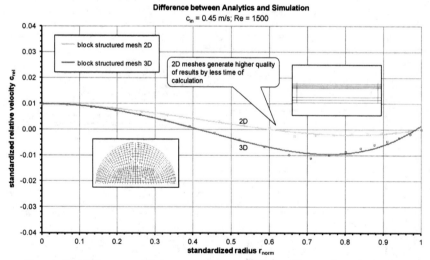

Figure 8: Differences between two- and three dimensional meshes

4 MODELLING PULSATION IN SCREW COMPRESSORS

In this case the modelling of the screw compressor's discharge area is simplified by three rotationally symmetric parts, shown in Figure 9. The working chamber is modelled by a cylinder which size changes just like in the case of a piston moving inside the cylinder. The moving condition is predefined by the volume curve of the screw compressor. The discharge port is modelled by a simple cylinder with fixed sized and the measuring site defined inside. The connection between both, the discharge area changes its size depending on the size of the real discharge area but with a circular shape. Therefore the radius of the working chamber is defined as the maximal radius of the discharge area.

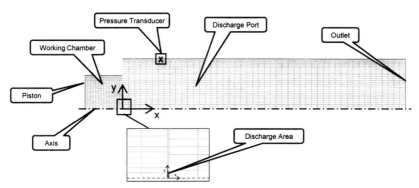

Figure 9: Simulation model for pulsation of a screw compressor

The course of the chamber volume and the discharge area are shown in figure 10 as values as well as illustration for several points in time. The cross-sectional rise in this model is not identical as reality but sufficient similar. The neglect of the shape's influence has to be considered especially in the case of the redirection of the flow in real discharge ports. The investigation of it is the next step of analysis, but for the time these simplifications accomplish the purpose.

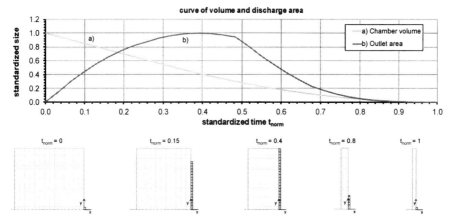

Figure 10: Characteristics of the simulation model

5 SIMULATION RESULTS

To compare the simulation results with real data the measurement sides are positioned similarly. As an example figure 11 shows the sound pressure over time of four discharge procedures. The main courses proceed very similar with minor deviation in the middle of each discharging procedure. The maxima and minima and therefore the pulsation value matches very well. This leads to the conclusion that the simulated results accomplish a sufficient quality despite of the simplifications.

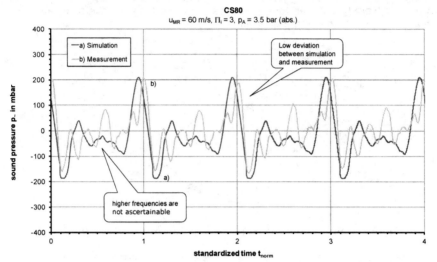

Figure 11: Measured and simulated sound pressure over time

Figure 12 gives an overview about the pulsation for several discharge pressures and rotor tip speeds while comparing the results of the simulation and the measurements. It can be seen that the correlation between experiment and theory is well enough to get a first impression about pulsation impact during design phase of a screw compressor.

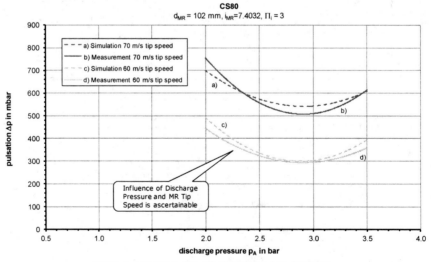

Figure 12: Results of simulated pulsation

6 CONCLUSION

This paper describes the influence of the operating parameters against pulsation and therefore noise emission in screw compressors. To predict pulsation effects computational fluid dynamics can be used but requires the right settings e.g. boundary conditions or mesh types to get accurate results. Compromises have to be made to find the right balance between accurate results and acceptable calculation time as well. But even if there are minor deviations for the pulsation of simulation and measurements, it is great to see that real measurement data can be reproduced via CFD with the help of a simple approach.

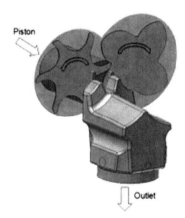

Figure 13: Three-dimensional model of discharge port

Usind rotary symmetrically meshes the calculation time will be reduced significantly which allows to analyze the influence of different volume ratios on pressure pulsation in future. For these investigations the discharge port can be integrated as a real three-dimensional body with representative meshes as you can see in figure 13. While the three-dimensional meshes of the discharge port could be created of CAD-files the chambers can be modelled as pistons circling the pivot while using polar coordinates. An additional software-module has to be developed to create the connection between the moving and the stationary mesh.

REFERENCE LIST

(1) Lerch R: Technische Akustik, Springer Verlag, Berlin, 2009
(2) Oertel H: Strömungsmechanik, Vieweg+Teubner, Wiesbaden, 2009
(3) Fujiwara A & Sakurai N: Experimental analysis of screw compressor noise and vibration, in the 1986 international compressor engineering conference in Perdue
(4) Koai K L & Soedel W: Gas pulsations in twin screw compressors – part I and II, in the 1990 international compressor engineering conference in Perdue
(5) Kovacevic A: An integrated model for the performance calculation of screw machines, in the 2007 international conference on compressors and their systems in London
(6) Mujic E, Kovacevic A, Stosic N & Smith I K: Analysis and measurement of discharge port influence upon screw compressor noise, TMT 2005, Antalya, 2005
(7) Sigloch H: Technische Fluidmechanik, Springer Verlag, Berlin, 2009

CFD analyses of a radial inflow turbine

M Cerdoun [1], A Ghenaiet [2]
[1] Laboratory of Thermal Power Systems, Applied Mechanics, Ecole Militaire Polytechnique, Algeria
[2] Faculty of Mechanical Engineering, University of Sciences and Technology USTHB, Algeria

ABSTRACT

The operating characteristics of a turbo-compressor are tightly depending on the steady and unsteady operating characteristics of its powering turbine. The present study is aimed to highlight the flow structures through the twin-entry volute and the rotor of a radial turbine. The influence of tongue wake is depicted by a low momentum toward the rotor entry, but the effect of this critical region does not extend beyond a tangential angle of 90 deg. The unsteady simulations at different operating conditions and the FFT analyses of the pressure fluctuations due components interactions have revealed a space-time periodic behaviour that may be described by the double Fourier decomposition. The time mode analysis has permitted to determine the different frequencies and the prevailing modes. Different results were obtained based on the type of computational domains; accordingly the one-blade simulation did not reveal all the flow features as was the case of full-rotor simulation.

NOTATION

BPF	blade passing frequency	Φ	Aerodynamic property
H	total enthalpy	λ	conductibility
h	static enthalpy	μ_{eff}	effective dynamic viscosity
m	interaction spatial mode	$\vec{\Omega}$	vector rotation
n	interaction temporal mode	$\vec{U}$	peripheral speed
N_R	number of rotor blades	Ω_R	rotor speed of rotation
N_s	number of stator blades	Ω_m	speed of rotation m^{th} space mode
P	pressure	ρ	density
Pr	Prandtl number	θ	circumferential position
r	radial coordinate	$\Delta\theta$	spatial period
t	time		
S	source term		
v	volume		
u_i	flow velocity component		
x	axial coordinate		

1 INTRODUCTION

Turbo-compressors used to boost the power of diesel engines are operated by radial turbines which have larger capacity to absorb or deliver power. Improving their design requires a better understanding of their turbulent and unsteady flows characterised by complex 3D structures owing to the relative motion between rotor and stator. In the last two decades, flow simulations in the design of turbomachinery have grown to a considerable extent as consequence of rapid evolution of the computing power. Descombes et al (1) carried out 2.5D simulation near the optimum running point in order to construct the local aerothermodynamic flow field of a radial turbine. By using commercial code Fluent, Putra and Joos (2) studied the flow through a radial turbine stator vane and a segment of asymmetric volute. Their results showed a good agreement with the measured flow angle distribution, but the secondary flow vortices were weaker. Full stage calculations with CFX commercial code were carried out by Simpson et al (3) on a series of vaned and vaneless volutes. The obtained results showed that the vaneless volutes exhibit lower levels of losses, yielding improved turbine efficiency. According to Simpson et al. (3,4) the non-uniformity of the flow field for a vaneless volute is prominent just downstream of the tongue. Spence et al. (5) conducted an experimental investigation to assess the differences in performance for vaned and vaneless volutes, and showed that the vaneless volute produces the best efficiency at all tested conditions. A survey presented by Baines (6) has given insights into an up-to-date understanding of pulse flows in radial turbines and the factors influencing their performance. CFD analyses of a vaneless turbine by Hellström and Fuchs (7,8,9) were without supporting test data, nevertheless their results showed that in pulse flows the large scales of turbulence should be handled by the large eddy simulations. Although, numerical studies for single-entry radial turbines exist, but the twin-entry turbines have not been considered, despite they are extensively used in turbochargers. Some of researchers conducted experimental investigations of the effects of partial and unequal admission on the radial turbine performance. Pischiger and Wunsche (10) studied the twin-entry turbines under steady conditions and showed that unequal inlet conditions have a significant effect on the turbine flows patterns and performance. Also the measurements by Dale and Watson (11) have indicated significant effects of partial and unequal admission on both the swallowing capacity and efficiency of a twin-entry turbine, and indicated that the maximum turbine efficiency does not necessarily occur under equal admission. These conclusions are also supported by the experiments of Capobianco and Gambrotta (12). Yeo and Baines (13) extended the study of Dale and Watson using laser two-focus velocimetry techniques to examine the flow inside a turbine. Further investigations by Baines and Yeo (14) showed that for different operating points and equal admission the incidence angle is close to $-30°$. Performances of twin-entry radial turbine were predicted by Aghaali and Hajilouy (15) and Hajilouy et al. (16), under steady state for full and partial admission conditions. The present paper is concerned with the steady and unsteady flows computations and the analyses of flow structures and volute/rotor interactions for a twin-entry radial inflow turbine, operating a type of a radial compressor schwitzer. This paper is a continuity of a previous study that concerned the unsteady flows through the centrifugal compressor's components and the impeller/diffuser/scroll interactions phenomena, carried out by Khalfallah and Ghenaiet (17). The spectral analyses of pressure fluctuations based on the chorochronic periodicity model of Teyler and Sofrin (18) were performed by means of the fast Fourier transform (FFT), applied successively to the signals of pressure obtained when considering one-blade and full-rotor simulations. The obtained results allowed assessing the complex flows and the interactions phenomena through this type of a radial turbine depending on the type of simulations. The turbine operation under such unsteady conditions is directly influencing the centrifugal compressor.

2 CFD MODELLING

The studied twin-entry radial turbine has a rotor with 12 blades and an inlet diameter of 96 mm and an external outlet diameter of 86.3 mm with a blade angle at hub and shroud of 54 deg and 37 deg, respectively. The intake area of each entry is 2.263×10^{-3} m^2. The complex geometry of the rotor and volute required using a measuring machine equipped with a contact-less optical feeler which has three translations guidance allowing an easy displacement of the beam of light in three orthogonal directions over the mapped surfaces. The volute was split into several parts to make accessibility of the spot of light easier. For the steady flow solution, a computation domain with one blade passage and an entire volute was considered, as shown by Fig.1. The grid of the volute has hexahedral meshing refined in areas of high curvatures and gradients, particularly around the tongue and the rotor interface (Fig.2). For the rotor domain, an H-grid topology was adopted with an O-grid wrapped around blade for minimizing skew angles and ensuring boundary layer resolution. In the tip clearance about 0.205 mm, twelve uniform cells were used to capture the tip leakage flows (Fig.3). By varying the rotor grids around the blade surfaces and along the streamwise and spanwise directions, five mesh sizes were generated. The predicted turbine performance tended to stabilize for a grid size equal to 670000 nodes.

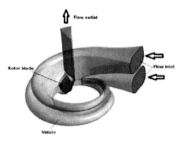
Figure 1 Full computational domain

Figure 2 Volute grid

(a) Blade and hub

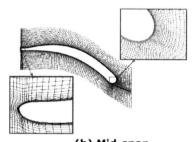
(b) Mid-span

Figure 3 Rotor grid

The CFX-Solver solves three-dimensional Reynolds-stress-averaged Navier-Stokes, with mass-averaged velocity and time-averaged density and pressure and energy equations. The mean form of the governing equations, expressed in a finite-volume formulation that is fully conservative, include the following:

$$\frac{\partial \rho}{\partial t} + \frac{\partial}{\partial x_j}(\rho u_j) = 0 \quad (1.a)$$

$$\frac{\partial}{\partial t}(\rho u_i) + \frac{\partial}{\partial x_j}(\rho u_j u_i) = -\frac{\partial P}{\partial x_i} + \frac{\partial}{\partial x_j}\left[\mu_{eff}\left(\frac{\partial u_i}{\partial x_j} + \frac{\partial u_j}{\partial x_i}\right)\right] + S_{ui}$$

$$\vec{S}_u = -2\vec{\Omega} \times \vec{U} - \vec{\Omega} \times (\vec{\Omega} \times \vec{r}) \quad (1.b)$$

$$\frac{\partial}{\partial t}(\rho H) - \frac{\partial P}{\partial t} + \frac{\partial}{\partial x_j}(\rho u_j H) = \frac{\partial}{\partial x_j}\left(\lambda \frac{\partial T}{\partial x_j} + \frac{\mu_t}{Pr_t}\frac{\partial h}{\partial x_j}\right) + \frac{\partial}{\partial x_j}\left\{u_i\left[\mu_{eff}\left(\frac{\partial u_i}{\partial x_j} + \frac{\partial u_j}{\partial x_i}\right) - \frac{2}{3}\mu_{eff}\frac{\partial u_l}{\partial x_l}\delta_{ij}\right] + \mu\frac{\partial k}{\partial x_j}\right\} + S_E$$

$$H = h + \frac{u_i u_i}{2} + k \quad (1.c)$$

To account for the frame of reference transfer and pitch change between volute exit and rotor inlet, the frozen interface was selected. The flow was solved through the volute in a stationary frame and in a rotating frame for the rotor. The boundary conditions considered were total pressure and temperature at volute inlet and a mass flow rate at draft pipe outlet. The minimum and maximum values of y+ on the surfaces of volute and rotor were 0.39 - 92.2, but it was difficult to have small values of y^+ near hub and tongue. When y^+ exceeds 2, the solver automatically uses the wall function, whereas the low Reynolds method is used for y^+ less than 2 such as the major part of the blade. In the present flow simulations, the k-ω SST turbulence model with an automatic wall function is considered.

3 STEADY SIMULATIONS

In the steady simulations, a high resolution scheme is used in the discretization of the advection term and the k-ω SST turbulence model with an automatic wall function. The Turbulence numeric option is the first order. The flow simulations were carried out in several steps. First, the frozen rotor simulation was obtained using the local time step. Afterwards, the results of these simulations were used as an initial guess for the stage interface simulations to predict performance and operating range of this radial turbine.

3.1 Flow in the volute

The streamlines plotted at mid-span of hub side volute tend to follow perfectly the volute, but there are non-uniformities observed near the tongue and a part of the crossing flow that recovers the main flow. Figure 4 depicts the Mach number in the hub and shroud side of volute, showing that the flow accelerates uniformly till the tongue region. The lower momentum of fluid entering the rotor is due to the wake from the tongue and the flow mixing between that emerging from under the tongue with that circling the entire rotor. It seems that the influence of wake and mixing do not extend further than a tangential angle of 90 deg from the tongue.

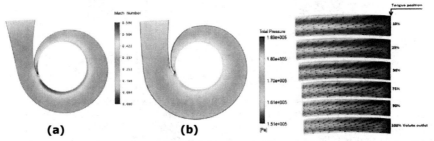

Figure 4 Mach number: a) hub entry volute, b) shroud entry volute

Figure 5 Interspace velocity vectors over total pressure

The downstream flow in the interspace between the shroud side exit and hub side exit of volute, as seen from several planes (Fig.5), affects the incidence of the flow to the rotor to cause high losses. In practice, the best efficiency occurs at an optimum incidence angle (16). At a distance of 10% from the lip, a decrease in total pressure is noted through the entire circumference (Fig.5), as caused by the mixing losses and the wakes due to jets from both sides that separate near the lip. Also, there is a decrease in total pressure corresponding to a recirculation zone between the tongue and inlet of interspace. A comparison between the distance of 10% and 25% shows the same distribution in total pressure with a reduction in the wake effect. At a distance of 50%, the effect of recirculation persists, but it is difficult to observe the mixing zone. The tongue effect is depicted by a proportion of infiltrating mass flow that takes part in feeding the rotor, thus creating mixing losses between the principal flow and that infiltrating. The wake is dumped by convection and the tongue effect is more illustrated by the lower stagnation pressure after a distance of 75%. The velocity vectors at different interspaces positions depict a mixing zone near the divider, which are largely inclined from the radial direction. Moreover, the flow entering from the shroud side has an axial component of velocity, but by approaching the volute exit, there is no axial velocity, except near the tip region.

3.2 Flow in the rotor

The vectors of relative flow velocities (Fig.6) show a clear vortex caused by the high negative incidence at rotor entry. Near the hub pressure corner there are very low velocities. The re-attaching flow onto the blade pressure side is seen at mid-span. The running clearance between blade and casing lets the fluid to traverse from high pressure to low pressure side. Over 70% blade length, the flow over blade tip is found to be more inclined to the streamwise direction and diverted towards the blade trailing edge, where there is a down turning of the flow.

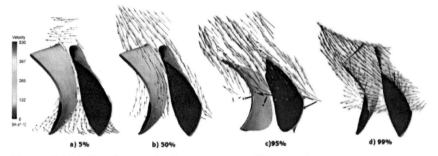

Figure 6 Relative flow velocity vectors at different blade to blade spans

The spanwise surfaces and cross sections of rotor exhibit high losses regions (Relative total pressure) at the inlet and in tip clearance that become stronger toward the rotor exit near shroud. At the mid of trailing edge, there is a wake due to difference in velocities of viscous layers emerging from both pressure side and suction side. Figure 7.b depicts a clear region of losses at rotor outlet due to wake extending from trailing edge. At 95% spanwise near shroud there are high losses on blade suction side (Fig.7.a) starting from the leading edge and growing up gradually along the flow, which are mainly caused by the secondary flow and migration of the low energy fluid from hub to shroud that interacts with the rotor tip leakage. From leading edge at pressure side, there is an area of high loss due to negative incidence. The flow incidence has strong influence on the rotor flow structure; a not suitable angle produces a vortex that occupies a large part of the channel reducing considerably the efficiency at off-design operating conditions.

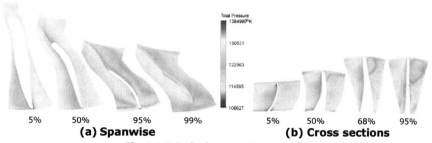

Figure 7 Relative total pressure

3.3 Predicted aerodynamic performance

The mass averaging of flow properties such as total enthalpy and other properties was with respect to the inlet and outlet of computational domain. The predicted total to total isentropic efficiency and expansion ratio as function of reduced mass flow parameter and rotational speeds are presented in Fig.8, revealing that the design point is close to a mass flow of 0.28 kg/s and a speed of 60000 rpm (Fig.8.a), for a maximum efficiency around 80.8%. The choked regime is recognized by the vertical constant portion of characteristics (Fig.8.b) corresponding to a reduced mass flow of 6.47×10^{-5} (mass flow=0.401kg/s). The maximum of efficiency at each rotational speed occurs close to a velocity ratio of 0.707 (Fig.8.c) which is near the theoretical value.

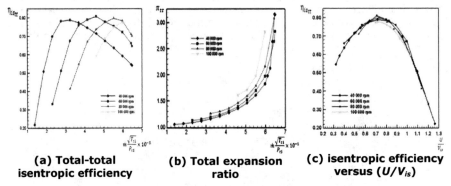

(a) Total-total isentropic efficiency (b) Total expansion ratio (c) isentropic efficiency versus (U/V_{is})

Figure 8 Radial turbine performance

4 UNSTEADY SIMULATIONS

A second order transient scheme is used for the transient term discretization. The transient rotor/stator interface has been used to account for the transient interaction effects at the sliding interface and to predict the true transient interaction of the flow between impeller and volute. The transient relative motion on each side of the general grid interface (GGI) connection was simulated, and the interface position is updated at each time step, as the relative position on each side of the interface varies with time. The transient simulations considering one-blade and full-rotor computational domains, were performed at the design point (60000 rpm, 0.28 kg/s), as well as near choke and at low mass flow. The total time simulation corresponds to 2 rounds of rotor. For a time step equal to 4.16 μsec, the rotor is rotated by the corresponding angular displacement equivalent to 1.5 deg of rotor rotation, which required substantial computation resources. Steady state computations were run first in order to produce initial results. The used transient scheme is the first-order backward Euler with an automatic time step initialization. The solver performs a number of 20 iterations for each time step which was sufficient to have a target residual inferior to 10^{-5}.

4.1 Chorochronic periodicity

The chorochronic model originates from the fact that a flow property is characterized by a double periodicity; that means it has a space periodicity at a given moment and presents a time periodicity at a fixed position:

$$\Phi(t,\theta,x,r) = \Phi(t,\theta + \Delta\theta,x,r) = \Phi(t + \Delta t,\theta,x,r) \tag{2}$$

The period $\Delta\theta$ is equal to 2π if N_R and N_S are not multiples of the same number. A particular case presents when blade numbers are multiples of same number $\Delta\theta/N$.

The angular frequencies of the modes in stator frame depend only on the rotor harmonics n. For n=1 gives the rotor BPF and higher values yield multiples of it:

$$\omega_n = nN_R\Omega_R \tag{3}$$

The frequency controls the speed at which the modes spin because one circumferential wave length must pass a fixed point during one period. Modes with higher values of "m" must therefore spin more slowly for the same frequency. The fundamental period should be a positive value and is calculated in stator frame as:

$$T_R = 2\pi/N_R\Omega_R \tag{4}$$

Any aerodynamic property can be described by the double Fourier decomposition as proposed by Teyler and Sofrin (18) as follow:

$$\Phi(x,r,t,\theta) = \sum_{n=0}^{+\infty} \sum_{m=-\infty}^{+\infty} A_{mn}(x,r) e^{i[m\theta - nN_R\Omega_R t + \varphi_{mn}]} \tag{5}$$

The sum is modelled as a superposition of infinity of rotating waves, characterised by a phase φ_{mn}, an amplitude $A_{mn}(x,r)$ and a rotating velocity corresponding to:

$$\Omega_m = nN_R\Omega_R/m \tag{6}$$

In the interaction between the volute tongue, which is equivalent to one vane, and the rotor blade rows, n represents the time harmonics and m the space harmonics.

$$m = nN_R + kN_S \qquad k = ..., -1, 0, 1,... \tag{7}$$

Consequently, the analysis of interactions is reduced to few parameters such as amplitude and rotational speed. Because it is hard to observe these properties in the time domain, an analysis in the frequency domain is performed. By applying FFT applied to the spatial signals of static pressure at different interfaces revealed different existing modes and the dominating ones and their speeds of rotation.

4.2 One-blade unsteady simulations

The results of one-blade simulations are provided at design and off-design points, based on FFT applied to pressure signals recorded along the circumferential lines. For the volute/rotor interaction, temporal signals were recorded along circumferential lines corresponding to mid-height of interface, and around the tongue. Figure 9 shows the pressure fluctuations when the turbine operates at the design point and the subsequent spectrums obtained from spatial FFT at a time t=2msec. The abscissa gives the harmonics and the ordinate their amplitudes. The spatial FFT applied to static pressure signal recorded along the circumferential line inside the volute and near the tongue (Fig.9.a) depicts a dominating mode m=1 representing the tongue potential effect. The influence of this latter is much more important than the rotor potential mode as defined by the harmonic 12 in vicinity of tongue. Amongst all the modes presented in Fig.9.b, there is a principal one defined by the harmonic 12 and its multiples that corresponds to the spatial modulation dividing the flow into 12 lobes. The rotational speed and the harmonic range of this lobed structure are calculated from equations (7) giving $\Omega_m = \Omega_R = 60000$rpm. This mode, which is dumped and not convected through the rotor passage, represents the potential effect of the rotor blade which would be the unique existing mode in an isolated rotor.

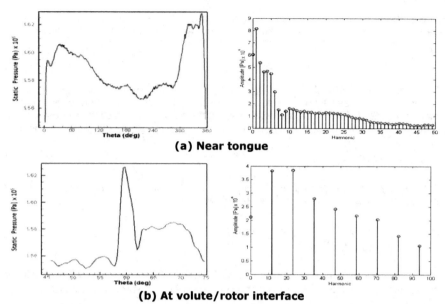

(a) Near tongue

(b) At volute/rotor interface

Figure 9 Spectrums of spatial pressure fluctuations (design point)

FFT spectrums of temporal pressure signals recorded at different positions (Fig.10) depict a peak corresponding to *BPF*=12000Hz, with harmonics smeared in the high frequency waveband. The fluctuations related to *BPF* are more dominating at the interspace, which are dumped through the tongue, and then their amplitudes increase again. FFT spectrum of temporal pressure at tongue depicts a dominating rotor effect. The multiples of *BPF* are considerable as compared to the other regions. Near the tongue, unsteadiness generated by the tongue is indicated by the existence of a weaker peak corresponding to 1000Hz, but in contrary to the *BPF*, the amplitude of this peak develops when passing by the tongue.

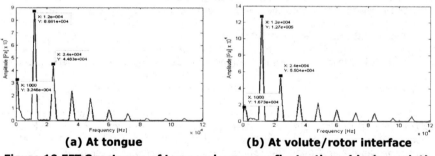

(a) At tongue (b) At volute/rotor interface

Figure 10 FFT Spectrums of temporal pressure fluctuations (design point)

In order to identify the interactions at off-design the spatial pressure signals were recorded. Near choke mass flow, Fig.11.a illustrates an existing dominating mode $m=1$, which when compared to design point has higher amplitude, which means the tongue effect is so important. Figure 11.b shows the existence of a mode $m=12$ corresponding to the rotor potential effect, but in comparison to design point, the choke condition seems to dump the principal harmonic 12 and increase its first multiple 24. When operating at low mass flow, the distribution of pressure around the rotor inlet shows a non-uniformity as compared to design point, and FFT spectrum (Fig.12.a) depicts a dominating harmonic $m=1$ corresponding to tongue effect. Figure 12.b shows spectrums at volute/rotor interface, where harmonic 12 is the dominating mode.

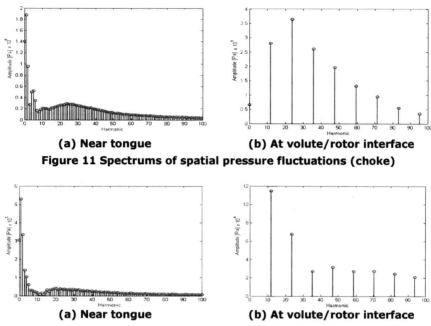

(a) Near tongue   (b) At volute/rotor interface
Figure 11 Spectrums of spatial pressure fluctuations (choke)

(a) Near tongue   (b) At volute/rotor interface
Figure 12 Spectrums of spatial pressure fluctuations (low mass flow)

Figure 13 presents FFT spectrums of temporal pressure fluctuations recorded near tongue and at volute/rotor interface when operating at choke. The frequency related to the tongue effect is 1000Hz. The peak that corresponds to $BPF=12000$ Hz is dominating around the tongue, but is dumped before the tongue. Also at the same position, a second frequency of 36000Hz becomes dominating. At the volute/rotor interface the first frequency is dumped and the multiple frequencies have practically the same amplitudes as BPF. The spectrums of temporal pressure (Fig.14) recorded near the tongue and at the volute/rotor interface exhibit a $BPF=12000$Hz as a dominating frequency around the tongue which is dumped farther and the frequency of 24000 Hz becomes dominating.

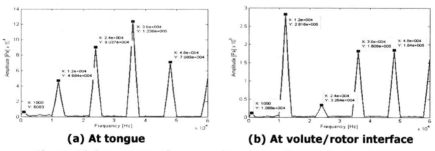

(a) At tongue   (b) At volute/rotor interface
Figure 13 Spectrums of temporal pressure fluctuations (choke)

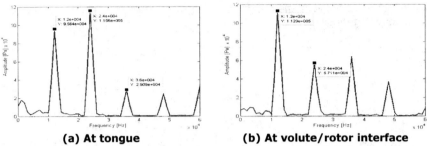

(a) At tongue   **(b) At volute/rotor interface**

Figure 14 Spectrums of temporal pressure fluctuations (low mass flow)

4.3 Full-blades unsteady simulations

The complete rotor simulations required a grid size of 0.95 million of nodes and were carried out using the same inlet and outlet boundary conditions and the solver parameters. The rotor blades number is equal to 12 and that of stator (one tongue) is equal to 1. The spatial signals were recorded along circumferential lines located at 25%, 50% and 75% span at volute/rotor interface and at a plane passing by the tongue. The temporal signals were recorded at points located at mid of blade passage illustrating highly fluctuating pressures forward and backward the tongue, about 90° and 30°, before and after the tongue. An example of pressure signal recorded close to tongue at mid span with FFT spectrum at an instant t=2$msec$ are presented by Fig.15, revealing that the interactions are composed by harmonic 1 and its multiples related to tongue. Also, there are dominating harmonic 2 and 3 in different spans near the tongue, which are a combination between rotor blades and tongue. The harmonic 12 is dominated by the tongue potential effect, and it is difficult to make a distinction between this harmonic and the multiples of harmonic 1. A comparison with one-blade simulations shows that only full-rotor simulations may detect the tongue effect.

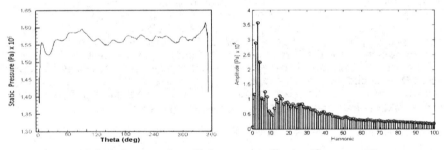

Figure 15 Spectrums of spatial pressure fluctuations near tongue

The pressure signals were recorded at volute/rotor interface along circumference lines at 25%, 50% and 75% span. Figure 16 related to pressure signal at mid span reveals that the interactions are composed of many dominating harmonics, such as the rotor potential effect defined by harmonic 12 and that related to tongue defined by harmonic 1 which is more dominating near shroud owing to the tip effect. According to one-blade simulations, harmonic 1 is not detected, and that simulations failed to detect harmonics lower than the rotor blades number.

The temporal pressure fluctuations were recorded before, at and after tongue and at volute-rotor interface. According to Fig.17 the blade periodic component corresponds to the flow events undergone by the blades at the fundamental *BPF* equal to 12000 Hz corresponds to the number of rotor blades multiplied by the rotor frequency. The first harmonic is 24000 Hz which is the multiple of *BPF*. The time spectrum depicts a first peaked value of 1000 Hz corresponding to the frequency of tongue.

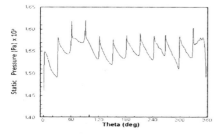 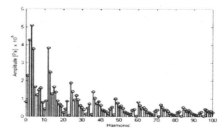

Figure 16 Spectrums of spatial pressure at rotor/volute - 50% span

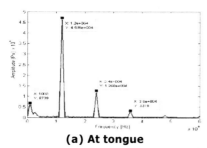 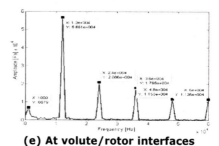

(a) At tongue   (e) At volute/rotor interfaces

Figure 17 Spectrums of temporal pressure fluctuations

4.4 Comparison

Table 1 summarizes the principal dominating modes, showing clearly that the one-blade passage simulation could not detect important modes, and there is decay in amplitudes for many harmonics. The one-blade simulation seems to detect only the modes defined by harmonics greater than the number of rotor blades in the volute/rotor interface, and overestimates the amplitudes of all modes inside the volute and rotor. Also, the effect of tongue is not detected when considering the volute/rotor interface. On the other hand, the full-rotor simulation may detect all modes even with harmonics less than the number of rotor blades. Inside the volute it detects the same modes as in the case of one-rotor passage simulation and the amplitudes of different modes seem to be more reasonable. The effect tongue is revealed to be more dominating near the volute tongue and at the volute/rotor interface.

Table 1 Principal modes

	Mode	One-blade simulations				Full-rotor simulations			
		n	m	$\Omega_m (rpm)$	Amplitude (kPa)	n	m	$\Omega_m (rpm)$	Amplitude (kPa)
Throuh tongue	1	*	1	*	820	*	1	*	120
	2	1	2	360000	530	1	2	360000	270
	3	1	3	240000	460	1	3	240000	355
	12	1	12	60000	180	1	12	60000	90
Volute/ rotor	1	-	-	-	-	*	1	*	22
	2	-	-	-	-	1	2	360000	43
	3	-	-	-	-	1	3	240000	51
	12	1	12	60000	37.5	1	12	60000	38

5 CONCLUSION

The aerothermodynamic performance of a twin-entry radial turbine has been assessed, and the maximum efficiency is shown to occur when the mass flow at the shroud side of volute is more than hub side. The influence of the tongue wake is depicted by a low momentum toward the rotor entry, but the effect of this critical region does not extend beyond a tangential angle of 90 deg. Also, the interspace region is characterized by flow mixing between the flow at exit of shroud side and hub side of volute. The spectral analyses have revealed the presence of a space-time periodic behaviour according to Teyler and Sofrin(18), and depicted the different space modes in the flow unsteadiness and their rotational speeds and amplitudes characterizing these volute/rotor interactions. The simulations based on one-passage did not predict all the flow features and failed to detect harmonics lower than the number of blades including the tongue effect. Only full-rotor transient simulations can produce accurate results for this type of radial machinery. The transient behaviours including the pulstile gas flow conditions at the volute entries is being considered for another study of the turbine and compressor matching.

REFERENCES

1. **Descombes, G, Maroteaux, F, Moreno, N, Jullien, J**, "Analysis of energy conversion within a radial turbine stage", *Int.J. Thermodynamics*, Vol.6 (No.1), pp. 41-48, March-2003.
2. **Putra, M A and Joos, F**, "Investigation of the Secondary Flow Behaviour in a Radial Turbine Nozzle," ASME Paper GT2006-90019. (2006).
3. **Simpson A T, Spence, S W T, Watterson, J K**, "A Comparison of the Flow Structures and Losses Within Vaned and Vaneless Stators for Radial Turbines" *ASME Journal of Turbomachinery* JULY 2009, Vol. 131 / 031010-1.
4. **Simpson, A, Spence, S W T and Early, J**, "A numerical and experimental study of the rotor inlet flow fields of radial turbines using vaned and vaneless stators", ASME Paper No.GT2009-59998, 2009.
5. **Spence, S, Rosborough, R, Artt, D and McCullough, G**, "A direct performance comparison of vaned and vaneless stators of radial turbines", *Journal of Turbomachinery* 129, 53–61, 2007.
6. **Baines, N C**, "Turbocharger turbine pulse flow performance and modelling – 25 years on" Concepts NREC, USA.
7. **Hellström, F and Fuchs, L**, "Numerical computations of pulsatile flow in a turbo-charger". AIAA-2008-073, 46th AIAA paper, (2008).
8. **Hellström, F and Fuchs, L**, "Effects of inlet conditions on the turbine performance of a radial turbine", ASME, paper GT2008-51088, (2008).
9. **Hellström, F and Fuchs, L**, "Numerical computation of the pulsatile flow in a turbocharger with realistic inflow conditions from an exhaust manifold", ASME, paper GT2009-59619, 2009.
10. **Pischinger, F and Wunsche, A**, "The characteristic behavior of radial turbines and its influence on the turbocharging process". *CIMAC Conference*, Tokyo, pp. 545-568, 1977.
11. **Dale, A and Watson, N**, "Vaneless radial turbocharger turbine performance", IMechE, C110/86, pp. 65-76 1986.
12. **Capobianco, M and Gambarotta, A**, "Performance of a twin-entry automotive turbocharger turbine", ASME Paper, 93-ICE-2 (1993).
13. **Yeo J H and Baines N C**, "Pulsating flow behaviour in a twin-entry vaneless radial-inflow turbine", IMechE Paper no. C405/004/90, 1990.
14. **Baines N C and Yeo J H**, "Flow in a radial turbine under equal and partial admission conditions", IMechE Paper no. C423/002/91, 1991.

15. **Aghaali, H, and Hajilouy-Benisi, A**, "Experimental Modelling of Twin-Entry Radial Turbine". IJST Iranian Journal of Science & Technology, Transaction B, Engineering, Vol. 32, No. B6, pp 571-584. 2008.
16. **Hajilouy, A, Rad, M, and Shahhosseini, M R**, "Modelling of Twin-Entry Radial Turbine Performance Characteristics Based on Experimental Investigation under Full and Partial Admission Conditions" Mechanical Engineering Vol. 16, No. 4, pp. 281-290, August 2009.
17. **Khalfallah, S and Ghenaiet, A**, "Impeller-Vaneless-Diffuser-Scroll Interactions and Unsteady Flow Analysis in a Centrifugal Compressor" IMechE, Compressors and Their Systems 2009 paper L20/C682/020, Chandos Publishing Oxford 2009.
18. **Tyler, J M and Sofrin T G**, "Axial flow compressor noise studies", SAE transaction, Vol.70, pp 309-332, 1962.

Numerical and experimental investigation of the efficiency of vaned diffuser of centrifugal compressor

O Obukhov [1], A Smirnov [1], O Gysak [2]
[1] PJSC «Sumy Frunze NPO», Ukraine
[2] Sumy State University, Ukraine

ABSTRACT

The results of the numerical and experimental investigation of vaned diffuser of centrifugal compressor stage are presented in the paper. Experimental investigations were carried out on an aerodynamic test rig of the JSC «Sumy Frunze NPO». Numerical simulations were performed using commercial CFD-software. Pressure recovery and loss coefficients were used as efficiency criteria for a wide range of mass flow rate. The velocity distributions along the vane surfaces of the diffuser were estimated numerically for the range of operating conditions of the compressor. As a result of the numerical investigation the recommendations on diffuser geometry modification to improve compressor stage efficiency have been stated.

NOMENCLATURE

$\bar{m}$ mass flow rate, kg/s;
Φ_0 flow coefficient;
ρ density, kg/m³;
D_2 impeller outlet diameter, m;
U_2 impeller tip speed, m/s;
H_{pol} polytropic head, J/kg;
R specific gas constant, J/(kg·K);
T temperature, K;
T^* total temperature, K;
σ gas dynamic function;
Π pressure ratio;
P pressure, Pa;
P^* total pressure, Pa;
c velocity, m/s;
ψ_{pol} polytropic head coefficient;
η_{pol} polytropic efficiency;
c_p specific heat at constant pressure, J/(kg·K);
ζ total pressure loss coefficient;
C_p static pressure recovery coefficient;
ω angular velocity, rad/s;
f cross-section area, m².

Subscripts
in stage inlet;
out stage outlet;
ps pressure surface of the vane;
ss suction surface of the vane.

INTRODUCTION

Long-term hydrocarbons production at one gas field may result in decrease of formational pressure. Drop of pressure at the compressor inlet necessitates the replacement of the compressor stages with the new ones which provide higher pressure ratio without significant modification of the baseline compressor parts, such as a casing, end covers, seals and bearings. This complex engineering problem can be solved using the stages with vaned diffusers, which enable the required pressure increase with minimal radial dimensions

Centrifugal compressor stage performance depends strongly on gas-dynamic characteristics of the stage elements, i.e. an impeller, diffuser, U-bend and return channel. Because of three-dimensional flow structure with absolute velocity Mach numbers $M_{c2}=0.5$-0.9 at the impeller exit, it is obviously that unfavorable geometry of the vaned diffuser can significantly decrease efficiency of compressor. Taking into account, that gas transportation centrifugal compressors are usually very powerful machines, decreasing energy costs of gas compression and transportation is of great practical importance.

EXPERIMENTAL INVESTIGATION

The experimental investigation of the centrifugal compressor model stage was carried out at the test rig of the JSC «Sumy Frunze NPO» (Sumy, Ukraine). The air was taken from the open air duct (Figure 1). The picture of the test rig is shown in Figure 2. The principal elements of the rig are suction pipeline 1, orifice meter 2, pipeline 3, axial inlet channel 4, centrifugal compressor model stage 5, D.C. motor 7, gear box 6, discharge pipeline 8 and throttle valve 9.

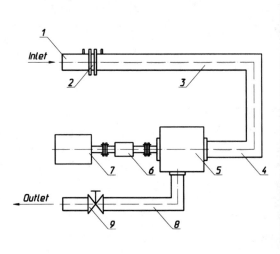

Figure 1 – Test rig scheme

Figure 2 – Test rig

The control cross-sections of the investigated stage, shown in Figure 3, are:
- 0-0 – stage inlet, located at the axial channel upstream of the impeller.
- 2-2 – impeller exit D_2=460 mm.
- 3-3 – vaned diffuser inlet D_3=1.076D_2.
- 4-4 – diffuser exit D_4=1.39D_2.
- 6-6 – return channel exit D_6=0.49D_2.

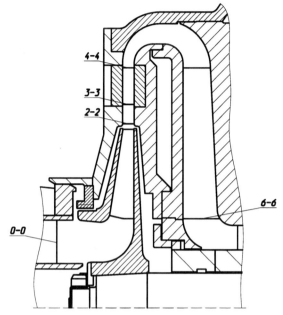

Figure 3 – Sectional view of the tested stage

Flow parameters were measured according to the recommendations [1]. Static pressure was measured in cross-sections 0-0, 3-3, 4-4, 6-6 using taps of 0.8 – 1 mm in diameter. Total pressure in cross-sections 0-0, 3-3, 4-4 was measured with total pressure probe [1, 2, 3]. The view of the total pressure probe and static pressure taps placement is shown in Figure 4.

Figure 4 –Total and static pressures measurement at cross-section 0-0

As a result of experimental research the integral gas-dynamic characteristics of the stage, such as polytrophic efficiency and polytrophic head coefficient as functions of the flow coefficient ($\eta_n = f(\Phi_0)$, $\psi_n = f(\Phi_0)$) were obtained. The measured distributions of the static pressure at the diffuser inlet and outlet are shown in Figures 5, 6.

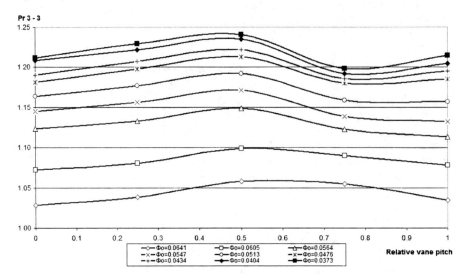

Figure 5 – Static pressure distributions at different flow rates (cross-section 3-3)

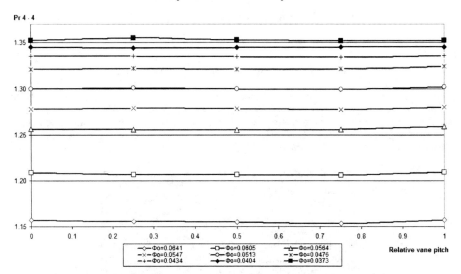

Figure 6 – Static pressure distributions at different flow rates (cross-section 4-4)

Strong flow non-uniformity has been indicated at the diffuser inlet. This is due to the jet-wake flow structure at the impeller outlet and blockage of the flow by diffuser vanes. When the flow enters the vane channel of the diffuser it decelerates and becomes more uniform at the outlet.

Gas-dynamic characteristics of the centrifugal compressor stage were calculated using following equations.

Flow coefficient:

$$\Phi_0 = \frac{4 \cdot \overline{m}}{\pi \cdot \rho_0 \cdot D_2^2 \cdot U_2} \tag{1}$$

Pressure ratio:

$$\Pi = \frac{P_{out}}{P_{in}} \tag{2}$$

Impeller tip velocity:

$$U_2 = \frac{\pi \cdot D_2 \cdot n}{60} \tag{3}$$

Gas dynamic function:

$$\sigma = \frac{\ln(\Pi)}{\ln\left(T_{out}/T_{in}\right)} \tag{4}$$

Polytropic head:

$$H_{pol} = R \cdot T_0 \cdot \sigma \cdot \left(\Pi^{\frac{1}{\sigma}} - 1\right) \tag{5}$$

Polytropic head coefficient:

$$\psi_{pol} = \frac{H_{pol}}{U_2^2} \tag{6}$$

Polytropic efficiency:

$$\eta_{pol} = \frac{\ln(\Pi)}{\left(\frac{k}{k-1}\right) \cdot \ln\left(\frac{T_{out}}{T_{in}}\right)} \tag{7}$$

NUMERICAL INVESTIGATION

Nowadays CFD software such as Ansys CFX, Fluent, FlowVision HPC, FlowER etc. is widely used for numerical modeling of gas flow in centrifugal compressors.

The FlowVision HPC software, which is used by engineers of the JSC «Sumy Frunze NPO» has some advantages over the other ones. The main one is a module of movable body. This module allows to take into account impeller disc friction, seal leakages and ambient heat exchange.

Reynolds-averaged Navier–Stokes equations are solved numerically by finite volume method. The numerical solving of these equations in the near-wall regions is a practical challenge, which requires high-performance computer clusters. Turbulence models are useful approach for that simulation. The SST-turbulence model was used for the presented research as the most recommended for centrifugal compressor application [4].

GEOMETRY

Compressor stage 3-D model geometry was created using ProE software. This is STL facet model format with chord height of 0.01 and angle control 0. According to [4] such detailed geometry is an optimal representation of the real shape of the flow path.

COMPUTATIONAL GRID

Set of conservation equations is solved numerically (by finite volume method). Numerical methods are sensitive to size and shape of the cells of the computational grid. The main advantage of the grid generation process in FlowVision is simplicity.

It is recommended [4] to use cube cells. The preliminary grid is generated by dividing the calculation domain in three directions into equal parts. Then the mesh is refined to better resolve the flow at the areas of high pressure and velocity gradients. Such approach allows to make the calculation process less time-consuming. For detailed investigation of the flow structure the mesh was refined in the impeller and diffuser. The generated grids are shown in Figure 7.

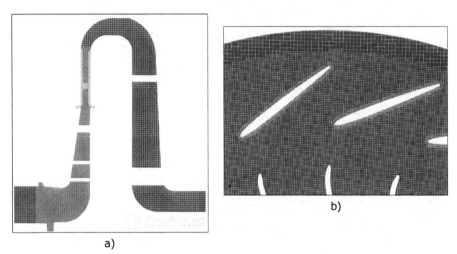

Figure 7 – Computational grid of the investigated stage
a) – meridional plane; b) – radial plane

ASSUMPTIONS

To perform the numerical simulation of the compressor stage the following assumptions have been made:
- Gas flow in the tip between the impeller and stator is not simulated.
- Seal leakages are not simulated.

BOUNDARY CONDITIONS

Experimentally measured total pressure and total temperature were specified at the impeller inlet boundary and the specific mass flow rate at the diffuser outlet. Non-slip and adiabatic conditions were imposed all over the solid walls. The impeller rotation speed was taken from the experimental data.

Boundary conditions are given in the Table 1.

Table 1 – Boundary conditions specified in FlowVision HPC

	Boundary condition	Remark
Inlet	$T^* = 15.2\ °C$ $P^* = -517\ Pa$	The reference pressure is $P_{ref} = 98472\ Pa$
Stator walls heat exchange	adiabatic	---
Impeller	$\omega = 885.38\ \dfrac{rad}{s}$	---
Exit	$\overline{c} = -18.4\ \dfrac{kg}{m^2 \cdot s}$	Specific mass flow rate $\overline{c} = \dfrac{\overline{m}}{f}\ \dfrac{kg}{m^2 \cdot s}$, where f – cross-section area.

CONVERGENCE CRITERIA

As a convergence criteria the mass flow rate discrepancy at the stage inlet equal to 0,001 ks/s and temperature discrepancy at the exit equal to 0,1 °C were used. As a result of numerical computation the averaged flow data at the interested cross-sections and visualization of the flow pattern may be obtained. Also, it is important to evaluate the qualitative structure of the flow. An example of such adequacy criteria of the obtained computations is mapping of high and low energy zones at the impeller exit (jet-wake flow). Figures 8 and 9 represent visualization of the flow in the investigated compressor stage.

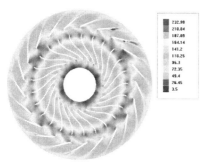
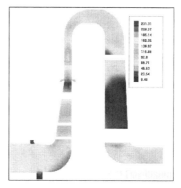

Figure 8 – Distribution of absolute velocity at the midspan

Figure 9 – Distribution of absolute velocity in meridional cross-section

COMPARISON OF RESULTS

The gas dynamic characteristics of the stage ($\eta_n = f(\Phi_0)$ and $\psi_n = f(\Phi_0)$) obtained numerically, were compared to experimental ones (Figure 10).

The pressure variations at the diffuser inlet and outlet related to the pressure at the stage inlet are shown in Figure 11.

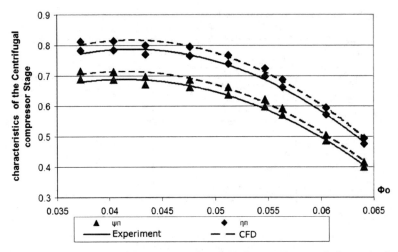

Figure 10 – Comparison of computational and experimental characteristics of the centrifugal compressor stage

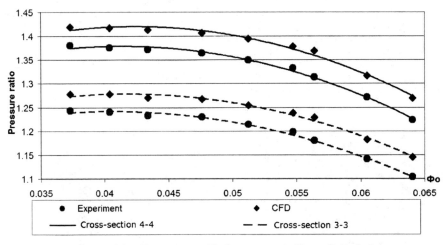

Figure 11 – Pressure ratio in cross-sections 3-3, 4-4

Maximal discrepancy between the calculated and the experimental values does not exceed 5%. Numerically obtained stage performance was overstated due to the used assumptions.

Acceptable agreement between calculated and experimental characteristics of the stage allows to perform numerical simulations instead of experiments.

VANED DIFFUSER EFFICIENCY

Due to the complex three-dimensional flow structure at the impeller outlet incorrectly selected geometry of the vaned diffuser can significantly decrease gas-dynamic efficiency of both the impeller and the centrifugal stage as a whole.

Criteria of the vaned diffuser efficiency are the loss and static pressure recovery coefficients. Static pressure recovery coefficient is calculated as follows:

$$c_p = \frac{P_4 - P_3}{P_3^* - P_3} \qquad (8)$$

Loss coefficient of the diffuser is calculated using the following equation:

$$\xi = \frac{P_3^* - P_4^*}{P_3^* - P_3} \qquad (9)$$

The aforementioned coefficients as a function of the flow coefficient are presented in the Fig. 12.

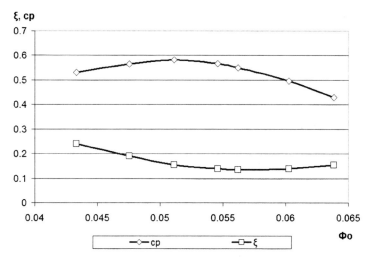

Figure 12 – Gas dynamic characteristics of the diffuser

Moreover, the velocity distributions along the vane surfaces were obtained. Absolute velocity distributions on the pressure and suction side of the vanes are shown in Fig. 13.

Low static pressure recovery and high velocity level along the vane surfaces of the diffuser at the design point of the compressor indicates that unfavorable geometry of the diffuser has been selected. The outlet diffuser velocity should not exceed 60 m/s for the tested stage [3]. The higher values of the absolute velocity indicate insufficient kinetic energy conversion.

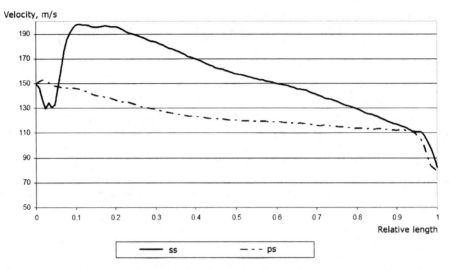

Figure 13 – Absolute velocity distributions along the pressure and suction vane surfaces at the compressor design point

CONCLUSIONS

Detailed study showed that the width of the vaned diffuser should be changed to $b_3=1,2b_2$ with corresponding correction of the vane leading edge angle. It is also recommended to shape the diffuser vanes according to [5]. Moreover, it may be reasonable to replace the single-row diffuser by the double-row one.

Simultaneous application of experimental and computational studies allows to obtain detailed flow patterns in centrifugal compressor stage.

Computational and experimental investigations have revealed some faults in vaned diffuser geometry. The diffuser pressure recovery is not sufficient to provide high efficiency of the compressor stage.

REFERENCE LIST

1. Galerkin U.B., Rekstin F.S. 1969. Methods of experimental investigation of centrifugal compressor machines [Metodi eksperimentalnogo isledovaniya centrobejnih kompresornix mashin (in Russian)]. - L.: Mashinostroenie. – 304 p.
2. Povh I.L. 1965. Aerodynamic experiment in machinery [Aerodinamicheskiy eksperiment v mexanike (in Russian)]. M.- L. Mashgiz. 480 p.
3. Kalinkevych N.V. 2011. Theory of turbocompressors: study book [Teoriya turbomashin: ychebnoe posobie (in Russian)] – Sumy: Sumy State University. - 221 p.
4. FlowVision HPC 3.08.00. Users' guide.
5. Kalinkevych M., Obukhov O., Smirnov A., Skoryk A. The Design of Vaned Diffusers of Centrifugal Compressors Based on the Given Velocity Distribution. International Conference on Compressors and their Systems 2011: City University London, UK, 5–6 September 2011. P. 61-69.

Simulation and validation of the compressor stage of a turbocharger using OpenFOAM

M Heinrich, R Schwarze
Institute of Mechanics and Fluid Dynamics,
Technical University Bergakademie Freiberg, Germany

ABSTRACT

A segregated pressure-based solver for compressible flows for turbomachinery applications has been developed using the open source CFD library OpenFOAM. The solver has been validated using two test cases. The first one is the simulation of one blade passage of the NASA low speed centrifugal compressor (LSCC). The second case is the computation of the complete compressor side of a turbocharger for light commercial vehicles including ported shroud and volute. Among others, the total pressure ratio and isentropic efficiency are chosen for comparison. In both cases, the numerical results and the test bench measurements show a good agreement.

NOMENCLATURE

c_p	J/(kg K)	specific heat at constant pressure
h	J	enthalpy
k_{eff}	W/(m K)	effective thermal conductivity
$\dot{m}_{corr}$	kg/s	corrected mass flow rate
p	Pa	pressure
p_r	Pa	reduced pressure
t	s	time
T	K	temperature
$\boldsymbol{u}$	m/s	velocity
$\boldsymbol{u}_{rel}$	m/s	relative velocity
y^+	-	dimensionless wall distance
η_{ad}	-	total/total adiabatic efficiency
η_{is}	-	total/total isentropic efficiency
μ	Pa·s	kinematic viscosity
ν	m^2/s	dynamic viscosity
κ	-	isentropic expansion factor
π_{tt}	-	total/total pressure ratio
ρ	kg/m^3	density
τ	Pa	shear stress tensor
ω	1/s	speed of rotation
Ω	kg/(m s^3)	dissipation function

© The author(s) and/or their employer(s), 2013

1. INTRODUCTION

The present work was conducted as part of the project called *innoSTREAM*. Its goal is the development and design of a lighter, yet stable and reliable turbocharger with thinner housing components made of modern materials such as TRIP steels. However, the superior material properties are highly temperature dependent (1). For this reason, they can only be used for the compressor side of the turbocharger with a maximum outlet temperature of approximately 200° C. This strong temperature dependency makes it vital to obtain detailed information about the flow field inside the turbocharger and the temperature distribution of the housing. Thus, detailed simulations using computational fluid dynamics (CFD) have to be conducted. The first part of the project focuses on the development and validation of the numerical model using a reference case and test bench measurements.

CFD is widely used in designing turbomachinery applications. Also, limitations and possible errors are well known and understood, see Denton (2). However, the use of OpenFOAM as an open source library for turbomachinery is less investigated. In recent years, additional models and methods were implemented in OpenFOAM for for incompressible and compressible turbomachinery flow simulation.

In this paper, a pressure-based solver for compressible flows of turbomachinery applications is presented based on the open source CFD library OpenFOAM. The solver is validated against the NASA low speed centrifugal compressor. Finally, the compressor side of a turbocharger is modelled to investigate the flow and temperature fields. The corresponding experimental data is used to assess the simulation results.

2. NUMERICAL SETUP

The numerical model was described in detail in (3). For this reason, the presentation of the governing equation and the turbulence model will be shortened.

2.1 Conservation Equations

The three dimensional Favre averaged compressible Navier-Stokes equations are solved using the open source CFD library OpenFOAM. The mass, momentum and energy conservation equations for a rotating reference frame are given as follows:

$$\frac{\partial \rho}{\partial t} + \nabla \cdot (\rho \boldsymbol{u}_{rel}) = 0, \tag{1}$$

$$\frac{\partial (\rho \boldsymbol{u})}{\partial t} + \nabla \cdot (\rho \boldsymbol{u}_{rel} \boldsymbol{u}) + \rho(\omega \times \boldsymbol{u}) = -\nabla p + \nabla \cdot \boldsymbol{\tau}, \tag{2}$$

$$\frac{\partial (\rho h)}{\partial t} + \nabla \cdot (\rho \boldsymbol{u}_{rel} h) = \frac{Dp}{Dt} + \Omega + \nabla \cdot (k_{eff} \nabla T) \tag{3}$$

where ρ denotes the density, p the pressure, h the enthalpy, $\boldsymbol{\tau}$ the shear stress tensor, and Ω the dissipation function, and $\boldsymbol{u}$ the absolute velocity which can be transformed into the relative velocity $\boldsymbol{u}_{rel}$ with respect to the rotating frame of reference by

$$\boldsymbol{u}_{rel} = \boldsymbol{u} - \omega \times \boldsymbol{r} \tag{4}$$

with the prescribed angular velocity ω. The effective thermal conductivity k_{eff} consists of a laminar and turbulent part.

The Spalart-Allmaras turbulence model is employed in the simulations (5). An all y^+ wall function approach is used based on Spaldings law of the wall to model the boundary layer at the wall (6).

2.2 Assumptions and Numerical Solution Control

For the simulations, a pressure-based, compressible solver has been developed using the OpenFOAM version 2.1.1 (4). It is designed for sub- and transonic flows. The multiple reference frame approach has been implemented to model the rotation of the compressor wheel. The flow conditions are considered to be steady-state. Local time stepping is utilized with a local Courant number of 0.3.

The convective terms in the model equations are discretized with second order linear upwind interpolation for $\boldsymbol{u}$, h and first order upwind interpolation for the modified turbulent viscosity $\tilde{\nu}$. Gradients are discretized with the central differencing scheme. The time integration for the local time stepping is performed using implicit first order Euler scheme. See (3) for a more detailed description.

For the evaluation of the CFD model, the total pressure ratio π_{tt}, the total/total adiabatic efficiency η_{ad} and the total/total isentropic efficiency η_{is} are evaluated. The mass flow rates are normalized with a reference pressure of 101,325 Pa.

3. NASA CENTRIFUGAL COMPRESSOR

3.1 Geometry and Boundary Conditions

The NASA low speed centrifugal compressor (LSCC) was experimentally investigated by Hathaway et. al (7) (8). Detailed measurements were conducted including laser anemometry, blade and surface static pressure tabs. The LSCC was designed to operate at low speeds to duplicate the flow fields of high-speed subsonic centrifugal compressors.

The impeller consists of 20 full blades with a backsweep of 55°. The inlet shroud and outlet diameter is 0.870 m and 1.524 m, respectively (see Figure 1). The tip clearance is constant with 2.54 mm. The compressor is followed by a vaneless diffusor designed to avoid backflow. The rotational speed is 1,862 min^{-1}, which results in a relative Mach number of 0.29 at the trailing edge of the blade at design point. The design point is defined at $\dot{m}_d = 30$ kg/s, $\pi_{tt} = 1.17$ and $\eta_{is} = 0.922$.

The computational domain consists of one blade passage. A block-structured hexahedral mesh is used with 1.25 x 10^6 cells and a blade passage resolution of 160 x 65 x 56 in meridional, spanwise and pitchwise direction, respectively. The tip clearance is resolved with 21 x 7 cells in pitchwise and spanwise direction.

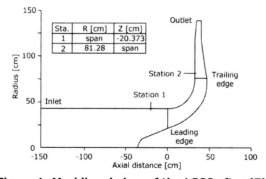

Figure 1: Meridional view of the LSCC after (7).

Three different simulations are performed for the design and off-design points of 23.7 kg/s and 30 kg/s and 38 kg/s. Since turbulent quantities are not known in detail at the inlet, the turbulent intensity is assumed to be 3 % with a turbulent length scale of 25 mm. The inlet pipe is extended to obtain a developed velocity profile at the first measuring station. The total pressure ratio as well as the adiabatic efficiencies are calculated between station 1 and 2 (see Figure 1).

3.2 Results

The impeller performance characteristics are shown in Figure 3 and in Figure 4. The simulation results are in very good agreement with the experimental results. The maximum deviation for the adiabatic efficiency is observed at the off-design points with an offset of about 1 %.

The absolute flow angle at station 2 is shown in Figure 4. The simulation agrees well from close to the hub up to the core of the flow. However, the flow angle is overpredicted by up to 8 % near the shroud surface and the slopes do not match. This indicates, that the turbulent near-wall behaviour is not predicted correctly by the numerical model. Therefore, more complex turbulence model like the k-ω SST model should be employed in future investigations.

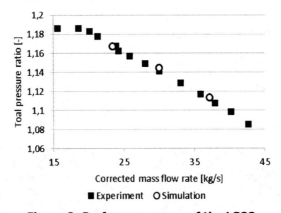

Figure 2: Performance map of the LSCC.

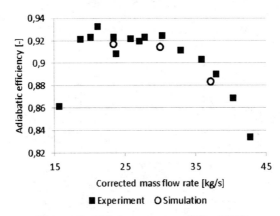

Figure 3: Efficiency map of the LSCC.

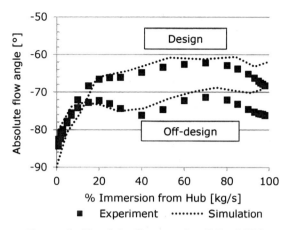

Figure 4: Absolute flow angle of the LSCC.

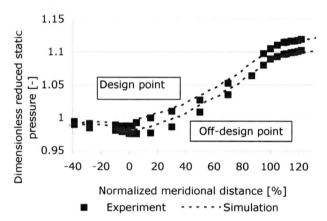

Figure 5: Dimensionless reduced static pressure of the LSCC.

The absolute flow angle at station 2 is shown in Figure 4. The simulation agrees well close to the hub up to the core of the flow. However, the flow angle is overpredicted by up to 8 % near the shroud surface and the slopes do not match. This indicates, that the turbulent near-wall behaviour is not predicted correctly by the numerical model. A more complex turbulence model like the k-ω SST model should be investigated.

Figure 5 shows the circumferential averaged normalized reduced static pressure defined as

$$p_r = \frac{p}{\left[1 + \dfrac{\frac{1}{2} u_{rel}^2 / c_p}{T + \frac{1}{2} u^2 / c_p}\right]^{\frac{\kappa}{\kappa-1}}} \quad (5)$$

with the specific heat at constant pressure c_p. A normalized meridional distance of 0 % is located at the leading edge of the blade while a value of 100 % stands for the trailing edge. For both design and off-design conditions, the simulation and experimental results match good. Considerable differences can be found at about 20-60% meridional distance with a deviation of up to 0.02.

In general, the presented solver shows a good performance in the NASA low speed centrifugal compressor test case. Global variables are predicted very well. However, local pressure and flow distributions reveal minor errors.

4. COMPRESSOR SIDE OF A TURBOCHARGER

4.1 Geometry and Grid

The investigated turbocharger is designed for light commercial vehicles. The numerical model is focused on the compressor side of the machine. The design point is at a rotational speed of 100,000 min^{-1}, which is equivalent to a circumferential velocity of 340 m/s at the trailing edge of the compressor wheel. This corresponds to a mass flow rate of 0.12 kg/s of air, a total pressure ratio of 2.05 and an isentropic efficiency of the compressor side of 0.736.

A cross section view of the turbocharger is shown in Figure 6. The blade consists of 7 main and 7 splitter blades. The hub and shroud leading edge diameter are 16 mm and 46 mm, respectively. The diameter of the trailing edge is 65 mm. The constant tip clearance has a value of 0.3 mm. The vaneless diffuser has a tapering span with a minimum height of 4 mm. The diffusor outlet-to-inlet diameter ratio is 1.66. Geometry simplifications are made to reduce the meshing expense and obtain accurate results in a reasonable time frame. Firstly, the fillet radius between the blades and the hub of the compressor wheel is neglected. Secondly, the gap between the compressor wheel and the diffuser is not modelled.

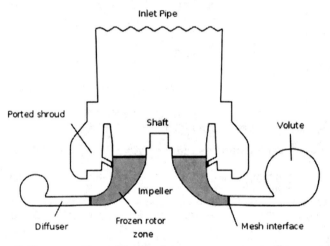

Figure 6: Cross section of turbocharger compressor flow passage.

The computational domain is divided into three different reference frames. The inlet and volute domain are meshed using a tetrahedral/prism hybrid mesh with five prism layers at the wall. The impeller is meshed using a block structured hexahedral mesh. All three domains are connected using arbitrary mesh interfaces. The mesh is locally refined at crucial regions, such as leading edge of the impeller blades, the tip clearance and the tongue of the volute.

A mesh dependency study was conducted with varying mesh sizes between (2.93 - 7.43) x 10^6 cells (9). The best compromise between accuracy and computational cost was achieved at a mesh resolution of 5.43 x 10^6 cells. The average dimensionless wall distance y^+ is about 7 for the inlet region, 12 in the impeller and 22 in the volute region. One blade passage consists of 104 x 42 x 30 cells in meridional, spanwise and pitchwise direction, respectively. The tip clearance gap is resolved with 8 x 14 cells in pitchwise and spanwise direction.

The turbocharger inlet and outlet has a diameter of 65 mm and 42 mm, respectively. The inlet pipe upstream the ported shroud and the outlet pipe downstream the volute are extended using an extrusion mesh in order to reduce the influence of the inflow and outflow boundary conditions. In this configuration, the inlet and outlet patches match the measurement positions of the test bench. Therefore, the boundary conditions can easily be adopted from the experiments.

4.2 Boundary Conditions and Initialization

The boundary conditions have to be varied to model the complete characteristic performance curves. At high mass flow rates, the pressure difference between the inlet and outlet is defined. In these cases, the total pressure at the inlet is set to 101,325 Pa. The static outlet pressure is varied, so that different mass flow rates are obtained. In contrast to that, the slopes at low mass flow rates are flat. For this reason, it is necessary to define the mass flow rate at the inlet itself. At the same time, the expected static pressure at the outlet is set accordingly.

The temperature at the inlet is set to 293.15 K. Other variables at the outlet are treated with zero gradient. The walls are defined as adiabatic and surface roughness is not considered.

Four angular speeds n are analysed: 53,000 min^{-1}, 100,000 min^{-1}, 123,000 min^{-1}, and 147,000 min^{-1}. The air is defined as perfect gas with with a constant heat capacity of c_p = 1004.5 J/(kg K) and an isentropic expansion factor of κ = 1.4. The temperature dependent heat conductivity and viscosity are modelled using Sutherland's law.

4.3 Results

Industrial test bench measurements, which are not documented in detail, are used for assessing the numerical results. The measurement uncertainty is unknown. The total pressure ratio is evaluated at the inlet and outlet boundaries. The mass flow rate is the integrated mass flux at the outlet patch.

The performance map for the four different rotational speeds is presented in Figure 7. The numerical results show a good accordance with the measurement data. The slopes are met very well and choke limit is predicted correctly. In general, the total pressure ratio is overpredicted by up to 5-10 %.

Due to the complex geometry and numerical model, there are various error sources for the deviations. One reason is the pressure-based algorithm in the proposed solver. This method causes errors when the flow approaches transonic conditions, when losses are not resolved properly. Another reason is the RANS-based simulation using wall functions. According to Baris (10), it is difficult to predict flow separation with this approach. This is particularly the case at flow conditions near stall, where the flow is dominated by unsteady flow separation at the rotor and reversed flow in the inlet region and the ported shroud (11). In order to resolve these effects, an unsteady simulation with a more complex turbulence model has to be conducted. Another source of error is the lack of real mesh movement where blade passage effects are not resolved.

Figure 8 shows the isentropic efficiency as a function of the mass flow rate for four different rotational speeds. At higher rotational speeds, η_{is} is predicted well with an offset between 1-4 %. The lower the rotational speed, the higher the overprediction of the isentropic efficiency.

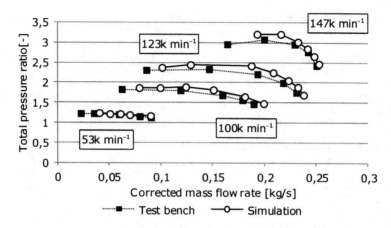

Figure 7: Total pressure ratio for different rotational speeds.

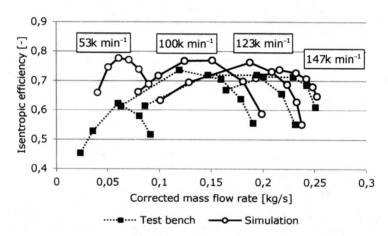

Figure 8: Isentropic efficiency for different rotational speeds.

The reason for these differences is the assumption of adiabatic walls. This suppresses the heat transfer to the surroundings, implying a higher ambient temperature. Furthermore, the influence turbine side is completely neglected. This is particularly noticeable at lower angular velocities. At these speeds, there is an significant heat flux from the turbine side to the compressor side of the turbocharger through the housing. This leads to an increased compressor outlet temperature, which lowers the experimentally measured isentropic efficiency. Since this effect is not captured in the simulations, the computed efficiency is considerably higher.

5. CONCLUSION

A pressure-based solver for turbomachinery application was presented using the open source library OpenFOAM. The solver utilizes the multiple reference frame approach and local time stepping. The solver was first validated against the NASA low speed centrifugal compressor. Secondly, the compressor side of a turbocharger was modelled including ported shroud and volute.

In the first case, both the impeller total pressure ratio and the adiabatic efficiency showed a very good agreement between simulation and experimental results. The deviations were less than 1 %. However, the flow angle distribution revealed minor errors compared to the measured data.

The results of the turbocharger simulation showed an overestimation of up to 10 % of the total pressure ratio for all operation points. The slopes and the choke limit were predicted well, though. The isentropic efficiency is met quite well for 147,000 min^{-1}. However, lower speeds result in a significant overestimation compared to the experimental measurements.

Reasons for the errors lie in geometry simplifications and the numerical model. Additionally, the accuracy is limited by defining the walls as adiabatic. Therefore, future work will concentrate on extending the numerical model to include the conjugate heat transfer in the housing of the turbocharger. This will improve the accuracy of the simulations and will give an insight of the temperature distribution of the housing. This information is necessary for the design of the turbocharger using new materials.

REFERENCES

1. Weiß, A., 2011. "Spannungs- und verformungsinduzierte Martensitbildungen in metastabilen austenitischen CrNi-Stählen". Habilitation, TU Bergakademie Freiberg.
2. Denton, J.D., 2010. "Some Limitations of Turbomachinery CFD". In ASME Turbo Expo 2010: Power for Land, Sea, and Air (GT2010) : GT2010-22540.
3. Heinrich, M., and Schwarze, R., 2013. "Simulation of the Compressor Stage of a Turbocharger - Validation of the Open Source Library OpenFOAM". In ASME Turbo Expo 2013: Power for Land, Sea, and Air (GT2013). GT2013-94511.
4. OpenFOAM, 2012. Version 2.1.1. http://www.openfoam.com.
5. Spalart, P., and Allmaras, S., 1994. "A One-Equation Turbulence Model for Aerodynamic Flows". *Recherche Aerospatiale*, pp. 5-12.
6. White, F. "Viscous Fluid Flow", McGraw-Hill Mechanical Engineering, United Kingdom.
7. Hathaway, M.D., Chriss, R.M., Strazisar, A.J., and Wood, J.R., 1995. "Laser Anemometer Measurements of the Three-Dimensional Rotor Fow Field in the NASA Low-Speed Centrifugal Compressor". NASA Technical Paper 3527.
8. Hathaway, M.D., Chriss, R.M., Strazisar, A.J., and Wood, J.R., 1993. "Experimental and Computational Investigation of the NASA Low-Speed Centrifugal Compressor Flow Field". NASA Technical Memorandum 4481.
9. Heinrich, M. and Schwarze, R., 2013. "Simulation of the compressor Sage of a Turbocharger - Validation of the Open Source Library OpenFoam". In ECCOMAS 2012, Vienna, Austria, 2012.
10. Baris, O., and Mendonca, F., 2011. "Automotive Turbocharger Compressor CFD and Extension Towards Incorporating Installation Effects". In ASME Turbo Expo 2011: Power for Land, Sea, and Air (GT2011). GT2011-46796.
11. Hellstrom, F., Gutmark, E., and Fuchs, L., 2012. "Large Eddy Simulation of the Unsteady Flow in a Radial Compressor Operating Near Surge". *Journal of Turbomachinery*, **134**(5), pp. 051006.1-051006.10.

Influence of the suction arrangement and geometry of the inlet port on the performance of twin screw compressors

M Pascu, M Heiyanthuduwage, S Mounoury, G Cook
Howden Compressors Ltd, UK

ABSTRACT

In the present paper, the influence of the port shape and suction arrangement on the overall performance of a twin screw compressor is investigated. Two suction variants were investigated by means of CFD for a compressor characterized by a rotor diameter of 163 mm, which allowed in-depth analyses and flow visualizations, confirmed by the experimental investigation carried out on the actual compressor.

NOMENCLATURE

CFD	computational fluid dynamics
SC	screw compressors
L/D	length of the rotors/ diameter of the rotors
GGI	general grid interface
PR	pressure ratio
ORI_PR=5	original compressor at pressure ratio 5
MOD_PR=5	modified compressor at pressure ratio 5
VF	velocity factor

1. INTRODUCTION

The trend in modern engineering in recent years has been to optimize existing technologies rather than to implement new ones. In the field of screw compressors, the manufacturing techniques have become so advanced that, nowadays, the rotor manufacturing can be done to very tight tolerances and the interlobe clearances created by the rotor meshing is in the order of few microns. Although the basic operation of such machines is well known and the analytical methods for their performance prediction are well established, only few attempts at investigating the flow in screw compressors by means of CFD can be identified in the available literature. Nevertheless, there are many advantages in considering CFD as integrated part of the design and optimization process of screw compressors. This is mostly because CFD complements the experimental and analytical efforts by providing an alternative cost-effective mean of simulating real fluid flows and substantially reduces lead times and costs of designs and production compared with an experimental based approach, Tiu and Liu [1].

Probably the most noticeable efforts in the field of numerical analysis of SC were made by Kovacevic et. al. [2] and [3], where in addition to establishing a mesh procedure specific to such flow machines, the author also explains adequate boundary calculations to encourage good convergence and minimal numerical errors.

Similar efforts were made by Sauls and Branch [4], where the commercial code ANSYS-CFX was used for the detailed analysis of a refrigeration screw compressors designed for use with R134a in air- and water-cooled chillers. Also benefiting from the mesh technique documented in, Steinmann [5] reported results from the modeling of a helical-lobed pump and a SC using ANSYS-CFX.

While the available literature includes several references on improvements of the compressor performance based on the analysis of the discharge port and discharge chamber, Mujic et. al. [6], Huagen et. al. [7] and Pascu et. al. [8], the investigation of the suction arrangement and inlet port remains fairly unexplored.

This is the area of concern for the present paper, where the influence of the port shape and suction arrangement on the overall compressor performance is investigated.

2. EXPERIMENTAL BACKGROUND

While its theoretical background has never really been the focus of research in the available literature, the shape of the suction port in a twin screw compressor is often the subject of experimental investigations.

A standard Howden SC assembly is depicted in Figure 1, where the three visible components are:
- the main casing, which hosts the two rotors in the working chamber, as well as the gas flange (suction).
- the inlet casing. The axial suction port is created at the interface between the main and the inlet casing.
- the discharge casing.

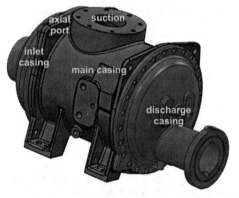

Figure 1 Compressor assembly for a standard WRV163145 compressor

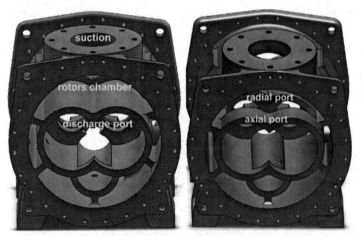

Figure 2 Compressor models – left: original (suction with axial port) right: modified (suction with axial port and machined radial port)

The general belief is that by opening the gas admission through a radial port at the suction will have a positive effect on the compressor performance, as it will reduce the throttling losses and improve the chamber filling process. In order to determine the influence of the radial suction port on the WRV163145 compressor performance, two suction scenarios were investigated: one with axial port at the compression chamber entry (the original compressor Figure 2-left), the second including the same axial port, as well as a radial port machined-off from the rotors chamber wall (modified), as shown in Figure 2-right. Both compressor models are characterized by equal rotor diameters 165 mm, L/D=1.45 and "N" rotor profile with 4/6 lobes.

Figure 3 Detail of the test bed (without compressor)

These compressors were tested on a fixed speed set up consisting of a 250hp main drive running at 1500rpm coupled to a speed increasing gearbox which in turn is connected to the compressor. The speed of the compressor is approximately 3500rpm. Lubrication of the compressor is provided by an oil system which takes oil from a pressure vessel and pumps this oil through a heat exchanger and filter to the compressor bearings, injection port and capacity control. Air is drawn into the compressor via a suction pipe and compresses with oil to exit at the discharge port. The air/oil mix is separated out, with oil returning to the system and the air exiting through the factory roof. A detail of the test rig is provided in Figure 3.

In Figure 4-top the measured mass flow is presented and it can be concluded that for small PRs both compressors are characterized by very similar behaviours. Obvious differences can be observed at higher PRs, where the modified compressor is characterized by smaller mass flow, 2-3% at PR=20.

The actual flow through a compressor is given by the difference between the theoretical flow and the leakages. With the theoretical flow identical for the two compressors, the only reason behind the measured differences lies with the leakage flow. Smaller mass flow in the modified compressor therefore implies an increase of the leakages.

When analyzing the power measurements (Figure 4-bottom), the two compressors seem to

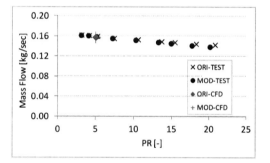

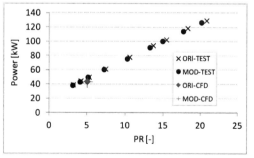

Figure 4 Experimental results

absorb similar values for small PRs, while at higher PRs the modified compressor consumes slightly more power. In conclusion, contrary to the initial assumption, this compressor has not benefited in any way from the introduction of the radial port, as test results have revealed no improvement in the compressor performance at smaller pressure ratios and a slight deterioration at higher pressure ratios, in terms of both power and flow. This is most likely due to an increase in the gas leakages in the modified compressor and a numerical analysis will be carried out next to reveal the source of these additional leaks.

3. NUMERICAL MODELS

3.1 Numerical mesh

Block structured grids are preferred for the mesh generation of complex geometries. The grid generation process is very much simplified as the domain is subdivided into a number of simpler blocks. For screw compressors, the preferred topologies include polyhedral and O-grids.

During the mesh generation process, the spatial domain of a screw compressor is replaced by a grid with discrete finite volumes and a composite grid, made of several structured and unstructured grid blocks patched together and based on a single boundary fitted coordinate system. The number of these volumes depends on the domain dimension and the required accuracy. The domain consists of four sub-domains, two of which refer to the compressor casing and the other two to the male and female rotors. The critical sub-domains in this setup are the two rotors as they contain the working chamber as well as the clearances and leakage paths (radial, axial, interlobe and blow-hole area).

Generating the grids for these domains is by far the most challenging part of the entire meshing procedure, as both micro- and macro- scales elements have to be solved. In this case, a technique dedicated to screw compressor rotors was employed, as described by Kovacević [3], included in SCORG© (Screw COmpressor Rotor Geometry grid generator). This procedure is fully explained in several publications included in the reference list and therefore, will not be repeated here. A simplified representation of the rotors mesh (cross-sectional view) is presented in Figure 5. This technique resulted in 143,000 nodes for the male and approximately 140,000 for the female, of structured mesh.

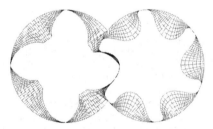

Figure 5 Mesh over rotors cross-section before 3D interpolation

The numerical model includes three more domains: the inlet casing, the main casing, which in turn includes the suction into the compressor and the discharge casing, including the discharge port. ANSYS ICEM v14.5 was used for the mesh generation process and special mesh refinement techniques were employed for sensitive flow areas, i.e. the interfaces with the rotors).

Figure 6 Representative mesh of the working chamber

The overall mesh statistics typically used for the compressor simulations are:
- Main casing approximately 82K nodes
- Discharge casing approximately 86K
- Inlet approximately 176K nodes

3.2 Adaptive meshing and timestep calculation

The compression process in a screw compressor occurs due to the rotation of the male rotor at a preset motor speed, which in turn drives the female rotor. To be able to accurately simulate this scenario, an adaptive mesh, which is modified at the beginning of each timestep, is required. In the previous paragraph, the procedure for generating the mesh for both the rotor and the casing domains, was presented. There is no need for interpolation of the casing mesh and therefore, it is kept constant through the simulation process.

Because of the rotation of the two rotors, the shape of the working chamber is changed constantly, as the gas particles enter the domain through the suction port, are compressed within the rotors and then discharged at the desired pressure. An adaptive meshing technique is utilized to capture all the changes which occur within the working chamber during the compression process.

The number of time changes required by the rotors mesh is 120 for the full rotation of the male rotor, with the number of nodes kept constant across the timesteps. This is calculated as:

$$No\ of\ grids = No\ of\ elements\ per\ lobe \cdot No\ of\ lobes\,(male) = 120$$

The rotor mesh is replaced with the updated lobe position at time intervals equal to the calculated timestep:

$$timestep = \frac{1}{n[rpm] \cdot No\ of\ grids} \qquad (1)$$

3.3 Boundary conditions

The numerical model includes the stationary domain with the major casing components (inlet, main and discharge) and the rotating domain, depicted by the two rotors, see Figure 7. Various interfaces were applied to each of these domains to ensure the flow transition between the different domains. All these interfaces were considered to be General Grid Interfaces (GGI). GGI connections refer to the class of grid connections where the grid on either side of the two connected surfaces does not match.

In general, GGI connections permit non-matching of node location, element type, surface extent, surface shape and even non-matching of the flow physics across the connection.

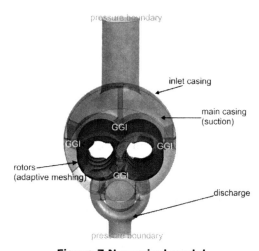

Figure 7 Numerical model

Both the suction and the discharge were simulated by pressure boundary conditions. The pressure boundary conditions are similar to the inlet or outlet boundaries, firstly because they couple pressure and velocity directly and secondly because for all equations, apart from the momentum equation, the boundary properties are calculated from the velocity. This procedure may cause instability in the compressor cycle especially when the flow changes its direction at the boundary. This is compensated by adding a boundary domain, in which an amount of mass is added or subtracted to maintain a constant pressure, is natural and gives a stable and relatively fast solution, [3]. Both compressor models were simulated for air ideal gas and ran for twelve full rotations (1440 timesteps), with the first six allowing a natural pressure build-up in the system and used as initial conditions for the last six, where wall roughness was added to the rotors on top of the previous settings.

The following operating conditions were simulated:

Table 1 Boundary conditions for the numerical models

P_1=1 bar	P_r=5	N=3500 rpm	V_i=5.0	T_1=18°C
				T_2=80°C

4. NUMERICAL RESULTS

The numerical simulations in the present paper were carried out in an attempt to reveal the flow phenomena behind the slight underperformance of the compressor model with a combined axial and radial suction port, as this was expected to improve the overall compressor performance.

Firstly, the comparison between the experimental and numerical results will be carried out, in order to validate the proposed CFD setup. For PR=5 a good agreement between the experimental results and the CFD predictions was calculated, as indicated in Figure 4.

The CFD models delivered very similar values in terms of power and mass flow, with no differences from one compressor model to the other. The absolute differences in the numerical and experimental mass flows are of approximately 4% and about 10% for the consumed power. The simulations were carried out with no consideration for the oil injection.

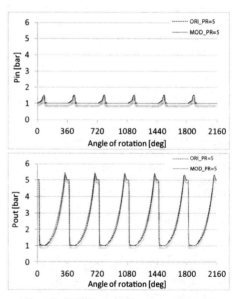

Figure 8 CFD results: pressure at the rotor-suction and rotor-discharge interfaces

Other values of PR are not included in this work, due to the extensive computational times, but simulations for PR=10 and PR=15 are currently in progress. It is assumed that due to the good validation of the numerical strategy, the rest of the experimental results will be repeatable through CFD analysis.

Next, the differences between the two compressor models will be assessed and comparative pressure plots will be analyzed first. A probe point was placed on one grid point of the male rotor at the axial interface with the suction (P_{in}) and discharge ports (P_{out}), respectively. The results are presented in Figure 8 and very little difference between the behaviour of the two compressors can be observed.

On the suction port side, both compressors are characterized by a stable pressure variation dictated by the boundary value of 1 bar. A peak to approximately 1.5 bar can be observed at the end/ start of each compression cycle (marked every 360 deg), when the admission into the compression chamber is closed and the system "sees" the higher pressure end of the working chamber at discharge. In a similar fashion, the second plot in Figure 8 captures the variation of the pressure on the discharge side during a full compression cycle, from the atmospheric pressure value at the suction to the 5 bar prescribed by the boundary condition at the discharge. Again, very little difference between the two compressors can be observed.

So far, the pressure plots have revealed no differences between the two compressors and thus, the flow field in the two main casings will be analyzed next. A detailed flow analysis was carried out for both compressor models under each set of the investigated parameters. In a first attempt to understand the flow mechanism in the two compressors models, the velocity vectors are plotted in the overall numerical model depicted in Figure 7.

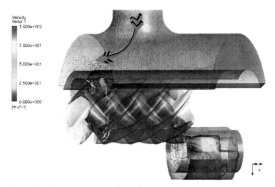

Figure 9 Velocity vectors in the original compressor model

The scale of the plots starts at 0 m/s and has a maximum of 100 m/s. Figure 9 depicts the velocity vectors in the original compressor; from suction (placed on the main casing) all the way to the axial port (rotors interface between the rotors and the inlet casing), the flow is steady and characterized by an average velocity of 6 m/s.

Figure 10 shows the same vector plot in the modified compressor, keeping the scale magnitude. It can be readily observed that the flow in this compressor is characterized by two streams: one, the incoming flow from the suction to the axial port; the second one (depicted in red because the vectors are outwith the chosen scale) is highly disturbed flow coming back into the main casing through the radial port. In this case, the average velocity on the rotors interface (from the inlet casing) is about 18 m/s.

In order to accentuate the differences between the flow paths in the suction casings of the compressors, two further snapshots of the same velocity vectors were taken from the front view. The still images are captured at the end of the rotation cycle, but are representative for any intermediate timestep after the internal pressure build-up was achieved.

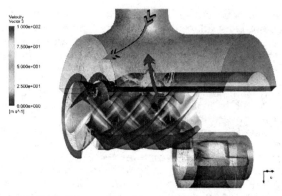

Figure 10 Velocity vectors in modified compressor model

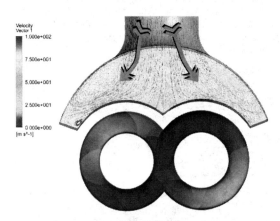

Figure 11 Velocity vectors in the original suction casing (front view)

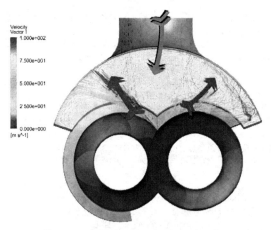

Figure 12 Velocity vectors in the modified suction casing (front view)

While Figure 11 reveals no other interferences to the main stream, Figure 12 shows the highly disturbed flow field, where the main flow stream encounters high-velocity flow returning from the radial interface with the rotors. Visually, it appears as if the high-velocity leakage through the rotors acts as a resistance to the main stream, as part of the incoming flow is redirected back into the suction area.

This is due to the atmospheric flow coming in direct contact with portions of the working chamber exposed to high pressure fields and sets the scene for increasing gas leakages. The CFD simulation was carried out on a single-phase fluid case with air ideal gas, but in reality, this compressor is an oil-injected machine, and a multiphase simulation would have revealed a combination of working gas and oil returning from the working chamber back into suction through the radial opening.

Figure 13 shows contour plots of the velocity on the axial port, both plotted on the same scale with a maximum at 20 m/s. While the original compressor is characterized by a fairly uniform velocity distribution, which averages out at 6 m/s, the axial port in the modified compressor averages at 18 m/s and peaks with a maximum of 33 m/s. This yields to a non-uniform high-velocity flow field at the axial interface with the rotors and is the main source for increased gas recirculation and therefore, increased leakages in the modified compressor.

Figure 13 Velocity contour plot on rotors interface: left-original suction casing; right-modified suction casing

CONCLUSIONS

The shape of the suction port in a twin screw compressor is often the subject of experimental investigations and the general belief is that by opening the gas admission through a radial port at the suction will have a positive effect on the compressor performance, as it would improve the filling process by reducing the throttling losses. This hypothesis was tested against a standard WRV163145 compressor and two suction scenarios were analyzed, both experimentally and numerically: one with axial port at the compression chamber entry (the original compressor), the second including the same axial port, as well as a radial port machined-off from the inner casing wall (the modified compressor).

Tests on the compressor models did not reveal any improvement in the modified casing and furthermore, for higher pressure ratios, a slight deterioration in the performance of this compressor was observed in terms of power and flow.

CFD simulations for PR=5 delivered results well validated by the experimental curves, i.e. no substantial difference between the two compressors at this pressure were noticeable.

The visual analysis at PR=5 for the two compressor models allowed the detailed analysis of the flow path in the main casing. It was concluded that the flow field in the original compressor was steady and uniform from suction all the way to the axial port. However, the flow in the modified compressor had two streams: one, the incoming flow from the suction to the axial port; the second one was highly disturbed high-velocity flow coming back into the main casing through the radial port. In this case, it appeared as if the high-velocity leakage through the rotors acted as a resistance to the main stream. This was due to the atmospheric flow coming in direct contact with portions of the working chamber exposed to high pressure fields and set the scene for increased gas leakages.

The CFD simulation was carried out on a single-phase fluid case with air ideal gas, but in reality, this compressor is an oil-injected machine, and a multiphase simulation would have revealed a combination of working gas and oil returning from the working chamber back into suction through the radial opening. It is expected that for the higher pressure ratios, this instability will increase and cause a further increase in the leakage flow, resulting thus in the deterioration of the compressor performance.

The combined experimental and numerical investigation revealed that there is no improvement to be expected by adding a radial port to the suction arrangement of the investigated SC and furthermore, at higher pressure ratios, a slight deterioration of the compressor performance can be expected. It is therefore recommended that, for this twin SC, the suction arrangement includes only an axial port. Though only one point on the PR axis was simulated by means of CFD, higher PRs are underway. However, due to the good agreement at PR=5, a full validation of the experimental curves is expected. The results presented in this paper emphasize the importance of including CFD in the early stages of the design process of SC, as they are very complex flow systems and the experimental analysis alone would have not revealed the full effect of introducing a radial suction port in the design of the WRV163145 compressor. With a CFD procedure fully validated such investigations will be carried out for all design modifications on other Howden compressor sizes/ranges. This will facilitate in-depth flow analyses and virtual DOE (design of experiments) ahead of the prototyping stage and will provide full compressor performance predictions where test results are not always available.

BIBLIOGRAPHY

[1] Tu, J., Yeoh, G. H., and Liu, C., 2008, "Computational fluid dynamics", Butterworth – Heinemann
[2] Kovacević, A., Stosic, N., and Smith, I., "CFD analysis of screw compressor performance", Centre for positive displacement compressor technology, City University, London
[3] Kovacević, A., Stosic, N., and Smith, I., "Screw compressors. Three dimensional computational fluid dynamics and solid fluid interaction", Springer Verlag, 2007
[4] Sauls, J., and Branch, S., 2009, "CFD analysis of refrigeration screw compressors", Ingersoll Rand
[5] Steinmann, A., 2006, "Numerical simulation of fluid flow in screw machines with moving mesh techniques in ANSYS CFX", Schraubenmaschinen 2006; VDI Verlag GmbH
[6] Mujic, E., Kovacević, A., Stosic, N., and Smith, I., "The influence of port shape on gas pulsations in a screw compressor discharge chamber", Centre for positive displacement compressor technology, City University, London
[7] Huagen, W., Xing, Z., Peng, X., and Shu, P., "Simulation of discharge pressure pulsation within twin screw compressors", Journal of Power and Energy, IMechE, 2004
[8] Pascu M., Kovacevic A., Udo N. (2012) Performance Optimization of Screw Compressors Based on Numerical Investigation of the Flow Behaviour in the Discharge Chamber. Proc Int Compressor Conf at Purdue, Purdue, pp. 1145
[9] Stosic, N., Smith, I., and Kovacević, A., "Screw compressors. Mathematical modelling and performance calculation", Springer Verlag, 2005
[10] Menter, F.R., "Zonal two-equation turbulence models for aerodynamic flows", AIAA Paper 96-2906, 1993
[11] Jovanović, J., "The statistical dynamics of turbulence", Springer Verlag, 2004
[12] Guerrato, D., Nouri, J. M., Stosic, N., Arcoumanis, C., and Smith, I., "Flow development in the discharge chamber of a screw compressor", Centre for positive displacement compressor technology, City University, London
[13] http://www.ansys.com/Products/Simulation+Technology/Fluid+Dynamics/ANSYS+CFX

MODELLING

A new dynamic heat pump simulation model with variable speed compressors under frosting conditions

N Park, J Shin, B Chung
SAC team, HAE Research Center, LG Electronics, Korea

ABSTRACT

A dynamic heat pump simulation model under frosting condition is proposed. Toward this end, a simple frosting model is proposed based on the perfect analogy between heat and mass transfer including heat conduction. The proposed frosting model is first validated against experimental data for the flat plate boundary layer at various temperatures and relative humidity. Then, the model is incorporated into a quasi-dynamic heat pump simulation which adopts segment-by-segment local heat exchanger model and variable speed compressors. It is shown that the proposed model can naturally represent non-uniform frosting on the heat exchanger surface, and the interaction between frosting and heat pump cycle.

1. INTRODUCTION

The present paper aims at proposing a simple and accurate air-source heat pump cycle model under frosting conditions. The impact of frosting on the performance of the evaporator mostly comes from the blockage of air passage as shown in Figure 1, which significantly reduces the front air velocity and, thus, the evaporating capacity. Under this situation, heat pump adjusts the evaporating temperature to secure suction superheat by throttling expansion valves. This may further accelerate frosting on the heat exchanger and the air passage blockage. Such a nonlinear interaction between heat pump and frosting has not yet been fully explored, which is the topic of the present paper.

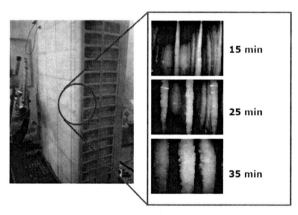

Figure 1 Typical frosting on the surface of outdoor unit heat exchanger

Another important feature to cover in this paper is the non-uniform frosting on the heat exchanger surface, which is caused by mal-distribution of refrigerant and/or non-uniform air velocity profiles. This requires the combination of frosting and dynamic heat pump simulation with localized heat exchanger model, as summarized in Figure 2 as a flow chart.

The present paper is organized as follows. In section 2, a quasi-dynamic heat pump model adopting variable-speed compressors is developed with emphasis given on evaporator modeling including refrigerant mal-distribution. In section 3, a new frosting model on a flat plate is developed and validated against experimental data. In section 4, proposed frost model is incorporated into the heat pump model with modification of heat transfer correlations and fan velocity to account for the frost effect. Then, the model runs to mimic chamber experiments under standard frosting conditions to see the interaction between heat pump cycle and frosting. Also given is the visualization of local frosting on the heat exchanger. Important conclusions and future works are summarized in the last section.

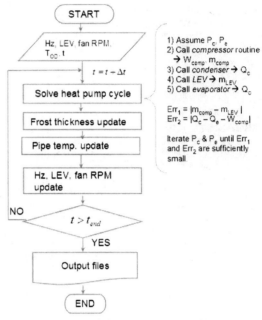

Figure 2 Flow chart of heat pump-frosting simulation

2. HEAT PUMP SIMULATION MODEL

Heat pump considered in the present study has 32 kW nominal heating capacity driven by one variable-speed and one fixed-speed scroll compressors. As shown in Figure 3, the outdoor unit has a C-shaped heat exchanger with top discharge of incoming air. Two 16 kW, cassette type indoor units are installed as the condensers. Modeling each component is briefly discussed in the following subsections.

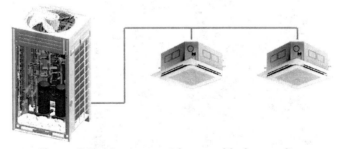

Figure 3 Heat pump outdoor and indoor units considered in the present study

2.1 Compressor model

Models of refrigerant mass flow rate and power consumptions from both fixed-speed and variable speed compressors are as follows:

$$\dot{m}_{ref} = V_{dis} \rho_{ref,s} Hz \cdot \eta_{vol} / 10^6 \; (kg/s),$$

$$W_{comp} = \frac{\dot{m}_{ref}}{\eta_{comp}} \left[\frac{k}{k-1}\right] \frac{P_s}{\rho_{ref,s}} \left[\left(\frac{P_d}{P_s}\right)^{k-1/k} - 1\right], \qquad (1)$$

where V_{dis}, $\rho_{ref,s}$, Hz, P_d, P_s and k are, respectively, compressor displacement volume (cc/rev), density at suction (kg/m³), compressor rotation speed in Hertz, discharge pressure (kPa), suction pressure (kPa), and polytropic constant. η_{vol} and η_{comp} are volumetric and total compressor efficiencies, which are closed by regression fits in terms of pressures and the compressor speeds. Correlations between the predictions and available experimental model are shown in Figure 4, where reference mass flow rate and power consumption are, respectively, 0.1 kg/s and 8 kW. As shown, good agreements within 10% error are observed for both refrigerant mass flow rate and compressor power, although some deviations are observed at some points mainly due to unsteady nature of collected experimental data during initial transient periods.

2.2 Evaporator model

The evaporator adopted is a fin and tube heat exchanger which consists of 28-passed refrigerant pipes in two rows with louvered fins. Since the outdoor unit has top discharge air flow, there should be non-uniform air mass flux distribution along the vertical direction, which is modeled as

$$\dot{m}_{air}(j, f_{RPM}) = \left(a_0 + a_1 z_j + a_2 z_j^2\right)\left(b_0 + b_1 f_{RPM}\right), \; z_j = \left(\frac{j-1}{N_{pass}-1}\right), \qquad (2)$$

where j is the index of heat exchanger pass $(1,...,N_{pass})$ and f_{RPM} is the fan rotation speed in RPM. Coefficients a's and b's are determined from the experimental data. The change of air mass flux due to frosting will be addressed in Section 4.

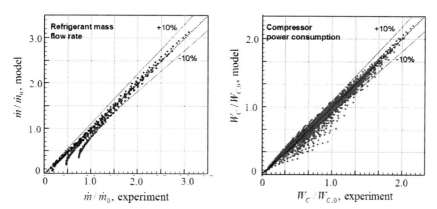

Figure 4 Validation of adopted variable speed compressor model for mass flow rate and power consumption against experimental data at various operating conditions

Due to non-uniform velocity profile, exit refrigerant inside each pipe has different level of superheat or exit quality. Toward minimizing this non-uniformity, capillary tubes attached to distributors with different lengths control the mass flow rates. Being a 28-degree-of-freedom problem, numerical solution of this intentional mal-distribution problem during the entire cycle simulation is highly time consuming job.

In order to make the simulation tractable, a linear mass distribution model $\dot{m}_j = cz_j + d$ is considered, where z_j is the non-dimensional pass number defined in Eq. (2) and $\dot{m}_j$ is the mass flow rate at pass index j. Coefficients c and d are chosen such that pressure loss differences among passes are minimized in the least square sense:

$$\frac{\partial J}{\partial c} = 0, \quad J = \sum_{j=1}^{N_{pass}} \left(\Delta P_j - \overline{\Delta P} \right)^2,$$

$$\sum_{j=1}^{N_{pass}} \left(cz_j + d \right) = \dot{m}_R, \quad (3)$$

where $\dot{m}_R$ is total mass flow rate, $\overline{\Delta P}$ is the mean pressure loss, and ΔP_j is the pressure loss at pass j. Pressure losses are modeled by

$$\Delta P = \frac{f_{lo} \cdot \rho \cdot v_f^2 L}{2 \cdot D} \Phi_{lo}^2,$$

$$\Phi_{lo}^2 = (1-x)^2 + 2.87 x^2 \left(\frac{P}{P_c} \right)^{-1} + 1.68 x^{0.8} (1-x)^{0.25} \left(\frac{P}{P_c} \right)^{-1.64}, \quad (4)$$

where x is the quality, P_c the critical pressure, f_{lo} the friction factor, v_f the refrigerant velocity, and L is the pipe length. Now the mass flux distribution problem is only single degree of freedom problem. Figure 5 shows the pressure losses along the passes predicted by the proposed linear model together with those from the uniform mass distribution assumption. It is shown that ΔP_j from the linear model has smaller variation, which means that the present linear model will predict more realistic refrigerant distribution.

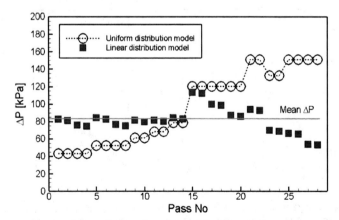

Figure 5 Prediction of pressure loss at each pass of evaporator predicted by the proposed linear distribution model (filled square) and the uniform distribution assumption (dotted line with hollow circles)

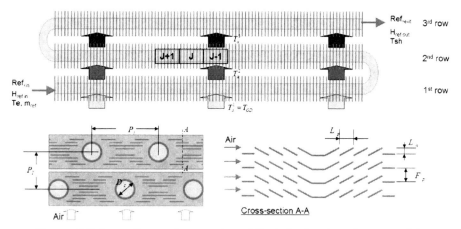

Figure 6 Schematics and geometry of fin-tube heat exchanger with louvered fins

Heat transfer inside each refrigerant pass is computed by ε-NTU method after dividing it into multiple control volumes as shown in Figure 6, yielding

$$\delta q_J = \varepsilon \left(\dot{m}C_p\right)_{J,\min} \left(T_a - T_{ref,J-1/2}\right) = \dot{m}_{ref}\left(h_{J+1/2} - h_{J-1/2}\right) \qquad (5)$$

for each control volume, where δq_J is the heat transfer at control volume J, h's denotes specific enthalpies at cell interface, and ε is the effectiveness given by

$$\varepsilon = \begin{cases} 1-\exp(-NTU), & \text{for two-phase region} \\ 1-\exp\left[\dfrac{\exp(-NTU^{0.78}C_r)-1}{NTU^{-0.22}C_r}\right] & \text{otherwise} \end{cases} \qquad (6)$$

Here, C_r and NTU are defined as

$$C_r = \frac{\left(\dot{m}C_p\right)_{\min}}{\left(\dot{m}C_p\right)_{\max}}, \quad \left(\dot{m}C_p\right)_{\min} = \begin{cases} \dot{m}_{air}C_{p,air}, & \text{for two-phase region} \\ \min\left[\dot{m}_{air}C_{p,air}, \dot{m}_{ref}C_{p,ref}\right] & \text{otherwise} \end{cases}, \qquad (7)$$

$$NTU = \frac{(UA)_J}{\left(\dot{m}C_p\right)_{J,\min}}, \quad \frac{1}{UA_J} = \frac{1}{h_r \pi d_i \delta_J} + \frac{\ln(d_o/d_i)}{2\pi k_p \delta_J} + \frac{1}{\eta h_a A_{a,J}}, \qquad (8)$$

where η is the fin efficiency modeled by Schmidt (1949), and J-factor model of the form

$$J = 14.3117 \, \mathrm{Re}_{D_c}^{J_1} \left(\frac{F_p}{D_c}\right)^{J_2} \left(\frac{L_h}{L_p}\right)^{J_3} \left(\frac{F_p}{P_l}\right)^{J_3} \left(\frac{P_l}{P_t}\right)^{-1.724}, \qquad (9)$$

is used for air-side heat transfer correlation (Chang and Wang 1997). Here Re_{Dc} is the Reynolds number based on pipe diameter and J's are parameters defined in Chang and Wang (1997). See Figure 6 for geometrical parameters of louver fin. For refrigerant side heat transfer, Dittus-Boelter model (McAdams 1942) and Gungor & Winterton model (1987) are used for single and two phase regions, respectively. The change of heat transfer correlation and fin efficiency due to frosting is discussed in Section 4.

Developed evaporator model is validated against experimental data as shown in Figure 7 yielding a reasonable agreement within 10% error for most operating conditions.

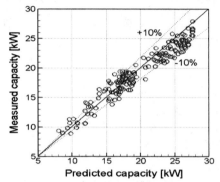

Figure 7 Validation of evaporator model in terms of capacity with experimental data at various operation conditions

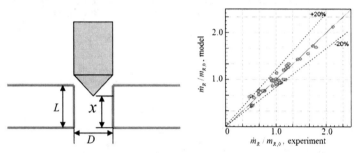

Figure 8 Geometrical configuration of electronic expansion valve (left) and validation of adopted model against experimental data (right)

2.3 Condenser model
As the condenser, cassette indoor units are considered as shown in Figure 3. Each indoor unit has a fin-tube heat exchanger which consists of 11-passed refrigerant pipes in two rows. We assume uniform refrigerant distribution and constant air mass flux at 33 m³/min for each indoor unit.

Similarly to evaporator model, ε-NTU based approach is used in a section-by-section manner for each refrigerant pipe. For air side heat transfer, J-factor model (Chang and Wang, 1997) is again used. For refrigerant side heat transfer, the correlation due to Dobson and Chatto (1998) is used for two-phase region, while Dittus-Boelter model (McAdams 1942) is used for single phase regions.

2.4 Electronic expansion valve model
The main role of electronic expansion valve (denoted as EEV hereinafter) is to control refrigerant mass flow rate and to secure desired suction superheat under various load conditions. Refrigerant mass flow rate at the exit of EEV is modeled as

$$\dot{m}_{LEV} = C_D A_{th} \sqrt{2\rho_l \left(P_{in} - P_{exit}\right)} \approx C_D A_{th} \sqrt{2\rho_l \left(P_{cond} - P_{evap}\right)}, \tag{10}$$

$$A_{th} = \frac{\pi D^2}{4L^2}\left[L^2 - (x-L)^2\right], \quad x = \frac{L}{PLS_{Max} - PLS_{min}}(PLS - PLS_{min}),$$

where PLS denotes the pulse of EEV, and 2000 pulse model is considered in the present study. See Figure 8 for geometry of EEV. C_D is an empirical constant, and $C_D = 1.43$ is used. Figure 8 also shows the validation of EEV model against existing experimental data. As shown the agreement between the experiment and the model prediction is acceptable.

2.5 Control logic implementation and cycle simulation

The heat pump model considered has three control parameters, or the speed of compressor, EEV step, and fan rotation speed. The speed of the inverter compressor is controlled to meet desired target condensing pressure $p_{c,targ}$, and the fixed-speed compressor is turned on when the inverter compressor alone fails to meet the target. On the other hand, EEV step varies to meet suction superheat within 3~5°C, and fan RPM varies between 250 to 1,200 RPM to keep desired gap between evaporating and outdoor temperature.

The actual heat pump system, or variable-refrigerant-flow (VRF) type system air conditioner, has highly complicated algorithm mainly based on Fuzzy control to meet these requirements as well as other ones including the oil recovery and fault detection, which are not considered here. In this study, a simpler P-D controller is adopted, which takes the form for the inverter speed, for example,

$$e = P_c - P_{c,\text{targ}},$$
$$dHz = k_p e + k_d \Delta t \frac{de}{dt}, \qquad (11)$$
$$Hz^{new} = Hz^{old} + dHz,$$

where k_p and k_d are gains chosen to closely mimic the actual Fuzzy controller. In addition to (11), limitations on compressor speed are imposed to manage the discharge temperature, compression ratio, high and low pressure, and the operating current within acceptable ranges. P-D controllers of the same form are adopted for EEV pulse and fan speed control as well.

For given control parameters, condensing and evaporating pressures are assumed. Then, refrigerant mass flow and compressor work are computed from Eq. (1). Also computed is mass flow from EEV by Eq. (10). Condenser and evaporator models are solved to compute capacities as well as exit refrigerant states. Finally, heat balance and mass flow rates from compressors and EEV are checked. These errors are driven to zero by nonlinear iteration for condensing and evaporating pressures. Thus, this simulation can be referred to as quasi-dynamic one in the sense that the system remains steady for given load condition and control parameters. This procedure is summarized in the shaded bracket of the flow chart in Figure 2.

2.6 Heat pump model validation

Developed heat pump model is used to compute heating cycle at standard outdoor temperature, or 7°C (DB) / 6°C (WB), and low temperature, or -10°C, conditions. As shown in Figure 9, simulated cycle data closely mimics actual cycle even during initial phase. Sharp changes of inverter speed at around 7 minute and 25 minute correspond to turning on and off of the constant speed compressor.

Qualitative comparison with experimental data shows that the present model underestimates heating capacity by 11~13%, while COP predictions are within 3% error. Reminding the level of prediction accuracy for all cycle components as shown in the previous subsections, this level of error is not surprising. Note also that experimental data used for the comparison represent the best performance from manually tuned cycle, while the simulation results are from automatic runs with P-D controller mentioned earlier. Thus, no further tuning is pursued here to closely match the experimental data.

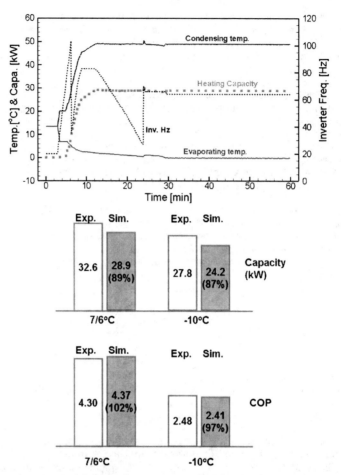

Figure 9 Cycle simulation data at standard heating condition (upper) and the validation of simulation against experimental data (lower)

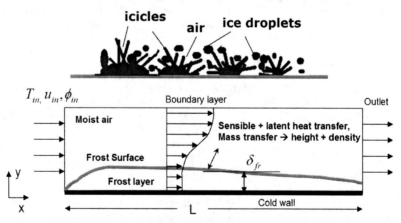

Figure 10 Sketch of actual frost formation on a flat plate (upper) and its macroscopic modeling (lower)

3. FROSTING MODEL DEVELOPMENT

In this section, we propose a simple and novel frosting model on flat plate boundary layer, which will be combined with the heat pump model developed in the previous section.

3.1 Frosting model based on perfect analogy

Although actual frosting consists of icicles, ice droplets and air in complex ways, we treat it as continuum with varying height and density as a result of heat and mass transfer at the interface as shown in Figure 10. Thus, main focus is given on predicting accurate frost thickness and density. Combining thickening and densification effect, one can compute the frost mass evolution:

$$\frac{dm_{fr}}{dt} = \rho_f A_T \frac{d\delta_f}{dt} + \delta_f A_T \frac{d\rho_f}{dt} = \dot{m}_{air}(w_{in} - w_{out}), \tag{12}$$

where m_{fr}, δ_f, ρ_f, A_T, w_{in} and w_{out} are frost mass, thickness, density, plate area, inlet and exit humidity ratio, respectively. Arranging the above equation leads to the following:

$$\frac{d\delta_f}{dt} = \frac{\dot{m}_{air}(w_{in} - w_{out})}{A_T\left[\rho_f + \delta_f \frac{d\rho_f}{dT_{fs}}\frac{dT_{fs}}{d\delta_f}\right]}, \tag{13}$$

where T_{fs} is the frost surface temperature, and the chain rule $d\rho_f/dt = (d\rho_f/dT_{fs})(dT_{fs}/d\delta_f)(d\delta_f/dt)$ is applied. It is common approach that the numerator of (13) is replaced by $h_m(w_{in} - w_s)$ (see, e.g., Kandula 2011) by exploiting that the frost surface is saturated, where h_m is the mass transfer coefficient. Aside from the criticism given by Na and Webb (2004) that the frost surface is super-saturated, humidity at frost surface is uncertain quantity as one can readily imagine by the actual frost surface shown in Figure 10. Instead, we rely on the surface temperature and humidity to derive humidity at the exit of the flat plate:

$$w_{out} = w_{p,sat} - (w_{p,sat} - w_{in})\exp\left(-\frac{UA}{\dot{m}_a C_{p,a} Le^{2/3}}\right), \quad UA_T = \left(\frac{1}{h_c A_T} + \frac{\delta_f}{k_f A_T}\right)^{-1}, \tag{14}$$

where $w_{p,sat}$ is the saturated humidity ratio at the plate surface, and Le is the Lewis number assumed to be 1.0. Here, h_c and k_f denote convective heat transfer coefficient and thermal conductivity of frost to be defined later in this section. Note that Eq. (14) is strictly valid at the initial stage with thin frost thickness due to the saturation assumption at the surface. Obviously, Eq. (14) has the same form as the heat transfer with the plate surface in the sense that even the effect of conduction is included in the mass transfer. Thus, this model can be referred to as the `perfect analogy model' between heat and mass transfer.

Frost surface temperature required to close Eq. (13) obeys the following governing equation and boundary conditions (Kandula 2011):

$$k_f \frac{d^2 T}{dy^2} = -L_{sv}\left(\frac{d\rho_f}{dt}\right),$$

$$T(0,t) = T_p, \quad k_f \frac{dT}{dy}\bigg|_{\delta_f} = h_c(T_a - T_f) + \rho_f L_{sv}\frac{d\delta_f}{dt} \tag{15}$$

Where L_{sv} is the latent heat of sublimation. From the solution of Eq. (15), the frost surface temperature is given as

$$T_{fs} = \frac{T_p + \dfrac{L_{sv}}{2k_f}\left(\dot{m}_{fr} + \rho_f \dfrac{d\delta_f}{dt}\right) + \dfrac{\delta_f}{k_f}h_c T_a}{1 + \dfrac{h_c}{k_f}\delta_f}. \tag{16}$$

Using (14) and (16), Eq. (13) or the frost thickness evolution is readily computed once the frost density and thermal conductivity are known, which are defined in the next subsection.

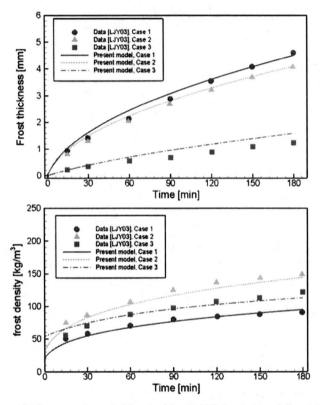

Figure 11 Temporal evolutions of frost thickness and frost density on a flat plate. Lines denote results from the proposed model and symbols are the experimental data by Lee *et al.* (2003)

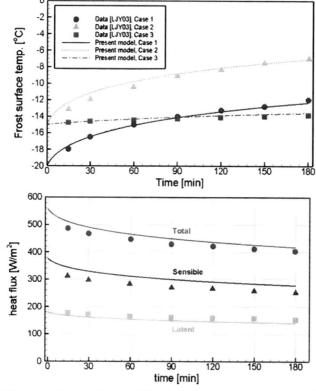

Figure 12 Temporal evolutions of frost surface temperature and heat flux on a flat plate for case 2. Lines denote results from the proposed model and the symbols denote experimental data by Lee et al. (2003)

3.2 Frost properties model
In the present study, constant frost thermal conductivity is assumed, which is approximated by the model due to Lee et al. (1997):

$$k_f = 0.132 + 3.13 \times 10^{-4} \rho_f + 1.01797 \times 10^{-4} \rho_f^2 \quad (W/m \cdot K). \tag{17}$$

For the frost density, we propose the following modification of da Silva et al. (2011) model:

Table 1 Summary of experimental conditions by Lee et al. (2003)

	T_{air} (°C)	Humidity, w	U (m/s)	T_p (°C)
Case 1	10	0.00528	1.75	-20
Case 2	15	0.00633	2.5	-15
Case 3	5	0.00322	1	-15

$$\rho_f = \rho_{f,ini} + \left(\rho_{f,ast} - \rho_{f,ini}\right)\left(1 - \exp\left(-\frac{\delta_f}{\delta_{ref}}\right)\right) \; (kg/m^3),$$

$$\rho_{f,ast} = 494 \exp(0.11 T_{fs} - 0.06 T_{dew}), \rho_{f,ini} = 0.5 \rho_{f,ast}, \delta_{ref} = 1mm, \tag{18}$$

Where T_{dew} is the dew point, and $\rho_{f,ast}$ is the original da Silva et al. (2011) model as the asymptotic solution. The model is designed to approach the asymptotic value when the frost is thicker than the reference value. The modification of the model aims at mimicking the solution of elliptic diffusion equation, which was introduced by previous studies (Lee et al. 1997; Na and Webb 2004).

3.3 Validation of proposed frosting model

The proposed model is applied to laminar flat plate problem and compared with experimental data by Lee et al. (2003), in terms of frost thickness, density, surface temperature and heat flux (Figures 11 and 12). Details on the experimental conditions are summarized in Table 1, where U is the airflow velocity, and T_p is the surface temperature. The proposed model shows an excellent agreement with experimental data for frost thickness, density, surface temperature and heat flux under various conditions.

Next, another flat plate frosting problem due to Hermes et al. (2009) is tackled with the proposed model as shown in Figure 13 with variable surface temperatures. The present model predicts frost thickness and density reasonable well, although some over-prediction of frost thickness up to 20% is observed. Nevertheless, the above results are sufficient to convince the overall validity of the proposed frosting model based on the perfect analogy.

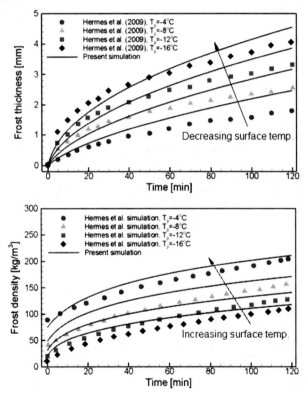

Figure 13 Temporal evolutions of frost thickness and frost density on a flat plate. Lines denote results from the proposed model and symbols denote the experimental data by Hermes et al. (2009). For all cases air temperature is 16°C, relative humidity is 80%, and frontal velocity is 0.7 m/s

4. HEAT PUMP SIMULATION UNDER FROSTING CONDITION

In this section, developed heat pump model in Section 2 ad frosting model in Section 3 are consolidated, and applied to heating operation under frosting conditions.

4.1 Fin efficiency and air mass flow
The existence of frost layer changes the fin efficiency. Here, we adopt the one-term approximation of the analytic solution by Xia and Jacobi (2012):

$$\eta = \eta_f \frac{A_{fin}}{A_{tot}} + \left(\frac{k_{fr}/\delta_{fr}}{h_{air}+k_{fr}/\delta_{fr}}\right)\frac{A_{tot}-A_{fin}}{A_{tot}}, \qquad (19)$$

$$\eta_f = \frac{2\lambda}{h_{air}H_{fin}\delta_{fr}}\tanh\left(\frac{\lambda H_{fin}}{2\delta_{fr}}\right)(k_{fin}t_{fin}+k_{fr}\delta_f),$$

$$\lambda = \delta_f\sqrt{\frac{h_{air}}{k_{fin}t_{fin}+\delta_f(k_{fin}h_{air}t_{fin}/k_{fr}+k_{fr})}}. \qquad (20)$$

where t_{fin} and H_{fin} are, respectively, fin thickness and fin height. It is shown, however, that increased thermal resistance due to frosting has only minor impact on the heat exchanger capacity (Ye and Lee, 2012). It is the air passage blockage that plays the most decisive role in the capacity degradation. In the present study, a simple working fit (Mader and Thybo 2012) is used to model this effect:

$$\frac{\dot{m}_{air}}{\dot{m}_{air,0}} = 1 - 0.06(1-\tilde{A}) - 0.94(1-\tilde{A})^2,$$

$$\tilde{A} = \frac{(H_{fin}-2\delta_{fr})(P_{fin}-2\delta_{fr})}{H_{fin}P_{fin}}, \qquad (21)$$

where $m_{air,o}$ is the air mass flow rate before frosting, and P_{fin} is the fin pitch.

4.2 Simulation results
Developed heat pump and frosting models are consolidated as mentioned in Figure 2, and is applied to simulate heating at 84% (2°C DB) and 100% (1°C DB) humidity outdoor conditions. Figure 14 visualizes temporal evolution of frosting on the heat exchanger surface at 84% humidity condition, which shows non-uniform frost growth.

It is shown that frost builds up from upper part of heat exchanger and spreads down mainly due to non-uniform air velocity. Also shown is that frosting on the second or the inner row starts after finishing of frosting on the first row. Although not shown here, the same trend is true to 100% humidity condition at 1°C DB.

Figure 15 shows temporal evolution of frost thickness at various locations from first and second rows of the heat exchanger. As shown in figure, frost thickness grows at completely different speeds depending on local air velocity and refrigerant temperature. It is also clear that higher relative humidity accelerates frosting. Frosting saturates at around 50 minute for 84% humidity condition, and 35 minute for 100% humidity condition, which correspond to average defrosting time at the same conditions in the experiments. As will be shown later, unnatural inflexion points in Figure 15 are associated with rapid changes of cycle with inverter compressors to match target pressures and suction superheat.

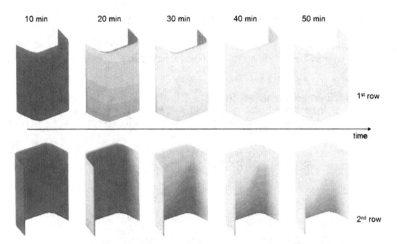

Figure 14 Visualization of frost formation on the outdoor heat exchanger at 84% humidity condition at 2°C DB

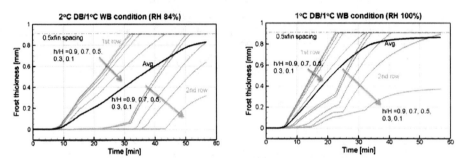

Figure 15 Frost formations on the outdoor heat exchanger under 84% (2°C DB, left) and 100% (1°C DB, right) humidity conditions monitored at various locations from top to bottom sides. Here, H denotes the height of the heat exchanger

Shown in Figures 16 and 17 are detailed investigation into cycle data, where the heat exchanger pipe temperature, air flow rate, condensing and evaporating temperatures, and capacity are plotted together with frost mass evolution. Here, the local pipe temperature is obtained by solving

$$\rho_p C_p A_p \frac{dT_p}{dt} = k_p A_p \frac{\partial^2 T_p}{\partial x^2} + h_{ref} dA_{ref}\left(T_{ref} - T_p\right) + \eta h_a dA_{air}\left(T_{air} - T_p\right), \qquad (22)$$

where ρ_p and C_p are, respectively, density and thermal capacity of pipe.

As shown in Figure 16, the gradients of pipe temperature become steeper at 40 minute and 30 minute, respectively, at 84% and 100% humidity conditions. This implies that monitoring heat exchanger pipe temperature is a good measure to determine when to defrost. Obviously, it is associated with fan air volume change as also shown in Figure 16. Although fan speed keeps 1200 RPM, the maximum, air volume reduces to 30% of non-frosting conditions. The same trend is true with condensing, evaporating temperatures and heating capacity as shown in Figure 17, where steep decrease of evaporating temperature occurs when frost mass is about to saturate.

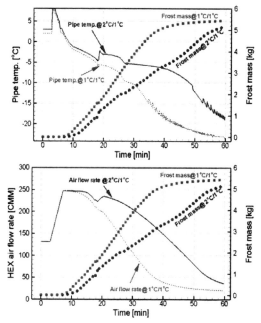

Figure 16 Evolutions of mean pipe temperature and fan air flow rate with respect to frost mass at 84% (2°C DB) and 100% (1°C DB) humidity conditions

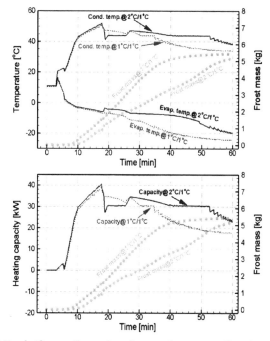

Figure 17 Evolutions of condensing and evaporating temperatures (left) and heating capacity (right) with respect to frost mass at 84% (2°C DB) and 100% (1°C DB) humidity conditions

5. CONCLUSIONS AND FUTURE WORKS

A dynamic heat pump model is developed, which is combined with frosting model on the heat exchanger. Toward this end, a new frosting model based on the perfect analogy between heat and mass transfer is proposed and validated against experimental data.

The heat pump model is based on quasi-dynamic algorithm adopting localized heat exchanger model with a linear mal-distribution model for the evaporator, and is also validated with experimental data.

Finally, heating simulation under frosting conditions are performed using the consolidated model. The model clearly represents realistic features of actual heat pump cycle as well as non-uniform frosting. Although predicted cycle shows good qualitative agreement with experimental results, a more detailed and quantitative validation is warranted, which is the topic of our subsequent research together with the development of defrosting model.

REFERENCES

[1] Chang YJ and Wang A, "Generalized heat transfer correlation for louver fin geometry", *Int. J Heat Mass Trans.*, Vol. 40 (1997), pp. 533-544.
[2] Da Silva L, Hermes CJL, and Melo C, "First-principles modeling of frost accumulation on fan-supplied tube-fin evaporators", *Appl. Thermal Eng.*, vol. 31 (2011), pp. 2616-2621.
[3] Dobson MK and Chato JC, "Condensation in smooth horizontal tubes", ASME J of Heat Transfer, Vol. 120 (1998), pp. 193-213.
[4] Gungor KE and Winterton RHS, "Simplified general correlation for saturated flow boiling and comparisons with data", *Chem. Eng. Res. Des*. Vol. 65 (1987), pp. 148-156.
[5] Hermes CJL, Piucco RO, Barbosa JR, and Melo C, "A study of frost growth and densification on flat surfaces", *Exp. Thermal Fluid Sci.*, Vol. 33 (2009), pp. 371-379.
[6] Kandula M, "Frost growth and densification in laminar flow over flat surfaces", *Int. J. Heat Mass Trans.*, Vol. 54 (2011), pp 3719-3731.
[7] Lee KS, Kim WS, and Lee TH, "A one-dimensional model for frost formation on a cold flat plate", *Int. J. Heat Mass Trans.*, Vol. 40 (1997), pp. 4539-4365.
[8] Lee KS, Jhee S, and Yang DK, "Prediction of the frost formation on a cold flat surface", *Int. J. Heat Mass Trans.*, Vol. 46 (2003), pp 3789-3796.
[9] Mader G and Thybo C, "A new method of defrosting evaporator coils", *Appl. Thermal Eng.*, Vol. 39 (2012), pp 72-85.
[10] McAdams WH, *Heat Transmission*, 2nd Ed., McGraw-Hill, New York, 1942.
[11] Na B and Webb RL, "New model for frost growth rate", *Int. J. Heat Mass Trans.*, Vol. 47 (2004), pp. 925-936.
[12] Schmidt TE, "Heat transfer calculations for extended surfaces", *Refrigeration Engineering*, April (1949) pp. 351-357.
[13] Xia Y and Jacobi AM, "An exact solution to steady heat conduction in a two-dimensional slab on a one-dimensional fin: application to frosted heat exchangers", *Int. J. Heat Mass Trans.*, Vol. 47 (2004), pp. 3317-3326.
[14] Ye HY and Lee KS, Private Communication, Hanyang University, Seoul, Korea, Sep. 6, 2012.

Developing simulation tools for design of low charge vapour compression refrigeration systems

G L Ding[a], T T Wang[a], J D Gao[a], Y X Zheng[b], Y F Gao[b], J Song[b]
[a] Institute of Refrigeration and Cryogenics, Shanghai Jiao Tong University, China
[b] International Copper Association Shanghai Office, China

ABSTRACT

Due to the environmental concerns of ozone depletion and global warming caused by refrigerants, techniques of reducing the refrigerant charge in vapour compression systems become quite important. The use of smaller diameter tubes is beneficial to reduce refrigerant charge inventory, but the performances of heat exchangers might decrease. In order to improve the efficiency of vapour compression system, simulation tools for heat exchangers and the entire refrigeration systems are needed. The room air conditioner optimized by the simulation tools are validated by experiments. It is shown that the refrigerant charge can be obviously decreased when using smaller diameter tubes.

NOMENCLATURE

A_o	Heat transfer area on air side (m²)	m	Mass flow rate (kgs⁻¹)
A_i	Inside surface area of tube (m²)	P	Pressure (Pa)
f	Friction factor	Q	Heat exchange (W)
h	Specific enthalpy (kJkg⁻¹)	T	Temperature (K)
j	Colburn factor	x	Quality

Greek

Δp	Pressure drop (Pa)	ε	Void fraction
α	Heat transfer coefficient (kWm⁻²K⁻¹)	η	Fin surface efficiency
ρ	Density (kgm⁻³)		

Subscripts

a	Air	out	Outlet
acc	Acceleration	r	Refrigerant
$back$	Back	top	Top
$bottom$	Bottom	l	Liquid
f	Friction, fin	tot	Total
$front$	Front	v	Vapour
in	Inlet	$wall$	Tube wall

1 INTRODUCTION

Fin-and-tube heat exchangers are widely used in vapour compression refrigeration system, in which R22 and R410A with high ODP or high GWP are commonly used as refrigerants. Due to the environmental concerns of ozone depletion and global warming caused by refrigerants, techniques of using natural refrigerants in vapour compression systems become quite important. Natural refrigerant R290 is a flammable working fluid, and using it as refrigerant in air conditioners may result in the risk of firing. In order to decrease the risk, the refrigerant charge needs to be decreased. And using smaller diameter (e.g. 5 mm) copper tubes to replace original larger diameter (7 mm or 9.52 mm) copper tubes in air conditioners is one of the effective solutions to decrease the refrigerant charge.

However, using smaller diameter tubes to replace the original larger diameter tubes will affect the heat exchange and pressure drop behavior of heat exchangers, and the performances of heat exchangers might decrease. In order to improve the efficiency and the heat exchange of the vapour compression system using smaller diameter tubes, simulation tools for design of heat exchangers and the entire refrigeration systems are needed. A simulation tool with three-dimensional distributed parameter model for design of fin-and-tube heat exchanger based on graph theory is developed. In addition, a simulation tool for the entire cycle, including the models for compressor, evaporator, condenser and capillary tube is also developed.

In this study, the software development ideas as well as the detailed mathematical model and algorithm of the simulation tools are introduced. And the air conditioner using 5 mm diameter copper tubes designed by these simulation tools is validated by experiment data.

2 DESIGN METHOD TO OPTIMIZE ROOM AIR CONDITIONER

The design method of room air conditioner system consists of two parts:
a. Using heat exchanger simulation tool to optimize the tube circuit of indoor unit and outdoor unit of room air conditioner system;
b. Using room air conditioner system simulation tool to calculate the system performance with optimal indoor unit and outdoor unit.

The detailed flow chart is shown in Figure 1.

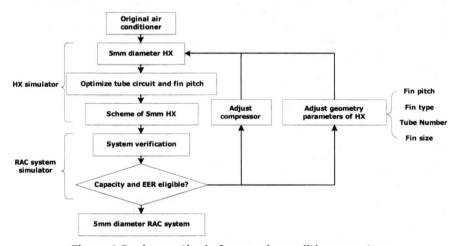

Figure 1 Design method of room air conditioner system

2.1 Heat exchanger simulation tool

In order to precisely simulate the heat transfer and pressure drop performance of both air side and refrigerant side for a fin-and-tube heat exchanger, a three-dimensional distributed parameter model for steady state performance needs to be established. Besides, the tube circuit of fin-and-tube heat exchanger is complicated and how to describe the complex tube connection in source codes becomes a problem. Liu et al. (2004) developed a general model based on graph theory to describe tube connection, which is used in this small diameter copper tube fin-and-tube heat exchanger simulation tool.

2.2 Room air conditioner simulation tool
In order to calculate the COP of the entire cycle of room air conditioner, the room air conditioner system simulator needs to be developed and the models of all the components, including evaporator, condensor, compresser and capilary, need to be established. For the system simulator, in order to satisfy the requirement of the calculation speed, the three-dimensional distributed parameter model is not suitable. Therefor, the lumped parameter system model needs to be established in system simulator for room air conditioner.

3 HEAT EXCHANGER SIMULATION TOOL FOR 5 MM DIAMETER TUBES

3.1 Mathematical model for fin-and-tube heat exchanger
The fin-and-tube heat exchanger is widely used both in evaporator and condenser of the vapour compression refrigeration system. As shown in Figure 2, the whole fin-and-tube heat exchanger is divided into some control volumes.

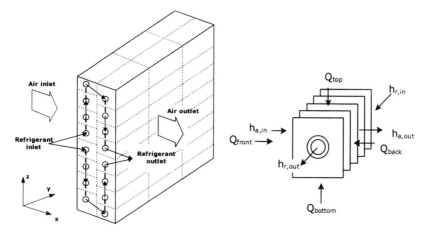

Figure 2 Control volumes of fin-and-tube heat exchanger

In order to simulate the steady-state performance of fin-and-tube heat exchanger, the heat transfer along the tube length direction is neglected, because it is much smaller than the heat transfer between the tube wall and refrigerant. The heat transfer characteristics of each control volume can be obtained by solving the governing equations. The schematic diagram of energy balance in a control volume is shown in Figure 2.

a. Refrigerant side:
Energy equation for refrigerant side is as follows:

$$Q_r = m_r(h_{r,in} - h_{r,out}) = \alpha_r \cdot A_r \cdot \left(\frac{T_{r,in} + T_{r,out}}{2} - T_{wall}\right) \quad (1)$$

where, the heat transfer coefficient α_r is calculated by empirical correlation of Hu et al. (2009) and Huang et al. (2010).

The momentum equation for refrigerant is as follows:

$$\Delta p_{total} = \Delta p_f + \Delta p_{acc} \tag{2}$$

where, Δp_{total}, Δp_{acc} and Δp_f represent total pressure drop, acceleration pressure drop and friction pressure drop of refrigerant, respectively. In two-phase region, the acceleration pressure drop is calculated as follows:

$$\Delta p_{acc} = m_r^2 \left[\frac{x_{r,out}^2}{\rho_v \cdot \varepsilon_{r,out}} + \frac{(1-x_{r,out})^2}{\rho_l \cdot (1-\alpha_{r,out})} \right] - m_r^2 \left[\frac{x_{r,in}^2}{\rho_v \cdot \varepsilon_{r,in}} + \frac{(1-x_{r,in})^2}{\rho_l \cdot (1-\alpha_{r,in})} \right] \tag{3}$$

b. Air side:
Energy equation for air side is as follows:

$$Q_a = m_a(h_{a,in} - h_{a,out}) = \alpha_a \cdot A_o \eta_o \cdot \left(\frac{T_{a,in} + T_{a,out}}{2} - T_{wall} \right) \tag{4}$$

where, $T_{a,in}$ and $T_{a,out}$ are inlet and outlet air dry bulb temperature of the control volume, respectively. The heat transfer coefficient α_a is calculated by empirical correlation of Wu et al. (2012).

c. Fin-and-tube: the energy equation is expressed as follows:

$$Q_r + Q_a + Q_{front} + Q_{back} + Q_{top} + Q_{bottom} = 0 \tag{5}$$

where, Q_{front}, Q_{back}, Q_{top}, and Q_{bottom} are heat conductions through fins from front row, back row, upper column, and bottom column, respectively. They are calculated by the temperature difference between the tube wall temperature of current control volume and corresponding temperature of neighbor tube, as shown in Figure 2.

3.2 Description of tube connection based on graph theory

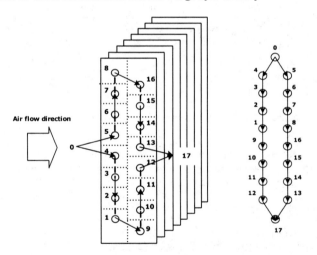

Figure 3 The directed graph corresponding to the tube connection

In order to simulate a heat exchanger with complex refrigerant circuitry, a special approach based on graph theory is used to link all the control volumes. The adjacent matrix is required to trace the confluence and branch of refrigerant flow. As shown in Figure 3, the directed graph is a kind of conceptualized hierarchy, depicted as a set of vertices connected by edges and each edge is endowed with certain direction. As far as heat exchanger considered, one vertex denotes a heat exchange tube, and the edge denotes the relationship between two tubes. Because the refrigerant flow direction has been considered, the edge is also endowed with flow direction. Thus the adjacent graph can be used to describe any flexible refrigerant circuits. The adjacency matrix is a mathematic data structure used to express above directed graph. The value of matrix element is expressed as follows, and the corresponding adjacent matrix is shown in Figure 4.

$$m_{i,j} = \begin{cases} 0 & (No.j \text{ is not connected to } No.i) \\ 1 & (No.j \text{ is connected to } No.i) \end{cases}$$

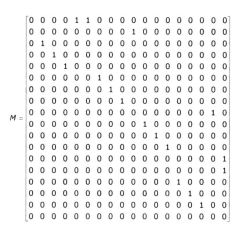

Figure 4 The adjacent matrix corresponding to Figure 3

3.3 Algorithm

In order to simplify the process of heat and mass transfer, the general simulation process is divided into two major different process: one is the heat transfer process, which is used to calculate the refrigerant enthalpy and refrigerant temperature with assumption that the refrigerant pressure is unchanged; the other is the pressure drop process, which is used to calculate the refrigerant pressure drop and adjust the refrigerant flow rate with the assumption that the refrigerant enthalpy is unchanged. The general simulation process is shown in Figure 5.

3.4 Functionality of heat exchanger simulation tool

In the fin-and-tube heat exchanger simulation tool, the 3D graphic user interface is also developed for users to define the size of the whole heat exchanger and the detailed geometry parameters of fin and enhanced tube. The input dialogs are presented in Figure 6. In addition, the output module is developed to output the performance parameters of the heat exchanger, which is shown in tables and charts as well as 3D colored graph, as shown in Figure 7.

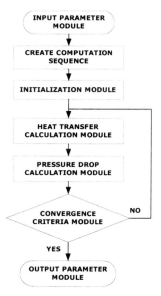

Figure 5 Logic flow chart of simulation algorithm

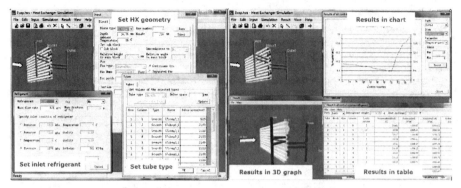

Figure 6 Data input windows   **Figure 7 Result output windows**

4 SYSTEM SIMULATION TOOL FOR AIR CONDITIONER

4.1 Mathematical model for system simulaion tool
A cycle of room air conditioner consists of four components, including evaportor, condensor, compressor and capilary, which need to be modeled respectively.

a. Compressor sub-model
The most important output parameters of the compressor sub-model are mass flow rate and input power, which can be calculated by Eq.(6)~(9).

$$m_{com} = \lambda V_{th} / v_{suc} \qquad (6)$$

Where, V_{th} is compressor theoretical exhaust volume, v_{suc} specific volume of suction line, λ is volumetric coefficient.

$$N_{in} = \frac{\frac{N_{th}}{\eta_i} + N_m}{\eta_{mo}} \qquad (7)$$

Where, the theoretical power of compressor N_{th} can be calculated by Eq. (8); the indicated efficiency is calculated by Eq. (9); N_m is friction power; η_{mo} is electrical efficiency.

$$N_{th} = V_h \cdot \lambda \cdot \frac{p_e \cdot m}{m-1}[(\frac{p_c}{p_e})^{(m-1)/m} - 1] \qquad (8)$$

$$\eta_i = \lambda_T + 0.0025(T_e - 273.15) \qquad (9)$$

b. Heat exchanger sub-model for condenser and evaporator
The heat exchanger sub-model is used to predict the refrigerant pressure, heat capacity and outlet air temperature at given conditions of inlet and outlet refrigerant mass flow rate, inlet refrigerant enthalpy and inlet air temperature. The heat transfer and pressure drop can be calculated by Eq. (10) ~ (12).

$$Q_r = m_r(h_{r,out} - h_{r,in}) = U_{i,ref}A_i(\frac{T_{a,in} + T_{a,out}}{2} - T_{rm}) \qquad (10)$$

Where, m_r is mass flow rate of refrigerant; $h_{r,in}$ and $h_{r,out}$ are the enthalpy of inlet and outlet; $U_{i,ref}$ is the overall average heat transfer coefficient of refrigerant side; $T_{a,in}$ and $T_{a,out}$ are the dry-bulb temperature of air inlet and outlet; T_{rm} is the average qualitative temperature.

$$\Delta p / L_1 = \left[f_m + \left(\frac{x_o - x_i}{x_m} \right) \frac{D_i}{L_1} \right] G_r^2 \frac{v_m}{D_i} \tag{11}$$

Where, f_m is coefficient of friction; L_1 is tube length of each path; D_i is inner diameter of tube; x_o, x_i and x_m are the refrigerant quality of inlet, outlet and average; G_r is the mass flux of refrigerant; v_m is average specific volume, which can be calculate by following equation.

$$v_m = x_m v_{gm} + (1 - x_m) v_{lm} \tag{12}$$

Where, v_{gm} and v_{lm} are specific volume of saturated gas and liquid.

c. Capillary sub-model
A distributed parameter model for adiabatic capillary tube is likely suitable for predicting the capillary tube characteristics accurately. In the model, the capillary tube is divided to some control volumes. For each control volume, the equation of continuity, energy and momentum are shown as follows:

$$m_1 = m_2 \tag{13}$$

$$h_1 + \frac{1}{2} G^2 v_1^2 = h_2 + \frac{1}{2} G^2 v_2^2 \tag{14}$$

$$p_1 - p_2 = G^2 (v_2 - v_1) + \frac{f_m v_m G^2}{2D} \Delta L \tag{15}$$

Where, subscript 1, 2 and m represent the inlet, outlet and average parameter.

4.2 Algorithm
The system simulator is used to predict the steady-state performance of room air conditioner and the coupling characteristic of all the components. To meet the requirement of real working condition of air conditioner, there are two algorithms for system simulator, shown as follows:

a. Algorithm 1:
The super heat degree of evaporator and sub cooling degree of condenser are known, while the refrigerant charge, the capillary tube length, the heat exchange and compressor power are predicted, as shown in Figure 8.

b. Algorithm 2:
The refrigerant charge and the capillary tube length are known, while the heat exchange, compressor power, the super heat degree of evaporator and sub cooling degree of condenser are predicted, as shown in Figure 9.

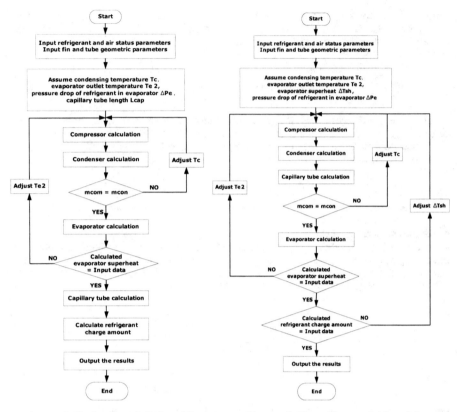

Figure 8 Flow chart of Algorithm 1   **Figure 9 Flow chart of Algorithm 2**

4.3 Functionality of system simulation tool

The graphic user interface is also developed. For input module, the dialogs for users to define the geometry parameters of evaporator, condensor, capillary tube and compressor are shown in Figure 10. For output module, the dialogs to output the COP and other system performance are shown in Figure 11.

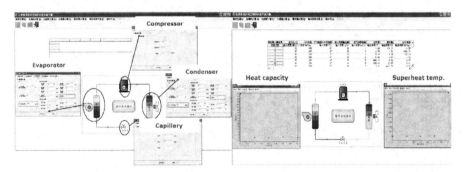

Figure 10 Input dialogs   **Figure 11 Output dialogs**

5 VALIDATION FOR SIMULATION TOOLS

In this study, room air conditioners with the cooling capacity of 2600 W, which possess more than 30% percent of the entire room air conditioner market, are optimized by the simulation tools introduced above. Smaller diameter tubes (5 mm diameter tube for indoor unit and 7 mm diameter tube for outdoor unit) are used to replace the original larger diameter tubes (7 mm diameter tube for indoor unit and 9.52 mm diameter tube for outdoor unit) in order to reduce refrigerant charge. The detailed geometry parameters of original and optimal room air conditioner are shown in Table 1.

Table 1 Structural parameters of the original and designed air conditioner

Structural parameters	Original air conditioner		Designed air conditioner	
	Indoor unit	Outdoor unit	Indoor unit	Outdoor unit
Tube diameter, mm	6.50	8.96	4.60	6.50
Length/Width/Height, mm	228/22/320	708/43.3/480	228/27.2/320	706/36/462
Row space/Column space, mm	11.0/19.0	21.6/25.4	13.6/19.0	18/21
Fin Thickness/Fin Pitch, mm	0.105/1.6	0.105/1.8	0.105/1.4	0.105/1.4
Fin type	Louver fin	Wavy fin	Louver fin	Wavy fin
Tube circuit	(diagram)	(diagram)	(diagram)	(diagram)

The system performance of the optimal air conditioner is validated by experiment data, as shown in Table 2.

Table 2 Exp. data and simulation results of designed air conditioner

Items	Simulation	Exp. data	Deviation
Refrigerant charge	275 g	285 g	3.5%
Cooling capacity	2403 W	2439 W	1.5%
COP	3.05	3.08	1.0%
Condensing temperature	46.5 °C	45.8 °C	1.5%
Evaporating temperature	7.9 °C	7.8 °C	1.3%

The refrigerant charge of optimal air conditioner with smaller copper tubes is much less than that of the original air conditioner, as shown in Figure 12.

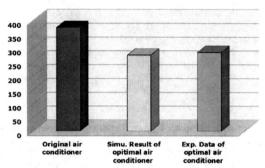

Figure 12 Refrigerant charge of optimal air conditioner

6 CONCLUSION

A general steady state distributed parameter model based on graph theory for the fin-and-tube heat exchanger with small diameter copper tubes is developed to analyze the performance of heat exchangers with complex refrigerant circuits. The system simulation tool with sub-model of each component for room air conditioner is also developed to predict the whole system performance of vapour compression refrigeration system. The deviations of the heat exchange and COP of simulation results are less than 5% compared with the experimental data. The results show that using 5 mm tubes to replace 7 mm or larger diameter tubes can obviously decrease the refrigerant charge in room air conditioners while the performance keeps well.

REFERENCE LIST

(1) Ding GL. 2007, Recent developments in simulation techniques for vapour-compression refrigeration systems, *Int. J. Refrig.* 30 (7): 1119-1133
(2) Hu HT, Ding GL, Huang XC, Deng B, Gao YF. 2009, Measurements and correlation of flow-boiling heat transfer of a R410A/Oil mixture inside a 4.18mm straight smooth tube. *HVAC&R Reserch* 15 (2): 287-314
(3) Huang XC, Ding GL, Hu HT, Zhu Y, Gao YF, Deng B. 2010, Condensation heat transfer characteristics of R410A-oil mixture in 5 mm and 4 mm outside diameter horizontal microfin tubes, *Experimental Thermal and Fluid Science* 34 (7): 845-856
(4) Liu J, Wei WJ, Ding GL, Zhang CL, Fukaya M, Wang KJ, Inagaki T. 2004, A general steady state mathematical model for fin-and-tube heat exchanger based on graph theory, *Int. J. Refrig.* 27 (8): 965-973
(5) Wei WJ, Ding GL, Hu HT, Wang KJ. 2007, Influence of lubricant oil on heat transfer performance of refrigerant flow boiling inside small diameter tubes. Part II: Correlations, *Experimental Thermal and Fluid Science* 32 (1): 77-84
(6) Wu, W., Ding, G.L., Gao, Y.F., et al, 2012, Experimental investigation of fin-and-tube heat exchanger with smaller diameter tubes. *The 6th Asian Conference on Refrigeration and Air Conditioning*, August 26-28, 2012, Xian, China
(7) X.Jia, C.P.Tso, P.K.Chia, 1995. A distributed model for prediction of the transient response of an evaporator. *Int. J. of Refrig*, 18 (5):336-342
(8) John.Judge, Reinhard Radermacher, 1997. A heat exchanger model for mixtures and pure refrigerant cycle simulations. *Int. J. of Refrig*, 20 (4): 244-255

Modeling of small-size turbocharger compressors' performance curves

K V Soldatova
Compressor Dept., TU Saint-Petersburg, Russia

ABSTRACT

Non-adiabatic performance curves of a small-size turbocharger compressor, which was tested at the Institute of Turbo machines (Hanover, Germany) in a range of periphery Mach numbers 0,73 – 1,44, were modeled using software developed at the Compressor Dept. TU Saint-Petersburg (Russia). The computer models and software, recently updated (the new 6th generation), were applied to small-size high Mach number compressors, along with this several improvements of basic algorithms and iterative processes in thermodynamic calculations were made. The head loss model demonstrated its validity after empirical coefficients' correlation according to new test data. The practical result of the work is that corrected algorithms and software could be used to model performances curves of small-size turbocharger compressors designed for periphery Mach numbers up to 1,44 rather satisfactorily in a wide range of RPM.

NOTATION

c – absolute velocity, m/s;
c_r – radial velocity, m/s;
c_u – tangential velocity, m/s;
C_p – specific heat;
D_0 – impeller eye diameter, m;
D_1 – impeller inlet diameter, m;
D_2 – impeller diameter, m;
k – isentropic coefficient;
$\dot{m}$ – mass flow rate;
M – Mach number;
$M_u = \dfrac{u_2}{\sqrt{kRT_{inl\,tot}}}$;
p – pressure, Pa;
R – gas constant, j/kg/K;
R_{bl} – radius of blade curvature, m;
Re – Reynolds number;

$\mathrm{Re}_u \dfrac{u_2 D_2}{\mu_{inl}} \dfrac{p_{inl\,tot}}{RT_{inl\,tot}}$;
T – temperature, K;
u_2 – impeller periphery speed, m/c;
w – relative velocity;
z – number of blades;
Φ – flow rate coefficient;
η_t – total polytropic efficiency;
π_t – total pressure ratio;
$\psi_T = c_{u2}/u_2$ - Euler work coefficient;

Subscripts
1 – blade row inlet;
des – design regime;
ex – compressor exit;
h – hub;
inl – compressor inlet;
s – blade suction side, shroud;

© The author(s) and/or their employer(s), 2013

ANNOTATION

Fuel saving of an internal combustion engine depends in particular on efficiency and operation range of a supercharger compressor. For proper design of latter we need the application of specific modeling tools to choose better variant and calculate expected performance map in a range of application. The Universal Modeling Method developed at TU SPb Compressor Dept. (1, 2, 3) has successful record of application to solve design problems of big industrial centrifugal compressors. The idea of the presented work is to apply this modeling technology to test data obtained at IT University Hanover. Problems appeared as CD modeling technology was not aimed to trans- and supersonic compressors tested non-adiabatically, i.e. with strong influence of heat transfer. The temporary solution of problems is presented. The Author is grateful to the Director of the Institute of Turbomachine and Fluid Dynamics TU Hannover Prof. J. R. Seume and Dipl.-Ing. T. Sextro who provided test data and gave necessary consultation for data reduction.

TEST METHODOLOGY AND RESULTS

The main data on the tested compressor is given in Table 1.

Table 1. Main parameters of tested compressor

1	2	3	4	5	6	7
Name\Parameter	D_2 mm	D_0 mm	$Z_{1,2}$	β_{bl2}^0	$\beta_{bl1s}^0 / \beta_{bl1h}^0$	M_u range
TC-1	48	36	6/12	50	15/62	1.442-0,735

Cross section of a compressor part of a small-size turbocharger is presented in Fig. 1.

Fig. 1. Cross section of a compressor part of a small-size turbocharger

Tests were conducted at the test facility shown in Fig 2.

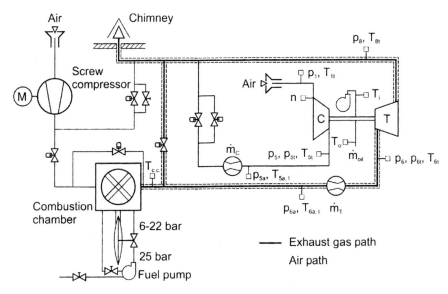

Fig. 2. Schematic of the rig for turbocharger tests (7)

The results of testing are presented and analyzed in a dimensionless form:

- flow rate coefficient: $\Phi = \dfrac{\dot{m}}{0{,}785 D_2^2 u_2} \dfrac{P_{inl\,t}}{RT_{inl\,t}}$, (1)

- work input coefficient: $\psi_i = \dfrac{C_p\left(T_{ext} - T_{inl\,t}\right)}{u_2^2} = \dfrac{(k-1)}{M_u^2}\left(1 - \dfrac{T_{ext}}{T_{inl\,t}}\right)$, (2)

- pressure ratio: $\pi_t = \dfrac{P_{ext}}{P_{inl\,t}}$, (3)

- efficiency: $\eta_t = \dfrac{k-1}{k} \dfrac{\ln \pi_t}{\ln \dfrac{T_{ext}}{T_{inl\,t}}}$. (4)

Flow temperatures at a compressor inlet and exit depend on active heat transfer as shown in Fig. 3:

Values of ψ_i and η_t calculated by the formulae (2), (3) reflect work input and head loss correctly if the compression process is adiabatic. Non-adiabatic performances $\eta_t, \psi_i = f(\Phi)$ defined by the formulae above are not correct. The objects of modeling below were performance curves $\pi_t = f(\Phi)$ that are measured correctly. Then with the help of Universal Modeling Method following performance curves $\eta_t, \psi_i = f(\Phi)$ are calculated, and after that measured and calculated performance curves $\pi_t = f(\Phi)$ were matched.

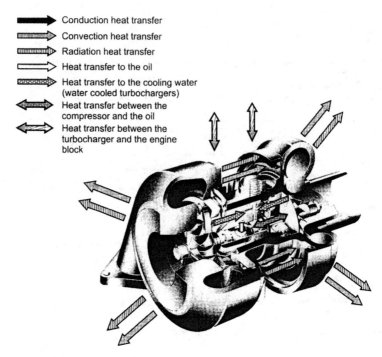

Fig. 3. Mechanism of heat transfer in/from a turbocharger (7)

MODELING METHOD

To model compressor performance following items are necessary: a model of mechanical work input, a model to calculate head loss in a flow path and an algorithm of gas parameter calculation in control planes of a flow path.

In design practice of CD TU SPb the model of work input performance is based on experimentally proven assumption that a function $\psi_T = f(\varphi_2)$ is linear (4, 5). It is independent of Mach criteria for impellers of big subsonic industrial compressors tested at $M_u \leq 0.9$. It means that pressure ratio performance of a compressor can be calculated for different RPM on a base of a single function $\psi_T = f(\varphi_2)$.

Two values of an Euler work coefficients define a work input performance $\psi_T = f(\varphi_2)$:

- a value at zero flow rate $\psi_{T0} < 1$. There is an empirical correlation to calculate it for big subsonic impellers. To match ψ_{T0} with exact test result an empirical coefficient $K\psi_{t0}$ can be chosen;
- a value at a design flow rate ψ_{Tdes}.

To calculate ψ_{Tdes} the scheme and formulae developed by prof. Y.Galerkin (6) is applied – Fig. 4.

In accordance with the scheme:

$$\psi_T = 1 - \varphi_2 ctg\beta_{bl2} - \Delta\overline{c}_{u2}. \qquad (5)$$

$$\Delta\overline{c}_{u2} = K_\mu \frac{\psi_T}{z \cdot \overline{l}_{m2}}, \qquad (6)$$

$\overline{l}_{m2}$ - normalized meridian distance from a gravity center a velocity diagram to an impeller exit, $K_\mu > 1$ - an empirical coefficient.

Loss calculation procedure is reduced to definition of friction drag force coefficients on all surfaces of a flow path, mixing loss coefficients where flow separation occurs and incidence losses at off- design regimes. Several dozens of empirical coefficients correlate loss coefficients with velocity level and gradients and with similarity criteria (2, 3). To extend the method to transonic and supersonic stages Prof. Y. Galerkin has added inductive losses calculation and takes into account negative influence of a choke on a boundary layer. More detailed description of 3D impellers and several improvements of iterative processes in thermodynamic calculations are added too. MSc Eng A. Drozdov added new algorithms to our updated software (6th generation). The Authors are grateful to both of them for possibility to apply new programs in this work. Fig. 5 presents an input menu and meridian shape of a 3D impeller of the 6th generation program for a stage performance map calculation.

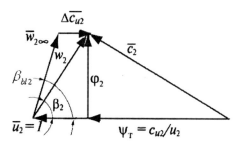

Fig. 4. Velocity triangle at an impeller exit

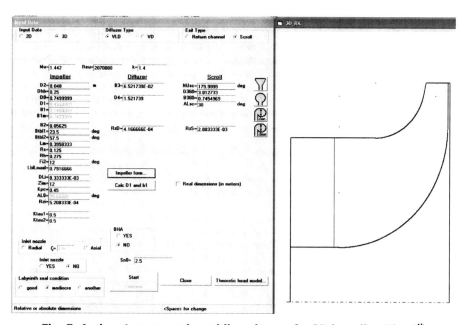

Fig. 5. An input menu and meridian shape of a 3D impeller. The 6th generation program for a stage performance map calculation

Several problems arouse in calculation of small-size turbocharger compressors due to simplified description of stage geometry in applied programs. The problems and ways of solution:
- the modeled impeller has split blades. The applied program calculates impellers with all full length blades. For approximate modeling 12 full length blades were applied with the half of a real thickness;
- there is no open impellers calculation in the programs at the moment as open impellers are practically not used in industrial compressors. Calculations are made for closed impellers. The influence of disc friction and labyrinth seal leakage is rather small as the impeller is of high flow rates ($\Phi_{des} \approx 0.13$);
- 1D calculation of flow at an impeller inlet is most important to define non-incidence flow rate. There is no problem for a 2D impeller where a blade inlet angle β_{bl1} is close to constant along the leading edge height. There is a method to choose necessary value of β_{bl1} for impellers designed by CD SPbTU method. Influence of blade blockage and blade load on critical streamline direction is calculated by formula presented in [5]. For presented compressors the values were found by series of calculation with different β_{bl1}. The value with better matching between calculated and measured flow rates was chosen for final modeling process.

MODELING PROCESS AND RESULTS

Non-adiabatic compression process adds problems to performance modeling. When adiabatic tests are reduced by the IDENT program, the work input performance is modeled by choice of K_μ and ψ_{T0} coefficients. Then an automatic process of empirical coefficients in the loss model takes place with an aim of best correlation of measured and calculated efficiencies. But there are no reliable measured values of work coefficient and efficiency at non-adiabatic tests.

The IDENT program at the moment can not find empirical coefficients for efficiency and work input performance matching pressure ratio performances - that is measured objectively at non-adiabatic tests. Some artificial ways were applied by the Author.

There is an opinion expressed in (7) that heat transfer deteriorate performances minimally at the highest RPM. The performance $\psi_i = f(\Phi)$ of the compressor was approximated most precisely by choice of K_μ and ψ_{T0} coefficients. The set of loss model empirical coefficients that well describes efficiency of subsonic industrial compressor stages was applied to calculate compressor efficiency curve. There is no information on roughness of the flow path surfaces. The photo in Fig. 1 shows that the roughness is sufficient in all elements of the compressor. The values 20 -25 micrometers for the impeller and the diffuser were applied arbitrary. Anyway, it was not enough to predict efficiency and pressure ratio correctly. For better matching the empirical coefficient that controls friction losses was increased by 40% in comparison with a value applied to big industrial compressors. The result of modeling is presented in Fig. 6.

The correlation of performances is rather good though only one empirical coefficient was changed for it. Three dozens of other empirical coefficients that participated at this exact calculation are the same as in well-proven calculations of industrial compressors.

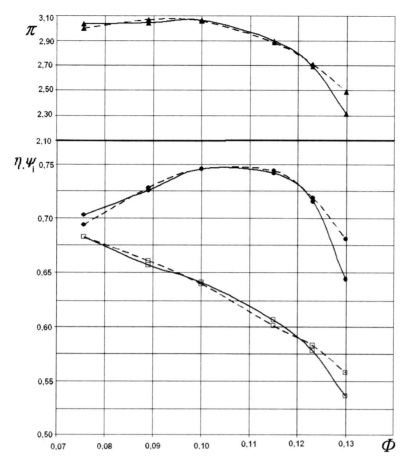

Fig. 6. Modeling of TC-1 compressor performances by the IDENT program

The independence of an empirical function $\psi_T = f(\varphi_2)$ of Mach criteria (4, 5) does not take place in modeling of non – adiabatic tests. To match pressure patio at different RPM it was necessary to choose individual values of K_μ and ψ_{T0} coefficients for different RPM, i.e. for different Mach numbers. The set of empirical coefficients was the same that was used for modeling of the compressor at highest RPM.

The results are presented in Fig. 7. The individually chosen values of K_μ and ψ_{T0} are presented in columns 3, 4 in Table 2.

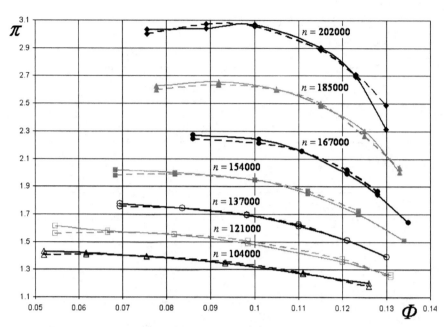

Fig. 7. Pressure ratio performances of TC-1 compressor in a range of RPM. Solid – test, stroke – modeling. Above – individual, below – approximated values of K_μ and ψ_{T0}

For systematization of results the individually chosen values of $K_\mu, \psi_{T0} = f(M_u)$ were approximated by the equations:

$$K_\mu = 3(1,45 - M_u)^{3,2} + 1,7, \qquad (7)$$

$$\psi_{T0} = 0,5 + 3,5(M_u - 0,735). \qquad (8)$$

The approximated values are presented in columns 5, 6 in Table 2. Graphic representation of the table 2 content is shown in Fig. 8. The empirical coefficients are calculated for test data in a range of M_u =0.735-1.442.

Pressure ratio performances prediction with approximated values of $K_\mu, \psi_{T0} = f(M_u)$ in Fig. 7 demonstrates acceptable result. Let us notice that $K_\mu, \psi_{T0} = f(M_u)$ values are practically constant for range of $M_u > 1,15$. It means that performance $\psi_{T0} = f(\varphi_2)$ is independent of M_u at high Mach numbers – as it is independent in case of subsonic stages tested adiabatically. It is possible to propose that deviation of the independence rule does not reflect flow character under different M_u but is due to strong heat transfer processes at non-adiabatic tests.

The performance map for different RPM of the compressor was calculated by the computer program CSPM-G6E with the mentioned above set of empirical coefficients and approximated $K_\mu, \psi_{T0} = f(M_u)$ values from the column 5, 6 of Table 2. The performance map is shown in Fig. 9.

Table 2. Values of K_μ and ψ_{T0} coefficients used for performance modeling of TC-1 compressor in a range of RPM

1	2	3	4	5	6
$n \cdot 10^{-3}$ RPM	M_u	K_μ	ψ_{T0}	$K_{\mu\,approx}$	$\psi_{T0\,approx}$
202	1.442	1.800	2.100	1.700	2.100
185	1.313	1.650	2.100	1.705	2.100
167	1.192	1.800	2.100	1.739	2.100
154	1.089	1.900	1.700	1.815	1.739
137	0.971	2.200	1.200	1.985	1.326
121	0.857	2.200	1.200	2.264	0.927
104	0.735	2.700	0.500	2.725	0.500

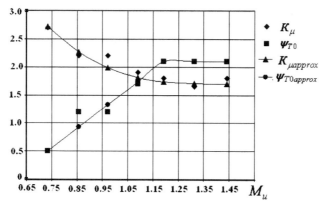

Fig. 8. Graphic representation of $K_\mu, \psi_{T0} = f(M_u)$ for individually chosen values and for their approximation by formulae (7), (8)

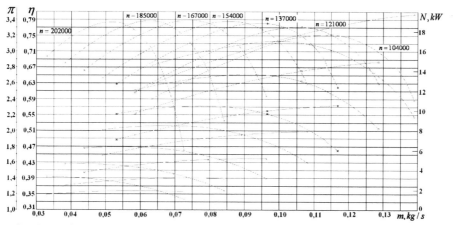

Fig. 9. Performance map of the compressor TC-1 calculated with modified set of empirical coefficients and approximated values of $K_\mu, \psi_{T0} = f(M_u)$

The configuration of performances seems quite logical. Up to the time when more universal ways of modeling would be available the described above methods of modeling could be recommended for use.

CONCLUSION

Turbine – compressor heat exchange influences exit temperature at different level at different RPM in turbochargers. Compressor performance is especially difficult to model because of indefinite mechanical work input that is measured by temperature rise in a compressor. If there are test data for a supercharger compressor at wide range of RPM the described modeling method can be applied for more or less reliable performance modeling. The Universal modeling computer programs [2] serve as a basic tool for modeling. The following should be done:
- the empirical coefficient that controls friction losses must be increased by 40% of the value in a standard set of Universal modeling method,
- the pressure ratio performance curves matching must be achieved by variation of function $\psi_T = f(\varphi_2)$ for each RPM,
- the empirical formulae (7), (8) are a key matter to define function $\psi_T = f(\varphi_2)$ for each RPM. For any other compressor the exact equations can differ.

The application of the modeling procedure to other compressor test data could demonstrate ivalidity or inconsistence of the former.

REFERENCES

1. Galerkin, Y. B. Turbo compressors. // LTD information and publishing center. – Moscow. – 2010 (In Russian).
2. Galerkin Y., Danilov K., Popova E. Design philosophy for industrial centrifugal compressor. // International Conference on Compressors and their systems. – London: City University, UK. – 1999.
3. Galerkin Y., Mitrofanov V., Geller M., Toews F. Experimental and numerical investigation of flow in industrial centrifugal impeller. // International Conference on Compressors and their systems. – London: City University, UK. – 2001.
4. Galerkin Y., Karpov A. Development of method for centrifugal compressor impeller pressure characteristic modeling on results of model stages tests. // Compressors & Pneumatics. – Moscow. – 2011. – No. 6. – Page 13-17 (In Russian).
5. Galerkin Y., Kozhukhov Y. Pressure performance modeling of a centrifugal compressor impeller on experimental data and calculations of a nonviscous three -dimensional flow. // Compressors & Pneumatics. – Moscow. - 2007. – No. 4. – Page 32-37 (In Russian).
6. Seleznev K., Galerkin Y. Centrifugal compressors. // Leningrad. – 1982 (In Russian).
7. Shaaban S. Experimental investigation and extended simulation of turbocharger non-adiabatic performance. // Doctor-Engineering Dissertation. – Hanover. - Germany. – 2004.

Development of a generalized steady-state simulation framework for positive displacement compressors and expanders

I H Bell, V Lemort
University of Liège, Belgium

E A Groll, J E Braun, W T Horton
Purdue University, USA

ABSTRACT

In this paper, a generalized framework is presented that can be used to simulate the steady-state performance of a wide range of positive displacement compressors and expanders (scroll, piston, screw, rotary, spool, etc.). The complete simulation algorithm is described, and an emphasis is placed on the numerical methods required to obtain robust behavior of the simulation. This formulation has been implemented into an open-source software package entitled PDSim written in the Python language. PDSim is the first open-source generalized compressor and expander simulation package. The source code can be freely obtained on the internet.

1 INTRODUCTION

The detailed simulation of positive displacement compressors and expanders is of great interest, both academically and industrially. These simulation programs are typically developed in order to conduct computer-based prototyping. That is, with a properly tuned and validated model, it is possible to accurately predict the performance of the machine at conditions that the machine has not been tested at previously. Additionally, by using a model that is fully mechanistic, the actual performance can be predicted at operating conditions quite different than those at which the machine had been previously tested.

One of the shortcomings of many of the existing simulation codes for positive displacement machines is that they are highly specific, generally having been developed for a particular type of machine (scroll, reciprocating, etc.) and are typically structured in a way that makes the simulations very difficult to modify. Also, many of the simulation codes that have been developed are highly proprietary, which means that there is a large amount of simulation code development that is duplicative amongst industrial entities and academic researchers.

In this paper, a generalized framework for the simulation of positive displacement compressors and expanders is proposed, and the full description of the framework is provided. The full source code for the framework is also provided[1].

---
[1]The PDSim website is at http://pdsim.sf.net, the code of PDSim as of publication can be found at http://hdl.handle.net/2268/147968

NOMENCLATURE

Symbol	Description	Symbol	Description
A_{tube}	Cross-sectional area of tube (m^2)	p_{out}	Outlet pressure (kPa)
A_{valve}	Frontal area of valve (m^2)	p_{in}	Inlet pressure (kPa)
c_p	Spec. heat at const. pressure (kJ kg^{-1} K^{-1})	Pr	Prandtl Number (-)
		$\dot{Q}$	Heat transfer rate (kW)
		$\vec{r}$	Residual array
c_v	Spec. heat at const. volume (kJ kg^{-1} K^{-1})	Re	Reynolds number (-)
		T_{out}	Outlet temperature (K)
C_D	Valve drag coefficient (-)	T_{in}	Inlet temperature (K)
D	Diameter (m)	T_{wall}	Wall temperature (K)
D_{port}	Diameter valve port (m)	U	Total internal energy (kJ)
D_{valve}	Diameter valve plate (m)	V	Volume (m^3)
f	Friction factor (-)	$\dot{W}$	Power (kW)
$\vec{f}_0, \vec{f}_1, ...$	Time derivative of $\vec{x}_0$..	$\dot{W}_{electrical}$	Electrical Power (kW)
G	Mass flux (kg m^{-2} s^{-1})	$\dot{W}_{mechanical}$	Mechanical Power (kW)
h	Specific enthalpy (kJ kg^{-1})	$\vec{x}_0, \vec{x}_1, ...$	State variable array
h	Step size (s,radian)	y	Valve position (m)
J	Jacobian	$\dot{y}$	Time deriv. valve disp. (m s^{-1})
k	Thermal conduct. (kW m^{-1} K^{-1})		
		y_{tr}	Transit. valve disp. (m)
k_{valve}	Valve spring coefficient (N m^{-1})	a	Heat trans. coeff. (kW m^{-2} K^{-1})
L	Length (m)	η_{motor}	Motor efficiency (-)
$\dot{m}_i$	Mass flow rate flow path (kg/s)	μ	Viscosity (Pa s)
		ω	Rotational speed (rad s^{-1})
$\dot{m}_{tube}$	Mass flow rate in tube (kg/s)	θ	Crank angle (rad)
		ρ	Density (kg m^{-3})
Δp	Change in pressure (Pa)	ε	Surface roughness (m)
p	Pressure (kPa)		

1.1 Existing Simulation Codes

There are a number of steady-state simulation codes that are described in literature that have been tailor-made for a single type of positive displacement machine. A selection of the simulation codes available are summarized here:

- Scroll compressor and expander: (1; 2; 3; 4; 5)
- Bowtie compressor: (6)
- Linear compressor: (7; 8)
- Two-stage Rotary compressor: (9)
- Z-compressor: (10)
- Reciprocating compressor: (11)
- Screw compressor: (12; 13)

There are also a number of simulation codes in use in industry, though they tend to be more rudimentary in capability than the tools developed by academic researchers. In any case, the details of the simulation tools developed in industry are not in the public domain.

To the author's knowledge, there has never been a generalized simulation code developed for all types of positive displacement compressors and expanders.

2 COMPONENT MODELS

In this section all of the component models that form the simulation code are described.

2.1 Thermophysical Properties

Thermophysical properties of the working fluid are used in all the other component models and constitute nearly all the running time of the simulation code. Thus the efficiency of thermophysical property evaluation is of the utmost importance. The CoolProp property library (14) is used to obtain the necessary thermodynamic and transport properties.

2.2 Tube

A tube is a component of the simulation that allows for pressure drop and heat transfer. It is modeled as being quasi-steady, and either the outlet or the inlet state is fixed. The following assumptions are employed:

- Quasi-steady - tube is treated as operating at steady state, and the mean mass flow rate is employed for the tube
- No mass storage - tube is considered to be operating at steady-state, thus no mass is added to or removed from the tube
- Fixed wall temperature, fixed wall heat transfer flux, or user-defined heat transfer model
- Pressure drop calculated
- Single-phase flow enforced through the tube

The flow through the tubes is generally turbulent, and thus the friction factor and heat transfer coefficient and can be given by the following forms from Churchill (15)

$$f = 8\left(\left(\frac{8}{Re}\right)^{12} + (A+B)^{-1.5}\right)^{\frac{1}{12}} \tag{1}$$

$$A = \left(2.457 \ln\left(\left(\frac{7}{Re}\right)^{0.9} + 0.27\frac{\varepsilon}{D}\right)^{-1}\right)^{16}, \quad B = \left(\frac{37530}{Re}\right)^{16} \tag{2}$$

and Gnielinski (16)

$$a = \frac{k}{D}\frac{(f/8)(Re - 1000)Pr}{1 + 12.7\sqrt{f/8}(Pr^{2/3} - 1)} \tag{3}$$

where the properties are evaluated at the known state and the Reynolds number is given by $Re = GD/\mu$ where G is the mass flux given by $G = \dot{m}_{tube}/A_{tube}$ and A_{tube} is the cross-sectional flow area of the tube given by $A_{tube} = \pi D^2/4$. If the inlet state of the tube and is given and the wall temperature is known, the outlet temperature can be calculated from

$$T_{out} = T_{wall} - (T_{wall} - T_{in})\exp\left(-\frac{\pi DLa}{\dot{m}_{tube}c_p}\right) \tag{4}$$

and the change in pressure by

$$\Delta p = -\frac{fG^2 L}{2\rho D} \tag{5}$$

which yields the outlet pressure of $p_{out} = p_{in} + \Delta p$.

If instead the outlet state is known, the inlet temperature can be calculated from Equation 4. The inlet pressure is calculated from $p_{in} = p_{out} - \Delta p$.

2.3 Flow Path

A flow path connects two flow nodes. These flow nodes could be control volumes, or they could be the inlet or an outlet for a tube section. In a flow path, the upstream state is fully specified and in general the downstream pressure is known. The model for the flow path must then calculate the mass flow rate through the flow path.

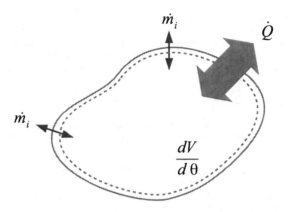

Figure 1: Schematic of control volume

There are a wide range of flow models that can be employed for a flow path. The flow model selected depends on the geometry, flow type, and other parameters. It is also important to use different flow models if the flow can be treated as being compressible or not. Table 1 presents a list of single phase flow models that have been used in the past in literature.

Table 1: List of some flow models used in literature

Type	Reference	Compressible	Friction	Notes
Isentropic Nozzle	(17)	Yes	No	Most common model, tends to overpredict flow rate though simple to implement and use
Incompressible pipe flow	(18)	No	Yes	
Fanno flow	(19)	Yes	Yes	
Nozzle/Fanno combination	(20)	Yes	Yes	
Superposition method	(21)	No	Yes	
Hybrid leakage	(17)	Yes	Yes	Very computationally efficient, intended only for leakage

2.4 Valve

There are a number of different types of valves that exist in positive displacement machines, the most common type are reed valves as they are simple to fabricate. While reed valves are dynamic valves, there are also other types of valves that are driven by the kinematics of the machine like poppet valves. Kinematic valves are more straightforward to analyze as their opening and closing are deterministic and are not governed by dynamics.

Reed valves can be analyzed as a one-degree-of-freedom vibrational system, as described in literature (6; 7), and a module for their analysis is included in PDSim.

2.5 Control Volume

A control volume is the fundamental component that allows for the analysis of compressors and expanders. The assumptions employed for the control volume are:

- Uniform temperature and pressure throughout the control volume
- Negligible kinetic energy of the control volume
- Negligible gravitational effects
- Thermal interaction possible through heat transfer
- Mass flow into or out of control volume possible

Two thermodynamic properties are needed to fix the thermodynamic state for pure fluids and fluid mixtures that are treated like pure fluids (see Lemmon(22)). The equations of state used are all of the Helmholtz energy explicit formulation and are explicit in temperature and density. Therefore, temperature and density are the most straightforward state variables to use in the analysis that follows.

The derivatives of the temperature and density as a function of crank angle are needed. These derivatives will then be implemented into the solver that is used to calculate the properties throughout the course of a cycle.

Other sets of state variables are also possible. For instance, temperature and pressure are a physically consistent and easily-comprehensible set of state variables, but they are not well suited for positive displacement machine simulation. For instance, if the fluid passes into the two-phase region during the working process due to pressure drop or wet expansion, temperature and pressure can no longer be used to uniquely define the state.

The conservation of mass for the control volume yields the following form for the derivative of the mass in the control volume

$$\frac{dm}{dt} = \sum_i \dot{m}_i \tag{6}$$

where for rotational machines working in the crank-angle domain, the derivative of the mass with respect to the crank angle can instead be given by

$$\frac{dm}{d\theta} = \frac{1}{\omega}\sum_i \dot{m}_i \tag{7}$$

where $\dot{m}_i$ is the mass flow rate corresponding to the i-th flow path and ω is the rotational speed. The sign of $\dot{m}_i$ is positive if flow is into the control volume. The conservation of energy is given by

$$\frac{dU}{dt} = \dot{Q} + \dot{W} + \sum_i \dot{m}_i h_i \tag{8}$$

where the boundary work term $\dot{W}$ is given by $\dot{W} = -p \cdot (dV/dt)$ and $\dot{m}_i h_i$ is the enthalpy flow term corresponding to the product of the upstream enthalpy and the mass flow rate for the i-th flow path. Expansion of the derivative of the total internal energy like in Bell (1) allows for the solution for the time derivative of the temperature, yielding

$$\frac{dT}{dt} = \frac{-T\left(\frac{\partial p}{\partial T}\right)_\rho \left[\frac{dV}{dt} - \frac{1}{\rho}\frac{dm}{dt}\right] - h\frac{dm}{dt} + \dot{Q} + \sum_i \dot{m}_i h_i}{mc_v} \tag{9}$$

and the crank-angle derivative of temperature can be obtained from

$$\frac{dT}{d\theta} = \frac{-T\left(\frac{\partial p}{\partial T}\right)_\rho \left[\frac{dV}{d\theta} - \frac{1}{\rho}\frac{dm}{d\theta}\right] - h\frac{dm}{d\theta} + \frac{\dot{Q}}{\omega} + \frac{1}{\omega}\sum_i \dot{m}_i h_i}{mc_v} \tag{10}$$

The conversion from dU/dt to dT/dt is carried out because all the terms in Eqn. 9 are explicit in temperature and density, the state variables upon which the equation of state is based.

Mass and specific internal energy are a viable pair of state variables, but difficulty arises from the fact that the equations of state are explicit in temperature and density rather than density and specific internal energy and it is therefore necessary to numerically solve for the temperature given the internal energy and the density which involves significant computational effort.

The thermophysical properties are evaluated using the CoolProp Thermophysical Property library (see section 2.1).

The heat transfer term $\dot{Q}$ is machine-specific and depends on the geometry as well as the operating conditions. It is not possible to provide a sufficiently general model for the heat transfer. It should be noted that the sign of $\dot{Q}$ has been selected such that it is positive if the heat transfer is into the control volume.

The geometric terms V and $dV/d\theta$ (or dV/dt) are given by the geometry of the machine. In the case of a piston machine, the geometry is trivial, while in the case of other more complex machines like screw and scroll compressors, calculation of the necessary geometric terms is very involved. For that reason, no further analysis of the geometry is presented here, though characteristic plots of volume versus crank angle for a scroll compressor can be seen in Bell (1).

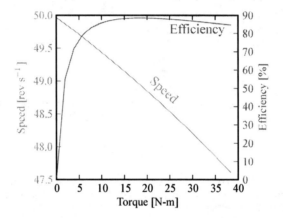

Figure 2: Motor curves for a hermetic compressor's motor

2.6 Motor
A motor is generally characterized by its slip speed, its electrical input power and its delivered mechanical torque. Motor manufacturers typically provide motor efficiency and slip speed curves or tabulated data as a function of the mechanical torque delivered. The slip speed is always less than the frequency of the applied electrical signal. The mechanical power output of the motor is given by $\dot{W}_{mechanical} = \omega \tau$ and the efficiency is given by

$$\eta_{motor} = \frac{\dot{W}_{mechanical}}{\dot{W}_{electrical}} \qquad (11)$$

The fraction of the electrical power that is not converted to mechanical power is lost as heat.

When the motor is run in reverse as an electrical generator, similar analysis applies.

2.7 Bearings
There are two primary families of bearings in positive displacement machines - dry bearings and oil-lubricated bearings. For oil-lubricated bearings, their performance

can be characterized by numerically solving Reynold's equation (23). For finite-length journal bearings, it is possible to utilize the numerical solutions of Raimondi and Boyd (24; 25; 26), or to use the Sommerfeld analytic solutions (23). The two most common types of bearings found in positive displacement machines are journal bearings and thrust bearings. Journal bearings are typically employed to support the radial load on a shaft, and can be quite efficient load-carrying devices when dimensioned properly.

3 CYCLE SOLVER

When the machine has reached steady-state operation, the state variables for each control volume (as well as for any valves) will be equal at the beginning and end of the cycle. In the simulation, this continuity must be enforced numerically, by active or passive means. In a following section the cycle solver is presented which enforces the continuity. Here we have assumed that an array of state variables $\vec{x}_0$ is available at the beginning of the cycle. This state variable array will arrive from the initialization as described below.

3.1 Differential Equation Integrator

The condition of the machine is known at the beginning of the cycle from the initialization protocol, and a system of tightly-coupled differential equations describes the future behavior of the machine. In order to obtain the behavior of the machine over the course of the cycle, it is necessary to integrate the coupled, non-linear system of equations. The set of differential equations that arise can be expressed in functional form as

$$\vec{f} = f(t, \vec{x}) \qquad (12)$$

where t is the variable of integration, and $\vec{x}$ is array of state variables and other variables that are involved in the integration. As an example, for a machine with two control volumes A and B that employ two dynamic valves 1 and 2, assuming the thermodynamic state variables that are used are T and ρ, the array $\vec{x}$ and the time-derivative of the array $\vec{x}$ are given by

$$\vec{x} = \{ \; T_A \quad T_B \quad \rho_A \quad \rho_B \quad y_1 \quad \dot{y}_1 \quad y_2 \quad \dot{y}_2 \; \} \qquad (13)$$

$$\vec{f} = \dot{\vec{x}} = \left\{ \frac{dT_A}{dt} \quad \frac{dT_B}{dt} \quad \frac{d\rho_A}{dt} \quad \frac{d\rho_B}{dt} \quad \frac{dy_1}{dt} \quad \frac{d\dot{y}_1}{dt} \quad \frac{dy_2}{dt} \quad \frac{d\dot{y}_2}{dt} \right\} \qquad (14)$$

Equivalently, the derivatives with respect to the crank angle may be used in place of the derivatives with respect to time.

There are a multitude of numerical integrators that may be used to integrate a system of non-linear differential equations. Numerical integrators for systems of differential equations can be generally classified based on the following taxonomy:

- Explicit/Implicit
- Single-Step/Multi-Step
- Fixed-Step-Size/Adaptive-Step-Size
- Order of Truncation Error

The selection of the best integrator for the simulation of a given machine is largely problem-dependent and iterative. If no further knowledge is available, it is in general recommended to begin with the Simple Euler Integrator with 5,000 to 10,000 steps per cycle. If the numerical instabilities remain too large, the number of steps should be increased. Three of the most-successful methods are presented below.

3.1.1 Simple Euler Integrator

In the simple Euler integrator, the derivatives are evaluated at the current step $(t_{old}, \vec{x}_{old})$, and new values of the state variable $(\vec{x}_{new})$ are predicted based on a linear extrapolation of the derivatives at this point. This can be expressed as

$$\vec{x}_{new} = \vec{x}_{old} + hf(t_{old}, \vec{x}_{old}) \tag{15}$$

which is a fixed step-size method, and exhibits error that is on the order of the step size to the first power. This method is quite simple to implement although it often struggles with severe numerical instabilities due to the stiffness of the system of differential equations.

3.1.2 Heun's Integrator

Heun's Integrator, also known as the Improved Euler Rule, or the Modified Euler Rule, is a predictor-corrector method. The new state variable array $\vec{x}_{new}$ is predicted using the same first step as in the Simple Euler Integrator, and the corrected step is obtained using an average of the derivatives at the beginning and predicted values:

$$\begin{aligned}\vec{x}^*_{new} &= \vec{x}_{old} + hf(t_{old}, \vec{x}_{old}) \\ \vec{x}_{new} &= \vec{x}_{old} + \frac{h}{2}\left[f(t_{old}, \vec{x}_{old}) + f(t_{old} + h, \vec{x}^*_{new})\right]\end{aligned} \tag{16}$$

which yields an error that is on the order of the step size to the second power.

3.1.3 Adaptive Runge-Kutta Integrator

The simple integrators (Euler and Heun) are not robust enough to handle some positive displacement machines. For instance, in scroll compressors it is common for the system of differential equations to be very stiff, requiring a very large number of steps in order to avoid numerical oscillations. Adaptive methods can more easily solve stiff system of differential equations, particularly when the the stiffness varies over a rotation.

For a given step size h, the adaptive Runge-Kutta method (27) is employed with the coefficients from Cash and Carp (28). This yields the following algorithm:

$$\vec{f}_1 = f(t_{old}, \vec{x}_{old}) \tag{17}$$

$$\vec{x}_1 = \vec{x}_{old} + h\left(\frac{1}{5}\vec{f}_1\right) \tag{18}$$

$$\vec{f}_2 = f\left(t_{old} + \frac{1}{5}h, \vec{x}_1\right) \tag{19}$$

$$\vec{x}_2 = \vec{x}_{old} + h\left(\frac{3}{40}\vec{f}_1 + \frac{9}{40}\vec{f}_2\right) \tag{20}$$

$$\vec{f}_3 = f\left(t_{old} + \frac{3}{10}h, \vec{x}_2\right) \tag{21}$$

$$\vec{x}_3 = \vec{x}_{old} + h\left(\frac{3}{10}\vec{f}_1 - \frac{9}{10}\vec{f}_2 + \frac{6}{5}\vec{f}_3\right) \tag{22}$$

$$\vec{f}_4 = f\left(t_{old} + \frac{3}{5}h, \vec{x}_3\right) \tag{23}$$

$$\vec{x}_4 = \vec{x}_{old} + h\left(-\frac{11}{54}\vec{f}_1 + \frac{5}{2}\vec{f}_2 - \frac{70}{27}\vec{f}_3 + \frac{35}{27}\vec{f}_4\right) \tag{24}$$

$$\vec{f}_5 = f(t_{old} + h, \vec{x}_4) \tag{25}$$

$$\vec{x}_5 = \vec{x}_{old} + h\left(\frac{1631}{55296}\vec{f}_1 + \frac{175}{512}\vec{f}_2 + \frac{575}{13824}\vec{f}_3 + \frac{44275}{110592}\vec{f}_4 + \frac{253}{4096}\vec{f}_5\right) \quad (26)$$

$$\vec{f}_6 = f\left(t_{old} + \frac{7}{8}h, \vec{x}_5\right) \quad (27)$$

The new values for the state variables $\vec{x}_{new}$ are given by

$$\vec{x}_{new} = \vec{x}_{old} + h\left(\frac{37}{378}\vec{f}_1 + \frac{250}{621}\vec{f}_3 + \frac{125}{594}\vec{f}_4 + \frac{512}{1771}\vec{f}_6\right) \quad (28)$$

and an estimate of the error is given by

$$\frac{\vec{\varepsilon}}{h} = -\frac{277}{64512}\vec{f}_1 + \frac{6925}{370944}\vec{f}_3 - \frac{6925}{202752}\vec{f}_4 - \frac{277}{14336}\vec{f}_5 + \frac{277}{7084}\vec{f}_6 \quad (29)$$

If the maximum absolute error given by $\varepsilon_{max} = \max(|\vec{\varepsilon}|)$ is less than the allowed error per step of $\varepsilon_{allowed}$, the step is accepted, and the step size used was unnecessarily small, so the step size for the next step is increased using

$$h_{next} = 0.9h\left(\frac{\varepsilon_{allowed}}{\varepsilon_{max}}\right)^{0.2} \quad (30)$$

If the error is too large ($\varepsilon_{max} > \varepsilon_{allowed}$), the step must be tried again, but this time with a smaller step. The new step size is obtained from

$$h_{next} = 0.9h\left(\frac{\varepsilon_{allowed}}{\varepsilon_{max}}\right)^{0.3} \quad (31)$$

The parameter 0.9 is a safety factor that forces the method to be conservative in its step resizing.

The Adaptive Runge-Kutta Integrator handles stiff systems of equations with relative ease, though it is more complex to implement than Heun's Integrator or the Simple Euler Integrator. In addition, the adaptive Runge-Kutta method can tailor the step-size to the instantaneous requirements in order to maintain the required error-per-step with the minimum of computational effort.

3.2 Continuity Solver

There are two basic families of methods available to enforce the continuity of the state array $\vec{x}$ at the end of one cycle and the beginning of the following cycle.

The first category of methods are passive methods that are based on beginning the next cycle at the state given by the end of the prior cycle. In this method, state variable array continuity is enforced by repetitively resetting the initial values for the state variables. If $\vec{x}_{end}$ is the end of one cycle, and $\vec{x}_{start}$ is the state array at the start of the following cycle, $\vec{x}_{start}$ is set to be equal to $\vec{x}_{end}$ until $|\vec{x}_{start} - \vec{x}_{end}|$ is less than the desired tolerance.

The second category are active methods that use numerical solvers of multi-dimensional non-linear system of equations in order to enforce continuity between cycles. In this case, the residual vector is given by $\vec{r} = \vec{x}_{start} - \vec{x}_{end}$ and the numerical solver is used to approach the solution $\vec{r} = 0$. The simplest variant of this method is the extension of the Newton-Raphson method to multiple dimensions, given by

$$\mathbf{J}\vec{v} = -\vec{r} \quad (32)$$

where the Jacobian matrix is given by

$$\mathbf{J} = \begin{bmatrix} \frac{\partial r_1}{\partial x_1} & \frac{\partial r_1}{\partial x_2} & \cdots & \frac{\partial r_1}{\partial x_n} \\ \frac{\partial r_2}{\partial x_1} & \frac{\partial r_2}{\partial x_2} & \cdots & \frac{\partial r_2}{\partial x_n} \\ \vdots & \vdots & \ddots & \vdots \\ \frac{\partial r_n}{\partial x_1} & \frac{\partial r_n}{\partial x_2} & \cdots & \frac{\partial r_n}{\partial x_n} \end{bmatrix} \quad (33)$$

and the solution for $\vec{x}_{start}$ is updated by

$$\vec{x}_{start} = \vec{x}_{start} + \vec{v} \qquad (34)$$

until $\max(|\vec{v}|)$ is less than the convergence criterion.

The Jacobian matrix **J** is evaluated numerically by columns using first-order finite differences. The increment in x_i for the finite difference should be selected with care to ensure that it is simultaneously a) small enough to limit truncation error in the finite difference and b) large enough to be at least a few orders of magnitude greater than the uncertainty in the outlet state vector from the integration of the state variables over the course of the cycle.

For this application, construction of the Jacobian matrix is computationally very expensive, so Jacobian updating methods can be used that update the Jacobian matrix using information about the shape of the solution surface rather than rebuilding the Jacobian matrix at every iteration of the solver. There are a number of methods in this family, of which Broyden's method (29) is the best known. Broyden's method can save a significant amount of computational effort, but as the accumulated error in the cycle integrator can be rather large if the step size is not conservative enough, it is sometimes necessary to rebuild the Jacobian matrix entirely as it no longer faithfully represents the shape of the functional surface due to numerically noisy Jacobian updates.

Selection of the active or passive method is problem-specific. Building the Jacobian matrix in Equation 33 takes $n+1$ cycles, where in each cycle one of the inputs of $\vec{x}_{start}$ is changed. If there are a large number of control volumes, building the Jacobian matrix is extremely computationally expensive and it is preferable to use the passive method, even though it takes quite a few cycles to reach adequate convergence. For instance, in scroll compressors there can be as many as 10 or 15 control volumes (20-30 state variables in total), and convergence is usually achieved in fewer than 30 cycles.

If on the other hand, there is a large disparity in thermal inertia between control volumes, the active method is preferable as it short-cuts the convergence process of the chamber with the slow time constant. This disparity in thermal inertia is commonly experienced in the case of a residential refrigerator compressor where the shell volume is much greater than the volume of the working chamber. As a result, the time constant of the shell volume is large due to its relatively large mass, and the passive method experiences very slow convergence.

4 OVERALL SOLVER

The outermost solver enforces energy balances on each of the lumped masses that form the machine and updates the outlet temperature of the machine. This solver also includes preconditioning and post-processing steps. Figure 3 provides a flowchart of the entire simulation.

4.1 Preconditioning
In the first level of the preconditioner, the guess value for the outlet temperature can be obtained by assuming a constant adiabatic efficiency on the order of the expected machine efficiency. In the second level of the preconditioner, the prediction from stage one is improved by running just one cycle without running the full continuity solver. Using the results from this cycle it is therefore possible to obtain a new estimate for the discharge temperature, the value of which is quite close to the discharge temperature at model convergence.

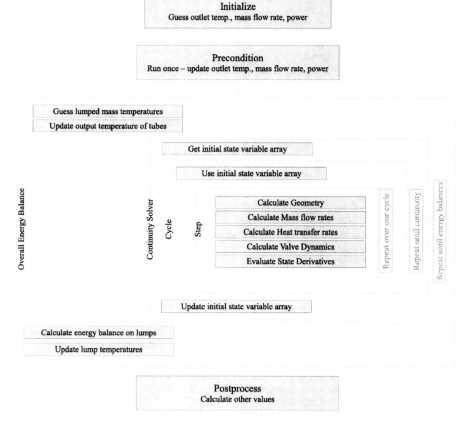

Figure 3: Flow chart of simulation

4.2 Solving

Once the preconditioning is complete, guesses are made for the lumped mass temperatures. These guesses are based on operator experience and are problem specific.

With the initial guess for the lump temperatures, the tubes are calculated in order to determine the tube pressures and enthalpies that the control volumes will interact with.

The overall solver is then executed. It enforces energy balances on each of the lumped masses by adjusting the lumped mass temperatures using a non-linear system of equations solver.

At each step of the overall solver, the continuity solver is executed. Within the continuity solver, the cycle is run numerous times. Within the cycle, the integrator is used to calculate the values over one cycle. The integrator integrates the values obtained from one step.

The hierarchical and nested behavior of the solvers result in a simulation algorithm that is fairly complex, though it is straightforward to simulate simple machines like a one-cylinder reciprocating compressor.

5 CONCLUSIONS

The above analysis has been implemented within a software package entitled PDSim (Positive Displacement SIMulation), which is an open-source code used to simulate the steady-state performance of positive-displacement compressors and expanders. The code is implemented in the programming language Python, with performance-critical components implemented in Cython(30). In this way, the high-level nature of Python can be maintained while achieving computational speeds equivalent to low-level programming languages like C++[2].

The code is highly object-oriented, and Python is a modern programming language that allows for easy modification and extension. To date, PDSim has been used to simulate reciprocating compressors and expanders, as well as scroll compressors.

REFERENCES LIST

[1] Ian Bell. *Theoretical and Experimental Analysis of Liquid Flooded Compression in Scroll Compressors*. PhD thesis, Purdue University, 2011.

[2] Y. Chen, N.P. Halm, E.A. Groll, and J.E. Braun. Mathematical Modeling of Scroll Compressor. Part I- Compression Process Modeling. *Int. J. Refrig.*, 25:731--750, 2002.

[3] Y. Chen, N.P. Halm, E.A. Groll, and J.E. Braun. Mathematical Modeling of Scroll Compressor. Part II- Overall scroll compressor modeling. *Int. J. Refrig.*, 25: 751--764, 2002.

[4] Vincent Lemort. *Contribution to the Characterization of Scroll Machines in Compressor and Expander Modes*. PhD thesis, University of Liège, 2008.

[5] Vincent Lemort, Ian Bell, Eckhard Groll, and James E. Braun. Analysis of liquid-flooded expansion using a scroll expander. In *19th Compressor Engineering Conference at Purdue University*, number 1323, 2008.

[6] Jun-Hyeung Kim. *Analysis of a Bowtie Compressor with Novel Capacity Modulation*. PhD thesis, Purdue University, 2005.

[7] Craig Bradshaw. *A Miniature-Scale Linear Compressor for Electronics Cooling*. PhD thesis, Purdue University, 2012.

[8] Craig R. Bradshaw, Eckhard A. Groll, and Suresh V. Garimella. A comprehensive model of a miniature-scale linear compressor for electronics cooling. *International Journal of Refrigeration*, 34(1):63 -- 73, 2011.

[9] M. M. Mathison, James E. Braun, and Eckhard A. Groll. Modeling of a two-stage rotary compressor. *HVAC&R Research*, 14:719--748, 2008.

[10] Miguel Jovane. *Modeling and analysis of a novel rotary compressor*. PhD thesis, Purdue University, 2007.

[11] Pawan J. Singh. A digital reciprocating compressor simulation program including suction and discharge piping. In *International Compressor Engineering Conference at Purdue 1984*, 1984.

[12] Thomas W. Bein and James F. Hamilton. Computer modeling of an oil flooded single screw air compressor. In *1982 International Compressor Engineering Conference at Purdue University*, pages 127--134, 1982.

[13] Pawan J. Singh and Ghanshyam C. Patel. A generalized performance computer program for oil flooded twin-screw compressors. In *1984 International Compressor Engineering Conference at Purdue University*, pages 544--553, 1984.

[14] Ian Bell. CoolProp: An open-source thermophysical property library - http://coolprop.sf.net, 2013.

---

[2]Examples are available in the source code for reciprocating compressors and expanders and scroll compressors

[15] S.W. Churchill. Friction factor equation spans all fluid flow regimes. *Chemical Engineering*, 84(24):91--92, 1977.

[16] V. Gnielinski. Neue gleichungen für den wärme-und den stoffübergang in turbulent durchströmten rohren und kanälen. *Forschung im Ingenieurwesen*, 41(1):8--16, 1975.

[17] Ian H. Bell, Eckhard A. Groll, James E. Braun, and W. Travis Horton. A computationally efficient hybrid leakage model for positive displacement compressors and expanders. *Int. J. Refrig.*, 2013.

[18] Noriaki Ishii, Kenichi Bird, Kiyoshi Sano, Mamoru Oono, Shinji Iwamura, and Takayuki Otokura. Refrigerant Leakage Flow Evaluation for scroll compressors. In *1996 Compressor Conference at Purdue University*, 1996.

[19] Kazutaka Suefuji, Masao Shiibayashi, and Kenji Tojo. Performance analysis of hermetic scroll compressors. In *1992 International Compressor Engineering Conference at Purdue University*, 1992.

[20] Kenji Tojo, Masato Ikegawa, Naoki Maeda, Sigeru Machida, Masao Shiibayashi, and Naoshi Uchikawa. Computer modeling of scroll compressor with self adjusting back-pressure mechanism. In *1986 International Compressor Engineering Conference at Purdue University*, 1986.

[21] Th. Afjei, P. Suter, and D. Favrat. Experimental analysis of an inverter-driven scroll compressor with liquid injection. In *1992 Compressor Engineering Conference at Purdue University*, 1992.

[22] E.W. Lemmon. Pseudo-Pure Fluid Equations of State for the Refrigerant Blends R-410A, R-404A, R-507A, and R-407C. *Int. J. Thermophys.*, 24(4):991--1006, 2003.

[23] Bernard J. Hamrock, Steven R. Schmid, and Bo O. Jacobson. *Fundamentals of Fluid Film Lubrication*. CRC Press, 2004.

[24] A.A. Raimondi and John Boyd. A Solution for the Finite Journal Bearing and its Application to Analysis and Design: I. *A S L E Transactions*, 1(1):159--174, 1958.

[25] A.A. Raimondi and John Boyd. A Solution for the Finite Journal Bearing and its Application to Analysis and Design: II. *A S L E Transactions*, 1(1):175--193, 1958.

[26] A.A. Raimondi and John Boyd. A Solution for the Finite Journal Bearing and its Application to Analysis and Design: III. *A S L E Transactions*, 1(1):194--209, 1958.

[27] Steven C. Chapra and Raymond P. Canale. *Numerical Methods for Engineers*. McGraw-Hill, 2006.

[28] J.R. Cash and Alan H. Karp. A Variable Order Runge-Kutta Method for Initial Value Problems with Rapidly Varying Right-Hand Sides. *ACM Transactions on Mathematical Software*, 16(3):201--222, 1990.

[29] C.G. Broyden. A class of methods for solving nonlinear simultaneous equations. *Mathematics of Computation*, 19:577--593, 1965.

[30] S. Behnel, R. Bradshaw, C. Citro, L. Dalcin, D.S. Seljebotn, and K. Smith. Cython: The best of both worlds. *Computing in Science Engineering*, 13(2): 31 --39, 2011.

A parallel object oriented code framework for numerical simulation of reciprocating compressors – introduction of solid parts modeling

J López[1], J Rigola[1], O Lehmkuhl[1,2], A Oliva[1]
[1] Universitat Politècnica de Catalunya (UPC),
    Centre Tecnològic de Transferència de Calor (CTTC), Spain
[2] Termo Fluids S.L., Spain

ABSTRACT

A partitioned coupled approach is employed to modeling a reciprocating compressor in a modular way. The approach allows the implementation of an object oriented parallel code framework for simulation of multiphysics systems in general and hermetic reciprocating compressors in particular.

Several works in compressor modeling have been presented before. Those works already addressed the resolution of the fluid flow by using this code framework. Now, a new model for simulation of solid components has been developed. In this way the thermal effect of the solid parts on the working fluid can be considered as well. Some numerical results are presented to show first achievements in this research line.

1 INTRODUCTION

In this work the simulation of a reciprocating compressor is addressed considering it as a thermal and fluid dynamic system. The simulation is done by using an in-house object oriented parallel code framework for simulation of multiphysics and multiscale systems. As stated by its name, this code is a generalist software solution to simulate problems involving mixed phenomena, commonly found in many engineering and science applications.

The single physics involved in a multiphysics problem usually have different mathematical and numerical properties, with distinct code implementation requirements. Therefore the implementation of monolithic schemes to solve the combined problem is software challenging and would lead to an inflexible tool. The main resolution algorithm of this code is based on a partitioned coupled approach (i.e. separate solvers are used for distinct physics). This provides a feasible and efficient way to simulate physical systems. Unlike monolithic approaches, partitioned strategies allow rapid setup of the simulated cases which is very useful when new system configurations must be tested. For example, consider the two compressor configurations in Figure 1. Although both schemes are quite similar, from the software point of view these few differences may involve annoying code changes when a monolithic approach is employed. The code framework used in this work makes easier to deal with these situations; the compressor components (or just elements) are treated as puzzle pieces and can be inserted, replaced or removed from the global computational model as appropriate.

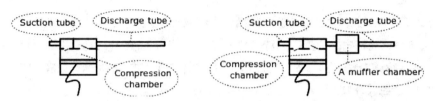

Figure 1. Two similar configuration schemes to show code modularity.

Numerical simulation of reciprocating compressors is not new (1-3). Some previous works (4-6) where addressed to the use of CFD&HT models (i.e. computational fluid dynamic and heat transfer models or just high-level models) in compressor simulations. Concretely, in those works a new CFD&HT model for the resolution of the recirculation gas into the shell was coupled with other zero and one-dimensional models for the resolution of the other parts of the compressor. Although those studies illustrated the possibilities of such type of simulations, the main contribution of those researches was the modular and efficient code being used here. Thanks to those investigations it is now possible to run system simulations by modeling each involved phenomena with the desired level of detail (CFD&HT, SIMLEC based (10), zero-dimensional models, etc.).

A detailed CFD&HT solution of the whole domain of a reciprocating compressor is now unfeasible and probably not necessary. The use of multiple models allows an efficient management of the computational resources by taking more attention on the critical parts of the compressor. For example, detailed studies such as in (4-6) can reveal important information about the recirculation gas into the shell, which could contribute in better geometry designs improving the compressor efficiency. Nevertheless, other compressor parts such as internal tubes or chambers can be solved by means of simpler low-level models (with low computational expenses) without put in risk the reliability of the numerical results.

Up to now, an integrated solution of the fluid dynamic phenomenology into the compressor system has been successfully addressed. However the solid parts of the compressor (tube walls, motor and crankcase body, the shell, etc.) were not modeled and constant temperatures should be assumed. The aim of the current paper is the development and coupling of a new model to take into account the thermal effect of the solid parts on the global compressor behavior. In order to check the reliability of the code after those new changes, some detailed and global illustrative numerical results from the simulation of a domestic R134a reciprocating compressor are presented at the end of the document.

2 COMPRESSOR MODELING APPROACH

Viewed as a thermal and fluid dynamic system, a reciprocating compressor is composed of several elements such as tubes, chambers, valves, the compression chamber and so on (see Figure 1 and Figure 2). Each element has some physical entity representing one or several phenomena (fluid flow, heat transfer, etc.) occurring inside its domain.

The partitioned approach employed for the numerical simulation of the compressor assumes that each element composing the system must be able to solve itself by giving the appropriate boundary conditions. Therefore each component in a simulation must be assigned with a specific computational model. This model can be queried in order to know the state of the element and/or update the boundary conditions of its surrounding elements.

This section provides a brief explanation of the computational models (zero and one-dimensional models) employed in the simulation of reciprocating compressors. Some details about the global resolution algorithm are also commented at the end. Deeper understanding on mathematical formulation of the models can be found in (1-3). For details in the CFD&HT models one must refer to (7-9).

2.1 Computational models for simulation of reciprocating compressors

As commented above, an appropriate computational model is assigned to each of the identified compressor elements. These can be zero or one-dimensional models, here referred as low-level models, or can be detailed CFD&HT models, referred as high-level models.

The general one-dimensional conservation equations for fluid flow are semi-discretized as:

$$\frac{\partial m}{\partial t} + \Sigma \dot{m}_e - \Sigma \dot{m}_w = 0 \tag{1}$$

$$\frac{\partial m \bar{v}}{\partial t} + \Sigma \dot{m}_e v_e - \Sigma \dot{m}_w v_w = F_s \tag{2}$$

$$\frac{\partial m(\bar{h} + \bar{e}_c)}{\partial t} + \Sigma \dot{m}_e (h_e + e_{c,e}) - \Sigma \dot{m}_w (h_w + e_{c,w}) = \dot{Q}_{wall} + \bar{V}\frac{\partial \bar{p}}{\partial t} \tag{3}$$

Tube elements are solved according to pressure-based methods. The SIMPLEC algorithm is used with staggered mesh for velocity map. Upwind criteria are used for convective terms (10).

In the same way, pressure correction approach for the compression chamber is used as well. Chambers treatment is similar to that of compression chamber with the volume at the current time step V equal to that at the previous instant V_0, as the volume of chamber remains constant.

Valve orifices elements are always connected between a chamber and a compression chamber. These elements implement both a multidimensional model based in modal analysis as well as a simplified model based on a mass-spring system. The resolution of the valve provides the instantaneous valve position and an effective flow area. After sending these parameters to the neighbor Chamber elements they are employed into the momentum equation. The presented numerical results are obtained by using the simplified model.

There is a specific element called Fixed Value used only to fix boundary conditions. It provides for example pressure, temperature, humidity, etc. These types of elements are usually employed in the inlet and outlet of the compressor domain in order to fix the input temperature and pressure and the output pressure.

Regarding to the solid elements the following fully implicit discrete energy equation can be applied. This equation assumes possible heat conduction between solid parts, convection heat exchanges with the fluid flow and heat transport by radiation.

$$\frac{\rho_k^{n-1} \bar{c}_p (T_k^n - T_k^{n-1}) V_k}{\Delta t} = \sum_j \frac{T_k^n - T_j^n}{R_{kj}} A_{kj} + \sum_i \dot{Q}_{ki}^{conv,n} + \sum_j \dot{Q}_{kj}^{rad,n} \tag{4}$$

Convection heat exchanges are evaluated using empirical information. Heat transfer coefficients employed inside chambers, tubes and the compressor chamber, as well as the one used to evaluate heat exchange between the shell and the environment, have been evaluated in the same way as (2). On the other side, radiation heat exchanges can be considered between the shell, the environment and the motor. In such case it can be assumed that radiation transport take place between concentric spheres.

Finally when some elements are modeled by using CFD&HT the TermoFluids code (7) is used. The modular strategy of the code framework makes possible to integrate the computational models implemented into the TermoFluids code (high-level models) with the rest of models (low-level models).

2.2 Global resolution algorithm

The global resolution algorithm is aimed to obtain an integrated solution of the compressor system by coupling the computational models used to solve each of its elements.

As mentioned before, all compressor elements are capable of solving itself for given the necessary boundary conditions. These boundary conditions are taken from the neighbor elements at each system iteration in order to determine the state of the element by solving its governing equations (e.g. momentum, pressure correction, energy, etc.). At the same time, the solution of such equations provides the set of boundary conditions for neighbors' resolution at the next iteration. Iterations continue until convergence is reached at a given time step and then the next time step calculation starts after updating the variables.

The above strategy to solve the system of equations posed by the whole compressor problem is based on a Block-Jacobi method. It has been observed that this algorithm performs well when the fluid flow in the compressor is solved assuming constant solid temperatures. However, introducing the thermal effect of the solid parts makes the global resolution algorithm more challenging. Although the compressor simulation does still converge to stationary cyclic conditions, the Block-Jacobi strategy may not be the best choice. For this reason, the introduction of alternative solvers such as based in Newton-Raphson strategies or genetics based solvers will be object of study in the future. Anyway, as already commented, the current implemented methodology is enough to solve the problem with the prescribed conditions.

Regarding the time integration scheme, note that the time step used in the solids resolution can be the same used in the fluid flow resolution. However, time scales for the solid elements are usually much larger than for the fluid flow. Hence, larger time steps are employed for the time integration scheme of solid parts in order to reduce computational cost. In the simulations presented in the following sections the time step is of the same order as the cycle time (i.e. $\Delta t = 1/f_n$, where f_n stands for the nominal frequency of the compressor).

3 CASE DESCRIPTION

3.1 Compressor configuration

In Figure 2 it is shown the scheme of the compressor configuration simulated in the present work. As indicated, the scheme is mostly composed by chamber elements connected by tube elements. Note that the most relevant parts of a reciprocating compressor have been represented by one or several elements. For example, the suction muffler is represented by a set of resonator chambers and tubes, while the compression chamber is represented by a single chamber element. Several solid elements are also included in the simulation although they are not present in this picture (see Figure 4). Note that the motor, the crankcase and the cylinder head have been grouped into one single solid element.

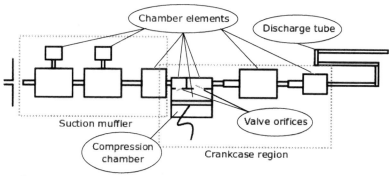

Figure 2. Configuration scheme of the simulated compressor.

3.2 Working conditions

The results presented in the following section are obtained from the simulation of a domestic hermetic reciprocating compressor working with R134a as refrigerant. The compressor capacity is 7.5 cm³, the evaporation temperature -10°C and the condenser temperature is 55°C. The inlet compressor temperature is set to 32°C and the environment temperature is set to 30°C. The motor electrical efficiency is assumed to be 0.7 in order to considerate motor heating due to electrical and mechanical losses. The compressor geometry parameters are presented in Table 1.

Table 1. Compressor geometry parameters

Compressor chamber		Motor	
Cylinder diameter	24.29 mm	Crank radius	7.95 mm
Clearance ratio	1.69 %	Connecting rod length	30.00 mm
Suction valve		Discharge valve	
Valve stop	-	Valve stop	1.194 mm
First frequency	269.2 Hz	First frequency	325.2 Hz
Suction port		Discharge port	
Diameter	6.3 mm	Diameter	5.4 mm

4 NUMERICAL RESULTS

In this section the results obtained from the numerical simulation of the above reciprocating compressor are presented. These results are qualitatively contrasted with experimental data from (2). It must be highlighted that the aim of this comparison is not the validation of the employed code but to verify the numerical predictions are in good agreement with experimental data.

Some global comparative results are detailed in Table 2 considering mass flow rate, power consumption, volumetric efficiency as $\eta_v = \dot{m}/(\rho_{in} V_{cl} f_n)$ and COP as $Q_{ev}/\dot{W}_e$ on the prescribed working conditions. It can be observed that the volumetric efficiency is overestimated by approximately 12%. The power consumption and the COP are also overestimated although in a smaller percentage. These discrepancies could probably diminish by considering more solid regions; in this sense the cylinder head could be defined separately from the motor and the crankcase since it is a singular region due to compression chamber proximity. Furthermore, thermal effects due to the oil pool should also be considered. The absence of the oil could clearly explain the overheating in the discharge part of the compressor observed in Figure 4. In Figure 3 some instantaneous illustrative numerical data is presented in order to show some detailed results available by means of the simulation.

Table 2. Global comparative results with R134a at $T_{evap}=-10°C$

	$\dot{m}$ (kg/h)	η_v %	$\dot{W}_w$ (W)	COP %
Experimental	7.18	63.6	207	1.74
Numerical	8.08	71.64	218	1.80
Discrepancy %	12.5	12.5	5.0	3.5

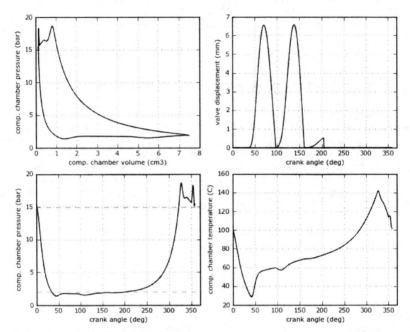

Figure 3. Some detailed instantaneous numerical illustrative results.

The temperature of the main solid parts (i.e. shell, suction muffler, motor, cylinder head and crankcase) in the stationary cyclic conditions of the compressor are compared with experimental values in Figure 4. As can be observed most fluid and solid temperature predictions are similar to the experimental values. However it must be highlighted the overheating of the fluid in the discharge point (F8). As mentioned above it could be explained by the absence of the oil pool in the simulation.

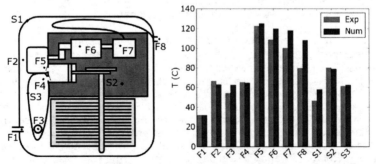

Figure 4. Comparison between experimental and numerical temperatures at different positions of the reciprocating compressor.

5 CONCLUSIONS

A parallel object oriented code framework for numerical simulation multiphysics systems have been used to simulate a domestic reciprocating compressor working with R134a as fluid refrigerant. The computational models for resolution of the fluid flow and the solid elements have been briefly explained.

Some global and detailed illustrative numerical results obtained from the simulation have been presented. These results are in good agreement with experimental data from (2), although some discrepancies are observed probably to the absence of the oil pool or other simplifications in the solid part resolution.

The resolution of the solid parts and its coupling with the fluid dynamic models is a step forward to the achievement of an efficient and modular tool for numerical simulation of reciprocating compressors. The modularity provided by the code framework made possible the development and integration of the solid computational model in a rapid and easy way. It must be highlighted that no changes where necessary in the code kernel to introduce this new model. Moreover nor the efficiency of the code neither the reliability of the results is affected by this code extension. These points are of great significance in software development.

A global experimental validation has been carried out (Table 2) while a first detailed thermal analysis has also been presented (Figure 4). Deeper analysis of the numerical simulations must be done in order to understand better the current discrepancies and reduce them. After that, the code will be able to use in parametric studies aimed to improve compressors efficiency.

REFERENCES

(1) Pérez-Segarra C.D., Rigola J., Oliva A., 2003, Modeling and numerical simulation of the thermal and fluid dynamic behavior of hermetic reciprocating compressors. Part 1: Theoretical basis, Int. J. Heating, Ventilating, Air-Conditioning Refrigerating, 9: p. 215-236.
(2) Pérez-Segarra C.D., Rigola J., Oliva A., 2003, Modeling and numerical simulation of the thermal and fluid dynamic behavior of hermetic reciprocating compressors. Part 2: Experimental investigation, Int. J. Heating, Ventilating, Air-Conditioning Refrigerating, 9: p. 237-249.
(3) Damle R., Rigola J., Pérez-Segarra C.D., Oliva A., 2008, An object oriented program for the numerical simulation for hermetic reciprocating compressors behavior, 19th International Compressor Engineering Conference, 1402, pp.1-8, Purdue University, IN, USA.
(4) López J., Rigola J., Lehmkuhl O., Oliva A., 2009, Numerical study of suction gas flow in the shell of hermetic refrigeration compressors, Int. Conference on Compressors and their Systems, City University, London, UK: C682/037 p. 395-404.
(5) López J., Rigola J., Lehmkuhl O., Oliva A., 2010, Introduction of CFD&HT Analysis into an Object Oriented One Dimensional and Transient Program for Numerically Simulate Hermetic Refrigeration Compressors, Int. Compressor Engineering Conference at Purdue, USA: p. 1382-1390.
(6) López J., Lehmkuhl O., Rigola J., Pérez-Segarra C.D., 2012, Use of a Low-Mach Model on a CFD&HT Solver for the Elements of an Object Oriented Program to Numerically Simulate Hermetic Refrigeration Compressors. Int. Compressor Engineering Conference at Purdue, USA.
(7) Lehmkuhl O., Borrell R., Pérez-Segarra C.D., Soria M., Oliva A., A new parallel unstructured CFD code for the simulation of turbulent industrial problems on

low cost PC cluster, Parallel Computational Fluid Dynamics, Ankara, Turkey, 2007.
(8) Lehmkuhl O., 2012, Numerical Resolution of Turbulent Flow on Complex Geometries, Ph.D. thesis, Universitat Politècnica de Catalunya.
(9) Borrell R., 2012, Parallel Algorithms for Computational Fluid Dynamics on Unstructured Meshes, Ph.D. thesis, Universitat Politècnica de Catalunya.
(10) S.V. Patankar, Numerical heat transfer, Hemisphere Publishing Corp., 1980.

Analysis of the basic geometrical parameters influence on the efficiency of the Roots-type compressor on the basis of thermodynamic processes simulation

A M Ibraev, S V Vizgalov, I G Khisameev
Kazan National Research Technological University, Russian Federation

ABSTRACT

Roots compressors are widely used in various industrial processes for compression and transportation of gas streams, which can contain additional inclusions in the form of liquid or solid particles (1, 2). These compressors have a number of advantages connected with rotor manufacturing and machine operation in the range of pressure ratio from 1.2 to 1.6. In this range Roots compressors are successfully competitive with the other types of rotary machines, including screw compressors, in mass, sizes and power efficiency.

The authors analyze the influence of rotor geometrical parameters on the integral characteristics of the compressor based on the mathematic model.

The paper provides the recommendations for the ranging of compressor geometrical parameters.

1 INTRODUCTION

Nowadays the solution of current problems such as decrease of power consumption of the compressor and increase of its efficiency is possible due to the rotor profile improvement, the rational selection of its geometrical parameters and parameters of compressor housing depending on its operation modes, including the input and output pressures, rotor rotational speed, and the parameters of gas to be compressed.

Fig. 1 depicts the Roots compressor, which is a two-lobe machine with spur rotors. The rotors have equal outside diameter and equal number of teeth of the male and female rotors, which is equal to 2 or 3.

Rotor profile defines the theoretical capacity of the compressor V_T, the function of volume change of working space according to the angle of rotor turning, and sealing properties of the clearances between working spaces at given outside diameter D, rotor length L, and the distances between axes A. The parameters of the rotor profile influence the effective capacity V_d and the efficiency parameters such as the volumetric efficiency η_v, the indicated efficiency η_{ind}, and the specific indicated capacity $\overline{N}_{ind}$.

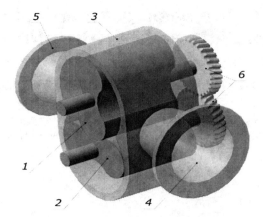

Fig. 1 Diagram of Roots type compressor:
1 – male rotor; 2 – female rotor; 3 – housing;
4 – suction pipe; 5 – discharge pipe; 6 – gears

2 GEOMETRICAL ANALYSIS

The ideal compressor is considered to be a unit without gas back streaming through the clearances between the working spaces, without heat exchange between the gas and the walls, and without gasdynamic losses during gas flow. In this case the theoretical capacity of the Roots type compressor is defined as

$$V_T = \frac{\pi D^3}{2} \cdot k_s \cdot k_L \cdot n_0, \ m^3/s \qquad (1)$$

where k_S is the volume utilization factor, k_L is the relative rotor length ($k_L = L/D$), n_0 is the rotational speed of the rotor, rps.

Value k_S is defined by the area of the rotor face F_{tor}:

$$k_S = 1 - \frac{4 \cdot F_{tor}}{\pi \cdot D^2}. \qquad (2)$$

The theoretical capacity is a function of rotor profile, which is characterized by the parameter k_S. The k_S depends on the type of the curve defining the profile of face cross-section, the rotor face angle ψ_p and the relative distance between axes $\overline{A} = A/D$.

A few main types of profiles are applied in Russian and Ukrainian industry. They are involute, elliptic (circumferential), and linear profiles (3). Fig. 2 shows these types of profiles, as well as their characteristic sections (Fig. 2 illustrates the ¼ part of the whole image of two-lobe rotor). A rotor lobe is divided into two parts (the peak and the root with the radius a=D/2). Each profile has the facing area A-B on the peak which represents a circular arc with the radius R=D/2 (its boundaries are defined by the facing angle ψ_p), and also the facing area E-F on the rotor root in the form of a circular arc with the radius $c = D \cdot (\overline{A} - 0,5)$. Facing areas are intended for the formation of deep clearances with equidistant walls and for the reduction of leakages through the radial clearances along the rotor outside diameter.

The involute profile (Fig. 2a) includes the areas B-C and C-D, formed by the involute circle and the area D-E which is an extended involute. Facing angle on the root is a little larger than that on the peak: $\psi_p' > \psi_p$. This profile is obtained by spinning.

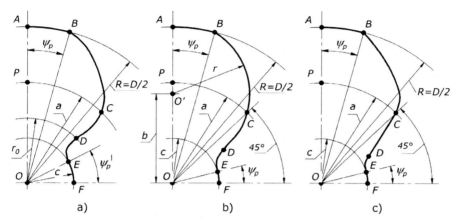

Fig. 2 Profiles of a rotor of the Roots type compressor
a - involute, b - circumferential, c - linear

Circumferential profile (fig. 2b) is formed by the following sections: B-C is a circular arc, with the radius r and the centre which is in the point O', located on the profile axis at the distance b from the centre of rotation – point O; section C-D is the curve of spinning obtained from a condition of engagement of two rotors, and D-E is a normal epicycloid, described by the point B of the other rotor. Independent parameters for specifying the profile of this rotor are the three following parameters - $D, \overline{A}, \psi_p$. The other two parameters can be written as:

$$b = \frac{D \cdot (1 - \overline{A}^2)}{4 \cdot (\cos\psi_p - \overline{A}\sqrt{2}/2)}, \quad r = 0{,}5\sqrt{D^2 + 4 \cdot b^2 - 4 \cdot D \cdot b\cos\psi_p} \;. \quad (3)$$

Linear profile (fig. 2c) has a section C-D on the rotor root having the form of a straight line, tangent to the epicycloid D-E, section B-C on the rotor peak is obtained by spinning, its coordinates are defined subject to the condition of the rotor gearing. Analysis of the linear profile and equations for the definition of coordinates of its sections are given in work (4).

According to the conditions of the rotors gearing and of the rotor root strength, the parameter $\overline{A}$ can not be less than 0.6 ... 0.62. Its upper limit usually makes 0.72 ... 0.73, and the facing angle ψ_p does not exceed 15 deg. Analysis of the influence of parameters $\overline{A}$ and ψ_p also has shown that with the increase of $\overline{A}$ value the area, occupied by the rotor body, increases. It essentially reduces the volume utilization factor, while the dependence of k_s on $\overline{A}$ is linear (Fig. 3). The profile type makes considerably smaller impact on the k_s and theoretical capacity. The circumferential profile, having the largest tooth peak end area among the profiles being considered, has the smallest k_s in the characteristic. The involute profile, possessing the least full peak, has a higher value of k_s. The linear profile occupies the intermediate position. With the growth of $\overline{A}$ the effect of profile type decreases and the curves converge.

Profile facing in much smaller extent influences the magnitude of k_s. With the growth of ψ_p from 0 deg. (the profile without facing) to 18 deg., the k_s decreases no more than by 2...3 %, and at $\psi_p > 10$ deg. the influence of profile type has no effect on the magnitude of k_s. Insignificant decrease of k_s shows the expediency of increase of the rotor facing angle. It can be used as a method for increasing compressor efficiency by better sealing of radial clearances. Limitation factors of the ψ_p angle magnitude are the simultaneous growth of transfer and jammed volumes arising in the engagement and the decrease of tightness of the profile slot in the engagement of the rotors that can be defined by calculations using the mathematical model.

Thus, the geometrical analysis of three types of rotor profiles showed that it is more preferable to use rotors with a smaller value of the $\overline{A}$ parameter at the level of 0.615...0.63, and facing at the level of $\psi_p = 5 ... 10$ deg.

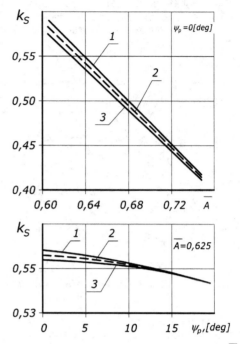

Fig. 3 Dependence of k_s parameter on $\overline{A}$ and ψ_p: 1 - involute profile, 2 - linear profile, 3 - circumferential profile

3 MATHEMATICAL MODEL

More complete investigation of the influence of the rotor parameters on compressor performance was carried out by means of mathematical model. At any angle of rotation of the male rotor there are three chambers in compressor with different pressures, which contact with each other through the clearances: δ_R is the radial clearance, δ_P is the profile clearance and δ_T is the face clearance. Fig. 4 presents the lay-out diagram of the chambers. A is the suction chamber connected with the suction port. Its volume increases during the rotor rotation due to the rotor drops out of a gear. The chamber B is the

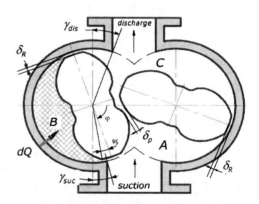

Fig. 4 Diagram for mathematical model of Roots compressor

isolated chamber, where gas is transferred from the suction side to the discharge side without volume change. C is the discharge chamber, the volume of which decreases due to the rotor engages a gear. Each chamber has a certain law of the volume variation depending on the angle of the male rotor rotation $V = f(\varphi)$.

Mathematical model of the Roots type compressor was developed on the basis of energy equilibrium equations in the working chamber and the ideal gas law. It takes into account the chamber volume change according to the rotor rotation angle, the gas leak-in during the suction period, the gas mass transfer through the clearances between the adjacent chambers having various pressures, the gas heat exchange with the chamber walls, and the gas back-streaming into the working chamber at the discharge side. In this case the gas thermodynamic parameters in the volume of the working chamber are homogeneous at any period of time. The pressure and temperature gradients are zero in the chamber.

The initial equations of the mathematical model are the equation of the first law of thermodynamics for gas mass in the working chamber and the equation of state

$$dQ + \sum (dM_{inf.j} \cdot h_{inf.j}) - \sum (dM_{lks.j} \cdot h) = dU + dL ; \qquad (4)$$

$$p \cdot V = M \cdot R \cdot T ; \qquad (5)$$

where dQ is the elementary change of heat entered the working chamber as a result of convective heat transfer through the walls; $\sum (dM_{inf.j} \cdot h_{inf.j})$ is the total power interchange in the working chamber occurred due to the gas leakage through the suction port, discharge port and the clearances; $\sum (dM_{lks.j} \cdot h)$ is the power interchange occurred due to the gas leaks from the chamber through the clearances; M is the gas mass in the working chamber, $M = M_0 + M_{inf} - M_{lks}$; $h_{inf.j}$ is the gas enthalpy of j-th inflow, h is the gas enthalpy in the working chamber; dU is the elementary change of the gas intrinsic energy in the working chamber; dL is the elementary change of work done by gas; R is the gas constant; p, T are pressure and temperature of gas in the working chamber. Indices: "inf" relates to the gas inflows into the chamber, "lks" relates to the leaks of gas from the chamber, "0" is the starting value.

After some transformations (3) we obtained the following differential equations describing the gas pressure and temperature change in the working chamber, depending on the angle of rotor rotation

$$\begin{cases} \dfrac{dp}{d\varphi} = \dfrac{k-1}{\omega \cdot V} \cdot \left(\dfrac{\omega \cdot dQ}{d\varphi} + \sum \dot{m}_{inf} \cdot h_{inf} - \sum \dot{m}_{lks} \cdot h - \dfrac{k}{k-1} \cdot \omega \cdot p \dfrac{dV}{d\varphi} \right) \\ \dfrac{dT}{d\varphi} = \dfrac{(k-1) \cdot T}{P \cdot V \cdot \omega} \cdot \left(\dfrac{\omega \cdot dQ}{d\varphi} + \dfrac{k-1}{k} \cdot \left(\sum \dot{m}_{inf} - \sum \dot{m}_{lks} \right) \cdot h + \sum \dot{m}_{inf} \cdot h_{inf} - \sum \dot{m}_{inf} \cdot h - p \dfrac{\omega \cdot dV}{d\varphi} \right) \end{cases} \qquad (6)$$

where k is the gas adiabatic exponent (k=1,4 for air); ω is the rotor angular velocity, rad·s^{-1}; m is the mass flow through the ports and clearances. The gas enthalpy in this case can be defined as for the ideal gas $h = C_p \cdot T$, where C_p is the isobaric heat capacity of gas.

The system of equations (6) is supplemented with the dependence of the working chamber volume on the angle of the rotor rotation in the form of $V = f(\varphi)$, which for the Roots type compressor was defined by the numerical method for each chambers: the suction chamber, isolated chamber and discharge chamber.

There are several working stages of the Roots type compressor, which differ from each other by the character of volume chamber change and the leakage arrangement:
- suction into the double chamber with the increasing volume;
- selection of a working chamber at the suction;

- carrying over the "isolated" working chamber of fixed volume to the discharge side;
- disclosure of the working chamber at the discharge;
- double chamber with the decreasing volume at the discharge.

Simultaneously there are several chambers which contact with each other trough the clearances. Here various stages of the working process take place, displaced from each other according to the angle of the rotor rotation, by the angles, divisible by 90 deg. It is sufficient to carry out the simulation for only one working chamber with the volume V as the rest are the same chambers for the rotation angles φ±90 deg, φ±180 deg and etc.

The member of equations (6), taking into account the convective heat transfer between the gas and the walls of the chamber was determined by the function

$$\frac{dQ}{d\tau} = \omega \cdot \frac{dQ}{d\varphi} = \overline{\alpha}(\varphi) \cdot f_w \cdot (T_w - T), \qquad (7)$$

where the heat transfer coefficient $\overline{\alpha}$ averaged over the surface of the walls f_w was determined according to (5), where the experimental data were summarized by the equation of Nusselt for the periods of suction, transfer of the isolated chamber, and discharge:

$$Nu(\varphi) = B \cdot Re(\varphi) + A_1 \cdot Pr + A_2, \qquad (8)$$

where the coefficients A_1, A_2, and B depend on the parameter П; Re, Pr are the Reynolds and Prantdl numbers. Characteristic velocity was the linear velocity of the chamber centre $v=\omega \cdot D/4$, characteristic length was the hydraulic diameter of the working chamber. For the period of the opening of the working chamber at the discharge the period average values of the heat transfer coefficient were determined explicitly using an empirical equation:

$$\overline{\alpha} = 836,451 \cdot П + 0,518 \cdot n_0 - 1734,75 \qquad (9)$$

where n_0 is the rotational speed of the rotors, expressed in rpm. This equation has a range of applicability of П = 1,2 ... 1,6 and n_0 = 1500 ... 3000 rpm.

The mass flow of gas through the suction and discharge ports was determined as a function of their area f:

$$\dot{m}_w = \mu \cdot f \cdot W_{ad} \cdot \rho_1 \qquad (10)$$

where μ is the flow coefficient of the port, determined experimentally by the method of static blowing for various positions of rotors; ρ_1 is the gas density before the port, W_{ad} is the speed of adiabatic gas flow through the port, which is defined as

$$W_{ad} = \sqrt{2RT_2 \frac{k}{k-1}\left[1 - \left(\frac{p_1}{p_2}\right)^{\frac{k-1}{k}}\right]} \quad - \text{when} \quad \frac{p_1}{p_2} \geq \beta = \left(\frac{2}{k+1}\right)^{\frac{k}{k-1}}; \qquad (11)$$

$$W_{ad} = \sqrt{2RT_2 \frac{k}{k+1}} \quad - \text{when} \quad \frac{p_1}{p_2} < \beta, \qquad (12)$$

where k is the adiabatic exponent of gas.

Gas flow through the narrow clearances between the rotors and the housing compressor was determined by the method proposed by S.E. Zakharenko (6):

$$\dot{m} = \sqrt{\frac{\rho_1 \cdot p_1 \cdot \ell^2 \cdot \delta^2 \left[\left(\frac{p_1}{p_2}\right)^2 - 1\right]}{\ln(p_1/p_2) + \xi + \lambda \cdot \Delta}} \text{ ,kg/s} \qquad (13)$$

where ℓ, δ - clearance sizes, the front length and the minimum height of clearance; ξ - coefficient of local resistances at the inlet and the outlet of the clearance due to a sudden contraction and expansion; λ - friction coefficient of the gas in the clearance as a function of the Reynolds number; Δ - shape factor of the clearance.

The system of equations of the mathematical model was solved numerically using the method of Runge - Kutt - Feldberg (8). Diagrams of pressure and temperature according to the angle of the rotor rotation resulted from solving the equations of mathematical model. Indicated power and indicated efficiency of the compressor were determined by integrating the equation $p = f(\varphi)$

$$N_{ind} = 4 \frac{\omega}{2\pi} \int_{\varphi_0}^{\varphi_k} p(\varphi) d\varphi \text{ ;} \qquad (14)$$

$$\eta_{ind} = N_{ad}/N_{ind} \text{ ,} \qquad (15)$$

where N_{ad} is the adiabatic compressor power.

Volumetric efficiency was determined using two formulas to control the accuracy:

$$\eta_v = 1 - \frac{\Sigma V_{inf.j}}{V_{is.cham}} \text{ ;} \quad \eta_v = \frac{\Sigma V_{inf.win}}{V_{is.cham}} \text{ ,} \qquad (16), (17)$$

where $\Sigma V_{inf.j}$ is the total volume of gas inflows into the suction chamber from the chamber V_{+180}, V_{+90} during the period of suction, referred to the suction conditions; $\Sigma V_{inf.win}$ is the amount of gas inleakage into the suction chamber only through the suction port; $V_{is.cham}$ is the volume of the isolated working chamber separated from the suction chamber at the angle, equal to $(90°-\gamma_{suc}-\psi_p)$.

The actual capacity of compressor can be defined as

$$V_d = 4 \cdot V_{is.cham} \cdot \eta_v \cdot n_0 \text{, m}^3/\text{s,} \qquad (18)$$

where n_0 is the rotor speed, rps.

Specific indicated power:

$$\overline{N}_{ind} = N_{ind}/V_d. \qquad (19)$$

The calculated results were compared with the data obtained from the experimental investigation of three types of Roots compressors with different profiles of the rotors. The discrepancy between the model and the experiment was not more than the error of experimental data. This confirms the adequacy of the proposed model.

4 RESULTS AND DISCUSSION

The developed mathematical model for Roots type compressor was used for determining the effect of the rotor facing angle ψ_p and relative spacing on centers $\overline{A}$ on the parameters of compressor effectiveness ($\overline{N}_{ind}$, η_v, η_{ind}). Calculations were carried out with constant values of the installed clearances between rotors and housing. The changes of working clearances due to thermal deformation of compressor rotors and housing were taken into account. Air was considered as the working gas. The dependences obtained are presented in Fig. 5, 6.

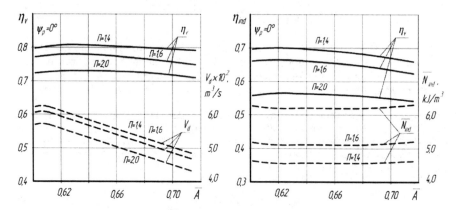

Fig. 5 Influence of the $\overline{A}$ parameter on compressor characteristics

Fig. 5 shows the characteristics of the Roots type compressor depending on the value of $\overline{A}$ at different Π for the circumferential profile (for other types of profiles the values are quite close). As expected, the capacity of compressor is reduced with the growth of the $\overline{A}$. The linear decrease of the V_d allows us to conclude that the decrease of the k_s value and, respectively, the theoretical capacity with the growth of $\overline{A}$ have the significant influence on the parameter $\overline{A}$.

In the rest investigated area the energy parameters $\overline{N}_{ind}$, η_{ind} decrease, and volumetric efficiency was observed with the growth of the $\overline{A}$. The highest values of η_{ind}, η_v are in the area of lower values of this parameter. However, the decrease of the $\overline{A}$ below 0.62 is unacceptable due to the risk of the reduction rigidity of rotor and the damage of mechanical design. For some of the profiles, in particular, for involute ones, tightness of gearing is lost. Depending on the mode of operation of the proposed compressor and on the design of its supports, it is recommended to choose the value of $\overline{A}$ ranged between 0.62 and 0.64.

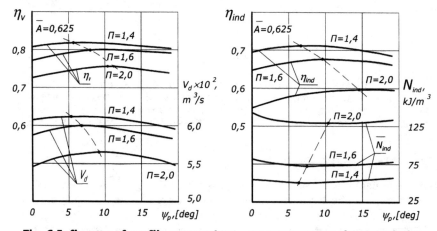

Fig. 6 Influence of profile parameter ψ_p on compressor characteristics

Fig. 6 shows the influence of the facing angle ψ_p on the power and volume characteristics. There is a noticeable improvement of performance due to the increase of depth of the radial slot and the reduce of leakages through these slots. However, at ψ_p = 5 ... 7 deg. the parameters reach the values which are close to the limit ones, and then under further increase of ψ_p they may slightly decrease. This can be explained by the fact that, even the low facing angles result in the reduction of gas leakage through the radial slots, and they do not influence the total leakage through all slots. Therefore, their further increase does not have the impact on performance. On the other hand, facing angle increase changes the influence of the other parameters. There is a fall of the volume utilization factor (Fig. 3) and theoretical capacity, as well as tightness of the profile clearance is damaged due to the changes in its shape, the values of crossover and jammed volumes increase.

For small values of ψ_p the influence of these parameters is marginal, but at ψ_p > 10 deg., this influence is significant for the power and volumetric characteristics (Fig. 6), resulting in dropping.

Thus, there are the certain values the facing angle ψ_p at which the volume and power parameters reach their maximum values. The magnitude of these values depends, primarily, on the operation mode of the compressor, as well as on the geometrical parameters which determine the amount of air leakage through the slots, and depends less on the profile type.

Influence of the operating mode on the position of extremum on the performance curves can be clearly observed on the graphs in Fig. 6. For small values of П (П= 1.4 or less) gas leakage through the clearances is small and reduction in leakage through the radial clearances in connection with the facing of the rotor ceases to have a significant impact on the performance even at small values of ψ_p. With the growth of П gas leakages increase and their effect on the indices and extremums on the performance characteristics shift to the higher values of ψ_p. Since with the growth of П the influence of jammed volumes and changes in the profile tightness is also growing, the characteristics become steeper and the maximum grows.

As seen from the graphs in Fig. 6, the maximum values of the different indices are observed at different values of ψ_p. So, for the circumferential profile at П = 1.4 the η_v reaches the maximum value at ψ_p = 6 deg., and V_d - at ψ_p =5 deg. This is due to the fact that the actual performance depends not only on the volumetric efficiency, but also on the theoretical capacity, which decreases with the growth of the ψ_p. For η_{ind} the extremum is slightly shifted to the higher ψ_p which, apparently, is connected with the increase of the angle of transfer of the isolated working chamber with the growth of the facing angle. For the specific indicated power $\overline{N}_{ind}$ the position of the extremums almost coincide with that on the η_v curves.

5 CONCLUSIONS

Computer analysis of two-lobe Roots compressor showed that the maximum values of the volumetric efficiency η_v and indicated efficiency η_{ind} can be observed at values of $\overline{A}$ = 0.62 ... 0.64. The lower values are generally not permitted under the terms of the strength and stiffness of the rotor.

Analysis of the impact of the rotor face angle ψ_p shows that for small values of П =1.4...1.6 at values ψ_p = 5 ... 7 deg. reaches maximum η_v, η_{ind}. But the compressor design at the values of П = 1.8 ... 2.0, which are relatively higher for this type of compressor, requires the large facing angle ψ_p = 10 ... 15 deg., as in this case, the leaks through the radial clearance are more significant. The results of this study can be used for designing Roots-type compressors operated under air conditions.

REFERENCES

(1) Golovintsev, A.G., Rumyantsev, V.A., Ardashev, V.I., 1964, Rotatsionnye kompressory (Rotary compressor (in Russian)). Moskva: Mashinostroenie, 315 p.
(2) Vizgalov, S.V., 2004, Vliyanie vnutrennego okhlazhdeniya na effektivnost rabochego protsessa shesterenchatogo kompressora (Influence of internal cooling to efficiency of working process of Root's compressor (in Russian)). Dissertation, Kazan national research technology university. 212 p.
(3) Ibraev, A.M., 1987. Povyshenie effektivnosti raboty rotornykh nagnetatelei vneshnego szhatiya na osnove analiza vliyaniya geometricheskikh parametrov na ikh kharakteristiki (Higher efficiency of rotary blower by analysis of geometry parameters to them characteristics (in Russian)). Dissertation, St. Petersburg State Polytechnical University. 208 p.
(4) Ibraev, A.M., Chekushkin, G.N., Khamidullin, M.S. 1999. Raschet i analiz lineinogo profilya shesterenchatykh kompressorov. Sbornik nauchnykh trudov: Proektirovanie i issledovanie kompressornykh mashin (Calculation and analysis of linear profile of Root's compressor. Collection of scientific papers: Design and research compressors (in Russian)). Kazan. 193-198 p.
(5) Sharapov, I.I., 2009. Razrabotka metodiki izmereniya i rascheta parametrov protsessa teploobmena v shesterenchatom kompressore s tselyu povysheniya tochnosti rabochego protsessa (Development of the method of measurement and calculation of parameters of the heat transfer process in Root's compressor in order to improve the accuracy of the workflow (in Russian)). Dissertation, Kazan national research technology university. 145 p.
(6) Zakharenko S. E. K voprosu o protechkakh gaza cherez shcheli (On the question of gas leakages through the gaps ((in Russian)) Trudy LPI. - №2. - M.-L., Mashgiz, 1953. - 144-160 p.
(7) Vizgalov, S.V., Ibraev, A.M., Khisameev, I.G., Sharapov, I.I., 2010. Analiz koeffitsienta podachi shesterenchatogo kompressora (Analysis of volume efficiency of Root's compressor (in Russian)). Moskva.: Kompressornaya tekhnika i pnevmatika, 2010, №8, p.2-8.
(8) Forsythe G., Malcolm, M., Moler C., 1977. Computer methods for mathematical computations. Prentice-Hall, Inc., Englewood cliffs, 1977.

A comprehensive simulation model of the dynamics of the revolving vane machine

A Subiantoro[1], K T Ooi[2]
[1] TUM CREATE (Technische Universität München –
  Campus for Research Excellence And Technological Enterprise), Singapore
[2] School of Mechanical and Aerospace Engineering,
  Nanyang Technological University, Singapore

ABSTRACT

A new and comprehensive mathematical model to simulate the revolving vane (RV) machine has been developed. The model allows the dynamics of the components to be fully dictated by forces and torques interactions, instead of restriction provided by their geometries alone. In general, the new model produces very similar kinematics predictions with that from the old model which is geometrically dictated. However, the new model is able to show some features which are previously not observable. These include the intermittent contact between the vane and the slot and the effects of the collisions between the vane and the slot.

NOMENCLATURE

C_d	coefficient of discharge (-)		v	velocity (m/s)
d	distance (m)		V	volume (m^3)
e	eccentricity (m)		w	width (m)
F	force (N)		z	height (m)
g	gravitational acceleration (m/s^2)			
h	specific enthalpy (J/kg)		**Greek symbols**	
I	rotational inertia (kg·m^2)		α	angular acceleration (rad/s^2)
l	length (m)		δ	gap (m)
m	mass (kg)		η	friction coefficient (-)
P	pressure (Pa)		φ	angle (rad)
r	radius (m)		μ	viscosity (Pa·s)
Q	heat flux (J)		ρ	density (kg/m^3)
t	time (s)		Σ	summation
T	torque (N·m); temperature (K)		θ	angle (rad)
u	specific internal energy (J/kg)		ω	angular velocity (rad/s^2)

Subscripts

0	at center of vane or vane slot		disc	discharge tank
1	at the right of vane or vane slot		f	friction
2	at the left of vane or vane slot		imp	impact
b	bearing; at the base of vane		m	motor
c	cylinder		n	normal
C	cylinder center		r	rotor
cont	vane contact		P	pressure
cv	control volume		R	rotor center
DCV	discharge chamber		SCV	suction chamber

s	isentropic	V	point of interest at vane
S	point of interest at vane slot	ve	exposed vane
suct	suction reservoir	VS	vane slot chamber
v	vane	VS0	vane slot without vane

1 INTRODUCTION

Ever since its invention, rotary machines have rapidly become more popular due to their superior vibration characteristics, simplicity and compactness. However, frictions between the rubbing components have been one of the limitations (1). In response, Teh and Ooi (2) introduced a novel mechanism called the Revolving Vane (RV) machine (Figure 1). Unlike the conventional rotary machines, the cylinder of the RV machine rotates together with the rotor, which reduces the relative velocities and frictional losses between the rubbing components.

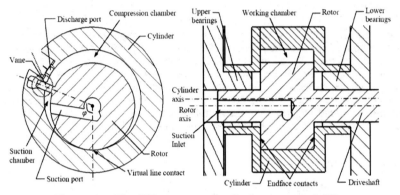

Figure 1 – The RV compressor design concept (3)

Many works have been carried out to model various aspects of the machine (2-7). The models have also been shown to be in good agreement with the experimental results (8). However, the existing simulation models have a lot of limitations, mainly because they assume that the dynamics of the components are completely geometrically defined. With these assumptions, a proper analysis of the effects of the vane geometry is not possible. Neither can it properly model the vane dynamics, the leak through the vane gaps, nor the impact forces on the vane slot.

In this paper, a new modeling method to simulate the RV machine more comprehensively is presented. The dynamics of the components are no longer assumed to be solely geometrically dependent. Instead, they are fully dictated by Newtonian mechanics, i.e. the forces and torques acting on each component.

2 GEOMETRICAL MODELS

The geometries of the RV machine are modeled based on those shown in Figure 2. Points C and R are the cylinder and rotor centers, respectively. Point V_1 is a point at the right hand side of the vane that is of a distance of the cylinder radius, r_c, from the cylinder center. This is also the vane contact point when the vane is in contact with the vane slot at its right hand side. Point S_1 is the tip of the right hand side of the vane slot opening. The distance between this point to point C is always equal to the radius of cylinder, r_c. In Figure 2, points V_1 and S_1 are shown to be at the same position, indicating that the contact point of the vane and the vane slot is currently at the right hand side of the vane.

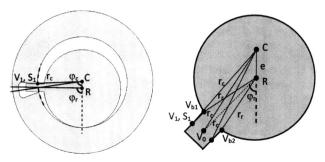

Figure 2 – The RV machine and the rotor model

The angle of the rotor, φ_r, is defined as the angle between a vertical line, which passes through both points C and R, and a line that is the midline of the vane, which passes through the rotor center, R, and point V_0. Point V_0 itself is a point that lies at the midline of the vane and is of a distance of the cylinder radius, r_c, from the cylinder center, C. Point V_2 is a point at the left hand side of the vane that is of a distance of the cylinder radius, r_c, from the cylinder center, C. This is the contact point when the vane contact is at the left hand side of the vane. Points V_{b1} and V_{b2} are points at the right hand and left hand sides of the base of the vane. They are of distances of the rotor radius, r_r, from the rotor center, R.

The angle of the cylinder, φ_c, is defined as the angle between the vertical line, which passes through C and R, and a line that connects point C and the midpoint of the vane slot opening, S_0 (see Figure 3). Point S_2 is the tip of the left hand side of the vane slot opening. Point R_1 is the projection of point S_1 at the midline of the vane, while point R_2 is the projection of point S_2 at the midline of the vane. When the vane is in contact with the vane slot at its right hand side as illustrated in Figure 2, the distance between points R_1 and S_1 is equal to half of the vane width, w_v. Point C_1 is a point at the line that connects point C with the midpoint of the vane slot opening, S_0. The distance between point C_1 and point S_1 is equal to half of the width of the vane slot opening, w_{vs}. There is also point C_2 (not shown in the figures) that is of the same distance from point S_2. However, because this point always coincides with point C_1, point C_2 will be merged with point C_1 in the subsequent discussions. Figure 3 also shows that points V_1, V_0, V_2, S_1, S_0 and S_2 are all of the distances of the cylinder radius, r_c, from the cylinder center, C.

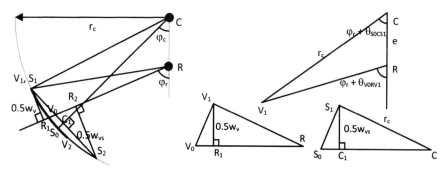

Figure 3 – Geometrical model #1 (vane is in contact with vane slot at V_1)

2.1 The angle of the cylinder

To determine the angle of the cylinder, the triangles CS_0S_1, CV_1R and RV_0V_1 shown in Figure 3 are used. By observing the triangles, the relationship between the various geometrical parameters according to Equation (1) can be derived. If the contact point is at the left hand side of the vane, the equation should be adjusted accordingly. Equation (1) can then be used to find the angle of the cylinder, φ_c, when the vane is in contact with the vane slot.

$$r_c \sin(\varphi_c + \theta_{S_0CS_1}) = \sin\varphi_r \sqrt{e^2 + r_c^2 - 2er_c \cos(\varphi_c + \theta_{S_0CS_1}) - 0.25w_v^2} + \frac{w_v}{2}\cos\varphi_r \quad (1)$$

2.2 The exposed length of the vane

During the operation, the vane extends and retracts in and out of the vane slot. To find the lengths of the vane at either sides of the vane that are exposed to the fluids inside the suction and discharge chambers, Equation (2) is needed.

$$\sin\theta_{V_0RV_{b1}} = \sin\theta_{V_0RV_{b2}} = \frac{w_v}{2r_r} \quad (2)$$

From the triangle RV_1V_{b1}, the expression for the length of the exposed vane, l_{ve}, at its right hand side is obtained as expressed in Equation (3). If the contact point is at the left hand side of the vane, the equation should be adjusted accordingly.

$$l_{ve1} = \left\{ r_r^2 + l_{RV_1}^2 - 2\left(\sqrt{\left(r_r^2 - \frac{w_v^2}{4}\right)\left(l_{RV_1}^2 - \frac{w_v^2}{4}\right)} + \frac{w_v^2}{4} \right) \right\}^{0.5} \quad (3)$$

2.3 The gap between the vane sides and the tips of the vane slot opening

The gap between the sides of the vane to the tips of the vane slot opening is necessary to check if the vane is in contact with the vane slot opening. These gaps may also cause the working fluid to leak from the high pressure chamber to the low pressure chamber. To help the understanding of the procedure, a geometrical model when the vane is not in contact with the vane slot is shown in Figure 4.

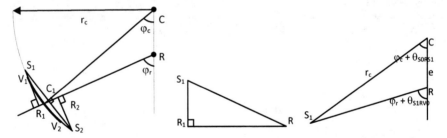

Figure 4 – Geometrical model #2 (vane is not in contact with the vane slot)

The vane is not in contact with the vane slot when neither the distance between R_1 and S_1 nor between R_2 and S_2 is equal to half of the vane width. The triangles CRS_1 and R_1RS_1 are used and give the gap between the right hand side of the vane and the right hand tip of the vane slot opening as expressed in Equation (4). A similar approach can be performed to find the gap between the left hand side of the vane and the left hand tip of the vane slot opening.

$$d_{S1} = l_{R_1S_1} - \frac{w_v}{2} \quad (4)$$

3 DYNAMICS MODEL

To model the dynamics of the RV machine, the free body diagrams for the cylinder and the rotor must be observed separately. The forces and torques at the cylinder when the vane is in contact with the vane slot at V_1 are shown in Figure 5.

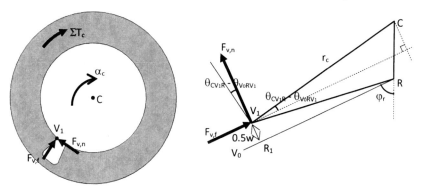

Figure 5 – The forces and torques at the cylinder (vane is in contact at V_1)

The torque balance equation of the cylinder is expressed in Equation (5).

$$I_c \alpha_c = F_{v,n} r_c \cos\left(\theta_{CV_1R} - \theta_{V_0RV_1}\right) - \frac{dl_{vel}/dt}{|dl_{vel}/dt|} F_{v,f} r_c \sin\left(\theta_{CV_1R} - \theta_{V_0RV_1}\right) + \Sigma T_c \tag{5}$$

Where: $F_{v,f} = |F_{v,n}| \eta_v$ (6)

The ΣT_c term consists of all the torques at the cylinder which are not caused by the vane contact force nor the vane side friction. Rearranging and generalizing Equation (5) to be also applicable when the vane is in contact with the vane slot at V_2 and using Equation (6) result in Equation (7). In this equation, the contact points V_1 and V_2 have been replaced by a general V_{cont}. If the vane contact point is at V_1, then $F_{v,n}/|F_{v,n}| = 1$. If the vane contact point is at V_2, then $F_{v,n}/|F_{v,n}| = -1$, and if the vane is not in contact with the slot, there is no contact force and therefore $F_{v,n}/|F_{v,n}| = 0$.

$$F_{v,n} = \frac{I_c \alpha_c - \Sigma T_c}{r_c \cos\left(\theta_{CV_{cont}R} - \frac{F_{v,n}}{|F_{v,n}|}\theta_{V_0RV_{cont}}\right) - \frac{F_{v,n}}{|F_{v,n}|}\frac{dl_{ve}/dt}{|dl_{ve}/dt|}\eta_v r_c \sin\left(\theta_{CV_{cont}R} - \frac{F_{v,n}}{|F_{v,n}|}\theta_{V_0RV_{cont}}\right)} \tag{7}$$

The forces and torques acting on the rotor are shown in Figure 6.

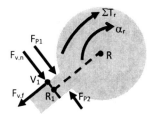

Figure 6 – The forces and torques at the rotor (vane is in contact at V_1)

Based on Figure 6, and assuming that the areas of the vane exposed to P_1 and P_2 are the same, the torque balance equation of the rotor when the vane is in contact with the vane slot at V_1 can be expressed as shown in Equation (8).

$$I_r \alpha_r = (F_{P2} - F_{P1})\left(r_r + \frac{l_{ve1}}{2}\right) - F_{v,n} l_{RR_1} - \frac{dl_{ve1}/dt}{|dl_{ve1}/dt|} F_{v,f} \frac{w_v}{2} + \Sigma T_r \qquad (8)$$

where: $F_p = Pl_{ve}l_{cham}$ \qquad (9)

The ΣT_r term consists of all the torques acting at the rotor which are not caused by the vane contact force, the vane side friction nor the fluid pressures acting on the vane. Rearranging and generalizing Equation (8) to be also applicable when the vane contact is at V_2 and using Equation (6) give Equation (10). Points R_1 and R_2 have been generalized as R_{cont} in this equation.

$$F_{v,n} = \frac{-I_r \alpha_r + (F_{P2} - F_{P1})\left(r_r + \frac{l_{ve}}{2}\right) + \Sigma T_r}{l_{RR_{cont}} + \frac{dl_{ve}/dt}{|dl_{ve}/dt|} \eta_v \frac{w_v}{2}} \qquad (10)$$

To simplify the subsequent discussions, let us introduce the new parameters shown in Equations (11)-(14).

$$R_{v,n,c} = r_c \cos\left(\theta_{CV_{cont}R} - \frac{F_{v,n}}{|F_{v,n}|}\theta_{V_0RV_{cont}}\right) - \frac{F_{v,n}}{|F_{v,n}|}\frac{dl_{ve}/dt}{|dl_{ve}/dt|}\eta_v r_c \sin\left(\theta_{CV_{cont}R} - \frac{F_{v,n}}{|F_{v,n}|}\theta_{V_0RV_{cont}}\right) \qquad (11)$$

$$T_c = \Sigma T_c \qquad (12)$$

$$R_{v,n,r} = l_{RR_{cont}} + \frac{dl_{ve}/dt}{|dl_{ve}/dt|}\eta_v \frac{w_v}{2} \qquad (13)$$

$$T_r = -F_{P1}\left(r_r + \frac{l_{ve1}}{2}\right) + F_{P2}\left(r_r + \frac{l_{ve2}}{2}\right) + \Sigma T_r \qquad (14)$$

Equations (7) and (10) calculate the contact force at the cylinder and the rotor, respectively. They are action and reaction forces. These forces are zero when the vane is not in contact with the vane slot. Substituting Equations (11)-(14) into Equations (7) and (10) and equating them give Equation (15).

$$\frac{I_c \alpha_c - T_c}{R_{v,n,c}} = \frac{T_r - I_r \alpha_r}{R_{v,n,r}} \qquad (15)$$

The conservation of energy equation before and after the vane hits the vane slot is expressed in Equation (16) with t_{imp} is the impact time between the vane and the vane slot. This parameter should be obtained experimentally. If necessary, coefficient of restitution can be introduced in Equation (16) by adding a term into T_r and T_c in Equations (12) and (14).

$$\frac{1}{2}I_r\omega_{r1}^2 + \frac{1}{2}I_c\omega_{c1}^2 + \frac{1}{2}(\omega_{r1} + \omega_{r2})t_{imp}T_r + \frac{1}{2}(\omega_{c1} + \omega_{c2})t_{imp}T_c = \frac{1}{2}I_r\omega_{r2}^2 + \frac{1}{2}I_c\omega_{c2}^2 \qquad (16)$$

where: $\omega_{r2} = \omega_{r1} + \alpha_r t_{imp}$

Finally, combining Equations (15) and (16) give the accelerations of the rotor and the cylinder as expressed in Equations (17) and (18).

$$\alpha_r = \frac{-2\left(\omega_{r1} - \frac{R_{v,n,c}}{R_{v,n,r}}\omega_{c1}\right) + \frac{t_{imp}}{I_c}\frac{R_{v,n,c}}{R_{v,n,r}}\left(T_r\frac{R_{v,n,c}}{R_{v,n,r}} + T_c\right)}{t_{imp}\left(1 + \frac{I_r}{I_c}\frac{R_{v,n,c}^2}{R_{v,n,r}^2}\right)} \quad (17)$$

$$\alpha_c = -\frac{I_r}{I_c}\alpha_r\frac{R_{v,n,c}}{R_{v,n,r}} + \frac{T_r}{I_c}\frac{R_{v,n,c}}{R_{v,n,r}} + \frac{T_c}{I_c} \quad (18)$$

4 THERMODYNAMICS MODEL

There are three chambers of interest in a RV machine. These are the suction, discharge and vane slot chambers. The working fluid inside each chamber experiences thermodynamics processes. These processes can be modeled according to the equation of the 1st law of thermodynamics as shown in Equation (19).

$$\frac{d\left(m_{cv}\left(u_{cv} + \frac{v_{cv}^2}{2} + gz_{cv}\right)\right)}{dt} = \frac{dQ_{cv}}{dt} - P_{cv}\frac{dV_{cv}}{dt} + \sum_j\left(h_j + \frac{v_j^2}{2} + gz_j\right)\frac{dm_j}{dt} \quad (19)$$

The working fluids flow into or out of each chamber through various flow paths. For the suction chamber, these include the suction port, the vane gap, the radial clearance and the endface gaps. For the discharge chamber, these include the discharge port, the vane gap, the radial clearance and the endface gaps. While for the vane slot, these include the vane gaps and the endface gaps. All of these flows can be modelled using the isentropic flow model as expressed in Equation (20).

$$\frac{dm}{dt} = C_d \rho_1 A \sqrt{2(h_1 - h_{2s})} \quad (20)$$

5 SIMULATION PROCEDURES

To carry out the simulation, a programming code was written in the FORTRAN programming language. Air was arbitrarily used as the working fluid. The fourth order Runge-Kutta method was employed to solve the differential equations. In the analysis, some assumptions were made. These include the adiabatic process assumption and that all the potential energy and the kinetic energy of the control volume terms are negligible. We also considered only the flows through the suction port, the discharge port and the gaps between the vane and the vane slot. Flows and leakages through other paths are neglected. The components are rigid and the effect of lubricants to leakage is negligible.

The simulation was for a RV compressor, but the model can also be adapted easily for a RV expander. Following our integration-step independence test, a rotor angle step of 0.001 rad has been chosen. As for the impact time, to our knowledge, there has been no reported study about the impact time of a vane and a vane slot. Therefore, a test was conducted using the RecurDyn V7R3 (9) commercial software to estimate the appropriate impact time. To verify the accuracy of the software, it was first benchmarked by testing it to simulate the collisions of two steel balls (10-11). The simulation results are found to be in good agreement with their experimental results, with differences of less than 20%. The software is then used to simulate the collision between the RV compressor vane and the vane slot. It was found that for each contact, it lasts for around 0.1 ms. Therefore, an impact time of 0.1 ms is used in this simulation. Lastly, the dimensions of the RV compressor and the constant parameters used are listed in Table 1 while the initial parameters used in our simulation are shown in Table 2.

Table 1 – Major dimensions and parameters of the RV compressor model

Parameter	Value	Parameter	Value
Rotor radius	29 mm	Rotor bearing radial clearance	15 μm
Cylinder radius	35 mm	Cylinder bearing length	38 mm
Compressor chamber length	50 mm	Cylinder bearing radius	19 mm
Rotor rotational inertia	0.334 g m^2	Cylinder bearing radial clearance	15 μm
Cylinder rotational inertia	1.63 g m^2	Suction pressure	1 bar
Motor torque	2.5 Nm	Suction temperature	25 °C
Vane total length	19.6 mm	Discharge pressure	8 bar
Vane width	4 mm	Discharge temperature	25 °C
Vane slot opening width	4.1 mm	Coefficient of discharge	0.58
Vane slot initial volume	12,054 mm^3	Friction coefficient	0.15
Rotor bearing length	64 mm	Endface gap	10 μm
Rotor bearing radius	11 mm	Lubricant viscosity	0.0034 Pa s

Table 2 – List of initial parameters

Parameter	Value	Parameter	Value
Rotor angle	0°	Suction chamber temperature	25 °C
Cylinder angle	0°	Discharge chamber pressure	8 bar
Rotor angular velocity	3,000 rev/min	Discharge chamber temperature	25 °C
Cylinder angular velocity	2,485.7 rev/min	Vane slot pressure	1 bar
Suction chamber pressure	1 bar	Vane slot temperature	25 °C

6 RESULTS AND DISCUSSIONS

The angular velocities of the rotor and the cylinder obtained from the old and the new models are shown in Figure 7. Both simulation models give very similar angular velocities, both in trends and values. This suggests that the effect of the vane thickness to the kinematics of the components is small. A closer look at Figure 7 shows that some sharp changes in the angular velocities are observed with the new model. These are not observable in the old model. These changes refer to the points where collisions between the vane and the vane slot occur, affecting the velocities of both components. This is also why all the sharp changes of both the rotor and the cylinder occur at the same rotor angles.

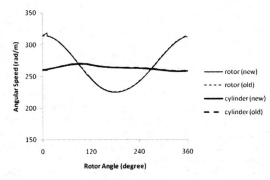

Figure 7 – Angular velocities of the rotor and the cylinder

The vane contact forces obtained from the old and the new models are shown in Figure 8. The general trends of the forces from both models are somewhat similar. They are positive until around a rotor angle of 100°, then drops until its minimum at around the 120° angle. It then remains in the negative region until it becomes positive again near the end of the cycle. However, while the old model produced a continuous and smooth vane contact force profile, the new model gives a fluctuating profile. This fluctuation indicates that the vane oscillates during operation, different from the assumption of the old model. The magnitudes of the forces are also different. This difference is thought to be due to the fact that the old model does not include the collision process between the vane and the vane slot, unlike the new model. This also indicates that the old model may have underestimated the vane contact forces as it neglects of the impact of the collision between the vane and the slot. Interestingly, Figure 8 shows that the vane only flips sides twice during the operation, although it oscillates during the whole operation. This is in agreement with the old model, which shows that the vane contact force flips its sign twice during a complete cycle.

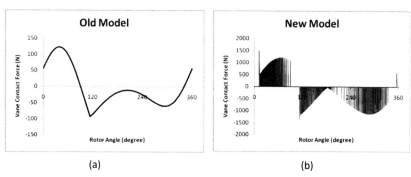

Figure 8 – Vane contact forces of: (a) the old and (b) the new models

From the results presented here, it can be seen that with the proposed model, various new features that are not possible to be investigated using the established models are now observable. In the future, comparison with experimental data needs to be carried out to verify the accuracy of the model.

7 CONCLUSIONS

A new and comprehensive mathematical model to simulate the RV machine has been developed. The model does not assume that the dynamics of the components are only geometrically dependent. Instead, it uses the forces and torques interactions of the components.

The comparison of the kinematics data of the new and the old simulation models indicate that while the general trends and values obtained from both models are very similar, some unique features are only observable with the new model. These include the findings that vane thickness does not affect the kinematics behavior of the compressor significantly. Some sharp changes in the angular speeds of the components caused by the collision between the vane and the slot previously unobserved with the old model are now observable with the new model.

Dynamically, unlike the old model, the new model is able to show the relative movements between the vane and the vane slot. It was found that the vane is not always in contact with the vane slot. Rather, it is only intermittently in contact with the vane slot during operation, as shown by the fluctuating vane contact force and the vane gaps. It is also found that the old model underestimated the contact force acting on the vane because it does not consider the impact force at the vane. The observation shows that while the vane oscillates inside the vane slot, for most of the time, it oscillates very close to either one of the tips of vane slot. In fact, the vane only flips sides twice in one complete cycle.

Experimental investigations will be carried out in the future to verify the accuracy of the new simulation model.

ACKNOWLEDGMENT

This work was financially supported by the Singapore National Research Foundation under its Campus for Research Excellence And Technological Enterprise (CREATE) program. The views expressed herein are solely the responsibility of the authors and do not necessarily represent the official views of the Foundation.

REFERENCE LIST

(1) Ozu, M., Itami, T., 1981, Efficiency analysis of power consumption in small hermetic refrigerant rotary compressors. International Journal of Refrigeration, 4(5):265-270.
(2) Teh, Y.L., Ooi, K.T., 2009, Theoretical study of a novel refrigeration compressor - Part I: Design of the revolving vane (RV) compressor and its frictional losses. International Journal of Refrigeration, 32(5):1092-1102.
(3) Teh, Y.L., Ooi, K.T., 2008, Design and friction analysis of the improved Revolving Vane (RV-i) compressor. International Compressor Engineering Conference at Purdue, 1233.
(4) Teh, Y.L., Ooi, K.T., 2009, Theoretical study of a novel refrigeration compressor. Part III: leakage loss of the revolving vane (RV) compressor and a comparison with that of the rolling piston type. International Journal of Refrigeration, 32(5):945-952.
(5) Teh, Y.L., Ooi, K.T., Djamari, D.W., 2009, Theoretical study of a novel refrigeration compressor - Part II: Performance of a rotating discharge valve in the revolving vane (RV) compressor. International Journal of Refrigeration, 32(5):1103-1111.
(6) Subiantoro, A., Ooi, K.T., 2009, Introduction of the Revolving Vane expander, HVAC&R Research, 15(4):801-816.
(7) Subiantoro, A., Ooi, K.T., 2010b, Design Analysis of the novel Revolving Vane Expander in a transcritical carbon dioxide refrigeration system. International Journal of Refrigeration, 33(4):675-685.
(8) Teh, Y.L., Ooi, K.T., 2009, Experimental study of the Revolving Vane (RV) compressor. Applied Thermal Engineering, 29(14-15):3235-3245.
(9) FunctionBay, Inc., 2010, RecurDyn V7R3 (www.functionbay.co.kr).
(10) Karpushin, V.B., Suhanov, I.I., 2000, The contact area-the impact time relationships for the steel balls. Siberian Russian Student Workshops and Tutorials on Electron Devices and Materials (Cat. No. 00EX416):167-171.
(11) Hessel, R., Perinotto, A.C., Alfaro, R.A.M., Freschi, A.A., 2006, Force-versus-time curves during collisions between two identical steel balls. American Journal of Physics, 74(3):176-179.

AUTHOR INDEX

Alves, M 555
Ancel, C 613
Angel, B 613
Arjeneh, M 511
Asal, W 103

Beinert, M 247, 625
Bell, I H 87, 453, 717
Bianchi, G 173, 183
Biswas, A 57
Bradshaw, C R 341
Branch, S 591
Brasz, J J 467
Braun, J E 87, 717
Brümmer, A 197, 407
Buckney, D 237

Calvi, T 173
Cerdoun, M 635
Chukanova, E 129
Chung, B 681
Cipollone, R 27, 173, 183
Contaldi, G 173, 183
Cook, G 669
Cremaschi, L 57

Declaye, S 431
Deokar, P 57
Deschamps, C J 301, 555
Ding, G L 697
Diniz, M C 301

Diny, M 453
Drozdov, A 533

Estruch, O 577
Etemad, S 511

Feng, Q K 219, 265

Galerkin, Y B 477, 533
Gao, J D 697
Gao, Y F 697
Ghenaiet, A 523, 635
Ginies, P 613
Groll, E A 15, 87, 717
Gross, D 285, 613
Guillaume, L 431, 453
Guo, J 397
Gysak, O 649

Harrison, D K 77
Hauser, J 247, 625
He, Z 397
Heidari, M 323
Heinrich, M 659
Heiyanthuduwage, M 669
Herlemann, S 247
Heyder, J 353
Hokey, D 113
Holmes, C S 363
Hori, K 257
Horton, W T 87, 717
Hossain, M A 257
Hrnjak, P S 277, 311
Huang, P X 113
Huang, R 219
Hütker, J 407

Ibraev, A M 227, 739

Ignatiev, K M 445

Ihnatenko, V 501

Inoue, T 257

Jia, Z 141

Junaidi, A Z 67

Kalinkevych, M 489, 501

Karnaz, J 163

Kethidi, M 601

Khalfallah, S 523

Khamidullin, M S 209, 227

Khisameev, I G 209, 227, 739

Kim, H-J 565

Köhler, A 353

Kovacevic, A 129, 237, 417, 601

Kurtulus, O 15

Lajús Junior, F C 555

Lee, G-H 565

Lee, S-W 565

Lee, T-J 565

Legros, A 431, 453

Lehmkuhl, O 577, 731

Lemort, V 431, 453, 717

Li, H 141

Li, L S 151

Li, T 219

Lin, S 397

Linkamp, A 197

Liu, F 265

López, J 731

Ma, T 141

Machu, E H 375, 545

Masuda, M 257
Matsuo, K 363
Milligan, W J 77
Mounoury, S 669
Muir, D I 77
Murgia, S 173, 183
Mustafin, T N 209

Nalimov, V N 209

Obukhov, O 649
Oguz, E 385
Olenick, D 113
Oliva, A 577, 731
Onbasioglu, S 385
Ooi, K T 67, 749
Ozdemir, A R 385

Park, N 681
Pascu, M 669
Pearson, A B 5
Pereira, E L L 301
Perevozchikov, M M 445
Pérez-Segarra, C D 577
Pullen, K R 511

Qian, Z G 151
Quoilin, S 431, 453

Rane, S 417, 601
Rigola, J 577, 731
Rufer, A 323

Sachs, R 103
Saifetdinov, A G 227
Sauls, J 591

Schwarze, R 659
Sharapov, I I 227
Shcherbakov, O 501
Shin, J 681
Skoryk, A 489
Smirnov, A 649
Smith, I 1
Soldatova, K V 477, 533, 707
Song, J 697
Stosic, N 129, 237, 601
Subiantoro, A 67, 749

Tang, Y 335

Ueno, H 257

Valenti, G 173
Vizgalov, S V 739

Wang, T T 697
Wang, Z L 219, 265
Wölfel, F 353
Wu, W F 219, 265
Wu, X 57

Yakupov, R R 209
Yang, Q C 151
Yonkers, S 113

Zhang, X 141
Zhao, X 141
Zhao, Y Y 151
Zheng, Y X 697
Zimmermann, A J P 277, 311